D0216267

DATE DUE

MAY 1 9 1998	

DEMCO, INC. 38-2931

Molecular Genetics of Bacterial Pathogenesis

Molecular Genetics of Bacterial Pathogenesis

A TRIBUTE TO STANLEY FALKOW

Editors:

Virginia L. Miller
Department of Microbiology and Molecular Genetics
University of California, Los Angeles
Los Angeles, California

James B. Kaper
Center for Vaccine Development
University of Maryland School of Medicine
Baltimore, Maryland

Daniel A. Portnoy
Department of Microbiology
University of Pennsylvania School of Medicine
Philadelphia, Pennsylvania

Ralph R. Isberg
Department of Molecular Biology and Microbiology
Tufts University School of Medicine
Boston, Massachusetts

ASM PRESS
Washington, D.C.

Library of Congress Cataloging-in-Publication Data

Molecular genetics of bacterial pathogenesis/editors, Virginia L.
   Miller . . . [et al.].
      p.   cm.
   "A tribute to Stanley Falkow".
   Includes bibliographical references and index.
   ISBN 1-55581-082-9
   1. Bacterial diseases—Pathogenesis.   2. Virulence (Microbiology).
3. Pathology, Molecular.   I. Miller, Virginia L.
QR201.B34M65   1994   94-19193
616'.014—dc20   CIP

Cover figure: Salmonella typhimurium invasion of the M cells of murine Peyer's patch 30 min after infection of a ligated intestinal loop. The uptake of *S. typhimurium* by these specialized epithelial cells is associated with a pronounced membrane ruffling. At this time period the bacteria are exclusively associated with M cells which are selectively killed in the early host-parasite interaction. The M cell shown here has swollen and distorted mitochondria as well as the contraction of its cytoplasmic contents from the adjacent cells. The adjacent cells seen here are enterocytes that contain no internalized bacteria and appear to be completely unaffected by the presence of the invasive organism. $\times 10,000$. (Transmission electron micrograph courtesy of Stanley Falkow.)

Contents

v

PART 6. OTHER ASPECTS OF BACTERIAL PATHOGENESIS

Contributors

Patrik M. Bavoil
Department of Microbiology and Immunology, University of Rochester
Medical Center, Box 672, Rochester, NY 14642

James B. Bliska
Department of Molecular Genetics and Microbiology, State University of
New York at Stony Brook, Stony Brook, NY 11794-5222

Jim Brunton
Departments of Microbiology and Medicine, The Toronto Hospital, and
Samuel Lunenfeld Research Institute at Mount Sinai Hospital, Toronto,
Ontario, Canada M5G 2C4

Carleen M. Collins
Department of Microbiology and Immunology, University of Miami School
of Medicine, Miami, FL 33101

Jorge H. Crosa
Department of Molecular Microbiology and Immunology, L-220, Oregon
Health Sciences University, 3181 S.W. Sam Jackson Park Road, Portland,
OR 97201

Walter S. Dallas
Division of Cell Biology, The Wellcome Research Laboratories, Burroughs
Wellcome Company, 3030 Cornwallis Road, Research Triangle Park,
NC 27709

Gordon Dougan
Department of Biochemistry, Imperial College of Science, Technology and
Medicine, London SW7 2AZ, United Kingdom

Lynn P. Elwell
Division of Molecular Genetics and Microbiology, Wellcome Research
Laboratories, 3030 Cornwallis Road, Research Triangle Park, NC 27709

Stanley Falkow
Department of Microbiology and Immunology, Stanford University School of
Medicine, Sherman Fairchild Building, Stanford, CA 94305

B. Brett Finlay
Biotechnology Laboratory and Departments of Biochemistry and
Microbiology, University of British Columbia, Vancouver, British Columbia,
Canada V6T 1Z3

Ronald E. Gill
Department of Microbiology and Immunology, Campus Box B175, 4200 East
Ninth Avenue, Denver, CO 80262

Patricia Guerry
Enteric Diseases Program, Naval Medical Research Institute, Bethesda,
MD 20889

Fred Heffron
Department of Microbiology and Immunology, Oregon Health Sciences
University, 3181 S.W. Sam Jackson Park Road, Portland, OR 97201-3098

Richard A. Hull
Department of Microbiology and Immunology, Baylor College of Medicine, Houston, TX 77030

Sheila I. Hull
Department of Microbiology and Immunology, Baylor College of Medicine, Houston, TX 77030

Ralph R. Isberg
Howard Hughes Medical Institute and Department of Molecular Biology and Microbiology, Tufts University School of Medicine, 136 Harrison Avenue, Boston, MA 02111

James B. Kaper
Center for Vaccine Development, University of Maryland School of Medicine, 10 South Pine Street, Baltimore, MD 21201

Michael Koomey
Department of Microbiology and Immunology, M 6791, Medical Science Building II, University of Michigan Medical School, Ann Arbor, MI 48109

Donald J. LeBlanc
Department of Microbiology, University of Texas Health Science Center at San Antonio, 7703 Floyd Curl Drive, San Antonio, TX 78284

Catherine A. Lee
Department of Microbiology and Molecular Genetics, Harvard Medical School, 200 Longwood Avenue, Boston, MA 02115

David A. Low
Division of Cell Biology and Immunology, Department of Pathology, Room 5B310MC, University of Utah Medical School, Salt Lake City, UT 84132

Carl F. Marrs
Department of Epidemiology, School of Public Health, University of Michigan, Ann Arbor, MI 48109

John J. Mekalanos
Department of Microbiology and Molecular Genetics, Harvard Medical School, 200 Longwood Avenue, Boston, MA 02115

Jeff F. Miller
Department of Microbiology and Immunology, University of California, Los Angeles, Los Angeles, CA 90024

Virginia L. Miller
Department of Microbiology and Molecular Genetics, University of California, Los Angeles, 405 Hilgard Avenue, Los Angeles, CA 90024

Frits R. Mooi
National Institute of Public Health and Environmental Protection, P.O. Box 1, 3720 BA Bilthoven, The Netherlands

Paul E. Orndorff
Department of Microbiology, Pathology, and Parasitology, College of Veterinary Medicine, North Carolina State University, 4700 Hillsborough Street, Raleigh, NC 27606

Dorothy E. Pierson
Department of Microbiology, University of Colorado Health Sciences Center, Denver, CO 80262

Daniel A. Portnoy
Department of Microbiology, University of Pennsylvania School of Medicine, Philadelphia, PA 19104-6076

David A. Relman
Departments of Medicine and of Microbiology and Immunology, Stanford University School of Medicine, Stanford, CA 94305, and Department of Veterans Affairs Medical Center, Palo Alto, CA 94304

Craig E. Rubens
Division of Infectious Diseases, Children's Hospital and Medical Center, and Department of Pediatrics, University of Washington, Seattle, WA 98105

David B. Schauer
Division of Toxicology and Division of Comparative Medicine, Massachusetts Institute of Technology, 77 Massachusetts Avenue, Room E18-564, Cambridge, MA 02139

Richard P. Silver
Department of Microbiology and Immunology, University of Rochester Medical Center, 601 Elmwood Avenue, Rochester, NY 14642

Pamela L. C. Small
Rocky Mountain Laboratories, Hamilton, MT 59840

Magdalene So
Department of Microbiology and Immunology, Oregon Health Sciences University, 3181 S.W. Sam Jackson Park Road, L220, Portland, OR 97201-3098

Joseph W. St. Geme III
Department of Molecular Microbiology, Washington University School of Medicine, 660 South Euclid Avenue, Box 8230, St. Louis, MO 63110

Scott Stibitz
Division of Bacterial Products, Center for Biologics Evaluation and Research, Food and Drug Administration, Bethesda, MD 20892

Lucy S. Tompkins
Department of Medicine, Division of Infectious Diseases and Geographic Medicine, and Department of Microbiology and Immunology, Stanford University Medical Center H 1537J, Stanford, CA 94305

Alison Ann Weiss
Molecular Genetics, Biochemistry, and Microbiology, University of Cincinnati, 231 Bethesda Avenue, ML 524, Cincinnati, OH 45267

Rodney A. Welch
Department of Medical Microbiology and Immunology, University of Wisconsin–Madison, Madison, WI 53706

Foreword

The history of infectious disease research can be divided into several major periods. Prior to the 19th century, even great thinkers such as Hippocrates were aware of the communicable nature of many diseases but suggested that their origins had most to do with poisonous vapors arising from the earth or the heavens. Jenner's work leading to the development of vaccination was certainly the crowning achievement of the late 18th century, providing the foundation for the fields of immunology and virology. Throughout the 19th century, research focused on proving the germ theory of disease and its implications for the budding field of epidemiology. This led the way for the "golden age of microbiology," during which dozens of bacterial species, both pathogenic and nonpathogenic, were isolated and characterized. The late 19th and early 20th centuries saw research efforts shift in the direction of immunology with the identification of attenuated live vaccines, protective bacterial antigens (e.g., diphtheria toxin), and therapeutic antisera as well as a variety of other fundamental discoveries in humoral and cellular immune mechanisms. During the mid-20th century, the floodgates opened with the development of methods for tissue culture, methods in enzymology, protein and nucleic acid structural analysis, bacterial genetic manipulation, and methods in virology. It was during this exciting period that Stanley Falkow joined the "microbe hunters" and began his scientific career. Over the last four decades, and no less today, Stan Falkow, together with his many students, postdoctoral fellows, and collaborators, has continued to astonish us with his innovative approaches to infectious disease research. This book is truly a testimony to the profound influence that Stan has had on our field.

I am sincerely pleased to introduce this book to the reader from the perspective of a friend, fan, and colleague of Stan Falkow. Although I have never formally been Stan's student, I did work in his lab for about 2 weeks during the course of a collaboration with him in 1980. Several of the contributors to this book were lab members at that time (including Richard and Sheila Hull, Jim Kaper, Mike Koomey, Dan Portnoy, Rod Welch, and Alison Weiss). One needs only look at this group of superb investigators to appreciate the impact that Stan has had on our field. However, they are simply typical of both past and present Falkow alumni and the quality of investigator Stan has consistently trained and attracted through his research efforts. In seeding the field, Stan laid the foundation of modern molecular genetic analysis of infectious diseases, particularly analysis of bacterial pathogens. Clearly, from my perspective, much of the most exciting research in medical bacteriology currently under way throughout the world can be traced to Stanley or his trainees.

There are numerous research areas that I associate with Stan's influence. Rising prominently among them is the use of genetic manipulations to define virulence determinants in what Stan accurately called "molecular Koch's postulates." It is not widely known that Stan Falkow was a participant at the famous 1972 midnight delicatessen meeting on Waikiki Beach in Honolulu at which Herb Boyer and Stan Cohen hatched their collaboration to develop recombinant DNA technology. Stan generously contributed the R factor that he had characterized (RSF1010) to the Boyer-Cohen plasmid slicing experiments with only one stipulation: if they got the method to work, then Stan wanted to apply it to the isolation of the gene encoding *Escherichia coli* heat-labile enterotoxin. This historical note provides yet another insight into Stan's devotion to his goal over the next two decades: the molecular genetic analysis of virulence factors.

A natural extension of Stan's early success in identifying genes that encode virulence determinants was the application of these genes to the problem of the diagnosis and epidemiology of infectious diseases. He is certainly one of the founders of the field of molecular diagnosis and molecular epidemiology, having been among the first to apply hybridization methods to the characterization and detection of pathogenic bacteria in clinical samples. When I visited Stan's lab in 1980, this effort was well under way, with Steve Moseley spearheading the effort to detect enterotoxigenic *E. coli* in stool specimens with various heat-labile and heat-stable enterotoxin gene probes. Then, as it is today, Stan's famous axiom that "s--- is my bread and butter" was as true as ever. Steve, Stan, and myself all have continued to work on diarrheal diseases and occasionally have gotten a little too close to our work. Indeed, all three of us have contracted shigellosis in one way or another.

Stan learned an early lesson in experiencing an unusual "mode of communication" of shigellosis during his research at Walter Reed Army Institute of Research with Sam Formal. Stan and Sam had infected a group of monkeys with *Shigella flexneri,* and as they were walking to the animal room to check on their patients, Stan became aware of a strange sound of something hitting the inside of the door as they turned the key. Sam, being older, wiser, and shorter than Stanley, had no problem avoiding the projectiles that the monkeys were accurately directing at the investigators. Alas, Stan was not so fortunate. This was perhaps the first time that fecal-oral transmission occurred via a "smart bomb."

Rather than attempt in this short forward to outline the many other marvelous achievements of Falkow and his scientific "children and grandchildren," I would simply ask the reader to read on and enjoy this fine collection of chapters.

Stan and I share an unflinching enthusiasm for what lies ahead in our field. We also share a keen devotion to fine Scottish malt whiskey. So, as I end this forward, let me raise my glass in celebration of Stan's 60th birthday. Stanley, thank you for the inspiration, perspiration, and intoxication you have brought to science. *L' Chaim!*

John Mekalanos

Research and Training Legacy of Stanley Falkow

Stanley Falkow was born in Albany, New York, in 1934. Rather than follow his parents into the clothing trade, Stanley set his sights on a scientific career after reading Paul de Kruif's book *Microbe Hunters* (3) at the age of 11. He graduated cum laude with a B.S. in bacteriology from the University of Maine in 1955. His initial graduate studies with Allan Campbell at the University of Michigan were disrupted by an attack of infectious mononucleosis, and he subsequently spent 1956 to 1957 in the Department of Bacteriology and Serology at Newport Hospital, Newport, Rhode Island. These years as a clinical microbiologist provided a solid foundation for his later work in developing novel methods of identifying and classifying bacteria. One long-lived product of this period was a diagnostic decarboxylase medium that Stan developed and that is still produced today (Bacto Decarboxylase Medium; Difco Laboratories, Detroit, Mich.). The early experience in clinical microbiology was a fitting beginning for an individual who was later acclaimed as "the first and still the most significant figure to eliminate the barriers between clinical and basic research in microbiology" (1).

In 1957, Stan entered Brown University in Providence, Rhode Island, where he studied with Charles Stuart and earned his Ph.D. in biology in 1961. He received a President's Fellowship from the American Society for Microbiology to study with Julius Marmur at Brandeis University, where he worked on DNA for the first time. One outcome of his postdoctoral work was the first direct evidence that the F factor was composed of DNA (11). Another product of his short time at Brandeis was a highly cited review article entitled "New Approaches to Bacterial Taxonomy" that applied molecular genetic techniques to bacterial classification (10).

From 1961 to 1966, Stan worked at the Walter Reed Army Institute of Research, where he became assistant chief of the Department of Bacterial Immunology in 1963. There he worked with Lou Baron and Sam Formal (the latter had also trained with Charles Stuart). Sam recalls that he received a phone call from Dr. Stuart one afternoon saying that he had a student who might benefit from spending some time at Walter Reed and would Sam please arrange it. When Sam asked for a recommendation for Stan to take to his superiors, Dr. Stuart replied "Oh, don't worry. He's halfway decent." Sam relayed this assessment of Stan to his superiors, and, because they knew Dr. Stuart, Stan had a job the next day.

The years at Walter Reed were extremely productive and established Stan as a gifted and versatile young researcher. At Walter Reed he learned bacterial genetics and started his work on sex factors, R factors, and virulence factors. He worked on the transfer of episomes and chromosomal genes between bacterial species, thereby establishing genetic hybrids with which to assess the effects of various genes on virulence. He also worked on the detection and characterization of sex factors and R factors and developed many new mating systems with which to study pathogenicity and virulence in enteric bacteria (1).

From Walter Reed, Stan went to Georgetown University Medical School as associate professor of microbiology in 1966. There he trained the first of many graduate students and postdoctoral fellows, in addition to developing his courses in microbiology for graduate and medical students. An important focus of his work at Georgetown was the characterization of plasmids which encode antibiotic resistance (R factors). Stan's group was the first to physically isolate an R factor (6) and demonstrate that it was extrachromosomal DNA. Early studies on plasmid replication and conjugation are described in this volume by Don LeBlanc and Rich Silver (see chapter 1), two of Stan's trainees at Georgetown. Stan's studies on the characterization of plasmids encoding enterotoxins and intestinal adherence factors of enterotoxigenic *Escherichia coli* were also initiated at Georgetown. DNA hybridization techniques were used to study genetic relatedness among R factors, bacteriophages, and enteric pathogens. Collaborative studies with Don Brenner on the genetic relatedness among several groups of enteric bacteria provided a definitive basis for classifying these organisms. Although the majority of the work at Georgetown involved pure molecular biology rather than the study of bacterial pathogenicity, the latter aspect was never far from Stan's thoughts. Don Brenner recalls Stan's first television interview, which occurred in a rowboat on the Potomac River. He was trying to answer questions about fecal pollution in the Potomac while holding onto the boat for dear life. He brought the hazards of pollution in the Potomac to the public's attention, warning that "on a particularly bad day, one cup of Potomac River water is equivalent to a half gram of human feces." This vivid metaphor is typical of the facility with which Stan conveys his message to his listeners, whether they are students, faculty, or the general public.

In 1972, Stan moved across the country to Seattle to become professor of microbiology and immunology at the University of Washington School of Medicine. During the years in Seattle, Stan's impact expanded beyond microbiology to issues of national and international science policy, because this was the dawn of the era of recombinant DNA. Stan's involvement in this field arose naturally from his research on R factors, and the U.S.-Japan Conference on Bacterial Plasmids held in Honolulu in November 1972 proved to be a watershed event in science. During a midnight meal at a kosher delicatessen on Waikiki Beach, Stanley Cohen, Herbert Boyer, Charles Brinton, Ginger Brinton, and Stan discussed the possibility of creating hybrid DNA molecules from plasmids. Cohen and Boyer agreed to collaborate on constructing such hybrid molecules using the *Eco*RI restriction enzyme purified by Boyer and a small plasmid that Cohen had derived from the plasmid R6-5 using sheared DNA. Stan offered the use of the RSF1010 R factor that he had characterized for their experiments in return for the use of this system to clone genes encoding the heat-stable enterotoxin (ST) of enterotoxigenic *E. coli* (ETEC) (9). Thus it was that the first hybrid molecule containing DNA from two different species harbored the streptomycin resistance gene from RSF1010. Subsequently, Maggie So, then a graduate student with Stan, used this system to clone the ST gene of ETEC, a process described elsewhere in this volume by Maggie and Walt Dallas (see chapter 4); this was the first virulence gene to be cloned. The onset of the recombinant DNA era engendered great controversy about the use of this technology. As the leading expert on the diseases and plasmids of *E. coli*, Stan was named to the first Recombinant DNA Advisory Committee and was heavily involved in writing the initial guidelines on recombinant DNA research for the National Institutes of Health.

While the great debate about recombinant DNA was proceeding on the national and international stages, the scientific productivity of the Falkow lab in Seattle was reaching new heights. Antibiotic resistance studies were being expanded to include *Haemophilus influenzae* and *Neisseria gonorrhoeae* (see chapter 2 by Lynn Elwell). The ability of small, discrete elements called transposons to transfer resistance between species that normally did not exchange genetic material was discovered (see chapter 3 by Fred Heffron and Ron Gill). A simple technique for visualizing plasmids by agarose gel electrophoresis was developed (12) (previous methodologies used laborious equilibrium density ultracentrifugation gradients). Such studies led to the establishment of a new branch of clinical microbiology called molecular epidemiology. This field of study uses molecular genetic techniques to establish similarities and differences among various bacterial isolates and is described in detail by Lucy Tompkins in this volume (see chapter 5). While these studies were proceeding, fundamental insights into the genetic control of virulence factors were being generated for a variety of pathogens including ETEC, uropathogenic *E. coli*, *N. gonorrhoeae*, *Bordetella pertussis*, *Yersinia enterocolitica*, *Yersinia pestis*, *Chlamydia trachomatis*, and *Vibrio cholerae*. The ultimate blend of molecular genetics and clinical microbiology, however, was to be found in the development of diagnostic DNA probes for the identification of pathogenic bacteria directly from clinical specimens. Steve Moseley, a graduate student in Stan's laboratory, used a colony DNA hybridization technique and cloned genes encoding the heat-stable and heat-labile enterotoxins of ETEC to detect homologous sequences in *E. coli* isolated from patients with diarrhea (13). This technique revolutionized the identification of bacterial pathogens, since laborious and expensive phenotypic tests such as animal toxin assays could be replaced by a simple colony hybridization. To confirm the utility of this technique, Stan sent Steve to Bangladesh to test *E. coli* freshly isolated from patients with diarrhea. The limits of this technology were pushed even further when homologous sequences were detected directly in clinical specimens, such as feces (13) and urethral exudates (14), inoculated directly onto filter paper placed on agar plates. Thus, it was now possible to identify bacterial pathogens by detecting specific DNA sequences without the need to obtain isolated colonies. DNA probes for the detection of *N. gonorrhoeae*, *C. trachomatis*, *Yersinia* spp., *Salmonella* spp., *Shigella* spp., *B. pertussis*, *Campylobacter* spp., and other pathogens were soon developed in Stan's lab, and it became possible to identify virtually any pathogenic bacterial species by this technique.

At the height of this period of productivity, Stanley left Seattle for a sunnier climate and the chair of the Department of Medical Microbiology at Stanford University School of Medicine. He rebuilt that department by recruiting a number of outstanding young faculty. During that period, the productivity in his own lab continued unabated and two major new trends emerged. One trend was the blending of bacterial pathogenesis with cell biology as Stan and his students and postdoctoral fellows began to examine the responses of host epithelial cells to an infecting bacterium. Two examples from a single species are illustrative of this trend, which is now a standard approach for studying bacterial pathogenesis. Ralph Isberg, a postdoctoral fellow with Stan, showed that a 103-kDa outer membrane protein of *Yersinia pseudotuberculosis,* called invasin, could confer upon *E. coli* K-12 the ability to enter epithelial cells (8) (see chapter 15 by Dorothy Pierson). Subsequent work demonstrated that binding of invasin to the host cell receptor (a subset of β_1 integrins) is sufficient to trigger the phagocytic response of the host cell. *Y. pseudotuberculosis* also produces an excreted protein, YopH, which is a phospho-tyrosine phosphatase. Jim Bliska, another postdoctoral fellow in Stan's lab, showed that the interaction of *Y. pseudotuberculosis* expressing YopH and macro-phages results in dephosphorylation of several macrophage proteins (2) that are not affected by a *yopH* mutant strain (see chapter 24 by Jim Bliska). The response of the macrophage to this bacterial activity is to *not* phagocytize adherent *Y. pseudotuberculosis.* The second major trend to emerge was the coordinate regula-tion of multiple virulence factors in a pathogenic species by a common regulatory factor. Alison Weiss, a graduate student who had moved from Seattle to Stanford with Stan, constructed a series of transposon mutants of *B. pertussis* defective in different virulence factors. Although the majority of the mutants were deficient in only one virulence factor, such as pertussis toxin or the filamentous hemagglutinin, some mutants contained a single insertion of Tn5 which eliminated the expression of multiple virulence factors (15). Alison and Stan proposed a model of a *trans*-acting gene product which acts as a positive regulator for multiple virulence factors and which acts in response to changes in environmental conditions such as temper-ature and ion concentration. Further work showed that this model is indeed correct (see chapter 27 by Scott Stibitz and Jeff Miller) and that similar regulatory mecha-nisms are crucial for the virulence of nearly all bacterial pathogens.

The accumulating volumes of information concerning the pathogenesis of di-verse bacterial species led Stan to crystallize his thoughts on the subject in several highly cited reviews (4, 5, 7). He formulated his "molecular Koch's postulates" to examine the potential roles of genes and their products in the pathogenesis of infection and disease in a manner similar to the way that the original Koch's postulates were formulated to identify the causal relationship between an organism and a specific disease (4). These postulates were defined as follows:

> 1) The phenotype or property under investigation should be associated with patho-genic members of a genus or pathogenic strains of a species. 2) Specific inactiva-tion of the gene(s) associated with the suspected virulence trait should lead to a measurable loss in pathogenicity or virulence. 3) Reversion or allelic replacement of the mutated gene should lead to restoration of pathogenicity (4).

In another article (5), Stan referred to his view of pathogenesis as the "Zen" of bacterial pathogenicity. All Zen sects of the Buddhist faith believe that true enlightenment cannot be achieved by the study of books or through the teaching of others. Rather, enlightenment results only from meditation, thought, and insight rather than from extrinsic information. Thus, Stan's view of bacterial pathogenicity results largely from his own insights deduced from fragmentary data rather than from the "cold analysis of results from careful experiments." One example of such enlightenment is Stan's prediction a year before the actual event that *N. gonorrhoeae* would soon acquire R factors (see chapter 2 by Lynn Elwell).

Stan's extraordinary talents as a researcher are matched by his exceptional skills as a teacher. These skills are displayed whether the audience numbers 1 or 1,000. He won outstanding teacher awards given by medical students at the University of Washington and Stanford numerous times. He has trained over 25 graduate students and 60 postdoctoral fellows in his laboratories at Georgetown, the University of Washington, and Stanford. About Stanley as a teacher, Gordon Dougan writes, "I regard him as a model on how to train and encourage young scientific workers. This characteristic combined with this unmatched scientific ability makes him untouchable and unique as a human being." Lucy Tompkins also writes of Stanley as a mentor and role model, particularly for women:

> He has actively supported the career development of all of us, showing that he is truly gender-blind. Now that I have become deeply involved in directing a program to develop and support women in medicine and medical science, I am even more cognizant of my great fortune to be one of the few women in medicine who were guided and supported by a mentor, and who have also had a role model to emulate.

Stan has been extremely supportive of the people in his lab, always generously giving them credit. In fact, in the words of one former student, "He is sort of the perennial proud father . . . you sort of know your work can't really be as remarkable as all that, but it's tempting to believe it, and this in turn is an inspiration to work harder." He creates an atmosphere of excitement and urgency. Somehow he made you feel your work was important and that you were the best. Above all, though, he made it fun, and no matter how discouraged you felt, when you went in to talk to Stan you always left filled with enthusiasm and confidence. Perhaps Stan is a great teacher because he never stopped being a student himself; he often says that it is "the fortunate professor who learns from his students."

It is difficult to define what makes someone special, in part because what is "special" depends on the observer. However, Pam Small comes close to defining this quality in Stanley Falkow:

> In considering a successful scientist it is perhaps natural to stress things like creativity and intelligence. However, these things are largely accidents of genetics, and they certainly do not guarantee success or productivity. The things about Stan that really stand out are his enthusiasm for pathogenesis, his enjoyment of life and people, and the fact that he is so annoyingly human and lets that humanness show.

Like anyone who makes a special impact on the world around them, Stanley Falkow's contributions to science are greater than the sum of his individual achievements. He had the foresight to see that molecular biology and microbial pathogenesis are two fields made for each other, and time has certainly proven him right. More recently, he saw a similar attraction between cell biology and pathogenesis. Undoubtedly, he is not the only person to have made these observations, but he had a tremendous impact for two reasons. First, Stan is the great persuader—he has extraordinary charisma and communication skills, and basically could sell just about anything. He instills a sense of excitement in those around him, and also projects that in the seminars he gives and the articles he writes. The second reason for his overwhelming impact is that Stan is a great teacher who has trained and inspired numerous undergraduates, graduate students, infectious disease fellows, and postdoctoral fellows, many of whom have gone on to make significant contributions of their own.

The importance of mentorship in the training of scientists has been noted repeatedly. Stan counts among his own mentors such giants of molecular biology as Allan Campbell and Julius Marmur. We, the graduate students, postdoctoral fellows, and sabbatical visitors who have passed through his laboratory, consider ourselves extremely fortunate to count Stan as our mentor. Therefore, to celebrate his 60th birthday, we have compiled these various chapters on bacterial pathogenesis not only to disseminate the latest information in the field but to pay tribute to Stan's insights, contributions, and training legacy in the field of microbial pathogenesis.

James B. Kaper and Virginia L. Miller

REFERENCES

1. **Anonymous.** 1986. Becton Dickinson award. *ASM News* **52:**98–99.
2. **Bliska, J. B., K. Guan, J. E. Dixon, and S. Falkow.** 1991. Tyrosine phosphate hydrolysis of host proteins by an essential *Yersinia* virulence determinant. *Proc. Natl. Acad. Sci. USA* **88:**1187–1191.
3. **de Kruif, P.** 1926. *Microbe Hunters.* Harcourt Brace and World, New York.
4. **Falkow, S.** 1988. Molecular Koch's postulates applied to microbial pathogenicity. *Rev. Infect. Dis.* **10:**S274–S276.
5. **Falkow, S.** 1990. The "Zen" of bacterial pathogenicity, p. 3–9. *In* B. H. Iglewski and V. L. Clark (ed.), *Molecular Basis of Bacterial Pathogenesis.* Academic Press, Inc., San Diego.
6. **Falkow, S., R. V. Citarella, J. A. Wohlhieter, and T. Watanabe.** 1966. The molecular nature of R-factors. *J. Mol. Biol.* **17:**110–116.
7. **Finlay, B. B., and S. Falkow.** 1989. Common themes in microbial pathogenicity. *Microbiol. Rev.* **53:**210–230.
8. **Isberg, R. R., and S. Falkow.** 1985. A single genetic locus encoded by *Yersinia pseudotuberculosis* permits invasion of cultured animal cells by *Escherichia coli* K12. *Nature* (London) **317:**262–264.
9. **Lear, J.** 1978. *Recombinant DNA—The Untold Story,* p. 59–60. Crown Publishers, Inc., New York.
10. **Marmur, J., S. Falkow, and M. Mandel.** 1963. New approaches to bacterial taxonomy. *Annu. Rev. Microbiol.* **17:**329.
11. **Marmur, J., R. Rownd, S. Falkow, L. S. Baron, C. Schildkraut, and P. Doty.** 1961. The nature of intergenic episomal infection. *Proc. Natl. Acad. Sci. USA* **47:**972–979.
12. **Meyers, J. A., D. Sanchez, E. P. Elwell, and S. Falkow.** 1976. Simple agarose gel electrophoretic method for the identification and characterization of plasmid deoxyribonucleic acid. *J. Bacteriol.* **127:**1529–1537.
13. **Moseley, S. L., I. Huq, A. R. M. A. Alim, M. So, M. Samadpour-Motalebi, and S. Falkow.** 1980. Detection of enterotoxigenic *Escherichia coli* by DNA colony hybridization. *J. Infect. Dis.* **142:**892–898.

14. **Totten, P. A., K. K. Holmes, H. H. Handsfield, J. S. Knapp, P. L. Perine, and S. Falkow.** 1983. DNA hybridization technique for the detection of *Neisseria gonorrhoeae* in men with urethritis. *J. Infect. Dis.* **148:**462–471.

15. **Weiss, A. A., and S. Falkow.** 1984. Genetic analysis of phase change in *Bordetella pertussis. Infect. Immun.* **43:**263–269.

A Look through the Retrospectoscope

I have been asked to provide an article about the evolution of my thoughts concerning bacterial pathogenesis. The retrospectoscope is a marvelous instrument, particularly if it is supplied with rose-colored lenses. I cannot pretend to reconstruct objectively the events that affected my career. I can provide an autobiographical view about how my thinking changed and, more importantly, the people who affected those changes.

I will not bore the reader with facts about my early life except to say that I was born in Albany, New York, on 24 January 1934. Suffice it to say, my life was influenced by my family's very limited financial resources, my Orthodox Jewish upbringing, the Depression, and World War II. The latter two effects are quite standard fare for those of my age. A good deal of Yiddish was spoken at home. The more outrageous phrases still pepper my conversation to this day. Neither of my parents was highly educated. My mother always felt superior since she had completed the sixth grade before starting work, while my father, who sold women's shoes, never completed grammar school. Consequently, there was no family profession, nor was there much in the way of great expectations for me to pursue a professional career of any kind.

My family moved to Newport, Rhode Island, in 1942, and my father was promptly drafted into the U.S. Army. This led to citizenship for my father, who had emigrated from Kiev in the Ukraine. Also, a family crisis resulted since our meager financial reserves were stressed even further. My mother worked selling women's corsets, my sister, Jean, worked part-time in a variety of jobs, and I secured a paper route. This is only important in the sense that I was alone much of the time, and I evolved an extensive fantasy world for myself in books and the radio serials of the times. Although I was attracted to science, particularly astronomy, it is an understatement to say that I was not a particularly good student. When I was about 11 years old, I happened to see a book in the library called *Microbe Hunters* by Paul de Kruif. Why I selected this volume, I do not know. After reading it, there was never another doubt in my mind that I wanted to be a bacteriologist—more precisely, a medical bacteriologist. Psychiatrists might say that I became fixated at an early age. Fine. In subsequent years, I have met many people who were also affected by that book. Most of them were drawn to study medicine. I was enchanted with the stories about Leeuwenhoek, Pasteur, and Koch, and I fantasized about a life of laboratory research.

Because of my fascination with the stories in *Microbe Hunters,* I longed to own a microscope. I worked out a deal with a local toy store owner to barter my work for a genuine Gilbert Hall of Science microscope. Of course, this instrument, as dear to me in my memory as the finest microscope available today, was turned on all sorts of things. I spent hours in a microscopic world seeing my first bacteria in soured milk and marvelous animals like rotifers in the "hay infusions" I had prepared with pond water. It is probably not an accident that 50 years later my personal research efforts involve microscopy.

GOING TO COLLEGE

This visual excitement and intellectual enthusiasm, alas, was not evident in my schoolwork. In my junior year at Rogers High School in Newport, Rhode Island, my adviser counseled me to "save myself and my parent's money and set my sights on a career in the military." Faced with this alternative and given that the Korean "conflict" would be my destination should I embark on a military career, I got all A's in my senior year of high school. In the fall of 1951, at the age of 17, I was admitted to the University of Maine as a bacteriology major in the College of Agriculture. Years later, I found out that *all* out-of-state students were accepted by the University of Maine during those years; the state needed the tuition monies. As good luck would have it, I was one of only two entering bacteriology majors in the class that year. The other was Peter Pattee, now a professor of microbiology at Iowa State University. Peter and I were nurtured by a group of professors who loved to teach and who cared about us. Frank Dalton, Charles Buck, and the chairman, Elmer Reeve Hitchner, together with Fred Radke, a biochemist, and Marvin Meyer, a parasitologist, introduced me to science and the laboratory. They were the first to show me the importance of teaching and let me in on the secret that the teacher also learns from the student.

As my first year in college came to an end, I began to think about summer employment. I wrote to the Newport Hospital to ask if there were any jobs available in the laboratory. I explained that I was a bacteriology major and added, dramatically, that I would do any menial task if they would only let me work there. To my considerable amazement and delight, I received a letter from the pathologist, William Freeman, stating that they would indeed take me on as a summer volunteer. I would be taught aspects of medical technology, including bacteriology and serology. I was also expected to act as an assistant during post-mortem examinations. They couldn't pay me, and I was expected to provide my own "whites." It was very exciting except, perhaps, the autopsy part. Early in June 1952, I found myself sitting across from a middle-aged woman, Alice Schaeffer Sauzette (an author of the Schaeffer and Fulton spore stain), who gravely held up a dish of an attractive red gel and said, "This is a blood plate." Thus, my apprenticeship in bacteriology began. Each day at 7:00 a.m. I went to the wards to draw preoperative blood samples. By 8:00 a.m. I was at Alice's side as she began to "read the plates." I learned to recognize the common bacteria, indigenous and otherwise, that could be isolated from humans. My passion became Gram staining bacterial colonies and doing direct smears on every patient excretion, secretion, and other effluvia within my grasp. These were and are the roots of my thinking about bacterial pathogenicity. Even then it was clear that "pathogens" were not always associated with clinical disease. Like all bacteriologists of that era, the identification of enteric bacteria was a challenge because they were biochemically and serologically so alike. Why did *Salmonella* organisms cause disease, while the seemingly closely related *Citrobacter* organisms did not? It was clear even in those early days that *Escherichia coli* and *Shigella* spp. were closely related. Why was one pathogenic and the other not? That summer (the experience of which was to be repeated on every vacation period and summer holiday during my college days) was the foundation of experience that still serves me to this day. It is why I choose to work with so many different pathogens. Quite simply, during those clinical days and later for the 2 years as a full-time clinical microbiologist, I worked with most pathogenic bacterial species mentioned in contemporary textbooks. Much of my life then was immersed in bacterial infection from the bedside to the laboratory and even sometimes to the autopsy table. It is easier to understand better the biology of the host-parasite relationship if one has seen its consequences. To the distress of my family and many friends, the thought of becoming a physician was never a serious option for me; rather, the idea of understanding how microbes worked was the driving force of my curiosity.

Back at Maine, under Dr. Hitchner's guidance, I also began to learn about the elements of research. While he was clearly pleased with my progress in the real world of medical bacteriology, he made it absolutely clear to me that I was, at best, a novice scientist. I learned from him that there is a difference between technical skill and discovery science. He taught me about controls and about keeping careful records, and he taught me to study carefully the experiments that seemed to fail because they often proved to be the most informative. At Maine, a bacteriology major was required to pursue an independent research project and to write a senior thesis as a prerequisite for graduation. Mine concerned the L forms of *Proteus* spp. and formed the basis of my first publication in 1956. Hitchner insisted that I learn as much as possible about the science of bacteriology and the biology of bacteria. Hence, I took courses in soil and dairy bacteriology, as well as in bacterial physiology. To this day, I am grateful to Hitchner for introducing me to soil bacteriology (his personal love), for it introduced me to the interdynamics of microbial populations and the interactions between microbes and the interactions of microbes with more complex coresident living forms.

GRADUATE SCHOOL, ILLNESS, GRADUATE SCHOOL

In my last year at Maine, I applied to several graduate programs in bacteriology and was offered a teaching assistantship at the University of Michigan. I began there in the fall of 1955. Allan Campbell had just joined the faculty as an assistant professor. My assignment was to assist him in the general bacteriology course. Allan gave me a copy of the newly published book *The Microbial World* by Stanier, Douderoff, and Adelberg. It was a revelation to me. With Allan's guidance, my mind began to slowly absorb this ''new'' form of bacteriology. This lasted only a few months, however. I became seriously ill with infectious mononucleosis. I was incapable of continuing my graduate work. Nevertheless, the short period of time with Allan introduced me to the idea that the fundamental basis of bacterial pathogenicity might be found at the genetic and physiological levels.

The physical illness caused by the viral illness was one matter, but I also suffered from emotional difficulties. Coincident with the fever, swollen lymph nodes, and clinically apparent hepatitis came emotional turmoil—anxiety and panic attacks. I feared crowded places. As strange as it seems to people who know me now, I was too fearful to fly and it was virtually impossible for me to talk to a group of people. These symptoms stayed with me at some level for the next few decades. With the help of understanding friends and with professional guidance, I have learned to sublimate many of these fears and symptoms.

Because of my illness, I returned to Newport and the relative security of the clinical laboratory. For the next year my life became immersed in the technical aspects of medical technology and I earned certification as a medical technologist (American Society of Clinical Pathology). My education was extended to the technical and theoretical clinical aspects of hematology, blood chemistry, serology, and, of course, more medical bacteriology. I continued to perform as a "diener" in the autopsy room. The pathologists helped me learn morbid pathology as well as normal histology. I tried to continue with research, as fundamental as it might be. My interest in *Proteus* and *Providencia* spp. as an undergraduate led me to develop a tryptophan deaminase test for their rapid identification in clinical isolates. Because I identified mistakenly a *Citrobacter* sp. as a *Salmonella* sp., my ego drove me to develop a simplified lysine decarboxylase test to differentiate one from the other. Both of these research efforts were published in the *American Journal of Clinical Pathology*. The important lesson learned from this experience was that it is possible to perform satisfactory and satisfying research with the most simple of experimental tools.

Dr. C. A. Stuart, a noted expert on enteric bacteria, worked at Brown University, about an hour's drive from Newport. We met and he set up a time each week to discuss with me different aspects of bacteriology and immunology. He also arranged for me to take a tutorial course with Herman Chase, a professor of genetics. Stuart was a remarkable microbiologist and served as the president of the Society for American Bacteriologists (the forerunner of today's American Society for Microbiology [ASM]). The microorganism he called 2991 was later called *Providencia stuartii* in his honor. Chase, who was called by his students "Uncle Herman," assigned me readings in the book *The Chemical Basis of Heredity,* edited by William McElroy and Bentley Glass. It was my first introduction to molecular genetics. I still remember the intellectual stimulation I received when I read the paper by Seymour Benzer in that volume; it can still be read with profit. Each week I met with Chase and we studied the papers of Watson and Crick, Lederberg, and the school of microbial genetics emerging at the Pasteur Institute. Meanwhile, Stuart took me into the intricacies of the family *Enterobacteriaceae.* In the spring of 1957, after meeting with Stuart and Chase for about 5 months, J. Walter Wilson, the chairman of biology, offered me a marvelous opportunity. Was I interested in becoming a graduate student in biology under the newly created predoctoral fellowship program of the National Cancer Institute? They offered me a tax-free stipend of $2,500 with full tuition paid. My salary at Newport Hospital was then $4,700 before taxes. It was too good to be true, and I accepted in a heartbeat. In retrospect, my good fortune was part of the response to the Russian *Sputnik,* which accelerated the U.S. commitment to basic science. Would that the same commitment to nontargeted research were with us today.

BROWN UNIVERSITY

I became a full-time graduate student in the Biology Department at Brown University in the fall of 1957. Of course, I wanted to concentrate on microbiology, but the department had other ideas. The requirements for a Ph.D. demanded courses in genetics, histology, embryology, cell biology, and physiology (really biochemistry). The histology, embryology, and cell biology courses were year-long, with four 1-hour lectures each week. In each course, it was expected that the student would do a small research project to be reported at the end of the year. I examined the histochemistry of *Tetrahymena* spp., did my initial bacterial conjugation experiments for genetics, and was introduced to electron microscopy by Elizabeth LeDuc. Mac Edds, a marvelous teacher and research embryologist, coaxed me into working with *Dictyostelium* spp. The *Dictyostelium* experiments were simply feeding these predatory creatures different bacteria as a food source. I felt somewhat like a traitor. In any event, the amoeba form of *Dictyostelium* could clear a petri dish of bacterial growth with extraordinary dispatch prior to aggregating into "slugs." However, I was, as one might guess, interested in the bacterial survivors of the phagocytic carnage. The answer played a role in my thinking throughout the rest of my career. The enteric bacterial survivors, *E. coli* and *Salmonella* spp., as well as *Pseudomonas aeruginosa,* were mucoid variants. I had seen remarkably similar strains isolated from the sputum of cystic fibrosis patients. Cystic fibrosis patients had always suffered from serious lung infections. Prior to the antibiotic era, they were largely caused by staphylococci (reportedly quite mucoid variants), but with the advent of antimicrobial therapy in the 1950s, one increasingly saw gram-negative infections, particularly with *E. coli* and later with *P. aeruginosa.* These student experiments with *Dictyostelium* spp., therefore, led me to begin to understand that the bacterial genetic apparatus selected to resist phagocytosis by mammalian phagocytes had its roots much, much earlier in evolution in the early battles against natural microbial predators. It is less difficult for me to think about the determinants of pathogenicity like that of botulinum toxin when I consider that it was likely evolved to act on some microscopic predatory bacteriophagic creature like nematodes in the soil rather than as a deliberate strategy to poison some poor soul who ate a spoiled can of food. Yet, the life of microbes outside of the human disease state gets relatively little space in textbooks, or in conversation among bacteriologists for that matter. I tease my friends who work on *Pseudomonas* spp. that it is not a "proper" pathogen because it causes disease only in compromised human hosts. Yet, *P. aeruginosa* comes armed with an impressive array of toxins and other virulence attributes and it regulates them exquisitely. What is the real target of these virulence determinants in the natural aquatic environment inhabited by *P. aeruginosa?*

My graduate research initially focused on the *Proteus* and *Providencia* group. Stuart had isolated from a patient an unusual *Proteus* isolate that, in his words, turned into a *Providencia* isolate. He thought that this was a natural genetic change, but the original "*Proteus*" isolate could not be stabilized and was fully "converted" into a *Providencia* isolate in vitro. "Doc" Stuart loved to speculate

on the evolution of the enteric bacteria on the basis of his serological and biochemical studies, but in the context of the times, most bacteriologists knew little about
microbial genetics. Sal Luria rightly characterized microbiology as the last strong-
hold of Lamarckism. Now, almost 40 years later, my suspicion is that Stuart had
isolated a *Providencia* strain that possessed a urease plasmid and that this plasmid
was unstable and was spontaneously lost in subculture. However, in 1958 there
was no clear view of bacterial extrachromosomal elements and microbial genetics
was still in its infancy. I tried all kinds of genetic manipulations involving UV
mutagenesis, the search for temperate bacteriophage, and DNA transformation.
I learned a lot of techniques, I read a lot of papers, and I was happily ingesting
information in the classroom about a number of different new disciplines. My
favorites were genetics, evolution, and cell biology.

Then, as now, graduate students bonded together and talked passionately late
into the night of science, experiments to do, and dreams and hopes. My memories
of these days in the laboratory still bring a smile and a warm feeling. I was Stuart's
last student; he was retiring because he had serious heart disease. Yet, he worked
every day at the bench and his rich, booming voice (he was virtually deaf and
didn't know he was shouting) could be heard throughout the building. I worked
at a bench next to him. He watched my every move and noted every error with
ear-splitting relish. We would talk as we worked, although it was more like I would
listen while he talked. (I am accused by my own students today of the same thing.)
He had a game that he played with me. "Stan, boy," he would yell, "did you
ever know Karl Landsteiner?" Of course I knew who Landsteiner was, but I
couldn't know him because he had died years earlier. He would "hrumph" loudly
and then go on to tell marvelous stories about the greats of bacteriology and
immunology that he had known. He peppered these stories with all kinds of spicy
gossip. Throughout all of this, he also led me through his thoughts about bacteria
as evolving populations and what he had learned about their serological properties.
He knew that I wanted to study their genetics. He would say with a touch of
sadness in his voice that he wished he could learn genetics and biochemistry but
that it was too late for him. "You do it for me," he ordered.

That year Seymour Lederberg joined the Brown faculty fresh from his postdoctoral work with Luria at the Massachusetts Institute of Technology (MIT). I
was assigned to assist Seymour in teaching a course on phages. At that time there
were relatively few opportunities to learn about phages outside of the famous
Cold Spring Harbor course. During those days, the course was more likely to be
populated by mature scientists than by graduate students and postdoctoral students. This was Seymour's first job teaching. Each week we did the class experiment before it was given to the students. In essence, Seymour was my personal
instructor. It was my good fortune. Not unexpectedly, I was awed by the precise
nature of the experiments. For the first time, I truly realized what the term "genetic selection" meant. The lessons learned from Herman Chase about the evolution of populations began to take on new meaning for me. Stuart and Lederberg
were my wellsprings of information, but oh, what different information! Lederberg asked me to accompany him to several evening sessions in Boston at MIT
where I met and could hear wonderful scientists like Luria, Magasanik, and Cy

Levinthal. Meanwhile, I drove Stuart to Boston to meet with medical microbiologists and with epidemiologists from the state laboratories to discuss the latest outbreaks of this or that infectious disease. Stuart and Lederberg were both dedicated scientists and teachers, but they came from different worlds. They did not really understand each other, but far from being caught in the middle of their philosophical differences, I was the beneficiary of their focus as teachers.

Having completed sufficient experiments on *Proteus* spp. to earn a master's degree, Stuart permitted me to change my focus to another area of the enteric organisms. I had always been fascinated by the typhoid bacillus. When Lou Baron published a paper on the conjugative transfer of genes from *E. coli* to *Salmonella typhi,* I was interested in following his lead. Shortly thereafter, Norton Zinder published a paper describing genetic recombination between an *E. coli* Hfr strain and *Salmonella typhimurium.* My request for strains was honored by both investigators, and I began examining the genetic organization of *Salmonella* spp. in the fall of 1958. I wanted to identify the genes associated with *Salmonella* pathogenicity. I guess I still do.

The Hfr $\times$ *S. typhi* crosses went well. It is difficult to convey the delight we felt in 1959 when the first Lac$^+$ *Salmonella* colonies appeared on the genetic selection plates. Stuart was possibly more excited than I was. When some of the recombinants showed inheritance of other sugar markers, he was beside himself with enthusiasm. "By God," he yelled, "they're turning into *E. coli*! Get some antisera and see if the antigens are changing too." They were. Baron, with Charlie Brinton, was getting ready to publish information indicating that common pili were inherited in the crosses, but at the time Stuart and I didn't know it. When the typhoid recombinants agglutinated in *E. coli* antisera, Stuart was ecstatic and he dropped his own experiments to devote his time to analyzing the recombinants. As I write this pleasant memory, I understand how he may have felt, for I have shared the same enjoyment from my students' work.

I had been at Brown only for about 18 months when there was a change in my status. Stuart was officially retired, and as an emeritus professor he could not mentor a graduate student. It was also quite clear that most of the faculty felt Stuart could not possibly direct my research on bacterial genetics and that I should be working with Lederberg. Seymour thought the work was all right, but he was really interested in restriction-modification of phages. Sal Luria suggested a compromise—why not work on *Shigella* spp. with Seymour or even with him. Stuart did not want me to "fall into the clutches" of anyone who was not going to permit me to work on the genetics of *Salmonella* pathogenicity. "You have to go somewhere else to complete your research," he said. "I fixed it so you can graduate from here and I'll still be your professor of record, but you need to go." How Stuart "fixed it" I'll never know. I called Norton Zinder at Rockefeller University. While he was initially interested, he contacted me a short time later to say that he was giving up the *Salmonella* work because he had become thoroughly absorbed in the F-specific RNA phages he had discovered. One day, a small, restless man was in the laboratory talking to Doc when I arrived. His opening greeting was, "*Nu!* Doctor? You've done a nice piece of preliminary work in a short period of time. How would you like to come to Walter Reed next year to work?" It was

Sam Formal, who had been greatly influenced by Stuart during his undergraduate years at Brown. He and Doc had somehow worked out a way for me to finish my research at the Walter Reed Army Institute of Research working under Lou Baron's direction. Baron and I only met several months later at an ASM meeting. Thus, I left Brown in the fall of 1960. I had been in graduate school only 2.5 years. I was actually hired as a GS-7 bacteriologist in the federal civil service, even though I was still a registered graduate student at Brown.

WALTER REED ARMY INSTITUTE OF RESEARCH

My years at Walter Reed were crucial in forming my ideas about bacterial pathogenicity. I worked closely with Lou Baron. He, like Allan Campbell, had been a student of Sol Spiegelman at the University of Illinois but had come into bacterial genetics through the course at Cold Spring Harbor and Josh Lederberg's influence. Baron would often say that if an experiment involved more than four plates and two pipettes, it was probably overdesigned. Baron was a master experimentalist. He did not agonize about all the reasons that an experiment would not work; he did not wait until he had everything ready for the perfect well-controlled experiment. Baron did simple genetic experiments that asked a single, relatively straightforward question. He depended on genetic selection and the resolving power of recombination to provide the answer.

Sam Formal was down the hall from me, and Richard Finkelstein was directly across the hall. Thus, there was a smorgasbord of enteric pathogens around. My first few months at Walter Reed took an unexpected turn, however. Baron had received a Lac$^+$ strain of *S. typhi* that had been isolated from the blood of a febrile patient. He had discovered that it transferred the Lac trait to other bacterial strains at a high frequency. A year earlier Ed Adelberg had reported the isolation of F-*lac* and the concept of episomes (the term "plasmid" was not yet in general use) had just become fixed. In the context of the time, the operon was just being described, and Jacque Monod had defined molecular biology as anything published in the new *Journal of Molecular Biology*. It appeared that the Lac trait of this strain, ST-2, was self-transmissible, a naturally occurring F-*lac*-like genetic element. Lou graciously offered me the chance of characterizing it in more detail.

Baron had made another key observation: the Lac element could be transferred to *Serratia marcescens*. In 1960, very few bacteriology laboratories knew how to isolate high-quality DNA. The manipulation of DNA molecules was virtually unheard of. However, the idea that bacterial species might have dramatic differences in their DNA base compositions was just becoming appreciated. One tool for investigating molecules of different base compositions was cesium chloride density gradient centrifugation, which had been described by Meselson and Stahl in their famous replication experiments. I knew enough then to understand that *S. typhi* had a G + C content of 50%, while *S. marcescens* had a G + C content of 56%. DNA molecules of *Serratia* and *Salmonella* should not overlap. Might it be possible that the extrachromosomal Lac element of ST-2 was actually visible as a separate band in a density gradient by using a model E ultracentrifuge? It is old-fashioned now, but it was heady stuff 33 years ago. I called Julius Marmur, then at Brandeis University, to ask if it was possible to do this experiment, and he agreed that it would be worth a try. I was given a President's Fellowship from ASM to go to Brandeis and work with Julius. The experiment worked. The *Serratia* strain with the transmissible Lac element had a perceptible extra band of DNA the same density as the band from *S. typhi*. This was an exciting experimental finding, but its greater influence was to get me thinking more about the role of extrachromosomal elements in the evolution of microorganisms. My experience with Julius and in his laboratory during little more than a 6-week period provided me with the expertise necessary to be able to isolate DNA from virtually any bacterial species—trivial now, but of extraordinary value then. Julius also introduced me to the idea of molecular evolution. I spent considerable time making melting curves of DNAs from various pathogenic species. Julius, together with Manley Mandel, was compiling a volume of information on the variations in DNA base composition among bacterial families. Marmur, Mandel, and I wrote a review article on the subject of genetic and molecular taxonomy. This experience initiated my serious thinking in molecular terms about the biology of enteric bacteria in particular and pathogens in general.

In the following years at Walter Reed, my scientific horizons continued to be expanded by being introduced to new experimental methods and ways of scientific thinking. So many people were willing to help me and other young scientists. Brian McCarthy and Ellis Bolton at the neighboring Carnegie Institute of Washington introduced me to DNA-DNA hybridization using DNA agar. This initiated a steady interaction with the people at Carnegie over the next decade. Thanks to people like Roy Britten, Dean Cowie, Don Brenner, David Kohne, and Bill Hoyer, I learned a good deal about nucleic acids and the evolutionary clues that could be found in their careful study. The National Institutes of Health (NIH) weren't far away, and I particularly remember attending the "Lambda Lunches," where I met and was helped enormously by people like Gary Felsenfeld, Schalom Hirschman, Bruce Ames, Gordon Tompkins, and Marty Gellert, as well as my fellow beginners John Roth and Jerry Fink. It was, as Luria called it, the "golden age of molecular biology," and every day seemed to bring a new revelation and a new way to study the genetic and molecular bases of living things, particularly phages and bacteria in those days, but not very many people worked on pathogens.

In my laboratory at Walter Reed, I continued with my efforts to examine the genetic organization of the typhoid bacillus. I was stymied by the lack of a suitable animal model to study *S. typhi* pathogenicity. Consequently, I began to look for alternative *Salmonella* spp. to study. *Salmonella* Hfr strains were just becoming available, but the restriction-modification in bacterial crosses outside of *E. coli* K-12 were horrendous. It was still the early years of microbial genetics. We had fine animal models for *S. typhimurium,* but the pathogenic strains did not mate at all like *S. typhi* with *E. coli* donors. The best *S. typhimurium* recipients were poor pathogens.

Sam Formal inveigled me to begin work on the genetics of *Shigella* spp. I started a series of experiments crossing *E. coli* genes into *Shigella flexneri.* Compared with *Salmonella* spp., it was easy work. Sam Formal and Herman Schneider then began to test these hybrid strains in an opium-starved "virulence" guinea pig model. We identified a number of chromosomal sites that seemed to be essential for *Shigella* virulence. To my continuing great surprise, the chromosomal regions detected by this quite indirect genetic approach have proved to be correct. Of course, in 1962 we didn't have a clue that *Shigella* spp. possessed an essential virulence plasmid. (Ah, the retrospectoscope—we could never isolate a virulent Hfr *Shigella* strain, likely because F and the *Shigella* resident virulence plasmid were incompatible.) Nevertheless, these experiments were very important to me because they established the idea that one could treat virulence as a phenotype. Yet, it was obvious that it was a multifactorial phenotype. I continued to be a collaborator with Sam and his associates on the work with *Shigella* spp. Perhaps, more importantly, Sam always served as a source of wise counsel throughout my career. I learned from Sam to take a multipronged approach to the study of pathogenicity. His work to understand how a pathogen worked was a combination of basic science, experimental pathology, and immunology. However, at that time most of my efforts were becoming focused on plasmids, particularly R factors, rather than pathogenicity.

My focus on plasmids was influenced by advice from well-meaning elders and the natural progression of my interests. It was also influenced by my increasing frustration in trying to find a practical way to look at the genetics of *Salmonella* pathogenicity. In contrast, the work on the transmissible "episome" of *S. typhi* went extremely well. My efforts with John Wohlheiter and Ron Citarella in the Baron laboratory focused on transferring this Lac element, as well as the classical F factor, into *Proteus* spp. *Proteus* spp. had the lowest G + C content of any enteric organism, and it seemed likely that it would be much easier to look at the extrachromosomal DNA of episomes if they could be transferred into *Proteus mirabilis*. A turning point for me came when I was invited to give a seminar during the summer course at Cold Spring Harbor. I presented the work with the Lac element, including its crude molecular characterization and F-*lac* in *Serratia* spp. That evening I was taken aside by several distinguished microbial geneticists who told me they thought the work was very interesting, but as one very succinctly put it, "Why typhoid? No one cares about typhoid anymore. If you want to study episomes, why not just study F? Your work will mean more."

Following the publication of the work demonstrating that extrachromosomal DNA could be seen in *Serratia* spp., I began to receive letters from people who wanted to look at different naturally occurring plasmids. It seemed likely that R factors would be very much like F-*lac* and the *S. typhi* Lac element except that they encoded for multiple antibiotic resistance. These were the days before electronic mail and before postdoctorals "networked" around the world by telephone. Investigators wrote letters to one another. I began a lively correspondence with Tsutomu Watanabe in Japan and Naomi Datta in England concerning the R factors that they had isolated from patient material. As this was evolving, it became clear that one could indeed visualize extrachromosomal DNA in *Proteus* spp. as very clear bands in a model E ultracentrifuge or that extrachromosal DNA could be isolated in a preparative ultracentrifuge. Thus, in 1964 I began to be seduced away from the direct study of pathogenicity in enteric bacteria to focus more on transferable antibiotic resistance in pathogenic enteric species. Increasingly, I found that I was able to employ the genetic and molecular tools I was learning to use to productively examine the nature of transmissible drug resistance. I spent less time on pathogenicity per se; it seemed to me that we did not have the genetic tools to dissect this multifactorial property properly, nor were the virulence assays either practical or revealing. The available tools were just adequate to look at a relatively small chromosome like that of a phage or an episome, but they were not sufficient to dissect the chromosome of anything other than *E. coli* K-12, and even this had limitations. Of course, in subsequent years I did return to the study of pathogenic mechanisms, and that is documented elsewhere in this volume.

In 1966, I began the transition from Walter Reed to an academic position at Georgetown University. I had arrived at a time in my life when I wanted my own laboratory and to have the opportunity to teach. I was a 31-year-old associate professor of microbiology. When I began as a professor, I had never mentored anyone, and, indeed, I was not that much older than my first students. I have tried to document here just how much of my professional development was molded by my teachers and advisers. My teachers at Maine, at Newport Hospital, at Brown, and at Walter Reed had been my advocates. They told me in so many ways that they cared about me and my development. Now I understand that my development was part of their own, just as the development of my students and coworkers has been part of the evolution of my own career. Over the years, I simply tried to help my students the way I had myself been helped all along the way.

I am told that the other chapters in this book are designed to tell the story of the evolution of research in my laboratory over the past 28 years. I have no idea what will be said. However, there are several other events that occurred in other arenas and that have had a most profound effect on my thinking, and which, in turn, affected the people working with me, whether they knew it or not.

At Georgetown, all of the laboratory effort was directed initially toward understanding the nature of R-factor transfer and replication. In September 1968, an important turning point in my thinking took place at the CIBA Foundation Symposium on Bacteriophage and Episomes in London. This was my first trip overseas. As I walked toward customs at Heathrow Airport, I heard the unmistakable voice of Allan Campbell behind me saying, "I thought you were going to work on the genetics of pathogenicity?" His first words to me since we parted in Michigan. "How did you get into the episome business?" This was a wonderful meeting for me in several ways. I had the opportunity to meet a number of scientists face to face, and I especially admired H. Williams Smith. "Willie" Smith, in my view, was the founding father of microbial pathogenesis. His papers on the role of adherence plasmids encoding K88 and K99, as well as the *E. coli* enterotoxins, are still marvelous in their simplicity of design and the scientific insight that they provided. Willie was a wonderful Welshman who came to the point. "If I had your genetics training," he said, "I'd work out the molecular basis of Ent, K88, and K99, but I'm just a simple vet with manure on my boots." Willie didn't want to collaborate; he simply wanted someone to do it and to let him know the data. He was always extraordinarily generous in supplying me with strains. It was not uncommon to receive a letter from Willie with a new piece of data and a strain that he thought we might like to look at. I had been unaware of his work before this meeting, but I instantly became his admirer. We enjoyed a lively, slanderous correspondence until his death several years ago.

This meeting prompted the extension of our studies of R plasmids (the term "plasmid" was now coming into play because of the CIBA meeting) to try to unfold the mechanism of plasmid-mediated pathogenicity. We made good progress, and it was satisfying to learn that the Ent plasmids were F-like, but we could not distinguish the genes for the two kinds of enterotoxins, the heat-labile or heat-stable enterotoxin. The Ent and the K plasmids fell right into line with the phylogeny of the plasmid world that we were pursuing in collaboration with Naomi Datta and her coworker Bob Hedges. The different incompatibility groups of plasmids were quite distinct at the DNA level, yet they carried seemingly common resistance determinants and, in some cases, enterotoxins, adhesins, or colicins, etc. The experimental methods at the time were still not sensitive enough to examine these carried genes in detail. I suppose it is difficult now for people to envision a world devoid of agarose gels, Southern blots, cloning, and sequencing. Our plasmid world was restricted to isolation by preparative cesium chloride density gradient centrifugation. Molecular mass was determined by sedimentation in sucrose gradients or by measuring the size of open circular DNA in the electron microscope. DNA-DNA hybridization was quantitative solution hybridization. Most of the work revolved around relatively large plasmids (>60 MDa) ("kilobases" was a rarely used term at the time). Thus, it was becoming painfully clear that the same technical limitations that limited our progress with understanding the genetic basis of *Salmonella* pathogenicity were limiting our precise understanding of plasmid-mediated pathogenicity. One could surely do genetic experiments with mutants that no longer expressed one or another pathogenic phenotype, but the precise nature of the genetic lesion, let alone the number of genes, was unknown. In most cases, the precise product of the genes was ill defined. We were learning, however, that the comparison of different isolates that varied in their pathogenicities was a bad idea. It was much more instructive to look at mutant variants of the same bacterial strain or the same plasmid. This is a given now, but it is important to see that many of the existing dogmas of bacterial pathogenicity at the time were based on experiments performed with different isolates of the same species that varied in their pathogenic potentials.

UNIVERSITY OF WASHINGTON

In June 1972, I left Georgetown University and took a position as a professor of microbiology and immunology at the University of Washington in Seattle. The majority of the research in the laboratory was directed toward R-plasmid replication, but I had just received a small grant from NIH to work on the *E. coli* enterotoxins on the basis of our observation that the human strains associated with traveler's diarrhea had plasmids virtually indistinguishable from those from piglets defined by Will Smith.

The move to Seattle was advantageous for my laboratory. First of all, there was a larger concentration of bacteriologists, particularly Helen Whiteley, Neal Groman, and Gene Nester, who used microbial genetics in their work. John Sherris, the chairman, was very supportive. The graduate student pool was larger than that at Georgetown. Second, since leaving Walter Reed, I had little contact with clinical infectious diseases except for a productive collaboration with the pediatric infectious diseases unit at the Washington Hospital Center in Washington, D.C., led by Sid Ross and his colleague, Waheed Kahn.

At Georgetown, I mainly taught microbial genetics to graduate students. There was little opportunity to teach medical bacteriology or to exploit my earlier clinical experience. At the University of Washington, however, John Sherris assigned me to teach most of the lectures on the clinically important bacteria to second-year medical students. Eventually, I coordinated the course. Teaching this subject each year meant rethinking the pathogenesis of infection for each microbial agent in light of the new information available both from the benches of basic scientists and from the clinical arena. Together with George Ray, Larry Cory, and Jim Plorde, we taught medical bacteriology, virology, and parasitology. We listened to each other's lectures over a 9-year period. Every year it was a new, enjoyable, and instructive experience. The lectures were integrated. For each microbe or disease syndrome, the basic biology of the microbe(s) in question was presented. Bill Kirby and Lucy Tompkins discussed the clinical features and treatment, Hjordis Foy discussed the epidemiology of the disease, and Tom Norris discussed the pathology of infection. I often hear younger faculty complain about teaching medical students; personally, I enjoy it. Each year I learn more and more new exciting insights into the pathogenesis of infectious diseases from teaching this subject. In Seattle, it was also possible to return to my roots by attending the "bacteriology plate rounds" held daily in the clinical microbiology laboratory to discuss the latest isolates from patients seen on the infectious diseases service. To this day, I very much enjoy attending infectious disease rounds. From these rounds I can gain an overview of the entire biology of the host-parasite relationship. Hence, when I arrived in Seattle, I was virtually met at the door by King Holmes and Marvin Turck, who led splendid infectious diseases units. I began a very close collaborative relationship that lasted my entire time in Seattle. That year was a turning point in another way, and once again it revolved around a scientific meeting.

The Second U.S.-Japan Symposium on Bacterial Episomes and Plasmids was held in Honolulu in November 1972. For several years, most workers in the plasmid field had been feeling frustrated and rather stuck in the same rut. Technically, there had been few advances. The genetic analysis of plasmid determinants was still primitive. The major question that plagued most people's minds concerned the mechanism(s) by which plasmids gained their accessory genes. It was clear, for example, that highly related F-like plasmids could easily carry colicins, antibiotic resistance genes (many different kinds), or enterotoxin genes. However, prior to arriving at this meeting, one could almost sense a strong wave of optimism that things were finally beginning to move ahead. The only sad note had been the news that "Tom" Watanabe, at the height of his career, had died of gastric cancer;

the meeting was dedicated to his memory. The night before the meeting began, Stanley Cohen, Herb Boyer, Charles Brinton, Ginger Brinton, and I went out in search of a late night snack and found ourselves in, of all places in Waikiki Beach, a Jewish deli. The story has been told before, and most versions are rather inaccurate. Some day I may even write my version of it. Our initial discussion consisted of the usual frenzied and almost breathless exchange of new data that occurs when scientists meet. Although our original focus was directed at understanding recombination mechanisms in R plasmids, the discussion suddenly gave way to the idea that one could link diverse fragments of DNA together into an autonomously replicating genetic element. It was a combination of the discovery of plasmid transformation and a variety of small (<12-kb) diverse plasmids encoding antibiotic resistance and of Boyer's discovery of *Eco*RI and its properties that led to this idea. Indeed, the first "recombinant" DNA experiment was the linking together of the plasmids pSC101 with RSF1010. The potential power of this experimental approach was obvious to all of us. Prior to leaving the meeting, Herb Boyer and I spent an afternoon at a Honolulu shopping center in search of the new rage, "crazy shirts." We agreed that if the experiment that he and Cohen were to perform worked, he would call and let me know. I told him that we wanted to "isolate" the fragment of DNA encoding the enterotoxin genes. Herb called me in the spring of 1973. Maggie So, then a graduate student, went to his laboratory in the fall of 1973 and we "cloned" the *E. coli* heat-stable enterotoxin gene at the beginning of 1974. From the moment I knew what has become known as recombinant DNA, it was clear to me that I was going to once more direct my attention toward the study of pathogenicity.

My initial plan was to leave the R-plasmid field as quickly as possible. "Quickly" took almost 10 years. I wrote a monograph on R plasmids as a kind of swan song—the only book I have written, and I expect the only book I will ever write. However, one could not anticipate that my talented graduate and undergraduate students, together with Naomi Datta and her coworkers, would be so extraordinarily successful in investigating plasmid-mediated ampicillin resistance gene transposition. The major mechanism of plasmid acquisition of resistance markers had finally been recognized virtually simultaneously in a half-dozen laboratories. Also, Jorge Crosa and his research associate, Linda Luttropp, spearheaded a group in the laboratory focused on R-plasmid replication. Jorge had joined me in Georgetown as a postdoctoral fellow. Neither of us realized that the political conditions in his native Argentina would keep us together for over a decade. Consequently, Jorge had a small island of relative independence in the laboratory.

During this transitional period, Crosa and I also worked with Herb Boyer and Paco Bolivar to construct new cloning vectors, the pBR series. An incidental observation made during these plasmid constructions led to the development of agarose gel analysis of plasmid DNA by my technical associate, Jane (Aldrich) Meyers. Our knowledge about gene transposition, together with the simplicity of agarose gel plasmid analysis, freed us from the tedious task of measuring plasmid molecular mass by microscopy or sucrose sedimentation. As a direct consequence, an approach that embraced clinical epidemiology and plasmid DNA analysis evolved into an experimental discipline that has become known as molecular epidemiology.

The use of agarose gels for plasmid analysis also came just as there was the spread of β-lactam resistance into the gonococcus, and this led to the synthesis of our work on Tn3 transposition with the evolution of resistance elements in pathogenic enteric bacteria, *Haemophilus influenzae,* and, subsequently, the gonococcus. It seemed almost perverse that my eagerness to study pathogenicity was being overshadowed by such provocative findings about plasmid biology. Yet, we did begin to expand our studies of virulence determinants, and the *E. coli* enterotoxin genes were cloned, as was the K88 adhesin.

Unfortunately, our cloning of the *E. coli* enterotoxin genes subsequently became part of the growing debate about the potential dangers of recombinant DNA technology. Of course, the impact of recombinant DNA technology on the scientific world is well known. The question of the potential hazards associated with recombinant DNA research continues today and initially centered around the question of whether the pathogenicities of bacteria might be modified by the insertion of foreign genes. At the Hawaii meeting, a committee had been established to consider the nomenclature of bacterial plasmids. This committee was composed of Richard Novick, Naomi Datta, Roy Curtiss, Roy Clowes, Stan Cohen, and me. We became a working party assigned to participate in a meeting held at Asilomar in Pacific Grove, California, to discuss the potential biohazards associated with recombinant DNA research and whether there should be a moratorium on recombinant DNA experiments. This was an extraordinary meeting that has been written about in considerable detail. I was appointed to the first NIH Recombinant DNA Advisory Committee, which met in San Francisco immediately following the Asilomar meeting. It was evident that scientists, politicians, and the public at large wanted guidelines to govern recombinant DNA research. It was the first time that I had to articulate what the term "pathogen" meant to me and what constituted a pathogenic determinant. The questions asked were without precedent. I recall very clearly being asked what I thought might be the pathogenic consequences of adding fruit fly DNA to *E. coli*. I really didn't know and said so, but declared with certainty that "the little sucker wouldn't fly." However, the more I reviewed the literature and pondered my own experience in the clinical laboratory and at the research bench, it became increasingly obvious that random DNA was very unlikely, indeed, to have any effect on pathogenicity. A pathogen was a highly evolved specialist and the product of long evolution. Pathogenicity was exquisitely regulated. My experience on the Recombinant DNA Advisory Committee was both fascinating and frustrating. The politics were dreadful. Yet, in retrospect, it was clearly an important time for me to think about the broader meaning of a pathogen beyond its capacity to cause disease. For the first time I began to focus more on the role of pathogenicity in the biology of the microbe than in its capacity to harm humans. These same questions came up a few years later when I served on a U.S. Food and Drug Administration panel to consider the use of antibiotics in animal feeds and their potential effects on human and animal health. I was strongly opposed to the practice of supplementing feeds with subtherapeutic levels of antimicrobial agents to stimulate animal growth. I viewed plasmids as a single interconnected gene pool and think that, for haploid bacteria, plasmids and phages are the most likely sources of bacterial diversity and specialization, including pathogenicity.

Ultimately, my laboratory in Seattle began to focus more and more on bacterial pathogens and pathogenic determinants. The laboratory's dedicated approach to bacterial pathogenicity began with the arrival of a new cadre of graduate students and postdoctoral fellows who, more or less simultaneously, began work on the virulence determinants of *Yersinia* spp., *Bordetella pertussis, Vibrio cholerae,* and the uropathogenic *E. coli* and the use of virulence genes as probes in molecular epidemiological studies. This work, beginning in about 1978, set the research tone of the laboratory for the next 7 years.

STANFORD UNIVERSITY

In 1980, I agreed to be chairman of the Department of Microbiology at Stanford University, and the laboratory made the transition in 1981. Those first years at Stanford were very difficult for me. Recruiting young faculty was a challenge, as was revamping the teaching program. These were expected and rewarding activities that for the first time took me away from the day-to-day activities of the laboratory. Finally, I concluded that the chairman's life was simply not for me. In 1984 my major focus again became my own laboratory.

Change also affected my personal life. Lucy Tompkins (who had been a graduate student in my laboratory at Georgetown University from 1967 to 1970) and I were married in December 1983. When asked about the evolution of my thoughts about pathogenicity, it is perhaps not usual to think of one's wife. However, because we have parallel careers, it is to be expected that we will share our ideas. Lucy is organized, logical, and factual; she has research grants and an independent research program, but she is also an infectious diseases physician who sees patients on a regular basis. I am disorganized, illogical, and inclined to be more intuitive than bothered by facts. Lucy does not suffer unsupported pronouncements well. She has forced me to think about the concepts of pathogenicity. Her questions and her counterpoints based on her own research experience, bolstered by the experiments of nature that she sees in the clinic, have played an enormous role in the evolution of my thinking. For example, in recent years when I have tended to paint pathogens as a misunderstood lot who really are trying to simply "make a living," Lucy has countered by simply saying, "That's nonsense and you know it. Just because most infections are asymptomatic doesn't mean that the host is not deliberately harmed as part of the successful pathogenic life cycle." And, of course, she is right. Pathogens probably always cause some degree of cellular disruption or at least sufficient disruption to muster the immune system. In fact, I have grown to believe that tissue damage may be a necessary regulatory prerequisite for pathogenic species to sense in order to achieve successful replication on or within the host.

In recent years, the focus of the laboratory has turned increasingly toward studying the cell biology of infection. The experimental findings with invasive

Yersinia organisms led to the inescapable conclusion that it would be necessary to become conversant with the tools and language of the cell biologist. This led to my spending the summer at the NIH Rocky Mountain Laboratory (RML) in Hamilton, Montana, in 1984. Claude Garon, a colleague of long standing and an electron microscopist who had helped me learn DNA microscopy, invited me to spend time in his laboratory. With help from Garon and his coworkers, Fred Hayes and Dan Corwin, I became a fledgling electron microscopist of animal cells infected by bacteria. My first photographic efforts were greeted with less than universal enthusiasm. Looking at some of my first efforts, Virginia Miller was moved to say, "Oh, Stanley, these will never do." I still do much of my laboratory's electron microscopy at RML. Thanks to the technical expertise of Nafisa Ghori, the quality of the work, if not the focus of the pictures, has become much better. Parenthetically, it should be noted that to someone interested in bacterial pathogenicity, RML is a wonderful place to work; the facility is dedicated to research on infectious agents. Claude as well as John Swanson, Bruce Chesbro, and, more recently, Harlan Caldwell lead marvelous research units of staff scientists and fellows in studies ranging from gonococcal pathogenesis, the biology of Lyme disease, and the determinants of chlamydial pathogenicity to the virology and immunopathology of scrapie. Marshall Bloom, an expert in Aleutian mink disease, is a source of wisdom for me and is my fishing buddy. I have returned to RML every summer since 1984 and visit throughout the year. The Bitterroot Valley, with its independent people and beautiful vistas, is the best place in the world for me to think about science.

I hope it is now clear that the evolution of my thoughts about pathogenicity is inextricably intermingled with my teachers, scientific colleagues, and, of course, those who worked with me in the laboratory. The reader no doubt will have noted that with very few exceptions, I have avoided naming any of my former students. This reflects nothing more than my anxiety that I might forget someone. I suppose they will tell their stories in the chapters that make up this volume. We might have different versions. For my part, I simply wished to have a chance to tell the factors that influenced my scientific life that could only be known or felt by me.

Marcel Marceau, the famous mime, said that he learned his trade because at the most emotional times in life, words fail. I feel similarly about expressing my thanks to my former students and those who helped me during my career. This book, I am told, is to celebrate my 60th birthday. I prefer to think of it as more a celebration of shared ideas.

ACKNOWLEDGMENT. I thank Sara Fisher for excellent editorial assistance. Sara's friendship and common sense also are gratefully acknowledged.

Stanley Falkow

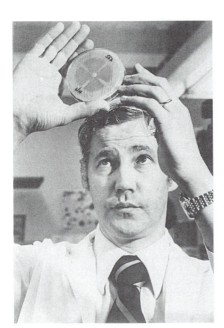

(Top) 1992 Gordon Conference on Microbial Toxins and Pathogenesis ([from left to right] *back row*: Jim Bliska, Scott Stibitz, Dan Portnoy, Staffan Normark, Jim Kaper; *middle row*: Mike Koomey, David Relman, Patrik Bavoil, Lucy Tompkins, Pam Small, Ralph Isberg, Rod Welch, Carleen Collins, Steve Moseley, Virginia Miller; *bottom row*: Carol Francis, Stanley Falkow [seated], Catherine Lee); (bottom left) Stanley Falkow in Seattle working on test of antibiotic resistance (Doug Wilson/*New York Times* Pictures, 29 June 1977; (middle) Charles A. Stuart, Ph.D. mentor; (bottom right) Stanley Falkow (right) with Allan Campbell.

Part 1. Retrospective Look at Early Advances

The majority of the chapters in this book are topical reviews of subjects currently under investigation by alumni of the Falkow laboratory. This part, however, takes a retrospective look at several topics that were important advances in the field but that are no longer being pursued by Stan and his current laboratory personnel. These chapters depict the state of the art at the time when Stan and his people were investigating these topics, describe how key experiments came about, and give a flavor of the laboratory environment at the time. Donald LeBlanc and Richard Silver (chapter 1) write about the investigations at Georgetown into the molecular nature of extrachromosomal elements. Lynn Elwell (chapter 2) then takes up the story at the University of Washington and discusses seminal studies on R factors and antibiotic resistance. These investigations led directly to the discovery of transposons, which is described by Fred Heffron and Ron Gill (chapter 3). As soon became apparent, plasmids could not only encode antibiotic resistance but also encode toxins and other virulence factors. Magdalene So and Walter Dallas (chapter 4) describe the first recombinant DNA experiments in Stan's lab and the cloning of genes encoding enterotoxins. The various studies on plasmids, antibiotic resistance genes, and virulence factor genes led to the development of a new field called "molecular epidemiology," which is discussed by Lucy Tompkins (chapter 5).

James B. Kaper

1

Molecular Genetics of Bacterial Pathogenesis
Edited by V. L. Miller, J. B. Kaper, D. A. Portnoy, and R. R. Isberg
© 1994 American Society for Microbiology, Washington, DC 20005

Chapter 1

Molecular Nature, Conjugal Transfer, and Replication of Extrachromosomal Elements, 1961 to 1973

Donald J. LeBlanc and Richard P. Silver

Any good poet, in our age at least, must begin with the scientific view of the world; and any scientist worth listening to, must be something of a poet, must possess the ability to communicate to the rest of us his sense of love and wonder at what his work discovers.

—Edward Abbey

We review in this chapter some of Stan Falkow's early contributions relative to the molecular nature, conjugative transfer, and replication of extrachromosomal elements, particularly R factors. The time frame to be covered is from 1961, 5 years before one of us (R.P.S.) joined his laboratory, to 1973, 1 year after the other (D.J.L.) left. It would be impossible to cover, in the space of a single chapter, all of Stanley's contributions during this period, because then, as has been true throughout his career, his research interests and contributions were many and varied. He has, indeed, made seminal contributions to many areas of microbiology. We will address only that work which was of primary interest to us at the time.

We hope to convey not only a feeling for the science of the time but also the affection and esteem in which the contributors hold the man to whom this book is dedicated. It also is somewhat difficult, after almost 30 years, to separate the science from images of Washington, D.C., in the late 1960s: the 1968 riots when R.P.S. watched the city burn from his apartment in Arlington, Virginia, and had to leave the lab before dark because of the curfew, tanks in the streets during the 1967 march on the Pentagon, and so on. So much was happening that shaped our generation and the generations to come.

In 1967, at the CIBA Foundation Symposium on Bacterial Episomes and Plasmids, S. E. Luria (23) divided the history of bacterial genetics into the "Stone

Donald J. LeBlanc • Department of Microbiology, University of Texas Health Science Center at San Antonio, 7703 Floyd Curl Drive, San Antonio, Texas 78284. **Richard P. Silver** • Department of Microbiology and Immunology, University of Rochester Medical Center, 601 Elmwood Avenue, Rochester, New York 14642.

Age or, as some people call it, the Luria-Delbrück Age, 1943–1946; the Bronze, or Lederberg Age, 1946–1953; the Golden, or Hayes-Wollman-Jacob Age, 1953–1961; and the Existentialist or Desperate Age, which is the present one, in which lots of things happen, lots of biochemists and ultracentrifugologists appear, and the philosophers get desperate.'' So, we found ourselves in the middle of the Desperate Age, a time in which bacterial genetics was looking for ''hidden unity in variety'' (23), a time in which the diversity of genetic elements in bacteria was being appreciated. These elements included the F factor, bacteriophage λ, resistance transfer factors, colicinogenic factors, and penicillinase plasmids (23), elements that differed in size and demonstrated various degrees of autonomy. It also is important to point out that the major tools for the molecular analyses performed during this period included equilibrium density gradient centrifugation in CsCl, and later (30) in ethidium bromide-containing CsCl, and rate zonal sedimentation through neutral and alkaline sucrose gradients. The earliest equilibrium gradients were run in a Spinco model E analytical ultracentrifuge which occupied nearly half of a wall in the laboratory and was affectionately referred to as the ''gray elephant''; later, for both isopycnic and rate zonal gradients, the Spinco model L preparative ultracentrifuge was used. Few methods were available for the effective enrichment of extrachromosomal DNA, and agarose and acrylamide gels were yet to be used for DNA analyses. Many techniques, such as restriction endonuclease mapping, dideoxy sequencing, molecular cloning, and PCR, so routine in today's laboratory environment, all were to emerge later. We often tell students that they can now do in 1 or 2 days what took us several years to accomplish. The work from the early 1960s, prior to Stanley's move to Georgetown University in 1966, was performed in Lou Baron's laboratory at the Walter Reed Army Institute of Research, in Washington, D.C., and included a brief stint in Julius Marmur's laboratory at Brandeis University, in Waltham, Massachusetts. A paper published by Baron and coworkers in 1959 (3), in which they reported the isolation of a culture of *Salmonella typhosa* that was able to utilize lactose, represents the starting point of the work described below.

CHOOSING A GRADUATE MENTOR

I (R.P.S.) arrived at Georgetown University in the fall of 1966. I had just completed a master's degree at the University of Houston, studying the growth of a phage in *Proteus mirabilis,* and was committed to continuing in bacterial genetics. Early in the semester I was told by the chairman of the department, Arthur Saz, that a young geneticist from Walter Reed would soon be joining the faculty. I recall standing in Arthur's office, never having met Stanley, deciding that I wanted to work in his lab. In retrospect, I have often thought how lucky I was, in a world that is governed so much by chance, to have found my mentor, someone who has probably influenced me more than any man I have known other than my father.

Molecular Nature of Extrachromosomal Elements

''The results presented in this study are certainly consistent with the hypothesis that the genetic component of the episomal element is deoxyribonucleic acid.''

This quotation was taken from a 1961 paper in the *Proceedings of the National Academy of Sciences* (24) that presented the first direct evidence that F was composed of DNA. The authors included J. Marmur, R. Rownd, S. Falkow, L. S. Baron, C. Schildkraut, and P. Doty. The studies described in that paper actually were made possible by information contained in a manuscript that had been submitted to *Genetics* 2 months prior to the submission of the manuscript to the *Proceedings of the National Academy of Sciences*. It was in the *Genetics* paper (10) that Stanley, along with Marmur, W. F. Carey, W. M. Spilman, and Baron, reported the transfer of the ability to utilize lactose from the strain of *S. typhosa* described in 1959 (3) not only to *Shigella dysenteriae, Salmonella typhimurium,* and *Escherichia coli* but also to two strains of *Serratia marcescens*. The importance of the latter transfers lay in the difference in the G+C contents between *S. marcescens* and the other enteric recipients.

The results described in the two papers published in 1961 (10, 24) actually constituted the beginning of a series of studies conducted in the early to mid-1960s in which Stanley and associates used what was then a relatively direct approach to demonstrate that F and various F' contained DNA. These factors were transferred by conjugation from what were considered the natural hosts, *Escherichia, Salmonella,* and *Shigella* species (50% G+C), to two different hosts with drastically different G+C contents, *S. marcescens* (58% G+C) and *P. mirabilis* (39% G+C). This permitted the separation in CsCl density gradients of the plasmids as satellite bands (10, 12, 24). The results showed that F, F', and P-*lac* were composed of double-stranded DNA with an average G+C content of 50%, the same as those of *Escherichia, Salmonella,* and *Shigella* species.

Stanley, with Ed Johnson and Lou Baron from Walter Reed, in a 1967 review on conjugation and extrachromosomal elements (9), cited the results of genetic experiments as well as theoretical arguments that had led to the general feeling among investigators that F and F' were circular molecules. The genetic map of R factors also had been reported to be circular (26). However, at the time that the review was written, there was no direct evidence for the physical circularity of any of these genetic elements. It was not until 1969 that Stanley's laboratory, including one of us (8), as well as S. Cohen's group (5), was able to provide physical evidence to show that the F-like R factor, R1, was isolated from *E. coli* as a covalently closed circular (ccc) molecule with a molecular weight (MW) of approximately 64×10^6 and was present at approximately one to two copies per chromosome equivalent.

It also was pointed out in the 1967 review that many investigators, including T. Watanabe (33) and N. Datta (6), believed on the basis of genetic evidence that R factors were composite genetic elements composed of up to seven drug resistance markers and a transfer and replication component referred to as the RTF, or resistance transfer factor. Data obtained by Haapala and Falkow (16) showed that when R1 was transferred to *P. mirabilis* and then isolated from this host, three species of extrachromosomal DNA were observed: the original R factor, with an MW of 64×10^6 and a 51.5% G+C content, present at approximately 5 copies; a molecule with an MW of 52×10^6 and a 48% G+C content, also present at 5 copies, designated the RTF; and a species with an MW of 12×10^6 and a 56%

G + C content, present at approximately 50 copies per chromosome equivalent, termed the r determinant. Three groups that had obtained these types of results (5, 8, 16, 27) generally agreed that the two smaller species were subunits of the larger molecule. Not all investigators, however, agreed with this assessment. Watanabe, who published the first comprehensive review on R factors (33), and another Japanese investigator (26), had shown that the R factors of *E. coli* and *Shigella* species invariably consisted of single genetic units composed of an RTF plus resistance genes. Anderson (1), on the other hand, had found that some R factors in *S. typhimurium* were dissociable into a component he called Δ (equivalent to the RTF) and individual resistance genes, each capable of autonomous replication. In a 1970 paper stemming from the thesis of R.P.S. (32), we proposed a hypothesis to unify the different findings, namely, that R factors represented recombinational assemblages of two or more replicons. This, of course, turned out to be the case for many R factors. In fact, some years later one of us (D.J.L.) showed that the first transmissible multiple antibiotic resistance plasmid to be described in a gram-positive bacterial species, *Enterococcus faecalis* (formerly *Streptococcus faecalis*) (18), was also a composite plasmid comprising at least two different replicons (2) and that one of these was highly related to one of two replicons, pAMα1Δ1, that made up the nontransmissible enterococcal plasmid pAMα1 (29).

The Freifelder Technique

In 1968 the Freifelders published information about a technique designed to specifically label extrachromosomal DNA following its transfer by conjugation (13). Stanley had been made aware of this procedure, the development of which constituted a large proportion of Dorothy Freifelder's Ph.D. thesis, prior to its publication. It was to become a major tool for both of us during our respective tenures in Stanley's laboratory. The procedure involved the use of a donor strain of *E. coli* that was prototrophic for thymine synthesis and resistant to 5-bromouracil in the presence of deoxyadenosine, and thus unable to incorporate exogenous thymine, and a recipient strain that was auxotrophic for thymine but that was unable to replicate its own DNA because of an inability to repair lesions caused by high doses of UV light. However, a conjugative plasmid could transfer from such a genetically blocked donor into the irradiated recipient at nearly normal frequencies. The transferred plasmid also could replicate in the recipient and was the only DNA into which label was incorporated when the mating was conducted in the presence of [^{3}H]thymidine. Initially, it was Stanley's intention to use the procedure to see if R factors dissociated in *E. coli* after transfer, which turned out not to be the case. Subsequently, however, the technique proved to be useful for the characterization of the R factor R1*drd-19* (31), its RTF (32), and its replication following conjugation (11) and for the analysis of superinfection immunity (21, 22). The reason that R1*drd-19* was chosen was because most naturally occurring R factors transfer at low frequencies and the amount of labeling would have been too low for detectability. Datta and associates (25) had described a mutant of plasmid R1, R1*drd-19*, which had been selected for high-frequency transfer.

Nature and Replication of R-Factor DNA Following Conjugation

Initial studies in which the Freifelder technique was used involved an examination of the labeled DNA after 60 min of mating in neutral sucrose gradients. The bulk of the label was found in two peaks with sedimentation values of 50S and 75S, a ratio which, if the same molecule was present under both peaks, would indicate open circular (oc) and ccc forms of the plasmid, respectively. Two approaches were used to confirm the ccc nature of the 75S species. First, limited DNase treatment produced a quantitative conversion of the 75S species to the 50S species. Second, when sedimented in an alkaline sucrose gradient, the 75S neutral sucrose gradient species migrated three to four times faster than the original 50S species, and the introduction of single-stranded breaks by the administration of controlled doses of X irradiation resulted in conversion of the fast-sedimenting species to the slower-sedimenting one in alkaline sucrose. Both the 75S and 50S species sedimented in CsCl equilibrium gradients at a buoyant density of 1.711 g/ml, corresponding to a G+C content of 51.5%. On the basis of the sedimentation values for both species of the labeled DNA and the kinetics of loss of the ccc form on X irradiation, the molecular weight of R1drd-19 was calculated to be approximately 65 × 10^6. Today, it is difficult to appreciate the number of gradients that were run for these experiments. We (this includes Stanley) were often in the lab through the night collecting fractions and preparing radioactive samples for counting.

Pulse-chase experiments were used to demonstrate that the ccc form of the plasmid was the end product of replication and that the oc form was an intermediate in the replication process and not just the result of nicking of some of the ccc molecules during cell lysis and manipulation. In addition, after a 15-min pulse, a third species of labeled DNA that sedimented at 44S was seen; this rate would be expected of a linear double-stranded 65-MDa molecule. Thus, the replicative progression was from linear to oc to ccc, which indicated that the plasmid was transferred from the donor to the recipient as a single-stranded molecule whose complementary strand was synthesized in the recipient.

An important question that needed to be addressed was whether the DNA synthesis in irradiated recipients was a true reflection of replication following normal conjugation conditions. It was assumed that, if the irradiated recipient could serve as a donor of R1drd-19 to a secondary recipient in a standard conjugation experiment, this would be an indication that the product of replication in the Freifelder system was likely the same as that synthesized in a normal mating situation. To test this assumption, an irradiated recipient resistant to lysis from without by T6 phage was mixed with a donor that was sensitive to lysis by the phage. At timed intervals the mating mixture was treated with T6 to kill the original donor cells. The phage-treated cells were then mixed with a second T6-resistant recipient that was also resistant to nalidixic acid. The mating with irradiated recipients produced transconjugants that were able to serve as donors of the R1drd-19-associated chloramphenicol and tetracycline resistance traits within 30 min.

It was presumed from the results presented above that R-factor DNA synthesis occurred on the entering strand in the recipient, since in the Freifelder system

DNA was labeled only in the recipient and labeled DNA was seen immediately after the donor and recipient cells were mixed. The rolling circle model of DNA replication (14) predicted that the 5' end would enter the recipient cell first and become associated with some structure in the recipient, most likely the cell membrane. The next logical step, then, was to show an association of R-factor DNA with a structural component of the cell, and in the Freifelder system, the only detectable R-factor DNA would be that labeled in the recipient. The nonionic detergent Brij 58 was used to lyse the cells in the mating mixture since this approach had been used to demonstrate the attachment of φX174 DNA to cell structures (20). The kinetics of appearance of cell structure-bound (presumably membrane-bound) DNA were very similar to the kinetics of appearance of 44S (linear monomer) DNA seen in the pulse-chase experiments. As the membrane-bound fraction decreased, there was a sequential appearance of free 50S (oc) and then 75S (ccc) molecules.

On the basis of the results of the combination of the pulse-chase and continuous-labeling experiments, it was concluded that immediately following the transfer of R-factor DNA from a donor cell, presumably as a single strand, a linear double-stranded molecule is synthesized on the bacterial membrane. The completed molecule then circularizes and one of the strands is immediately sealed and released into the cytoplasm, where the second strand is sealed, and a supercoiled ccc molecule is the end product (11). Thus, R1 replication after conjugation was analogous to that of viral replication in *E. coli* K-12 and was consistent with the Gilbert and Dressler (14) version of the rolling circle model for F-*lac*. Although much more is known today relative to strand specificity, *oriT*, enzymatic mechanisms, and the genetic determinants responsible for them, a general description of the immediate fate of single-stranded plasmid DNA following conjugation would not differ drastically from that just described (34). It was not known at that time whether vegetative replication of R1*drd-19* followed the same mode. Of course, we now know that vegetative replication of the vast majority of plasmids of gram-negative bacterial origin occurs via a theta mode (4).

When R1 was transferred to *P. mirabilis,* three independent replicons were observed. The largest was 65 MDa in size, with a buoyant density of 1.711 g/ml, the same as that seen in *E. coli.* The two other species, not seen in *E. coli,* were 55 and 12 MDa in size, with buoyant densities of 1.709 and 1.716 g/ml, respectively. The larger of these two molecules appeared to contain the transfer functions of R1 and was assumed to be the RTF, whereas the smaller species was assumed to contain the resistance determinants. E. Meynell had obtained an isolate of *E. coli* originally containing R1*drd-19* that had lost all of the resistance traits but that retained the high-frequency transfer function. Presumably, then, this strain contained the RTF that had been observed in *P. mirabilis.* Stanley obtained this isolate from Meynell, and we (R.P.S.) transferred its plasmid to the thymine-prototrophic, 5-bromouracil-resistant strain DF110 that was used as the donor for the Freifelder procedure. Since the transfer factor no longer encoded any resistance traits, there was no selection for its presence in the DF110 strain. However, it did encode an F-like pilus which rendered any host susceptible to the phage MS2, and since it was derived from R1*drd-19,* it still transferred at a high frequency. Thus, it was possible to simply select for recipient colonies and then to screen a reasonable number of these colonies for susceptibility to MS2. One such isolate was then used as the donor in mating experiments by the Freifelder technique. The labeled plasmid DNA was shown to exist as ccc and oc molecules that sedimented in neutral sucrose gradients with sedimentation values of 64 and 44, respectively. These sedimentation rates were consistent with a size of 50 MDa, similar to the size observed for the RTF in *P. mirabilis*. The buoyant density of this molecule was shown in CsCl gradients to be 1.709 g/ml, exactly the same as that determined for the RTF from *P. mirabilis*. Thus, the R1*drd-19* molecule had lost approximately 15 MDa of DNA in the loss of the resistance determinants, equivalent to approximately 23% of the original molecule. On the basis of the previously determined buoyant density of R1 (1.711), the buoyant density of the RTF of 1.709 g/ml, and the proportion of R1*drd-19* represented by the RTF, it was calculated that the lost r determinant should have had a buoyant density of 1.716 g/ml, the same as that observed for the smallest of the three DNA species observed with R1-harboring *Proteus* strains.

One of us (R.P.S.) also took a genetic approach to look at the conjugal transfer of R1*drd-19*. We attempted an interrupted mating experiment to determine the order of transfer of the antibiotic resistance markers. Although from an analysis of the selection data it appeared possible to successfully interrupt transfer, replica plating analysis of unselected markers revealed that all colonies resistant to one antibiotic also were resistant to the other three antibiotics. The data revealed the phenotypic expression of the antibiotic resistance markers and confirmed that R plasmids are indeed transmitted rapidly (within 5 min) and behave genetically as a single unit of transmission. Stanley described the experiment in his book (7) and noted that "one need only imagine the potential clinical implications of even a single similar event in the gut."

From the beginning, the clinical significance of his work has always been a driving force. In 1967, Stanley organized the Symposium on Infectious Multiple Drug Resistance: Genetics, Molecular Nature, and Clinical Implications of R Factors, which was held at the Georgetown University School of Medicine. The participants included Tsutomu Watanabe, Naomi Datta, P. A. M. Guinee, R. Rownd, and David Smith. To our knowledge, this was the first symposium in the United States to focus on the importance of R plasmids to human and veterinary medicine. When I (R.P.S.) think about the clinical implications of our work, however, I think mostly of 1968, when my student deferment had expired and I was about to be drafted. I requested a hearing with my draft board in Massachusetts. It was the height of the Vietnam conflict, and Stanley wrote a letter to the board describing the significance of our work to the control of infectious diseases in Southeast Asia. "After they read this letter, there won't be a dry eye in the house," he stated with pride. To this day Stanley thinks it was his letter, and not my age, that kept me from the jungles of Vietnam.

CHOOSING A POSTDOCTORAL MENTOR

Halfway through my (D.J.L.) third year as a graduate student at the University of Massachusetts at the beginning of the spring semester of 1969, it seemed appropriate to begin thinking of possible postdoctoral mentors. I had narrowed my choices to an extension of my training in microbial physiology or to making a complete shift into microbial genetics. My graduate mentor, Bob Mortlock, was very helpful with suggestions for further training in microbial physiology, but he confessed to being at a loss as far as microbial geneticists were concerned. Consequently, I turned to another member of my dissertation committee, Curt Thorne. When Curt asked me just what it was in the area of microbial genetics that I thought would be of interest to me, I remembered a lecture in which the concept of R factors had been discussed. I remembered becoming excited about these new genetic elements, especially after having read a recent review on the subject (33). Upon relating this to Curt, he said he knew of a young investigator at Georgetown University by the name of Falkow who was combining molecular and genetic approaches in studies on plasmids, including R factors. Solely on the basis of this information, I immediately wrote to Stanley to inquire as to the possibility of doing postdoctoral work in his laboratory beginning sometime in 1970. In his reply, Stanley suggested that since he was going to be presenting a seminar in our department later in the spring (having been invited by Curt Thorne), we should plan to talk more about a postdoctoral fellowship at that time. After hearing what amounted to the most lucid and exciting seminar presentation that I had ever attended, in which Stanley talked about some of the work of Haapala and Silver, I was sold. Later that afternoon Stanley asked me why I wanted to study plasmids. My answer was that I really wasn't interested in plasmids per se, but that I thought they might provide useful tools for genetic analyses. He indicated that he expected to have space by the summer of 1970 and that if I wanted to join his laboratory I should think about writing for a National Institutes of Health postdoctoral fellowship. The proposal was written, submitted, and funded. I have always suspected that the success of the proposal had much more to do with Stanley's emerging reputation than to my grantsmanship at that point in my career. When I arrived in Stanley's laboratory on 15 July 1970, Dan Haapala, Rich Silver, and Lucy Tompkins, Stanley's first three graduate students, had all completed their dissertations. Dan was at the Carnegie Institute in Washington, D.C., Rich had just left to do postdoctoral work in Stan Cohen's laboratory at Stanford, and Lucy was about to leave for Hanover, New Hampshire, to enter her first year of medical school at Dartmouth. Pat Guerry had completed a master's degree with Stanley and was staying in the laboratory as his technician, and Vickers Hershfield was the only graduate student, but she was not yet into her dissertation problem. The only other person in the laboratory besides Stanley was Carol-Sue, Stanley's personal secretary and the person responsible for typing each draft of "the book" (on a typewriter), which was finally published in 1975 (7).

Studies on Superinfection Immunity

It had been shown that a bacterial strain harboring a sex factor or an R factor could not accept the same, or a closely related, extrachromosomal element by conjugation, a phenomenon referred to as "exclusion," "entry exclusion," or "surface exclusion" (25). If such a plasmid was able to circumvent exclusion, it still may not be inherited by the recipient strain. This latter phenomenon was explained on the basis of the replicon hypothesis (19) in terms of a competition between the resident and superinfecting plasmid for a specific intracellular site (probably the membrane) at which replication could be initiated. Thus, "superinfection immunity," a term borrowed by Watanabe (33) from the phage literature, was a reflection of two distinct effects, entry exclusion and plasmid incompatibility, i.e., the inability of two closely related plasmids to coexist stably in the same host. Stanley suggested that I (D.J.L.) apply the Freifelder technique to an analysis of these phenomena.

The first set of experiments involved an examination of matings between a common donor strain, RS2, containing the F-like (fi$^+$) plasmid R1drd-19 and the Freifelder recipient strain AB2500 into which different plasmids had been introduced. The plasmids present in the different recipients included the homologous RTF of R1drd-19, the related fi$^+$ plasmid R222 or F, the unrelated fi$^+$ R factor N3, and no plasmid. Two types of matings were conducted with each mating pair, one in which the recipient had been irradiated and [^{3}H]thymidine incorporation was determined and a second 30-min standard mating with unirradiated cells, after which the conjugation frequency was determined. The incorporation of label and the frequency of R1drd-19 transfer for the matings between RS2 and AB2500 were taken to represent 100%, and the results obtained with the other mating pairs were judged relative to this control. When the recipient strain carried the unrelated N3 R factor, both incorporation of label and conjugation frequencies were approximately 60% relative to those of the AB2500 recipient, a slight decrease that was likely a result of restriction rather than either exclusion or incompatibility, since R1 and N3 were unrelated. When the recipient harbored F, the incorporation of label was 21% and the frequency of inheritance of R1drd-19 by the F$^+$ recipient was 4% relative to the control mating, a difference between the two types of matings of less than 1 log unit. The incorporation of label by recipients harboring plasmid R222 or RTF was between 10 and 19% of that by AB2500 and in the same range as that seen with the F$^+$ recipient. However, the frequencies of inheritance of R1drd-19 by the R222- and RTF-containing recipients were 0.24 and 0.001, respectively. The drops in relative labeling to between 10 and 21% that of the control were considered reflections of entry exclusion mediated by the related plasmids F and R222 and the homologous plasmid RTF. The nearly 100- and 10,000-fold decreases in the frequency of inheritance of R1drd-19 by the recipients harboring R222 and RTF, respectively, were attributed to incompatibility. Thus, the presence of N3 in a recipient had no effect on its ability to obtain R1drd-19 by conjugation. On the other hand, F, R222, and RTF all inhibited the entry of R1drd-19 during mating, but only R222 and RTF were incompatible with it.

The labeled DNA present in lysates of each of the mating mixtures described above was examined in neutral sucrose gradients. As expected (31), the DNA from the R1drd-19 × AB2500 mating sedimented in two major peaks at 50S and 75S, equivalent to an oc and a ccc form, respectively, of a 65-MDa plasmid. The labeled DNA from the R1drd-19 × F$^+$ mating also sedimented at 50S and 75S. Thus, although F was able to exert entry exclusion on the transfer of R1drd-19, it appeared to be compatible with it. On the other hand, the DNA synthesized in the recipients containing an incompatible plasmid sedimented very differently. Instead of the 50S and 75S species, after 60 min of mating there was a broad distribution of counts in the gradient at locations equivalent to sedimentation values ranging between 50 and 65. These results suggested that incompatibility either involved the synthesis of aberrant replicative forms or caused the accumulation of normal replicative intermediates that generally existed only transiently and in very small amounts. That the latter interpretation was correct was suggested by the results of pulse-chase experiments conducted during R1drd19 × AB2500 (compatible) and R1drd-19 × AB2500(RTF) (incompatible) matings. Each mating mixture was labeled with [³H]thymidine for 20 min, at which time an excess of unlabeled thymine was added and the incubation was continued. At different time intervals, beginning at the end of the pulse, the cells in aliquots of each mating mixture were lysed and the labeled DNA was subjected to neutral sucrose gradient analysis. At the end of the pulse in the compatible mating, most of the label was present in a 44S DNA species, the sedimentation rate expected of a linear double-stranded 65-MDa molecule. There also was a small amount of a 65S species. Forty minutes into the mating, at 20 min after the chase, nearly all of the label sedimented at 50S and 75S. After 30 min of mating with the incompatible recipient [AB2500(RTF)], at 10 min postchase, the label was distributed throughout the sucrose gradient, with most of it sedimenting at rates of between 44S and 65S. Samples taken after 60 and 90 min of mating, at 40 and 70 min postchase, also produced very broad distributions of label in sucrose gradients. The only differences seemed to be reflected in shifts toward the higher sedimentation rates, with a significant proportion of the label sedimenting at 75S by 90 min. The persistence of what appeared to be replicative intermediates in an incompatible mating could be due to an inability to process them once they are formed or, possibly, to a highly reduced rate of replication. The following experiment was conducted to determine if replicative intermediates could be processed in an incompatible recipient and to gain some insight into the difference in such processing between a compatible and an incompatible host. A pulse-chase experiment was performed with each mating, R1drd-19 × AB2500 and R1drd-19 × AB2500(RTF), and samples were taken for neutral sucrose gradient analysis at various intervals postchase. All sucrose gradients for each mating were made identically, the same number of fractions was collected from each gradient, and the percentage of the total counts in the gradient represented in each fraction was calculated and plotted. In order to assess the actual molecular species being processed during specific time intervals, the percentage of label in each fraction of one gradient was subtracted from the percentage of label in the corresponding fraction of a gradient from a later time postchase. Thirty minutes into the incompatible mating, at 10 min after the chase, most of the label that had been present at the time of chase had been converted to 44S and 50S species. After 60 min of mating, most of the DNA labeled by the 20-min pulse had been converted to 57S and 65S species. Only during the interval between 60 and 90 min was there a significant conversion to 65S DNA, and between 90 and 150 min of mating, most of the observed change was reflected in an accumulation of the 75S ccc monomer. Clearly, these results suggested that replication of a plasmid could proceed to a normal end product in a recipient harboring an incompatible genetic element, but that the rate at which such replication occurred was slowed considerably relative to the rate of its replication in a compatible recipient. The data from the experiment described above as well as those from a similar one involving a compatible mating were used to plot the percentages of 75S DNA present against the time of mating in order to assess the rate of R1drd-19 replication in the two types of matings. The results indicated that the plasmid was replicated approximately 11 times faster in a host with a compatible plasmid than it was upon transfer to a recipient with an incompatible plasmid. It was estimated that the actual times required to convert an entering single-stranded molecule to a double-stranded ccc form were approximately 3 and 33 min in compatible and incompatible hosts, respectively (22). The distribution of plasmids to progeny cells would, of course, also require a round of vegetative replication.

Replication of R-Factor DNA on the Cell Membrane Following Conjugation

The results of the preliminary experiments described above suggested that upon transfer by conjugation, single-stranded R1drd-19 attaches to the recipient cell membrane, where the complementary strand is synthesized to form a linear double-stranded monomer which is converted to an oc form and, subsequently, the ccc form. The two latter DNA species were not associated with the cell membrane and were assumed to be cytoplasmic. The following approach was used to refine those preliminary experiments. Brij 58 causes the release of cytoplasmic components from the bacterial cell but leaves cellular structural components (including the cell membrane) intact (15). The DNA attached to such cellular structures, presumably the membrane, can be separated from cytoplasmic DNA by centrifugation in a discontinuous sucrose gradient composed of a 60% sucrose shelf overlaid with a 20% sucrose solution. The rate and time of centrifugation can be adjusted such that membrane-associated DNA will migrate through the 20% sucrose solution onto the 60% sucrose shelf, whereas any DNA in the cytoplasm will remain near the top of the 20% sucrose layer. Brij 58 lysis and sedimentation in discontinuous sucrose gradients were coupled with the Freifelder mating technique to examine further the attachment of R1drd-19 DNA to the recipient cell membrane. As observed previously, the kinetics of appearance of membrane-bound DNA were very similar to the kinetics of appearance of 44S seen in pulse-chase experiments. As the membrane-bound fraction decreased, there was a corresponding appearance of free cytoplasmic molecules which, upon sedimentation in linear neutral sucrose gradients, first appeared as 50S oc forms and then 75S ccc molecules, the end products of R1drd-19 replication (11).

It was reasoned that the 11-fold slowing of the rate of R1 replication upon entrance into the RTF-containing recipient might have been related to an inability to associate with, or dissociate from, a cellular membrane site. The following experiments were conducted by me (D.J.L.) and Vickers Hershfield (17) in order to compare the kinetics of membrane association of R1 in a compatible mating versus that in an incompatible mating. Freifelder matings were set up between the R1drd-19-containing donor and a plasmid-free as well as an RTF-containing recipient. The transferred DNA was labeled, and samples were taken at various intervals after the initiation of mating. Each sample was diluted in KCN to inhibit further replication, the cells were lysed with Brij 58, and the sedimentation of labeled DNA in discontinuous 20 to 60% sucrose gradients was examined. After 20 min of mating, approximately 80% of the label in both types of matings was associated with the fast-sedimenting cell fraction (60% sucrose shelf). Seventy minutes later, 50% of the label from the compatible mating was found in the cytoplasmic fraction (top of the 20% sucrose layer), whereas only 13% of the label from the incompatible mating had been released into this fraction. The sedimentation properties in linear sucrose gradients of DNA from the cytoplasmic fractions and the membrane fractions (after solubilization with sodium dodecyl sulfate) were examined. The cytoplasmic fraction from the compatible mating sedimented in two well-defined peaks at 50S and 75S, whereas the DNA from the membrane fraction sedimented in an equally defined peak at 44S, with lesser amounts that included a 50S DNA species but no 75S DNA species sedimenting at faster rates. The sedimentation properties of the DNA from the incompatible mating were quite different. In this case, the distribution of the cytoplasmic fraction was very broad, with a hint of a peak at 75S. The membrane fraction from the incompatible mating sedimented in two broad peaks, a small amount at 44S and the majority with an even broader distribution at between 60S and 68S. These latter species had been seen as very transient replicative intermediates following compatible matings, but they accumulated following incompatible matings (22). Sedimentation in this region of a neutral sucrose gradient would be consistent with a partially replicated circular form of a 65-MDa plasmid. The results of this set of experiments suggested that in both compatible and incompatible matings, the transferred DNA attaches to a cellular component that is assumed to be the membrane. However, it was not possible to assess whether such attachment was specific in either mating. The major difference between a compatible and an incompatible mating, relative to attachment to the recipient cell surface, appeared to be the time required for the synthesis from entering single strands of double-stranded oc monomers and their release into the cytoplasm and subsequent conversion to double-stranded ccc monomers. In the incompatible cell, most of the DNA that enters remains fixed to the membrane and accumulates as replicative intermediates. The block or slow-down in replication could be due to the action of a repressor-like substance produced by the resident plasmid or to a competition for a specific replication site. It remains possible that the reason for the delay in the release of replicative intermediates from the membrane site in incompatible cells is that the site of attachment is not the R1-specific site, because that is already occupied by the RTF.

The precise role of the cell membrane in plasmid incompatibility has yet to be resolved. There seems to be no doubt as to its involvement in plasmid replication and the expression of incompatibility immediately following conjugation. However, whether the membrane plays a role in vegetative plasmid replication or even maintenance in the presence of the same or a related plasmid remains an open question. Partition systems (*par*) associated with low-copy-number plasmids, including R1*drd-19,* are responsible for the stable maintenance and distribution of these elements and are also incompatibility determinants. The so-called killer systems, which have also been described for R1*drd-19*, have been designated incompatibility determinants. For more information on these systems and their roles in plasmid maintenance and incompatibility, the reader is referred to the review by Nordstrom and Austin (28).

OUT OF SITE, BUT NOT OUT OF MIND, SOME FINAL THOUGHTS FROM AN EARLY GRADUATE STUDENT AND AN EARLY POSTDOC

It seems that the most significant contributions to science often come from those, like Stanley Falkow, who are somehow able to take something seen by all, see it differently, and give it additional meaning. This volume is a testament not only to his many contributions but to the many contributions he has influenced. For us it was indeed a pleasure and an honor to have been a small part of these accomplishments. Yet, we all leave Stanley's laboratory eventually, some after less time than others. However, the influence and the memories are never gone. Perhaps the most valuable lesson to be taken from him, other than always to use big data points when plotting your results, is the concept that some of the most elegant and informative experiments are the simplest, those that provide a single answer to a single question. The need to simplify even the most complex of problems and to ask one single but answerable question at a time has influenced our entire research careers. We also learned the folly of being consumed by our own self-importance. As Stanley repeatedly told us, "One hundred years from now it will be reported that Silver and Falkow, or LeBlanc and Falkow, incorrectly reported . . ."

Stanley has remained an important sounding board and adviser for us both. He has been the first person to whom we have turned for advice on major career moves. We cherish his insight and friendship. Perhaps, in the final analysis, the finest lesson that a true teacher reveals is that you can never truly be rid of him. For this we are grateful.

REFERENCES

1. **Anderson, E. S.** 1968. The ecology of transferable drug resistance in the *Enterobacteriaceae. Annu. Rev. Microbiol.* **22:**131–180.
2. **Banai, M., M. A. Gonda, J. M. Ranhand, and D. J. LeBlanc.** 1985. *Streptococcus faecalis* R plasmid pJH1 contains a pAMα1Δ1-like replicon. *J. Bacteriol.* **164:**626–632.
3. **Baron, L. S., W. F. Carey, and W. M. Spilman.** 1959. Characterization of a high frequency of recombination (Hfr) strain of *Salmonella typhosa* compatible with *Salmonella, Shigella,* and *Escherichia* species. *Proc. Natl. Acad. Sci. USA* **45:**1752–1757.
4. **Cairns, J.,** 1963. The chromosome of *Escherichia coli. Cold Spring Harbor Symp. Quant. Biol.* **28:**43–45.
5. **Cohen, S. N., and C. A. Miller.** 1969. Multiple molecular species of R factor DNA isolated from *Escherichia coli. Nature* (London) **224:**1273–1277.
6. **Datta, N.** 1965. Infectious drug resistance. *Br. Med. Bull.* **21:**254–259.
7. **Falkow, S.** 1975. *Infectious Multiple Drug Resistance.* Pion Ltd., London.
8. **Falkow, S., D. Haapala, and R. P. Silver.** 1969. Relationships between extrachromosomal elements, p. 136–158. *In* G. E. W. Wolstenholme and M. O'Connor (ed.), *Bacterial Episomes and Plasmids.* Churchill, Ltd., London.

9. Falkow, S., E. M. Johnson, and L. S. Baron. 1967. Bacterial conjugation and extrachromosomal elements. *Annu. Rev. Genet.* **1**:87–116.

10. Falkow, S., J. Marmur, W. F. Carey, W. M. Spilman, and L. S. Baron. 1961. Episomic transfer between *Salmonella typhosa* and *Serratia marcescens. Genetics* **46**:703–706.

11. Falkow, S., L. S. Tompkins, R. P. Silver, P. Guerry, and D. J. LeBlanc. 1971. The replication of R-factor DNA in *Escherichia coli* K12 following conjugation. *Ann. N. Y. Acad. Sci.* **182**:153–171.

12. Falkow, S., J. A. Wohlhieter, R. V. Citarella, and L. S. Baron. 1964. Transfer of episomal elements to *Proteus.* I. Transfer of F-linked chromosomal determinants. *J. Bacteriol.* **87**:209–219.

13. Freifelder, D. R., and D. Freifelder. 1968. Studies on *E. coli* sex factors. I. Specific labeling of F'lac DNA. *J. Mol. Biol.* **32**:15–23.

14. Gilbert, W., and D. Dressler. 1968. DNA replication: the rolling circle model. *Cold Spring Harbor Symp. Quant. Biol.* **33**:473–484.

15. Godson, G. N., and R. L. Sinsheimer. 1967. Lysis of *E. coli* with a neutral detergent. *Biochim. Biophys. Acta* **149**:476–488.

16. Haapala, D. K., and S. Falkow. 1971. Physical studies of the drug-resistance transfer factor in *Proteus. J. Bacteriol.* **106**:294–295.

17. Hershfield, V., D. J. LeBlanc, and S. Falkow. 1973. Membrane attachment of R-factor deoxyribonucleic acid in compatible and incompatible cell pairs following conjugation. *J. Bacteriol.* **115**: 1208–1211.

18. Jacob, A. E., and S. J. Hobbs. 1974. Conjugal transfer of plasmid-borne multiple antibiotic resistance in *Streptococcus faecalis* var. *zymogenes. J. Bacteriol.* **117**:360–372.

19. Jacob, F., S. Brenner, and F. Cuzin. 1963. On the regulation of DNA replication in bacteria. *Cold Spring Harbor Symp. Quant. Biol.* **28**:329–348.

20. Knippers, R., and R. L. Sinsheimer. 1968. The process of infection with bacteriophage φX174. XX. Attachment of the parental DNA of bacteriophage φX174 to a fast-sedimenting cell component. *J. Mol. Biol.* **34**:17–29.

21. LeBlanc, D. J., and S. Falkow. 1972. Effects of superinfection immunity on plasmid replication following conjugation, p. 309–318. *In* V. Krcmery, L. Rosival, and T. Watanabe (ed.), *Bacterial Plasmids and Antibiotic Resistance.* Springer-Verlag, New York.

22. LeBlanc, D. J., and S. Falkow. 1973. Studies on superinfection immunity among transmissible plasmids in *Escherichia coli. J. Mol. Biol.* **74**:689–701.

23. Luria, S. E. 1969. Introduction, p. 1–3. *In* G. E. Wolstenholme and M. O'Connor (ed.), *Bacterial Episomes and Plasmids.* Churchill, Ltd., London.

24. Marmur, J., R. Rownd, S. Falkow, L. S. Baron, C. Schildkraut, and P. Doty. 1961. The nature of intergeneric episomal infection. *Proc. Natl. Acad. Sci. USA* **47**:972–979.

25. Meynell, E., G. G. Meynell, and N. Datta. 1968. Phylogenetic relationships of drug-resistance factors and other transmissible bacterial plasmids. *Bacteriol. Rev.* **32**:55–83.

26. Mitsuhashi, S. 1965. Transmissible drug-resistance factor R. *Gunma J. Med. Sci.* **14**:169–209.

27. Nisioka, T., M. Mitani, and R. Clowes. 1969. Composite circular forms of R factor DNA. *J. Bacteriol.* **97**:376–385.

28. Nordstrom, K., and S. J. Austin. 1989. Mechanisms that contribute to the stable segregation of plasmids. *Annu. Rev. Genet.* **23**:37–69.

29. Perkins, J., and P. Youngman. 1983. *Streptococcus* plasmid pAMα1 is a composite of two separable replicons, one of which is closely related to *Bacillus* plasmid pBC16. *J. Bacteriol.* **155**:607–615.

30. Radloff, R., W. Bauer, and J. Vinograd. 1967. A dye-buoyant-density method for the detection and isolation of closed circular duplex DNA: the closed circular DNA in HeLa cells. *Proc. Natl. Acad. Sci. USA* **57**:1514–1521.

31. Silver, R. P., and S. Falkow. 1970. Specific labeling and physical characterization of R-factor deoxyribonucleic acid in *Escherichia coli. J. Bacteriol.* **104**:331–339.

32. Silver, R. P., and S. Falkow. 1970. Studies on resistance transfer factor deoxyribonucleic acid in *Escherichia coli. J. Bacteriol.* **104**:340–344.

33. Watanabe, T. 1963. Infective heredity of multiple drug resistance in bacteria. *Bacteriol. Rev.* **27**: 87–115.

34. Willets, N., and B. Wilkins. 1984. Processing of plasmid DNA during bacterial conjugation. *Microbiol. Rev.* **48**:24–41.

Molecular Genetics of Bacterial Pathogenesis
Edited by V. L. Miller, J. B. Kaper, D. A. Portnoy, and R. R. Isberg
© 1994 American Society for Microbiology, Washington, DC 20005

Chapter 2

R Plasmids and Antibiotic Resistances

Lynn P. Elwell

Truth lies at the end of a circle.

—Elbert Hubbard

My theory is perfectly correct, it is the facts that are misleading.
  —Dialogue from the Alfred Hitchcock movie *The Lady Vanishes*

IN THE BEGINNING

I joined Stanley Falkow's laboratory in the summer of 1974 as a postdoctoral fellow. Stanley and King Holmes had just received a grant to study *Neisseria gonorrhoeae* and I was to work with Leonard Mayer, who was a graduate student in the lab. Leonard was interested in the genetic analysis of virulence determinants in the gonococcus, and my first job was to subculture these organisms on a daily basis in order to keep their pili intact; that was easy enough. My second job was to cut rectangular windows over the main blood vessels in 11-day-old chicken embryos with a dentist's drill and then inject suspensions of various gonococcal cultures into the embryos without killing them; this was not so easy. Although Leonard showed an inordinate amount of patience and good humor (albeit, a tad forced as time passed and unusable embryos piled up), I never quite got the hang of it.

Part of the problem, I think, was that my heart really wasn't in it; I had hoped to work with antibiotic resistance and R plasmids in Falkow's lab. The next best thing, I decided, was to spend some time in the clinical microbiology laboratory that was associated with the University of Washington Medical School and teaching hospital. Professor John Sherris, who was the chairman of the Department of Microbiology, suggested that I attend plate rounds and get to meet folks like Fritz Schoenknecht and Marie Coyle. This surprised me a bit. Although I got my degree at a medical school microbiology department (a mere 150 miles [241 km] down the road), the prevailing attitude there was that there is basic research and there is clinical microbiology, and never the twain should meet. It was not until much

---

Lynn P. Elwell • Division of Molecular Genetics and Microbiology, Wellcome Research Laboratories, 3030 Cornwallis Road, Research Triangle Park, North Carolina 27709.

later that I fully appreciated Professor Sherris's enlightened attitude and instincts in this regard. In any event, during the first session I attended, the subject of high-level ampicillin resistance in *Haemophilus influenzae* was discussed. As soon as I got back to the lab I asked Stanley whether it would be worthwhile looking into this problem; after all, three infants had died from meningitis caused by drug-resistant *H. influenzae* (35, 47). Stanley said that a medical student in the lab, David Seibert, was already attempting to find R plasmids in drug-resistant clinical isolates but, so far, to no avail. Over lunch at a pub on the shore of North Lake, I asked David Seibert if he would mind if I joined him in the *Haemophilus* plasmid quest; to my everlasting gratitude, he said he wouldn't mind.

H. INFLUENZAE ACQUIRES R PLASMIDS

Since David's original lysates were devoid of visible extrachromosomal DNA in isopycnic gradients, we reasoned that the copy number of the putative plasmid might be extremely low. Therefore, we grew cells in a makeshift minimal medium containing [^{3}H]thymidine and were ecstatic to see a peak of counts some 20 fractions below the linear, chromosomal DNA peak (Fig. 1). An examination of this peak under the electron microscope revealed a plasmid species of approximately 30 MDa in mass which we dubbed RSF007 (Fig. 2). Three ampicillin-resistant *H. influenzae* isolates harbored RSF007, but a fourth isolate (*Haemophilus parainfluenzae*) contained a much smaller plasmid species (about 4 MDa); we called it RSF0885 (13). We were able to transform a drug-susceptible *Haemophilus* recipient strain to ampicillin resistance by using purified RSF0885 plasmid DNA but not by using plasmid RSF007. At about this time, Grace Thorne and Edmund Farrar told us that they had successfully transferred ampicillin resistance between *H. influenzae* isolates immobilized on membrane filters, presumably by conjugation. The larger size of RSF007 suggested to us that this plasmid might encode transfer functions, and Thorne and Farrar subsequently proved this to be so (74).

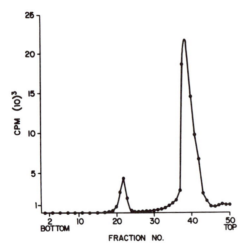

FIGURE 1. Ethidium bromide-CsCl gradient of a cleared lysate of *H. influenzae* G32(RSF007) grown in minimal medium in the presence of [^{3}H]thymidine.

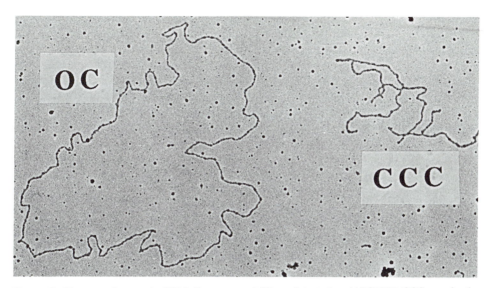

FIGURE 2. Electron micrograph of *H. influenzae* ampicillin-resistant plasmid RSF007. CCC, covalently closed circular form; OC, open circular form. Final magnification, × 13,874.

Incidentally, while all of this was going on, Jorge Crosa taught me how to lyse gram-negative cells in such a way as to preserve their extrachromosomal elements; apparently I was vortexing them to death.

Of course, the question that haunted us was, where did these R plasmids come from? After all, *H. influenzae* had been exquisitely susceptible to ampicillin for over a decade, and suddenly strains that harbored plasmids specifying β-lactamase were emerging. One obvious possibility was that these plasmids were donated en bloc from another bacterial species. Although there must be certain limits to the range of plasmid transfer in nature, the host range of some plasmids is surprisingly broad. In fact, Darryl Reanney (54) estimated that at least 356 extrachromosomal elements can introduce genetic information into *Escherichia coli,* an amount of exogenously contributed DNA, in the form of plasmids, equal to or greater than the DNA content of the *E. coli* chromosome. The other possibility involved an important and intriguing variation on the basic theme of extrachromosomal genetic exchange, namely, the phenomenon of transposition. Transposition is an event by which genetically and physically discrete segments of DNA are able to move to new sites (usually leaving an original copy behind), to invert, and to undergo deletion or amplification without the extensive DNA sequence homology required for classical recombination. Stanley went on record favoring the transpositional event alternative to explain the sudden and unexpected accretion of R plasmids by pathogenic *Haemophilus* strains.

About this time, Johannes (Hans) de Graaff, on sabbatical from the Free University, Amsterdam, came to the lab and enthusiastically joined the *Haemophilus* project. We knew that both RSF007 and RSF0885 encoded a TEM-type β-lactamase characteristic of the enzyme found in certain enteric strains, but this was a far cry from proving an enteric origin for these plasmids. Perhaps the mole fraction G + C values would provide a clue to their origin(s) since the *E. coli* chromosome has a G + C ratio of 0.50, in comparison with a G + C ratio of 0.38 for the *H. influenzae* chromosome. We found mole fraction G + C values of 0.39 and 0.40 for plasmids RSF007 and RSF0885, respectively, values closer to those for *H. influenzae* than *E. coli*. A related question that we asked ourselves was, are RSF007 and RSF0885 disparate plasmid species or are they closely related? The relatedness of these plasmids was of more than trivial epidemiological importance. If we had observed only a single plasmid species within the resistant *Haemophilus* strains, a case might have been made that we were dealing with a single "clone" of drug-resistant *H. influenzae* that had become widely disseminated throughout the world. The finding of the much smaller RSF0885 plasmid in a clinical isolate of *H. parainfluenzae* indicated either that there had been several independent genetic events leading to the development of ampicillin resistance or that a single plasmid had already undergone a series of recombinational interactions. We prepared radiolabeled RSF0885 plasmid DNA and did hybridization experiments with whole-cell DNA extracted from *H. influenzae*(RSF007) and *E. coli*(RP4). RP4 was chosen since it is a plasmid of the P incompatibility group, originating in *Pseudomonas aeruginosa* with a mole fraction G + C of about 0.62. Furthermore, RP4 mediates ampicillin resistance by virtue of the TEM-type β-lactamase contained within the well-characterized transposon TnA. (The term "TnA" actually represents three closely related transposons designated Tn*1*, Tn*2*, and Tn*3*. DNA sequence analysis by Chen and Clowes [6] suggests that the *Haemophilus* and *Neisseria* R plasmids contain DNA sequences most closely related to, if not identical to, Tn*2*. However, for simplicity, I have chosen to use the generic designation TnA throughout this chapter.) RSF0085 shared approximately 27% of its polynucleotide sequences with RSF007 as well as with the totally unrelated plasmid RP4 (Table 1). This result strongly suggested that the region of homology between the two *Haemophilus* R plasmids and between RSF0885 and

TABLE 1

Hybridization between ³H-labeled RSF0885 and RSF1030 plasmid DNA and whole-cell DNA

Source of unlabeled DNA	% Relative DNA sequence homology with ³H-labeled plasmid DNA[a]	
	RSF0885	RSF1030
H. influenzae G32(RSF007)	27.3	48.3
H. influenzae G32(RSF0885)	100	21.2
H. influenzae G32(R⁻)	0	1
E. coli J5(RP4)	26.2	NT[b]
E. coli J5(R⁻)	0.8	0
E. coli C600(RSF1030)	NT	100

[a] The degree of DNA-DNA duplex formation was assayed by the S1 endonuclease method (7). The actual extent of binding of ³H-labeled RSF0885 and RSF1030 plasmid DNA with whole-cell *H. influenzae* (RSF0885) and *E. coli* (RSF1030) DNA was approximately 87%. All other reactions were normalized to these values taken as 100%. Each value shown is the average of three separate reactions.

[b] NT, not tested.

RP4 was confined to the TnA DNA segment. Of course, we were quite fortunate to have in the adjoining laboratory folks like Maggie So, Fred Heffron, Ron Gill, and Craig Rubens who were busy unraveling the molecular nature, origin(s), and insertion specificity of TnA (28, 29, 69) (see chapter 3 on Tn3). They provided us with plasmid RSF1030, a 5.6-MDa replicon originally isolated from *E. coli* and harboring the entire TnA sequence of enteric origin. Table 1 shows the DNA-DNA hybridization values when radiolabeled RSF1030 DNA was reacted with various whole-cell DNA preparations. DNA extracted from *H. influenzae* (RSF007) hybridized 48% with RSF1030. The arithmetic indicated that the 30-MDa *Haemophilus* plasmid shares a region with RSF1030 which is approximately 2.6 MDa in mass, a number reasonably close to the reported 3.2-MDa mass of TnA. On the other hand, DNA extracted from strains harboring RSF0885 shared a lower level of sequence homology (21%) with RSF1030. In terms of the molecular mass, the region of homology was 1.2 MDa, which was close to the estimated mass of the structural gene for β-lactamase.

To visualize in more detail the exact nature of the TnA sequence in both *Haemophilus* plasmids, we heteroduplexed each plasmid with R648, a replicon containing the entire ampicillin transposon with the added feature of one *Eco*RI cleavage site that served to orient TnA in heteroduplex analyses. Figure 3A shows a single-stranded molecule of R648 prepared after cleavage with *Eco*RI. The TnA region is identified by virtue of the inverted repeat base sequences that are located at both ends of the ampicillin transposon and that anneal in a single-stranded molecule to form a stalk. Figure 3B shows a heteroduplex of RSF007 and a linear single-stranded R648 molecule. It is clear that the double-stranded region corresponds to the TnA sequence by virtue of its orientation to the short arm of R648. In addition, the double-stranded region measures 3.2 MDa in mass, which is in good agreement with the reported size of TnA. Figure 3D shows a heteroduplex of RSF0885 and *Eco*RI-cut R648. Here again, the orientation of the short, single-stranded arm of R648 shows that the double-stranded region of homology is within the TnA segment. The fact that the double-stranded region between R648 and RSF0885 starts at the same point as the double-stranded region between R648 and RSF007 argues that only one of the inverted repeats is present in RSF0885. In addition, the region of homology measures approximately 1 MDa in mass, confirming the DNA-DNA hybridization data (Table 1) that indicated that only one-third of the TnA DNA segment is present in RSF0885 (8).

This particular set of heteroduplex molecules provided a great deal of support to the hypothesis that an indigenous, phenotypically "silent" *Haemophilus* plasmid had become the unwitting recipient of TnA from an anonymous donor R plasmid. In the not too distant past, perhaps an enteric plasmid or one of the more promiscuous plasmids of the P incompatibility group carrying ampicillin resistance in the form of TnA might have been introduced into *H. influenzae*. Presumably, this "visitor" could not be stably maintained in *H. influenzae*, but it did reside in its host long enough for TnA to transpose into an indigenous *Haemophilus* plasmid where it could be stably maintained. If this speculation is essentially correct, at least two separate events must have transpired, since RSF007 (30 MDa) and RSF0885 (3 MDa) are essentially unrelated and share base sequence homology

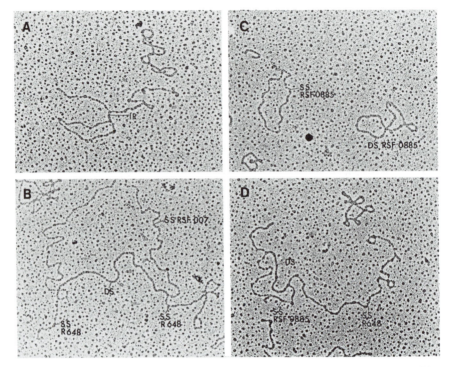

FIGURE 3. Heteroduplexes of RSF007 and RSF0885 plasmid DNA with *Eco*RI-cut R648 plasmid DNA. (A) *Eco*RI endonuclease-cut single-stranded molecule of R648. The TnA region is marked by the inverted repeats that reanneal in a single-stranded molecule, forming a double-stranded region (arrow). (B) Heteroduplex of RSF007 and *Eco*RI-cut R648. (C) Single- and double-stranded molecules of RSF0885. Note the absence of an inverted repeat nucleotide base sequence in the single-stranded molecule. (D) Heteroduplex of RSF0885 and *Eco*RI-cut R648. SS, single-stranded DNA; DS, double-stranded DNA; IR, inverted repeat nucleotide base sequence.

only within the confines of their TnA segments. We suspected this to be the case on the basis of DNA-DNA hybridization analysis results and later confirmed it by way of heteroduplex analysis. A heteroduplex molecule formed between RSF007 and RSF0885 clearly showed that these two plasmids shared one continuous region of homology that constituted approximately 1 MDa of the RSF0885 strand (8).

The plot thickened considerably when Mark Richmond and Jon Saunders found a *Haemophilus* R plasmid (pUB701) specifying tetracycline resistance in the United Kingdom (64) and van Klingeren et al. (76) described an R plasmid (pR1234) encoding both tetracycline and chloramphenicol resistance in a *Haemophilus* strain from the Netherlands (Table 2). RSF007 and pUB701 were found to share about 65% of their polynucleotide sequences with the isolate from the Netherlands. Heteroduplex analysis with single strands of RSF007 and pUB701 underscored the close relationship between these R plasmids (Fig. 4). RSF007 has the characteristic single-stranded loop of about 3.2 MDa indicative of TnA.

TABLE 2
Properties of selected, naturally occurring, large-molecular-weight *Haemophilus* R plasmids

Plasmid	Resistance(s) specified[a]	Molecular mass (MDa)	Mol fraction G + C content	Origin[b]	Reference
RSF007	Ap	30	0.39	USA	13
pUB701	Tc	31	0.39	UK	64
pR1234	Tc, CAM	38	?	NL	76
pKRE5367	Ap	30	?	FRG	40
pFR16017	Tc	33	?	FRG	40
pHK539	Ap, Tc	36	?	DK	37
RSF017	None	26	0.36	USA	20
pW266	None	27	?	FRG	40

[a] Ap, ampicillin; Tc, tetracycline; CAM, chloramphenicol acetyltransferase.
[b] DK, Denmark; NL, Netherlands; FRG, West Germany; UK, United Kingdom; USA, United States.

Plasmid pUB701 lacks this structure and, in turn, contains a most prominent (1,400-bp) inverted repeat sequence and a 4-MDa single-stranded loop, characteristic of the tetracycline resistance determinant Tn*10*. It is also clear from Fig. 4 that polynucleotide differences between RSF007 and pUB701 are not strictly confined to the drug resistance determinants. For example, a small substitution loop near the Tn*A* sequence and three additional, closely grouped substitution loops are clearly visible in the heteroduplex molecule. Therefore, although the "cores" of RSF007 and pUB701 are highly related, in terms of their polynucleotide sequences, they are not identical (20). A couple of additional things are worth men-

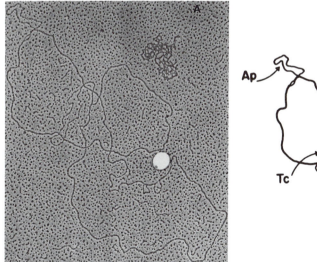
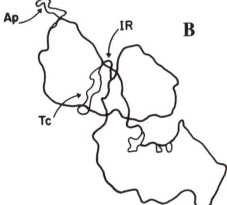

FIGURE 4. (A). Heteroduplex of RSF007 and pUB701 plasmid DNA. (B) Diagram of RSF007/pUB701 heteroduplex. Thick line, double-stranded DNA; thin line, single-stranded DNA; Ap, ampicillin resistance transposon (Tn*A*); Tc, tetracycline resistance transposon (Tn*10*); IR, inverted repeat sequence.

tioning about plasmid pUB701. First, it is not self-transmissible, which might be explained by the insertion of Tn*10* into the transfer operon of the plasmid, thus inactivating transfer functions (64). Second, the tetracycline resistance determinants found on Tn*10* are inducible in all species examined to date except *Haemophilus* spp., in which there appears to be constitutive expression of resistance. The reason for this unregulated expression appears to be an inactive repressor molecule (42).

If, in fact, certain *Haemophilus* isolates do harbor a common core plasmid that has served as a target for various transposable elements bearing drug resistance determinants, then it should be possible to find it (we thought). We screened over 200 drug-susceptible *H. influenzae* and *H. parainfluenzae* clinical isolates and were highly encouraged to find one strain, *H. influenzae* CL-17, that harbored a 26-MDa cryptic plasmid (20). This would have been a perfect candidate for the putative indigenous plasmid, since a translocation of either Tn*A* or Tn*10* into it would result in R plasmids of 30 to 31 MDa in mass. Unfortunately, this particular plasmid had insufficient nucleotide sequence homology with either pUB701 or RSF007 to merit further study (20). The apparent rarity of the appropriate cryptic plasmid may provide a clue as to why it took so long for R plasmids to appear among clinical isolates of *H. influenzae*. Perhaps *Haemophilus* spp. have been relatively late in acquiring high-level drug resistance because of the low probability of both the initial intergeneric transfer event coupled with the equally low probability of the establishment and stable maintenance of foreign or newly created R plasmids. Once established and stably maintained, however, these R plasmids presumably were able to spread throughout the bacterial population, especially in response to the selective pressure of antibiotic use (64).

The quest for the indigenous target plasmid ultimately had a happy ending; it occurred in the Federal Republic of Germany. During our work on the tetracycline resistance plasmid from the United Kingdom, Rainer Laufs took a hiatus from the University of Göttingen and came to Stanley's lab to learn about infectious drug resistance. Subsequent to his return to the Federal Republic of Germany, Rainer and colleagues managed to isolate and characterize a series of conjugative *Haemophilus* R plasmids containing various permutations of transposable drug resistance determinants (34, 36). In addition, they examined 699 *H. influenzae* strains isolated from healthy people, and one strain appeared to have harbored the "phantom" replicon (40). Plasmid pW266 (Table 2) is 27 MDa in mass, is phenotypically silent, and shares approximately 82% nucleotide sequence homology with the cores of other characterized, high-molecular-weight *Haemophilus* R plasmids; in short, it was an ideal candidate. The coup de grace, of course, would have been to rig a strain that harbors both pW266 and an R plasmid containing Tn*A* or Tn*10*, or both, and monitor the acquisition of drug resistance determinants by pW266 by the mechanism of transposition. As far as I know, this has never been accomplished, perhaps because of incompatibility or instability problems or because pW266 appears to have a predilection to integrate into its host chromosome, as do many *Haemophilus* extrachromosomal elements (73). In theory, at least, the discovery of pW266 presented a rare and provocative opportunity to catch Mother Nature "in the act."

GONOCOCCI ACQUIRE R PLASMIDS

One spring afternoon in 1975, Hans de Graaff and I were huddled in the green glow of the JEOL instrument panel gazing at DNA heteroduplex molecules when Stanley walked into the electron microscopy room. "*Nu*," he says to me, "it's time we got back to gonorrhoeae." I found this to be a less than tempting proposal, but I assumed he shared my vague sense of guilt about working on *Haemophilus* plasmids under the aegis of a grant to study gonococci. However, he assured me that guilt (actually, he used a Yiddish word that I forget or that I should not repeat) played no role in his decision. He wanted me to do a more comprehensive survey of the indigenous plasmids in a variety of *Neisseria* species, pathogenic and "nonpathogenic," because he was certain that it was only a matter of time before the gonococcus would be blessed with R plasmids. In fact, he told us that he was going off to a conference on sexually transmitted diseases in London and was planning to make that very prediction in front of God and everyone. And he did (23). He suggested to his London audience that gonococci would acquire R plasmids in much the same manner as *Haemophilus* species did. A few strains of *N. gonorrhoeae* had already been shown to harbor a phenotypically cryptic, 24-MDa plasmid, an attractive recipient for Tn*A* or other transposable resistance determinants. He further suggested that commensal *Neisseria* species might serve initially as a kind of reservoir or breeding ground for newly created R plasmids. Olsen and Shipley (48) had already shown that a P incompatibility group plasmid, R1822, could be transmitted by conjugation from *P. aeruginosa* and *Pseudomonas fluorescens* to a strain of *Neisseria perflava* and that the resultant transconjugants stably maintained R1822 if ampicillin pressure was applied. On the basis of this datum, as well as other clues and hunches, Stanley proposed that commensal *Neisseria* species might ultimately serve as unwitting hosts for the creation and maintenance of R plasmids which would eventually penetrate not-so-benign recipients like *N. gonorrhoeae* and/or *Neisseria meningitidis*.

Hans and I were a bit surprised that he would risk such a prediction in public—in England, no less—and we asked him whether he was sure he knew what he was doing. He seemed unmoved by our pusillanimous pleas, and as he walked to the door, I suggested that instead of wearing a tweed sport coat to the conference he wear a long, flowing black gown embossed with gold crescent moons and shooting stars and shoes that turned up at their ends. He seemed unamused by my suggestion and quietly closed the door behind him.

Cryptic Plasmids in Neisseriae

Originally, the search for plasmids in *N. gonorrhoeae* was stimulated by the fact that many strains possessed phenotypic properties that have been shown to be plasmid encoded in other bacterial species. These include drug resistance, pilus biosynthesis, and bacteriocin production. The in vitro and in vivo transition from a piliated, virulent state to a nonpiliated, avirulent form was a most suggestive switch, in that the apparent irreversible loss of the ability to produce pili which accompanied a change in colonial morphology was certainly compatible with the loss of a plasmid species. Moreover, the role of piliation, in terms of its ability to drastically modify gonococcal pathogenicity, was reminiscent of the role played by plasmid-mediated adhesins, which have been well characterized in a variety of diarrheagenic *E. coli*.

Maness and Sparling (41) and Engelkirk and Schoenhard (21) were the first groups to demonstrate the presence of extrachromosomal DNA in *N. gonorrhoeae*, using density equilibrium centrifugation. While those studies affirmed the presence of plasmid DNA in the gonococcus, the first biophysical characterization was reported by Mayer, Holmes, and Falkow (43). Five of the six gonococcal isolates that they examined harbored a plasmid of 2.6 MDa in mass that was present as a multicopy pool of 24 to 32 copies per chromosome equivalent, with a mole fraction G + C content of 0.50 (identical to that of *N. gonorrhoeae* chromosomal DNA). Most significant was the observation that there was no loss or alteration of the plasmid pool during the in vitro type 1 to type 4 colony transition associated with the loss of pilus biosynthesis and virulence. Stiffler et al. (72) found a 24.5-MDa cryptic plasmid residing with the 2.6-MDa plasmid species in both ampicillin-susceptible and -resistant isolates and showed that this larger plasmid was not related to the amino acid requirements of the strains that they studied. Although of little interest at the time of its discovery, this 24.5-MDa plasmid would ultimately play a central role in the dissemination of drug resistance among strains and species of *Neisseria* (56).

One of the difficulties in negotiating any large-scale survey of plasmid incidence is the fact that the technology for isolating and characterizing extrachromosomal DNA was time-consuming and relatively costly, thus limiting the number of isolates that could conveniently be examined. Jane Meyers and David Sanchez, in Stanley's laboratory, were working very hard on developing a method by which the number of extrachromosomal DNA species as well as their molecular masses and configurations could be quickly screened by agarose gel electrophoresis. Their efforts paid off handsomely; the method that they developed proved to be applicable to virtually all gram-negative clinical isolates and laboratory strains and was suitable for the detection and enumeration of plasmids in partially purified whole-cell lysates (45). Plasmids with molecular masses in the range of 0.6 to 95 MDa were easily visualized by this convenient and relatively quick method. This procedure soon became an important adjunct for survey work as well as an important aid in the epidemiological investigation of plasmid dissemination, and the timing could not have been better!

We initially applied the Meyers and Sanchez method to 21 clinical isolates of *N. gonorrhoeae:* 12 strains isolated from classical urethral infections and 9 strains from disseminated gonococcal infections. Twenty of the 21 *N. gonorrhoeae* strains examined contained plasmids; the 2.6-MDa species appeared in 19 isolates (90%), whereas the 24.5-MDa plasmid was found in only 1 strain. Interestingly, none of the 11 meningococcal isolates harbored extrachromosomal DNA (14). The apparent absence of plasmid DNA in *N. meningitidis* and the high incidence in gonococci posed interesting evolutionary questions. It seemed likely that the gonococcus and meningococcus evolved from the same ancestral species. Did this ancestral strain contain a plasmid(s) and later diverge into two species, namely,

N. gonorrhoeae, which to this day retains an extrachromosomal element, and *N. meningitidis*, which lost this accessory element? Or did the acquisition of plasmid DNA by the gonococcus occur after the divergence of a plasmid-negative ancestral strain into two species?

The commensal *Neisseria* species comprised a heterogeneous group which could arbitrarily be subdivided into three general categories. These groups were (i) *N. subflava*, *N. perflava*, and *N. flava*, which did not contain plasmid DNA; (ii) *N. lactamica*, a species in which the presence of extrachromosomal DNA is not common but exceptions occur; and (iii) *N. sicca* and *N. flavescens*, species which generally possess multiple plasmids with large molecular masses. However, since only a few commensal strains were analyzed in that survey, it was somewhat perilous to make firm conclusions regarding the incidence of cryptic plasmids solely on the basis of these data (14).

R Plasmids in Neisseriae

Stanley's London prediction became a reality in early 1976, but not exactly by the mechanism that he had envisioned. Gonococcal isolates with high-level ampicillin resistance and producing β-lactamase were isolated in the Far East (3), London (50), and Liverpool (49). During the following 6 months, similar resistant gonococcal isolates were identified in 11 countries; 94 strains were isolated from patients in 16 states in the United States (33). At about this time, a graduate student in Stanley's laboratory, Marilyn Roberts, joined the *Neisseria* project.

We analyzed a selection of these clinical isolates and found that strains originating from men returning from the Far East and their sexual contacts harbored a common 4.6-MDa plasmid (pMR0360; Fig. 5) (19). Transformation studies and the isolation of spontaneous ampicillin-susceptible segregants devoid of extrachromosomal DNA strongly suggested that the β-lactamase gene resided on pMR0360. In addition, a resistant strain isolated in London as well as three strains originating

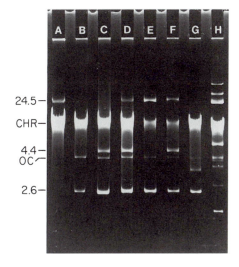

FIGURE 5. Agarose gel electrophoresis of ethanol-precipitated DNA from cleared lysates of penicillin-susceptible (Pen^s) and penicillin-resistant (Pen^r) *N. gonorrhoeae* strains. Lanes: A, strain KH45 (Pen^s) containing the 24.5 × 10^6-Da cryptic plasmid; B, strain F62 (Pen^s) containing the 2.6 × 10^6-Da cryptic plasmid; C, strain CDC66 (Pen^r), Far East isolate; D, strain CDC67 (Pen^r), Far East isolate; E, spontaneous Pen^s derivative of strain CDC01; F, strain CDC01 (Pen^r), from a case contact of a patient infected in the Far East; G, strain IPL (Pen^r) from London, England; H, standard plasmid DNAs ranging in size from 62 × 10^6 (uppermost band) to 1.9 × 10^6 (lowest band) Da. OC refers to the open circular form of the 2.6 × 10^6-Da cryptic plasmid. Numbers indicate molecular masses (in megadaltons). CHR indicates chromosomal DNA.

TABLE 3
Molecular characteristics of small β-lactamase-specifying plasmids found in *N. gonorrhoeae* and *Haemophilus* species

Plasmid	Host (source)	Molecular mass (MDa)	Mol fraction G + C content	TnA (%)
pMR0200	*N. gonorrhoeae* (West Africa)	3.4	0.41	40
pMR0360	*N. gonorrhoeae* (Far East)	4.6	0.41	40
RSF0885	*H. parainfluenzae* (United States)	4.1	0.40	34
pJB1	*H. ducreyi* (Canada)	5.7	?	100
pHD131	*H. ducreyi* (Kenya)	7.0	?	100

in Liverpool harbored a slightly smaller (3.2-MDa) R plasmid designated pMR0200 (Fig. 5). Further analysis showed both gonococcal R plasmids to have a mole ratio G + C content of 0.40 and to contain approximately 40% of the TnA sequence (Table 3) (19).

The clinical situation with respect to drug-resistant gonococci seemed dire; these R plasmids were soon endemic in isolates from North America, the Caribbean and Europe, as well as Africa and Asia (59). What was the primary mode of dissemination of the gonococcal R plasmids? Transformation as an important mechanism was initially ruled out by the failure to transform the gonococcus with covalently closed circular DNA (57). But there was an important clue; whereas an examination of drug-susceptible clinical isolates of *N. gonorrhoeae* revealed that 1 of 30 contained an indigenous 24.5-MDa plasmid (14), about one-half of the ampicillin-resistant gonococcal strains isolated in the Far East harbored this large plasmid. This suggested the possibility that this plasmid might possess sex factor activity and that R-plasmid transmission between gonococci was mediated by conjugation. Marilyn Roberts negotiated a series of mating experiments using a modified filter method and showed that the transfer of R plasmids occurred only when the gonococcal strain possessed the 24.5-MDa plasmid as well as the R plasmid (57). Cell-free filtrates of the donor strains could not transfer the R plasmid to *N. gonorrhoeae* or *E. coli* recipient cells, lessening the likelihood that phage-mediated transduction could account for these results. Simultaneous with Marilyn's finding, Eisenstein et al. (12) and Kirven and Thornsberry (38) also reported the conjugal transfer of penicillinase-producing R plasmids between *N. gonorrhoeae* strains.

Of course, the epidemiological implications of these results were quite ominous; not only must one be concerned about widespread dissemination among gonococci but one must also be concerned about the extension of high-level ampicillin resistance to strains of *N. meningitidis* which were historically quite drug susceptible. This possibility was made more likely by the finding that in overnight mating experiments, significant numbers of gonococcal recipients received the 24.5-MDa plasmid, and thus retained the potential of transmitting the R plasmid to new strains and species (57).

Furthermore, we were struck by the gross similarity between the two gonococcal R plasmids and the *H. parainfluenzae* plasmid, RSF0885, which also had a mole ratio G + C content of 0.40 and harbored approximately 40% of the TnA sequence (Table 3). Subsequent DNA-DNA duplex studies showed that RSF0885 shared over 90% of its nucleotide base sequence with pMR0360 (4.6-MDa species) and was 70% related to pMR0200, the 3.4-MDa gonococcal R plasmid (18), hence, a denouement of sorts: the β-lactamase-specifying R plasmids in *N. gonorrhoeae* were virtually indistinguishable from the small R plasmid originating in *H. parainfluenzae*. Of course, this did not necessarily mean that a *Haemophilus* species served as the direct source of the replicon now residing in the gonococcus. Our hunch was that this event represented an extension of the enteric R-plasmid pool across what appeared to be an increasingly vulnerable species barrier (18, 24).

In any event, the emergence of R plasmids in *Haemophilus* species and *N. gonorrhoeae* ushered in an ominous chapter in the story of drug resistance plasmids. Historically, plasmids have played a critical clinical role, mainly in specific epidemic diseases such as typhoid or dysentery and in isolated hospital environments. Hence, it could have been argued that antibiotic resistance plasmids really only affected a rather limited portion of the population. However, with the emergence and dissemination of penicillinase-producing strains of *H. influenzae* and the gonococcus, R plasmids gained the medically important potential to affect a much broader segment of the population since both of these organisms can cause serious or even fatal infections in individuals living in the community at large (15).

TOBRAMYCIN RESISTANCE IN A BURN UNIT

My "last hurrah" in Seattle was accompanied by a new cast of characters, namely, Erling Ordal, Barbara Minshew, and Julia Inamine. Stanley had a habit of saying to anyone who would listen, "Go down the hall and visit Erling; he's a treasure of microbiology." So, on occasion, I would hang out for awhile in Dr. Ordal's lab, a magnificently cluttered place awash in exotic smells and sights. Professor Ordal was in the process of retiring and he claimed that he was "slowly getting things in order," although to the casual observer, at least, there was no visible evidence of this. One day, he appeared at my door in one of his stained and rumpled lab coats—the scientific equivalent of Columbo. "I've got a student I want you to be a mentor to," he said. "She's about to make a terrible mistake and get involved in a 'Mickey Mouse' project—so, she should work with you." With that, he went back down the hall to his lab, and the next morning Julia Inamine came by and we agreed to do a project together.

Barbara Minshew had just joined the Department of Surgery at the University of Washington, and she brought along with her a few tobramycin-resistant strains of *Klebsiella pneumoniae* and *Enterobacter cloacae*. In 1974 investigators (46) had conducted a survey of resistance to gentamicin and tobramycin among strains of the family *Enterobacteriaceae* in cultures of samples obtained from over 1,000 patients with burn wounds at Parkland Memorial Hospital, Dallas, Texas. In the course of this retrospective study, tobramycin-resistant isolates of *K. pneumoniae*, *E. cloacae*, *E. coli*, *Citrobacter* spp., and *P. aeruginosa* were collected. These resistant gram-negative organisms appeared at various intervals over an 11-month period when tobramycin was used as an investigational drug in patients with burns over more than 40% of their body surface areas and who exhibited persistent signs

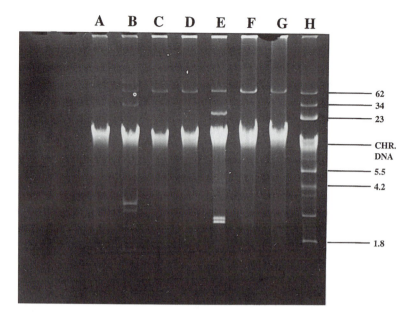

FIGURE 6. Agarose gel electrophoresis of ethanol-precipitated DNA from cleared lysates of tobramycin-resistant (TOB^r) and tobramycin-susceptible strains. Plasmid pIE098 is a 64-MDa conjugative R plasmid specifying resistances to tobramycin, kanamycin, neomycin, streptomycin, and ampicillin. Lanes: A, *E. coli* 185(R$^-$); B, TOB^r *E. cloacae* clinical isolate; C and D, *E. coli* 185(pIE098), derived from mating with the TOB^r *E. cloacae* strain; E, TOB^r *K. pneumoniae* clinical isolate; F and G, *E. coli* 185(pIE098), derived from mating with the TOB^r *K. pneumoniae* isolate; H, purified standard plasmids for molecular weight determinations. Numbers indicate megadaltons. CHR. DNA, chromosomal DNA.

of sepsis after treatment with gentamicin and carbenicillin. One *K. pneumoniae* and two *E. cloacae* strains were found to be capable of enzymatically acetylating tobramycin with an activity closely resembling that of aminoglycoside-6'-acetyltransferase, a hallmark of plasmid involvement (16).

We carried out a series of mating experiments and successfully cotransferred the resistances to tobramycin, kanamycin, and neomycin to *E. coli* transconjugants. Osmotic lysates from all of the transconjugant strains acetylated tobramycin. The two tobramycin-resistant clinical isolates (*E. cloacae* 264 and *K. pneumoniae* 355) were examined for the presence of extrachromosomal DNA by agarose gel electrophoresis (Fig. 6). All of the resistant strains contained a common high-molecular-mass (approximately 64-MDa) plasmid species which was absent from the tobramycin-susceptible strain. DNA-DNA hybridization studies and restriction endonuclease analyses definitively proved that these R plasmids were identical (17).

The incidence of R-plasmid-mediated drug resistance among clinical isolates is most evident in highly selective environments such as intensive care units and burn wards, where the daily administration of high-potency antibiotics is often a necessity. In this investigation we proved, by a variety of molecular techniques, the identities of two conjugative R plasmids isolated from bacterial strains from two gram-negative genera originating in a hospital burn unit. This study was limited by its retrospective nature; over an 11-month period, tobramycin resistance emerged in gram-negative bacterial species representing five different genera (46), but unfortunately, only the *Enterobacter* and *Klebsiella* spp. were saved. Had the other three isolates been available for analysis, it might have been possible to confirm that a single R plasmid "took a romp" through a burn unit within less than a year after tobramycin therapy had been approved for experimental use. (For an update on this important subject, see chapter 5 on molecular epidemiology.)

My tenure at the University of Washington had come to a close. Now, as I look back, there was a marvelous symmetry to the whole affair; hanging out in the clinical microbiology lab had led me to drug-resistant *Haemophilus* species, and shooting the bull with the late Professor Ordal led me to tobramycin-resistant clinical isolates. Perhaps Woody Allen was right after all when he said, "In life, half of the battle is merely showing up."

A DECADE AND A HALF LATER

Proposed Origin of the Large R Plasmids

Molecular analyses, time, and the clarity afforded by hindsight have revealed two main groups of antibiotic resistance plasmids in *Haemophilus* species and *N. gonorrhoeae* (Table 2 and 3). The large (30- to 38-MDa), conjugative *Haemophilus* plasmids are the most important cause of resistance in clinical strains of *H. influenzae*. These plasmids, isolated throughout the world, carry various permutations of transposable genes specifying β-lactamase, tetracycline resistance, and chloramphenicol acetyltransferase (Table 2). Most importantly, these large R plasmids also appear to constitute a worldwide "family" of sorts, in that they all appear to be highly related but have individual differences in terms of their exact nucleotide sequences. Thus, the earlier findings of our group (20, 64) and Laufs and colleagues (34, 36) have been extended and confirmed by other investigators. Arnold Smith, Marilyn Roberts, and their group (60) found 70 to 90% homology between the prototypical ampicillin resistance plasmid RSF007 and nine epidemiologically distinct plasmids specifying resistances to tetracycline and chloramphenicol. Albritton et al. (2) have described similar conjugative tetracycline and tetracycline-chloramphenicol resistance plasmids in *Haemophilus ducreyi* originating in Kenya and Seattle, respectively.

Faced with such an imposing collection of highly related plasmids, it is tempting to embrace the original hypothesis posited to account for their origins. That is, a series of transpositional events targeted a common core plasmid in the form of an indigenous *Haemophilus* extrachromosomal element. There is, however, no compelling reason to believe that this putative cryptic plasmid originated in *H. influenzae*. Quite the opposite; Scheifele and Fussell (65) have shown that β-lactamase-specifying strains of *H. parainfluenzae* are far more prevalent in the oropharynxes of children than are similar strains of *H. influenzae*. In addition, the former strains can conjugally transfer their R plasmids conferring ampicillin resistance to *H. influenzae* recipients (66). Brunton et al. (5) have proposed that *H. parainfluenzae* strains may have served as convenient in vivo reservoirs for the assembly and maintenance of plasmids that eventually ended up in pathogenic species of *Haemophilus*.

Proposed Origins of the Small R Plasmids

How were the smaller (3.4- to 7.0-MDa; Table 3) β-lactamase-specifying plasmids generated? The original idea that these R plasmids might have evolved by transposition of Tn*A* into an ancestral *Haemophilus* plasmid was first proposed on the basis of the finding that the mole ratio G + C content for the *Haemophilus* chromosome and for all of the small R plasmids (including those found in *Neisseria* species) was 0.40, in contrast to a ratio of 0.50 for *Neisseria* chromosomal DNA. This hypothesis has essentially withstood the test of time. In fact, subsequent experimental evidence suggests that they did indeed evolve from a single ancestral plasmid and that they differ from each other by virtue of small deletions or insertions within their Tn*A* regions and sometimes in their non-Tn*A* regions (5, 59). For example, the two small *H. ducreyi* R plasmids pJB1 and pHD131 are identical except that the larger one (pHD131) contains a 1.3-MDa insertion in its non-Tn*A* region. Both plasmids, in turn, are highly related to the gonococcal R plasmids pMR200 and pMR0360 (Table 3). Heteroduplex analysis showed that pHD131 is identical to pMR0360, except that the *H. ducreyi* R plasmid harbors the entire Tn*A* sequence. Similarly, pJB1 is identical to pMR0200 except for the fact that pJB1 also contains an intact ampicillin resistance transposon (5).

If our assumptions are correct regarding the evolution of these elements, the question remains, whence this ancestral plasmid? Brunton and colleagues (5), once again, favor *H. parainfluenzae* as the breeding ground for their origin. They base this conclusion on the identification of a number of small cryptic plasmids (e.g., pHPA300) originally isolated from drug-susceptible strains of *H. parainfluenzae*. These cryptic plasmids can form heteroduplexes with all of the plasmids listed in Table 3. Thus, the heteroduplex molecules formed between pHPA300 and pJB1 showed complete homology with the exception of the Tn*A* segment of pJB1. In addition, heteroduplexes between the cryptic plasmid and pHD131 (7.0-MDa R plasmid found in *H. ducreyi*) were identical except for an extra 1.8-MDa insertion loop contributed by the pHD131 strand. Moreover, heteroduplexes formed between pHPA300 and the smaller (3.2-MDa) gonococcal plasmid showed similar homology, except that a 300-bp segment adjacent to a smaller (1.7-kb) Tn*A* loop was absent from the gonococcal plasmid. Finally, heteroduplexes of pHPA300 and the larger (4.6-MDa) gonococcal plasmid were identical to former molecules, apart from a similar 1.8-kb insertion loop contributed by the gonococcal plasmid strand.

Thus, a strong case can be made that most, if not all, of these small β-lactamase-specifying plasmids originated by the insertion of Tn*A* into phenotypically cryptic plasmids commonly residing in *H. parainfluenzae*. Presumably, these plasmids underwent independent insertions or deletions, or both, within and outside of their Tn*A* regions as they migrated into and through strains of *N. gonorrhoeae*, *H. influenzae*, and *H. ducreyi* (5, 9, 10, 55, 75, 78). However, a critical experimental link in the proposed evolutionary pathway for these small R plasmids is missing. Brunton and colleagues (5) were unable to create a β-lactamase-specifying plasmid by the insertion of Tn*A*, of enteric origin, into *H. parainfluenzae* cryptic replicons. This is reminiscent of the apparent inability of Laufs et al. (40) to reconstitute an R plasmid using the 27-MDa indigenous plasmid pW266 as a transposon acceptor.

The 24.5-MDa Gonococcal Plasmid

The large (24.5-MDa) indigenous gonococcal plasmid pLE2451 was first described in 1974 (14). Flett et al. (25) subsequently showed that this plasmid is capable of efficiently mobilizing β-lactamase plasmids between gonococci and from gonococci to *H. influenzae* and *E. coli*. However, pLE2451 was not detectable in either *E. coli* or *Haemophilus* transconjugants, suggesting that it is inherently unstable in these hosts. In fact, its host range is quite limited and appears to be stably maintained only in the gonococcus and *Neisseria cinerea* (26). The key epidemiological role played by this plasmid is discussed in more detail in other sections of this chapter.

Pathway from Commensal Organisms to Pathogens

Many key questions remain unanswered; what pathway did these plasmids follow as they migrated from commensal organisms into various pathogenic organisms, and what were the modes of plasmid transfer along their journey? In his 1975 London speech, Stanley postulated that when the gonococcus eventually acquired R plasmids, commensal *Neisseria* spp. would play a critical role. There is now solid experimental support for this hypothesis. Genco et al. (26) examined a number of commensal and pathogenic *Neisseria* species for their ability to engage in the conjugal exchange of β-lactamase-specifying plasmids with *N. gonorrhoeae*. They found that R-plasmid stability varied significantly, from quite stable in *N. flava* and *N. cinerea* to highly unstable in isolates of *N. perflava-N. sicca*. Moreover, the large mobilizing plasmid pLE2451 was stable only in *N. cinerea*, which was, in turn, capable of retransferring pLE2451 and the R plasmid back into *N. gonorrhoeae* recipients at a high frequency. These observations are significant since they represent the first consistently successful attempt to transfer the ampicillin-resistant R plasmid from another bacterial species into *N. gonorrhoeae*. In a related finding, McNicol et al. (44) successfully demonstrated the conjugal transfer of R plasmids from *Haemophilus* spp. to the gonococcus only when *N. cinerea* was used as a transfer intermediate. Thus, *N. cinerea* stably maintained R plasmids and facilitated R-plasmid transfer to *N. gonorrhoeae* in a triparental mating system. Related to these findings, it should be noted that Sparling et al. (71) reported the transfer of a β-lactamase-specifying R plasmid from *H. parainfluenzae* to an *N. gonorrhoeae* recipient. However, this experiment has not been repeated.

Or did these R plasmids enter the gonococcus by way of an enteric organism? Piffaretti et al. (51) reported the mobilization of pMR0360 (4.4-MDa gonococcal R plasmid) from *E. coli* to *N. gonorrhoeae* using the incompatibility group P plasmid pUB307. They also showed that when a complete Tn*A* sequence carried on a pMR0360 derivative was introduced into *N. gonorrhoeae*, it underwent a deletion. The nature of the deletion was reminiscent of that found in the Tn*A* sequences present in the original gonococcal β-lactamase plasmids. More recently, Piffaretti and Soldati (52) introduced pHD131, an element originating in *H. ducreyi* and harboring an intact Tn*A* segment, into the gonococcus from *E. coli* by both transformation and conjugative mobilization. Plasmids with no detectable deletions in their Tn*A* sequences were recovered. Thus, the absence in *N. gonorrhoeae* of R plasmids containing complete Tn*A* segments cannot be ascribed to incompatibility between the genetic element and the host, but apparently can be ascribed to a barrier to the introduction of certain foreign DNA into the gonococcus.

Transformation as a Mechanism of R-Plasmid Dissemination

Whereas Roberts and Falkow (57) had originally ruled out transformation as an important mechanism in the dissemination of the gonococcal R plasmids, Sox et al. (70), using plasmid DNA purified from *N. gonorrhoeae*, showed that the 4.6-MDa plasmid could be transformed into competent gonococcal recipients. Twenty percent of the transformed plasmids were deleted derivatives. The most common deleted plasmid had a molecular mass of 3.4 MDa and a restriction digestion fingerprint identical to that of the 3.4-MDa gonococcal R plasmid pMR0200. Those workers postulated that transformation might account for the evolution of this plasmid (70).

R Plasmids in *N. meningitidis*

Our original report (14) that *N. meningitidis* strains were bereft of extrachromosomal DNA has long since gone by the boards. Verschueren et al. (77) have purified cryptic plasmids of 3.3 and 4.8 MDa, respectively, from two different isolates of *N. meningitidis,* and Ison et al. (32) have shown that 11 of the 16 isolates of *N. meningitidis* that they examined harbored indigenous plasmids showing a high degree of homology with the 2.6-MDa cryptic plasmid found in the majority of *N. gonorrhoeae* strains. A few cryptic plasmids which do not share nucleotide sequences with the small gonococcal plasmid have also been identified in strains of *N. meningitidis* (77). In addition, Dillon et al. (11) identified a strain of *N. meningitidis* harboring both the 4.6-MDa β-lactamase-producing R plasmid and the 24.5-MDa mobilizing plasmid that is found in a significant number of gonococcal isolates. Those authors could not, however, demonstrate the mobilization of this R plasmid from its *N. meningitidis* host to either a gonococcal or an *E. coli* recipient. More recently, Ikeda et al. (31) successfully used the 24.5-MDa plasmid to transfer the 4.6-MDa gonococcal R plasmid to *N. meningitidis*. Three of 20 donor strains (*N. gonorrhoeae* harboring both the 24.5-MDa and the 4.6-MDa R plasmids) showed detectable conjugation frequencies ($>10^{-5}$) in mating experiments with *N. meningitidis*. On the other hand, all 20 strains were capable of mobilizing the small R plasmid to *N. gonorrhoeae* and *E. coli* recipients. The β-lactamase-specifying plasmid was stably maintained in *N. meningitidis*, whereas the large, mobilizing plasmid was not detected in any of the transconjugants (31). In a related study, Roberts and Knapp (61), using strains harboring both the β-lactamase plasmid and the 25.2-MDa tetracycline resistance conjugative plasmid, mobilized both the 4.6- and 3.2-MDa R plasmids into a variety of *Neisseria* spp., including *N. meningitidis*.

Small Cryptic Plasmid in *N. gonorrhoeae*

The 2.6-MDa cryptic gonococcal plasmid remains just that, cryptic. Its irrelevance with respect to virulence and pilus production in *N. gonorrhoeae* has been known for a long time (43). Subsequently, it has been shown not to be associated with the gonococcal outer membrane protein PI, PII, or PIII, the immunoglobulin A1 proteases, the 37-kDa iron-regulated protein, or the receptor protein for the iron-sequestering siderophores (59). No function has been assigned to this extrachromosomal element, even though its complete nucleotide sequence has been determined (39). Since approximately 96% of the clinical isolates of *N. gonorrhoeae* harbor this plasmid, it is hard to believe that such a highly conserved replicon is utterly superfluous.

R Plasmids Conferring Sulfonamide Resistance in *Haemophilus* and *Neisseria* Species

Finally, a new group of plasmids ranging in size from 4.9 to 9.4 MDa has been described for *N. meningitidis*, *H. ducreyi*, and a variety of commensal *Neisseria* species (22, 53, 63). These plasmids are genetically related to the incompatibility group Q plasmid RSF1010, a plasmid of enteric origin specifying linked resistance to streptomycin and sulfonamides. In addition, RSF1010 has an extremely broad host range, including strains of *E. coli*, *P. aeruginosa*, *Proteus mirabilis*, and several *Salmonella* species (4). Albritton et al. (1) characterized a 4.9-MDa plasmid, designated pHD148, that specified sulfonamide resistance in clinical isolates of *H. ducreyi*. DNA-DNA heteroduplex analysis showed that this plasmid was 80% related to RSF1010. Some of these plasmids specify only sulfonamide resistance, whereas others encode resistance to penicillin, streptomycin, and sulfonamide (22, 62, 63). Plasmids of the latter variety have been detected in isolates of *Neisseria mucosa*, *N. subflava*, *N. sicca*, and *Eikenella corrodens* (62). Data obtained by several investigators strongly suggest that the multiresistance plasmids carrying streptomycin and sulfonamide resistance and the TEM β-lactamase were created by the transposition of TnA into RSF1010 (59). The fact that this precise transpositional event occurs under laboratory conditions lends credence to this hypothesis. However, Heffron and colleagues (28) found that their laboratory-generated RSF1010::TnA derivatives contained the entire TnA segment, whereas the naturally occurring counterpart did not.

Finally, the cumulative experimental and epidemiological evidence strongly suggests that these sulfonamide resistance plasmids were disseminated from enteric bacteria directly to *Haemophilus* species (5). This is in marked contrast to the β-lactamase plasmids, which are thought to have been generated by a series of recombinational events within *Haemophilus* species.

AT THE END

As luck (or whatever) would have it, the penetration of R plasmids into the genera *Neisseria* and *Haemophilus* happened at a most fortuitous time for many members of the Falkow laboratory. While Mother Nature was engineering new and medically threatening microscopic life forms, a timely convergence of opportunities, ideas, and techniques was occurring in suite G305 of the Health Sciences Building of the University of Washington: specific advances like the S1 endonuclease method for determining DNA relatedness (Jorge Crosa and Don Brenner), a rapid method for plasmid detection and preliminary characterization (Jane Meyers and David Sanchez), the molecular analysis of the transposition process and transposon origins (Magdalene So, Fred Heffron, Ron Gill, and Craig Rubens, as well as Bob Hedges in London), open access and dialogue with a fine clinical microbiology laboratory (Professor John Sherris), and the aspiration of taking R-plasmid research from a strictly epidemiological role in post facto statistics to a level where it might be possible to predict (perhaps intervene with?) plasmid-associated resistance before it is encountered in the clinic and the community at large. Of course, none of this happened by mere chance. In large part, we had Stanley's uncanny scientific insight as well as his good-humored (occasionally profane) prodding and cajoling to thank for this most gratifying and productive confluence.

As it turned out, we and all of the other research groups who joined in the fray became more or less bystanders as the prevalence of drug-resistant isolates of *H. influenzae* in U.S. hospitals went from 5 to 30% by 1983 (68) and the incidence of penicillinase-producing *N. gonorrhoeae* increased to 30 to 70% of isolates in the Philippines and 10 to 30% in East Asia and parts of Africa (27). Despite Stanley's prescience and bold predictions, we had to be content with passively tracking and monitoring the meandering paths of newly created R plasmids and their rearranged derivatives as they emerged, one by one, in clinical isolates of *H. ducreyi, N. meningitidis,* and gonococci. In retrospect, it probably was a bit naive to seriously consider that the goal of prospective intervention would ever be realized. After all, the dynamic of infectious drug resistance dissemination outside of the laboratory is inherently complex and involves a host of variables. And where should one look? Schaberg et al. (67) have reported the transfer of R plasmids from *Serratia marcescens* to *E. coli* in urine! Those investigators suggest that the urinary catheter bag is a potential site for extraintestinal R-plasmid transfer among members of the family *Enterobacteriaceae.* Thus, in situ transfer of resistance factors in the hospital environment may contribute significantly to the incidence of nosocomial infections.

Regardless of whether the dreaded exchange takes place in a contaminated operating room sink or in a patient's oropharynx or whether it takes place by conjugal transfer or by transformation, there are many other hurdles that must be overcome. For example, most organisms that acquire R plasmids do not persist in the ecological niche in which they find themselves. For a newly transferred R plasmid to survive, it must find a host endowed with exceptional survival value when antibiotics are present as well as when they are not. Finally, to make matters even more complex and unpredictable, human-to-human transmission is undoubtedly the most important factor in the spread of drug-resistant bacteria (5).

Emerge and spread they do, however; antibiotic-resistant pathogenic bacteria currently pose a very serious public health problem for the United States and the world. Examples abound; tuberculosis, once thought to be under control thanks to an arsenal of effective therapeutic drugs, has risen like a phoenix from the ashes of defeat. Drug-resistant *Pneumococcus* strains cause otitis media meningitis among children in day-care centers, and in hospitals, clinicians grapple with multidrug-resistant *Staphylococcus* and *Enterococcus* species. All of this occurs against the backdrop of pharmaceutical companies deemphasizing their anti-infective research programs. So, one wonders, where will the next generation of antibiotics come from?

One final consideration: the high-molecular-weight "ancestral" *Haemophilus* R plasmid was found only after examination of 699 drug-susceptible *Haemophilus* isolates, an incidence of 0.014% (40). On the other hand, 10 to 53% (depending on the geographical origin) of *N. gonorrhoeae* isolates harbor a large, cryptic plasmid that possesses both mobilizing and transfer functions. Furthermore, this indigenous replicon has been around since the 1940s (58). So, why hasn't this plasmid served as a transposon target within *N. gonorrhoeae* in an analogous fashion to that proposed for *Haemophilus* species? It would seem to be a sitting duck, and the sheer numbers are overwhelmingly in favor of this scenario for the gonococcus. Back in 1946, the noted British chemist Cyril Hinshelwood proposed a nongenetic theory of drug adaptation in his book *The Chemical Kinetics of the Bacterial Cell* (30). Dr. Hinshelwood rejected the idea that resistance arises by way of spontaneous mutation and was so convinced that his kinetic theory of adaptation was correct that he wrote, "Adaptive changes should so easily occur in ways generally similar to those suggested by the [kinetic] models which have been studied that if they do not, then it is hard to evade the question why not."

Perhaps history is trying to tell Cyril and the rest of us something rather important. Maybe it is merely trying to tell us to take to heart Paul Berg's timely dictum, "Mother Nature always bats last."

REFERENCES

1. **Albritton, W. L., J. L. Brunton, L. Slaney, and I. MacLean.** 1982. Plasmid-mediated sulfonamide resistance in *Haemophilus ducreyi*. *Antimicrob. Agents Chemother.* **21:**159–165.
2. **Albritton, W. L., I. W. MacLean, L. A. Slaney, A. R. Ronald, and H. Deneer.** 1984. Plasmid-mediated tetracycline resistance in *Haemophilus ducreyi*. *Antimicrob. Agents Chemother.* **25:** 187–190.
3. **Ashford, W. A., R. G. Golash, and V. G. Hemming.** 1976. Penicillinase-producing *Neisseria gonorrhoeae*. *Lancet* ii:657–658.
4. **Barth, P. T., and N. J. Grinter.** 1974. Comparison of the deoxyribonucleic acid molecular weights and homologies of plasmids conferring linked resistance to streptomycin and sulfonamides. *J. Bacteriol.* **120:**618–630.
5. **Brunton, J., D. Clare, and M. A. Meier.** 1986. Molecular epidemiology of antibiotic resistance plasmids of *Haemophilus* species and *Neisseria gonorrhoeae*. *Rev. Infect. Dis.* **8:**713–724.
6. **Chen, S.-T., and R. C. Clowes.** 1987. Nucleotide sequence comparisons of plasmids pHD131, pJB1, pFA3, and pFA7 and β-lactamase expression in *Escherichia coli*, *Haemophilus influenzae*, and *Neisseria gonorrhoeae*. *J. Bacteriol.* **169:**3124–3130.
7. **Crosa, J. H., J. Brenner, and S. Falkow.** 1973. Use of a single-strand specific nuclease for analysis of bacterial and plasmid deoxyribonucleic acid homo- and heteroduplexes. *J. Bacteriol.* **115:** 904–911.

8. **De Graaff, J., L. P. Elwell, and S. Falkow.** 1976. The molecular nature of two beta-lactamase specifying plasmids isolated from *Haemophilus influenzae* type b. *J. Bacteriol.* **126:**439–446.

9. **Dickgiesser, N.** 1984. A molecular characterization of *H. influenzae* plasmid pVe445: a comparison with *N. gonorrhoeae* plasmids pNG10 and pNG18. *Plasmid* **11:**99–101.

10. **Dickgiesser, N., P. M. Bennett, and M. N. Richmond.** 1982. Penicillinase-producing *Neisseria gonorrhoeae* and molecular comparison of 5.3- and 7.4-kilobase β-lactamase plasmids. *J. Bacteriol.* **151:**1171–1175.

11. **Dillon, J. R., M. Pauze, and K. H. Yeung.** 1983. Spread of penicillinase-producing and transfer plasmids from the gonococcus to *Neisseria meningitidis. Lancet* **i:**779–781.

12. **Eisenstein, B. I., T. Sox, G. Biswas, E. Blackman, and P. F. Sparling.** 1977. Conjugal transfer of the gonococcal penicillinase plasmid. *Science* **195:**998–1000.

13. **Elwell, L. P., J. de Graaff, D. Siebert, and S. Falkow.** 1975. Plasmid-linked ampicillin resistance in *Haemophilus influenzae* type b. *Infect. Immun.* **12:**404–410.

14. **Elwell, L. P., and S. Falkow.** 1977. Plasmids of the genus *Neisseria,* p. 134–154. *In* R. Roberts (ed.), *The Gonococcus.* John Wiley & Sons, Inc., New York.

15. **Elwell, L. P., and S. Falkow.** 1977. Genetic loose change. *Sciences* **17:**8–11.

16. **Elwell, L. P., and S. Falkow.** 1986. The characterization of R plasmids and the detection of plasmid-specified genes, p. 683–721. *In* V. Lorian (ed.), *Antibiotics in Laboratory Medicine,* 2nd ed. The Williams & Wilkins Co., Baltimore.

17. **Elwell, L. P., J. Inamine, and B. Minshew.** 1978. Common plasmid specifying tobramycin resistance found in two enteric bacteria isolated from burn patients. *Antimicrob. Agents Chemother.* **13:**312–317.

18. **Elwell, L. P., M. Roberts, and S. Falkow.** 1978. Common β-lactamase specifying R plasmid isolated from the genera *Haemophilus* and *Neisseria,* p. 255–256. *In* D. Schlessinger (ed.), *Microbiology—1978.* American Society for Microbiology, Washington, D.C.

19. **Elwell, L. P., M. Roberts, L. Mayer, and S. Falkow.** 1977. Plasmid-mediated beta-lactamase production in *Neisseria gonorrhoeae. Antimicrob. Agents Chemother.* **11:**528–533.

20. **Elwell, L. P., J. R. Saunders, M. H. Richmond, and S. Falkow.** 1977. Relationships among some R plasmids found in *Haemophilus influenzae. J. Bacteriol.* **131:**356–362.

21. **Engelkirk, P. G., and D. E. Schoenhard.** 1972. Physical evidence of a plasmid in *Neisseria gonorrhoeae. J. Infect. Dis.* **127:**197–200.

22. **Facinelli, B., and P. E. Varaldo.** 1987. Plasmid-mediated sulfonamide resistance in *Neisseria meningitidis. Antimicrob. Agents Chemother.* **31:**1642–1643.

23. **Falkow, S., L. P. Elwell, J. de Graaff, F. Heffron, and L. Mayer.** 1976. A possible model for the development of plasmid-mediated penicillin-resistance in the gonococcus, p. 20–133. *In* R. D. Catterall and C. S. Nicol (ed.), *Sexually Transmitted Diseases.* Academic Press Publishers, London.

24. **Falkow, S., L. P. Elwell, M. Roberts, F. Heffron, and R. Gill.** 1977. The transposition of ampicillin resistance: nature of ampicillin resistant *H. influenzae* and *N. gonorrhoeae,* p. 115–125. *In* J. Drews and G. Hogenauer (ed.), *R-Factors: Their Properties and Possible Control.* Springer-Verlag, New York.

25. **Flett, F., G. O. Humphreys, and J. R. Saunders.** 1981. Intraspecific and intergeneric mobilization of non-conjugative resistance plasmids by a 24.5 Mdal conjugative plasmid of *Neisseria gonorrhoeae. J. Gen. Microbiol.* **125:**123–129.

26. **Genco, C. A., J. S. Knapp, and V. L. Clark.** 1984. Conjugation of plasmids of *Neisseria gonorrhoeae* to other *Neisseria* species: potential reservoirs for the β-lactamase plasmid. *J. Infect. Dis.* **150:**397–401.

27. **Handsfield, H. H., E. G. Sandstrom, J. S. Knapp, P. Perine, W. L. Whittington, D. E. Sayers, and K. K. Holmes.** 1982. Epidemiology of penicillinase-producing *Neisseria gonorrhoeae* infections: analysis by auxotyping and serotyping. *N. Engl. J. Med.* **306:**950–954.

28. **Heffron, F., C. Rubens, and S. Falkow.** 1975. Translocation of a plasmid DNA sequence which mediates ampicillin resistance: molecular nature and specificity of insertion. *Proc. Natl. Acad. Sci. USA* **72:**3623–3627.

29. **Heffron, F., R. Sublett, R. W. Hedges, A. Jacob, and S. Falkow.** 1975. Origin of the TEM beta-lactamase gene found on plasmids. *J. Bacteriol.* **122:**250–256.

30. **Hinshelwood, C.** 1946. *The Chemical Kinetics of the Bacterial Cell.* Oxford University Press (Clarendon), London.

31. **Ikeda, F., A. Tsuji, Y. Kaneko, M. Nishida, and S. Goto.** 1986. Conjugal transfer of beta-lactamase-producing plasmids of *Neisseria gonorrhoeae* to *Neisseria meningitidis. Microbiol. Immunol.* **30:** 737–742.

32. **Ison, C. A., C. M. Beillinger, and J. Walker.** 1986. Homology of cryptic plasmid of *Neisseria gonorrhoeae* with plasmids from *Neisseria meningitidis* and *Neisseria lactamica. J. Clin. Pathol.* **39:**1119–1123.

33. **Jaffe, H. W., J. W. Biddle, S. R. Johnson, and J. J. Wiesner.** 1981. Infections due to penicillinase-producing *Neisseria gonorrhoeae* in the United States: 1976–1980. *J. Infect. Dis.* **144:**191–197.

34. **Jahn, G., R. Laufs, P.-M. Kaulfers, and H. Kolenda.** 1979. Molecular nature of two *Haemophilus influenzae* R factors containing resistances and the multiple integration of drug resistance transposons. *J. Bacteriol.* **138:**584–597.

35. **Kahn, W., S. Ross, W. Rodriguez, G. Controri, and A. K. Saz.** 1974. *Haemophilus influenzae* type B resistant to ampicillin. A report of two cases. *JAMA* **229:**298–301.

36. **Kaulfers, P.-M., R. Laufs, and G. Jahn.** 1978. Molecular properties of transmissible R factors of *Haemophilus influenzae* determining tetracycline resistance. *J. Gen. Microbiol.* **105:**243–252.

37. **Kilian, M.** 1976. A taxonomic study of the genus *Haemophilus* with the proposal of a new species. *J. Gen. Microbiol.* **93:**9–62.

38. **Kirven, L. A., and C. Thornsberry.** 1977. Transfer of beta-lactamase genes of *Neisseria gonorrhoeae* by conjugation. *Antimicrob. Agents Chemother.* **11:**1004–1006.

39. **Korch, C., P. Hagblom, H. Ohman, M. Goransson, and S. Normark.** 1985. Cryptic plasmid of *Neisseria gonorrhoeae:* complete nucleotide sequence and genetic organization. *J. Bacteriol.* **163:** 430–438.

40. **Laufs, R., F.-C. Riess, G. Jahn, R. Fock, and P.-M. Kaulfers.** 1981. Origin of *Haemophilus influenzae* R factors. *J. Bacteriol.* **147:**563–568.

41. **Maness, M. J., and P. F. Sparling.** 1973 Multiple antibiotic resistance due to a single mutation in *Neisseria gonorrhoeae. J. Infect. Dis.* **128:**321–326.

42. **Marshall, B., M. C. Roberts, A. Smith, and S. B. Levy.** 1984. Homogeneity of transferable tetracycline-resistance determinants in *Haemophilus* species. *J. Infect. Dis.* **149:**1028–1029.

43. **Mayer, L. W., K. K. Holmes, and S. Falkow.** 1974. Characterization of plasmid deoxyribonucleic and from *Neisseria gonorrhoeae. Infect. Immun.* **10:**712–717.

44. **McNicol, P. J., W. L. Albritton, and A. R. Ronald.** 1986. Transfer of plasmid-mediated ampicillin resistance from *Haemophilus* to *Neisseria gonorrhoeae* requires an intervening organism. *Sex. Transm. Dis.* **13:**145–150.

45. **Meyers, J., D. Sanchez, L. P. Elwell, and S. Falkow.** 1976. A simple agarose gel electrophoretic method for the identification and characterization of plasmid deoxyribonucleic acid. *J. Bacteriol.* **127:**1529–1537.

46. **Minshew, B. H., R. K. Holmes, J. P. Sanford, and C. R. Baxter.** 1974. Transferable resistance to tobramycin in *Klebsiella pneumoniae* and *Enterobacter cloacae* associated with enzymatic acetylation of tobramycin. *Antimicrob. Agents Chemother.* **6:**492–497.

47. **Nelson, J. D.** 1974. Should ampicillin be abandoned for treatment of *Haemophilus influenzae* disease? *JAMA* **229:**322–324.

48. **Olsen, R. H., and P. Shipley.** 1973. Host range and properties of the *Pseudomonas aeruginosa* R-factor R1822. *J. Bacteriol.* **113:**772–779.

49. **Percival, A., J. E. Corkill, P. O. Arya, S. Rowlands, C. D. Alergant, and E. H. Anrels.** 1976. Penicillin-producing gonococci in Liverpool. *Lancet* **ii:**1379–1382.

50. **Phillips, I.** 1976. Beta-lactamase-producing penicillin-resistant gonococcus. *Lancet* **ii:**656–657.

51. **Piffaretti, J. C., J. Frey, and A. Arini.** 1986. A new conjugative system for transferring plasmids into *Neisseria gonorrhoeae,* abstr. H-98, p. 143. *Abstr. 86th Annu. Meet. Am. Soc. Microbiol. 1986.* American Society for Microbiology, Washington, D.C.

52. **Piffaretti, J. C., and L. Soldati.** 1990. TnA transposons can be introduced and maintained in *Neisseria gonorrhoeae. Res. Microbiol.* **141:**519–528.

53. **Pintado, C., C. Salvador, R. Rotger, and C. Nombela.** 1985. Multiresistant plasmid from commensal *Neisseria* species. *Antimicrob. Agents Chemother.* **27:**120–124.

54. **Reanney, D.** 1976. Extrachromosomal elements as possible agents of adoption and development. *Bacteriol. Rev.* **40:**552–590.

55. **Roberts, M., L. P. Elwell, and S. Falkow.** 1977. Molecular characterization of two R-plasmids isolated from *Neisseria gonorrhoeae. J. Bacteriol.* **131:**557–563.

56. **Roberts, M., L. P. Elwell, and S. Falkow.** 1978. Introduction to the mechanisms of genetic exchange in the gonococcus: plasmids and conjugation in *Neisseria gonorrhoeae,* p. 38–43. *In* G. Brooks, E. Brooks, E. Gotschlich, K. Holmes, W. Sawyer, and F. Young (ed.), *Immunobiology of Neisseria gonorrhoeae.* American Society for Microbiology, Washington, D.C.

57. **Roberts, M., and S. Falkow.** 1977. Conjugal transfer of R plasmids in *Neisseria gonorrhoeae. Nature* (London) **266:**630–631.

58. **Roberts, M., P. Piot, and S. Falkow.** 1979. The ecology of gonococcal plasmids. *J. Gen. Microbiol.* **114:**491–494.

59. **Roberts, M. C.** 1989. Plasmids of *Neisseria gonorrhoeae* and other *Neisseria* species. *Clin. Microbiol. Rev.* **2:**S18–S23.

60. **Roberts, M. C., C. D. Swenson, L. M. Owens, and A. L. Smith.** 1980. Characterization of chloramphenicol-resistant *Haemophilus influenzae. Antimicrob. Agents Chemother.* **18:**610–615.

61. **Roberts, M. C., and J. S. Knapp.** 1988. Transfer of β-lactamase plasmids from *Neisseria gonorrhoeae* to *Neisseria meningitidis* and commensal *Neisseria* species by the 25.2-megadalton plasmid. *Antimicrob. Agents Chemother.* **32:**1430–1432.

62. **Rotger, R., E. Garcia-Valdes, and E. P. Trallero.** 1986. Characterization of a β-lactamase-specifying plasmid isolated from *Eikenella corrodens* and its relationship to a commensal *Neisseria* plasmid. *Antimicrob. Agents Chemother.* **30:**508–509.

63. **Rotger, R., F. Rubio, and C. Nombela.** 1986. A multiresistance plasmid isolated from commensal *Neisseria* species is closely related to the enterobacterial plasmid RSF1010. *J. Gen. Microbiol.* **132:**2491–2496.

64. **Saunders, J., L. P. Elwell, S. Falkow, R. Sykes, and M. Richmond.** 1978. β-Lactamase and R-plasmids of *Haemophilus influenzae. Scand. J. Infect. Dis. Suppl.* **13:**16–22.

65. **Scheifele, D. W., and S. J. Fussell.** 1981. Frequency of ampicillin-resistant *Haemophilus parainfluenzae* in children. *J. Infect. Dis.* **143:**495–498.

66. **Scheifele, D. W., S. J. Fussell, and M. C. Roberts.** 1982. Characterization of ampicillin-resistant *Haemophilus parainfluenzae. Antimicrob. Agents Chemother.* **21:**734–739.

67. **Schaberg, D. R., A. K. Highsmith, and I. K. Wachsmuth.** 1977. Resistance plasmid transfer by *Serratia marcescens* in urine. *Antimicrob. Agents Chemother.* **11:**449–450.

68. **Smith, A. L.** 1983. Antibiotic resistance in *Haemophilus influenzae. Pediatr. Infect. Dis.* **2:**352–355.

69. **So, M., R. Gill, and S. Falkow.** 1975. The generation of a ColE1-Ap^r cloning vehicle which allows detection of inserted DNA. *Mol. Gen. Genet.* **142:**239–249.

70. **Sox, T. E., W. Mohammed, and P. F. Sparling.** 1979. Transformation-derived *Neisseria gonorrhoeae* plasmids with altered structure and function. *J. Bacteriol.* **138:**510–518.

71. **Sparling, P. F., T. Sox, W. Mohammed, and L. Guymon.** 1978. Antibiotic resistance in the gonococcus: diverse mechanisms of coping with a hostile environment, p. 44–52. *In* G. Brooks, E. Brooks, E. Gotschlich, K. Holmes, W. Sawyer, and F. Young (ed.), *Immunobiology of Neisseria gonorrhoeae.* American Society for Microbiology, Washington, D.C.

72. **Stiffler, P. W., S. A. Lerner, M. Bohnhoff, and J. A. Morello.** 1975. Plasmid deoxyribonucleic acid in clinical isolates of *Neisseria gonorrhoeae. J. Bacteriol.* **122:**1293–1300.

73. **Stuy, J. H.** 1980. Chromosomally integrated conjugative plasmids are common in antibiotic-resistant *Haemophilus influenzae. J. Bacteriol.* **142:**925–930.

74. **Thorne, G. M., and W. E. Farrar.** 1975. Transfer of ampicillin resistance between strains of *Haemophilus influenzae* type B. *J. Infect. Dis.* **132:**276–281.

75. **van Embden, J. D. A., M. Dessens-Kroon, and B. van Klingeren.** 1985. A new β-lactamase plasmid in *Neisseria gonorrhoeae. J. Antimicrob. Chemother.* **15:**247–258.

76. **van Klingeren, B., J. van Embden, and M. Dessens-Kroon.** 1977. Plasmid-mediated chloramphenicol resistance in *Haemophilus influenzae. Antimicrob. Agents Chemother.* **11:**383–387.

77. **Verschueren, H., M. Dekegel, D. Dekegel, C. Gilquin, and S. De Mayer.** 1982. Plasmids in *Neisseria meningitidis. Lancet* **i:**851–852.

78. **Yeung, K.-H., J. R. Dillon, M. Pauze, and E. Wallace.** 1986. A novel 4.9-kilobase plasmid associated with an outbreak of penicillinase-producing *Neisseria gonorrhoeae. J. Infect. Dis.* **153:**1162–1165.

Molecular Genetics of Bacterial Pathogenesis
Edited by V. L. Miller, J. B. Kaper, D. A. Portnoy, and R. R. Isberg
© 1994 American Society for Microbiology, Washington, DC 20005

Chapter 3

Transposon Tn*3*, 1973 to 1980

Fred Heffron and Ronald E. Gill

In this chapter we review the work on transposable drug resistance, specifically, the ampicillin resistance transposon Tn*3*, that was carried out in Stanley
Falkow's laboratory. This chapter starts by describing the events that occurred
in the fall of 1973 and ends with the departure of Ron Gill from the laboratory in
1980. Like many other successful laboratory directors, Stanley had tremendous
foresight in choosing timely research areas and was a great motivator of his students. When this work was started, Stanley still did an occasional experiment in
the laboratory, but later, he limited himself to microscopy. He was terrific to be
around in the lab, singing ribald songs, telling Jewish stories, and trying to teach
us a smattering of Yiddish.

The genesis of the work on Tn*3* was several small plasmids that we received
from Bob Hedges and Naomi Datta encoding various antibiotic resistance markers
(8, 9). Jorge Crosa and Stanley had just developed a simple method of carrying
out DNA-DNA hybridization using S1 nuclease (7). One of the central questions
of the time was how antibiotic resistance spread so quickly in bacteria. One possibility was that each drug resistance plasmid, or R factor, encoded a different
resistance gene. A second possibility was that the same gene somehow spread from
plasmid to plasmid. Recall that this question arose long before DNA sequencing,
restriction enzymes, and gel electrophoresis of DNA were available. To distinguish these two possibilities, we hybridized a small β-lactamase-encoding plasmid,
now called RSF1030 (about 5.5 MDa), against a large number of ampicillin-resistant (Ap^r) clinical isolates that were otherwise unrelated. We assumed that the
gene for Ap^r would show observable cross-hybridization, if there were any DNA
identity. In fact, there was much stronger hybridization than anticipated (about
30% of the plasmid hybridized, corresponding to almost 3 MDa). This was perplex-
ing because we expected only 10 to 15% hybridization on the basis of the size of
the β-lactamase gene. The results of these experiments were ultimately published
in the *Journal of Bacteriology* in 1975 (20), and although our experiments were

Fred Heffron • Department of Microbiology and Immunology, Oregon Health Sciences University,
3181 S.W. Sam Jackson Park Road, Portland, Oregon 97201-3098.   **Ronald E. Gill** • Department
of Microbiology and Immunology, Campus Box B175, 4200 East Ninth Avenue, Denver, Colorado
80262.

completed first, they were published out of chronology with the results of those of Hedges and Jacob (16). The observation that the same β-lactamase gene and some of its flanking DNA was moving from plasmid to plasmid seemed clear, but one could imagine many different mechanisms. At that time it was not clear whether this phenomenon was peculiar to Tn3 or could be generalized to other antibiotic resistance genes.

Stanley's laboratory was one of the first to grasp the generality of these findings and its implication for the spread of antibiotic resistance. One reason for this was the close tie between N. Datta's laboratory and Stanley's lab. Datta pioneered the study of infectious drug resistance, including the classification of plasmids into groups on the basis of incompatibility. More importantly, for the study of Tn3, her laboratory showed that R factors could evolve quickly (9) and that a similar penicillinase was expressed from many isolates (10, 11). Datta as well as A. E. Jacob and R. W. Hedges provided us with a wealth of antibiotic-resistant isolates. We compared several of the isolates by DNA-DNA heteroduplex analysis. The isolates all contained related plasmids, but the plasmids specified resistance to different antibiotics. This experiment gave a striking result, although it was not published for several years (15). The simplest interpretation of the electron micrographs was that most of the antibiotic resistance genes were located on insertions into a common precursor plasmid. From Stanley's standpoint, the collaboration with Datta's lab was ideal because his laboratory could provide the newest technology but did not have the necessary isolates.

A second reason why we realized the importance of the findings was the proximity of Russell Chan, who had discovered Tn10 with David Botstein (5, 35). Russell was carrying out his postdoctoral work with Lee Hartwell at the University of Washington. Chan and Botstein first suggested that Tn10 was analogous to bacteriophage Mu and that this family would probably include only a few genetic elements. Chan wrote to Botstein about our findings, and as told by Chan, this led Nancy Kleckner to start her beautiful work.

By 1974 it was clear that many antibiotic resistance genes, and probably other genes as well, were present on transposons and that these sequences were likely related to bacterial insertion sequences that had been described earlier (31, 33). There were many reasons to be excited about this result. For one thing, the dogma was still that DNA sequences were fixed and the idea that DNA could move around was new. The possible relationship between these bacterial transposons and the controlling elements in corn that McClintoch had described 20 years earlier was also clear (16, 20). Controlling elements underwent rearrangement in a developmentally controlled pattern. This suggested that perhaps all development might be linked to transposable elements. Development is unidirectional; programmed rearrangements could provide unidirectionality. The early work on bacterial transposable elements may have influenced thinking among immunologists as well and provided credence for rearrangement theories before there was any solid evidence. Thus, the suggestion that there may be many of these movable genetic elements and that they did something important, namely, spread drug resistance in bacteria, was very exciting.

The next breakthrough on Tn3 came in the summer of 1974 when Craig Rubens came to the laboratory (18, 30). Craig Rubens was one of many undergraduates always working in Stan's lab. They usually started in the summer but continued in the fall if everything went well. The experience was positive for many, and they continued in science; some are well known. Craig is now a pediatric infectious disease specialist at Children's Hospital in Seattle and has made many discoveries about drug resistance in bacteria. Together, we successfully developed a more defined way to study transposition. The method relied on the use of a large plasmid as a Tn3 donor and a smaller plasmid as a target. The two could be grown together in the same cell and then separated by centrifugation on the basis of their different molecular masses. Different fractions from the gradient were then used to transform *Escherichia coli* by the CaCl$_2$ method (6), and recombinants were found at an intermediate density. This method is too cumbersome to use now, but it provided a way to isolate and study a large number of insertions generated in strains with different genetic backgrounds. We mapped the insertions by heteroduplex analysis and electron microscopy following cleavage with *Eco*RI provided by Herb Boyer (*Eco*RI cleaved the target molecule once). Stanley has always been interested in importing the newest technology to his laboratory. He had sent Maggie So to Herb Boyer's and Stanley Cohen's laboratories to learn cloning. She brought back agarose gels as well as cloning techniques that we immediately used for our work. The simplest way to map the insertions would have been to use multiple restriction enzymes, but few were available. The advantage of mapping insertions by hetero-duplex analysis following cleavage with *Eco*RI was that the results were visually beautiful and provided better insight into what was going on. Another advantage was that Stan did much of the work. For example, he noticed that there was a short palindrome at either end of the transposon. The results of these studies have generally held up. They showed that Tn3 transposed independently of host functions, that Tn3 insertions inactivated genes, that Tn3 insertions were not ran-domly spaced around a target molecule, and that insertions could take place in either orientation relative to a target gene. These results mirrored those of Ko-pecko and Cohen (26), Kleckner and colleagues for Tn10 (21, 22), and Berg and colleagues for Tn5 (3).

MOLECULAR DISSECTION OF Tn3

A big thrill at this time (1974) was when I (F.H.) presented this work in front of other graduate students, faculty, and Stan. Normally, this was not a high point on anyone's weekly schedule. Craig Rubens and I had made a 10-foot-long (ca. 3-m-long) map of all the insertions, and we unfurled it in front of a packed room that included many faculty members from other departments. The target plasmid (RSF1010) encoded resistance to streptomycin and sulfonamides. It was clear from the work that these two genes formed an operon in which the first gene encoded sulfa resistance. Insertions in this gene simultaneously reduced or elimi-nated resistance to streptomycin, showing that Tn3 insertions were polar on expression of downstream genes. Several of the faculty at the University of Wash-ington at this time started talking about implications—all of this adding to the excitement.

The next project, in chronological order, was to try to reproduce in the laboratory the evolution of R factors (19). This work was almost entirely Stanley's invention and was again based on the collaboration with Datta's lab and clinical isolates sent to us from her laboratory (1, 2). Stan had received a small broad-host-range R factor from Datta's lab, RSF1010, that encoded streptomycin and sulfonamide resistance and that was originally isolated from a *Salmonella* sp. In the same package as RSF1010 were several clinical isolates that also encoded Ap^r as well as streptomycin resistance (Str^r) and sulfonamide resistance (Sul^r), as was found on RSF1010 (27). Eventually, the Ap^r isolates were compared with RSF1010 and were found to contain a Tn*3* insertion in a target plasmid; these Tn*3* insertions were identical to those that we isolated in RSF1010. Of the clinical isolates examined by heteroduplex analysis, one even contained an insertion at the same position as that in one of our laboratory constructs. The results certainly suggested that R factors could evolve by transposition and that more and more of these drug resistance genes were likely to spread to pathogenic microbes. Stanley was very excited about this work, and it formed the basis for an article that predicted the appearance of penicillin-resistant *Neisseria gonorrheae* just shortly before the strains were first documented in U.S. sexually transmitted disease clinics (12).

Rather than investigate the types of events that Tn*3* catalyzed as other groups did (25, 28) or develop Tn*3* as a genetic tool, as was so elegantly done for Tn*10* by Kleckner and colleagues (23, 24, 29), Stanley encouraged me (F.H.) to try to understand the transposition functions that Tn*3* might encode. This seemed the most important question to me as well. Jim Champoux, a faculty member in the adjoining laboratory, pointed out that a site-specific recombination enzyme, like bacteriophage λ Int, was likely to be encoded near the site on which it acted. Therefore, it was likely that Tn*3* encoded its own transposition enzyme(s). Pat Bedinger, now a faculty member at the University of North Carolina, worked with me on the project. She had terrific hands and the biochemical skill to make the project work. There were two things that were necessary to analyze the functions encoded by Tn*3*: a way of making mutations, preferably small deletions that could be easily mapped, and a way of quickly assaying whether the mutant transposons were defective for transposition. Jim Champoux made the critical suggestion of using in vitro mutagenesis to make the mutations. Recall that at this time (winter of 1975 and 1976), a description of in vitro mutagenesis had only just been published (4). Champoux had the idea independently but had been scooped. We chose to use a mutagenesis method similar to that developed in Paul Berg's laboratory but added some novel twists. Our protocol differed in that we made single-strand nicks in the plasmid with DNase I rather than double-strand breaks as in Berg's protocol (4). Exonuclease III was used to remove various amounts of target DNA; this was followed by treatment with S1 single-strand nuclease to create a single-stranded gap and ligation to reseal the plasmid. Following these steps, the DNA was transformed into bacteria, and three pools of transformants that differed in the size of the deletion generated were collected. For some inexplicable reason, a description of the ligation step, mentioned above, was omitted from the final published paper, making this method irreproducible, although the results were certainly correct (17). This was unfortunate because it was one of the first papers on in vitro mutagenesis published, but our method was never used because it could not be reproduced.

In order to quickly and easily assay for transposition of our mutant Tn*3* elements, we made use of a new conjugation assay for transposition. The assay was an adaptation of a newly described method to genetically label cryptic plasmids (34). In that assay, the conjugation donor strain contained the small Tn*3* donor plasmid, and the conjugative F factor served as a transposition recipient. Transposition of Tn*3* into the F factor was then detected by selecting for the transfer of Ap^r into an appropriate recipient strain. The recipient strain was a *polA* mutant, which does not allow replication of the Tn*3* donor plasmid if it is mobilized into the recipient. Therefore, we reasoned that the only way that Ap^r could be transferred to the recipient was as a result of transposition of Tn*3* to form an F::Tn*3* recombinant. In the absence of transposition, no Ap^r transconjugants would form.

We assayed 680 transformants from the mutagenesis to see whether Tn*3* could still transpose. By the way in which the assay was done, we could tell at a glance by the low number of colonies on some selective plates that the experiment had worked and that we had identified transposition-defective mutants. This was certainly a high point for all of us, because these were the first mutants identified to effect transposition. Stanley, in an uncharacteristically modest mood, said that he felt lucky to be part of the study. I (F.H.) threw some of the plates up in the air so hard that they stuck to the ceiling, making red goopy circles on the ceiling tiles.

The greatest stroke of luck in that work was the serendipitous isolation of a Tn*1*-containing plasmid (ΔAp) which could be used in complementation studies to determine whether any of the Tn*3* mutations identified *trans*-acting products required for transposition, i.e., transposase. It started as a Tn*1* insertion into RSF1010 located in the Sul^r gene that was polar on the downstream streptomycin resistance (Str^r) gene. That procedure selected an Str^r derivative in which the polar effect on Str^r was relieved. This isolate (ΔAp) retained its sulfonamide susceptibility (Sul^s) but was now Str^r and ampicillin susceptible (Ap^s). We determined that ΔAp contained a fairly large deletion within the Tn*1* sequences, which we now know removed portions of both the β-lactamase and the Tn*3* repressor-resolvase genes. Fortunately, the deletion did not include the transposase gene and, in fact, resulted in the overexpression of transposase. The fact that ΔAp was Ap^s and was compatible with the RSF1050 plasmid meant that it could be used to determine whether any of the nontransposable Tn*3* mutants could be complemented in *trans*. On the basis of the results of complementation with ΔAp, the nontransposable mutants were divided into three classes—those that did not transpose even when complemented (class I), those that transposed at nearly normal frequency when complemented (class II), and some that transposed at a much higher frequency than anything we had seen, even wild-type Tn*3* (class III). The first class of deletion mutants removed the end of the transposon required in *cis* for transposition. The second class included deletions in a large region later shown to encode the transposase. The third class of deletions was later shown to remove portions of both the transposase and the repressor-resolvase genes; their behaviors were inexplicable at the time. Not only did complementation result in an abnormally high frequency of transposition but the transposition product also appeared to be some sort of plasmid fusion and so did not appear to be the product of typical transposition events. Stanley even looked at the plates to see that this was not a fabrication (I suspect).

I (F.H.) left the laboratory in 1976 and moved to Brian McCarthy's laboratory at the University of California, San Francisco. I continued to work on Tn3, a decision that Stanley supported, and to maintain close ties with the Falkow laboratory. Stanley has always been very generous with students, and if they wanted to continue working on a project after leaving his lab, he felt that this was to everyone's benefit. The collaboration between the two labs proved synergistic, although probably none of us thought so at the time.

FINAL ERA OF Tn3

I (R.E.G.) began my thesis research on transposition the summer of 1976 and marked the beginning of the final era of the Tn3 saga in the Falkow lab. As a graduate student, I had actually come to the University of Washington with the intent of working for Stanley on some aspect of bacterial pathogenesis. This never came to be. In retrospect, I have a vivid recollection of my first meeting with Stanley, a meeting which certainly had a considerable influence on my choice of a thesis project. I arrived in the lab in the summer preceding my entry into graduate school. Having been an undergraduate at a small liberal arts college and having never seen or experienced "real" research, I had planned to work during the summer in the Falkow lab to be sure that research was an appropriate career choice.

When I arrived, I was told that Stan was busy finishing up some assays on heat-stable enterotoxin. Not knowing what to expect, I was led to the back of the lab where Stan was sitting hunched over the balance, surrounded by small weighing cups containing pink mice in various states of evisceration. He flashed a warm smile, and before I could react, I found myself shaking his slimy, mouse blood-stained hand. Needless to say, this first impression of bacterial pathogenesis research was enough to ensure that I chose to work on Tn3.

As I began my thesis work in earnest in late 1976, there were two questions which remained to be answered from Heffron's analysis of his Tn3 mutants: (i) what are the sequences which transpose as a result of complementation of the transposition-deficient Tn3 mutants, especially the class III mutants, and (ii) what is the identity of the Tn3 gene(s) and its protein products which are required for transposition? For these studies, we used a modification of Heffron's conjugation assay for transposition in which the relatively small conjugation plasmid, R388, was used as a transposition target. This greatly facilitated analysis of the transposed sequences since the plasmid was so much easier to isolate than the F plasmid and it contained unique restriction sites which were particularly useful in heteroduplex analysis. In these and subsequent studies, we relied heavily on heteroduplex analysis. Although these seemed to be personally turbulent times for Stanley, he always looked forward to "going on the scope" in search of the always elusive heteroduplexes. He seemed to be able to use it as a quiet time for escape and reflection, and often emerged from the electron microscopy suite renewed and invigorated (especially when he had data!). It was undoubtedly a good thing that Stanley enjoyed the electron microscope; quite frankly, scanning of the grids most often gave me severe motion sickness, so that I would emerge from the electron microscope suite dizzy and nauseous.

The data that we obtained confirmed that the class II transposition-deficient mutants undergo bona fide transposition in the presence of either a complementing Tn3 or a Tn1 transposon (13). Restriction enzyme digestion and heteroduplex analysis indicated that the transposed DNA consisted of just the mutant transposon and did not contain sequences derived from either the pMB8 portion of the donor RSF1050 plasmid or sequences from the complementing plasmid. These results would, of course, indicate that the class II mutant Tn3 elements were defective in one or more *trans*-acting functions which are required for transposition. Pairwise complementation tests between individual type I mutants revealed evidence for only one transposition function—designated *tnpA* or transposase (14).

The class III mutants proved to be a far more challenging problem, and the findings for the class III mutants were a much more significant contribution to understanding the mechanism of transposition (13). These mutants were peculiar in two ways: the frequency of transposition when complemented by ΔAp was much higher than even wild-type transposition, and the product of transposition was abnormal.

We first wanted to rule out the possibility that this behavior was due in some way to the ΔAp complementing plasmid. ΔAp was, after all, a derivative of Tn1, which is similar, although not identical, to the Tn3 element found on RSF1050. Furthermore, the effects of the ΔAp deletion on its own transposition had not been fully characterized, and, in fact, results obtained by Dougan had indicated that ΔAp failed to produce at least one of the Tn1-specified proteins in minicells (11a).

Two new complementing plasmids were made to resolve these uncertainties. The new plasmids were Ap^s derivatives of RSF1010::Tn1 and RSF1010::Tn3; each one had point mutations in the β-lactamase gene, but they were shown to transpose normally. Our complementation results with these plasmids clearly demonstrated that the two components of the class III mutant phenotype were functionally independent. In contrast to the very high transposition frequency when complemented by ΔAp, complementation of class III mutants by Tn1 or Tn3 elements with point mutations affecting only Ap^r resulted in a frequency of transposition comparable to that of class II mutants, about 20% of the wild-type frequency. However, the product of transposition from the class III mutants remained aberrant, regardless of the complementing transposon or the frequency at which transposition occurred. These results suggested that Tn1 and Tn3 encode a gene, designated *tnpA*, which is required for transposition and which is defective in both the class II and class III mutants; a second gene, designated *tnpR*, which down regulates the frequency of transposition and which is defective in both the class III mutants and ΔAp; and an additional locus which apparently acts in *cis* and which is required for a normal transposition product.

As was the case for the class II mutants, we were able to determine the structures of the sequences transposed by complementation of the class III mutants, thanks in large part to the seemingly endless hours that Stanley spent working on the electron microscope. The electron microscopic data and the results of restriction enzyme analysis were both consistent with the interpretation that complementation of these mutants resulted in a single insertion into the target plasmid. The insertion contained two copies of the mutant Tn*3* element and a single copy of the pMB8 portion of RSF1050. The Tn*3* elements flank each side of the pMB8 sequences and are directly repeated. The endpoints of the transposed DNA appeared to correspond to the inverted repeats of the Tn*3* elements and so, in this respect, resembled the Tn*3* transposition products that we normally observed.

The similarity of the class III transposition product to that of normal Tn*3* invited us to speculate that the nature of the class III product is a consequence of the inability of these mutants to process an otherwise normal, although transient, intermediate of transposition. This processing step would normally require a site within the transposon, deleted in the class III mutations, which serves as a site of recombination to resolve these cointegrates into their final products. In 1979, Shapiro (32) incorporated these observations into his elegant molecular model for replicative transposition in which he showed how cointegrates, identical to those which we described, could be intermediates in transposition. The basic tenets of Shapiro's model have stood the test of time, and many of the individual steps have been validated in vitro.

In order to complete our genetic description of Tn*3*, we needed to observe the proteins which were encoded by the genes predicted by our mutant analysis. The method of choice for identifying plasmid-encoded proteins was the use of minicells. Minicell analysis was already in use in the lab, but when Gordon Dougan joined the lab as a postdoctoral fellow, he brought with him new minicell-producing strains and introduced to our procedure a few critical modifications that greatly improved the quality of our results. The experimental strategy was a simple one—look at the proteins produced from RSF1050 and a variety of its deletion and linker insertion mutants (19a) in order to correlate certain proteins with their positions on the Tn*3* map and the phenotypes of the respective mutants.

It had previously been determined that the β-lactamase gene, encoding resistance to ampicillin, was located at the far right end of the transposon and was transcribed in a rightward direction. All of the mutations which affected transposition frequency mapped to the part of the transposon next to Ap^r and abolished the production of a 19-kDa protein labeled in minicells. These included frameshift mutations (designated B° mutations) introduced at the unique *Bam* site within Tn*3*, located within the *tnpR* gene. Initially, however, we were not able to identify anything consistent with the product of the *tnpA* gene. It was of some note that the B° mutations resulted in the production of a truncated peptide, a peptide that was apparently produced at a much higher level than the wild-type protein. This suggested to us that the *tnpR* gene could be autoregulatory, as is the *c*I repressor of bacteriophage lambda, a finding confirmed by Stanley Cohen's group by using *lac* fusions. If *tnpR* regulated its own expression, then perhaps it could also regulate the level of transposase and thereby regulate transposition frequency. Therefore, we examined a variety of strains into which we had introduced the B° mutation to determine whether a new protein product could be identified. It took some time to recognize that the high-molecular-mass "junk" in some of the lanes was not junk at all. Rather, it corresponded to a 120-kDa protein (precisely the predicted coding capacity of the *tnpA* region of Tn*3*) which was present in the B° lanes or a truncated form in the strains with both a B° mutation and an additional mutation in the left-hand portion of the transposon—the *tnpA* region. When I think back to how long these data were in my notebook before I recognized the correlation, I am embarrassed. It is something that Stanley has never forgotten; it seems like he takes every opportunity to remind me of the importance of critical unbiased observation and evaluation of data.

The genetic organization of Tn*3* was verified with the publication of the DNA sequence by Heffron et al. (17) in 1980. Their sequence contained three open reading frames, consistent with *tnpA*, *tnpR*, and *bla* (encoding β-lactamase). Furthermore, they found evidence for a site (now designated *res*) embedded within the *tnpR* gene which is required for resolution of the cointegrate transposition intermediates. The *tnpA* gene product is required for transposition, whereas the *tnpR* product is now known to function both as a transcriptional repressor of *tnpA* and itself and as the resolvase enzyme which is required for resolution of the cointegrate intermediates.

CONCLUSION

In almost all respects Stanley Falkow was and is a wonderful role model. He had so many sayings that kept you going late at night, such as "I want you to work hard and make me famous." Stan has a very keen nose for what is interesting and what is new. He taught us to have respect for anyone who had done good work—that is unusual in molecular biology. We are deeply indebted to him for the mark he has made on our lives and our careers.

REFERENCES

1. **Anderson, E. S., and M. J. Lewis.** 1965. Drug resistance and its transfer in *Salmonella typhimurium*. *Nature* (London) **206:**579–583.
2. **Barth, P., and M. J. Grinter.** 1974. Comparison of the deoxyribonucleic acid molecular weights and homologies of plasmids conferring linked resistance to streptomycin and sulfonamide. *J. Bacteriol.* **120:**618–630.
3. **Berg, D. E., J. Daviues, B. Allet, and J. D. Rochaix.** 1975. Transposition of R factor genes to bacteriophage lambda. *Proc. Natl. Acad. Sci. USA* **72:**3628–3632.
4. **Carbon, J., T. E. Shenk, and P. Berb.** 1975. Biochemical procedure for production of small deletions in simian virus 40 DNA. *Proc. Natl. Acad. Sci. USA* **72:**1392–1396.
5. **Chan, R. K., D. Botstein, T. Watanabe, and U. Ogata.** 1972. Specialized transduction of tetracycline resistance by phage P22 in *Salmonella typhimurium*. II. Properties of a high-frequency-transducing lysate. *Virology* **50:**883–898.
6. **Cohen, S. N., A. C. Chang, and L. Hsu.** 1972. Nonchromosomal antibiotic resistance in bacteria:

genetic transformation of *Escherichia coli* by R-factor DNA. *Proc. Natl. Acad. Sci. USA* **69:** 2110–2114.

7. Crosa, J. H., D. J. Brenner, and S. Falkow. 1973. Use of single-strand specific nuclease for analysis of bacterial and plasmid deoxyribonucleic acid homo- and heteroduplexes. *J. Bacteriol.* **115:** 904–911.

8. Datta, N. 1971. Prevalence of extrachromosomal drug resistance. R factors in *Escherichia coli*. *Ann. N. Y. Acad. Sci.* **182:**59–64.

9. Datta, N., and R. W. Hedges. 1971. Compatibility groups among fi-R factors. *Nature* (London) **234:**222–223.

10. Datta, N., and P. Kontomichalou. 1965. Penicillinase synthesis controlled by infectious R factors in Enterobacteriaceae. *Nature* (London) **208:**239–241.

11. Datta, N., and M. H. Richmond. 1966. The purification and properties of a penicillinase whose synthesis is mediated by an R-factor in *Escherichia coli*. *Biochem. J.* **98:**204–209.

11a. Dougan, G., M. Saul, A. Twigg, R. Gill, and D. Sherratt. 1979. Polypeptides expressed in *Escherichia coli* K-12 minicells by transposition elements Tn*1* and Tn*3*. *J. Bacteriol.* **138:**48–54.

12. Falkow, S., L. P. Elwell, J. de Graaff, F. Heffron, and L. Mayer. A possible model for the development of plasmid-mediated penicillin resistance in the gonococcus, p. 120–134. *In* R. D. Catterall and C. S. Nicol (ed.), *Sexually Transmitted Diseases*.

13. Gill, R. E., F. Heffron, G. Dougan, and S. Falkow. 1978. Analysis of sequences transposed by complementation of two classes of transposition-deficient mutants of Tn*3*. *J. Bacteriol.* **136:** 742–756.

14. Gill, R. E., F. Heffron, and S. Falkow. 1979. Identification of the protein encoded by the transposable element Tn*3* which is required for its transposition. *Nature* (London) **282:**797–801.

15. Gorai, A. P., F. Heffron, S. Falkow, R. W. Hedges, and N. Datta. 1979. Electron microscope heteroduplex studies of sequence relationships among plasmids of the W incompatibility group. *Plasmid* **2:**485–492.

16. Hedges, R. W., and A. E. Jacob. 1974. Transposition of ampicillin resistance from RP4 to other replicons. *Mol. Gen. Genet.* **132:**31–40.

17. Heffron, F., P. Bedinger, J. J. Champoux, and S. Falkow. 1977. Deletions affecting the transposition of an antibiotic resistance gene. *Proc. Natl. Acad. Sci. USA* **74:**702–706.

18. Heffron, F., C. Rubens, and S. Falkow. 1975. Translocation of a plasmid DNA sequence which mediates ampicillin resistance: molecular nature and specificity of insertion. *Proc. Natl. Acad. Sci. USA* **72:**3623–3627.

19. Heffron, F., C. Rubens, and S. Falkow. 1977. Transposition of a plasmid deoxyribonucleic acid sequence that mediates ampicillin resistance: identity of laboratory-constructed plasmids and clinical isolates. *J. Bacteriol.* **129:**530–533.

19a. Heffron, F., M. So, and B. J. McCarthy. 1978. In vitro mutagenesis of a circular DNA molecule by using synthetic restriction sites. *Proc. Natl. Acad. Sci. USA* **75:**6012–6016.

20. Heffron, F., R. W. Sublett, R. W. Hedges, A. Jacob, and S. Falkow. 1975. Origin of the TEM-beta-lactamase gene found on plasmids. *J. Bacteriol.* **122:**250–256.

21. Kleckner, N., D. F. Barker, D. G. Ross, and D. Botstein. 1978. Properties of the translocatable tetracycline-resistance elements Tn*10* in *Escherichia coli* and bacteriophage lambda. *Genetics* **90:** 427–461.

22. Kleckner, N., R. K. Chan, B. K. Tye, and D. Botstein. 1975. Mutagenesis by insertion of a drug-resistance element carrying an inverted repetition. *J. Mol. Biol.* **97:**561–575.

23. Kleckner, N., K. Reichardt, and D. Botstein. 1979. Inversions and deletions of the Salmonella chromosome generated by the translocatable tetracycline resistance element Tn*10*. *J. Mol. Biol.* **127:**89–115.

24. Kleckner N., J. Roth, and D. Botstein. 1977. Genetic engineering in vivo using translocatable drug-resistance elements. New methods in bacterial genetics. *J. Mol. Biol.* **116:**125–159.

25. Kopecko, D. J., J. Brevet, and S. N. Cohen. 1976. Involvement of multiple translocating DNA segments and recombinational hotspots in the structural evolution of bacterial plasmids. *J. Mol. Biol.* **108:**333–360.

26. Kopecko, D. J., and S. N. Cohen. 1975. Site specific recA-independent recombination between

bacterial plasmids: involvement of palindromes at the recombinational loci. *Proc. Natl. Acad. Sci. USA* **72:**1373–1377.

27. **Mathew, M., and R. W. Hedges.** 1976. Analytical isoelectrofocusing of R factor-determined β-lactamases: correlation with plasmid compatibility. *J. Bacteriol.* **125:**713–718.

28. **Nisen, P. D., D. J. Kopecko, J. Chou, and S. N. Cohen.** 1977. Site-specific DNA deletions occurring adjacent to the termini of a transposable ampicillin resistance element (Tn*3*). *J. Mol. Biol.* **117:** 975–978.

29. **Ross, D. G., J. Swan, and N. Kleckner.** 1979. Physical structures of Tn*10*-promoted deletions and inversions: role of 1400 bp inverted repetions. *Cell* **16:**721–731.

30. **Rubens, C., F. Heffron, and S. Falkow.** 1975. Transposition of a plasmid deoxyribonucleic acid sequence that mediates ampicillin resistance: independence from host *rec* functions and orientation of insertion. *J. Bacteriol.* **128:**425–434.

31. **Saedler, H., J. Besemer, B. Kemper, B. N. Rosenwirth, and P. Starlinger.** 1972. Insertion mutations in the control region of the Gal operon of E. coli. I. Biological characterization of the mutations. *Mol. Gen. Genet.* **115:**258–265.

32. **Shapiro, J. A.** 1979. Molecular model for the transposition and replication of bacteriophage Mu and other transposable elements. *Proc. Natl. Acad. Sci. USA* **76:**1933–1937.

33. **Shapiro, J. A.** 1969. Mutations caused by the insertion of genetic material into the galactose operon of *Escherichia coli. J. Mol. Biol.* **40:**93–105.

34. **So, M., F. Heffron, and S. Falkow.** 1978. Method for the genetic labeling of cryptic plasmids. *J. Bacteriol.* **133:**1520–1523.

35. **Tye, B. K., R. K. Chan, and D. Botstein.** 1974. Packaging of an oversize transduction genome by Salmonella phage P22. *J. Mol. Biol.* **85:**485–500.

Molecular Genetics of Bacterial Pathogenesis
Edited by V. L. Miller, J. B. Kaper, D. A. Portnoy, and R. R. Isberg
© 1994 American Society for Microbiology, Washington, DC 20005

Chapter 4

Enterotoxigenic *Escherichia coli*, 1971 to 1979

Magdalene So and Walter S. Dallas

It may be s--- to you, but it's been my bread and butter!

—Stanley Falkow

This chapter is intended to provide a brief review of the work on enterotoxigenic *Escherichia coli* that was undertaken in Stanley's lab in the 1970s. It is mostly anecdotal and is not intended to be a literature review of the field. There will probably be omissions because of the advancing ages of the authors of this chapter, so for this reason, we beg for mercy beforehand.

Stanley's interest in the extrachromosomal nature of enterotoxin genes was undoubtedly a result of the interplay between his clinical microbiology background and his episome work in a lab studying bacterial gastroenteritis (at the Walter Reed Army Institute of Research). Working on enterotoxin plasmids was therefore a natural extension of these interests. Stanley's work in this field and in the field of plasmid-encoded drug resistance genes (see chapter 3) eventually increased the awareness of the public and the research community that plasmids were not a curiosity of geneticists but, rather, were important elements in both bacterial pathogenesis and antibiotic resistance.

CHARACTERIZATION OF ETEC PLASMIDS

I (M.S.) entered the Microbiology Graduate Program at the University of Washington in 1971. By the early spring of 1972 it was clear that I was not interested in throwing in my lot with any of the professors then present in the department. John Sherris, chair of the department at that time, took me into his office and expressed the concern of members of the graduate committee that I did not have the interest, commitment, or drive of a successful student. I told him my problem: I had not found a lab in which I wanted to work. Sherris then told me I had one more chance to prove myself. The department had just hired a new

Magdalene So • Department of Microbiology and Immunology, Oregon Health Sciences University, 3181 S.W. Sam Jackson Park Road, L220, Portland, Oregon 97201-3098. *Walter S. Dallas* • Division of Cell Biology, The Wellcome Research Laboratories, Burroughs Wellcome Company, 3030 Cornwallis Road, Research Triangle Park, North Carolina 27709.

faculty member. This person, Stanley Falkow, was arriving in Seattle in the summer, and maybe I should speak to him. The next thing I remember was sitting in Stanley's office and asking if I could do a rotation with him. Stanley said something about allowing me to work in his lab for the summer and that my performance would be reassessed after that period. As a general rule, faculty are loathe to let problem students rotate in the lab—the reason being that if these students haven't worked out in three rotations, they would most likely not work out in a fourth. So to this day, I am grateful to Stanley for having given me a second chance. Passing on this tradition of being merciful to the misfits, my lab now has two students from other departments who couldn't find a good match after a year of rotations. They're both doing very well.

Stanley began his work on enterotoxigenic *E. coli* shortly after he arrived in Seattle. His interest in these pathogenic *E. coli* was aroused by the elegant work of H. W. Smith, who in the late 1960s showed that toxin production in entertoxigenic strains of *E. coli* that cause diarrhea in piglets was dependent on a plasmid (14). Smith and M. A. Linggood called these plasmids Ent and further observed that the expression of the K88 antigen, which conferred on *E. coli* the ability to colonize the small bowels of piglets, also depended on a plasmid (16). It is now known that the K88 colonization factor is fimbriae and that different fimbrial types determine host range (infectivity for humans, piglets, or calves).

Smith and Carlton Gyles initiated characterization of enterotoxins from these diarrheagenic *E. coli,* including demonstration of two distinct classes of toxins (10, 15). One class was heat stable (and therefore named ST), of low molecular weight, and not immunogenic; the second class was heat labile (named LT), of high molecular weight, and antigenic. Some strains produced only ST, but all strains that produced LT also produced ST. Joan Skerman, a postdoctoral fellow in Stanley's lab, used the "triple-cross" technique of E. S. Anderson (1) to show that enterotoxin production in a strain of *E. coli* that produces diarrhea in humans is also dependent on a plasmid (13).

Gyles, on sabbatical leave from the University of Guelph, and I worked on characterizing the plasmids of several toxigenic *E. coli* strains. He used the triple-cross method to look at the prevalence of transmissible plasmids in enterotoxigenic *E. coli,* and I estimated plasmid size by sucrose sedimentation rates and contour length measurements of the DNA after it was visualized electron microscopically. Luckily for us, several toxigenic strains contained only one plasmid. Some strains, on the other hand, such as H10407, contained numerous plasmids, and none of the Ent$^+$ exconjugants from a cross of H10407 with a plasmidless recipient contained only one plasmid. We were able to assign a particular-sized plasmid with toxin production in this strain only by comparing the plasmid profiles of the different Ent$^+$ and Ent$^-$ exconjugants from this cross. Our conclusion from those studies was that the plasmids encoding ST tended to be small (30 MDa; although there were exceptions) and those encoding ST + LT tended to fall into a narrow range of 80 to 90 mDa (9).

My next effort was to use DNA hybridization to determine the extent of relatedness among the Ent plasmids at the genetic level. These experiments, done with Jorge Crosa, who had just started his postdoctoral fellowship with Stanley, showed that the ST + LT plasmids from *E. coli* toxigenic for piglets were very related to each other (~88% homology) and had significant sequence identity (55%) with the ST + LT plasmids from an *E. coli* strain isolated from humans. However, the ST + LT plasmids were not related at all to the two ST plasmids examined. Further hybridization experiments showed that the ST + LT plasmids belonged to the F incompatibility group and that the sequences held in common by Ent and F plasmids were those of the transfer and pilus genes (18). Those studies indicated that the ST + LT plasmids had their origins in a common F factor. They suggested that the enterotoxin plasmids were the products of an evolutionary process in which plasmid replicons in an *E. coli* strain, under pressures to survive in an animal host, acquired the appropriate virulence genes—a novel concept at the time.

Also at about this time, other labs were studying restriction-modification systems in bacteria. Their work indicated that restriction enzymes could be used to modify plasmids so that pieces of foreign DNA could be cut with an enzyme (such as *Eco*RI) and cloned into a plasmid that was also suitably cut with the same enzyme. DNAs from prokaryotic and eukaryotic sources were "cloned" into plasmids by this technique. Two plasmids were used as vehicles for the initial cloning experiments. One was pSC101, which had a single *Eco*RI site and a useful antibiotic marker, Tc^r (tetracycline resistance) (4). However, there was no way to detect the presence of an insert in the cloning site. Furthermore, pSC101 was a relatively low-copy-number plasmid. The second plasmid, ColE1 (2, 12), was multicopied, and the copy number could be amplified with chloramphenicol (a trait which was very useful at a time when plasmid DNA isolation was not the perfunctory process it has now become and expression systems were only in our dreams). Furthermore, the *Eco*RI site was in the colicin gene, so recombinant plasmids could be identified by testing recombinants for the loss of colicin production. However, it lacked a marker for the easy selection of transformants.

Stanley, being good friends with Herb Boyer and Stan Cohen, knew about this cloning technique and realized that it would be most useful for isolating virulence genes, such as the ST and LT genes. In order to learn the newfangled cloning technique, Stanley allowed me to spend some time in the Cohen and Boyer labs. It was in Boyer's lab that I learned how to do *Eco*RI digestions of B41, an ST plasmid, to run an agarose gel, and to perform ligations of the B41 fragments into the pSC101 plasmid. Since there was no way to detect the presence of an insert in this plasmid, we had to test 72 clones in the suckling mouse assay before we found one that was ST$^+$ (17).

The low frequency of detection of inserts encoding hard-to-assay phenotypes gave us the impetus to construct a cloning vehicle which would incorporate the best features of the existing vehicles. We wanted a plasmid that had a selectable marker, a high copy number, and a cloning site that inactivated a gene which could be assayed easily. ColE1 had the last two features; we had only to place an antibiotic resistance gene into the plasmid. This was not an easy task in those days, but we were in the perfect position to do it. Fred Heffron, a graduate student in Stanley's lab at the time, was working on antibiotic resistance plasmids. In the course of his studies, he found that the β-lactamase gene, encoding ampicillin resistance, was part of a transposable element, which he named Tn*A* (subsequently renamed Tn*3*; see chapter 3). Together with Ron Gill, he devised a method to label plasmids with Tn*3*, so Gill and I hopped the Tn*3* transposon onto ColE1, and one such recombinant, RSF2124 (20), was to serve as the vehicle for cloning the LT gene (19). Subsequently, Paco Bolivar and Ray Rodriguez in Herb Boyer's lab generated the popular cloning vehicle pBR322, which Jorge Crosa tested for transmissibility.

As part of the ST cloning experiments, I had made heteroduplexes between pSC101 and the recombinant in order to give visual impact to the presence of the ST-containing insert in the vector. One of the micrographs I had taken, which never made it into the cloning paper, was a picture of single-stranded DNA of the recombinant plasmid. This picture showed a very obvious stem-loop structure. Single-stranded pSC101 DNA did not yield such a structure, so the stem-loop must have come from the ST-containing DNA. (At the time, I did not know the significance of these data, but this observation was to prove to be a useful guide in my future experiments while I was a postdoctoral fellow at the University of California at San Francisco. It subsequently led me to studies that eventually showed that the ST gene is a part of a transposable element.)

My first impression of working in Stanley's lab was that I was working in a foreign country. Growing up as a Roman Catholic Chinese immigrant in California did not prepare me in the least for the cultural and linguistic change of pace in my new home. I quickly learned Yiddish, and to this day, my command of the numerous words describing incompetent people has stood me in good stead. Also helpful to me in my role as department head today was the irreverent attitude Stanley often had toward administrators, especially those in the dean's office. I remember his attitude well, and this puts my job as administrator into proper perspective. Most of all, I am thankful to Stanley for helping me get through this time of change. He made graduate school fun and exciting for me, and by bringing the field of microbial pathogenesis into the spotlight, he made my journey into this scientific discipline so much easier.

MOLECULAR NATURE OF LT

I (W.D.) arrived in Seattle fresh from completing my master's degree in microbiology from North Carolina State University with Walt Dobrogosz. I had been working on cyclic AMP receptor protein (CRP) and had isolated and characterized a CRP mutant (7). I was interested in the medical aspects of microbiology and within a month of my arrival had been accepted into Stanley's lab, where I worked with Maggie So on subcloning the LT gene. All my knowledge of genetics until that time had been self-taught, and I quickly realized how limited it was in the environment into which I had cast my lot. Stan essentially left me with Maggie until she departed about 6 months later for her postdoctoral fellowship. It took me a while to get into the swing of things and to keep track of all the plasmids with their unusual names. Maggie was an excellent teacher, and I think I caught on fairly quickly, although I never could get experiments to work as cleanly and as quickly as she could.

Maggie had just cloned the LT gene, and my interest was quickly drawn to defining the molecular nature of LT. This aspect had not been suggested by Stan, but he encouraged me to continue. Stanley saw the new molecular genetic techniques as they might be applied to the study of microbial pathogenesis and epidemiology, and he always had plenty of suggestions about what could be done and what it might tell us. My problem was manifold. I had to remember all those plasmid names and the significance of each plasmid, I had to master the techniques already used in the lab, and then I had to get these new methodologies to work. At times it seemed too much, but it was not long before I had my own ideas and time to test them. When Stan was in town, which seemed to be about half the time (his involvement with drafting the recombinant DNA guidelines added to a busy schedule), he would make the rounds late in the afternoon and drop by to chat. Stan would enter the lab and say "*Nu*, Walter," at which point my experiment would be put on hold and we would discuss any recent developments. Next Stan would put his feet up on my clean lab bench (the only one in his labs) and he would start telling me what he had heard through his network of scientific contacts and what he was thinking, and quickly a group would gather to listen. Stan could easily move the discussion between molecular genetics and microbial pathogenesis, and what he spoke about was so exciting that after he left, your battery was charged and you couldn't wait to do some more experiments; classwork and other details necessary for fulfilling degree requirements had to be fit around experiments as best they could.

Stan recognized that minicells might be useful for determining the protein products encoded by the LT plasmid (8). Bob Tait, then in Herb Boyer's lab at the University of California at San Francisco, was using this method, and I was fortunate to be able to work in the Boyer lab for 2 weeks learning about minicells and lots of molecular genetics. I returned to Seattle holding a precious Styrofoam cooler with about eight different restriction enzymes as well as T4 polymerase and ligase, all purified in the Boyer lab. At that time few enzymes were commercially available, and the quality of those enzymes was quite variable. The Boyer lab experience opened my eyes even more to what could be done with molecular genetics. I also learned from them the importance of being open and giving to people who seek your help. While I was there I gave a *very* informal talk to the lab group about my work. Also in the lab was a certain freelance writer who was composing a story about recombinant DNA technology. In my opinion, the article that she eventually published was not accurate, and I must confess that I am the unnamed, cavalier young scientist mentioned in the article who could not recite the recombinant DNA guidelines.

Back in Seattle I got the minicell technology up and going. Using minicells, we found that a number of proteins were expressed from the recombinant plasmid with the cloned LT gene. Our next task was to determine which was LT. The classic approach was to isolate plasmid mutations and then correlate a particular protein with an LT phenotype (the expression or absence of LT in cells harboring a mutated plasmid). The approach was analogous in a way to the one Maggie had used to identify Ent plasmids in mating experiments when a number of plasmids were present. When Fred Heffron had been in Stan's lab, he used a random deletion mutagenesis technique in his seminal study on transposon Tn*3* (11). I tried but could never get the method to work. Therefore, I used restriction enzymes to make deletion derivatives of the recombinant LT plasmid and then analyzed the proteins encoded in each plasmid in minicells. At that time techniques were new and there was no such thing as "rapid." The analysis took quite a lot of effort, but we were successful in defining the molecular nature of LT. We found the toxin to be analogous to cholera toxin in structure: a pentameric B subunit with a larger catalytic A subunit that, unlike cholera toxin, was infrequently processed in *E. coli*. We were fortunate in being able to collaborate with two other labs to confirm and extend our observations. Mike Gill used our mutants to prepare extracts for adenylate cyclase activation assays, and with the data that he collected, we were able to correlate catalytic activity with the A subunit (6). Dick Finkelstein was gracious enough to give us some antibody directed against cholera toxin and also antiserum that neutralized LT. With these sera we were able to immunoprecipitate the B subunit (5). It was especially generous of them to share the reagents with us, since Dick's group was also attempting to define the molecular nature of LT by a protein purification approach (3). Ironically, we were competing while at the same time we were helping each other (we gave Dick recombinant strains that produced more toxin compared with the amount produced by the clinical isolates he had been using).

Later I learned Maxam-Gilbert DNA sequencing from Ben Hall's group in the Genetics Department at the University of Washington, and we determined the LT B-subunit cistron sequence. There were no computer sequence analysis programs at that time, but Stan had purchased a Radio Shack computer, and while I labored at sequencing he played with his computer, writing some programs to help analyze the sequence as it was determined. The computer was at Stan's house, so he would take the data home and analyze them there. I would have to call him at home to get the results right away or wait until the morning. Of course, he could not just tell me what the results were. He first had to explain what changes he had made to upgrade his program, and then how the program worked, and then how he planned to make improvements, and finally, at long last, the results. He also wrote other programs, a favorite being KAPOW. This program showed a cannon that traced a path across the screen. When it finally stopped, it belched the letters KAPOW (it was as corny as I have described, but Stan was so proud of it I could only praise the fruit of his great effort).

Initially, all our recombinant DNA work had to be approved by a local advisory committee. I isolated a nalidixic acid-resistant K-12 strain, and we used this strain for all of our work, which had to be done in a single room (my lab) with the door closed. On the door was a bright sign which proclaimed that recombinant DNA experiments were in progress. What the sign did was attract attention, and since the top half of the door was glass, I soon learned to empathize with fish in an aquarium. Every week those of us doing recombinant experiments would take rectal swabs (of themselves) and streak the swabs onto MacConkey agar with nalidixic acid to determine if someone had become a carrier of (infected with) a recombinant strain. No one ever recovered a recombinant *E. coli* strain by this method, but some of us learned the right and wrong ways to swab.

The pace in Stan's lab was fast and furious. We realized that we were doing cutting-edge science, and most of us fed off that energy. At the same time, I never felt that Stan was putting pressure on me to produce. It was a great place to learn and to grow as a scientist. Stanley had a clinical microbiology side to him that always kept us aware of the practical side of our work. A unique part of the lab environment was the infectious disease fellows (M.D.'s getting lab experience) who were in the lab and who were always willing to expound—at length—on pathogenic bacteria and how the techniques and approaches used in the lab could be applied to the diagnosis, treatment, and understanding of pathogenesis. As I worked on my project I would think about vaccine development (I eventually generated a recombinant strain that expressed only the B subunit of LT, a toxoid that is devoid of toxin activity but that is immunogenic), and Stan was very interested in using DNA fragments as diagnostic tools (I did the initial Southern blots to screen clinical isolates for the presence of LT genes).

As I reflect on the time spent in Stan's lab I think of Chinese circles—rings that are separate but linked together. Nature actually did our first cloning experiment by placing the toxin genes on plasmids, and we used the knowledge of plasmids gained by previous students to study these virulence determinants; eventually, we did our own recombinant DNA experiments and cloned the genes to smaller plasmids. The transposon work with Tn*3* led to the development of a plasmid to clone the ST gene, and later, the ST gene was shown to be part of a transposon. In Stan's lab we did serious work with serious effort, but most of the time we did not take ourselves too seriously.

REFERENCES

1. **Anderson, E. S.** 1965. A rapid screening test for transfer factors in drug sensitive Enterobacteriaceae. *Nature* (London) **208:**1016–1017.
2. **Bazaral, M., and D. R. Helinski.** 1972. Characterization of multiple circular DNA forms of colicinogenic factor E₁ from Proteus mirabilis. *Biochemistry* **7:**3513–3520.
3. **Clements, J. D., R. J. Yancey, and R. A. Finkelstein.** 1980. Properties of homogeneous heat-labile enterotoxin from *Escherichia coli. Infect. Immun.* **29:**91–97.
4. **Cohen, S. N., A. C. Y. Chang, H. W. Boyer, and R. B. Helling.** 1973. Construction of biologically functional bacterial plasmids *in vitro. Proc. Natl. Acad. Sci. USA* **70:**3240–3244.
5. **Dallas, W. S., and S. Falkow.** 1979. The molecular nature of heat-labile enterotoxin (LT) of *Escherichia coli. Nature* (London) **277:**406–407.
6. **Dallas, W. S., D. M. Gill, and S. Falkow.** 1979. Cistrons encoding *Escherichia coli* heat-labile toxin. *J. Bacteriol.* **139:**850–858.
7. **Dallas, W. S., Y.-H. Tseng, and W. J. Dobrogosz.** 1976. Regulation of membrane functions and fatty acid composition in *Escherichia coli* by cyclic AMP receptor protein. *Arch. Biochem. Biophys.* **175:**295–302.
8. **Dougan, G., and D. Sheratt.** 1977. The transposon Tn1 as a probe for studying ColE1 structure and function. *Mol. Gen. Genet.* **151:**151–160.
9. **Gyles, C., M. So, and S. Falkow.** 1974. The enterotoxin plasmids of *Escherichia coli. J. Infect. Dis.* **130:**40–49.
10. **Gyles, C. L.** 1971. Heat-labile and heat-stable forms of the enterotoxin from *E. coli* strains enteropathogenic for pigs. *Ann. N. Y. Acad. Sci.* **176:**314–322.
11. **Heffron, F., P. Bedinger, J. J. Champoux, and S. Falkow.** 1977. Deletions affecting the transposition of an antibiotic resistance gene. *Proc. Natl. Acad. Sci. USA* **74:**702–706.
12. **Hershfield, V., H. W. Boyer, C. Yanofsky, M. A. Lovett, and D. R. Helinski.** 1974. Plasmid ColE1

as a molecular vehicle for cloning and amplification of DNA. *Proc. Natl. Acad. Sci. USA* **71:** 3455–3459.

13. **Skerman, F. J., S. B. Formal, and S. Falkow.** 1972. Plasmid-associated enterotoxin production in a strain of *Escherichia coli* isolated from humans. *Infect. Immun.* **5:**622–624.

14. **Smith, H. W.** 1968. The transmissible nature of the genetic factor in *Escherichia coli* that controls enterotoxin production. *J. Gen. Microbiol.* **52:**319–334.

15. **Smith, H. W., and C. L. Gyles.** 1970. The relationship between two apparently different enterotoxins produced by enteropathogenic strains of *Escherichia coli* of porcine origin. *J. Med. Microbiol.* **3:**387–401.

16. **Smith, H. W., and M. A. Linggood.** 1971. Observations on the pathogenic properties of the K88 Hly and Ent plasmids of *Escherichia coli* with particular reference to porcine diarrhoea. *J. Med. Microbiol.* **4:**467–485.

17. **So, M., M. Betlach, H. Boyer, and S. Falkow.** 1976. The molecular cloning of an *Escherichia coli* plasmid determinant which encodes the production of heat stable (ST) enterotoxin. *J. Bacteriol.* **128:**463–472.

18. **So, M., J. H. Crosa, and S. Falkow.** 1975. Polynucleotide sequence relationships among Ent plasmids and the relationship between Ent and other plasmids. *J. Bacteriol.* **121:**234–238.

19. **So, M., W. S. Dallas, and S. Falkow.** 1978. The molecular cloning of an *Escherichia coli* plasmid determinant which encodes the production of heat labile (LT) toxin. *Infect. Immun.* **21:**405–411.

20. **So, M., R. Gill, and S. Falkow.** 1975. Generation of a ColE1-Ap[r] cloning vehicle which allows detection of inserted DNA. *Mol. Gen. Genet.* **142:**239–249.

Molecular Genetics of Bacterial Pathogenesis
Edited by V. L. Miller, J. B. Kaper, D. A. Portnoy, and R. R. Isberg
© 1994 American Society for Microbiology, Washington, DC 20005

Chapter 5

Molecular Epidemiology: Development and Application of Molecular Methods To Solve Infectious Disease Mysteries

Lucy S. Tompkins

I met Stanley in 1967 upon my entrance to graduate school in the Department of Microbiology at Georgetown University, having worked as a research technologist for 6 years after graduation from college. Although I thought I wanted to become an immunologist, I was quickly dissuaded from this by the chairman and rather accidentally wound up in Stanley's laboratory, where the emphasis was on the molecular biology of resistance plasmids, about which I knew virtually nothing. When I arrived in the lab, Rich Silver, Stan's first graduate student, had worked out a system, based on prior work by the Freifelders, to analyze the replication of R plasmids during conjugation (see chapter 1). My project was to describe the topological forms of R plasmids which were produced during transfer from the host and synthesis of the complementary strand in the recipient. At that time, the only methods we had available to "visualize" plasmid DNA were either to determine its molecular mass by ultracentrifugation in cesium chloride, using the model E ultracentrifuge, or to incorporate radiolabeled precursors into replicating plasmids during conjugation. The latter method involved extracting DNA from cells at different stages in conjugation, separating linear, nicked circular, and covalently closed plasmid DNA molecules by neutral sucrose gradient sedimentation, and collecting fractions which were then precipitated with trichloroacetic acid and counted in the scintillation counter, which took up a good portion of our small laboratory. Stan would hang over my shoulder waiting for the data, reading the counts off the scintillation counter screen; sometimes, he analyzed the printout even before I could do it myself. In retrospect, he was still in a phase of his development where he viewed himself as one of us, instead of as our mentor, and felt that he had a perfect right to examine the data first. In those days, Stanley also liked to work at the bench with us, especially if there was an experiment that required that many samples be processed rapidly. It was clear that he loved "get-

Lucy S. Tompkins • Department of Medicine, Division of Infectious Diseases and Geographic Medicine, and Department of Microbiology and Immunology, Stanford University Medical Center H 1537J, Stanford, California 94305.

63

ting the primary data,'' and his enthusiasm for each day's new discoveries undoubtedly influenced the nature of my own scientific career. Both in the clinical laboratory and in the research lab one can get ''data'' every day.

Following a long hiatus during which I received medical training, I reentered the Falkow laboratory as an infectious diseases postdoctoral fellow in 1976, the first of many infectious diseases physicians to follow. My goal was to learn the latest molecular techniques that had been developed during the initial recombinant DNA era and to apply them to a clinically relevant problem. My project was undertaken at the beginning of the recombinant DNA era, and the potential applications of molecular methods to diagnostic microbiology and infectious diseases seemed limitless. Stanley's laboratory was already in the forefront of this revolution. A year or two before I arrived at the University of Washington, his laboratory had developed agarose gel electrophoresis to separate plasmid DNA molecules (16), and Lynn Elwell et al. (4) had discovered resistance plasmids in *Neisseria gonorrhoeae* isolates by using this technique (see chapter 2). At about the same time Elwell, Julia Inamine, and Barbara Minshew had examined the plasmids which had been isolated from two enteric species cultured from patients in a burn unit and the endonuclease restriction profiles of those plasmids (4). They discovered plasmids in both species which had similar relative molecular masses, as determined by agarose gel electrophoresis. Furthermore, they also demonstrated that the plasmids were genotypically identical by comparing the patterns of plasmid DNA fragments obtained after restriction endonuclease digestion and separation by agarose gel electrophoresis. Meanwhile, Jorge Crosa and his colleagues in Mexico had previously demonstrated the applicability of this technique for studying ampicillin-resistant *Shigella dysenteriae* isolated during a Central American epidemic of dysentery (1).

Because of my interest in R plasmids and in multiply resistant bacteria causing nosocomial infections, Stan and I decided that I should investigate the basis of multiple antibiotic resistance in a collection of *Serratia marcescens* isolates and other enteric species sent to him by Jim Plorde from the Seattle Veterans Administration Hospital (VAH). Using a lysis and DNA extraction protocol given to me by Dan Portnoy, I analyzed the plasmids contained within the VAH *S. marcescens* strains and quickly discovered that a single 45-MDa plasmid was present in most of the isolates. Subsequent conjugation studies showed that this plasmid encoded a novel pattern of aminoglycoside resistance and also expressed β-lactamase and other resistance markers (35). Analysis of *Escherichia coli* transconjungants produced by conjugation with each *Serratia* donor by restriction endonuclease analysis confirmed the identities of the 45-MDa plasmids. Analysis of other enteric bacteria that were isolated from Seattle VAH patients and that expressed similar aminoglycoside resistance patterns by using the same techniques demonstrated that the 45-MDa plasmid had been mobilized in situ from *Serratia* donors to other species of the family *Enterobacteriaceae* (35). These data supported our hypothesis that the Seattle VAH had experienced a plasmid epidemic which we thought was probably mediated by in vivo conjugation, which occurred either in the gastrointestinal tract or the urinary tract of patients or in the urinary catheter drainage bags. This study was one of several early examples of the application of molecular methods to tracing the natural history and epidemiology of pathogenic bacteria causing nosocomial infections. Stanley, Jim Plorde, and I published this study in the *Journal of Infectious Diseases* and coined the phrase "molecular epidemiology" to describe the application of molecular biology to the study of infectious disease epidemiology (35). Later, Fred Tenover, Jim Plorde, and I began to investigate the mechanism by which the 45-MDa plasmid pLST1000 had acquired aminoglycoside resistance, and subsequently, Fred and his colleagues at the Seattle VAH fully characterized the molecular basis for the novel phenotype (33).

During this period, Stanley had a great interest in understanding plasmid evolution and suggested that perhaps pLST1000 might have evolved and been disseminated to other geographical areas long before we became aware of it in Seattle. Therefore, we planned to determine whether molecularly similar clones of this plasmid were present in other enteric species outside of our geographical area, but I wasn't able to pursue this idea, having just accepted a faculty position at the University of Washington and assumed other responsibilities. Subsequently, Hopkins et al. (6) in Boston showed that pLST1000 was indeed an endemic plasmid in several species isolated throughout the world. Their group has continued to trace the evolution of antibiotic resistance and has discovered that resistance and transmission of resistance markers evolve in very complex ways, demonstrating in a very contemporary way what Louis Pasteur said more than 100 years before the discovery of plasmids and transposons: "The bacteria will always have the last word."

The more practical observation that we made from this study of VAH *Serratia* isolates was that the plasmid profile correlated with the phenotype. Thus, all the *S. marcescens* isolates that were obtained from patients who were epidemiologically linked and that contained a single 45-MDa plasmid expressed the same colicin type pattern and antibiogram, whereas strains isolated from VAH patients who did not have an association with the case patients contained different numbers and sizes of plasmids, and they had distinct colicin patterns and slightly different antibiograms. Shortly after this discovery, Dennis Schaberg, the next infectious diseases postdoctoral fellow to work in Stan's lab, began to analyze a collection of multiply resistant enteric bacteria which he had collected from several hospitals experiencing outbreaks of infection during his tenure at the Centers for Disease Control (CDC). Dennis compared the plasmid profiles of *Klebsiella, Pseudomonas,* and *Serratia* strains linked to outbreaks with those of control strains and obtained results which were similar to those obtained in the *Serratia* analysis, that is, that the plasmid profile could be used to distinguish among strains, even those which expressed similar phenotypes. Together we published these observations (26, 37) and suggested that this fingerprinting method would be a useful tool for distinguishing among strains obtained during outbreak investigations, even when the function of the plasmids might be cryptic.

Our suggestion about the utility of a genotyping system was based upon the way in which epidemiologists approach the task of solving an outbreak. One of the first steps taken by the epidemiologist is to consult the clinical laboratory. This is essential because the laboratory must first determine whether the outbreak was caused by a particular species and, more importantly, whether a single strain of the species was responsible for infection in the case group compared with the strains causing infection in a control group. Species identification is usually based upon standard phenotypic properties, including biochemical patterns of fermentation and enzymatic activity and antigenic analyses. Similarly, strain identification is also commonly determined by comparing serotypic markers, antibiotic susceptibility patterns, bacteriophage susceptibility, etc. Therefore, the laboratory's contribution to an epidemiological investigation is to determine whether the case patients (individuals exposed to similar risk factors) were infected by the same strain of a particular species. In some instances, a putative reservoir of the infection may also be identified, and thus, the laboratory would attempt to isolate and type microorganisms cultured from this source. For example, in order to confirm that an outbreak of salmonellosis is associated with a particular food ingested by a significant number of ill persons, the same strain of enteric *Salmonella* sp. (as detected by serotyping) would be expected to be isolated from the case patients and from the food (source) that had been epidemiologically linked to the infection. In many instances, phenotypic analyses will suffice to confirm the clinical and epidemiological data. However, many phenotyping methods are performed only by reference laboratories, and sometimes, isolates may appear to be phenotypically identical even though they are genotypically distinct, thus producing inaccurate data which may bias the subsequent investigation. Our studies suggested that molecular fingerprinting by plasmid profile analysis was a more specific method for identifying case-related isolates and to distinguish these from the isolates obtained from other patients.

Many others have subsequently applied plasmid fingerprinting to analyses of hospital- and community-acquired strains of gram-negative bacilli (11; reviewed by Mayer [15]) and gram-positive bacteria, including staphylococci, streptococci, and enterococci (12, 17, 20, 21). Plasmid profile analysis has been found to be quite useful for rapid analysis, and currently, this technique is viewed as one of the "gold standard" methods for distinguishing among strains obtained from outbreaks. Since plasmid acquisition and loss may occur over time, the plasmid profile provides a "snapshot" of the epidemic during a relatively short interval. However, the method is not particularly useful when isolates contain a single plasmid encoding a virulence gene(s).

Agarose gel electrophoresis of plasmid DNA is rapid, requires very little equipment, and can be performed in routine clinical laboratories. At Stanford University Medical Center (SUMC), the Clinical Microbiology Laboratory routinely analyzes a wide variety of bacterial species associated with nosocomial infections, and this service has become an integral component of the Infection Control Program. As an example of its utility, methicillin-resistant *Staphylococcus aureus* (MRSA) isolates suddenly appeared at SUMC at a time when the affiliated Palo Alto VAH had experienced a prolonged outbreak of MRSA infections. Since house staff rotate between the two medical centers, there was concern that the endemic Palo Alto VAH strains might have been spread to SUMC patients by cross-transmission via a house staff carrier. Plasmid profile analysis permitted a quick determination which showed that the MRSA strains were unique in each

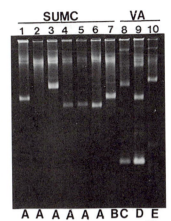

FIGURE 1. Agarose gel electrophoresis of plasmid DNA isolated from methicillin-resistant *S. aureus* isolates obtained from patients at SUMC and Palo Alto Veterans Affairs Medical Center (PAVAMC), indicated as VA in the figure. Unique patterns are identified as letters A, B, etc. None of the SUMC strains were genotypically identical to PAVAMC strains, as indicated by unique plasmid profiles, showing that cross-transmission of MRSA from PAVAMC to SUMC did not occur.

institution (Fig. 1). These results suggested that personnel were not responsible for the introduction of MRSA strains into SUMC. Subsequently, MRSA isolates disappeared from SUMC following appropriate treatment of infected patients and the institution of more meticulous infection control practices.

Plasmid profile analysis has been particularly useful for describing the epidemiology of community outbreaks and so-called sporadic cases of salmonellosis in the United States. Taylor and colleagues (32) traced an outbreak of *Salmonella* infections to contaminated marijuana from which they isolated a *Salmonella* strain with a plasmid profile identical to the pattern seen in isolates obtained from clinical cases. Similarly, using this method, Spika et al. (30) described the epidemiology of chloramphenicol-resistant *Salmonella* isolates transmitted to humans by contaminated hamburger which was traced to a particular abattoir. The comparison of plasmid profile analysis with phenotypic methods for distinguishing *Salmonella typhimurium* strains has clearly shown that this genotypic method is more specific, and strains expressing identical phenotypes can sometimes be differentiated by this technique (5). Plasmid analysis may not always disclose differences among strains. For example, Wachsmuth et al. (40) have shown that restriction endonuclease digestion of genomic DNA extracted from food-associated strains of enteric pathogens, including enterohemorrhagic *E. coli,* demonstrated differences that were not revealed by plasmid profile analysis. Thus, other methods of producing a genotype may be necessary in selected instances.

Before discussing further developments in molecular fingerprinting, it is important to note the pioneering studies of Moseley, Falkow, and their colleagues to develop DNA probes as epidemiological tools to study the natural history and epidemiology of enterotoxigenic *E. coli* strains in Bangladesh (18, 19). Their seminal studies subsequently provided the foundation for the development and application of DNA probes for the detection of virtually any pathogen, either directly or indirectly, in clinical materials. Thus, many of the genotyping schemes currently used were directly derived from studies from Stan's laboratory to isolate and characterize genes encoding pathogenic determinants.

Two genotyping methods now in use can be directly traced to molecular studies that Jim Kaper began in Stanley's laboratory at the University of Washington to examine the distribution of enterotoxin genes in *Vibrio cholerae*. He and his colleagues demonstrated that individual strains of non-O1 *V. cholerae* each contained a unique pattern of endonuclease restriction sites that could be displayed by simply separating all genomic DNA fragments produced by restriction endonuclease digestion by electrophoresis, creating a genomic fingerprint (7). The technique is commonly referred to as restriction endonuclease analysis (REA). As an example of the utility of REA, we compared this technique with another genotyping method originally developed by Selander and colleagues (27), called alloenzyme analysis, to explore the epidemiology of nosocomial *Legionella* infections at SUMC. Several aspects of this continuing outbreak were unusual, including these features: (i) many immunocompetent patients were infected; (ii) the majority of the infections were extrapulmonary, including prosthetic valve endocarditis, sternal wound infections, pleuritis, and pericarditis; and (iii) infections were caused by *Legionella pneumophila* and *Legionella dumoffii*, a species rarely implicated in Legionnaire's disease (36). In conjunction with investigators at CDC, our group analyzed *L. pneumophila* isolates obtained from patients at SUMC and those isolated from the potable water supply of the hospital, using REA and alloenzyme analysis (39). The REA results, which can be produced very quickly compared with the time needed to obtain results by alloenzyme typing, allowed us to conclude that the patient isolates of *L. pneumophila* were genotypically identical to those cultured from the hot water system, thereby showing that the potable water was a probable reservoir of infection. However, the reservoir of *L. dumoffii* was not apparent, nor did we understand how organisms present in the water could be transmitted to patients to establish these unusual extrapulmonary infections. Then, in 1989, the mystery of *Legionella* sp. transmission was solved by the discovery that topical application of contaminated tap water near fresh surgical wounds was the likely, and novel, mode of transmission (14). We also cultured *L. dumoffii* from the water supplied to two patient rooms in the hospital; all isolates had identical REA patterns, identifying the hospital hot water supply system as the definitive reservoir of *Legionella* in the medical center.

While we were continuing to track the outbreak at SUMC, investigators at CDC observed that the SUMC potable water supply contained amoebae. Later epidemiological studies by the group at CDC showed a high correlation between disease and exposure to water sources containing protozoans, which serve as the natural reservoir for the amplification and growth of amoebae in fresh water. Therefore, we postulated that amoebae might serve as the relevant reservoir for the replication of both *Legionella* species at SUMC and that the culture methods that we had used to isolate *L. dumoffii* from the water would not have had sufficient sensitivity to disclose their intracellular location. This hypothesis then led us to develop species-specific primers to amplify *Legionella* species from water, including those inside amoebae, by PCR (polymerase chain reaction) (31). It was later shown that a few culture-negative water samples were positive for *L. pneumophila* by PCR amplification; however, this technique failed to disclose any culture-negative, PCR-positive samples containing *L. dumoffii* (13). We subsequently realized that the two patient rooms from which we successfully isolated *L. dumoffii* were infrequently used, and the water pipes in those rooms contained large amounts of organic material, including a biofilm composed of *L. dumoffii* and other microorganisms. Thus, the patients who received care in those rooms were exposed to higher concentrations of bacteria when the tap was turned on because of the release of a bolus of water in comparison with the concentrations of bacteria to which patients who received care in other rooms that were used on a regular basis were exposed. Over a period of 8 years in which we investigated nosocomial legionellosis at SUMC, we found that REA and PCR amplification were sensitive and specific methods for analyzing hospital water systems. In addition, we proposed that it was more cost-efficient to prospectively survey hospital environmental samples for *Legionella* species than to assess the infection rate, especially since *Legionella* infections are relatively uncommon.

The aforementioned study by Kaper et al. (7) also contributed a second molecular epidemiological technique, called restriction fragment length polymorphism (RFLP) or Southern hybridization. Those investigators demonstrated that unique strains of *V. cholerae* could be distinguished by the pattern obtained by reacting endonuclease restriction fragments which were electrophoretically separated on agarose gels with radiolabeled cholera toxin genes. Compared with REA, in which hundreds of small fragments are produced by endonuclease digestion, the advantage of this technique is that the pattern is greatly simplified, and it is easier to visually compare patterns from a large collection of isolates. Specific RFLP fingerprint "types" can be generated, much as serotypes are produced. Ribosomal gene probes are widely used for RFLP analysis (ribotyping) because they contain highly conserved sequences which react with all bacterial species and are commercially available. The disadvantages of this technique are that the additional step for probing the restriction fragments is time-consuming and the method does not lend itself easily to routine clinical laboratory methods.

Essentially any genetic sequence which will hybridize to only a few genetic sequences contained within the genomic DNA can be used to develop an RFLP fingerprint. Thus, random DNA sequences can also be used for RFLP analysis. In collaboration with Mitchell Cohen, who obtained his infectious diseases postdoctoral fellowship research training in Stanley's laboratory and who knew firsthand the power of genetic typing methods, our group used random genomic *Salmonella* DNA fragments cloned in bacteriophage lambda as probes to develop RFLP patterns of endonuclease-digested genomic DNAs that had been isolated from *Salmonella enteritidis, S. typhimurium,* and other enteric *Salmonella* spp. during outbreak investigations. The resulting hybridization patterns clearly showed differences among the isolates, thus demonstrating that the RFLP technique could yield useful epidemiological data even in the absence of plasmids or well-characterized virulence genes (38).

Pulsed-field gel electrophoresis of large chromosomal fragments provides a simple alternative to RFLP analysis. In this method, "rarely" cutting restriction endonucleases are used to produce a small number of large fragments which can then be separated by pulse-gel electrophoresis to produce a pattern similar to the

RFLP pattern. This technique is now widely used in many reference laboratories and in some diagnostic labs as well.

PCR amplification technology, often in combination with REA or RFLP analysis, has recently been exploited to produce molecular fingerprints. For example, *Rickettsia* and *Rochalimaea* species can be identified by comparison of fragment patterns with the species-specific patterns produced by PCR amplification and endonuclease restriction of the citrate synthetase gene (23). The combination of PCR amplification with "random" primers and endonuclease restriction digestion may produce molecular fingerprints which are even more complex and discriminating than those produced by other means. Thus, random primer PCR fingerprinting can be applied to differentiate isolates which may appear to be clonal by other methods, thereby further clouding the definition of a "strain" or a "clone."

While the term "molecular epidemiology" is frequently applied to genotyping methods, it can also be used in another context to describe molecular techniques that have been adapted to the detection of uncultured microorganisms or those whose presence is not readily detected by conventional methods. David Relman, Jeff Loutit, Stanley, and I applied eubacterial 16S rRNA primers ("universal" primers) to amplify the prokaryotic 16S rDNA of the unknown bacillus embedded in the spleen and lymph nodes of a patient with disseminated bacillary angiomatosis (8), a newly described infectious disease primarily affecting AIDS patients (28, 41). As has often been the case, the strategy that we used was first proposed by Stanley during an infectious diseases grand rounds at which Jeff and I presented a patient with disseminated bacillary angiomatosis. Stanley suggested that we could use eubacterial 16S rRNA primers (universal primers) to amplify the prokaryotic 16S rDNA of the unknown bacillus embedded in the patient's spleen and lymph nodes. This approach incorporated two molecular methods: (i) amplification of target 16S rRNA by using eubacterial 16S rRNA primers and (ii) DNA sequence analysis of the amplified 16S rDNA to make a taxonomic assignment. In his typical fashion, Stanley noted that this was a perfectly straightforward application of two molecular methods and should be a "piece of cake." Two years later, we had identified the new bacillus as a previously unidentified member of the *Rochalimaea* genus most closely related to *Rochalimaea quintana* (24). Subsequent work by several other groups has now shown that this species, now called *Rochalimaea henselae*, not only causes bacillary angiomatosis but also causes cat scratch disease (2, 22). Furthermore, it has recently been demonstrated by PCR fingerprinting methods that bacillary angiomatosis may be caused by both *R. henselae* and *R. quintana* (10). Most recently, Jane Koehler and her colleagues have obtained evidence that asymptomatic, bacteremic cats may serve as a reservoir of *R. henselae*. Using PCR amplification with *R. henselae*-specific primers, those investigators also demonstrated that cat fleas contain *R. henselae*, implying that fleas might be vectors of infection to humans (9), and thereby strongly suggesting that cat scratch disease and bacillary angiomatosis may be zoonoses (9, 34).

Since the discovery of the agent of bacillary angiomatosis, two other uncultured bacterial pathogens have been identified by PCR amplification by using eubacterial 16S rDNA primers (25, 29) (see chapter 34).

In summary, molecular methods are very useful tools for solving epidemiological problems and for detecting and identifying both "old" and "new" pathogens. In addition to their utilities in solving infectious disease "mysteries," in some instances the data derived from these techniques have provided a foundation upon which to approach the study of microbial pathogenesis. This is well illustrated by current studies in our laboratory and in others to investigate the nature of the genes in *Rochalimaea* species encoding pathogenic determinants; these were spawned by molecular epidemiological studies to detect and characterize the infectious agents that cause bacillary angiomatosis and cat scratch disease.

As much as this chapter concerns molecular epidemiology, it also reflects my exposure to Stanley's approach to the pathogenesis of infectious diseases. I have worked with Stanley in various capacities throughout my career, first as a student, then as a fellow, and now as a faculty colleague and spouse. My training with him has enabled me to take a variety of approaches to the study of infectious diseases that I encountered on the wards or in the clinical microbiology laboratory. The results of some of those studies have returned to the clinical setting where they have benefited patients. It was Stan's love of infectious diseases and his ability to deduce a pathogenic pathway by imagining himself as a bacterial pathogen that drew me toward the field. (Ironically, such an imagination convinced me that perhaps I should head toward medicine, rather than science, since I believed at the time that I could never be a good investigator if that sort of fantasizing was a requirement for success.) Many of the applications of molecular methods that we used at the beginning of what is now a burgeoning field of molecular epidemiology were derived from studies initially performed by Stanley and his students in search of pathogenic determinants. Stanley's insights into the pathogenesis and natural history of infectious diseases have transformed the way in which pathogenesis studies are performed, and he has refocused the rather narrow molecular view to a broader perspective of the cell biology underlying the interaction between the host and the parasite. The development of unique approaches to elucidate the epidemiology and natural history of the infectious diseases caused by pathogenic bacteria epitomizes his creative approach to science.

REFERENCES

1. **Crosa, J. H., J. Olarte, L. J. Mata, L. K. Luttrop, and M. E. Penarando.** 1977. Characterization of an R plasmid associated with ampicillin resistance in *Shigella dysenteriae* type 1 isolated from epidemics. *Antimicrob. Agents Chemother.* **11:**553–558.
2. **Dolan, M. J., M. T. Wong, R. L. Regnery, J. H. Jorgensen, M. Garcia, J. Peters, and D. Drehner.** 1993. Syndrome of *Rochalimaea henselae* adenitis suggesting cat scratch disease. *Ann. Intern. Med.* **118:**331–336.
3. **Elwell, L. P., J. M. Inamine, and B. H. Minshew.** 1978. Common plasmid specifying tobramycin resistance found in two enteric bacteria isolated from burn patients. *Antimicrob. Agents Chemother.* **13:**312–317.
4. **Elwell, L. P., M. Roberts, L. W. Mayer, and S. Falkow.** 1977. Plasmid-mediated beta-lactamase production in *Neisseria gonorrhoeae. Antimicrob. Agents Chemother.* **11:**528–533.
5. **Holmberg, S. D., I. K. Wachsmuth, F. W. Hickman-Brenner, and M. L. Cohen.** 1984. Comparison of plasmid profile analysis, phage typing, and antimicrobial susceptibility testing in characterizing *Salmonella typhimurium* isolates from outbreaks. *J. Clin. Microbiol.* **19:**100–104.
6. **Hopkins, J. D., T. F. O'Brien, and M. Syvanen.** 1988. Functional and structural map of pLST1000: a multiresistance plasmid widely distributed in *Enterobacteriaceae. Plasmid* **20:**163.

7. **Kaper, J. B., H. B. Bradford, N. C. Roberts, and S. Falkow.** 1982. Molecular epidemiology of *Vibrio cholerae* in the U.S. Gulf Coast. *J. Clin. Microbiol.* **16:**129–134.

8. **Kemper, C. A., C. M. Lombard, S. C. Deresinski, and L. S. Tompkins.** 1990. Visceral bacillary epithelioid angiomatosis: possible manifestations of disseminated cat scratch disease in the immunocompromised host: a report of 2 cases. *Am. J. Med.* **89:**216–222.

9. **Koehler, J. E., C. A. Glaser, and J. W. Tappero.** 1994. *Rochalimaea henselae* infection: a new zoonosis with the pet cat as reservoir. *JAMA* **271:**531–535.

10. **Koehler, J. E., F. D. Quinn, T. G. Berger, P. E. LeBoit, and J. W. Tappero.** 1992. Isolation of *Rochalimaea* species from cutaneous and osseous lesions of bacillary angiomatosis. *N. Engl. J. Med.* **327:**1625–1631.

11. **Lee, S. C., D. N. Gerding, and P. P. Cleary.** 1984. Plasmid macroevolution in a nosocomial environment: demonstration of a persistent molecular polymorphism and construction of a cladistic phylogeny on the basis of restriction data. *Mol. Gen. Genet.* **194:**173–178.

12. **Locksley, R. M., M. L. Cohen, T. C. Quinn, L. S. Tompkins, M. B. Coyle, J. M. Kirihara, and G. W. Counts.** 1982. Multiply antibiotic resistant *Staphylococcus aureus:* introduction, transmission, and evolution of nosocomial infection. *Ann. Intern. Med.* **97:**317–324.

13. **Loutit, J. S., and L. S. Tompkins.** 1993. Evaluation of a DNA amplification procedure for detection of *Legionella pneumophila* and *Legionella dumoffii* in water, p. 176–178. *In* J. M. Barbaree, R. F. Breiman, and A. P. Dufour (ed.), *Legionella: Current Status and Emerging Perspectives.* American Society for Microbiology, Washington, D.C.

14. **Lowry, P. W., R. J. Blankenship, W. Gridley, N. J. Troup, and L. S. Tompkins.** 1991. A cluster of *Legionella* sternal-wound infections due to postoperative topical exposure to contaminated tap water. *N. Engl. J. Med.* **324:**109–112.

15. **Mayer, L. W.** 1988. Use of plasmid profiles in epidemiologic surveillance of disease outbreaks and in tracing the transmission of antibiotic resistance. *Clin. Microbiol. Rev.* **1:**228–243.

16. **Meyers, J. A., D. Sanchez, L. P. Elwell, and S. Falkow.** 1976. Simple agarose gel electrophoretic method for the identification and characterization of plasmid deoxyribonucleic acid. *J. Bacteriol.* **27:**1529–1537.

17. **Mickelsen, P. A., J. J. Plorde, K. P. Gordon, C. Hargiss, J. McClure, F. D. Schoenknecht, F. Condie, F. C. Tenover, and L. S. Tompkins.** 1985. Instability of antibiotic resistance in a strain of *Staphylococcus epidermidis* isolated from an outbreak of prosthetic valve endocarditis. *J. Infect. Dis.* **152:**50–58.

18. **Moseley, S. L., P. Echeverria, J. Seriwatana, C. Tiripat, W. Chaicumpa, T. Sakuldaipeara, and S. Falkow.** 1982. Identification of enterotoxigenic *Escherichia coli* by colony hybridization using three enterotoxin gene probes. *J. Infect. Dis.* **145:**863–869.

19. **Moseley, S. L., S. L. I. Huq, A. R. M. A. Alim, M. So, M. Sampour-Motalebi, and S. Falkow.** 1980. Detection of enterotoxigenic *Escherichia coli* by DNA colony hybridization. *J. Infect. Dis.* **142:**892–898.

20. **Murray, B. E., D. A. Church, A. Wagner, K. Zacheck, M. E. Levison, M. J. Ingerman, E. Abrutyn, and B. Mederske-Sameraj.** 1986. Comparison of two β-lactamase-producing strains of *Streptococcus faecalis. Antimicrob. Agents Chemother.* **30:**861–864.

21. **Parisi, J. T., and D. W. Hecht.** 1980. Plasmid profiles in epidemiologic studies of infections by *Staphylococcus epidermidis. J. Infect. Dis.* **141:**637–643.

22. **Regnery, R. L., J. G. Olson, B. A. Perkins, and W. Bibb.** 1992. Serological response to *Rochalimaea henselae* antigen in suspected cat-scratch disease. *Lancet* **339:**1443–1445.

23. **Regnery, R. L., C. E. Spruill, and B. D. Plikaytis.** 1991. Genotypic identification of rickettsiae and estimation of intraspecies sequence divergence for portions of two rickettsial genes. *J. Bacteriol.* **173:**1576–1589.

24. **Relman, D. A., J. S. Loutit, S. Falkow, and L. S. Tompkins.** 1990. The agent of bacillary angiomatosis: an approach to the identification of uncultured pathogens. *N. Engl. J. Med.* **323:**1573–1580.

25. **Relman, D. A., T. M. Schmidt, R. P. MacDermott, and S. Falkow.** 1992. Identification of the Whipple's disease bacillus. *N. Engl. J. Med.* **327:**293–301.

26. **Schaberg, D. R., L. S. Tompkins, and S. Falkow.** 1981. Use of agarose gel electrophoresis of

plasmid DNA to "fingerprint" bacterial strains of gram negative bacilli. *J. Clin. Microbiol.* **13:** 1105–1108.

27. **Selander, R. K. D. A. Caugant, H. Ochman, J. M. Musser, M. N. Gilmour, and T. S. Whittam.** 1986. Methods of multilocus enzyme electrophoresis for bacterial population genetics and systematics. *Appl. Environ. Microbiol.* **51:**873–874.

28. **Slater, L. N., D. F. Welch, D. Hensel, and D. W. Coody.** 1990. A newly recognized fastidious gram-negative pathogen as a cause of fever and bacteremia. *N. Engl. J. Med.* **323:**1587–1593.

29. **Solnick, J., J. O'Rourke, A. Lee, B. J. Paster, F. E. Dewhirst, and L. S. Tompkins.** 1993. An uncultured gastric spiral organism is a newly identified species of *Helicobacter* in humans. *J. Infect. Dis.* **168:**379–385.

30. **Spika, J. S., S. H. Waterman, G. W. SooHoo, M. E. St. Louis, R. E. Pacer, S. M. James, M. L. Bissett, L. W. Mayer, J. Y. Chie, B. Hall, K. Greene, M. E. Potter, M. L. Cohen, and P. A. Blake.** 1987. Chloramphenicol resistant *Salmonella newport* traced through hamburger to dairy farms—a major persisting source of human salmonellosis in California. *N. Engl. J. Med.* **516:**565–570.

31. **Starnbach, M. N., S. Falkow, and L. S. Tompkins.** 1989. Species specific detection of *Legionella pneumophila* in water by DNA amplification and hybridization. *J. Clin. Microbiol.* **27:**1257–1261.

32. **Taylor, D. N., I. K. Wachsmuth, Y. Shangkuon, E. V. Schmidt, T. J. Barrett, J. S. Schrader, C. S. Scherach, H. B. McGee, R. A. Feldman, and D. J. Brenner.** 1982. Salmonellosis associated with marijuana—a multistate outbreak traced by plasmid fingerprinting. *N. Engl. J. Med.* **306:** 1249–1253.

33. **Tenover, F. C., T. D. Gootz, K. P. Gordon, L. S. Tompkins, S. A. Young, and J. J. Plorde.** 1984. Development of a DNA probe for the ANT-2″-aminoglycoside modifying enzyme structural gene. *J. Infect. Dis.* **15:**678–687.

34. **Tompkins, L. S.** 1994. *Rochalimaea* infections: are they zoonoses? *JAMA* **271:**553–554. (Editorial.)

35. **Tompkins, L. S., J. J. Plorde, and S. Falkow.** 1980. Molecular analysis of R-factors from multiresistant nosocomial isolates. *J. Infect. Dis.* **141:**625–636.

36. **Tompkins, L. S., B. Roessler, S. C. Redd, L. Markowitz, and M. L. Cohen.** 1988. *Legionella* prosthetic valve endocarditis. *N. Engl. J. Med.* **318:**530–535.

37. **Tompkins, L. S., D. R. Schaberg, and S. Falkow.** 1979. Use of agarose gel electrophoresis of plasmid DNA as an epidemiologic marker of bacterial strains, abstr. 637. *Program Abstr. 19th Intersci. Conf. Antimicrob. Agents Chemother.*

38. **Tompkins, L. S., N. Troup, A. Labigne, and M. L. Cohen.** 1986. Cloned, random chromosomal sequences as probes to identify *Salmonella* species. *J. Infect. Dis.* **154:**156–162.

39. **Tompkins, L. S., N. Troup, T. Woods, W. Bibbs, and R. M. McKinney.** 1987. Molecular epidemiology of *Legionella* species by restriction endonuclease and alloenzyme analysis. *J. Clin. Microbiol.* **25:**1875–1880.

40. **Wachsmuth, I. K., J. A. Kiehlbauch, C. A. Bopp, D. N. Cameron, N. A. Strockbine, J. G. Wells, and P. A. Blake.** 1991. The use of plasmid profiles and nucleic acid probes in epidemiologic investigations of foodborne, diarrheal disease. *Int. J. Food Microbiol.* **12:**77–90.

41. **Welch, D. F., D. A. Pickett, L. N. Slater, A. G. Steigerwalt, and D. J. Brenner.** 1992. *Rochalimaea henselae* sp. nov., a cause of septicemia, bacillary angiomatosis, and parenchymal bacillary peliosis. *J. Clin. Microbiol.* **30:**275–280.

Part 2. Adhesins

The years in Seattle marked the shift in Stan's research from studying plasmid replication and antibiotic resistance to studying the actual virulence factors of bacteria. The initial virulence factors to be cloned were genes encoding extracellular toxins, specifically, the heat-labile and heat-stable enterotoxins of enterotoxigenic *Escherichia coli* (ETEC). After the initial cloning of these enterotoxin genes, described in this volume by Maggie So and Walt Dallas (chapter 4), genes encoding adhesins of extracellular pathogens were cloned and characterized. The first adhesin to be studied was the K88 adhesin of ETEC associated with animal diarrhea. The cloning of genes encoding K88ac by Pat Shipley, Gordon Dougan, and colleagues (10) was a natural progression from the characterization of the ENT plasmids of ETEC, which had been pursued for several years in the Falkow lab. After the initial focus on diarrheagenic *E. coli,* attention then shifted to *E. coli* strains that cause a distinctly different disease, namely, urinary tract infections (UTIs). Richard Hull, a postdoctoral fellow in the lab, and Ron Gill constructed a cosmid gene library of an *E. coli* strain associated with UTIs. Ron was finishing his Ph.D. work on Tn*3* (see chapter 3) and had developed a protocol for constructing cosmid gene libraries using the pHC79 cosmid vector that had recently been developed by Hohn and Collins in Germany (4). This highly detailed protocol, dubbed ''The Gospel According to Gill,'' was used for constructing cosmid gene libraries of several pathogens in the Falkow lab. The uropathogenic *E. coli* library was highly productive, yielding clones for type 1 pili, P pili, and hemolysin (5, 12). After Ron left the laboratory, Richard and Sheila Hull, who was also a postdoctoral fellow, continued to work on P and type 1 pili, respectively. The Hulls have continued their research with P pili and review the various bacterial adhesins associated with UTI in this section (chapter 6).

At Stanford, work continued on the adhesins of UTI *E. coli,* and studies on adhesins of other pathogens were initiated. Staffan Normark arrived for a sabbatical in Stan's lab and began to work on the P pili. Staffan's work on P pili continued for many years after that sabbatical, and the research of Staffan and colleagues resulted in an elegant story for P pili which has become a paradigm of bacterial adhesins. David Low, a postdoctoral fellow in the laboratory, investigated the regulation of pilus expression, a topic he reviews elsewhere in this volume (chapter 28). Paul Orndorff entered the laboratory as a postdoctoral fellow with the intent of working on the heat-labile enterotoxin of ETEC. However, upon Paul's arrival, Stan said that Walt Dallas and other workers had the enterotoxin story well in hand and suggested that he should work on the type 1 pili. This was the beginning of numerous papers by Paul and colleagues on this widely distributed adhesin, which he reviews here (chapter 7).

One of the graduate students who had moved from Seattle with Stan was Mike Koomey. Mike's Ph.D. project concerned *Neisseria gonorrhoeae*, particularly the immunoglobulin A protease of this species. He has continued to study the genetics of *N. gonorrhoeae*, particularly the mechanisms of pilus antigenic variation, a topic which he discusses here (chapter 8). The pili of *N. gonorrhoeae* are important members of the class of adhesins known as type IV pili. Carl Marrs was a postdoctoral fellow with Stan at Stanford, and during this time he cloned the genes encoding the type IV pili of *Moraxella bovis* (6). Carl has continued to study type IV pili in *Moraxella*, *Eikenella*, and *Kingella* species and reviews this topic in this section (chapter 9). Type IV pili are also produced by agents of diarrhea, including enteropathogenic *E. coli* (EPEC) (3).

Another graduate student who moved from Seattle was Alison Weiss, who studied *Bordetella pertussis* for her Ph.D. thesis. Alison constructed a Tn5 library of *B. pertussis* and isolated mutants for a variety of virulence factors of this species, including pertussis toxin, adenylate cyclase, and filamentous hemagglutinin (FHA) (11). Alison reviews the secretion of *B. pertussis* virulence factors elsewhere in this volume (chapter 22). The FHA of *B. pertussis*, an important adherence factor in this species, has been extensively characterized by David Relman, a postdoctoral fellow in Stan's lab, and colleagues (8). Frits Mooi, who spent a sabbatical with Stan at Stanford, reviews the roles of FHA and fimbriae in the adhesion of *B. pertussis* in this section (chapter 10). Interestingly, at the time that Pat Shipley and colleagues were cloning genes encoding the K88ac adhesin, Frits was in The Netherlands cloning genes encoding K88ab, an antigenic variant of K88ac (7). Joe St. Geme investigated *Haemophilus influenzae* as a postdoctoral fellow in Stan's lab, focusing on the adherence of nontypeable strains. He describes here the hemagglutinating pili synthesized by both type b and nontypeable *H. influenzae* (chapter 11); these pili appear to play a role in nasopharyngeal colonization. In addition to these pili, nontypeable *H. influenzae* can synthesize high-molecular-weight adhesins which are similar to FHA from *B. pertussis*.

In addition to filamentous structures such as pili, other bacterial components such as outer membrane proteins can mediate adherence to eukaryotic cells. Invasin, a 103-kDa outer membrane protein produced by *Yersinia pseudotuberculosis* and other *Yersinia* species, was discovered by Ralph Isberg while he was a postdoctoral fellow in Stan's lab at Stanford. As reviewed by Dorothy Pierson elsewhere in this volume (chapter 15), invasin mediates the entry of *Yersinia* species into mammalian cells by binding to β_1 integrins on the cell surface. EPEC strains produce a protein related to invasin called "intimin" which mediates the intimate adherence of EPEC to epithelial cells. Intimin was discovered in the laboratory of Jim Kaper, who initiated studies on EPEC after finishing his postdoctoral fellowship with Stan in Seattle. Jim reviews intimin, type IV pili, and other virulence factors of EPEC in this section (chapter 12). A homolog of intimin is crucial for the pathogenesis of colonic hyperplasia in mice caused by *Citrobacter freundii* biotype 4280, as reviewed by David Schauer (chapter 13). The gene encoding the intimin of *C. freundii* (*eae*) was cloned by David, then a postdoctoral fellow in Stan's lab, using a probe for the EPEC *eaeA* gene received from Jim Kaper (9). Carol Francis, then a graduate student with Stan, collaborated with Jim on studies of EPEC (2), as has Brett Finlay, a former postdoc of Stan's now in Vancouver (1). Such interactions and collaborations are not uncommon among those who trained with Stan, even many years after they leave the sheltered environment he provided.

James B. Kaper

REFERENCES

1. **Finlay, B. B., I. Rosenshine, M. S. Donnenberg, and J. B. Kaper.** 1992. Cytoskeletal composition of attaching and effacing lesions associated with enteropathogenic *Escherichia coli* adherence to HeLa cells. *Infect. Immun.* **60:**2541–2543.
2. **Francis, C. L., A. E. Jerse, J. B. Kaper, and S. Falkow.** 1991. Characterization of interactions of enteropathogenic *Escherichia coli* O127:H6 with mammalian cells in vitro. *J. Infect. Dis.* **164:**693–703.
3. **Girón, J. A., A. S. Y. Ho, and G. K. Schoolnik.** 1991. An inducible bundle-forming pilus of enteropathogenic *Escherichia coli*. *Science* **254:**710–713.
4. **Hohn, B., and J. Collins.** 1980. A small cosmid for efficient cloning of large DNA fragments. *Gene* **11:**291–298.
5. **Hull, R. A., R. E. Gill, P. Hsu, B. H. Minshew, and S. Falkow.** 1981. Construction and expression of recombinant plasmids encoding type 1 or D-mannose-resistant pili from a urinary tract infection *Escherichia coli* isolate. *Infect. Immun.* **33:**933–938.
6. **Marrs, C. F., G. Schoolnik, J. M. Koomey, J. Hardy, J. Rothbard, and S. Falkow.** 1985. Cloning and sequencing of a *Moraxella bovis* pilin gene. *J. Bacteriol.* **163:**132–139.
7. **Mooi, F. R., F. K. De Graaf, and J. D. A. van Embden.** 1979. Cloning, mapping and expression of the genetic determinant that encodes for the K88ab antigen. *Nucleic Acids Res.* **6:**849–865.
8. **Relman, D. A., M. Domenighini, E. Tuomanen, R. Rappuoli, and S. Falkow.** 1989. Filamentous hemagglutinin of *Bordetella pertussis*: nucleotide sequence and crucial role in adherence. *Proc. Natl. Acad. Sci. USA* **86:**2637–2641.
9. **Schauer, D. B., and S. Falkow.** 1993. Attaching and effacing locus of a *Citrobacter freundii* biotype that causes transmissible murine colonic hyperplasia. *Infect. Immun.* **61:**2486–2492.
10. **Shipley, P. L., W. S. Dallas, G. Dougan, and S. Falkow.** 1979. Expression of plasmid genes in pathogenic bacteria, p. 176–180. *In* D. Schlessinger (ed.), *Microbiology—1979*. American Society for Microbiology, Washington, D.C.
11. **Weiss, A. A., E. L. Hewlett, G. A. Myers, and S. Falkow.** 1983. Tn5-induced mutations affecting virulence factors of *Bordetella pertussis*. *Infect. Immun.* **42:**33–41.
12. **Welch, R. A., R. Hull, and S. Falkow.** 1983. Molecular cloning and physical characterization of a chromosomal hemolysin from *Escherichia coli*. *Infect. Immun.* **42:**178–186.

Molecular Genetics of Bacterial Pathogenesis
Edited by V. L. Miller, J. B. Kaper, D. A. Portnoy, and R. R. Isberg
© 1994 American Society for Microbiology, Washington, DC 20005

Chapter 6

Adherence Mechanisms in Urinary Tract Infections

Richard A. Hull and Sheila I. Hull

In 1971, Smith and Linggood (56) concluded that bacterial adherence to target tissue was a critical step in bacterial pathogenesis. This observation has spawned a generation of research into the significance and mechanisms of bacterial adherence in human disease, including urinary tract infections (UTIs). Understanding of the mechanisms of bacterial adherence, at least for *Escherichia coli,* has been advanced by the application of molecular genetic methods. Understanding of the role of bacterial adherence in promoting disease has also been advanced by thoughtful clinical investigations and extrapolation from animal models. Still, knowledge in both areas remains incomplete.

This chapter is intended as an overview of the bacterial adhesins that promote colonization of the human urinary tract. Particular attention is addressed to the adherence properties that may contribute to symptomatic UTI in patients with underlying urinary defects including urinary catheterization. Patients with complicated infection are more prone to severe disease that includes renal damage and urosepsis, and such patients may benefit most from preventative measures. From a practical standpoint, preventative therapies may also be more effective for this patient group because (i) the target group is well defined, for example, people who have indwelling catheters, and (ii) the likelihood of symptomatic infection is high, especially with an indwelling catheter in place. Many clinical and microbiological issues relevant to catheter-associated UTI are discussed in reference 67.

Recent progress regarding the adherence properties of bacteria associated with acute community-acquired UTI is also reviewed. For a more comprehensive treatment of the molecular basis of adherence of uropathogenic *E. coli* and for an overview of some clinical aspects of UTI, the reader is referred to several excellent recent reviews (17, 25, 32, 57).

SOURCE OF UTI BACTERIA

For all bacterial UTIs, there must first be a ready supply of infectious bacteria. In the majority of instances, bacteria colonize the urinary tract by ascending the

Richard A. Hull and Sheila I. Hull • Department of Microbiology and Immunology, Baylor College of Medicine, Houston, Texas 77030.

urethra. The primary source of infecting bacteria, whether gram positive or gram negative, for either community-acquired or nosocomial UTI is most likely the individual's own bowel flora. For women, the capacity of bacteria for vaginal colonization may also contribute to urinary tract colonization. In one study of *Klebsiella pneumoniae* UTI in spinal cord injury patients, the infecting strain was prospectively isolated from the bowel in 80% of patients (9). Those investigators also found that prior *K. pneumoniae* bowel colonization always preceded UTI with this species, suggesting that the individual's own flora was the only source of urinary tract colonization.

Prior colonization of a body site adjacent to the urethra is nearly always a prerequisite step for urinary tract invasion. In order to occupy this ecological niche, urinary tract strains must compete with indigenous flora. As a consequence, the first role of bacterial adherence in UTI may be to facilitate bowel colonization. Wold et al. (68) investigated bacterial factors associated with stable colonization of the bowel and found that resident *E. coli* frequently expressed P fimbriae and adhered to colonic epithelium by using these fimbrial adhesins. *E. coli* Dr pili also adhere to bowel tissue, but their contribution to bowel colonization has not been directly demonstrated. Type 1 pili have also been proposed as potential bowel colonization factors (59). Unlike other adhesins discussed in this chapter, type 1 pili are present in most *E. coli* and other members of the family *Enterobacteriaceae* (7). A more complete discussion of type 1 pili is presented in chapter 7.

In general, the non-*E. coli* members of the family *Enterobacteriaceae* and enterococci do not colonize the bowel well. They are usually minor components of the gram-negative bowel flora in humans and are found to be in the majority only when the level of competing flora is reduced, for example, following antibiotic therapy. Whether the uropathogenic bacteria in these genera are better at colonizing the bowel is not clear. The uroepithelial cell adhesin (UCA) of urinary *Proteus mirabilis* is a potential candidate as a bowel adhesin for this species. UCA pili share many structural features with F17 pili, which promote bovine bowel colonization (70). For other UTI bacteria such as *Staphylococcus saprophyticus* or for nosocomial uropathogens such as *Klebsiella*, *Providencia*, or *Enterococcus* species, little information is available regarding bowel tissue adherence mechanisms. It seems attractive to consider prospectively the elimination of specific uropathogenic strains or species from the bowel flora as a means of preventing UTIs. However, in the Donovan study, many more people experienced bowel colonization with *K. pneumoniae* than experienced symptomatic UTI, so that the net benefit of prophylaxis to eliminate *K. pneumoniae* from all patients would have been marginal. Tullus (64) examined the frequency of UTI in children prospectively colonized with piliated (P$^+$) bacteria. It was reported that P$^+$ *E. coli* colonization was a poor predictor of clinical outcome; most children with P$^+$ *E. coli* did not experience UTI with that organism.

INVASION OF THE URINARY TRACT

The mechanisms and associated adherence organelles by which bacteria colonize the urinary tract are varied and seem to reflect the site of colonization, whether or not the colonization is symptomatic, and also the method of urinary management of the afflicted individual (e.g., normal voiding, intermittent or short- or long-term indwelling catheterization, or use of an external urinary catheter). Research regarding the adherence of bacterial uropathogens during the past decade has focused primarily upon acute community-acquired UTI in women and girls with normal voiding. It began with the seminal discovery in 1976 (60) that *E. coli* associated with upper UTI was more likely to adhere to exfoliated uroepithelial cells than *E. coli* randomly isolated from stools.

The adherence organelles, identified as fimbrial structures that bound to a region of the P blood group antigen, were called P pili. The genes encoding P pili were cloned in 1981 and were called *pap* (pyelonephritis-associated pili). Genes for P pilus expression were found at increased frequency among UTI bacteria and among blood isolates, but they were restricted to the species *E. coli* (8, 20). With one recent exception (41), P pilus adherence has been shown to contribute to upper urinary tract colonization in numerous murine and primate model studies (25). It is now recognized that P pili consist of a family of adherence organelles that recognize different receptor regions of the P blood group antigen (23, 29). The substantial progress over the past years in understanding the genetic and physical properties of the P family of adhesins is presented in reference 17 and is summarized here. The pili consist of a shaft of polymerized *papA*-encoded protein that projects outward from the surface of the bacteria. They are assembled upon a protein base composed of the *papC* product. *papE*- and *papG*-encoded proteins reside at the tip of the shaft, and the *papE* product connects the *papA* product to the shaft and also adheres to fibronectin. The *papG* product is the receptor-specific protein that binds to the P blood group antigen. Synthesis and assembly of P pili also require accessory proteins encoded by *papF, papH, papJ,* and *papK;* their production is regulated by products of the *papB* and *papI* genes as well as at least three genes outside of the *pap* gene cluster. The *papG*-encoded protein alone is sufficient to recognize and bind the carbohydrate receptor. Assembly of the *papAEG* gene product complex depends upon the *papD*-encoded protein, a periplasmic chaperone.

More recent studies have begun to address the structural basis for the interaction of the *papG* product with its specific carbohydrate receptor and with the *papD*-encoded protein, which is required for pilus assembly. In one study, three separate regions of the *papG*-encoded protein were identified to be required for carbohydrate receptor binding. Missense mutations in any one of these regions abolished receptor binding but did not interfere with pilus assembly (30). In other studies, the *papD* gene product was shown to interact with amino acids at the carboxyl end of the *papG* product and other *pap* structural proteins. A combination of crystallography, mutagenesis, and in vitro experiments was used to show that the *papD* product interacts specifically with the COOH-terminal seven amino acids of the *papG*-encoded protein (31). As investigators move closer to understanding the structural requirements for recognition and binding of bacterial adhesins to tissue and for the assembly of functional adherence organelles, rational design of novel antiadherence or antiassembly agents will become possible.

The physical and genetic characteristics of other UTI-associated adherence

TABLE 1
UTI-associated adhesins for which receptors are known[a]

Adhesin name	Example	Receptor	Reference(s)
P family[b]	Pap, Pap-2, Pap-3, Prs, ONAP, P_{ia2}, Prf, Fso, Fst	P blood group antigen	17, 30
S family	Sfa, type 1C, Sfr	Sialyloligosaccharides of glycophorin A (for Sfa)	45, 47, 50
Dr family	Dr, Afa-I, Afa-III, F1485	Dr blood group antigen	42
M	M	Glycophorin A^M	52
G	G	N-Acetylglucosamine	52
Type 1	Type 1	Mannosides	11

[a] Comprehensive listings of other fimbrial and nonfimbrial adhesins associated with extraintestinal infections can be found in references 17 and 25.
[b] Families include genetically related adhesins.

organelles such as Dr, S, G, M, and type 1C pili are discussed in detail elsewhere (17, 25). UTI-associated adhesins with known tissue receptors are listed in Table 1.

Adherence Associated with Symptomatic Bladder Colonization

E. coli

Several *E. coli* fimbrial adhesins are associated with symptomatic bladder infections, including type 1, type 1C, Dr, and, to some extent, P. It was shown directly that type 1 pili increase the bladder-colonizing capacity of bacteria in a murine model system (18, 22). In correlation studies, type 1C and Dr pili are found to be more frequently associated with cystitis bacteria than bacteria from other sites (21, 43, 58). The mechanism by which type 1 pili contribute to colonization of the human bladder is unclear. Virkola et al. (66) found that while rodent bladder epithelium contains receptors for type 1 pili and binds type 1-piliated bacteria, human bladder epithelial cells do not; type 1 pilus-mediated binding directly to bladder cells of humans is unlikely. Instead, type 1 pilus attachment to Tamm-Horsfall protein in urinary tract mucus may contribute to bladder colonization. While bacteria attached to urinary tract mucus sloughed from the bladder wall will be carried away swiftly upon voiding, a portion of the uromucoid and the associated bacteria may remain attached to the bladder wall. Just as residual urine left in the bladder upon incomplete voiding may act as a reservoir for bacterial growth, the residual bacteria bound to retained uromucoid may provide seed for persistent colonization. The slime capsules produced by bacteria have been postulated in other systems to facilitate bacterial colonization; for example, enlarged glycocalyx produced by *Pseudomonas aeruginosa* isolated from lung infections in cystic fibrosis patients may impede clearance of the bacteria from the lung and protect the bacteria from host immunity. Perhaps *E. coli* has found a tool in type 1 pili for recruiting a slime capsule from the host to provide a method for persistence in the bladder.

Other gram-negative bacteria

While adhesins functionally similar to type 1 pili are found among non-*E. coli* members of the family *Enterobacteriaceae,* they share limited homology with *E. coli* type 1 pili at the DNA sequence level (7). There is little direct evidence describing a role for type 1 adherence in symptomatic bladder infections for other members of the family *Enterobacteriaceae* in community-acquired disease. However, a second fimbrial protein, designated *P. mirabilis* fimbriae (PMF) (4, 37), has recently been identified in two urinary *P. mirabilis* isolates. When the *pmf* gene was inactivated, the capacity of the *P. mirabilis* strain for bladder colonization in a murine UTI model was reduced; PMF adherence may also be a significant factor contributing to bladder colonization by this organism. Urease expression by some uropathogenic bacteria may also contribute to adherence; ammonia produced by urease has been shown to alter the bladder surface, resulting in greater bacterial adherence (24, 48).

Gram-positive bacteria

S. saprophyticus, which is a significant cause of community-acquired acute UTI (1), and *Enterococcus faecalis,* which primarily causes nosocomial UTI, are the two principal gram-positive uropathogenic bacteria. Urinary isolates of *S. saprophyticus* have been tested for their capacities to colonize mouse bladder epithelium and have been shown to preferentially attach to tight junctions of uroepithelial cells (38). The mechanism of adherence is unknown. Likely candidates include one or more of the various hemagglutinins expressed by urinary *S. saprophyticus* or teichoic acid (6, 14, 15, 63).

Little is known about the urinary tract adherence mechanisms of *E. faecalis,* although it is the primary or secondary cause of UTI in certain patient groups. Guzman et al. (16) have shown that urinary *E. faecalis* isolates attach to exfoliated uroepithelial cells more efficiently than isolates from other infection sites. They suggest that carbohydrates on the bacterial surface may promote adherence. Details of this adherence mechanism or of its significance in disease await discovery.

Adhesins Associated with Symptomatic Bladder Colonization in People with Indwelling Urinary Catheters

It is often assumed, perhaps correctly, that since urine is an excellent growth medium for a variety of bacteria, the catheterized bladder becomes an ecological niche suitable for colonization on a first-come, first-served basis. Often, these colonizations persist for weeks to months without adverse symptoms. At other times colonization will progress to symptomatic urinary disease. While the ultimate outcome of a colonization may depend to a large extent on the physical and immune status of the host individual, bacterial adherence may also promote bladder colonization by selected genera or individual strains. For example, type 1 pili may contribute to bladder colonization through their capacity to colonize the uromucoid coating urinary catheters (39). MR/K pili produced by *Providencia stuartii* were also correlated with persistence in catheterized urinary bladders (40).

Eddeland and Hedelin (10) found that *P. mirabilis*, unlike *E. coli*, was rarely found in the urine without simultaneously growing in the urethra. They proposed that the urethra may be the primary colonization site in patients with long-term indwelling catheters. Little is known about bacterial adherence to urethral epithelium, especially in relation to urethral catheterization.

Numerous studies of UTI in the catheterized patient group have attempted to correlate known hemagglutination phenotypes or uroepithelial cell adherence with bladder colonization and symptomatic disease. At one time or another, nearly every enterobacterial hemagglutinin, with the possible exception of P fimbriae, has been implicated. Each adhesin may contribute to colonization to some degree, or there may be mechanisms of adherence or colonization, or both, that are completely different from those identified previously. Physical trauma to the bladder wall or urethra resulting from catheterization may provide unique receptors for a specialized set of bacterial adhesins. For example, the *K. pneumoniae* type 3 pili and *E. coli* Dr pili which bind receptors in muscle and endothelial tissue may find new opportunities for colonization at wound sites (44, 62). Progress in this area has been impeded by poor understanding of how nosocomial uropathogens, such as *Proteus* species, *Klebsiella* species, non-P piliated *E. coli*, and *E. faecalis*, interact with the milieu of the bladder wall in individuals undergoing intermittent or short- or long-term catheterization.

Adhesins Associated with *E. coli* Asymptomatic Bladder Colonization

Certain *E. coli* strains occasionally colonize the bladders of urologically intact individuals for extended periods of time without producing overt undesirable symptoms and without causing a measurable decrease in renal function (19). Left untreated, these organisms seem to establish a stable colonization and allow the bladder to resist superinfection with more pathogenic bacteria. Asymptomatic colonizing strains often do not possess the adherence mechanisms commonly associated with urinary bacteria, such as pili of the P or Dr family. The adherence mechanisms used by such strains to persist in the bladder are unknown but are under investigation.

Adhesins Associated with Ureteral Colonization

The next anatomical step in the progression of UTI is movement through the ureters to the kidneys. Bacteria may ascend by simple diffusion against urine flow, or ascent may be assisted by adherence to ureter epithelium and colonization of the ureter walls. Among the uropathogenic *E. coli*, several fimbrial adhesins including P, type 1, and Dr pili adhere to ureteral tissue (12, 44). *S. saprophyticus* urinary isolates also adhere to ureteral epithelium, possibly using a pilus-related mechanism (13).

Adhesins Associated with Kidney Colonization and Urosepsis

The *E. coli* adherence proteins associated with renal colonization, primarily pili of the P adhesin family, have been the subjects of intense study over the past

14 years. This topic has received thorough attention in recent reviews (for example, see references 25 and 61). P pili facilitate the attachment of *E. coli* to kidney tissue and promote colonization of that organ. They may also contribute to the tissue inflammation associated with symptomatic pyelonephritis. P pili associated with either intact *E. coli* or endotoxin alone were sufficient to induce inflammation in a murine UTI model (33).

P pili appear to be a risk factor in the progression of kidney infection to urosepsis. Otto et al. (46) reported that 100% of 18 *E. coli* urosepsis patients without complicating medical conditions were infected with P$^+$ bacteria. The frequency of P$^+$ *E. coli* urosepsis isolates from patients with complicating factors was 57% of seven isolates. Johnson et al. (26) reported that 67% of 75 urosepsis *E. coli* isolates expressed P adherence. Their patient population included both uncompromised patients (37%) and those with one or more compromising conditions. In a previous study (27), those investigators found that 100% of 12 urosepsis isolates from uncomplicated infections were P$^+$. P pili appear to be more common among sepsis isolates than among urinary isolates (70 to 80%) in the uncomplicated pyelonephritis group but are underrepresented in urosepsis isolates from the complicated UTI group. How P pili act or how they may interact with other bacterial virulence factors to promote urosepsis is unclear.

Understanding of how other genera of uropathogens, such as *P. mirabilis*, which has a propensity toward kidney invasion and urosepsis, colonize the kidney is limited. Silverblatt (54) and Silverblatt and Ofek (55) demonstrated that some *P. mirabilis* strains had increased virulence in a rodent retrograde pyelonephritis model but found that P$^+$ strains were at a disadvantage in hematogenous infections. Wray et al. (69) described a fimbrial protein named UCA in a urinary *P. mirabilis* isolate that promoted adherence to exfoliated human uroepithelial cells. When a recombinant molecule containing this gene and adjacent genes were transferred to an *E. coli* host, the strain exhibited a uroepithelial cell adherence pattern typical of *P. mirabilis* (70). The contribution of this adhesin to kidney colonization and urosepsis by *P. mirabilis* is under investigation.

Two *P. mirabilis* hemagglutinating pili, MR/K and MR/P, have recently been suggested as virulence factors for kidney colonization. In one set of studies, MR/P pili were found to be associated with 65% of pyelonephritis bacteria (5). Both MR/K and MR/P pili have been shown to adhere to human kidney tissue (53). However, other studies have shown little correlation between hemagglutination patterns and urinary tract virulence or uroepithelial adherence (2, 36). These diverse results speak for the complexity of adherence properties in this species.

Strains of *S. saprophyticus* isolated from UTIs have been shown to adhere to laminin, a connective tissue protein found in the kidney (49). The molecular basis for laminin adherence or its contribution to urovirulence is unknown.

PROPHYLAXIS

Understanding of how bacteria interact with and colonize urinary tract epithelial cells has suggested methods for prevention that have proven to be both exciting and frustrating. P pili are a critical virulence factor for kidney colonization by *E. coli;* few P$^-$ strains are found in uncomplicated pyelonephritis, and P$^+$ strains are required for virulence in animal models of upper urinary tract colonization. This makes the P pilus a prime candidate as a vaccine antigen for the prevention of community-acquired UTI. But the major P pilus structural protein is also antigenically diverse, so that antipilus vaccines, while offering protection against challenge with similar strains, may have restricted application. Also, the ability to identify a target population that should be vaccinated is limited. Vaccination of all women and girls for the prevention of a nonepidemic disease that is rarely fatal may be impractical. Still, other components of P pili that have greater sequence conservation, such as the *papE* product, or that are directly involved in adherence, such as the *papG* product, may be effective vaccine antigens. Kaak et al. (28) have recently shown that anti-idiotypic antibodies specific for the receptor-binding domain of the *papG* product offer protection. Studies on host susceptibility may help to identify individuals at the greatest risk for acute UTI so that immunization can efficiently be provided to people who would receive the greatest benefit. For example, tissue from people with certain blood groups was shown to have a greater capacity to bind P$^+$ bacteria; these individuals may be more susceptible to UTI and would most likely receive greater benefit from a P pilus-based vaccine (34, 35).

The target population of a vaccine for the prevention of chronic, complicated UTI is more easily defined and includes individuals with renal reflux, renal scarring, neurogenic bladder, or urinary catheters. But the choice of an effective vaccine antigen for this patient population is not as clear as that for acute UTIs because chronic UTIs are caused by a diverse group of organisms. The mechanisms of urinary tract colonization by these organisms are different from those by *E. coli* and are only poorly understood. In one study, a multivalent killed whole-cell vaccine consisting of single strains of *Proteus, Providencia, Klebsiella,* and *Enterococcus* and six strains of *E. coli* was tested for efficacy in a murine model. It provided protection against challenge with *E. coli,* but its protective value versus the other genera, which are more often associated with chronic infection, was not reported (65).

Another nonantimicrobial strategy for preventing UTI involves the instillation of soluble receptor analogs into the bladder to prevent bacterial colonization (59). The effectiveness of this strategy has been demonstrated with P pilus receptor analogs in murine model systems. However, as a practical matter, instillation of analogs into the bladder has not yet proven useful. Individuals who have in place a urinary catheter that could be used for receptor instillation are not prone to infections by bacteria with known receptor-binding properties, such as the P type. Healthy young women who may experience UTIs with P$^+$ *E. coli* may be (rightly) disinclined to use a catheter for that purpose. Receptor analog therapy may become most effective when analogs for the prevention of infection in the catheterized patient are identified or improved means of delivery are developed.

A third preventive measure currently being investigated, called "bacterial interference," is targeted primarily toward preventing UTIs in individuals who experience recurrent UTIs and who have predisposing conditions. Harmless bacteria are used to colonize either the bladder or the periurethral region with the goal of preventing subsequent colonization with disease-causing organisms. Studies by Reid et al. (51) have shown that certain strains of *Lactobacillus* are able to establish stable colonization of the vagina of women subject to recurrent UTIs. The *Lactobacillus* colonization appeared to reduce the frequency of subsequent UTIs and vaginal colonization with uropathogens.

E. coli bacteria have also been investigated for use in preventing symptomatic UTI (3). Those experiments identified an *E. coli* isolate which displayed a capacity for prolonged asymptomatic colonization of the human urinary bladder. When instilled into the bladder of several individuals predisposed to UTIs, stable bladder colonization by the test strain was achieved with few undesirable symptoms. The effectiveness of the test strain for preventing subsequent symptomatic disease is under investigation.

ACKNOWLEDGMENTS. This work was supported by Public Health Service grants AI 21009 and AI 18462 from the National Institutes of Health.

REFERENCES

1. **Abrahamsson, K., S. Hansson, U. Jodal, and K. Lincoln.** 1993. *Staphylococcus saprophyticus* urinary tract infections in children. *Eur. J. Pediatr.* **152:**69–71.

2. **Adegbola, R. A., D. C. Old, and B. W. Senior.** 1983. The adhesions and fimbriae of *Proteus mirabilis* strains associated with high and low affinity for the urinary tract. *J. Med. Microbiol.* **16:**427–431.

3. **Andersson, P., I. Engberg, G. Lindin-Janson, K. Lincoln, R. Hull, S. Hull, and C. Svanborg.** 1991. Persistence of *Escherichia coli* bacteriuria is not determined by bacterial adherence. *Infect. Immun.* **59:**2915–2921.

4. **Bahrani, F. K., S. Cook, R. A. Hull, G. Massad, and H. L. T. Mobley.** 1993. *Proteus mirabilis* fimbriae: N-terminal amino acid sequence of a major subunit and nucleotide sequences of the genes from two strains. *Infect. Immun.* **61:**884–891.

5. **Bahrani, F. K., and H. L. T. Mobley.** 1993. *Proteus mirabilis* MR/P fimbriae: molecular cloning, expression, and nucleotide sequence of the major subunit gene. *J. Bacteriol.* **475:**457–464.

6. **Beuth, J., H. L. Ko, F. Schumacher-Perdreau, G. Peters, P. Heczko, and G. Pulverer.** 1988. Hemagglutinaton by *Staphylococcus saprophyticus* and other coagulase-negative staphylococci. *Microb. Pathog.* **4:**379–383.

7. **Buchanan, K., S. Falkow, R. A. Hull, and S. I. Hull.** 1985. Frequency among *Enterobacteriaceae* of DNA sequences encoding type 1 pili. *J. Bacteriol.* **162:**799–803.

8. **Denich, K., A. Craiu, H. Rugo, G. Muralidhar, and P. O'Hanley.** 1991. Frequency and organization of *papA* homologous DNA sequences among uropathogenic digalactoside-binding *Escherichia coli* strains. *Infect. Immun.* **59:**2089–2096.

9. **Donovan, W. H., R. Hull, D. X. Cifu, H. D. Brown, and N. J. Smith.** 1990. Use of plasmid analysis to determine the source of bacterial invasion of the urinary tract. *Paraplegia* **28:**573–582.

10. **Eddeland, A., and H. Hedelin.** 1983. Bacterial colonization of the lower urinary tract in women with long-term indwelling urethral catheters. *Scand. J. Infect. Dis.* **15:**361–365.

11. **Firon, N., I. Ofek, and N. Sharon.** 1984. Carbohydrate-binding sites of the mannose-specific fimbrial lectins of enterobacteria. *Infect. Immun.* **43:**1088–1090.

12. **Fujita, K., T. Yamamoto, and R. Kitagawa.** 1991. Binding sites for P and/or type 1 piliated *Escherichia coli* in human ureter. *J. Urol.* **146:**217–222.

13. **Fujita, K., T. Yokota, T. Oguri, M. Fujime, and R. Kitagawa.** 1992. In vitro adherence of *Staphylococcus saprophyticus, Staphylococcus epidermidis, Staphylococcus haemolyticus* and *Staphylococcus aureus* to human ureter. *Urol. Res.* **20:**399–402.

14. **Gatermann, S., R. Marre, J. Heesemann, and W. Henkel.** 1988. Hemagglutinating and adherence

properties of *Staphylococcus saprophyticus:* epidemiology and virulence in experimental urinary tract infections of rats. *FEMS Microbiol. Immunol.* **1:**179–185.

15. **Gatermann, S., H.-G. W. Meyer, and G. Wanner.** 1992. *Staphylococcus saprophyticus* hemagglutinin is a 160-kilodalton surface polypeptide. *Infect. Immun.* **60:**4127–4132.

16. **Guzman, C. A., C. Pruzzo, G. LiPiri, and L. Calegari.** 1989. Role of adherence in pathogenesis of *Enterococcus faecalis* urinary tract infection and endocarditis. *Infect. Immun.* **57:**1834–1838.

17. **Hacker, J.** 1990. Genetic determinants coding for fimbriae and adhesins of extraintestinal *Escherichia coli. Curr. Top. Microbiol. Immunol.* **151:**1–27.

18. **Hagberg, L., R. Hull, S. Hull, S. Falkow, R. Freter, and C. Svanborg-Éden.** 1983. Contribution of adhesion to bacterial persistence in the mouse urinary tract. *Infect. Immun.* **40:**265–272.

19. **Hanson, L. A., A. Fasth, U. Jodal, B. Kaijser, and C. Svanborg-Éden.** 1981. Biology and pathology of urinary tract infections. *J. Clin. Pathol.* **34:**695–700.

20. **Hull, R. A., S. I. Hull, and S. Falkow.** 1984. Frequency of gene sequences necessary for pyelonephritis associated pili expression among isolates of *Enterobacteriaceae* from human extraintestinal infection. *Infect. Immun.* **43:**1064–1067.

21. **Ikäheimo, R., A. Siitonen, U. Kärkkäinen, P. Kuosmanen, and H. Mäkelä.** 1993. Characteristics of *Escherichia coli* in acute community-acquired cystitis of adult women. *Scand. J. Infect. Dis.* **25:**705–712.

22. **Iwahi, T., Y. Abe, M. Nakao, A. Imada, and K. Tsuchiya.** 1983. Role of type 1 fimbriae in the pathogenesis of ascending urinary tract infection induced by *Escherichia coli* in mice. *Infect. Immun.* **39:**1307–1315.

23. **Johanson, I., R. Lindstedt, and C. Svanborg.** 1992. Roles of the *pap-* and *prs*-encoded adhesins in *Escherichia coli* adherence to human uroepithelial cells. *Infect. Immun.* **60:**3416–3422.

24. **Johnson, D. E., R. G. Russell, C. V. Lockatell, J. C. Zulty, J. W. Warren, and H. L. T. Mobley.** 1993. Contribution of *Proteus mirabilis* urease to persistence, urolithiasis, and acute pyelonephritis in a mouse model of ascending urinary tract infection. *Infect. Immun.* **61:**2748–2754.

25. **Johnson, J.** 1991. Virulence factors in *Escherichia coli* urinary tract infection. *Clin. Microbiol. Rev.* **4:**80–128.

26. **Johnson, J. R., I. Ørskov, F. Ørskov, P. Goullet, B. Picard, S. L. Moseley, P. L. Roberts, and W. Stamm.** 1994. O, K, and H antigens predict virulence factors, carboxylesterase B pattern, antimicrobial resistance, and host compromising factors among *Escherichia coli* strains causing urosepsis. *J. Infect. Dis.* **169:**119–126.

27. **Johnson, J. R., P. L. Roberts, and W. Stamm.** 1987. P fimbriae and other virulence factors in *Escherichia coli* urosepsis: association with patients' characteristics. *J. Infect. Dis.* **156:**225–229.

28. **Kaak, M. B., L. N. Martin, S. B. Svenson, G. Baskin, R. H. Steele, and J. A. Roberts.** 1993. Protective anti-idiotypic antibodies in the primate model of pyelonephritis. *Infect. Immun.* **61:**2289–2295.

29. **Karr, J. K., B. J. Nowicki, L. D. Truong, R. A. Hull, J. J. Moulds, and S. I. Hull.** 1990. *pap-2* encoded fimbriae adhere to the P blood group-related glycosphingolipid stage-specific embryonic antigen 4 in the human kidney. *Infect. Immun.* **58:**4055–4062.

30. **Klann, A. G., R. A. Hull, T. Palzkill, and S. I. Hull.** 1994. Alanine scanning mutagenesis reveals residues involved in binding of *pap-3* encoded pili. *J. Bacteriol.* **176:**2312–2317.

31. **Kuehn, M. J., D. J. Ogg, J. Kihlberg, L. N. Slonim, K. Flemmer, T. Bergfors, and S. Hultgren.** 1993. Structural basis of pilus subunit recognition by the PapD chaperone. *Science* **162:**1234–1241.

32. **Kunin, C. M.** 1994. Urinary tract infection in females. *Clin. Infect. Dis.* **18:**1–12.

33. **Linder, H., I. Engberg, I. Mattsby Baltzer, K. Jann, and C. Svanborg-Éden.** 1988. Induction of inflammation by *Escherichia coli* on the mucosal level: requirement for adherence and endotoxin. *Infect. Immun.* **56:**1309–1313.

34. **Lingstedt, R., G. Larson, P. Falk, U. Jodel, H. Leffler, and C. Svanborg.** 1991. The receptor repertoire defines the host range for attaching *Escherichia coli* strains that recognize globo-A. *Infect. Immun.* **59:**1086–1092.

35. **Lomberg, H., B. Cedergren, H. Leffler, B. Nilsson, A.-S. Carlström, and C. Svanborg-Éden.** 1986. Influence of blood group on the availability of receptors for attachment of uropathogenic *Escherichia coli. Infect. Immun.* **51:**919–926.

36. **Lomberg, H., P. Larsson, H. Leffler, and C. Svanborg-Édén.** 1982. Different binding specificities of *Proteus mirabilis* compared to *E. coli. Scand. J. Infect. Dis. Suppl.* **33:**37–42.

37. **Massad, G., C. V. Locatell, D. E. Johnson, and H. L. T. Mobley.** 1994. *Proteus mirabilis* fimbriae: construction of an isogenic *pmfA* mutant and analysis of virulence in a CBA mouse model of ascending urinary tract infection. *Infect. Immun.* **62:**536–542.

38. **McTaggart, L. A., R. C. Rigby, and T. S. J. Elliot.** 1990. The pathogenesis of urinary tract infection associated with *Escherichia coli, Staphylococcus saprophyticus,* and *S. epidermidis. J. Med. Microbiol.* **32:**135–141.

39. **Mobley, H. L. T., G. R. Chippendale, J. H. Tenney, R. A. Hull, and J. W. Warren.** 1987. Expression of type 1 fimbriae may be required for persistence of *Escherichia coli* in the catheterized urinary tract. *J. Clin. Microbiol.* **25:**2253–2257.

40. **Mobley, H. L. T., G. R. Chippendale, J. H. Tenney, A. R. Mayrer, L. J. Crisp, J. L. Penner, and J. W. Warren.** 1988. MR/K hemagglutination of *Providencia stuartii* correlates with adherence to catheters and with persistence in catheter-associated bacteriuria. *J. Infect. Dis.* **157:**264–271.

41. **Mobley, H. L. T., K. G. Jarvis, J. P. Elwood, D. I. Whittle, C. V. Lockatell, R. G. Russell, D. E. Johnson, M. S. Donnenberg, and J. Warren.** 1993. Isogenic P-fimbrial deletion mutants of pyelonephritogenic *Escherichia coli:* the role of αGal(1–4)βGal binding in virulence of a wild-type strain. *Mol. Microbiol.* **10:**143–155.

42. **Nowicki, B., A. Labigne, S. Moseley, R. Hull, S. Hull, and J. Moulds.** 1990. The Dr hemagglutinin, afimbrial adhesins AFA-I and AFA-III, and F1845 fimbriae of uropathogenic and diarrhea-associated *Escherichia coli* belong to a family of hemagglutinins with Dr receptor recognition. *Infect. Immun.* **58:**279–281.

43. **Nowicki, B., C. Svanborg-Édén, R. Hull, and S. Hull.** 1989. Molecular analysis and epidemiology of the Dr hemagglutinin of uropathogenic *Escherichia coli. Infect. Immun.* **57:**446–451.

44. **Nowicki, B., L. Truong, J. Moulds, and R. Hull.** 1988. Presence of the Dr receptor in normal tissues and its possible role in the pathogenesis of urinary tract infection. *Am. J. Pathol.* **133:**1–4.

45. **Ott, M., H. Hoschützky, K. Jann, I. Van Die, and J. Hacker.** 1988. Gene clusters for S fimbrial adhesin (*sfa*) and F1C fimbriae (*foc*) of *Escherichia coli:* comparative aspects of structure and function. *J. Bacteriol.* **170:**3983–3990.

46. **Otto, G., T. Sandberg, B.-I. Marklund, P. Ulleryd, and C. Svanborg.** 1993. Virulence factors and *pap* genotype in *Escherichia coli* isolates from women with acute pyelonephritis with or without bacteremia. *Clin. Infect. Dis.* **17:**448–456.

47. **Parkkinen, J., G. N. Rogers, T. Korhonen, W. Dahr, and Jukka Finne.** 1986. Identification of the O-linked sialyloligosaccharides of glycophorin A as the erythrocyte receptors for S-fimbriated *Escherichia coli. Infect. Immun.* **54:**37–42.

48. **Parsons, C. L., C. Stauffer, S. G. Mulholland, and D. Griffith.** 1984. Effect of ammonium on bacterial adherence to bladder transitional epithelium. *J. Urol.* **132:**365–366.

49. **Paulsson, M., A. Ljunge, and T. Wadstrom.** 1992. Rapid identification of fibronectin, vitronectin, laminin, and collagen cell surface binding proteins on coagulase-negative staphylococci by particle agglutination assays. *J. Clin. Microbiol.* **30:**2006–2012.

50. **Pawelzik, M., J. Heesemann, J. Hacker, and W. Opferkuch.** 1988. Cloning and characterization of a new type of fimbria (S/F1C-related fimbria) expressed by an *Escherichia coli* O75:NOK1: NOH7 blood culture isolate. *Infect. Immun.* **56:**2918–2924.

51. **Reid, G., A. W. Bruce, J. A. McGroarty, K.-J. Cheng, and J. W. Costerton.** 1990. Is there a role for lactobacilli in prevention of urogenital and intestinal infections? *Clin. Microbiol. Rev.* **3:**335–344.

52. **Rhen, M., P. Klemm, and T. Korhonen.** 1986. Identification of two new hemagglutinins of *Escherichia coli*, *N*-acetyl-D-glucosamine-specific fimbriae and a blood group M-specific agglutinin, by cloning the corresponding genes in *Escherichia coli* K-12. *J. Bacteriol.* **168:**1234–1242.

53. **Sareneva, T., H. Holthofer, and T. Korhonen.** 1990. Tissue-binding affinity of *Proteus mirabilis* fimbriae in the human urinary tract. *Infect. Immun.* **58:**3330–3336.

54. **Silverblatt, F. J.** 1974. Host parasite interaction in the rat pelvis. *J. Exp. Med.* **140:**1696–1711.

55. **Silverblatt, F. J., and I. Ofek.** 1978. Influence of pili on the virulence of *Proteus mirabilis* in experimental hematogenous pyelonephritis. *J. Infect. Dis.* **138:**664–667.

56. **Smith, H. W., and M. A. Linggood.** 1971. Observations on the pathogenic properties of the K88,

Hly, and Ent plasmids of *Escherichia coli* with particular reference to porcine diarrhea. *J. Med. Microbiol.* **4**:467–486.

57. Stamm, W. E., T. M. Hooton, J. R. Johnson, C. Johnson, A. Stapleton, P. L. Roberts, S. L. Moseley, and S. D. Fihn. 1989. Urinary tract infections: from pathogenesis to treatment. *J. Infect. Dis.* **159**: 400–406.

58. Stapleton, A., S. Moseley, and W. E. Stamm. 1991. Urovirulence determinants in *Escherichia coli* isolates causing first-episode and recurrent cystitis in women. *J. Infect. Dis.* **163**:773–779.

59. Svanborg-Éden, C., B. Anderson, G. Aniansson, R. Lindstedt, P. deMan, A. Neilsen, H. Leffler, and A. Wold. 1990. Inhibition of bacterial attachment: examples from the urinary and respiratory tracts. *Curr. Top. Microbiol. Immunol.* **151**:167–184.

60. Svanborg-Éden, C., L. A. Hanson, U. Jodel, U. Lindberg, and A. Sohl Akerlund. 1976. Variable adherence to normal human urinary tract epithelial cells of *Escherichia coli* strains associated with various forms of urinary-tract infection. Lancet **ii**:490–492.

61. Svanborg-Éden, C., S. Hausson, U. Jodal, G. Lidin-Jansson, K. Lincoln, H. Linder, H. Lomberg, P. de Mann, S. Marild, J. Martinell, K. Plos, T. Sandberg, and K. Stenqvist. 1988. Host parasite interaction in the urinary tract. *J. Infect. Dis.* **157**:421–426.

62. Tarkkanen, A.-M., B. L. Allen, B. Westerlund, H. Holthöfer, P. Kuusela, L. Risteli, S. Clegg, and T. K. Korhonen. 1990. Type V collagen as the target for type-3 fimbriae, enterobacterial adherence organelles. *Mol. Microbiol.* **4**:1353–1361.

63. Teti, G., M. S. Chiofalo, F. Tomasello, C. Fava, and P. Mastroeni. 1987. Mediation of *Staphylococcus saprophyticus* adherence to uroepithelial cells by lipoteichoic acid. *Infect. Immun.* **55**:839–842.

64. Tullus, K. 1986. Fecal colonization with P-fimbriated *Escherichia coli* in newborn children and relation to development of extraintestinal *E. coli* infections. *Acta Paediatr. Scand. Suppl.* **334**: 1–35.

65. Uehling, D. T., L. J. James, W. J. Hopkins, and E. Balish. 1991. Immunization against urinary tract infection with a multi-valent vaginal vaccine. *J. Urol.* **146**:223–226.

66. Virkola, R., B. Westerlund, H. Holthöfer, J. Parkkinen, M. Kekomaki, and T. K. Korhonen. 1988. Binding characteristics of *Escherichia coli* adhesins in the human urinary bladder. *Infect. Immun.* **56**:2615–2622.

67. Warren, J. W. 1987. Catheter-associated urinary tract infections. *Infect. Dis. Clin. North Am.* **1**: 823–854.

68. Wold, A., D. A. Caugant, G. Lidin-Janson, P. de Man, and C. Svanborg. 1992. Resident colonic *Escherichia coli* strains frequently display uropathogenic characteristics. *J. Infect. Dis.* **165**:46–52.

69. Wray, S. K., S. I. Hull. R. G. Cook, J. Barrish, and R. A. Hull. 1986. Identification and characterization of a uroepithelial cell adhesin from a uropathogenic isolate of *Proteus mirabilis*. *Infect. Immun.* **54**:43–49.

70. Wray-Cook, S. K., and R. Hull. Unpublished data.

Molecular Genetics of Bacterial Pathogenesis
Edited by V. L. Miller, J. B. Kaper, D. A. Portnoy, and R. R. Isberg
© 1994 American Society for Microbiology, Washington, DC 20005

Chapter 7

Escherichia coli Type 1 Pili

Paul E. Orndorff

The short review in this chapter concentrates upon what I think are the essential features of *Escherichia coli* type 1 pili as models of host-pathogen interactions, gene regulation, supramolecular assembly, and receptor binding. In these areas, I have been fortunate to have colleagues and collaborators of very high standards whose work I have greatly benefited from and whose work I hope I have represented accurately here. Also, my writing task was lightened by the existence of a number of recent reviews on various aspects of pilus expression and the possible roles of pili in disease (20, 43, 52, 55, 74, 101, 104). Finally, this volume itself contains excellent chapters that concern or touch upon several features of bacterial piliation.

Type 1 pili (fimbriae) of *E. coli* are filamentous, proteinaceous appendages approximately 5 nm wide and 1 to 2 μm long (18). They are principally composed of a single repeating polypeptide unit called pilin which is arrayed helically to form a hollow-cored fiber (18) (Fig. 1). Although these structures were thought for some time to be composed exclusively of pilin (18, 91), it is now known that at least three other minor components are incorporated into the pilus fiber (45, 71). One of these minor components (termed the adhesin) is responsible for the ability of type 1-piliated bacteria to bind to a variety of eukaryotic cells (28). Binding can be inhibited by mannose, alpha-linked mannose oligosaccharides, and certain mannose analogs (102), indicating that the receptor on the eukaryotic cell contains mannose or a closely related compound (38, 114).

Type 1 pili are the most common adhesive appendages found on members of the family *Enterobacteriaceae* (28). As such, they were likely first noticed at the turn of the century by Guyot (42), who reported hemagglutination by certain strains of *E. coli*. Subsequent observations documented agglutination of additional types of eukaryotic cells by *E. coli* (115). Again, these observations were consistent with type 1 pilus-mediated attachment. However, it was not until shortly after the development of the electron microscope that thin structures such as pili could be distinguished on bacterial cells.

The organelles that have become known as type 1 pili were probably first

Paul E. Orndorff • Department of Microbiology, Pathology, and Parasitology, College of Veterinary Medicine, North Carolina State University, 4700 Hillsborough Street, Raleigh, North Carolina 27606.

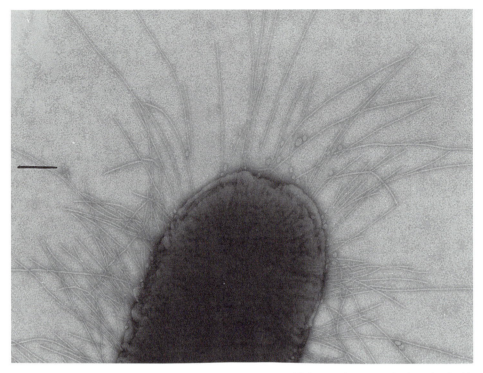

FIGURE 1. Electron micrograph of an *E. coli* cell with type 1 pili that has been negatively stained with 1.0% phosphotungstic acid. Bar, 0.25 μm.

seen by Houwink and VanItterson (49) in their early electron microscopic investigation of flagellation in bacteria. Subsequent studies by Duguid et al. (29) established the relationship of fimbriation to hemagglutination. Those authors classified four groups of pili by the criteria of morphology and by the ability of the piliated *E. coli* strains to agglutinate erythrocytes from different animal species. The majority of *E. coli* strains fell into group 1, which had the structures that were to become known as type 1 pili. Brinton (17) independently described various properties of the organelles, and Collier and DeMiranda (19) established the specific effect of mannose in inhibiting type 1-mediated hemagglutination of guinea pig erythrocytes. In 1959, Brinton (17) observed mannose-sensitive hemagglutination by purified type 1 pili, unequivocally establishing the role of these pili in the process of mannose-sensitive adherence. This observation has been confirmed independently many times (for reviews, see references 68 and 110).

ROLE OF TYPE 1 PILI IN INFECTION

Because most eukaryotic cells possess surface molecules containing mannose (28), it might be reasonable to assume that type 1 piliation would increase *E. coli* virulence by increasing the likelihood that type 1-piliated individuals would bind to epithelial cells, thus facilitating colonization—the first step in any infectious process. While this reasoning seems sound, there are reasons to doubt its veracity or at least its generality. Certainly one impediment to accepting type 1 pili as virulence factors is the ubiquity of the structures: 70% of all normal fecal isolates are type 1 piliated (21, 26, 109). Clearly, then, the property of piliation does not dedicate an individual to a virulent lifestyle. Nevertheless, in certain instances, type 1 piliation does influence the course of disease caused by *E. coli*. Some of these instances are reviewed below.

Early findings that type 1 pili bound to mucosal epithelial cells (27, 59, 100, 117) prompted investigations into the roles that these pili might play in *E. coli* colonization and pathogenesis (for a review, see reference 104). These studies can be summarized by saying that there is now good experimental evidence that type 1 pili promote colonization of the urinary tract (67) and oropharyngeal region (10), at least in certain laboratory animals. The role of these pili in colonization of the intestinal tract and their roles in nonopsonic engulfment and killing by phagocytic cells are more open to question. However, results from my own and other laboratories have indicated that type 1 pili play no role in facilitating colonization of the large or small bowel (10, 41, 90) but aid in protecting *E. coli* from phagocytic killing in vitro (39, 66, 126)—although the pili also promote binding to a variety of phagocytic cells (14; for a review, see reference 101). With regard to this latter apparent paradox, members of my laboratory and I have suggested that type 1 binding may disrupt or bypass the normal lytic pathway(s) in the phagocyte, resulting in reduced killing (66). This idea was based upon our observation that mutants lacking the minor adhesive component of the pilus were as susceptible to phagocytic killing as those lacking the whole pilus, indicating that protection from killing is related specifically to the binding process. More detailed independent biochemical tests have revealed that binding of the *fimH* product activates neutrophils (126), but the authors add that preliminary results support the observation of less killing by these cells (126). It may be that *fimH* product binding, while resulting in activation, causes the engulfment process to be mistimed or otherwise ineffective in killing piliated cells.

With regard to the role of type 1 pili in promoting *E. coli* extraintestinal colonization (10, 67), a good clinical correlation exists between type 1 piliation and the potential for causing cystitis and urethritis (44). This relationship has been supported by a variety of genetic and biochemical tests of colonization in experimental animals (2, 6, 67). In extraintestinal areas other than the urinary tract, the role of the pili has not been rigorously tested except for studies on their role in causing septicemia in neonatal rats (10, 11, 41). In these animals, type 1 pili appear to play a role in facilitating *E. coli* colonization of the oropharyngeal cavity and in the interindividual spread of *E. coli* K1 (10, 11). However, the effect of piliation upon oropharyngeal colonization is quite subtle, in that the effect of a mutation that eliminates piliation is not seen at doses that are high enough to give a standard 50% infectious dose (10). However, at much lower doses, there is a significant effect, with the piliated strains much more likely to colonize and go on to cause septicemia (11). This observation seems to make good sense if one sees colonization as a microorganism's way of increasing its infectious dose. That is, we (the experimenters) can increase the dose by giving larger inocula, or the bacteria can do it themselves—perhaps by attaching to the oral mucosa and replicating (10, 11, 104).

In contrast to the rather subtle conditional effect of type 1 piliation on virulence in inoculated animals, type 1 expression promotes (10-fold) the spread of *E. coli* in neonatal rat populations (11). The ability of type 1 piliation to promote interindividual transmission was demonstrated in a simple experiment in which two infected rat pups were introduced into a noninfected litter. If the two pups were infected with type 1-piliated *E. coli,* the microorganisms spread extensively among the littermates. If the pups were inoculated with an otherwise isogenic nonpiliated mutant, there was very little interindividual spread. Thus, it may be that type 1 piliation, while not making a given *E. coli* isolate dramatically more virulent in the sense of decreasing the 50% infectious dose, may make type 1-piliated individuals more prevalent in susceptible host populations (10, 11, 104) and could be the central reason that type 1-piliated strains are most often isolated from cases of extraintestinal disease caused by *E. coli*.

GENETICS OF TYPE 1 PILUS EXPRESSION

Type 1 and other bacterial pili have made good models for examining the control of gene expression (15, 40, 65), the assembly of supramolecular structures (55, 77, 86), and receptor-ligand interactions (47; for reviews, see references 20, 43, 74, and 97). Early work on the genetics of type 1 piliation established the location of the principal group of genes required for expression (82, 123) and described the peculiar property of phase variation—the high-frequency switching between piliated and nonpiliated states. The molecular mechanism of phase variation has now been investigated in detail (1, 30, 61) and continues to provide many interesting questions on the transcriptional regulation of gene expression. Problems involved in pilus biogenesis have been addressed only more recently, with progress being most rapid since the cloning of the genes for type 1 pili (50) and the development of techniques to examine the assembly process. A summary of the genes involved in type 1 pilus production and their organization, product size, product function, and mutant phenotypes are shown in Fig. 2.

In the area of the genetic control of type 1 pilus expression, most recent work has centered upon identifying genes whose products are involved in influencing the inversion of a small, ca. 300-bp segment of DNA encoding the *fimA* promoter (1, 69, 118, 122) (enlarged region, Fig. 2). The inversion of this promoter is the driving force behind phase variation (1). In the on promoter orientation, the *fimA* gene is transcribed, pilin is produced, and pili are expressed. In the off orientation, transcription is away from the *fimA* gene, pilin is not made, and pili are not expressed.

In most of the common laboratory strains of *E. coli* (e.g., *E. coli* B and *E. coli* K-12), the rate of inversion is slow enough so that populations of predominantly one phase or the other can be maintained by picking colonies composed principally of piliated or nonpiliated individuals (the hydrophobic nature of pili produces easily noticed differences in the colonial morphologies). However, in some strains, the inversion rate is quite rapid, and clonal populations of predominantly one piliation phenotype cannot be isolated. Some of the differences in strain-specific inversion rates can be attributed to the products of two genes (*fimB* and *fimE*), which are described below.

In the gene cluster shown in Fig. 2, there are two genes, *fimB* and *fimE* (69, 105), whose products influence promoter inversion. The *fimB* product is required for promoter inversion to take place, and the *fimE* product is involved in promoting a unidirectional (on to off) inversion (89). The *fimE* gene product also appears to suppress transcription of *fimA* (13, 89, 108). The requirement of the *fimB* product for inversion and the unusual unidirectional inversion activity of the *fimE* product were discovered recently by Bloomfield and colleagues (13, 36, 89). Their work corrected an inaccurate but widely accepted view that the *fimB* and *fimE* gene products each "flipped" the invertible segment one way or the other (69).

In *fimB* mutants, individuals are all in the off phase because of the unidirectional action of the *fimE* gene product. In *fimE* mutants, random on-to-off inversion occurs at a low rate (approximately 10^{-4} per cell per generation [30, 122]). This low rate allows the isolation of populations that are predominantly in one piliation phase or the other. It has been my experience that in the wild-type situation (*fimB*$^+$ *fimE*$^+$), the population is predominantly in the off orientation, possibly because of the naturally higher degree of activity of the unidirectional *fimB* product (13). However, the "natural" activity of the *fimB* product is probably subject to a variety of selective pressures, resulting in a wide assortment of *fimB* alleles, each of which produces a product with a distinct activity. Such allelic differences could be one possible reason that the range of inversion rates is quite wide between different *E. coli* isolates (36). Also, it may be useful to note that the products of the *fimB* and *fimE* genes have not been shown to act directly to effect inversion and suppression. That is, a direct interaction of the *fimB* and *fimE* products with the DNA of the *fimA* promoter region has not been demonstrated.

With regard to environmental factors influencing the inversion process, Bloomfield and colleagues (36) have concluded that certain environmental conditions (e.g., high or low temperature) influence the inversion rate and suggest that the different rates may be explained by a differential expression of the *fimB* and *fimE* products under a given environmental condition. According to this picture, the amount of time that the invertible promoter remains in the on or off orientation would depend upon the intracellular level of the *fimB* and *fimE* products. This would, in turn, affect the proportion of the population that is piliated under a particular environmental condition. However, not all observations agree with this type of regulatory control (12, 36). Nevertheless, it is clear that different environmental conditions do affect the inversion rate (17, 36). It remains to be seen if a given rate is influenced directly by the levels of the *fimB* and *fimE* gene products or if other currently unidentified factors play a more central role. There have also been reports that certain environmental conditions actually result in a controlled directional inversion which shifts the entire population to one promoter orientation or the other (57, 118, 119). However, this has been disputed (36), and the most conservative interpretation of the available data is that while environmental condi-

Gene	Molecular size of product	Proposed gene product function and brief description of the mutant phenotype
fimB	23 kDa	The fimB product is required for fimA promoter inversion; fimB mutants exhibit a nonpiliated phenotype in which the invertible fimA promoter (enlarged region) is essentially locked in the off orientation (transcription is away from the fimA gene). However, the fimB product is required for inversion in either direction. The reason for the off phenotype is because the adjacent fimE product catalyzes (directly or indirectly) the inversion of all on invertible segments to the off orientation (which can then not switch back in fimB mutants). A key to understanding this rather complicated mechanism is the phenotype exhibited by double mutants lacking both fimB and fimE. Such mutants are found to be locked in either the on or the off orientation.
fimE	22 kDa	The fimE product is required for unidirectional (on to off) inversion of the fimA promoter; fimE mutants exhibit a random on-off phase variation of fimA expression at a rate of approximately 10^{-4} switches per cell per generation. This phenotype is due to the lack of activity of the fimE gene product, which catalyzes (directly or indirectly) inversion of the fimA promoter from on to off. The naturally low activity of this gene in certain well-studied strains such as E. coli K-12 and E. coli B is responsible for the "normal" (infrequent) phase variation seen in these strains (in some cases, this low level of activity is due to IS1 insertions in or near the fimE gene [36]). Higher levels of fimE product result in populations that are predominantly in the off orientation (13). Some clinical E. coli isolates exhibit this phenotype.
fimA	17 kDa	The fimA gene product is the major structural subunit of pili; fimA mutants are nonpiliated because of the lack of the pilin monomer—the product of this gene (106, 107). The promoter for the fimA gene is invertible (enlargement above). When in the on orientation, fimA is transcribed and the cells are piliated. In the opposite (off) orientation, transcription is directed away from fimA, which is not transcribed, and the cell is not piliated.
fimC	30 kDa	The fimC product is a chaperone-like molecule required for pilus assembly; fimC mutants are nonpiliated, but pilin is still produced and processed (its signal sequence is cleaved) to its mature size. The fimC product is itself a periplasmic protein (70) necessary for the polymerization and assembly process. A product of similar size produced by a gene (papD) in a similar location in the pap pili gene cluster makes a chaperone required for proper folding of the pap pilin subunit. The fimC gene also encodes a chaperone-like molecule (70), and the papD gene product can substitute for the fimC product to produce functional type 1 pili, indicating a similar function for each (64).

Continued on following page

Gene	Molecular size of product	Proposed gene product function and brief description of the mutant phenotype
fimD	81 kDa	The *fimD* product is an outer membrane protein required for pilus assembly; *fimD* mutants are nonpiliated, but pilin is still produced and processed to its mature size. The *fimD* product is itself an envelope protein (106), and thus likely serves in the polymerization and assembly process. Similarly, large products from similarly situated genes are observed in most pilus gene clusters. Recent work on *fimD* suggests that it may integrate into the outer membrane and act as a polymerization channel for pilin monomers (72). Also, it may, by analogy with a similarly situated gene in the *pap* pilus cluster, serve as a molecular "usher" for pilin and other minor pilus components (23).
fimF	16 kDa	The *fimF* product is a minor pilus component thought to aid in initiating new pili; *fimF* mutants make approximately one-fourth the number of pili that the parent does. However, the pili are capable of binding erythrocytes and are morphologically indistinguishable from the parent.
fimG	14 kDa	The *fimG* product is a minor pilus component that regulates pilus length; *fimG* mutants make pili that are, on average, three times longer than the parent but have approximately the same number per cell. These pili bind receptor and, aside from their length, have a normal morphology.
fimH	30 kDa	The *fimH* product is a minor pilus component and the adhesive molecule (adhesin) responsible for mannose-sensitive adherence; *fimH* mutants make pili that are incapable of binding to and agglutinating guinea pig erythrocytes and fail to make a pellicle when grown in static broth (47, 85). However, the exact mutant phenotype produced depends upon the type of lesion.

FIGURE 2. The *fim* genes are located at approximately 98 min on the *E. coli* genetic map (7). All of the genes are transcribed from left to right as drawn (clockwise on the map [35]). The genetic mnemonics used throughout the text are from Bachmann (7) and are for *E. coli* K-12. In the initial descriptions of the genes (105–108), the *pil* and *hyp* designations were used (shown in parentheses in the illustration). The intercistronic regions are not drawn to scale in all cases. An additional open reading frame designated *fimI* has been reported (70); it is immediately to the right of *fimA*. Details of the function of the *fimI* gene product have not yet been communicated. More detailed explanations of gene function and additional references may be found in the text.

tions can influence the switching rate, selective outgrowth produces the final proportion of individuals in the on or off fimbriate state (103).

The ability of *E. coli* populations to maintain heterogeneity with regard to individual piliation phenotypes may be an important feature of the life cycle of the microorganisms. It is known, for example, that some laboratory conditions favor the growth of piliated individuals, whereas other conditions favor the growth of nonpiliated ones (103). Furthermore, in vivo, investigators have observed that piliated variants are more often recovered from certain host compartments, notably, the oropharynx (41) and bladder (56), whereas nonpiliated variants are more often associated with peritonitis (5, 87) and septicemia (41). Such observations suggest that maintenance of the potential for either state might permit exploitation of an expanded range of host niches (9). It thus remains of considerable interest to discern the factors affecting type 1 pilus expression.

Returning to the subject of the genetic factors that affect type 1 pilus expression, in addition to the *fimB* and *fimE* products—which appear to be dedicated to the *fimA* promoter inversion process—there are three other gene products that influence inversion as part of a more global regulatory activity. The first is the product of the *hip/himA* gene, integration host factor, which is required for inversion to take place (25, 31). This product is also required for the normal level of transcriptional activity of the *fimA* gene (25). The second is the product of the *lrp* gene. This gene product affects *fimA* expression and promoter inversion (12) as well as expression of other genes (16, 96, 98). The third product is encoded by the *hns* gene (111) (independently found and called *bglY* [79], *drdX* [40], *osmZ* [48, 58], *pilG* [65, 122], and *virR* [24, 84] by various investigators), which acts in repressing the rate of promoter inversion (65, 122). The *hns* gene encodes the histone-like protein HN-S, and work by Tom Kawula and me (65) has shown that mutants with lesions in this gene have a surprisingly high level of promoter inversion (ca. 100-fold faster than the parental strain). H-NS effects the transcriptional activity of a variety of genes as well as affects the rate of *fimA* promoter inversion. In particular, H-NS is implicated in influencing the normal temperature regulation of genes associated with virulence in *E. coli* (40) and *Shigella flexneri* (73, 74). H-NS has been suggested to act by influencing DNA supercoiling (48, 58). However, work from at least one laboratory has not supported H-NS-effected supercoiling as a plausible mechanism (65). Nevertheless, H-NS may act in a global fashion (92) to effect the formation of unique DNA substrates upon which other genes more intimately associated with transcription (or inversion) might act (e.g., the *fimB* and *fimE* gene products). Work on the transcriptional regulation of piliation genes is admirably reviewed in this volume by David Low (chapter 28).

PILUS BIOGENESIS

Type 1 pili grow from the base (22, 80). That is, new pilin monomers (*fimA* gene products) are added to the bottom of the growing fiber rather than at the tip. There are only two additional gene products required for this polymerization process. These are the products of the *fimC* and *fimD* genes (70–72, 106) (described in Fig. 2). These two gene products, while required for pilus expression, are not part of the pilus (106). There are, in fact, more genes encoding nonessential minor pilus components (products of the *fimF, fimG,* and *fimH* genes; Fig. 2) than are required for the actual assembly of the pilus fiber. I will hasten to add that while the minor pilus components are not required for the assembly of the pilus fiber, their absence does affect the number of pili per cell and pilus structure. Also, in the absence of one of the minor components, the *fimH* product (the adhesin), the ability of type 1 pili to carry out mannose-sensitive attachment is abolished (47, 85, 86, 116); these topics will be discussed later. However, the point to be taken here is that there are relatively few gene products (two) involved in producing the pilus fiber. This gives rise to the hope that the process of pilus biogenesis can be completely understood.

Regarding the function and properties of the *fimC* and *fimD* gene products, much of the most recent work suggests that the *fimC* gene product is a periplasmic chaperone-like protein (64) that binds to pilin and the minor components and escorts them to a polymerization center. A chaperone molecule is believed to be necessary in order to keep the pilin molecules from aggregating prematurely and to stabilize the minor pilus components during their time in the periplasm (63, 78). Evidence to support the role of the *fimC* product as a chaperone comes from its structural similarity to the chaperone protein from the morphologically similar pyelonephritis-associated pili (*pap* pili). For the *pap* pili, the structure and function of the chaperone, the *papD* gene product, has been carefully described (for a review, see reference 52). Further evidence of the chaperone-like nature of the *fimC* product comes from the ability of the *papD* gene product to substitute for the *fimC* product in the production of type 1 pili (64). In spite of the functional similarities of the two proteins, certain differences exist since the *fimC* product will not substitute for the *papD* product in assembling *pap* pili (64). Also, type 1 subunits show no obvious tendency to aggregate or degrade in vivo (106) in the absence of the *fimC* product. However, it is clear that at least one of the minor type 1 pilus components, the *fimH* gene product (the adhesin), is stabilized in vivo in the presence of the *fimC* gene product (53, 126). The documented role of the *fimC* product in directly binding other pilus components in vitro (64) is good circumstantial evidence of its function as a chaperone. However, its role in the polymerization process in vivo may involve alternative or additional activities that are not yet appreciated.

The other element required for pilus production, the *fimD* gene product, is an outer membrane protein that may serve as a channel for the emergence of the growing pilus fiber (72). This product also may be, by analogy with a protein of similar structure in *pap* pili (the *papC* gene product), a molecular "usher" (23). That is, in addition to serving as a conduit, the *fimD* product (usher) may serve to coordinate the initiation of a pilus. In the case of *pap* pili, this involves the initial formation of a tip structure composed of minor pilus components prior to the addition of the pilin monomers (77). It is thought that this is accomplished in *pap* pili by the usher having different affinities for the major and minor pilus components, producing a situation in which a pilus rod is unlikely to start without a tip structure first being bound to the usher (23, 63).

ROLES OF MINOR PILUS COMPONENTS

Turning now to the nonessential minor components of type 1 pili (the products of the *fimF*, *fimG*, and *fimH* genes) and their effects on pilus biogenesis, much work has focused upon the locations of these products in the pilus. In the morphologically similar *pap* pili, evidence is strong that all of the binding activity is at the pilus tip, where several minor pilus components act to present the adhesive component (the product of the *papG* gene product) on a thin flexible filament (77). For type 1 pili, such a high degree of structural organization at the tip has not emerged (for a review, see reference 55), and the most recently published biochemical evidence would tend to support the idea that the adhesin (the *fimH* gene product) is incorporated into the pilus shaft as well as being present at the tip (4, 45, 46, 55, 75, 76, 113). However, the adhesin may be active (capable of binding receptor) only when it is "exposed" at the tip. The evidence for this picture comes from an interesting experiment by Ponniah and colleagues (113), who showed that the more type 1 pili are broken, the more adhesive (per unit of protein) they become. In contrast, in another interesting experiment, other investigators found that mannose-coupled bovine serum albumin bound preferentially to the sides of type 1 pili (75). Still others have used antibody blocking to argue that the adhesive moiety resides exclusively at the tip (45). Also in favor of an exclusive tip location is the observation that pili behave, at low concentrations, as univalent structures (38), assuming multivalency only in the presence of $MgSO_4$ or at high concentrations when their hydrophobic interactions likely result in multimeric structures (17). On the other hand, another observation indicates that type 1 pili bind along their lengths to erythrocytes (125). The difficulty of the experiments needed to resolve the issue of single or multiple adhesin locations in the pilus has precluded a definitive answer.

One reason to favor the presence of the adhesive component in a specialized tip structure has come from preliminary evidence presented by Scott Hultgren and colleagues (64), who have shown, using a freeze-etching technique, a fibrillar tip structure on type 1 pili similar to that seen with *pap* pili. By analogy, it is likely that the minor type 1 pilus components (the *fimF*, *fimG*, and *fimH* gene products) make up the tip structure. Apart from the analogy, accumulated genetic evidence also tends to support a multicomponent tip structure for type 1 pili.

Genetic investigations into pilus structure have been devoted to an examination of the mutant phenotypes produced when cells have deletions of, or have mutant alleles of, one or more of the genes encoding the minor pilus components (71, 116). For some time, it was thought that whereas the *fimH* product actually interacted with the receptor on eukaryotic cells, at least one of the other minor pilus components was also required (71, 76) for that interaction. However, from the phenotypes of the *fimF* and *fimG* mutants, Walker Russell and I (116) were able to show that neither the *fimF* nor the *fimG* product was required for receptor binding (i.e., the loss of either one or both genes did not eliminate the ability of the *fimH* product to be incorporated into, or function in, the pilus fiber). Rather, each gene appeared to produce a product that affected normal pilus number per cell and normal pilus length (Fig. 3A): *fimF* mutants had fewer pili per cell but the pili had normal lengths (Fig. 3B), and *fimG* mutants had longer pili but normal numbers (Fig. 3C). Double mutants, which lacked both genes, had fewer and longer pili (71, 116) (Fig. 3D).

Interestingly, results from my laboratory had earlier indicated that cells with certain mutant alleles of *fimH* showed phenotypes strikingly similar to those seen for the *fimG* and *fimF-fimG* double mutants (47, 86, 116). In particular, *fimH* null mutants closely resemble *fimG* null mutants in terms of having longer pili. In fact,

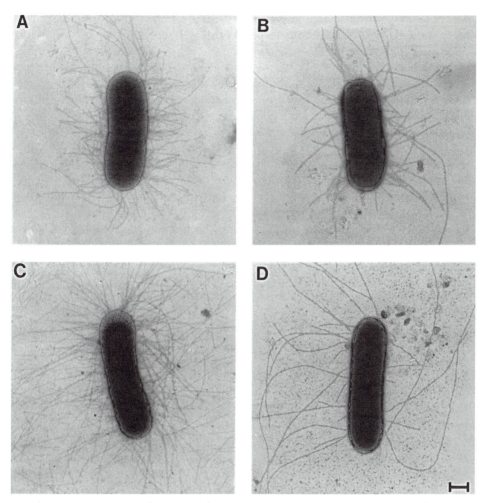

FIGURE 3. Electron micrographs of negatively stained *E. coli* cells that have lesions in genes producing two of the minor pilus components (after Russell and Orndorff [116]). (A) The parental stain showing pili of normal numbers per cell and length. (B) A *fimF* mutant that produces approximately one-fourth the number of pili produced by the parent. (C) A *fimG* mutant that produces pili that are two to three times longer than those of the parent. (D) A *fimF*-*fimG* double mutant that produces fewer and longer pili than the parent. Bar, 0.25 μm.

in our original description of *fimH* (then called *pilE*) (86), my colleagues and I mistakenly concluded that lesions in the 5′ portion of *fimH* were in an entirely different gene (termed *pilF*). We only later found that this morphological phenotype could be produced by *fimG* mutations as well as *fimH* mutations (47).

A MODEL FOR TYPE 1 PILUS ASSEMBLY

The above phenotypes suggest a way that the minor components may be involved in type 1 pilus biosynthesis. For example, one might suppose that the product of the *fimF* gene aids in pilus initiation: it serves as part of a nucleation center upon which other components coalesce to start a new pilus. Thus, mutants lacking this molecule make fewer pili because of a defect in this nucleation complex (71). In the case of *fimG,* the product of this gene may act as an inhibitor of pilus elongation (76, 86, 116). Thus, mutants lacking this product would have very long pili. In a mutant lacking both *fimF* and *fimG* products, the few pili that are initiated cannot be stopped, resulting in the characteristic phenotype of the double mutant (Fig. 3D).

The observation that certain *fimH* mutants mimic *fimG* or *fimF-fimG* double mutants may mean that all three minor components interact. For example, both *fimH* and *fimG* products may be present as a unit so that, in the absence of the *fimH* gene product, the *fimG* product is inactive, thus explaining how the *fimG* mutant phenotype (long pili) could be produced by *fimH* null mutants (86). If such a complex interacted with the *fimF* product, then one might easily imagine that certain *fimH* lesions would affect the normal interaction of all three components. This would explain why certain *fimH* mutants resemble *fimF-fimG* double mutants in having longer and fewer pili (47, 116).

One way to think of the initiation and elongation of type 1 pili that is consistent with the genetic evidence is to suppose that both the *fimH* product and the *fimG* product are normally present at the start and stop of each new pilus. A growing pilus would thus reach an average length before being stopped (or displaced) by a new one attempting to start on the same "base" (nucleation center). Average length would be determined by the ratio of the concentration of *fimG-fimH* product complexes to pilin molecules. That is, the higher the ratio, the more likely that a new complex will enter an occupied base per unit of time, and thus the shorter the pili would be before stoppage or displacement.

This model is inconsistent with the observation that the *fimH* product is intercalated into the pilus shaft (4, 75, 76). However, this inconsistency may be resolved by supposing that entry of the hypothetical *fimH-fimG* complex only weakens the pilus fiber but does not stop its growth (58, 71, 116). Thus, the pili would tend to break where these additions occurred, again creating an average length based upon the pilin-to-*fimG-fimH* product ratio. If one assumes that breakage would reveal an active *fimH* adhesin molecule, as some experiments suggest (113), then even broken pili may still be capable of binding. Salvaging the binding activities of broken pili could be very important for piliated microorganisms. That is, the energy expended to produce the fibers would seem to be prohibitively high if simple breakage would negate their usefulness—assuming, of course, that those of us who study pili are correct in categorizing binding as the "useful" function that pili serve.

Hultgren and colleagues have presented a very detailed biochemical model of the interactions of the minor *pap* pilus components that form the fibrillar tip structure (for a review, see reference 52). If one assumes a tip-like structure for type 1 pili, one might picture the *fimH* product as a fibrillar tip composed of *fimG* product molecules and with the *fimF* product being the "initiator" (60). Indeed, the phenotype of *pap* initiator mutants (*papF* mutants) is very similar to that of *fimF* mutants. However, *pap* pili appear to have more products at their tips (52), and inferring too much by analogy is probably ill advised.

Sorting out how pili are put together demands both biochemical and genetic approaches. Currently, members of my laboratory (127) and others (88) are employing strains in which we can rapidly and synchronously induce piliation de novo. Using pulse-chase and other biochemical techniques, we hope to establish an order of events in the construction of a pilus from transcription through assembly.

RECEPTOR BINDING

The adhesive component of type 1 pili is the product of the *fimH* gene. This product is capable of binding receptor without producing a pilus (3, 53, 73). However, the pilus structure likely plays a role in the sugar specificity (33, 34, 55, 113) and affinity (73) of the *fimH* product, probably by affecting its conformation. Evidence supporting this conclusion comes from the recent observation by Madison et al. (83) that the subtle differences in binding affinities exhibited by *fimH* products of closely related enterobacterial species are determined by the pilus fiber, composed of *fimA* gene products.

Investigations into the molecular nature of the adhesins of a number of different pili have revealed that specific parts of the adhesive molecules are responsible for different functions. For example, in *pap* and type 1 pili, the adhesin has distinct regions that are required for pilus incorporation, receptor binding, and, in the case of type 1 pili, other protein-protein interactions (47, 54, 120, 121). Also, the *pap* pili show a distinctive pattern of carbohydrate-binding preferences attributable, in part, to rather fine differences in the minor adhesive component (33, 51, 81, 93, 95, 121). For pili other than type 1 and *pap,* the receptor specificities of closely related adhesins (32, 94) have been mapped to discrete regions by using techniques of "domain swapping" and site-directed mutagenesis (8, 61, 62, 99, 124).

Type 1 pili bind the simple sugar mannose and a number of its structural analogs (102). This binding can be studied at a very fine level, and interactions of this type are crucial to the ability of many members of the family *Enterobacteriaceae* to colonize plant and animal tissues. However, little is known about the intermolecular factors that influence the ability of the adhesin to recognize and bind to receptor. It is known that the principal receptor on guinea pig erythrocytes is a 65-kDa glycoprotein (38) and that on leukocytes the receptors are the integrins CD11 and CD18 (37). In both cell types, the receptors contain mannose. However, it is not known if the receptors identified are the only ones on these cells. That is, other glycoproteins (or glycolipids) could be involved. In plant cells, yet other receptors are thought to be involved. For example, mannans (pure carbohydrate) are thought to be the molecules to which type 1 pili bind on yeasts (102).

It is easy to overstate the importance of the molecule or molecules that type 1 pili contact on eukaryotic cells since virtually all mannose-containing ligands could play a least a partial role in the events that follow binding. However, it does seem clear that binding triggers a strong reaction in B cells (112) and phagocytic cells (126). In the case of phagocytic cells, why this reaction does not result in greater killing of type 1-piliated microorganisms is unclear and intriguing (39, 66).

Regardless of the specific receptor for type 1 pili, all type 1-mediated interactions are inhibited by mannose, leading to the idea that this molecule (or one closely related to it) is the actual receptor. Interestingly, the affinity of mannose for type 1 pili can be enhanced by the addition of nonpolar groups to mannose. It is thought that such groups bind additional sites on the *fimH* molecule, increasing the affinity of the interaction (33, 34).

In addition to carbohydrate binding, type 1 pili can also bind to other proteins. In the two instances studied, these protein-protein interactions appear to involve the *fimH* gene product (47, 121). In the first case, genetic evidence indicates that *fimH* products on adjacent cells interact with each other to generate the surface film (pellicle) characteristic of type 1-piliated cells when they are grown in static broth (103). This interaction was found to be inhibited by mannose, suggesting that the site or sites of protein-protein binding overlap a region (or regions) on the *fimH* product involved in mannose binding. This notion was further supported by the finding that *fimH* mutants, isolated as individuals that could form a pellicle in the presence of the normally inhibitory mannose, were defective in erythrocyte binding (47). The interpretation of this result was that if these *fimH* mutants were producing an adhesin that was "blind" to the normally inhibitory effect of mannose on pellicle formation, they would also likely be "blind" to mannose on erythrocytes and thus unable to bind. The second instance in which the *fimH* product may participate in protein-protein interactions comes from the observation that type 1 pili from certain *E. coli* strains bind to immobilized fibronectin (121). Experiments showed that the binding was to a region of fibronectin that did not contain carbohydrate, but, as in the first instance, the binding was inhibited by mannose, suggesting that the pilus-fibronectin interaction likely involves the *fimH* product.

The significance of the protein-protein interactions carried out by the *fimH* product is unclear. However, in the case of pellicle formation, it is known that this property confers a selective advantage on cells growing in static culture in the laboratory (103). It would indeed be ironic if carbohydrate binding was of secondary (or incidental) importance to protein-protein interactions from the point of view of the bacterium.

Members of my laboratory and I have recently employed one additional genetic tactic to examine the finer points of receptor binding. This tactic has been to search for *fimH* mutants that bind to erythrocytes in the presence of mannose analogs that normally inhibit binding. The analog we have employed most successfully is fructose, a ketohexose that resembles mannose only around the two central anomeric carbons. Nevertheless, this compound inhibits binding. We have isolated a number of *fimH* mutants that can bind erythrocytes in the presence of fructose and have found that all of these mutants also bind erythrocytes in the presence of mannose and other normally inhibitory analogs. These data indicate that the receptor specificity of the *fimH* product has been changed rather dramatically.

Preliminary examination of the location of the *fimH* lesions that eliminate receptor binding (47) or alter receptor specificity has revealed no common theme—at least in terms of the primary amino acid sequence. However, several critical pieces of information needed to analyze the meaning of the location of such lesions are missing. For example, the crystalline structure of the *fimH* product or the receptor on guinea pig erythrocytes is not known. Furthermore, it is not known if the *fimH* product is active as a multimer or if its position (or amount) in the pilus alters the overall affinity or specificity for the receptor. Consequently, some of the *fimH* lesions may affect attributes other than tertiary structure. Improved means of detecting the *fimH* product in the pilus and data on the specific intermolecular interactions of the *fimH* product during the synthesis of a pilus may aid in understanding the process of receptor binding and receptor specificity.

CONCLUSIONS

Type 1 pili have provided paradigms for a number of central biological processes including host-pathogen interactions, gene regulation, assembly of supramolecular structures, and receptor-ligand interactions. In each case, I think that a great deal has been learned about how *E. coli* manages these complex processes. I hope that the review in this chapter has covered at least some of the interesting features of type 1 piliation and has highlighted areas that still present interesting challenges to understanding of these structures and their function.

ACKNOWLEDGMENTS. Work in my laboratory is supported by grants AI2223 and DK34987 from the National Institutes of Health.

REFERENCES

1. **Abraham, J. M., C. S. Freitag, J. R. Clements, and B. I. Eisenstein.** 1985. An invertible element of DNA controls phase variation of type 1 fimbriae of *Escherichia coli*. *Proc. Natl. Acad. Sci. USA* **82:**5724–5727.
2. **Abraham, S. N., J. P. Babu, C. S. Giampapa, D. L. Hasty, W. A. Simpson, and E. H. Beachey.** 1984. Protection against *Escherichia coli*-induced urinary tract infections with hybridoma antibodies directed against type 1 fimbriae or complementary D-mannose receptors. *Infect. Immun.* **48:** 625–628.
3. **Abraham, S. N., J. D. Goguen, and E. H. Beachey.** 1988. Hyperadhesive deletion mutant of type 1 fimbriated *Escherichia coli* associated with formation of *fimH* organelles (fimbriosomes). *Infect. Immun.* **169:**1023–1029.
4. **Abraham, S. N., J. D. Goguen, S. Sun, P. Klemm, and E. H. Beachey.** 1987. Identification of two ancillary subunits of *Escherichia coli* type 1 fimbriae by using antibodies against synthetic oligopeptides of *fim* gene products. *J. Bacteriol.* **169:**5530–5535.
5. **Alkan, M. L., I. Wong, and J. F. Silverblatt.** 1986. Change in degree of type 1 piliation of *Escherichia coli* during experimental peritonitis in the mouse. *Infect. Immun.* **54:**549–554.
6. **Aronson, M., O. Medalea, L. Schori, D. Mirelman, and I. Ofek.** 1979. Prevention of colonization of the urinary tract of mice with *Escherichia coli* by blocking of bacterial adherence with methyl-alpha-D-mannopyranoside. *J. Infect. Dis.* **139:**329–332.
7. **Bachmann, B. J.** 1990. Linkage map of *Escherichia coli* K-12, edition 8. *Microbiol. Rev.* **54:** 130–197.
8. **Bakker, D., P. T. J. Willemsen, L. H. Simons, P. G. Van Zijderveld, and F. K. deGraaf.** 1992. Characterization of the antigenic and adhesive properties of FaeG, the major subunit of K88 fimbriae. *Mol. Microbiol.* **6:**247–255.

9. **Beachey, E. H.** 1980. Bacterial adherence: adhesin-receptor interactions mediating the attachment of bacteria to mucosal surfaces. *J. Infect. Dis.* **143**:325–345.

10. **Bloch, C. A., and P. E. Orndorff.** 1990. Impaired colonization by and full invasiveness of *Escherichia coli* K1 bearing a site-directed mutation in the type 1 pilin gene. *Infect. Immun.* **58**:274–278.

11. **Bloch, C. A., B. A. D. Stocker, and P. E. Orndorff.** 1992. A key role for type 1 pili in enterobacterial communicability. *Mol. Microbiol.* **6**:697–701.

12. **Bloomfield, I. C., P. J. Callie, K. J. Eberhardt, M. S. McClain, and B. I. Eisenstien.** 1993. Lrp stimulates phase variation of type 1 fimbriation in *Escherichia coli* K-12. *J. Bacteriol.* **175**:27–36.

13. **Bloomfield, I. C., M. S. McClain, J. A. Princ, P. J. Calie, and B. I. Eisenstein.** 1991. Type 1 fimbriation of *fimE* mutants of *Escherichia coli* K-12. *J. Bacteriol.* **173**:5298–5307.

14. **Blumenstock, E., and K. Jann.** 1982. Adhesion of piliated *Escherichia coli* strains to phagocytes: difference between bacteria with mannose-sensitive pili and those with mannose-resistant pili. *Infect. Immun.* **35**:264–269.

15. **Blyn, L. B., B. A. Braaten, and D. A. Low.** 1990. Regulation of pap pilin phase variation by a mechanism involving differential dam methylation states. *EMBO J.* **9**:4045–4054.

16. **Braaten, B. A., J. V. Platko, M. W. vanderWonde, B. H. Simons, F. K. deGraaf, J. M. Calvo, and D. A. Low.** 1992. Leucine responsive regulatory protein (Lrp) controls expression of both the *pap* and *fan* pili operons in *Escherichia coli. Proc. Natl. Acad. Sci. USA* **89**:4250–4254.

17. **Brinton, C. C., Jr.** 1959. Non-flagellar appendages of bacteria. *Nature* (London) **183**:782–786.

18. **Brinton, C. C., Jr.** 1965. The structure, function, synthesis and genetic control of bacterial pili and a molecular model of DNA and RNA transport in gram negative bacteria. *Trans. N. Y. Acad. Sci.* **27**:1003–1054.

19. **Collier, W. H., and J. G. DeMiranda.** 1955. Inhibition of erythrocyte agglutination by fimbriate bacteria by alpha-D-mannose. *Antonie van Leeuwenhoek J. Microbiol. Serol.* **21**:133–140.

20. **De Graaf, F. K.** 1990. Genetics of adhesive fimbriae of intestinal *Escherichia coli. Curr. Top. Microbiol. Immunol.* **151**:29–54.

21. **Deneke, C. R., G. M. Thorne, and S. L. Gorbach.** 1979. Attachment pili from enterotoxigenic *Escherichia coli* pathogenic for humans. *Infect. Immun.* **26**:362–368.

22. **Dodd, D. C., and B. I. Eisenstein.** 1984. Kinetic analysis of the synthesis and assembly of type 1 fimbriae of *Escherichia coli. J. Bacteriol.* **160**:227–232.

23. **Dodson, K. W., F. Jacob-Dubuisson, R. T. Striker, and S. J. Hultgren.** 1993. Outer-membrane PapC molecular usher discriminately recognizes periplasmic chaperon-pilin subunit complexes. *Proc. Natl. Acad. Sci. USA* **90**:3670–3674.

24. **Dorman, C. J., N. N. Bhrian, and C. F. Higgins.** 1990. DNA supercoiling and environmental regulation of virulence gene expression in *Shigella flexneri. Nature*(London) **344**:789–792.

25. **Dorman, C. J., and C. F. Higgins.** 1987. Fimbrial phase variation in *Escherichia coli:* dependence on integration host factor and homologies with other site-specific recombinases. *J. Bacteriol.* **169**:3840–3843.

26. **Duguid, J. P., S. Clegg, and M. I. Wilson.** 1979. The fimbrial and nonfimbrial haemagglutinins of *Escherichia coli. J. Med. Microbiol.* **12**:213–227.

27. **Duguid, J. P., and R. R. Gillies.** 1957. Fimbriae and adhesive properties in dysentery bacilli. *J. Pathol. Bacteriol.* **74**:397–411.

28. **Duguid, J. P., and D. C. Old.** 1980. Adhesive properties of Enterobacteriaceae, p. 185–217. *In* E. H. Beachey (ed.), *Bacterial Adherence Receptors and Recognition*, vol. 6. Chapman & Hall, Ltd., London.

29. **Duguid, J. P., W. Smith, G. Dempster, and P. N. Edmunds.** 1955. Non-flagellar filamentous appendages ('fimbriae') and haemagglutinating activity in *Bacterium coli. J. Pathol. Bacteriol.* **70**:335–348.

30. **Eisenstein, B. I.** 1981. Phase variation of type 1 fimbriae in *Escherichia coli* is under transcriptional control. *Science* **214**:337–339.

31. **Eisenstein, B. I., D. S. Sweet, V. Vaughn, and D. I. Freidman.** 1987. Integration host factor is required for the DNA inversion that controls phase variation in *Escherichia coli. Proc. Natl. Acad. Sci. USA* **84**:6506–6510.

32. **Erickson, A. K., J. A. Willgohs, S. Y. McFarland, D. Benfield, and D. H. Francis.** 1992. Identifica-

tion of two porcine brush border glycoproteins that bind the K88ac adhesin of *Escherichia coli* and correlation of these glycoproteins with the adhesive phenotype. *Infect. Immun.* **60**:983–988.

33. **Firon, W., I. Ofek, and N. Sharon.** 1983. Carbohydrate specificity of the surface lectin of *Escherichia coli, Klebsiella pneumoniae,* and *Salmonella typhimurium. Carbohydr. Res.* **120**:235–249.

34. **Firon, W., I. Ofek, and N. Sharon.** 1984. Carbohydrate-binding sites of the mannose-specific fimbrial lectins of enterobacteria. *Infect. Immun.* **43**:1088–1090.

35. **Freitag, C. S., and B. I. Eisenstein.** 1983. Genetic mapping and transcriptional orientation of the *fimD* gene. *J. Bacteriol.* **156**:1052–1058.

36. **Gally, D. L., J. A. Bogan, B. I. Eisenstein, and I. C. Bloomfield.** 1993. Environmental regulation of the *fim* switch controlling type 1 fimbrial phase variation in *Escherichia coli* K-12: effects of temperature and media. *J. Bacteriol.* **175**:6186–6193.

37. **Gbarah, A., C. G. Gahmberg, I. Ofek, U. Jacobi, and N. Sharon.** 1991. Identification of the leukocyte adhesion molecules CD11 and CD18 as receptors for type 1-fimbriated (mannose-specific) *Escherichia coli. Infect. Immun.* **59**:4524–4530.

38. **Giampapa, C. S., S. N. Abraham, R. M. Chiang, and E. H. Beachey.** 1988. Isolation and characterization of a receptor for type 1 fimbriae of *Escherichia coli* from guinea pig erythrocytes. *J. Biol. Chem.* **263**:5362–5367.

39. **Goetz, M. B., S. M. Kuriyama, and J. J. Silverblatt.** 1987. Phagolysosome formation by polymorphonuclear neutrophilic leukocytes after ingestion of *Escherichia coli* that express type 1 pili. *J. Infect. Dis.* **156**:229–233.

40. **Goransson, M., B. Sonden, P. Nilsson, B. Dagberg, K. Forsman, K. Emanuelsson, and B. E. Uhlin.** 1990. Transcriptional silencing and thermoregulation of gene expression in *Escherichia coli. Nature* (London) **344**:682–685.

41. **Guerina, N. G., T. W. Kessler, V. J. Guerina, M. R. Neutra, H. W. Clegg, S. Langermann, F. A. Scannapieco, and D. A. Goldman.** 1983. The role of pili and capsule in the pathogenesis of neonatal infection with *Escherichia coli* K1. *J. Infect. Dis.* **148**:395–405.

42. **Guyot, G.** 1908. Über die bakterielle hämagglutination (bakterio-haemoagglutination). *Zentralbl. Bakteriol. Parasitenkd. Infektionskr. Hyg. Abt. 1 Orig.* **47**:640–653.

43. **Hacker, J.** 1990. Genetic determinants coding for fimbriae and adhesins of extra-intestinal *Escherichia coli. Curr. Top. Microbiol. Immunol.* **151**:1–28.

44. **Hagberg, L., U. Jodal, T. Korhonen, G. Lidin-Ganson, U. Lindberg, and C. Svanborg-Edén.** 1981. Adhesion, hemagglutination, and virulence of *Escherichia coli* causing urinary tract infections. *Infect. Immun.* **31**:564–570.

45. **Hanson, M. S., and C. C. Brinton, Jr.** 1988. Identification and characterization of *E. coli* type-1 pilus tip adhesion protein. *Nature* (London) **332**:265–268.

46. **Hanson, M. S., J. Hempel, and C. C. Brinton, Jr.** 1988. Purification of the *Escherichia coli* type 1 pilin and minor pilus proteins and partial characterization of the adhesin protein. *J. Bacteriol.* **170**:3350–3358.

47. **Harris, S. L., D. A. Elliott, M. C. Blake, L. M. Must, M. Messenger, and P. E. Orndorff.** 1990. Isolation and characterization of mutants with lesions affecting pellicle formation and erythrocyte agglutination by type 1 piliated *Escherichia coli. J. Bacteriol.* **172**:6411–6418.

48. **Higgins, C. F., C. J. Dorman, D. A. Stirling, L. Waddell, I. R. Booth, G. May, and E. Bremer.** 1988. A physiological role for DNA supercoiling in the osmotic regulation of gene expression in *S. typhimurium* and *E. coli. Cell* **52**:569–584.

49. **Houwink, A. L., and W. VanItterson.** 1950. Electron microscopical observations on bacterial cytology. II. A study on flagellation. *Biochim. Biophys. Acta* **5**:110–144.

50. **Hull, R. A., R. E. Gill, P. Hsu, B. H. Minshew, and S. Falkow.** 1981. Construction and expression of recombinant plasmids encoding type 1 or D-mannose-resistant pili from a urinary tract infection *Escherichia coli* isolate. *Infect. Immun.* **33**:933–938.

51. **Hull, S., S. Clegg, C. Svanborg Eden, and R. Hull.** 1985. Multiple forms of genes in pyelonephritogenic *Escherichia coli* encoding adhesins binding globoseries glycolipid receptors. *Infect. Immun.* **47**:80–83.

52. **Hultgren, S. J., S. N. Abraham, M. Caparon, P. Falk, J. W. St. Geme III, and S. Normark.** 1993. Pilus and nonpilus bacterial adhesins: assembly and function in cell recognition. *Cell* **73**:887–901.

53. **Hultgren, S. J., J. L. Duncan, A. J. Schaeffer, and S. K. Amundsen.** 1990. Mannose-sensitive haemagglutination in the absence of piliation in *Escherichia coli*. *Mol. Microbiol.* **4:**1311–1318.

54. **Hultgren, S. J., F. Lindberg, G. Magnusson, J. Kihlberg, J. M. Tennent, and S. Normark.** 1989. The PapG adhesin of uropathogenic *Escherichia coli* contains separate regions for receptor binding and for the incorporation into the pilus. *Proc. Natl. Acad. Sci. USA* **86:**4357–4361.

55. **Hultgren, S. J., S. Normark, and S. N. Abraham.** 1991. Chaperone-assisted assembly and molecular architecture of adhesive pili. *Annu. Rev. Microbiol.* **45:**383–415.

56. **Hultgren, S. J., T. N. Porter, A. J. Schaeffer, and J. L. Duncan.** 1985. Role of type 1 pili and effects of phase variation on lower urinary tract infections produced by *Escherichia coli*. *Infect. Immun.* **50:**370–377.

57. **Hultgren, S. J., W. R. Schwan, A. J. Schaeffer, and J. L. Duncan.** 1986. Regulation of production of type 1 pili among urinary tract isolates of *Escherichia coli*. *Infect. Immun.* **54:**613–620.

58. **Hulton, C. S. J., A. Seirafi, J. C. D. Hinton, J. M. Sidebotham, L. Waddell, G. D. Pavitt, T. Owen-Hughs, A. Spassky, H. Buc, and C. F. Higgins.** 1990. Histone-like protein H1 (H-NS), DNA supercoiling, and gene expression in bacteria. *Cell* **63:**631–642.

59. **Isaacson, R. E., P. C. Fusco, C. C. Brinton, Jr., and H. W. Moon.** 1978. In vitro adhesion of *Escherichia coli* to porcine small intestinal cells: pili and adhesive factors. *Infect. Immun.* **21:**392–397.

60. **Jacob-Dubuisson, F., J. Heuser, K. Dodson, S. Normark, and S. Hultgren.** 1993. Initiation of assembly and association of the structural elements of a bacterial pilus depend on two specialized tip proteins. *EMBO J.* **12:**837–847.

61. **Jacobs, A. A. C., B. Roosendaal, J. F. L. vanBreeman, and F. K. deGraaf.** 1987. Role of phenylalanine 150 of the receptor binding domain of the K88 fibrillar subunit. *J. Bacteriol.* **169:**4907–4911.

62. **Jacobs, A. A. C., B. H. Simons, and F. K. deGraaf.** 1987. The role of lysine-132 and arginine-136 in the receptor-binding domain of the K99 fimbrillar subunit. *EMBO J.* **6:**1805–1808.

63. **Jones, C. H., F. Jacob-Dubuisson, K. Dodson, M. Kuehn, L. Slonim, R. Striker, and S. J. Hultgren.** 1992. Adhesin presentation in bacteria requires molecular chaperons and ushers. *Infect. Immun.* **60:**4445–4451.

64. **Jones, C. H., J. Pinker, A. Nicholes, L. Slonim, S. N. Abraham, and S. J. Hultgren.** 1993. FimC is a periplasmic PapD-like chaperone which directs assembly of type 1 pili in bacteria. *Proc. Natl. Acad. Sci. USA* **90:**8397–8401.

65. **Kawula, T. H., and P. E. Orndorff.** 1991. Rapid site-specific DNA inversion in *Escherichia coli* mutants lacking the histonelike protein H-NS. *J. Bacteriol.* **173:**4116–4123.

66. **Keith, B. R., S. L. Harris, P. W. Russell, and P. E. Orndorff.** 1990. Effect of type 1 piliation on in vitro killing of *Escherichia coli* by mouse peritoneal macrophages. *Infect. Immun.* **58:**3448–3454.

67. **Keith, B. R., L. Maurer, P. A. Spears, and P. E. Orndorff.** 1986. Receptor-binding function of type 1 pili effects bladder colonization by a clinical isolate of *Escherichia coli*. *Infect. Immun.* **53:**693–696.

68. **Klemm, P.** 1985. Fimbrial adhesins of *Escherichia coli*. *Rev. Infect. Dis.* **7:**321–340.

69. **Klemm, P.** 1986. Two regulatory *fim* genes, *fimB* and *fimE* control the phase variation of type 1 fimbriae in *Escherichia coli*. *EMBO J.* **5:**1389–1398.

70. **Klemm, P.** 1992. FimC, a chaperone-like periplasmic protein of *Escherichia coli* involved in biogenesis of type 1 fimbriae. *Res. Microbiol.* **143:**831–840.

71. **Klemm, P., and G. Christiansen.** 1987. Three *fim* genes required for the regulation of length and mediation of adhesion of *Escherichia coli* type 1 fimbriae. *Mol. Gen. Genet.* **208:**439–445.

72. **Klemm, P., and G. Christiansen.** 1990. The *fimD* gene is required for cell surface localization of *Escherichia coli* type 1 fimbriae. *Mol. Gen. Genet.* **220:**334–338.

73. **Klemm, P., K. A. Krogfelt, L. Hedegaard, and G. Christiansen.** 1990. The major subunit of *Escherichia coli* type 1 fimbriae is not required for D-mannose-specific adhesion. *Mol. Microbiol.* **4:**553–559.

74. **Krogfelt, K. A.** 1991. Bacterial adhesion: genetics, biogenesis and role in pathogenesis of fimbrial adhesins of *Escherichia coli*. *Rev. Infect. Dis.* **13:**721–735.

75. **Krogfelt, K. A., H. Bergmans, and P. Klemm.** 1990. Direct evidence that the *fimH* protein is the mannose-specific adhesin of *Escherichia coli* type 1 fimbriae. *Infect. Immun.* **58:**1995–1998.

76. **Krogfelt, K. A., and P. Klemm.** 1988. Investigation of minor components of *Escherichia coli* type 1 fimbriae: protein, chemical, and immunological aspects. *Microb. Pathog.* **4:**231–238.

77. **Kuehn, M. J., J. Heuser, S. Normark, and S. J. Hultgren.** 1992. P pili in uropathogenic *E. coli* are composite fibers with distinct fibrillar adhesive tips. *Nature* (London) **356:**252–255.

78. **Kuehn, M. J., D. J. Ogg, J. Kihlberg, L. N. Slonim, K. Flemmer, T. Bergfors, and S. J. Hultgren.** 1993. Structural basis of pilus subunit recognition by the PapD chaperone. *Science* **262:**1234–1241.

79. **Lejeune, P., and A. Danchin.** 1990. Mutations in the *bglY* gene increase the frequency of spontaneous deletions in *Escherichia coli* K-12. *Proc. Natl. Acad. Sci. USA* **87:**360–363.

80. **Lowe, M. A., S. C. Holt, and B. I. Eisenstein.** 1987. Immunoelectron microscopic analysis of elongation of type 1 fimbriae in *Escherichia coli*. *J. Bacteriol.* **169:**157–163.

81. **Lund, B., B.-I., Marklund, N. Stromberg, F. Lindberg, K. A. Karlsson, and S. Normark.** 1988. Uropathogenic *Escherichia coli* can express serologically identical pili of different receptor binding specificities. *Mol. Microbiol.* **2:**255–263.

82. **Maccacaro, G. A., and W. Hayes.** 1961. The genetics of fimbriation in *Escherichia coli*. *Genet. Res.* **2:**394–405.

83. **Madison, B., I. Ofek, S. Clegg, and S. Abraham.** 1994. Type 1 fimbrial shafts of *Escherichia coli* and *Klebsiella pneumoniae* influence sugar-binding specificities of their FimH adhesins. *Infect. Immun.* **62:**843–848.

84. **Maurelli, A. T., and P. J. Sansonetti.** 1988. Identification of a chromosomal gene controlling temperature-regulated expression of *Shigella* virulence. *Proc. Natl. Acad. Sci. USA* **85:**2820–2824.

85. **Maurer, L., and P. E. Orndorff.** 1985. A new locus, *pilE*, required for the binding of type 1 piliated *Escherichia coli* to erythrocytes. *FEMS Microbiol. Lett.* **30:**59–66.

86. **Maurer, L., and P. E. Orndorff.** 1987. Identification and characterization of genes determining receptor binding and pilus length of *Escherichia coli* type 1 pili. *J. Bacteriol.* **169:**640–645.

87. **May, A. K., C. A. Bloch, R. G. Sawyer, M. D. Spengler, and T. I. Pruett.** 1993. Enhanced virulence of *Escherichia coli* bearing a site-targeted mutation in the major structural subunit of type 1 fimbriae. *Infect. Immun.* **61:**1667–1673.

88. **McClain, M. S., I. C. Bloomfield, K. J. Eberhardt, and B. I. Eisenstein.** 1993. Inversion-independent phase variation of type 1 fimbriae in *Escherichia coli*. *J. Bacteriol.* **175:**4335–4344.

89. **McClain, M. S., I. C. Bloomfield, and B. I. Eisenstein.** 1991. Roles of *fimB* and *fimE* in site-specific DNA inversion associated with phase variation of type 1 fimbriae in *Escherichia coli*. *J. Bacteriol.* **173:**5308–5314.

90. **McCormick, B. A., D. P. Franklin, D. C. Laux, and P. S. Cohen.** 1989. Type 1 pili are not necessary for colonization of the streptomycin-treated mouse large intestine by type 1 piliated *Escherichia coli* F-18 and *E. coli* K-12. *Infect. Immun.* **57:**3022–3029.

91. **McMichael, J. C., and J. T. Ou.** 1979. Structure of common pili from *Escherichia coli*. *J. Bacteriol.* **138:**969–975.

92. **Mekalanos, J. J.** 1992. Environmental signals controlling expression of virulence determinants in bacteria. *J. Bacteriol.* **174:**1–7.

93. **Morschhauser, J., H. Hoschutzky, K. Jann, and J. Hacker.** 1990. Functional analysis of the sialic binding adhesin SfaS of pathogenic *Escherichia coli* by site-specific mutagenesis. *Infect. Immun.* **58:**2133–2138.

94. **Mouricout, M., J. M. Petit, J. R. Carais, and R. Julien.** 1990. Glycoprotein glycans that inhibit adhesion of *Escherichia coli* mediated by K99 fimbriae: treatment of colibacillosis. *Infect. Immun.* **58:**98–106.

95. **Must, L. M.** 1989. Identification and characterization of *pilE*, a gene determining receptor binding of *Escherichia coli* type 1 pili. Ph.D. thesis. North Carolina State University at Raleigh.

96. **Newman, B. I., R. D'Ari, and R. T. Lin.** 1992. The leucine-Lrp regulon in *E. coli*: a global response in search of a raison d'etre. *Cell* **68:**617–619.

97. **Normark, S., M. Baga, M. Goransson, F. P. Lindberg, B. Lund, and B. E. Uhlin.** 1986. Genetics and biogenesis of *Escherichia coli* adhesins, p. 113–143. *In* D. Mirlman (ed.), *Microbial Lectins and Agglutinins: Properties and Biological Activity.* John Wiley & Sons, Inc., New York.

98. **Nou, X., B. Skinner, B. Braaten, L. Blyn, D. Hirsch, and D. Low.** 1993. Regulation of pyelonephri-

tis-associated pili phase-variation in *Escherichia coli:* binding of the PapI and the Lrp regulatory proteins is controlled by DNA methylation. *Mol. Microbiol.* **7:**545–553.

99. **Nowicki, B., A. Labigne, S. Moseley, R. Hull, S. Hull, and J. Moulds.** 1990. The Dr hemagglutinin, afimbrial adhesins AFA-I and AFA-III, and F1845 fimbriae of uropathogenic and diarrhea-associated *Escherichia coli* belong to a family of hemagglutinins with the Dr receptor recognition. *Infect. Immun.* **58:**279–281.

100. **Ofek, I., D. Mirelman, and N. Sharon.** 1977. Adherence of *Escherichia coli* to human mucosal cells mediated by mannose receptors. *Nature* (London) **265:**623–625.

101. **Ofek, I., and N. Sharon.** 1988. Lectinophagocytosis: a molecular mechanism of recognition between cell surface sugars and lectins in the phagocytosis of bacteria. *Infect. Immun.* **56:**539–547.

102. **Old, D. C.** 1972. Inhibition of the interaction between fimbrial haemagglutinins and erythrocytes by D-mannose and other carbohydrates. *J. Gen. Microbiol.* **71:**149–157.

103. **Old, D. C., and J. P. Duguid.** 1970. Selective outgrowth of fimbriate bacteria in static liquid medium. *J. Bacteriol.* **103:**447–456.

104. **Orndorff, P. E., and C. A. Bloch.** 1990. The role of type 1 pili in the pathogenesis of *Escherichia coli* infections: a short review and some new ideas. *Microb. Pathog.* **9:**75–79.

105. **Orndorff, P. E., and S. Falkow.** 1984. Identification and characterization of a gene product that regulates type 1 pili in *Escherichia coli. J. Bacteriol.* **160:**61–66.

106. **Orndorff, P. E., and S. Falkow.** 1984. Organization and expression of genes responsible for type 1 piliation in *Escherichia coli. J. Bacteriol.* **159:**736–744.

107. **Orndorff, P. E., and S. Falkow.** 1985. The nucleotide sequence of *pilA,* the gene encoding the structural component of type 1 pili in *Escherichia coli. J. Bacteriol.* **162:**454–457.

108. **Orndorff, P. E., P. A. Spears, D. Schauer, and S. Falkow.** 1985. Two modes of control of *pilA,* the gene encoding type 1 pilin in *Escherichia coli. J. Bacteriol.* **164:**321–330.

109. **Ørskov, I., F. Ørskov, A. Birch-Anderson, P. Klemm, and C. Svanborg-Éden.** 1982. Protein attachment factors: fimbriae in adhering *Escherichia coli* strains. *Semin. Infect. Dis.* **4:**97–103.

110. **Ottow, J. C. G.** 1975. Ecology, physiology and genetics of fimbriae and pili. *Annu. Rev. Microbiol.* **29:**79–108.

111. **Pon, C. L., R. Calogero, and C. Gualerzi.** 1988. Identification, cloning, nucleotide sequence and chromosomal map location of *hns,* the structural gene for *Escherichia coli* DNA-binding protein H-NS. *Mol. Gen. Genet.* **212:**199–202.

112. **Ponniah, S., S. N. Abraham, M. E. Dockter, C. D. Wall, and R. O. Endres.** 1989. Mitogenic stimulation of human B lymphocytes by the mannose-specific adhesin on *Escherichia coli* type 1 fimbriae. *J. Immunol.* **142:**992–998.

113. **Ponniah, S., R. O. Endres, D. L. Hasty, and S. N. Abraham.** 1991. Fragmentation of *Escherichia coli* type 1 fimbriae exposes cryptic D-mannose-binding sites. *J. Bacteriol.* **173:**4195–4202.

114. **Rodriquez-Ortega, M., I. Ofek, and N. Sharon.** 1987. Membrane glycoproteins of human polymorphonuclear leukocytes that act as receptors for mannose-specific *Escherichia coli. Infect. Immun.* **55:**968–973.

115. **Rosenthal, L.** 1943. Agglutinating properties of *Escherichia coli. J. Bacteriol.* **45:**545–550.

116. **Russell, P. W., and P. E. Orndorff.** 1992. Lesions in two *Escherichia coli* type 1 pilus genes alter pilus number and length without affecting receptor binding. *J. Bacteriol.* **174:**5923–5935.

117. **Salit, I. E., and E. C. Gotschlich.** 1977. Type 1 *Escherichia coli* pili: characterization of binding to monkey kidney cells. *J. Exp. Med.* **146:**1182–1194.

118. **Schwan, W. R., H. S. Seifert, and J. L. Duncan.** 1992. Growth conditions mediate differential transcription of *fim* genes involved in phase variation of type 1 pili. *J. Bacteriol.* **174:**2367–2375.

119. **Schwan, W. R., H. S. Seifert, and J. L. Duncan.** 1994. Analysis of the *fimB* promoter region involved in type 1 pilus phase variation in *Escherichia coli. Mol. Gen. Genet.* **242:**623–629.

120. **Simmons, B. L., P. Rathman, C. R. Malij, B. Oudega, and F. K. deGraaf.** 1990. The penultimate tyrosine residue of the K99 fibrillar subunit is essential for the stability of the protein and its interaction with the periplasmic carrier proteins. *FEMS Microbiol. Lett.* **67:**107–112.

121. **Sokurenko, E. V., H. S. Courtney, S. N. Abraham, P. Klemm, and D. L. Hasty.** 1992. Functional heterogeneity of type 1 fimbriae of *Escherichia coli. Infect. Immun.* **60:**4709–4716.

122. **Spears, P. A., D. Schauer, and P. E. Orndorff.** 1986. Metastable regulation of type 1 piliation in

Escherichia coli and isolation and characterization of a phenotypically stable mutant. *J. Bacteriol.* **168**:179–185.

123. **Swaney, L. M., Y. Lu, K. Ippen-Ihler, and C. C. Brinton, JR.** 1977. Genetic complementation analysis of *Escherichia coli* type 1 somatic pilus mutants. *J. Bacteriol.* **130**:506–511.

124. **Swanson, T. N., S. S. Bilge, B. Nowicki, and S. Moseley.** 1991. Molecular structure of the Dr adhesin: nucleotide sequence and mapping of receptor-binding domain by use of fusion constructs. *Infect. Immun.* **59**:261–268.

125. **Sweney, G., and J. Freer.** 1979. Location of binding sites on common type 1 fimbriae from *Escherichia coli*. *J. Gen. Microbiol.* **112**:321–328.

126. **Tewari, R., J. I. MacGregor, T. Ikeda, J. R. Little, S. J. Hultgren, and S. N. Abraham.** 1993. Neutrophil activation by nascent FimH subunits of type 1 fimbriae purified from the periplasm of *Escherichia coli*. *J. Biol. Chem.* **268**:3009–3015.

127. **Woodall, L. D., P. W. Russell, S. L. Harris, and P. E. Orndorff.** 1993. Rapid, synchronous and stable induction of type 1 piliation in *Escherichia coli* using a chromosomal *lacUV5* promoter. *J. Bacteriol.* **175**:2770–2778.

Molecular Genetics of Bacterial Pathogenesis
Edited by V. L. Miller, J. B. Kaper, D. A. Portnoy, and R. R. Isberg
© 1994 American Society for Microbiology, Washington, DC 20005

Chapter 8

Mechanisms of Pilus Antigenic Variation in *Neisseria gonorrhoeae*

Michael Koomey

The alteration of gene expression by DNA rearrangement appears to be an essential part of most biological systems. When an extremely high level of diversification for a set of gene products with a common function is advantageous, these rearrangements are achieved by the reassortment and recombination of repeated gene segments. Selective pressures have also led to the evolution of processes that maintain an uncompromised degree of plasticity despite repeated rearrangements. The generation of the variable domains in avian immunoglobulins (37, 57), antigenic variation of surface proteins in African trypanosomes (7, 59) and *Borrelia hermsii* (1, 35), and pilus antigenic variation in *Neisseria gonorrhoeae* each results from the transfer of genetic information from nonexpressed donor alleles or pseudogenes to an active expression locus, a process likened to gene conversion occurring in spore-forming fungi (16). Although significant advances have been made in the understanding of recombination mechanisms for all the organisms cited, it seems at this point in time that the gonococcal pilus system is most amenable to genetic dissection and analysis. However, the relative simplicity of the gonococcal system does not automatically translate into the presentation of a bright and shining paradigm. Several substantial reviews addressing this topic have been published (30, 39, 51). In this chapter, I wish to concentrate primarily on recent studies that have influenced understanding of variable pilus expression in gonococci. In particular, I hope to evaluate the evidence for alternative models which have been proposed to account for the asymmetric character of pilin gene recombination.

The expression of pili correlates very strongly with the capacity of the gonococcus to colonize its sole host, humans. This view is supported by the failure of a nonpiliated mutant to colonize the urethras of male volunteers (54) and retrospective interpretations of earlier studies with less well defined strains (24). In this sense, pilus expression appears to be a requisite determinant of infectivity, and as such, the organelle itself may be thought of as a colonization factor. However, the precise nature of these correlations remains unclear. By analogy with their

Michael Koomey • Department of Microbiology and Immunology, M 6791, Medical Science Building II, University of Michigan Medical School, Ann Arbor, Michigan 48109.

enterobacterial counterparts, gonococcal pili may function as adhesins mediating long-range binding interactions with epithelial cells. Conditions can be found under which piliated (P^+) organisms adhere more avidly than isogenic nonpiliated (P^-) mutants to tissue cell lines, but not all P^+ variants show this property (38). As yet, no specific cell receptor for piliated gonococci has been identified. The situation is further complicated by the possibilities that particular pilus antigenic variants might display different levels of piliation or might bind to a single receptor with various affinities or bind to altogether different receptors.

Alternatively, the correlation between pilus expression and infectivity may be less direct and may be a reflection of the importance of other pilus-associated properties such as competence for DNA transformation (47), bacterial cell-bacterial cell interactions (autoagglutination) (52), and twitching motility (19). The recent demonstrations of similarities and functional overlap between type IV pilus assembly components and proteins involved in two-step protein export raise the possibility that pilus expression may be critical for proper surface localization or membrane translocation of other molecules (36). This case is quite similar to that which exists for other bacterial pathogens which express highly related type IV pili as mucosal colonization factors and which include *Neisseria meningitidis* (20), *Pseudomonas aeruginosa* (34), *Moraxella* species (28, 58), strains of enteropathogenic *Escherichia coli* (11), *Dichelobacter nodosus* (8), and *Vibrio cholerae* (44). Gonococci and meningococci are unique among these species in their capacities to sequentially display a bevy of structurally distinct pili. These filaments are helically arrayed polymers of identical polypeptide subunits of ~18 kDa termed "pilin," and changes in the primary structure of pilin solely account for pilus antigenic variation. Structural heterogeneity was first noted as differences in pilus antigenicity between strains (6) and was subsequently documented for intrastrain variants (60, 61). While antigenic variation is generally considered to be a mechanism of host immune evasion and such a function for pilus variation fits well with the chronic nature of untreated gonococcal infection and the lack of demonstrable immunity, there is no direct evidence to support this idea. Pilus structural variation may also have important consequences for pilus functions (even though those functions are undefined) and may be important to different phases of infection and colonization of different hosts and niches within a host. The term "antigenic variation" is used here solely to denote changes in the primary structure of pilin and corresponding changes in pilus structure without implication for its biological function.

First hints at the basis for pilus variation came from early cloning studies which showed that the gonococcal genome contained a single complete pilin subunit gene (with one exceptional strain that contained two) and multiple copies of partial pilin gene copies lacking promoter- and amino-terminus-encoding sequences at loci designated *pilE* (expression) and *pilS* (silent), respectively (12, 29, 31). The sequence alignments of the *pilS* and *pilE* alleles demonstrated clearly that pilin genes consisted of conserved and variable segments corresponding to analogous domains of pilin. Further evidence that intragenic recombination might be the basis for pilus variability came next from sequencing studies of pilin mRNA from laboratory-derived variants (14). In subsequent studies, Southern blotting with allele-specific oligonucleotide probes demonstrated that changes in pilin expression resulted from the transfer of variant-encoding gene information from *pilS* loci to *pilE* (12, 50, 55). The apparent nonreciprocal nature of these rearrangements led to their characterization as gene conversion events. The observations that sequential rearrangements could involve different *pilS* alleles or different stretches of nucleotides within a single *pilS* allele indicated that substantial pilin variation can be generated with a relatively small repertoire of partial pilin genes. When combined with the fact that crossing over could occur within variable gene stretches, the degree of plasticity capable of being generated appears limitless. The conclusive evidence that these rearrangements are a result of homologous recombination came from studies demonstrating a lack of variation in recombination-deficient (*recA*) gonococcal strains (25). It is worth noting that pilus antigenic variants that occur (or that are selected) in male volunteers following infection with a defined gonococcal inoculum arise by recombination events indistinguishable from those observed for laboratory-derived variants (54). Therefore, it appears likely that the studies of in vitro variation are relevant to the processes operating in vivo.

PILUS ANTIGENIC VARIATION AND PHASE MUTATION

The lack of differentiating and selectable phenotypes that are associated with variable pilus expression and that can be used to score and to isolate variants has been a primary obstacle to studying the mechanisms of pilin gene rearrangements. The simplest solution to this problem takes advantage of the fact that alterations in pilus expression are accompanied by changes in colony morphology. Although its biophysical basis is poorly defined, it appears to reflect the influence of pili on bacterial cell-cell aggregation, with subtle differences in colony size, shape, and edge character occurring concomitantly with antigenic variation (48), while nonpiliated organisms display a total loss of aggregation and edge morphology (52). Early on, it was not widely appreciated that antigenic variation could be scored by alterations in colony morphology, so most studies focused on what appeared to be a related event, that being pilus phase (on-off) variation, wherein P^+ organisms give rise to P^- forms, and these in turn could sometimes spawn P^+ revertants. The findings that some P^- organisms had undergone gross rearrangement in their pilin gene locus (31, 40) and that subsequent P^+ revertants often expressed an altered variant pilin (14) provided the rationale for isolating antigenic variants. Fortuitously, it turned out that there is a direct relationship between antigenic and phase variation, since each is a manifestation of pilin gene recombination (25). P^+ to P^- transitions result from recombination events that create pilin gene alterations which result in a failure to express assembly-competent pilin. The two predominant classes of pilin gene lesions found are missense mutations, in which certain combinations of variable domains (in what appear to be otherwise normal pilin variants) appear to preclude assembly (2, 13, 50), and deletions whose endpoints lie in *pilE* and a *pilS* locus tandemly arrayed immediately upstream, resulting in the loss of the unique 5′ end of expression locus (49). Less frequently encountered alterations include small deletion (21) and frameshift and nonsense mutations (25) as well as exchanges resulting in the elaboration of grossly oversized L (long) pilins (10). The reverse process ($P^- \rightarrow P^+$ transitions) results from templated correction of the *pilE* mutations by recombination with *pilS* alleles (50, 55).

In this context, $P^- \rightarrow P^+$ transitions and direct variation of the $P^+ \alpha \rightarrow P^+$ β type are examples of what has been termed "programmed" or "productive" DNA rearrangements, while $P^+ \rightarrow P^-$ transitions are typical of unprogrammed or incidental rearrangements (5, 45). The latter view is supported by the stochastic nature of pilin gene mutations and the facts that P^- organisms are pilin mutants which appear during in vitro propagation and arise at rates significantly lower than those seen for $P^- \rightarrow P^+$ transitions (25). One unfortunate result of the focus on phase variations has been the misinterpretation that on-off transitions are in some way critical to pilus antigenic variation (with perhaps the nonpiliated state being an essential intermediate step in the process) or otherwise essential to the lifestyle of the organism. Nonetheless, the existence of $P^+ \rightarrow P^-$ phase variation, or perhaps more correctly, phase mutation, does raise two important points with regard to pilus antigenic variation. First, it appears to fulfill a tacit prediction of systems designed to generate a high degree of diversity, that is, that the process is not highly regulated and, as such, may generate molecules so diverse as to preclude their assembly and functionality. Second, it suggests that recombination may enhance the occurrence of nontemplated mutations and that those mutations may provide a less dramatic but still significant source for variation.

An alternate form of pilus phase variation has been proposed to occur by variable expression of PilC, a 110-kDa protein that copurifies with the gonococcal pilin subunit (22). On-off *pilC* expression results from frameshift mutations within a repetitive stretch of guanine residues early within the open reading frame, a mechanism predicted to be RecA independent. However, on $\rightarrow$ off pilus phase variation is reduced dramatically (1,000-fold) in a *recA* mutant (25). Therefore, the majority of P^- variants arising in wild-type strains appears to be due to pilin gene rearrangements (see below for further discussion of the influence of *pilC* expression on pilus variation). The role of *pilC* frameshifting in pilus phase variation has also been clouded by a recent report that PilC expression is not obligatory for pilus assembly.

GENE CONVERSION VERSUS TRANSFORMATION: INSIDE JOB OR EXTERNAL AFFAIR

There is a clear consensus that pilus antigenic variation and on-off expression are the results of recombination between the pilin expression locus and the multiple partial pilin gene alleles and that these exchanges appear to be nonreciprocal in character. Results of early studies were interpreted as indicating that these were examples of gene conversion events.

A novel explanation for the nonreciprocal character of pilin gene recombination had its genesis in observations that piliated cells are highly competent (able to take up DNA and be transformed by it) (47) and that gonococci undergo autolysis in culture (17). In this scenario, the source of new genetic information is exogenous DNA released by lysed cells in the population. The hypothesis was also fostered by a report that DNase I treatment reduced but did not completely inhibit the frequencies of pilus phase variation. The first report of this phenomenon indicated that $P^+ \rightarrow P^-$ transitions were reduced in the presence of DNase I (33). However, it was noted in the same study that colony morphology changes attributable to variable expression of Opa (opacity-associated) proteins were likewise reduced by that treatment. It is now clear that most if not all Opa on-off variation occurs by slipped-strand mispairing (32), a RecA-independent process, and so the results reported for the effect on pilus variation must be viewed with skepticism. In a later study, it was reported that DNase I reduces the frequency of reversion demonstrated by an uncharacterized pilin mutant 30-fold (42).

Seifert and colleagues (43) examined the potential role of transformation by scoring for transfer between cocultivated strains of a pilin gene construct bearing a minitransposon in *pilE*, and it was found that the marked pilin gene could be transferred to the recipient pilin expression locus in a DNase I-sensitive manner. By using a single strain with a slightly different construct featuring a promoterless *cat* gene marker inserted between the *pilE* promoter and open reading frame (of one of two pilin expression loci), it was then shown that strains carrying duplications of the marked gene could be recovered on the basis of their hyperresistance to chloramphenicol. Although DNase I had no effect on the rate at which hyperresistant colonies arose, the frequencies with which they arose was decreased almost 10,000-fold when examined in the background of a P^+ strain that was transformation defective because it harbored the *dud-1* (DNA uptake-deficient) mutation. This mutation was generated by chemical mutagenesis and has not yet been characterized genetically (3).

A second study examined the influence of DNase I on recombination events that led to the expression of an oversized, defective pilin and generation of what has been termed "L-pilin P^- phase variants" (10). These pilin mutants arise by translocation in toto to the pilin expression loci of two tandemly arrayed, in-frame partial pilin gene alleles, which accounts for the synthesis of an extended pilin. These defective partial pilin gene alleles normally reside in the *pilS6* locus, which maps approximately 20 kb downstream of the pilin expression locus. L-pilin mutants tolerate higher levels of antibiotics and therefore can be selected for and scored on the basis of their increased levels of resistance to kanamycin. The frequencies with which L-pilin phase variants arose was reduced fivefold in the presence of DNase I. When examined by Southern blotting with allele-specific oligonucleotide probes, the variants that arose under these conditions showed a pattern that was consistent with a reciprocal exchange rather than the nonreciprocal pattern seen for the L-pilin mutants that arose in the absence of DNase I. On the basis of these results, the authors proposed that two primary pathways existed for pilin variation, one being a transformation-mediated event which is sensitive to DNase I and the other being an intragenomic reciprocal recombination event which is observed only when the former pathway is blocked.

The overall impression engendered by these studies was that much of the data could be explained by transformation. It also appeared to provide raison d'etres for natural transformation and autolysis, the biological advantages of which have been the topic of much debate. There was, however, one clear inconsistency in the model that related to observations of the events responsible for $P^- \rightarrow$ P^+ reversions. Pilin gene mutants revert at a high frequency by an apparently nonreciprocal exchange that corrects the mutation in a templated fashion, although it had been known for some time that P^- variants display reduced competence with transformation frequencies down at least a 1,000-fold from those of isogenic P^+ organisms (47). As noted previously, most of the data for the early gene conversion models were in fact derived from studies of $P^- \rightarrow P^+$ reversion events. Only one study has examined the frequencies of direct pilus variation of the P^+ $\alpha \rightarrow$ P^+ β type (48). In that case, subtle changes in colony morphology were used to identify variants expressing pilins of altered relative mobility detected by immunoblotting, and the frequencies for these direct transitions were reported to range from 0.1 to 5% of progeny in single colonies.

These concerns just noted provided the impetus for two sets of studies that were based on the following premise. If transformation accounts for the observed nonreciprocity of exchanges between genophores, then mutants or variants that are transformation deficient or significantly reduced in their levels of competence should show quantitative and perhaps qualitative differences in pilin gene recombination. One test of this hypothesis (53) entailed the comparison of phase variation frequencies in a wild-type strain and an isogenic mutant that is transformation defective because it harbors the same *dud*-1 allele used in the studies of Seifert et al. (43). Despite its reduced competence for transformation, no significant differences in the frequencies of phase variation were found between the mutant strain and its parent. Moreover, the character of the recombination events seen in the mutant were nonreciprocal, as assessed by the hybridization patterns obtained by using allele-specific oligonucleotides.

In a study carried out by my group (62), strains carrying defined frameshift, missense, and nonsense mutations within the pilin expression locus were constructed. Reversion to a piliated state required correction of the lesions and provided a simple means of scoring productive recombination and antigenic variation. Examination of the mutants revealed a lack of correspondence between the frequencies with which they could be transformed (10^{-6} per recipient) and the incidence with which they gave rise to revertants ($>10^{-4}$/CFU per generation). Furthermore, the rates of reversion demonstrated by these mutants were not altered by growth in the presence of DNase I, conditions that abolished intercellular transfer of chromosomal markers during cultivation. By using a pilin mutant in which a frameshift mutation encompassed the introduction of a unique *Cla*I restriction site, the symmetry of recombination that resulted in reversion was scored by the Southern hybridization pattern. In all 65 independent events analyzed, the DNA rearrangements responsible for pilin variation were nonreciprocal in character. Although it was directly shown that reversion was RecA dependent, the degree of variation achieved by these events was not reported. This is an important concern, since it was possible that all revertants arose by recombination with one single partial pilin gene allele and that little diversity was actually generated by the process. To address this, the pilin expression loci of eight revertants arising in the presence of DNase I were cloned and sequenced, and the derived primary structures of the revertant pilins are shown in Fig. 1. The results of this limited

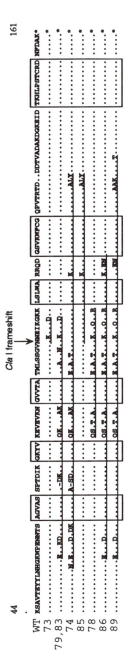

FIGURE 1. Pilin variation occurring in the absence of transformation. Pilin amino acid sequences (derived from DNA sequencing of *pilE*-containing plasmid clones) are of P⁺ revertants arising from the nonpiliated strain $N_{N99 - ClaI}$ during propagation in the presence of DNase I (26). The arrow indicates the location of the *ClaI* restriction site encompassing a frameshift mutation in *pilE* used to demonstrate the nonreciprocal character of the recombination events responsible for reversion. See the text and reference 62 for details. Amino acids are noted by single-letter code. Boxed residues indicate conserved pilin domains. Numbers on the left refer to revertant strain designations, while the numbering of the residues is based on those found in mature pilin. Revertants are ordered to emphasize the use of identical (or highly related) partial pilin gene copies as donors in productive recombination events. Revertants 73, 79, and 83 represent the use of one donor allele, 74 and 85 represent the use of another donor allele, and 78, 86, and 89 represent the use of a third donor allele. Solid underlining denotes the corresponding minimal nucleotide stretches introduced by recombination.

sampling indicate that a minimum of three partial pilin gene alleles were capable of acting as productive donors in this background. Another important feature of pilin variation seen in that study was the utilization of different-length tracts of the same partial gene copy. These findings indicated that both the extent of pilin diversification and the basic mechanisms manifest under these conditions were comparable to those found in earlier studies. These results and those found in spontaneous phase variants occurring in a transformation-defective mutant demonstrate that pilin gene exchanges can occur at a high frequency by an apparently nonreciprocal, transformation-independent recombination event.

In an effort to reconcile the contradictory natures of these sets of data and their interpretations, it seems prudent to define a specific point of contention. The term "gene conversion" is used here in the context that a partial pilin gene appears to have donated part of its sequence to the pilin expression locus without being altered itself. As in any system in which all products of a single event are not held together or otherwise recoverable, this outcome may represent one of three different forms of recombination. These are (i) nonreciprocal intrachromosomal exchanges, (ii) nonreciprocal exchanges between sister chromosomes, and (iii) reciprocal exchanges between sister chromosomes which then segregate, with one progeny expressing an altered pilus and the other carrying an altered partial gene copy with no change in pilus phenotype. The simplest distinction, then, is that gene conversion involves intracellular substrates, while in a transformation mechanism, the donor sequences are supplied by exogenous DNA derived from lysed cells.

I believe, and will attempt to make the case, that differences in the test systems themselves may account for many of the discrepancies. On the one hand, the rearrangements detailed by Seifert, Gibbs, and colleagues (10, 43) undoubtedly arose by transformation. On the other, there is strong evidence for the existence of an intracellular mechanism that could account for all pilus antigenic variation. Ideally, any test system should mirror the biologically relevant process. In my view, this should be productive pilin gene recombination, defined as those exchanges in which any one of multiple partial pilin genes undergoes recombination with the complete pilin gene, resulting in the expression of an altered pilin polypeptide that is capable of being assembled. In addition, the test system should involve native substrates and should not entail manipulations that might alter the cell physiology or metabolism. To what extent do the systems used adhere to these standards? Seifert and colleagues (43) examined recombination that involved translocation and duplication of the 5' end of the complete pilin gene and its associated *cat* gene marker. These events bear little resemblance to the stereotyped rearrangement in which the unique 5' end of the complete gene remains unaltered in position and sequence and only the 3' ends of *pilE* and *pilS* are involved. In fact, there was no evidence that the primary structure of pilin was varied as a consequence of any of those rearrangements. Gibbs and colleagues (10) showed that transfer of one particular defective partial pilin gene to the expression site resulted from transformation or from intragenomic reciprocal recombination when DNA uptake was blocked. In both of these sets of work, recombinants were selected on the basis of their increased resistance to antibiotics. It is not clear, therefore, if this selection unduly perturbed the system since the rearrangements may have occurred after exposure to the drug. This concern may be especially relevant to those situations, since such treatment would be either bacteriostatic or bactericidal to the majority of exposed organisms, possibly enhancing autolysis and the level of free DNA. In our study, the block of heterologous DNA at the 3' end of the pilin expression locus consisting of the *cat* gene (present to facilitate the introduction of the mutated constructs into the genome) may have altered or influenced recombination. Although each of these tests has its merits and weaknesses, I think a strong case could be made that the tests that led to the transformation-based model are not representative of productive pilin recombination. One point of dispute that cannot be readily explained by obvious methodological differences relates to the reported influence or lack thereof of DNase I treatment on pilin gene recombination (42, 62).

ANALOGIES IN OTHER PROKARYOTIC SYSTEMS

Some have argued for a transformation mechanism based on what was perceived to be the lack of documented gene conversion-like events in other prokaryotic systems. Granted, technical reasons make it difficult to recover all products of bacterial recombination, but in one case in which this is possible, conversion is the rule. The vast majority of recombination events between double-stranded DNA forms of the filamentous bacteriophage f1 appear to be a consequence of gene conversion (4), and the mechanism appears to be the formation of asymmetric heteroduplex molecules (9, 15).

The gonococcal situation is also reminiscent of the results of a study examining recombination between homologous sequences in direct orientation in the chromosome of *Salmonella typhimurium* (41). Two genetically marked *lac* operons were placed in direct order at the ends of a noninvertible chromosome segment, and Lac$^+$ recombinants were selected. The frequency that Lac$^+$ clones arose from this strain was 9×10^{-4}, and 73% of the events were classified as gene conversions because some information was duplicated while other information was lost. This frequency is only about fivefold lower than that seen for productive pilin gene recombination events, and the increased rates seen in gonococci could be due to the presence of multiple donor gene copies. In addition, a unidirectional bias in conversion polarity was seen in the *Salmonella* system, with only one of the *lac* operons acting as the recipient of information. Apparently, gene conversion-like events in *S. typhimurium* occur at a relatively high frequency and do so in the absence of transformation or intercellular exchanges.

It may also seem that reciprocal exchange and gene conversion demand fundamentally distinct recombination reactions, but mechanistically, these two outcomes may simply be alternative resolutions of the same basic event (16, 56). Several models of recombination invoke the formation of heteroduplex DNA that is corrected to complementarity by repair, and the choice of which strand to use as the template would have a profound effect on the symmetries of the resulting products.

FACTORS INFLUENCING PILIN GENE RECOMBINATION

What factors other than transformation might account for the observations? One simple explanation may be that the exchanges are in fact reciprocal but that the other product is never recovered. Alternatively, the net outcome of nonreciprocity may be influenced by events occurring specifically at *pilE*. Transcription may be one key factor by making this locus more accessible to exchange processes by virtue of its altered topology or supercoiling (18). By analogy with the mating-type switching of yeasts (27), one could envision that a double-stranded break within *pilE* might initiate recombination, but no evidence supportive of this possibility has been reported. Hill and colleagues (21) have proposed an analogous role for short direct and inverted oligonucleotide repeats within the pilin-encoding segment of *pilE*. They hypothesized that altered secondary structures and single-stranded nicks resulting from intrastrand interactions might target *pilE* for recombination-mediated repair. Since these repeats are also present in *pilS* gene copies, they suggested that transcription might increase the frequencies with which these structures form at *pilE* or stimulate their recognition as substrates in need of repair.

Another property unique to the pilin expression locus that might account for the situation may be its relative position within the genome. Certainly, any model of pilin gene recombination cannot avoid the curious organization of pilin genes, with all copies being oriented in the same 5' to 3' direction and the complete pilin gene being situated at the 3' end of the major pilin gene array (29). Curiously, the pilin gene organization is strikingly similar to that found for the pseudogenes and expression locus encoding avian immunoglobulins, whose diversification arises by gene conversion (37). The arrangement does not appear to be particularly relevant to a transformation-based system but might be critical to an intracellular event since the ratios and positioning of genes during active growth would depend on their orientations relative to the origin and terminus of chromosomal replication. It seems worth noting that while most partial pilin gene copies map upstream of the pilin expression locus, the specific recombination event scored by Gibbs et al. (10) "demanded" the use of a donor allele mapping downstream of the expression locus.

The PilC protein has been implicated in pilus biogenesis on the basis of the reduced piliation of mutants lacking this product. On the basis of the finding that some PilC mutants carry concomitant sequence alterations in the expressed pilin gene, it was proposed that PilC variation might provide a selective mechanism for the structural diversity of pili (23). In this model, a loss of PilC expression by frameshifting would lead to the accumulation of unassembled pilin subunits that are detrimental to cell growth. Variants that express altered pilins more rapidly processed into soluble secreted forms of pilin are relieved of growth inhibition, and the subsequent corrective frameshift mutation in *pilC* results in the assembly of the altered pilin. The frequency with which piliated variants are spawned by these events would be 10^{-13}, being the product of the frequencies of the individual steps (PilC$^+$ → PilC$^-$ $[10^{-5}]$ × pilin α → pilin β $[10^{-3}]$ × PilC$^-$ → PilC$^+$ $[10^{-5}]$). It is therefore difficult to envision what biological significance this process would have. Moreover, there is no evidence that this somewhat convoluted pathway generates any greater pilin diversity than that which arises during single recombination events occurring at frequencies of 10^{-3}. It is also worth noting here that the exact temporal order of *pilC* frameshifting followed by pilin gene recombination has not been confirmed and that the available data are also compatible with a reversed scheme of events (pilin recombination followed by selection for organisms bearing a *pilC* frameshift).

CONCLUSION

Little is known about the exact enzymatic machinery involved in pilin gene recombination. Even in the case of the *recA* gene product, it is not clear whether its requirement reflects a function in catalyzing homologous pairing and strand exchange, DNA repair, or activation of other factors. Although it is necessary, it alone is probably not sufficient and it is highly likely that gonococci express homologs of many of the recombination and mismatch repair proteins characterized in other species (46). The identification of equivalent gonococcal genes by interspecies complementation or by virtue of their DNA homologies and the construction of mutants altered in their expression should make it possible to define their role in the chemistry of the exchange reaction. Likewise, with the advent of methods for constructing strains that express pilin alleles manipulated in vitro (62), it is now possible to examine the influences of heterology, transcription, and genome position on recombination.

Despite the paucity of information on the biochemistry of recombination in gonococci, I come to the following conclusions with regard to the current understanding of pilin gene recombination. Strong evidence exists to show that productive pilin gene recombination can occur at high frequencies by a transformation-independent process and that the exchanges responsible appear to be nonreciprocal in character. This intracellular mechanism is compatible with all available data on productive pilin gene recombination. Furthermore, it has never been shown that uptake and integration of endogenous gonococcal DNA result in the expression of an altered pilin capable of being assembled, nor has it been demonstrated that such a system is capable of generating the pilin diversity displayed by the organism. Until convincing and direct evidence contradicting this current situation is presented, I will remain suspicious of the eager acceptance of transformation as an explanation for pilin gene conversion. Perhaps the question we should be seeking to answer at this time is whether or not any of the possible mechanisms can be disproved and, if so, what tests and results would be viewed as constituting conclusive proof.

ACKNOWLEDGMENTS. I am indebted to Stanley Falkow for his interest, support, and guidance in my studies of gonococcal pathogenesis and other favors too numerous to mention. I am grateful to Vic DiRita and Jim Kaper for critical comments on the manuscript and thank the anonymous reviewers of previous manuscripts for bringing some of the ideas presented here to my attention.

The work done in my lab was supported by Public Health Service grant AI-27837 and a grant from the Pew Scholars Program in the Biomedical Sciences.

REFERENCES

1. **Barbour, A.** 1989. Antigenic variation in relapsing fever *Borrelia* species: genetic aspects, p. 783–789. *In* D. E. Berg and M. M. Howe (ed.), *Mobile DNA*. American Society for Microbiology, Washington, D.C.

2. **Bergström, S., K. Robbins, J. M. Koomey, and J. Swanson.** 1986. Piliation control mechanisms in *Neisseria gonorrhoeae*. *Proc. Natl. Acad. Sci. USA* **83**:3890–3894.

3. **Biswas, G. D., S. A. Lacks, and P. F. Sparling.** 1989. Transformation-deficient mutants of piliated *Neisseria gonorrhoeae*. *J. Bacteriol.* **171**:657–664.

4. **Boon, T., and N. D. Zinder.** 1971. Genotypes produced by individual recombination events involving bacteriophage f1. *J. Mol. Biol.* **58**:133–151.

5. **Borst, P., and D. R. Greaves.** 1987. Programmed gene rearrangements altering gene expression. *Science* **235**:658–667.

6. **Buchanan, T. M.** 1975. Antigenic heterogeneity of gonococcal pili. *J. Exp. Med.* **141**:1470–1475.

7. **Donelson, J. E., and A. C. Rice-Ficht.** 1985. Molecular biology of trypanosome antigenic variation. *Microbiol. Rev.* **49**:107–125.

8. **Elleman, T. C., and P. A. Hoyne.** 1984. Nucleotide sequence of the gene encoding pilin of *Bacteroides nodosus*, the causal organism of ovine footrot. *J. Bacteriol.* **160**:1184–1187.

9. **Enea, V., G. F. Vovis, and N. D. Zinder.** 1975. Genetic studies of heteroduplex DNA of bacteriophage f1. Asymmetric segregation, base correction and implications for the mechanism of genetic recombination. *J. Mol. Biol.* **96**:495–509.

10. **Gibbs, C. P., B. Y. Reimann, E. Schultz, A. Kaufman, R. Haas, and T. F. Meyer.** 1989. Reassortment of pilin genes in *Neisseria gonorrhoeae* occurs by two distinct mechanisms. *Nature* (London) **338**:651–652.

11. **Giron, J. A., A. S. Y. Ho, and G. K. Schoolnik.** 1991. An inducible bundle-forming pilus of enteropathogenic *Escherichia coli*. *Science* **254**:710–713.

12. **Haas, R., and T. F. Meyer.** 1986. The repertoire of silent pilus genes in *Neisseria gonorrhoeae*: evidence for gene conversion. *Cell* **44**:107–115.

13. **Haas, R., H. Schwarz, and T. F. Meyer.** 1987. Release of soluble pilin antigen coupled with gene conversion in *Neisseria gonorrhoeae*. *Proc. Natl. Acad. Sci. USA* **84**:9079–9083.

14. **Hagblom, P., E. Segal, E. Billyard, and M. So.** 1985. Intragenic recombination leads to pilus antigenic variation in *Neisseria gonorrhoeae*. *Nature* (London) **315**:156–158.

15. **Hartman, N., and N. D. Zinder.** 1974. The effect of B specific restriction and modification of DNA on linkage relationships in f1 bacteriophage. II. Evidence for a heteroduplex intermediate in f1 recombination. *J. Mol. Biol.* **85**:357–369.

16. **Hastings, P. J.** 1988. Conversion events in fungi, p. 397–428 *In* R. Kucharlapati and G. R. Smith (ed.), *Genetic Recombination*. American Society for Microbiology, Washington, D.C.

17. **Hebeler, B. H., and F. E. Young.** 1975. Autolysis of *Neisseria gonorrhoeae*. *J. Bacteriol.* **122**:385–392.

18. **Helling, R. B.** 1967. The effect of arabinose-specific enzyme synthesis on recombination in the arabinose genes of *Escherichia coli*. *Genetics* **57**:665–675.

19. **Henrichsen, J.** 1983. Twitching motility. *Annu. Rev. Microbiol.* **37**:81–93.

20. **Hermodson, M. A., K. C. Chen, and T. M. Buchanan.** 1978. Neisseria pili proteins: amino-terminal amino acid sequences and identification of an unusual amino acid. *Biochemistry* **17**:442–445.

21. **Hill, S. A., S. G. Morrison, and J. Swanson.** 1990. The role of direct oligonucleotide repeats in gonococcal pilin gene variation. *Mol. Microbiol.* **4**:1341–1352.

22. **Jonsson, A. B., G. Nyberg, and S. Normark.** 1991. Phase variation of gonococcal pili by frameshift mutation in *pilC*, a novel gene for pilus assembly. *EMBO J.* **10**:477–488.

23. **Jonsson, A. B., J. Pfeifer, and S. Normark.** 1992. *Neisseria gonorrhoeae* PilC expression provides a selective mechanism for structural diversity of pili. *Proc. Natl. Acad. Sci. USA* **89**:3204–3208.

24. **Kellog, D. S., I. R. Cohen, L. C. Norins, A. L. Schroeter, and G. Reising.** 1968. *Neisseria gonorrhoeae*. II. Colonial variation and pathogenicity during 35 months in vitro. *J. Bacteriol.* **96**:596–605.

25. **Koomey, M., E. C. Gotschlich, K. Robbins, S. Bergstrom, and J. Swanson.** 1987. Effects of *recA* mutations on pilus antigenic variation and phase transitions in *Neisseria gonorrhoeae*. *Genetics* **117**:391–398.

26. **Koomey, M., and Q. Y. Z. Zhang.** Unpublished data.

27. **Kostriken, R., N. Strathern, A. J. Sklar, J. B. Hicks, and F. Heffron.** 1983. A site-specific endonuclease essential for mating-type switching in *Saccharomyces cerevisiae*. *Cell* **35:**167–174.

28. **Marrs, C. F., G. Schoolnik, J. M. Koomey, J. Hardy, J. Rothbard, and S. Falkow.** 1985. Cloning and sequencing of a *Moraxella bovis* pilin gene. *J. Bacteriol.* **163:**132–139.

29. **Meyer, T. F., E. Billyard, R. Haas, S. Storzbach, and M. So.** 1984. Pilus genes of *Neisseria gonorrheae:* chromosomal organization and DNA sequence. *Proc. Natl. Acad. Sci. USA* **81:**6110–6114.

30. **Meyer, T. F., C. P. Gibbs, and R. Haas.** 1990. Variation and control of protein expression in Neisseria. *Annu. Rev. Microbiol.* **44:**451–477.

31. **Meyer, T. F., N. Mlawer, and M. So.** 1982. Pilus expression in *Neisseria gonorrhoeae* involves chromosomal rearrangement. *Cell* **30:**45–52.

32. **Murphy, G. L., T. D. Connell, D. S. Barrit, M. Koomey, and J. G. Cannon.** 1989. Phase variation of gonococcal protein. II. Regulation of gene expression by slipped-strand mispairing of a repetitive DNA sequence. *Cell* **56:**539–547.

33. **Norlander, L., J. Davies, A. Norqvist, and S. Normark.** 1979. Genetic basis for colonial variation in *Neisseria gonorrhoeae*. *J. Bacteriol.* **138:**762–769.

34. **Pasloske, B. L., B. B. Finlay, and W. Paranchych.** 1985. Cloning and sequencing of the *Pseudomonas aeruginosa* PAK pilin gene. *FEBS Lett.* **183:**408–412.

35. **Plasterk, R. H. A., M. I. Simon, and A. G. Barbour.** 1985. Transposition of structural genes to an expression sequence on a linear plasmid causes antigenic variation in the bacterium *Borrelia hermsii*. *Nature* (London) **318:**257–263.

36. **Pugsley, A. P.** 1993. The complete general secretory pathway in gram-negative bacteria. *Microbiol. Rev.* **57:**50–108.

37. **Reynaud, C.-A., V. Anquez, H. Grimal, and J.-C. Weill.** 1987. A hyperconversion mechanism generates the chicken light chain gene preimmune repertoire. *Cell* **48:**379–388.

38. **Rudel, T., J. P. M. van Putten, C. P. Gibbs, R. Hass, and T. F. Meyer.** 1992. Interaction of two variable proteins (PilE and PilC) required for pilus-mediated adherence of *Neisseria gonorrhoeae* to human epithelial cells. *Mol. Microbiol.* **6:**3439–3450.

39. **Scocca, J. J.** 1990. The role of transformation in the variability of the *Neisseria gonorrhoeae* cell surface. *Mol. Microbiol.* **4:**321–327.

40. **Segal, E., E. Billyard, M. So, S. Storzbach, and T. F. Meyer.** 1985. Role of chromosomal rearrangement in *N. gonorrhoeae* pilus phase variation. *Cell* **40:**293–300.

41. **Segall, A. M., and J. R. Roth.** 1989. Recombination between homologies in direct and inverse orientation in the chromosome of Salmonella: intervals which are nonpermissive for inversion formation. *Genetics* **122:**737–747.

42. **Seifert, H. S., R. Ajioka, and M. So.** 1988. Alternative model for *Neisseria gonorrhoeae* pilin variation. *Vaccine* **6:**107–109.

43. **Seifert, H. S., R. S. Ajioka, C. Marchal, P. F. Sparling, and M. So.** 1988. DNA transformation leads to pilin antigenic variation in *Neisseria gonorrhoeae*. *Nature* (London) **336:**392–395.

44. **Shaw, C. E., and R. K. Taylor.** 1990. *Vibrio cholerae* O395 *tcpA* pilin gene sequence and comparison of the predicted protein structural features to those of type IV pilins. *Infect. Immun.* **58:**3042–3049.

45. **Simon, M., and I. Herskowitz.** 1985. Introduction, p. xv–xix. *In* M. Simon and I. Herskowitz (ed.), *Genome Rearrangement*. Alan R. Liss, Inc., New York.

46. **Smith, G. R.** 1988. Homologous recombination in prokaryotes. *Microbiol. Rev.* **52:**1–28.

47. **Sparling, P. F.** 1966. Genetic transformation of *Neisseria gonorrhoeae* to streptomycin resistance. *J. Bacteriol.* **92:**1364–1371.

48. **Swanson, J., and O. Barrera.** 1983. Gonococcal pilus subunit size heterogeneity correlates with transitions in colony piliation phenotype, not with changes in colony opacity. *J. Exp. Med.* **158:**1459–1472.

49. **Swanson, J., S. Bergström, O. Barrera, K. Robbins, and D. Corwin.** 1985. Pilus⁻ gonococcal variants. Evidence for multiple forms of piliation control. *J. Exp. Med.* **162:**729–744.

50. **Swanson, J., S. Bergström, K. Robbins, O. Barrera, D. Corwin, and J. M. Koomey.** 1986. Gene conversion involving the pilin structural gene correlates with pilus + ⇔ pilus − changes in *Neisseria gonorrhoeae*. *Cell* **47:**267–276.

51. **Swanson, J., and J. M. Koomey.** 1989. Mechanisms for variation of pili and outer membrane protein II in *Neisseria gonorrhoeae,* p. 743–761 *In* D. E. Berg and M. M. Howe (ed.), *Mobile DNA.* American Society for Microbiology, Washington, D.C.

52. **Swanson, J., S. J. Kraus, and E. C. Gotschlich.** 1971. Studies on gonococcus infection. I. Pili and zones of adhesion: their relation to gonococcal growth patterns. *J. Exp. Med.* **134:**886–906.

53. **Swanson, J., S. Morrison, O. Barrera, and S. Hill.** 1990. Piliation changes in tranformation-defective *Neisseria gonorrhoeae. J. Exp. Med.* **171:**2131–2139.

54. **Swanson, J., K. Robbins, O. Barrera, D. Corwin, J. Boslego, J. Ciak, M. Blake, and J. M. Koomey.** 1987. Gonococcal pilin variants in experimental gonorrhea. *J. Exp. Med.* **165:**1344–1357.

55. **Swanson, J., K. Robbins, O. Barrera, and J. M. Koomey.** 1987. Gene conversion variations generate structurally distinct pilin polypeptides in *Neisseria gonorrhoeae. J. Exp. Med.* **165:**1016–1025.

56. **Thompson, C. B.** 1992. Creation of immunoglobulin diversity by intrachromosomal gene conversion. *Trends Genet.* **8:**416–422.

57. **Thompson, C. B., and P. E. Neiman.** 1987. Somatic diversification of the chicken immunoglobulin light chain gene is limited to the rearranged variable gene segment. *Cell* **48:**369–378.

58. **Tønjum, T., C. Marrs, F. Rozsa, and K. Bøvre.** 1991. The type IV pilin of *Moraxella nonliquefaciens* exhibits unique similarities with pilins of *Neisseria gonorrhoeae* and *Bacteroides nodosus. J. Gen. Microbiol.* **137:**2483–2490.

59. **Vickerman, K.** 1978. Antigenic variation in trypanosomes. *Nature* (London) **273:**613–617.

60. **Virji, M., J. E. Heckels, and P. J. Watt.** 1983. Monoclonal antibodies to gonococcal pili: studies on antigenic determinants on pili variants of strain P9. *J. Gen. Microbiol.* **129:**1965–1973.

61. **Zak, K., J.-L. Diaz, D. Jackson, and J. E. Heckels.** 1984. Antigenic variation during infection with *Neisseria gonorrhoeae:* detection of antibodies to surface proteins in sera of patients with gonorrhea. *J. Infect. Dis.* **149:**166–173.

62. **Zhang, Q. Y., D. DeRyckere, P. Lauer, and M. Koomey.** 1992. Gene conversion in *Neisseria gonorrhoeae:* evidence for its role in pilus antigenic variation. *Proc. Natl. Acad. Sci. USA* **89:** 5366–5370.

Molecular Genetics of Bacterial Pathogenesis
Edited by V. L. Miller, J. B. Kaper, D. A. Portnoy, and R. R. Isberg
© 1994 American Society for Microbiology, Washington, DC 20005

Chapter 9

Type 4 Pili in the Families *Moraxellaceae* and *Neisseriaceae*

Carl F. Marrs

The genera *Moraxella, Acinetobacter,* and *Psychrobacter* have recently been removed from the family *Neisseriaceae* to form the new family *Moraxellaceae* (115). In this chapter I review studies on type 4 pili from the species *Moraxella bovis, Moraxella lacunata,* and *Moraxella nonliquefaciens* of the family *Moraxellaceae* and *Eikenella corrodens, Kingella denitrificans,* and *Kingella kingae* of the family *Neisseriaceae.* For an earlier review, see Marrs et al. (83). The type 4 pili of *Neisseria gonorrhoeae* are discussed by Michael Koomey elsewhere in this book (chapter 8).

The three species *M. bovis, M. lacunata,* and *M. nonliquefaciens* have been shown to be closely related by DNA-DNA hybridization studies with total genomic DNAs (138), pulse-RNA (mRNA)-DNA hybridization analysis (10), and transformation studies (9, 17) and have been referred to as "classical moraxellae" or as the "*M. lacunata* group" (17). Transformation and nutritional assays (17, 76, 77) and, more recently, multilocus enzyme electrophoresis (139) have been developed to distinguish between the three species. *M. lacunata* includes the strains previously described as *M. liquefaciens,* and *M. bovis* includes the strains which made up *M. equi* (17).

In 1972, Jackson and Goodman (71) proposed the removal of a set of facultative anaerobic bacterial strains from *Bacteroides corrodens* (strict anaerobic bacteria) into a new species called *E. corrodens.*

K. kingae was originally classified as *Moraxella kingii* (63) and was then transferred to the new genus *Kingella* within the family *Neisseriaceae* in 1976 (65). Later in 1976, the species *K. denitrificans* and *Kingella indologenes* were added to the genus *Kingella* (130). In 1990 *K. indologenes* was transferred out of the genus *Kingella* and was renamed *Suttonella indologenes* (30). In 1993, *Kingella orale* was added as a new species of the genus *Kingella* (29).

Carl F. Marrs • Department of Epidemiology, School of Public Health, University of Michigan, Ann Arbor, Michigan 48109.

PATHOGENESIS

Human Disease

M. lacunata (including strains previously called *M. liquefaciens*) appears to have been a significant causative agent of human conjunctivitis (and keratitis) and was frequently isolated in the past (5). It is only rarely isolated at present, and studies in the 1950s and 1960s in the United States and England reported the incidence of *Moraxella* conjunctivitis ranging from 0.1 to 1.0% of all forms of conjunctivitis (75, 91, 137, 144). The epidemiology of recent cases of *Moraxella* conjunctivitis and keratitis reveals two very different populations at highest risk: older male alcoholics (5, 39, 40) and teenaged females (52, 78, 113). In 1958, Henriksen (60) found that *M. nonliquefaciens* was present in 11.3% of nasal cultures from normal subjects, but he did not isolate any *M. lacunata* isolates from the upper respiratory tracts of normal individuals. Similarly, Bovre (11) in 1970 reported finding *M. nonliquefaciens* from 17.7% of nose specimens but only occasionally isolated *M. lacunata*. Thus, while *M. nonliquefaciens* may be considered part of the normal flora, *M. lacunata* does not appear to be. Despite this, ocular infections caused by *M. nonliquefaciens* are not more common than those caused by *M. lacunata* (146, 147). While most *Moraxella* infections have been associated with the diseases of the external eye, cases of both *M. lacunata* (26) and *M. nonliquefaciens* (33, 82) endophthalmitis have been reported. Additional rare diseases at nonocular sites include *M. lacunata* endocarditis (123, 126) and *M. nonliquefaciens* pulmonary diseases (8, 114), septicemia (125), and botryomycosis (41).

E. corrodens is part of the normal flora of the upper respiratory and gastrointestinal tracts of humans (34, 62). As a pathogen, it is involved in periodontal disease (23, 73, 74, 121, 135) and as a cause of abscesses in many areas of the body (20, 24, 32, 99, 101, 110, 131, 134).

K. denitrificans and *K. kingae* are commensal organisms of the human respiratory tract (69). *K. kingae* is infrequently reported as a pathogen in humans; however, a wide variety of diseases in which *K. kingae* as the etiologic agent have been described (1, 2, 7, 22, 25, 28, 93, 95, 143, 149). *K. denitrificans* is even less commonly found to cause disease, but it has been implicated in cases of endocarditis, bacteremia, and empyema (21, 46, 51, 133).

Infectious Bovine Keratoconjunctivitis

M. bovis is the primary cause of infectious bovine keratoconjunctivitis (IBK), a widespread, highly contagious ocular disease of cattle that causes temporary and, occasionally, permanent blindness (for reviews, see references 3, 45, 66, and 90). IBK ranks as one of the most important diseases of cattle throughout the world (45, 128, 129, 150) and results in substantial economic losses. The annual estimated incidence of IBK among all beef cattle in the United States is 5%, with half of all herds having infected cattle (150). IBK is also common among dairy cattle (105, 150), and epizootics occur, with morbidity approaching 90% (104, 105).

TYPE 4 PILI

Members of the type 4 class of pili are found on a variety of gram-negative bacterial species (27, 132). In addition to *M. bovis* (86), *M. lacunata* (117), *M. nonliquefaciens* (43, 141), *E. corrodens* (59, 61), *K. denitrificans* (152), and *K. kingae* (152), type 4 pili are also found on *N. gonorrhoeae* (89), *Neisseria meningitidis* (67), *Dichelobacter nodosus* (35, 68, 88), and *Pseudomonas aeruginosa* (94, 124). The main structural subunits of these pili, termed pilins, have *N*-methylphenylalanine at the start of the mature pilin and share a highly conserved amino-terminal domain of 25 to 30 amino acid residues (67), while the C-terminal part contains variable DNA regions. The functional significance of this conservation has been demonstrated by the ability of *P. aeruginosa* bacteria containing plasmids expressing the *D. nodosus* pilin gene to process and assemble the *D. nodosus* pilins into pili on the *P. aeruginosa* cell surface; these pili are structurally and immunologically indistinguishable from authentic *D. nodosus* pili (37, 87). Similar experiments in *P. aeruginosa* strains that contain an *M. bovis* pilin gene produced some strains that made only *M. bovis* pili (6, 36) and some that expressed chimeric pili containing both *M. bovis* pilin and *P. aeruginosa* pilin (36). *Vibrio cholerae* (136) and enteropathogenic *Escherichia coli* (31, 48) produce pilins related to type 4 pilins, but with less amino-terminal sequence homology and with different modified amino acids as the first residues of the mature protein. Patel et al. (96) noted shared antigenicities and immunogenicities of the type 4 pilins expressed by *P. aeruginosa*, *M. bovis*, *N. gonorrhoeae*, *D. nodosus*, and *V. cholerae*.

PHENOTYPES OF PILIATED VERSUS NONPILIATED BACTERIA

Colony Morphology and Pitting of Agar

Early investigators (4, 38, 66, 103) of *M. bovis* isolated from diseased cow eyes noted that on primary isolation the colonies were flat, umbonate with a glistening surface, firm, and dry with sharp, circular edges. Small depressions, or pits, in the agar surface occurred under the colonies. However, after a few rounds of subculturing on various types of blood agar plates, new colony types appeared; these were irregular in form, soft, and viscid, and no depressions in the agar surface were seen after removal of the colonies. These initial investigators termed the primary colony phenotype "smooth" and the secondary type "rough" (4, 66). Later investigators—first Pedersen (97) and then Sandhu et al. (122)—argued that on the basis of the physical and morphological properties of the colonies on blood agar plates, the primary isolate phenotype should be termed "rough" and the secondary type should be termed "smooth."

In 1969, Henriksen (61) described a colony type for *Moraxella kingii* (now *K. kingae*) on blood agar plates which he termed "corroding," emphasizing the pitting of the agar surface that occurred under the colonies. Henriksen and Bovre (64) then described a similar colony morphology for *M. nonliquefaciens* which they termed "spreading-corroding" (SC) and which they distinguished from a noncorroding (N) type. In 1970, Bovre and coworkers (12) examined *M. nonlique-*

faciens bacteria from either SC or N colonies in the electron microscope and determined that those from SC colonies had numerous fimbriae (pili), whereas the bacteria from N colonies lacked them. Further studies with *M. nonliquefaciens* (15), *M. bovis* (15, 98, 122, 127), and *M. kingii* (*K. kingae*) (59) confirmed the association between the SC colony type and piliation. There is some disagreement about the localization of the pili on the cell surface. Simpson et al. (127) studied thin sections of *M. bovis* in the electron microscope and concluded that the pili were distributed peritrichously around the cell. In contrast, Bovre and Froholm (15) show electron micrographs in which it appears that the pili are primarily polar in location. Henrichsen (58) has argued that when appropriately gentle techniques are used to place the bacteria on the grid and to stain the cells prior to electron microscopic examination, the polar localization of the pili is more apparent.

The SC versus N nomenclature is used throughout the rest of this chapter. The colony morphologies seen with the SC and N types vary dramatically with the makeup of the media used in the agar plates. Figure 1 shows the colony morphologies of *M. bovis* SC and N types on bovine or sheep blood agar plates (Fig. 1A) and on GC medium (Fig. 1B). When *M. bovis* Epp63 was propagated on 5% sheep blood agar, one N-type and six distinct piliated colony types were observed (119). As will be discussed in detail later, a single *M. bovis* strain can express two distinct pilin types (Q and I) differentiable by apparent molecular weight (85, 86). Colleagues and I thus examined whether any of the different piliated colony types were correlated to expression of a different pilin type. However, we discovered that individual colonies of each of the SC morphotypes could express either Q or I pilin (119). Thus, the different SC types observed must have factors other than pili type that determine their colony morphotype. Recently, colleagues and I have determined that for both *M. bovis* (54) and *K. denitrificans* (151) some N-type colonies which fail to pit the agar still contain bacteria which have pili on their

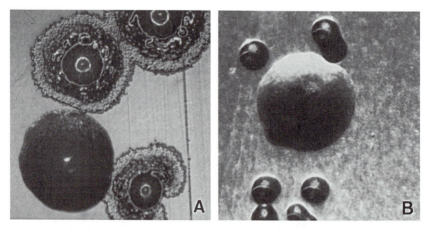

FIGURE 1. SC and N colony types of *M. bovis* on bovine blood agar (A) or GC agar medium (B). (A) The SC colonies exhibit a "fried egg" appearance, while the N colony is relatively featureless. (B) The SC colonies are small and domed, while the N colony is much larger.

surfaces. More research is necessary to determine what change is responsible for the N colony type in these bacteria.

How SC colonies pit the agar is still unclear. It may be due to an agarase enzyme which is coordinately regulated with the expression of pili. Alternatively, the physical characteristics of the SC colony, their tight clumping, and the hydrophobic nature imparted by the pili may exclude the semisolid agar surface in a manner which physically produces the pit below each colony.

Autoagglutination and Pellicle Formation

Some if not all of the colony morphological differences between SC and N colonies are due to the strong autoagglutination phenotype which occurs with piliated but not with nonpiliated bacteria. When SC colonies of *M. bovis* are suspended in distilled water or physiologic salt solution they autoagglutinate (122). Pugh and Hughes (102) reported that this autoagglutination could be inhibited by suspending the SC colonies in 10% magnesium chloride. Nonpiliated bacteria from N colonies can be evenly dispersed in water or broth (102, 122). The magnesium chloride treatment appears to break off the pili from the cell (122).

When SC variants of *M. nonliquefaciens, M. bovis,* or *K. kingae* were grown in broth without agitation, they were found to form an opaque rim that stuck to the glass wall close to the surface of the medium after 1 to 3 days of incubation (12, 15, 57, 122). This pellicle can be unstable, tending to detach and settle at the bottom when the tube is shaken or incubated for a prolonged period (15). While major pellicle formation is not observed upon static growth of cells from N-type colonies, when a slight surface rim does form, subcultivation from the pellicle results in SC colony types (15). Thus, pellicle formation can be used as an enrichment procedure to isolate the often rare variants switching from the nonpiliated to the piliated state.

Hemagglutination

Sandhu et al. (122) found that piliated, but not nonpiliated, *M. bovis* bacteria bound to sheep erythrocytes, resulting in a hemagglutination reaction within 3 to 4 h after the bacteria were mixed with the sheep erythrocytes at room temperature. Gil-Turnes and Ribeiro (47) studied the hemagglutination properties of eight different *M. bovis* strains with erythrocytes from chickens, rabbits, sheep, swine, and guinea pigs. Piliated, but not nonpiliated, bacteria hemagglutinated the erythrocytes from all but the guinea pigs.

Twitching Motility

Twitching motility is a special kind of bacterial surface translocation, first discovered in 1961 by Lautrop in *Acinetobacter calcoaceticus,* that may lead to the production of spreading zones on solid surfaces (78a). Twitching cells move predominantly singly, although smaller moving aggregates occur. The movement appears intermittent and jerky and does not regularly follow the long axis of the cell (55). Twitching motility occurs only when conditions allow a surface film of liquid of a suitable thickness to be formed and maintained around the bacterial colonies (57). Twitching is most pronounced on freshly poured thick plates with a low content of nutrients and with a relatively low concentration of agar only slightly dried before inoculation and incubated in a humid atmosphere (56). For example, *M. bovis* colonies give evidence of twitching motility only when incubated in a closed jar with a layer of water at the bottom; under normal atmospheric incubation conditions, spreading is not seen (55).

Twitching motility was originally reported in *M. lacunata* and *M. nonliquefaciens* in 1963 by Piechaud (100). Henrichsen et al. (59) found that for *M. nonliquefaciens, M. bovis,* and *M. kingii* (*K. kingae*), piliated bacteria from SC colonies had twitching motility, while nonpiliated N colony bacteria lacked twitching motility. In fact, it appears that the spreading colony phenotype used in part to define SC colonies is a result of the twitching motility (58). The mechanism by which twitching motility occurs and the role of pili in that mechanism are still unknown. Froholm and Bovre (42) suggested that some kind of contraction or retraction of pili after attachment of their ends might cause twitching motility in *Moraxella* bacteria. Bradley (18, 19) has proposed a similar, more detailed model for the retraction of pili as the basis for twitching motility. Henrichsen (57, 58) does not believe that the experimental data fit well with this model. He proposed that

> the movements of the individual cells could be the result of a combination of some or all of the following five factors: (a) interfacial tension phenomena; (b) local changes in the surface charge density of the growing bacteria; (c) affinity of the negatively charged cells for the water phase, primarily due to their hydrophilic carboxyl groups; (d) Brownian movements of the cells; and (e) long-range electrostatic repulsive forces between the surface of the agar gel and the electrical double layer of the bacteria.

Competence for DNA Transformation

In 1970, Bovre and Froholm (13) first noted a correlation between piliation in *M. nonliquefaciens* and competence to be genetically transformed. They later expanded these observations to include the association between piliation and transformation in *M. bovis* and *M. kingii* (*K. kingae*) (14, 16, 42). Tonjum et al. (140) have also demonstrated competence for DNA transformation in piliated *E. corrodens*. When pairs of colony variants (SC = piliated, N = nonpiliated) of the same strain of *M. bovis, M. nonliquefaciens,* or *K. kingae* were tested in streptomycin resistance transformation experiments with the same batch of DNA, Bovre and Froholm found that the SC-type cells yielded from about 10^2- to more than 10^6-fold more transformants than the N-type recipients (16, 42). Although transformants from N-type recipients were often found to have switched to SC colonies (probably because of the presence of undetected variants in the recipient population), pili are not essential for competence since streptomycin-resistant transformants do occur in bacteria which remain nonpiliated. Hackel (54) recently noted that unusual *M. bovis* N-type variants which still have pili observable by electron microscopy are transformed much better than N-type variants with no detectable pili. The role that pili play in transformation is unknown. It is not even clear if the pili are directly involved in facilitating transformation or whether their expression is coregulated with other factors directly involved in transformation.

ROLE OF TYPE 4 PILI IN DISEASE

While inoculations of human volunteers and rabbits with *M. lacunata* (91, 144) and guinea pigs with *M. nonliquefaciens* (145) can result in ocular disease, no experiments have been carried out comparing piliated versus nonpiliated bacteria. Therefore, the only evidence that pili are important in these infections is the finding that upon initial isolation from infected individuals the bacteria have an SC colony morphology (33, 113). Even less is known about the role of pili in disease caused by *Eikenella* and *Kingella* species.

In contrast, pili are clearly required for *M. bovis* to cause IBK. In 1972, Pedersen et al. (98) were the first to demonstrate the importance of pili in the establishment of IBK. In every calf they inoculated with piliated *M. bovis* it was possible to reisolate *M. bovis* from the eyes in subsequent days, and half of the calves went on to develop clinical disease, while calves inoculated with nonpiliated *M. bovis* never got disease and it was not possible to reisolate *M. bovis* from their eyes in subsequent days. Other workers then showed that isolated pili could be used as vaccines which would protect calves from disease after inoculation of homologous strains (72, 79, 106, 107). However, *M. bovis* strains in nature have a wide variety of antigenic differences in their pili (79, 80, 92). New antigenic types have been shown to appear during outbreaks of IBK (148), and vaccines that use one pilus type are not protective against heterologous pilus strain types (108, 109).

Additional studies demonstrated that a single *M. bovis* strain could switch from making a pilin subunit of one size to a pilin subunit of a different size (81, 86). Initially, my colleagues and I observed two sizes of pilins in strain Epp63 which we designated alpha for the higher-molecular-weight pilin and beta for the lower-molecular-weight pilin (86). Independently, Lepper and Power (81) in Australia identified three size groups of pilins which they named alpha, beta, and gamma by ascending size. They also observed that some strains could switch between having alpha and beta pilins and others could switch between having gamma and beta pilins (81). Unfortunately, our alpha was their beta, and our beta was their alpha. Therefore, both groups agreed to rename the pilin subunits as follows: Q (quick), I (intermediate), and S (slow) relative to their electrophoretic mobilities on sodium dodecyl sulfate-polyacrylamide gel electrophoresis. *M. bovis* strains expressing I pili are significantly less likely to colonize and cause disease in mice (81) and calves (81, 119, 120) than are *M. bovis* strains expressing either Q or S pili. However, after initial infection of calf eyes with an *M. bovis* strain expressing Q pili, the *M. bovis* bacteria isolated back out of the infected eye were as likely to express I pili as Q pili (120). It has also been observed that piliated *M. bovis* bacteria attach better to bovine corneas in vitro than nonpiliated bacteria of the same strain (70). Ruehl et al. (118) showed that Q-piliated *M. bovis* bound bovine corneas in vitro better than did I-piliated *M. bovis*. They also tested the binding of a recombinant *P. aeruginosa* strain which elaborates *M. bovis* pili (6) and showed that this Q-piliated *P. aeruginosa* strain bound bovine corneas in vitro better than the *P. aeruginosa* parent which does not make Q pili (118).

MOLECULAR GENETICS OF TYPE 4 PILI

M. bovis, *M. lacunata*, and *M. nonliquefaciens*

We originally cloned the *M. bovis* Q pilin gene (*tfpQ*) from strain Epp63 and placed it in an *E. coli* host background (86). In an *E. coli* host the *tfpQ* gene was expressed as pilin, but the pilin was not assembled into pili. Using this *tfpQ* gene as a probe in Southern hybridization analysis, we then determined that the transition between Q and I pilus types in Epp63 was the result of inversion of a 2.1-kb segment of chromosomal DNA (85). We noted a stretch of 36 bp within the conserved amino acid region of *tfpQ* which was 58% identical to the left inverted repeat (*hixL*) of the *Salmonella typhimurium* flagellar *hin* control region. Analysis of the sequence spanning the entire inversion region in orientation 1 (Q pilin expressed) compared with the sequence of PCR-amplified genomic DNA from orientation 2 (I pilin expressed) allowed the site-specific region of recombination to be localized to a 26-bp region within the previously noted site of similarity to *hixL* (44). The pilin gene on the side of the inversion opposite from the expressed pilin gene lacks not only a promoter but also the coding region for the start of the pilin gene (Fig. 2).

Using the *M. bovis tfpQ* gene as a probe, we then isolated the recombinant plasmid clone pMxL1, whose 5.9-kb insert contains the entire *M. lacunata* pilin gene inversion region (84). The inversion region of pMxL1 inverts in *E. coli* host

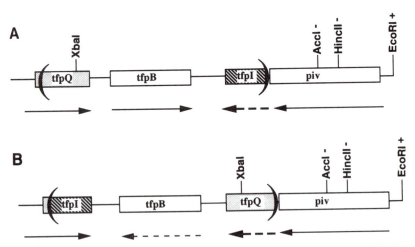

FIGURE 2. (A) Diagram of the *M. bovis* pilin gene inversion region in orientation 1, with *tfpQ* in the expression locus. Parentheses represent the sites of inversion, and lines beneath the map show the direction of transcription (dashed lines indicate genes for which we have no evidence of transcription). (B) The same region in orientation 2, with *tfpI* in the expression locus.

cells. Deletion derivatives of pMxL1 yielded some plasmids that still had the entire inversion region but were phase-locked into one or the other of the two potential orientations. Similarly, insertions of a 2-kb streptomycin resistance element (omega) within some regions outside of the inversion also resulted in phase-locked plasmids. These deletions and insertions thus localized a probable invertase necessary for the inversion event. Sequence analysis identified a gene that we named piv_{ml} (pilin inversion of *M. lacunata*). We then showed that the equivalent gene from *M. bovis* (piv_{mb}) could supply in *trans* the inversion function missing from one of the *M. lacunata* phase-locked inversion mutants (84, 116). Other deletion derivatives have shown that the *tfpB* gene can be removed and that inversion still occurs (49). Interestingly, the Piv invertases have no homology to the Hin family of invertases. However, the Piv invertases do have some amino acid homology to transposases of certain insertion sequence elements (49), being most related to those of IS*1000* in *Thermus thermophilus* (28% amino acid identity and 47% similarity with five gaps) and IS*492* in *Pseudomonas atlantica* (26% identity and 49% sequence similarity with five gaps).

Figure 3 shows a comparison of the pilin gene inversion regions of *M. lacunata* and *M. bovis*. The *tfpI* pilin gene of *M. lacunata* in plasmid pMxL1 contains a perfect tandem repeat of 19 bp at the middle of the recombination junction site (116, 117). This 19-bp insert causes a frameshift and disrupts the pilin gene; thus, pMxL1 makes pilin only in orientation 1, when the *tfpQ* gene is in the expression locus. One other difference is shown in Fig. 3: an AT deletion in pMxL1, in comparison with the sequence of Epp63, within an open reading frame (*tfpB*) which results in the pMxL1 *tfpB* open reading frame being one-third shorter than that in Epp63. PCR sequencing of the parent *M. lacunata* strain from which pMxL1 was derived showed no sequence differences between the genomic DNA and that of pMxL1.

Hackel et al. (53) looked at the effects on the inversion process in plasmids containing either the *M. bovis*(pMxB65) or the *M. lacunata*(pMxL40) pilin inversion region of mutations in the *E. coli* genes for the proteins HU, integration host factor (IHF), and factor for inversion stimulation (Fis). These three proteins are known to be involved in other bacterial inversion systems (50). Loss of the Fis protein did not affect the frequency of inversion of the pilin gene regions present on either pMxB65 or pMxL40, while the loss of HU decreased the rate of inversion on each plasmid. The involvement of IHF is still unclear. Loss of this protein did not affect the inversion rate on pMxL40, while the rate of inversion on pMxB65

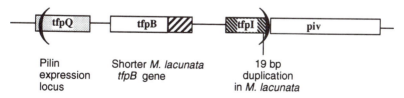

FIGURE 3. Comparison of the pilin gene inversion regions of *M. bovis* and *M. lacunata*. The hatched region of *tfpB* represents the portion of the gene present in *M. bovis* but absent from *M. lacunata*.

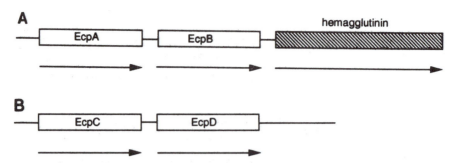

FIGURE 4. (A) Diagram of the arrangement of two pilin genes and a hemagglutinin gene from the *E. corrodens* type strain ATCC 23834. (B) Similar genes from *E. corrodens* 31745. The region past *ecpD* does not contain a gene equivalent to the hemagglutinin gene found in the type strain. The arrows below the genes show the directions of transcription.

was decreased with the loss of the beta subunit, but not the alpha subunit, of the IHF protein.

Tonjum et al. (141) used the *M. bovis tfpQ* gene as a probe to isolate a type 4 pilin gene from *M. nonliquefaciens* and named it *tfpA*. The TfpA pilin showed as much similarity in amino acid sequence to the type 4 pilins of *N. gonorrhoeae* and *D. nodosus* as it did to the type 4 pilins of *M. bovis* and *M. lacunata*. It is not yet known if *M. nonliquefaciens* pilins undergo gene inversion events.

E. corrodens

Type 4 pilin genes have been cloned from two different strains of *E. corrodens*. Rao and Progulske-Fox (111) reported the cloning of two tandemly arrayed pilin genes, *ecpA* and *ecpB* (Fig. 4A), from the type strain ATCC 23834 (originally an oral isolate). Downstream of these two pilin genes is a gene which encodes a 31.5-kDa hemagglutinin (112). Tonjum et al. (142) discovered that the clinical strain *E. corrodens* 31745 (isolated from spinal fluid) also contained two tandem type 4 pilin genes. However, the surrounding genetic region is different, and the genes are so different in sequence from *ecpA* and *ecpB* that they were given the unique names *ecpC* and *ecpD* (Fig. 4B).

K. denitrificans and K. kingae

Hybridization analysis by using either the *tfpQ* pilin gene or the *pilE1* gene from *N. gonorrhoeae* MS11 as the probe revealed that both *K. denitrificans* and *K. kingae* had multiple bands of hybridization (151). Weir (151) has now cloned three separately located, complete type 4 pilin genes from the *K. denitrificans* type strain ATCC 33394. Interestingly, genes *kdpA*, *kdpB*, and *kdpC* encode identical pilins at the amino acid sequence level.

SUMMARY

Type 4 pili are very interesting because of the wide range of gram-negative bacterial species which express type 4 pili, their functional conservation, their significant differences in genetic organization, and their possible importance in bacterial pathogenesis. Further studies on the function and genetics of type 4 pili may lead to a better understanding of the roles that they play in so many bacterial species.

REFERENCES

1. **Adachi, R., O. Hammerberg, and H. Richardson.** 1983. Infective endocarditis caused by *Kingella kingae. Can. Med. Assoc. J.* **128:**1087–1088.
2. **Amir, J., and P. G. Shockelford.** 1991. *Kingella kingae* intervertebral disk infection. *J. Clin. Microbiol.* **29:**1083–1086.
3. **Baptista, P. J. H. P.** 1979. Infectious bovine keratoconjunctivitis: a review. *Br. Vet. J.* **135:** 225–242.
4. **Barner, R. D.** 1952. A study of *Moraxella bovis* and its relation to bovine keratitis. *Am. J. Vet. Res.* **13:**132–144.
5. **Baum, J., H. B. Fedukowicz, and A. Jordan.** 1980. A survey of *Moraxella* corneal ulcers in a derelict population. *Am. J. Ophthalmol.* **90:**476–480.
6. **Beard, M. K. M., J. S. Mattick, L. J. Moore, M. R. Mott, C. F. Marrs, and J. R. Egerton.** 1990. Morphogenetic expression of *Moraxella bovis* fimbriae (pili) in *Pseudomonas aeruginosa. J. Bacteriol.* **172:**2601–2607.
7. **Bosworth, D. E.** 1983. *Kingella (Moraxella) kingae* infections in children. *Am. J. Dis. Child.* **137:** 650–653.
8. **Bottone, E., and J. Allerhand.** 1968. Association of mucoid encapsulated *Moraxella duplex* var. *nonliquefaciens* with chronic bronchitis. *Appl. Microbiol.* **16:**315–319.
9. **Bovre, K.** 1965. Studies on transformation in *Moraxella* and organisms assumed to be related to *Moraxella*. 3. Quantitative streptomycin resistance transformation between *Moraxella bovis* and *Moraxella nonliquefaciens* strains. *Acta Pathol. Microbiol. Scand.* **63:**42–50.
10. **Bovre, K.** 1970. Pulse-RNA-DNA hybridization between rodshaped and coccal species of the *Moraxella-Neisseria* groups. *Acta Pathol. Microbiol. Scand. Sect. B* **78:**565–574.
11. **Bovre, K.** 1970. Oxidase positive bacteria in the human nose. Incidence and species distribution as diagnosed by genetic transformation. *Acta Pathol. Microbiol. Scand. Sect. B* **78:**780–784.
12. **Bovre, K., T. Bergan, and L. O. Froholm.** 1970. Electron microscopical and serological characteristics associated with colony type in *Moraxella nonliquefaciens. Acta Pathol. Microbiol. Scand. Sect. B* **78:**765–779.
13. **Bovre, K., and L. O. Froholm.** 1970. Correlation between the fimbriated state and competence of genetic transformation in *Moraxella nonliquefaciens* strains. *Acta Pathol. Microbiol. Scand. Sect. B* **78:**526–528.
14. **Bovre, K., and L. O. Froholm.** 1971. Competence of genetic transformation correlated with the occurrence of fimbriae in three bacterial species. *Nature* (London) *New Biol.* **234:**151–152.
15. **Bovre, K., and L. O. Froholm.** 1972. Variation of colony morphology reflecting fimbriation in *Moraxella bovis* and two reference strains of *Moraxella nonliquefaciens. Acta Pathol. Microbiol. Scand. Sect. B* **80:**629–640.
16. **Bovre, K., and L. O. Froholm.** 1972. Competence in genetic transformation related to colony type and fimbriation in three species of *Moraxella. Acta Pathol. Microbiol. Scand. Sect. B* **80:** 649–659.
17. **Bovre, K., and N. Hagen.** 1981. The family Neisseriaceae: rod-shaped species of the genera *Moraxella, Acinetobacter, Kingella,* and *Neisseria,* and the *Branhamella* group of cocci, p. 1506–1529. *In* M. P. Starr, H. Stolp, H. G. Truper, A. Balows, and H. G. Schlegel (ed.), *The Prokaryotes: a Handbook on Habitats, Isolation, and Identification of Bacteria.* Springer-Verlag, New York.
18. **Bradley, D. E.** 1974. The adsorption of *Pseudomonas aeruginosa* pilus-dependent bacteriophages to a host mutant with nonretractile pili. *Virology* **58:**149–163.
19. **Bradley, D. E.** 1980. A function of *Pseudomonas aeruginosa* PAO polar pili: twitching motility. *Can. J. Microbiol.* **26:**146–154.

20. **Brill, C. B., L. S. Pearlstein, J. M. Kaplan, and E. L. Mancall.** 1982. CNS infections caused by *Eikenella corrodens. Arch. Neurol.* **39:**431–432.

21. **Brown, A. M., M. M. Rothburn, C. Roberts, and F. J. Nye.** 1987. Septicaemia with probable endocarditis caused by *Kingella denitrificans. J. Infect.* **15:**225–228.

22. **Chanal, C., F. Tiget, P. Chapuis, D. Campagne, M. Jan, and J. Sirot.** 1987. Spondylitis and osteomyelitis caused by *Kingella kingae* in children. *J. Clin. Microbiol.* **25:**2407–2409.

23. **Chen, C. K. C., R. G. Duford, H. S. Reynolds, and J. J. Zambon.** 1989. *Eikenella corrodens* in the human oral cavity. *J. Periodontol.* **60:**611–616.

24. **Cheng, A. F., J. R. South, and G. L. French.** 1988. *Eikenella corrodens* as a cause of brain abscess. *Scand. J. Infect. Dis.* **20:**667–671.

25. **Claesson, B., E. Falsen, and B. Kjellman.** 1985. *Kingella kingae* infections: a review and a presentation of data from 10 Swedish cases. *Scand. J. Infect. Dis.* **17:**233–243.

26. **Cooperman, E. W., and A. H. Friedman.** 1975. Exogenous *Moraxella liquefaciens* endophthalmitis. *Ophthalmologica* **171:**177–180.

27. **Dalrymple, B., and J. S. Mattick.** 1987. An analysis of the organization and evolution of type 4 fimbrial (MePhe) subunit proteins. *J. Mol. Evol.* **25:**261–269.

28. **de Groot, R., D. Glover, C. Clausen, A. L. Smith, and C. R. Wilson.** 1988. Bone and joint infections caused by *Kingella kingae:* six cases and review of the literature. *Rev. Infect. Dis.* **10:**998–1004.

29. **Dewhirst, F. E., C. K. C. Chen, B. J. Paster, and J. J. Zambon.** 1993. Phylogeny of species in the family *Neisseriaceae* isolated from human dental plaque and description of *Kingella orale* sp. nov. *Int. J. Syst. Bacteriol.* **43:**490–499.

30. **Dewhirst, F. E., B. J. Paster, S. LaFontaine, and J. Rood.** 1990. Transfer of *Kingella indologenes* (Snell and Lapage 1976) to the genus *Suttonella* gen. nov. as *Suttonella indologenes* comb. nov.; transfer of *Bacteroides nodosus* (Beveridge 1941) to the genus *Dichelobacter* gen. nov. as *Dichelobacter nodosus* comb. nov.; and assignment of the genera *Cardiobacterium, Dichelobacter,* and *Suttonella* to *Cardiobacteriaceae* fam. nov. in the gamma division of *Proteobacteria* on the basis of 16S rRNA sequence comparisons. *Int. J. Syst. Bacteriol.* **40:**426–433.

31. **Donnenberg, M. S., J. A. Giron, J. P. Nataro, and J. B. Kaper.** 1992. A plasmid-encoded type IV fimbrial gene of enteropathogenic *Escherichia coli* associated with localized adherence. *Mol. Microbiol.* **6:**3427–3437.

32. **Dupon, M., C. d'Ivernois, M. Malou, P. Tauzin-Fin, F. Boineau, and J. Y. Lacut.** 1991. Sacro-iliac joint infection caused by *Eikenella corrodens. Eur. J. Clin. Microbiol. Infect. Dis.* **10:**529–530.

33. **Ebright, J. R., J. R. Lentino, and E. Juni.** 1982. Endophthalmitis caused by *Moraxella nonliquefaciens. Am. J. Clin. Pathol.* **77:**362–363.

34. **Eiken, M.** 1958. Studies on an anaerobic, rod-shaped gram-negative microorganism: *Bacteroides corrodens* n. sp. *Acta Pathol. Microbiol. Scand.* **43:**404–416.

35. **Elleman, T. C., and P. A. Hoyne.** 1984. Nucleotide sequence of the gene encoding pilin of *Bacteroides nodosus,* the causal organism of ovine footrot. *J. Bacteriol.* **160:**1184–1187.

36. **Elleman, T. C., P. A. Hoyne, and A. W. D. Lepper.** 1990. Characterization of the pilin gene of *Moraxella bovis* Dalton 2d and expression of pili from *M. bovis* in *Pseudomonas aeruginosa. Infect. Immun.* **58:**1678–1684.

37. **Elleman, T. C., P. A. Hoyne, D. J. Stewart, N. M. McKern, and J. E. Peterson.** 1986. Expression of pili from *Bacteroides nodosus* in *Pseudomonas aeruginosa. J. Bacteriol.* **168:**574–580.

38. **Farley, H., I. O. Kliewer, C. C. Pearson, and L. E. Foote.** 1950. Infectious keratitis of cattle—a preliminary report. *Am. J. Vet. Res.* **11:**17–21.

39. **Fedukowicz, H., M. Alterman, and R. Newman.** 1971. Two decades of experience with *Moraxella. In* M. P. Solanes (ed.), *XXI International Congress of Ophthalmology. Mexico, 1970.* Excerpta Medica, Amsterdam.

40. **Fedukowicz, H., and H. Horwich.** 1953. The gram-negative diplobacillus in hypopyon keratitis. *Arch. Ophthalmol.* **49:**202–211.

41. **Feldman, S. R., and M. J. Petersen.** 1989. Botryomycosis caused by *Moraxella nonliquefaciens. Cutis* **43:**140–142.

42. **Froholm, L. O., and K. Bovre.** 1973. The fimbriated state and competence in genetic transformation of three *Moraxella* species, p. 125–139. *In* L. J. Archer (ed.), *Bacterial Transformation.* Academic Press, Inc., New York.

43. **Froholm, L. O., and K. Sletten.** 1977. Purification and N-terminal sequence of a fimbrial protein from *Moraxella nonliquefaciens. FEBS Lett.* **73:**29–32.

44. **Fulks, K. A., S. P. Stevens, M. R. Green, and C. F. Marrs.** 1990. Sequence analysis of the inversion region containing the pilin genes of *Moraxella bovis. J. Bacteriol.* **172:**310–316.

45. **George, L. W.** 1984. Clinical infectious bovine keratoconjunctivitis. *Comp. Cont. Ed. Pract. Vet.* **6:**712–724.

46. **Geraci, J. E., and W. R. Wilson.** 1982. Endocarditis due to gram-negative bacteria. Report of 56 cases. *Mayo Clin. Proc.* **57:**145–148.

47. **Gil-Turnes, C., and G. A. Ribeiro.** 1985. *Moraxella bovis* hemagglutinins: effect of carbohydrates, heating and erythrocytes. *Can. J. Comp. Med.* **49:**112–114.

48. **Giron, J. A., A. S. Y. Ho, and G. K. Schoolnik.** 1991. An inducible bundle-forming pilus of enteropathogenic *Escherichia coli. Science* **254:**710–713.

49. **Glasgow, A. C. (Emory University).** Personal communication.

50. **Glasgow, A. C., K. T. Hughes, and M. I. Simon.** 1989. Bacterial DNA inversion systems, p. 637–659. *In* D. E. Berg and M. M. Howe (ed.), *Mobile DNA.* American Society for Microbiology, Washington, D.C.

51. **Goldman, I. S., P. D. Ellner, E. L. Francke, G. J. Garvey, H. C. Neu, and N. Squilla.** 1980. Infective endocarditis due to *Kingella denitrificans. Ann. Intern. Med.* **93:**152–153.

52. **Grayson, M.** 1983. *Diseases of the Cornea,* 2nd ed., p. 122–123. C. V. Mosby Co., St. Louis.

53. **Hackel, M. A., F. W. Rosza, and C. F. Marrs.** Submitted for publication.

54. **Hackel, M. A. M.** 1993. Genetic analysis of *Moraxella bovis* pilus expression. Ph.D. thesis. University of Michigan, Ann Arbor.

55. **Henrichsen, J.** 1972. Bacterial surface translocation: a survey and a classification. *Bacteriol. Rev.* **36:**478–503.

56. **Henrichsen, J.** 1975. The occurrence of twitching motility among gram-negative bacteria. *Acta Pathol. Microbiol. Scand. Sect. B* **83:**171–178.

57. **Henrichsen, J.** 1975. The influence of changes in the environment on twitching motility. *Acta Pathol. Microbiol. Scand. Sect. B* **83:**179–186.

58. **Henrichsen, J.** 1983. Twitching motility. *Annu. Rev. Microbiol.* **37:**81–93.

59. **Henrichsen, J., L. O. Froholm, and K. Bovre.** 1972. Studies on bacterial surface translocation. 2. Correlation of twitching motility and fimbriation in colony variants of *Moraxella nonliquefaciens, M. bovis,* and *M. kingii. Acta Pathol. Microbiol. Scand. Sect. B* **80:**445–452.

60. **Henriksen, S. D.** 1958. *Moraxella duplex* var. *nonliquefaciens,* habitat and antibiotic sensitivity. *Acta Pathol. Microbiol. Scand.* **43:**157–161.

61. **Henriksen, S. D.** 1969. Corroding bacteria from the respiratory tract. I. *Moraxella kingii. Acta Pathol. Microbiol. Scand.* **75:**85–90.

62. **Henriksen, S. D.** 1969. Corroding bacteria from the respiratory tract. II. *Bacteroides corrodens. Acta Pathol. Microbiol. Scand.* **75:**91–96.

63. **Henriksen, S. D., and K. Bovre.** 1968. *Moraxella kingii* sp. nov., a haemolytic, saccharolytic species of the genus *Moraxella. J. Gen. Microbiol.* **51:**377–385.

64. **Henriksen, S. D., and K. Bovre.** 1969. Corroding and spreading colonies in *Moraxella nonliquefaciens. Acta Pathol. Microbiol. Scand.* **76:**459–463.

65. **Henriksen, S. D., and K. Bovre.** 1976. Transfer of *Moraxella kingae* Henriksen and Bovre to the genus *Kingella* gen. nov. in the family *Neisseriaceae. Int. J. Syst. Bacteriol.* **26:**447–450.

66. **Henson, J. B., and L. C. Grumbles.** 1960. Infectious bovine keratoconjunctivitis. I. Etiology. *Am. J. Vet. Res.* **21:**761–766.

67. **Hermodson, M. A., K. C. S. Chen, and T. M. Buchanan.** 1978. *Neisseria* pili proteins: amino-terminal amino acid sequences and identification of an unusual amino acid. *Biochemistry* **17:** 442–445.

68. **Hobbs, M., B. Dalrymple, P. T. Cox, S. P. Livingstone, S. F. Delaney, and J. S. Mattick.** 1991. Organization of the fimbrial gene region of *Bacteroides nodosus:* class I and class II strains. *Mol. Microbiol.* **5:**543–560.

69. **Hollis, D. G., G. L. Wiggins, and R. E. Weaver.** 1972. An unclassified gram-negative rod isolated from the pharynx on Thyer-Martin medium (selective agar). *Appl. Microbiol.* **24:**772–777.

70. **Jackman, S. H., and R. F. Rosenbusch.** 1984. *In vitro* adherence of *Moraxella bovis* to intact corneal epithelium. *Curr. Eye Res.* **9:**1107–1112.

71. **Jackson, F. L., and Y. Goodman.** 1972. Transfer of the facultatively anaerobic organism *Bacteroides corrodens* Eiken to a new genus *Eikenella. Int. J. Syst. Bacteriol.* **22:**73–77.

72. **Jayappa, H. G., and C. Lehr.** 1986. Pathogenicity and immunogenicity of piliated and non-piliated phases of *Moraxella bovis* in calves. *Am. J. Vet. Res.* **47:**2217–2221.

73. **Johnson, D. A., U. H. Behling, C.-H. Lai, M. Listgarten, S. Socransky, and A. Nowotny.** 1978. Role of bacterial products in periodontitis: immune response in gnotobiotic rats monoinfected with *Eikenella corrodens. Infect. Immun.* **19:**246–253.

74. **Johnson, D. A., U. H. Behling, M. Listgarten, and A. Nowotny.** 1978. Role of bacterial products in periodontitis: humoral immune response to *Eikenella corrodens. Infect. Immun.* **22:**382–386.

75. **Jones, B. R., B. E. Andrews, W. G. Henderson, and P. B. Schofield.** 1957. The patterns of conjunctivitis at Moorfields during 1956. *Trans. Ophthalmol. Soc. U.K.* **77:**291–305.

76. **Juni, E., G. A. Heym, M. J. Maurer, and M. L. Miller.** 1987. Combined genetic transformation and nutritional assay for identification of *Moraxella nonliquefaciens. J. Clin. Microbiol.* **25:**1691–1694.

77. **Juni, E., G. A. Heym, and R. D. Newcomb.** 1988. Identification of *Moraxella bovis* by qualitative genetic transformation and nutritional assays. *Appl. Environ. Microbiol.* **54:**1304–1306.

78. **Kowalski, R. P., and J. C. Harwick.** 1986. Incidence of *Moraxella* conjunctival infection. *Am. J. Ophthalmol.* **101:**437–440.

78a.**Lautrop, H.** 1961. *Bacterium anitratum* transferred to the genus *Cytophaga. Int. Bull. Bacteriol. Nomencl.* **11:**107–108.

79. **Lehr, C., H. G. Jayappa, and R. A. Goodnow.** 1985. Serologic and protective characterization of *Moraxella bovis* pili. Cornell Vet. **75:**484–492.

80. **Lepper, A. W. D., and L. R. Hermans.** 1986. Characterization and quantitation of pilus antigens of *Moraxella* bovis by ELISA. *Aust. Vet. J.* **63:**401–405.

81. **Lepper, A. W. D., and B. E. Power.** 1988. Infectivity and virulence of Australian strains of *Moraxella bovis* for the murine and bovine eye in relation to pilus serogroup sub-unit size and degree of piliation. *Aust. Vet. J.* **65:**305–309.

82. **Lobue, T. D., T. A. Deutsch, and R. M. Stein.** 1985. *Moraxella nonliquefaciens* endophthalmitis after trabeculectomy. *Am. J. Ophthalmol.* **99:**343–345.

83. **Marrs, C. F., K. A. Fulks, M. Hackel, F. W. Rozsa, W. W. Ruehl, S. P. Stevens, and S. K. Weir.** 1990. Studies on bacterial type 4 pili in the family *Neisseriaceae*, p. 69–75. *In* O. Olsvik and G. Bukholm (ed.), *Applications of Molecular Biology in Diagnosis of Infectious Diseases*. Norwegian College of Veterinary Medicine, Oslo.

84. **Marrs, C. F., F. W. Rozsa, M. Hackel, S. P. Stevens, and A. C. Glasgow.** 1990. Identification, cloning and sequencing of *piv*, a new gene involved in inverting the pilin genes of *Moraxella lacunata. J. Bacteriol.* **172:**4370–4377.

85. **Marrs, C. F., W. W. Ruehl, G. K. Schoolnik, and S. Falkow.** 1988. Pilin gene phase variation of *Moraxella bovis* is caused by an inversion of the pilin genes. *J. Bacteriol.* **170:**3032–3039.

86. **Marrs, C. F., G. Schoolnik, J. M. Koomey, J. Hardy, J. Rothbard, and S. Falkow.** 1985. Cloning and sequencing of a *Moraxella bovis* pilin gene. *J. Bacteriol.* **163:**132–139.

87. **Mattick, J. S., M. M. Bills, B. J. Anderson, B. Dalrymple, M. R. Mott, and J. R. Egerton.** 1987. Morphogenetic expression of *Bacteroides nodosus* fimbriae in *Pseudomonas aeruginosa. J. Bacteriol.* **169:**33–41.

88. **McKern, N. M., I. J. O'Donnell, A. S. Inglis, D. J. Stewart, and B. L. Clark.** 1983. Amino acid sequence of pilin from *Bacteroides nodosus* (strain 198), the causative organism of ovine footrot. *FEBS Lett.* **164:**149–153.

89. **Meyer, T. F., E. Billyard, R. Haas, S. Storzbach, and M. So.** 1984. Pilus genes of *Neisseria gonorrhoeae:* chromosomal organization and DNA sequence. *Proc. Natl. Acad. Sci. USA* **81:**6110–6114.

90. **Miller, R. B., and W. H. Fales.** 1984. Infectious bovine keratoconjunctivitis: an update. *Vet. Clin. North Am. (Large Anim. Pract.)* **6:**597–608.

91. **Mitsui, Y., S. Hinokuma, and C. Tanaka.** 1951. Etiology of angular conjunctivitis. *Am. J. Ophthalmol.* **34:**1579–1586.

92. **Moore, L. J., and J. M. Rutter.** 1987. Antigenic analysis of fimbrial proteins from *Moraxella bovis. J. Clin. Microbiol.* **25**:2063–2070.

93. **Morrison, V. A., and K. F. Wagner.** 1989. Clinical manifestations of *Kingella kingae* infections: case report and review. *Rev. Infect. Dis.* **11**:776–782.

94. **Pasloske, B. L., B. B. Finlay, and W. Paranchych.** 1985. Cloning and sequencing of the *Pseudomonas aeruginosa* PAK pilin gene. *FEBS Lett.* **183**:408–412.

95. **Patel, N. J., T. L. Moore, T. D. Weiss, and J. Zuckner.** 1983. *Kingella kingae* infectious arthritis: case report and review of literature of *Kingella* and *Moraxella* infections. *Arthritis Rheum.* **26**: 557–559.

96. **Patel, P., C. F. Marrs, J. S. Mattick, W. W. Ruehl, R. K. Taylor, and M. Koomey.** 1991. Shared antigenicity and immunogenicity of type 4 pilins expressed by *Pseudomonas aeruginosa, Moraxella bovis, Neisseria gonorrhoeae, Dichelobacter nodosus,* and *Vibrio cholerae. Infect. Immun.* **59**:4674–4676.

97. **Pedersen, K. B.** 1970. *Moraxella bovis* isolated from cattle with infectious keratoconjunctivitis. *Acta Pathol. Microbiol. Scand. Sect. B* **78**:429–434.

98. **Pedersen, K. B., L. O. Froholm, and K. Bovre.** 1972. Fimbriation and colony type of *Moraxella bovis* in relation to conjunctival colonization and development of keratoconjunctivitis in cattle. *Acta Pathol. Microbiol. Scand. Sect. B* **80**:911–918.

99. **Perez-Pomata, M. T., J. Dominguez, P. Horcajo, F. Santidrian, and J. Bisquert.** 1992. Spleen abscess caused by *Eikenella corrodens. Eur. J. Clin. Microbiol. Infect. Dis.* **11**:162–163.

100. **Piechaud, M.** 1963. Motility in the genus *Moraxella. Ann. Inst. Pasteur* **104**:291–297.

101. **Pollner, J. H., A. Khan, and C. U. Tuazon.** 1992. Severe soft-tissue infection caused by *Eikenella corrodens. Clin. Infect. Dis.* **15**:740–741.

102. **Pugh, G. W., and D. E. Hughes.** 1970. Inhibition of autoagglutination of *Moraxella bovis* by 10% $MgCl_2$. *Appl. Microbiol.* **19**:201–203.

103. **Pugh, G. W., and D. E. Hughes.** 1971. Infectious bovine keratoconjunctivitis induced by different experimental methods. *Cornell Vet.* **61**:23–45.

104. **Pugh, G. W., and D. E. Hughes.** 1972. Bovine infectious keratoconjunctivitis: *Moraxella bovis* as the sole etiologic agent in a winter epizootic. *J. Am. Vet. Med. Assoc.* **161**:481–486.

105. **Pugh, G. W., and D. E. Hughes.** 1975. Bovine infectious keratoconjunctivitis: carrier state of *Moraxella bovis* and the development of preventive measures against disease. *J. Am. Vet. Med. Assoc.* **167**:310–313.

106. **Pugh, G. W., and D. E. Hughes.** 1976. Experimental production of infectious bovine keratoconjunctivitis: comparison of serological and immunological responses using pili fractions of *Moraxella bovis. Can. J. Comp. Med.* **40**:60–66.

107. **Pugh, G. W., D. E. Hughes, and G. D. Booth.** 1977. Experimentally induced infectious bovine keratoconjunctivitis: effectiveness of a pilus vaccine against exposure to homologous strains of *Moraxella bovis. Am. J. Vet. Res.* **38**:1519–1522.

108. **Pugh, G. W., D. E. Hughes, V. D. Schulz, and C. K. Graham.** 1976. Experimentally induced infectious bovine keratoconjunctivitis: resistance of vaccinated cattle to homologous and heterologous strains of *Moraxella bovis. Am. J. Vet. Res.* **37**:57–60.

109. **Pugh, G. W., T. J. McDonald, and A. B. Larsen.** 1978. Experimentally induced infectious bovine keratoconjunctivitis: potentiation of a *Moraxella bovis* pilus vaccine immunogenicity by vaccination with *Mycobacterium paratuberculosis* bacterin. *Am. J. Vet. Res.* **39**:1656–1661.

110. **Raffensperger, J. G.** 1986. *Eikenella corrodens* infections in children. *J. Pediatr. Surg.* **21**: 644–646.

111. **Rao, V. K., and A. Progulske-Fox.** 1993. Cloning and sequencing of two type 4 (*N*-methylphenylalanine) pilin genes from *Eikenella corrodens. J. Gen. Microbiol.* **139**:651–660.

112. **Rao, V. K., J. A. Whitlock, and A. Progulske-Fox.** 1993. Cloning, characterization and sequencing of two haemagglutinin genes from *Eikenella corrodens. J. Gen. Microbiol.* **139**:639–650.

113. **Ringvold, A., E. Vik, and L. S. Bevanger.** 1985. *Moraxella lacunata* isolated from epidemic conjunctivitis among teen-aged females. *Acta Ophthalmol.* **63**:427–431.

114. **Rosett, W., D. M. Heck, and G. R. Hodges.** 1976. Pneumonitis and pulmonary abscess associated with *Moraxella nonliquefaciens. Chest* **70**:664–665.

115. **Rossau, R., A. van Lanschoot, M. Gillis, and J. DeLey.** 1991. Taxonomy of *Moraxellaceae* fam.

nov., a new bacterial family to accommodate the genera *Moraxella, Acinetobacter*, and *Psychrobacter* and related organisms. *Int. J. Syst. Bacteriol.* **41**:310–319.

116. **Rozsa, F. W., M. Hackel, S. P. Stevens, and C. F. Marrs.** 1991. Analysis of the pilin gene inversion region of *Moraxella lacunata*, p. 475–480. *In* M. Achtman, P. Kohl, C. Marchal, G. Morelli, A. Seiler, and B. Thiesen (ed.), *Neisseria 1990*. Walter de Gruyter & Co., Berlin.

117. **Rozsa, F. W., and C. F. Marrs.** 1991. Interesting sequence differences between the pilin gene inversion regions of *Moraxella lacunata* ATCC 17956 and *Moraxella bovis* Epp63. *J. Bacteriol.* **173**:4000–4006.

118. **Ruehl, W. W., C. F. Marrs, M. K. Beard, V. Shokooki, J. R. Hinojoza, S. Banks, D. Bieber, and J. S. Mattick.** 1993. Q pili enhance the attachment of *Moraxella bovis* to bovine corneas *in vitro*. *Mol. Microbiol.* **7**:285–288.

119. **Ruehl, W. W., C. F. Marrs, R. Fernandez, S. Falkow, and G. K. Schoolnik.** 1988. Purification, characterization, and pathogenicity of *Moraxella bovis* pili. *J. Exp. Med.* **168**:983–1002.

120. **Ruehl, W. W., C. F. Marrs, L. George, S. J. M. Banks, and G. K. Schoolnik.** 1993. Infection rates, disease frequency, pilin gene rearrangement, and pilin expression in calves inoculated with *Moraxella bovis* pilin specific isogenic variants. *Am. J. Vet. Res.* **54**:248–253.

121. **Samejima, Y., S. Ebisu, and H. Okada.** 1990. Effect of infection with *Eikenella corrodens* on the progression of ligature-induced periodontitis in rats. *J. Periodontal Res.* **25**:308–315.

122. **Sandhu, T. S., F. H. White, and C. F. Simpson.** 1974. Association of pili with rough colony type of *Moraxella bovis*. *Am. J. Vet. Res.* **35**:437–439.

123. **Sanyal, S. K., N. Wilson, K. Twum-Danso, A. Abomelha, and S. Sohel.** 1990. *Moraxella* endocarditis following balloon angioplasty of aortic coarctation. *Am. Heart J.* **119**:1421–1423.

124. **Sastry, P. A., B. B. Finlay, B. L. Pasloske, W. Paranchych, J. R. Pearlstone, and L. B. Smillie.** 1985. Comparative studies of the amino acid and nucleotide sequences of pilin derived from *Pseudomonas aeruginosa* PAK and PAO. *J. Bacteriol.* **164**:571–577.

125. **Sharma, D. L. B.** 1974. Fatal septicaemia due to *Moraxella nonliquefaciens*. *Arch. Dis. Child.* **49**:966–967.

126. **Silberfarb, P. M., and J. E. Lawe.** 1968. Endocarditis due to *Moraxella liquefaciens*. *Arch. Intern. Med.* **122**:512–513.

127. **Simpson, C. F., F. H. White, and T. S. Sandhu.** 1976. The structure of pili (fimbriae) of *Moraxella bovis*. *Can. J. Comp. Med.* **40**:1–4.

128. **Slatter, D. H., M. E. Edwards, C. D. Hawkins, and G. E. Wilcox.** 1982. A national survey of the occurrence of infectious bovine keratoconjunctivitis. *Aust. Vet. J.* **59**:65–68.

129. **Slatter, D. H., M. E. Edwards, C. D. Hawkins, and G. E. Wilcox.** 1982. A national survey of the clinical features, treatment and importance of infectious bovine keratoconjunctivitis. *Aust. Vet. J.* **59**:69–72.

130. **Snell, J. J. S., and S. Lapage.** 1976. Transfer of some saccharolytic *Moraxella* species to *Kingella* Henriksen and Bovre 1976, with descriptions of *Kingella indologenes* sp. nov. and *Kingella denitrificans* sp. nov. *Int. J. Syst. Bacteriol.* **26**:451–458.

131. **Stein, A., N. Teysseire, C. Capobianco, R. Bricot, and D. Raoult.** 1993. *Eikenella corrodens*, a rare cause of pancreatic abscess: two case reports and review. *Clin. Infect. Dis.* **17**:273–275.

132. **Strom, M. S., and S. Lory.** 1993. Structure-function and biogenesis of the type IV pili. *Annu. Rev. Microbiol.* **47**:565–596.

133. **Swann, R. A., and B. Holmes.** 1984. Infective endocarditis caused by *Kingella denitrificans*. *J. Clin. Pathol.* **37**:1384–1387.

134. **Tami, T. A., and G. S. Parker.** 1984. *Eikenella corrodens*: an emerging pathogen in head and neck infections. *Arch. Otolaryngol.* **110**:752–754.

135. **Tanner, A. C. R., C. Haffer, G. T. Brathal, R. A. Visconti, and S. S. Socransky.** 1979. A study of the bacteria associated with advancing periodontitis in man. *J. Clin. Periodontol.* **6**:278–307.

136. **Taylor, R. K., V. L. Miller, D. B. Furlong, and J. J. Mekalanos.** 1987. Use of *phoA* gene fusions to identify a pilus colonization factor coordinately regulated with cholera toxin. *Proc. Natl. Acad. Sci. USA* **81**:2833–2837.

137. **Thygeson, P., and S. J. Kimura.** 1963. Chronic conjunctivitis. *Trans. Am. Acad. Ophthalmol. Otolaryngol.* **67**:494–517.

138. **Tonjum, T., G. Bukholm, and K. Bovre.** 1989. Differentiation of some species of *Neisseriaceae* and other bacterial groups by DNA-DNA hybridization. *APMIS* **97:**395–405.
139. **Tonjum, T., D. A. Caugant, and K. Bovre.** 1992. Differentiation of *Moraxella nonliquefaciens, M. lacunata,* and *M. bovis* by using mutilocus enzyme electrophoresis and hybridization with pilin-specific DNA probes. *J. Clin. Microbiol.* **30:**3099–3107.
140. **Tonjum, T., N. Hagen, and K. Bovre.** 1985. Identification of *Eikenella corrodens* and *Cardiobacterium hominis* by genetic transformation. *Acta. Pathol. Microbiol. Immunol. Scand. Sect. B* **93:**389–394.
141. **Tonjum, T., C. F. Marrs, F. W. Rozsa, and K. Bovre.** 1991. The type 4 pilin of *Moraxella nonliquefaciens* exhibits unique similarities with the pilins of *Neisseria gonorrhoeae* and *Dichelobacter* (*Bacteriodes*) *nodosus. J. Gen. Microbiol.* **137:**2483–2490.
142. **Tonjum, T., S. Weir, K. Bovre, V. K. Rao, A. Progulske-Fox, and C. Marrs.** 1993. Sequence divergence in two tandemly located pilin genes of *Eikenella corrodens. Infect. Immun.* **61:** 1909–1916.
143. **Toshniwal, R., T. C. Draghi, F. E. Kocka, and C. A. Kallick.** 1986. Manifestations of *Kingella kingae* infections in adults: resemblance to neisserial infections. *Diagn. Microbiol. Infect. Dis.* **5:**81–85.
144. **van Bijsterveld, O. P.** 1967. Acute conjunctivitis and *Moraxella. Am. J. Ophthalmol.* **63:** 1702–1705.
145. **van Bijsterveld, O. P.** 1971. *In-vitro* phagocytosis in pyridoxine deficiency. *J. Med. Microbiol.* **4:**165–170.
146. **van Bijsterveld, O. P.** 1972. The incidence of *Moraxella* on mucous membranes and the skin. *Am. J. Ophthalmol.* **74:**72–76.
147. **van Bijsterveld, O. P.** 1973. Host-parasite relationship and taxonomic position of *Moraxella* and morphologically related organisms. *Am. J. Ophthalmol.* **76:**545–554.
148. **Vandergaast, N., and R. F. Rosenbusch.** 1989. Infectious bovine keratoconjunctivitis epizootic associated with area-wide emergence of a new *Moraxella bovis* pilus type. *Am. J. Vet. Res.* **50:** 1437–1441.
149. **Verbruggen, A.-M., D. Hauglustaine, F. Schildermans, L. van der Hauwaert, J. J. Rombouts, G. Wauters, and J. Vandepitte.** 1986. Infections caused by *Kingella kingae:* report of four cases and review. *J. Infect.* **13:**133–142.
150. **Webber, J. J., and L. A. Selby.** 1981. Risk factors related to the prevalence of infectious bovine keratoconjunctivitis. *J. Am. Vet. Med. Assoc.* **179:**823–826.
151. **Weir, S.** 1993. Studies of pili in *Branhamella catarrhalis, Eikenella corrodens,* and *Kingella denitrificans.* Ph.D. thesis. University of Michigan, Ann Arbor.
152. **Weir, S. K., and C. F. Marrs.** 1992. Identification of type 4 pili in *Kingella denitrificans. Infect. Immun.* **60:**3437–3441.

Molecular Genetics of Bacterial Pathogenesis
Edited by V. L. Miller, J. B. Kaper, D. A. Portnoy, and R. R. Isberg
© 1994 American Society for Microbiology, Washington, DC 20005

Chapter 10

Genes for the Filamentous Hemagglutinin and Fimbriae of *Bordetella pertussis*: Colocation, Coregulation, and Cooperation?

Frits R. Mooi

The view of bacterial adhesins has changed significantly in recent years. Until recently, adhesins were seen as passively mediating attachment to a single receptor molecule. It is now becoming clear that adhesins are much more complex structures which may have multiple binding abilities and whose functions are not always limited to passive adherence but may be extended to affect signal transduction in the target cell. To bring about these effects, adhesins mimic host molecules whose natural functions involve adherence and signal transduction. Another interesting theme that is emerging is the cooperation that may exist between different adhesins in establishing an infection. As will be discussed here, there is now evidence that the filamentous hemagglutinin (FHA) and fimbriae show many of these features. Most notably, FHA and fimbriae seem to act synergistically, a phenomenon which is reflected in the organization and regulation of their genes.

STRUCTURE OF FHA AND FIMBRIAL PROTEINS

FHA (molecular weight, 220,000) (see reference 13 for a recent review) is derived from a very large precursor (see below) and forms large complexes (2 by 40 nm) which are loosely associated with the cell surface (1). A family of surface-exposed high-molecular-weight proteins that are related to FHA has been found in nontypeable *Haemophilus influenzae* strains (35).

Bordetella pertussis fimbriae (see reference 19 for a recent review) have larger diameters (5 nm) and are much longer (up to several hundred nanometers long) than FHA (51). In contrast to FHA, fimbriae are firmly attached to the bacterial cell. *B. pertussis* produces two closely related fimbriae, the serotype 2 and serotype 3 fimbriae, which are composed of subunits, designated Fim2 and Fim3, with molecular weights of 22,500 and 22,000, respectively (10, 51). In addition to these

Frits R. Mooi • National Institute of Public Health and Environmental Protection, P.O. Box 1, 3720 BA Bilthoven, The Netherlands.

major subunits, *B. pertussis* fimbriae also contain a minor subunit (molecular weight, 40,000), designated FimD, which can be dissociated from the fimbrial structure by mild treatment with detergents (45). Hybridization experiments have shown that *B. pertussis* harbors only one copy of *fimD* (45); thus, while the major subunits of *B. pertussis* fimbriae may show antigenic variation, the minor component does not and is incorporated into antigenically diverse fimbriae. FimD exhibits similarity in amino acid sequence to a number of minor fimbrial subunits of other bacterial species which are known to contain receptor-binding sites. No evidence for additional minor fimbrial subunits has been found, suggesting that, compared with most other fimbriae, those of *B. pertussis* are simple in composition, comprising only two subunits: FimD and either Fim2 or Fim3.

GENE ORGANIZATION AND BIOSYNTHESIS

Like most *B. pertussis* virulence factors, the *fha* and *fim* genes are positively controlled by the *bvg* locus (see elsewhere in this volume). The *bvg, fha,* and *fim* genes are clustered on the chromosome and constitute three independent transcriptional units (Fig. 1).

The structural gene for FHA, *fhaB*, codes for a precursor protein (designated FhaB) with a molecular size of 367 kDa, of which approximately the first 2,000 N-terminal amino acid residues make up FHA (5, 6, 28). The 5′ end of *fhaB* codes for a putative signal peptide, suggesting that FhaB is transported across the cytoplasmic membrane by the signal peptide-dependent pathway. A second *fha* gene, *fhaC,* is transcribed independently from *fhaB*, together with the *fim* genes (47) (Fig. 1). Mutations in *fhaC* completely abolish FHA production but do not affect the synthesis of fimbriae (47). FhaC shows sequence similarities with ShlB and HpmB of *Serratia marcescens* and *Proteus mirabilis,* respectively (25, 40, 47). Both ShlB and HpmB are located in the outer membrane and are involved in the activation and secretion of the *S. marcescens* and *P. mirabilis* hemolysins ShlA and HpmA, respectively. ShlB interacts with the N terminus of ShlA, and it is significant that the N terminus of FhaB shows similarities with this region (5). Taken together, these observations suggest that FhaC is located in the outer membrane of *B. pertussis* and is involved in the transport of FhaB across this barrier. Presumably, transport requires an interaction between the N terminus of FhaB and FhaC.

The function of the approximately 1,600 C-terminal amino acids of FhaB (designated FhaB2) which are removed to form FHA is not clear. This region may have an accessory role in FHA synthesis like secretion and anchorage. It is also possible that FhaB2 has a distinct role in adherence, and it is interesting that, like FHA, FhaB2 contains an RGD sequence. FhaB2 is essential for FHA synthesis, since strains carrying a frameshift mutation in the part of *fhaB* which codes for FhaB2 do not contain FHA, suggesting that the truncated product is degraded (13). A function for FhaB2 in the anchorage of FHA was suggested by Domenighini et al. (6) on the basis of the fact that the C terminus of this molecule is proline

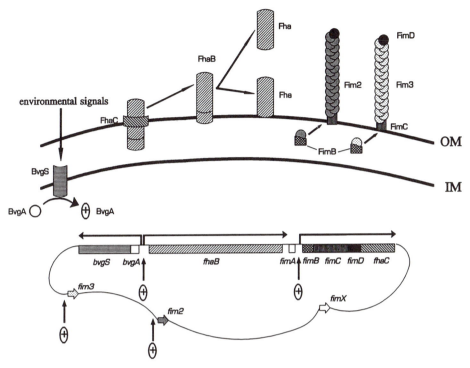

FIGURE 1. Organization and function of *bvg, fha,* and *fim* genes. Genes are represented by boxes, and arrowed lines indicate transcriptional units. The *bvgA* and *bvgS* genes code for a two-component sensory transduction system which translates environmental signals into transcriptional control of the *fha* and *fim* genes. BvgA is a positive regulator which binds to the promoter regions of the *fha* and *fim* genes. The *fhaB* gene codes for a large precursor molecule, which is transported across the inner (IM) and outer membrane (OM), after which it remains attached to the cell surface by means of its C-terminal region. This region is cleaved, and the resulting molecule (FHA) may be released from the cell or remain loosely associated with it. The *fhaC* gene codes for a protein which is required to transport FhaB across the outer membrane. The *fimA* gene probably represents the ancestral major fimbrial subunit gene. It contains a large deletion and is therefore inactive. The *fimB* gene codes for a periplasmic chaperon, which associates with fimbrial subunits to prevent nonproductive polymerization. The *fimC* gene codes for the anchorage protein, which is involved in the transport, assembly, and anchorage of fimbrial subunits. The *fimD* gene codes for the minor fimbrial subunit, which has been tentatively placed at the tip of the fimbrial structure. The *fim2, fim3,* and *fimX* genes code for the major fimbrial subunits and probably arose by duplication of *fimA*. The *fimX* gene is not expressed because of an inactive promoter.

rich, a feature found in a number of cell wall-associated proteins. This assumption is consistent with studies (2) which revealed a large precursor form of FHA, possibly the primary translation product of *fhaB,* that was anchored to the cell wall. This cell wall-associated molecule showed many of the properties of FHA such as binding to eukaryotic cells. Eventually, the large molecule was cleaved and FHA was released from the cell surface. It was suggested that this detachment of FHA would allow the bacteria to dissociate from the host cell and spread to other locations. FhaB2 was found to be associated only with bacterial cells and was eventually degraded. In conclusion, at present, a role of FhaB2 in anchorage is most consistent with the available data.

A cluster of four fimbrial genes, *fimABCD*, is located downstream of *fhaB* (14, 48) (Fig. 1). The *fimA* gene is homologous to the three major subunit genes located elsewhere on the chromosome (see below), but it contains a deletion of approximately 160 bp at its 5′ end and is therefore silent. It seems likely that *fimA* represents the primordial major subunit gene in *B. pertussis* and that the three major subunit genes were derived from *fimA* by gene duplication. This assumption is consistent with the observation that in many fimbrial operons, the major fimbrial subunit occupies the same relative position as *fimA*. Interestingly, complete *fimA* genes have been detected in *Bordetella parapertussis* and *Bordetella bronchiseptica* (22). Downstream of *fimA*, the *fimB* and *fimC* genes code for the chaperon and anchorage proteins, respectively. These accessory fimbrial proteins are involved in the transport, assembly, and anchorage of fimbrial subunits (9). Finally, *fimD* codes for the minor fimbrial subunit.

The observation that insertions in the *fim* cluster diminish or abolish FHA production has been taken as evidence that the *fimBCD* genes are involved in the biosynthesis of FHA (14). However, it is now clear that mutations in the *fim* cluster affect FHA production indirectly by decreasing the transcription or translation of *fhaC*, which is located downstream of *fimD*. Nonpolar, in-frame mutations in *fimBCD* affect only the production of fimbriae (47, 48). Insertion mutations in the region downstream of *fhaC* did not significantly affect FHA or fimbria production, suggesting that this region does not contain additional *fha* or *fim* genes (47).

Three major fimbrial subunit genes, *fim2*, *fim3*, and *fimX*, have been identified in *B. pertussis* (12, 21, 24) and are scattered around the chromosome (36). The *fim2* and *fim3* genes code for the serotype 2 and 3 fimbriae, respectively, while *fimX* is probably not expressed in most *B. pertussis* strains because of a deletion in its promoter region (see below). All three genes show a high degree of sequence similarity (approximately 60% identity).

REGULATION

By transferring the *fhaB* and *bvg* genes to *Escherichia coli*, Miller et al. (18) have shown that the *fhaB* promoter is directly activated by the *bvg* gene products. DNA retardation and DNase I protection experiments (30) have shown that the regions upstream of *fhaB* and *bvgA* are able to bind BvgA. Both of these regions contain the sequence TTTCCTA, which is part of an inverted repeat upstream of *fhaB* and a direct repeat upstream of *bvgA*. An oligonucleotide containing this sequence was shown to bind BvgA. Interestingly, very similar sequences are found in the promoter regions of the major fimbrial subunit genes and upstream of *fimB* (Fig. 2). In fact, binding of BvgA to the *fim3* promoter and the region located between *fimA* and *fimB* has been observed (3, 38). These observations suggest that *fhaB* and the *fim* cluster form distinct transcriptional units. Mutations in the *fim* cluster indicate that *fimBCD* and *fhaC* are part of a single transcriptional unit

```
                    ---------------------
fhaB   GCCGGACAGGGTTTGATGG-TTTGACTAAGAAATTTCCTACAAGTCTTGTATAAATATC   132 ATG
fim3   AG.C.C.C.CCGA.CTG.CGCGATTACCG.C..A....C...CAA.CATC.GCCCCCC.    69 ATG
bvg    C...CCGC....GCC.GC.CAGGA.T.C...CT.......TTTTGG.ATAGG.T.CTGT   143 ATG
fimB   A.G...TG.TCGGCTTGTCACGA..T.C-TTGT.....-.TTTC.T.CT.T.C.C.CGG   153 ATG
```

FIGURE 2. Nucleotide sequence similarities in the promoter region of *bvg*-activated genes. The *fhaB* sequence shown to bind to BvgA is overlined. Dots indicate sequence identities with the *fhaB* promoter region. Dashes refer to gaps introduced to increase the number of matches. The numbers refer to the distance to the initiation codons. The sequences of the following genes were obtained from previous reports: *fhaB* (30), *fim3* (21), *bvg* (30), and *fimB* (48).

and that *fimC*, *fimD*, and *fhaC* are translationally coupled (47). Thus, the production of FHA and fimbriae is intimately linked. The coregulation of these two adhesins suggests that they closely cooperate (see below).

In addition to the coordinate regulation by *bvg*, the major fimbrial genes are subject to a second type of control which occurs independently of the control of other virulence genes and involves random events (46). These random events involve insertion or deletion of cytosines in a run of 15 cytosine residues located in the promoter area. Presumably, these mutations affect the spacing between the site for a transcriptional activator, possibly BvgA, and the −10 box (46). The mutations in the reiterated sequence probably occur by a mechanism called slipped-strand mispairing (37). The *fimX* promoter region contains a relatively large deletion of approximately 9 bases in the C stretch (46), and this may explain why it is silent. Activation of *fimX* is an unlikely event, since it would require a single large insertion or many successive small insertions. The mutations in the *fim2* and *fim3* promoter region result in fimbrial phase variation, i.e., the random switch from one fimbrial serotype to another, and probably serve to evade the immune response of the host. Analogous mechanisms of transcriptional control have now been found in a number of other bacteria (29). Interestingly, fimbriae are the only *B. pertussis* surface molecules known to show intrastrain antigenic variation, suggesting that fimbrial antibodies may be more efficient than antibodies directed against other antigens in impeding the ability of *B. pertussis* to maintain itself in human populations.

After a shift from 22 to 37°C, *bvg*-regulated genes show different kinetics of induction: the *fha* and *fim* genes are activated within minutes, whereas the genes coding for pertussis toxin and adenylate cyclase toxin are activated after a lag of several hours (34). The early activation of the *fha* and *fim* genes suggests that they are involved in the initial steps in the pathogenesis of pertussis.

STRUCTURE AND FUNCTION

The FHA molecule has at least three distinct binding activities (i.e., for sulfated saccharides, the integrin CR3, and lactosylceramides) which have been located in different regions of the molecule (Table 1).

The region involved in binding to sulfated saccharides, such as heparin and heparan sulfate, has been mapped at the N terminus of the FHA molecule (16, 17). Since heparin is an inhibitor of FHA-mediated hemagglutination, it is conceiv-

TABLE 1
Binding features of the FHA molecule[a]

Location in FHA[b]	Receptor	Target	Function
1–400	Sulfated sugars	Mucus, epithelial cells, extracellular matrix	Attachment
1097–1099	CR3 integrin	Macrophages	Invasion
1141–1279	Lactosylceramides	Macrophages, ciliary cells	Attachment

[a] See text for references.
[b] Numbers refer to the positions of the amino acid residues in FhaB.

able that this activity is also mediated by the heparin-binding site of FHA. Sulfated saccharides are a major component of the extracellular matrix and the mucus of the respiratory tract and are also found on the surfaces of epithelial and many other cells (15, 50). Thus, recognition of sulfated saccharides may allow *B. pertussis* to adhere to respiratory mucus and epithelial cells. Indeed, Brennan et al. (4) have shown that binding of *B. pertussis* to epithelial cells can be inhibited by sulfated sugars. Adherence to the extracellular matrix may become important after it has become unmasked because of the extrusion of epithelial cells that have been damaged by *B. pertussis* toxins.

FHA binds to the integrin CR3 ($\alpha_m\beta_2$, CD11b/CD18) which is present on macrophages, and it has been shown that the RGD sequence present in FHA is important for this binding (28). Recently, it has also been shown that FHA is able to bind purified CR3 (41). Binding of *B. pertussis* to CR3 triggers uptake of the bacterium into the macrophage (33), and phagocytosis by means of CR3 may be beneficial to the bacterium since it does not generate an oxidative burst (49). The presence of CR3 on host cells as such is not sufficient for binding of FHA, and there is evidence that CR3 must be activated before FHA is able to bind (8). Two *B. pertussis* proteins have been implicated in the activation of CR3, pertussis toxin, and fimbriae. Exposure of macrophages to pertussis toxin increased the level of binding of *B. pertussis* to CR3 (42). This effect was localized to the B oligomer, and thus was not attributable to cellular intoxication by pertussis toxin. The monocyte receptor responsible for pertussis toxin-mediated activation of CR3 has not been identified. Studies performed by Hazenbos et al. (7) indicate that cross-linking of the integrin VLA-5 enhances the attachment of *B. pertussis* to monocytes by augmenting the binding activity of CR3. Those authors propose that the attachment to monocytes occurs in two steps: binding and cross-linking of VLA-5 by *B. pertussis* and then FHA-mediated binding to activated CR3. There is evidence that fimbriae mediate the first step in this process, i.e., binding to VLA-5 (see below). Thus, both pertussis toxin and fimbriae may act synergistically with FHA.

The carbohydrate recognition domain (CRD) of FHA binds to lactosylceramides and confers the ability to bind to ciliary cells and macrophages (27, 33). The CRD has been located between amino acid residues 1141 and 1279 (26). An 18-kDa polypeptide corresponding to this region was able to bind to lactosylceramide in a manner identical to that of native FHA. The binding to lactosylceramide is a property shared by FHA and the S2 subunit of pertussis toxin. The CRD domain of the S2 subunit has been mapped to a short segment near the N terminus (32), and this region shows sequence similarities with the CRD of FHA (26). Thus, the two proteins may use similar motifs to recognize their common target on ciliary cells and macrophages. The CRD domain is not directly involved in the uptake of *B. pertussis* by macrophages (33), but since it may bind to the same receptor as pertussis toxin, it is possible that binding of CRD to macrophages upregulates CR3 and thus contributes indirectly to uptake.

B. pertussis fimbriae bind to monocytes through the integrin VLA-5, and this binding may serve to facilitate uptake of *B. pertussis* in macrophages (7). Binding to monocytes could be inhibited by the minor fimbrial subunit FimD, suggesting that FimD recognizes VLA-5. It is not clear whether the major subunit is also involved in binding to VLA-5. Possibly, major and minor subunits are involved in binding to distinct receptors, as has been shown for a number of fimbriae (43).

BEHAVIOR OF MUTANTS IN ANIMAL MODELS

Information about the function of FHA and fimbriae has also been obtained by studying mutants in animal models. *B. pertussis* fimbrial mutants are less well able than the wild-type strain to colonize the tracheas of mice (20, 44). Interestingly, the mutant strain was not affected in its ability to colonize the nasopharynx or lungs (20). FHA mutants are impeded in their ability to persist in both the tracheas and nasopharynxes of mice (11, 20). No effect of FHA mutations was found on the persistence of *B. pertussis* in the lungs of mice (11, 20). The fact that both fimbrial and FHA mutants are affected in their ability to persist in the trachea suggests that fimbriae and FHA may cooperate not only in adherence to and invasion of macrophages but also in colonization of the trachea.

Studies in rabbits have shown that *B. pertussis* colonizes the lung as two approximately equal populations, one residing extracellularly and the other residing intracellularly, in pulmonary macrophages (33). Loss of the CR3 interaction, either by mutation of FHA or treatment with antibody to CR3, disrupted accumulation of viable bacteria inside pulmonary macrophages, confirming the crucial role that FHA plays in the invasion of macrophages. It has been suggested that this intracellular niche may contribute to the establishment of a carrier state (33).

MIMICRY

On the basis of the assumption that FHA may mimic the natural ligands of CR3, it was investigated whether there was a structural similarity between FHA and these molecules (31, 39). The natural ligands for CR3 include unknown determinants on endothelial cells which participate in leukocyte transmigration across endothelial barriers, complement component C3bi, and factor X of the coagulation cascade. Interestingly, anti-FHA monoclonal antibodies were found to cross-react with the vascular side of cerebral microvessels, and some of them were able to inhibit transmigration of leukocytes into the cerebral fluid. By using immunoblotting, two polypeptides were detected in cerebral capillary preparations with anti-FHA monoclonal antibodies. These observations indicate that FHA shares similarities in structure and function with endothelial proteins which interact with lymphocytes. Cross-reactivity of anti-FHA monoclonal antibodies was also observed with C3bi. A similarity in amino acid sequence between FHA and C3bi was found in a stretch of 11 amino acid residues, and this region was shown to be involved in the binding of a cross-reacting monoclonal antibody. Similarities in amino acid sequence were also observed between factor X and the C terminus of FHA.

Mimicry of natural receptor-ligand interactions may be beneficial for the pathogen in a number of ways. First, as illustrated by the interaction of FHA and fimbriae with CR3 and VLA-5, respectively, this mimicry may allow the pathogen to subvert host signal transduction to its own benefit. Second, mimicry may result in immunological tolerance. Third, if tolerance against the pathogen's determinant does not occur, an immunological response against it may still affect the ability of the host to clear the pathogen because of autoimmunity. And, finally, it may be expected that there is a strong selective pressure to conserve host structures which are involved in receptor-ligand interactions, because a mutation in one component (the receptor) would generally require a suppressor mutation in the other component (the ligand) to prevent dysfunction. Thus, in comparison with other surface molecules, it may be difficult to change the structure of natural receptors in order to escape from pathogens. Nevertheless, the host is probably subject to strong selective pressures to change the targets for pathogens. It has indeed been proposed that molecular mimicry is a major force in the generation of the diversity of host proteins which interact with pathogens (23).

CONCLUSIONS

It is intriguing that a region of the *B. pertussis* chromosome comprising 27 kb contains three linked loci involved in virulence: *bvg*, *fha*, and *fim*. The clustering of these loci is especially striking, because other virulence genes of *B. pertussis* are not found to be linked on the chromosome (36). The *fha* and *fim* genes are not only linked but also intermingled; one *fha* gene is located in the *fim* operon. The organization of these genes suggests that the fimbrial genes may have been inserted in the ancestral *fha* operon, between *fhaB* and *fhaC*. FHA production is coupled to fimbrial expression, and the intimate coregulation of *fha* and *fim* genes suggests that they cooperate in infecting the host. This assumption is corroborated by in vivo and in vitro studies which revealed that FHA and fimbriae may cooperate in colonization of the trachea and in invasion of macrophages. There is evidence that other virulence factors also cooperate with FHA. Like fimbriae, pertussis toxin may facilitate the uptake of *B. pertussis* into macrophages by upregulating the CR3 receptor. Also, it has been proposed that pertactin and FHA interact with each other and that binding to some eukaryotic receptors requires this interaction (2).

ACKNOWLEDGMENTS. I thank Jan van Embden, Cecile Geuijen, and Rob Willems for stimulating discussions and Camille Locht, Wouter Hazenbos, and Ralph van Furth for communicating unpublished work.

This work was supported by grant 900-514-106 and 900-514-135 from the Dutch Organization for Scientific Research.

REFERENCES

1. **Arai, H., and Y. Sato.** 1976. Separation and characterization of two distinct hemagglutinins contained in purified leukocytosis-promoting factor from *Bordetella pertussis. Biochem. Biophys. Acta* **444:**765–782.

2. **Arico, B., S. Nuti, V. Scarlato, and R. Rappuoli.** 1993. Adhesion of *Bordetella pertussis* to eukaryotic cells required a time-dependent export and maturation of filamentous hemagglutinin. *Proc. Natl. Acad. Sci. USA* **90:**9204–9208.

3. **Bertin, P., P. Boucher, M. C. Geoffroy, R. J. L. Willems, C. Geuijen, F. R. Mooi, and C. Locht.** 1993. Regulation of the genes involved in the biogenesis of FHA and fimbriae. *Biologicals* **21:** 31–32.

4. **Brennan, M. J., J. H. Hannah, and E. Leininger.** 1991. Adhesion of *Bordetella pertussis* to sulfatides and to the GalNAc beta4Gal sequence found in glycosphingolipids. *J. Biol. Chem.* **266:** 18827–18831.

5. **Delisse-Gathoye, A., C. Locht, F. Jacob, M. Raaschou-Nielsen, I. Heron, J. Ruelle, M. De Wolde, and T. Cabezon.** 1990. Cloning partial sequence, expression, and antigenic analysis of the filamentous hemagglutinin gene of *Bordetella pertussis. Infect. Immun.* **58:**2895–2905.

6. **Domenighini, M., D. Relman, C. Capiau, S. Falkow, A. Prugnoloa, V. Scarlato, and R. Rappuoli.** 1990. Genetic characterization of *Bordetella pertussis* filamentous haemagglutinin: a protein processed from an unusually large precursor. *Mol. Microbiol.* **4:**787–800.

7. **Hazenbos, W. L. W., C. Geuijen, B. M. van den Berg, F. R. Mooi, and R. van Furth.** Submitted for publication.

8. **Hazenbos, W. L. W., B. M. van den Berg, and R. van Furth.** 1993. Very late antigen-5 and complement receptor type 3 cooperatively mediate the interaction between *Bordetella pertussis* and human monocytes. *J. Immunol.* **151:**6274–6282.

9. **Hultgren, S. J., S. Abraham, M. Caparon, P. Falk, J. W. St. Geme III, and S. Normark.** 1993. Pilus and nonpilus bacterial adhesins: assembly and function in cell recognition. *Cell* **73:**887–901.

10. **Irons, L. I., L. A. E. Ashworth, and A. Robinson.** 1985. Release and purification of fimbriae from *Bordetella pertussis.* Proc. 4th Int. Symp. Pertussis, Geneva, 1984. *Dev. Biol. Stand.* **61:**153–163.

11. **Kimura, A., K. T. Mountzouros, D. D. Relman, S. Falkow, and J. L. Cowell.** 1990. *Bordetella pertussis* filamentous hemagglutinin: evaluation as a protective antigen and colonization factor in a mouse respiratory infection model. *Infect. Immun.* **58:**7–16.

12. **Livey, I., C. J. Duggleby, and A. Robinson.** 1987. Cloning and nucleotide sequence of the serotype 2 fimbrial subunit gene of *Bordetella pertussis. Mol. Microbiol.* **1:**203–209.

13. **Locht, C., P. Bertin, F. D. Menozzi, and G. Renauld.** 1993. The filamentous hemagglutinin, a multifaceted adhesin produced by virulent *Bordetella* spp. *Mol. Microbiol.* **9:**653–660.

14. **Locht, C., M. C. Geoffroy, and G. Renauld.** 1992. Common accessory genes for the *Bordetella pertussis* filamentous hemagglutinin and fimbriae share sequence similarities with the *papC* and *papD* gene families. *EMBO J.* **11:**3175–3183.

15. **Mawhinney, T. P., E. Adelstein, D. A. Morris, A. M. Mawhinney, and G. J. Barbero.** 1987. Structure determination of five sulfated oligosaccharides derived from tracheobronchial mucus glycoproteins. *J. Biol. Chem.* **262:**2994–3001.

16. **Menozzi, F. D., C. Gantiez, and C. Locht.** 1991. Interaction of the *Bordetella pertussis* filamentous hemagglutinin with heparin. *FEMS Microbiol. Lett.* **78:**59–64.

17. **Menozzi, F. D., R. Mutombo, G. Renauld, C. Gantiez, J. H. Hannah, E. Leininger, M. J. Brennan, and C. Locht.** 1994. Heparin-inhibitable lectin activity of the filamentous hemagglutinin adhesin of *Bordetella pertussis.* Infect. Immun. **62:**769–778.

18. **Miller, J. F., C. R. Roy, and S. Falkow.** 1989. Analysis of *Bordetella pertussis* virulence gene regulation by the use of transcriptional fusions in *Escherichia coli. J. Bacteriol.* **171:**6345–6348.

19. **Mooi, F. R.** *Bordetella pertussis* fimbriae. *In* P. Klemm (ed.), *Fimbriae, Aspects of Adhesion, Genetics, Biogenesis and Vaccines,* in press. CRC Press, Inc., Boca Raton, Fla.

20. **Mooi, F. R., W. H. Jansen, H. Brunings, H. Gielen, H. G. J. van der Heide, H. C. Walvoort, and P. A. M. Guinee.** 1992. Construction and analysis of *Bordetella pertussis* mutants defective in the production of fimbriae. *Microb. Pathog.* **12:**127–135.

21. **Mooi, F. R., A. ter Avest, and H. G. J. van der Heide.** 1990. Structure of the *Bordetella pertussis* gene coding for the serotype 3 fimbrial subunit. *FEMS Microbiol. Lett.* **66:**327–332.

22. **Mooi, F. R., and H. G. J. van der Heide.** Unpublished data.

23. **Murphy, P. M.** 1993. Molecular mimicry and the generation of host defense protein diversity. *Cell* **72:**823–826.

24. **Pedroni, P., B. Riboli, F. de Ferra, G. Grandi, S. Toma, B. Arico, and R. Rappuoli.** 1988. Cloning of a novel pilin-like gene from *Bordetella pertussis:* homology to the *fim2* gene. *Mol. Microbiol.* **2:**539–543.

25. **Pool, K., E. Schiebel, and V. Braun.** 1988. Molecular characterization of the hemolysin determinant of *Serratia marcescens. J. Bacteriol.* **170:**3177–3188.

26. **Prasad, S. M., Y. Yin, E. Rodzinski, E. T. Tuomanen, and R. Masure.** 1993. Identification of a carbohydrate recognition domain in filamentous hemagglutinin from *Bordetella pertussis. Infect. Immun.* **61:**2780–2785.

27. **Relman, D., E. Tuomanen, S. Falkow, D. T. Golenbock, K. Saukkonen, and S. D. Wright.** 1990. Recognition of a bacterial adhesin by an integrin: macrophage CR ($\alpha_M\beta_2$, CD11b/CD18) binds filamentous hemagglutinin of *B. pertussis. Cell* **61:**1375–1382.

28. **Relman, D. A., M. Dominighini, E. Tuomanen, R. Rappuoli, and S. Falkow.** 1989. Filamentous hemagglutinin of *Bordetella pertussis:* nucleotide sequence and crucial role in adherence. *Proc. Natl. Acad. Sci. USA* **86:**2637–2641.

29. **Robertson, B. D., and T. F. Meyer.** 1992. Genetic variation in pathogenic bacteria. *Trends Genet.* **8:**422–427.

30. **Roy, C. R., and S. Falkow.** 1991. Identification of *Bordetella pertussis* regulatory sequences required for transcriptional activation of the *fhaB* gene and autoregulation of the *bvgAS* operon. *J. Bacteriol.* **173:**2385–2392.

31. **Sandros, J., and E. Tuomanen.** 1993. Attachment factors of *Bordetella pertussis:* mimicry of eukaryotic cell recognition molecules. *Trends Microbiol.* **1:**192–196.

32. **Saukkonen, K., W. N. Burnette, V. L. Mar, H. R. Masure, and E. I. Tuomanen.** 1992. Pertussis toxin has eukaryotic-like carbohydrate recognition domains. *Proc. Natl. Acad. Sci. USA* **89:** 118–122.

33. **Saukkonen, K., C. Cabellos, M. Burroughs, S. Prasad, and E. Tuomanen.** 1991. Integrin-mediated localization of *Bordetella pertussis* within macrophages—role in pulmonary colonization. *J. Exp. Med.* **173:**1143–1149.

34. **Scarlato, V., B. Arico, A. Prugnola, and R. Rappuoli.** 1991. Sequential activation and environmental regulation of virulence genes in *Bordetella pertussis. EMBO J.* **10:**3971–3975.

35. **St. Geme, J. W., S. Falkow, and S. J. Barenkamp.** 1993. High molecular weight proteins of nontypable *Haemophilus influenzae* mediate attachment to human epithelial cells. *Proc. Natl. Acad. Sci. USA* **90:**2875–2879.

36. **Stibitz, S., and T. T. Garletts.** 1992. Derivation of a physical map of the chromosome of *Bordetella pertussis* Tohama I. *J. Bacteriol.* **174:**7770–7777.

37. **Streisinger, G., and J. E. Owen.** 1984. Mechanisms of spontaneous and induced frameshift mutations in bacteriophage T4. *Genetics* **109:**633–659.

38. **Top, J., P. Bertin, P. Boucher, F. R. Mooi, and C. Locht.** Unpublished data.

39. **Tuomanen, E. I., S. M. Prasad, J. S. George, A. I. M. Hoepelman, P. Ibsen, I. Heron, and R. M. Starzyk.** 1993. Reversible opening of the blood-brain barrier by anti-bacterial antibodies. *Proc. Natl. Acad. Sci. USA* **90:**7824–7828.

40. **Uphoff, T. S., and R. A. Welch.** 1990. Nucleotide sequencing of *Proteus mirabilis* calcium-independent hemolysin genes (*hpmA* and *hpmB*) reveals sequence similarity with the *Serratia marcescens* hemolysin genes (*shlA* and *shlB*). *J. Bacteriol.* **172:**1206–1216.

41. **Van Strijp, J. A. G., D. G. Rusell, E. Tuomanen, E. J. Brown, and S. D. Wright.** 1993. Ligand

specificity of purified complement receptor three (CD11b/CD18, $\alpha_m\beta_2$, Mac-1): indirect effects of an arg-gly-asp (RGD) sequence. *J. Immunol.* **151**:3324–3336.

42. **Van't Wout, J., W. N. Burnette, V. L., Mar, E. Rozdzinski, S. D. Wright, and E. I. Tuomanen.** 1992. Role of carbohydrate recognition domains of pertussis toxin in adherence of *Bordetella pertussis* to human macrophages. *Infect. Immun.* **60**:3303–3308.

43. **Westerlund, B., and T. K. Korhonen.** 1993. Bacterial proteins binding to the mammalian extracellular matrix. *Mol. Microbiol.* **9**:687–694.

44. **Willems, R. J. L.** 1993. Genetic and molecular studies on *Bordetella pertussis* fimbriae. Thesis. Rijks Universiteit, Utrecht, The Netherlands.

45. **Willems, R. J. L., C. Geuijen, H. G. J. van der Heide, M. Matheson, A. Robinson, L. Versluis, R. Ebberink, J. Theelen, and F. R. Mooi.** 1993. Isolation of a putative fimbrial adhesin from *Bordetella pertussis* and the identification of its gene. *Mol. Microbiol.* **9**:623–634.

46. **Willems, R. J. L., P. Paul, H. G. J van der Heide, A. ter Avest, and F. R. Mooi.** 1990. Fimbrial phase variation in *Bordetella pertussis:* a novel mechanism for transcriptional regulation. *EMBO J.* **9**:2803–2809.

47. **Willems, R. J. L., C. Geuijen, G. Renauld, P. Bertin, H. G. J. van der Heide, W. M. R. van de Akker, C. Locht, and F. R. Mooi.** Mutational analysis of the Bordetella pertussis *fim/fha* gene cluster: identification of a gene with sequence similarities to haemolysin accessory genes involved in export of FHA. *Mol. Microbiol.* **11**:337–347.

48. **Willems, R. J. L., H. G. J. van der Heide, and F. R. Mooi.** 1992. Characterization of a *Bordetella pertussis* fimbrial gene cluster which is located directly downstream of the filamentous hemagglutinin gene. *Mol. Microbiol.* **6**:2661–2671.

49. **Wright, S. D., and S. Silverstein.** 1993. Receptors for C3b and C3bi promote phagocytosis but not release of toxic oxygen from human phagocytes. *J. Exp. Med.* **158**:2016–2023.

50. **Yanagishita, M., and V. C. Hascall.** 1992. Cell surface heparan sulfate proteoglycans. *J. Biol. Chem.* **267**:9451–9454.

51. **Zhang, J. M., J. L. Cowell, A. C. Steven, P. H. Carter, C. C. McGrath, and C. R. Manclark.** 1985. Purification and characterization of fimbriae isolated from *Bordetella pertussis. Infect. Immun.* **48**:422–427.

Molecular Genetics of Bacterial Pathogenesis
Edited by V. L. Miller, J. B. Kaper, D. A. Portnoy, and R. R. Isberg
© 1994 American Society for Microbiology, Washington, DC 20005

Chapter 11

Early Events in the Pathogenesis of *Haemophilus influenzae* Disease

Joseph W. St. Geme III

Haemophilus influenzae is a fastidious, nonmotile, gram-negative bacterium. This organism was first isolated by Pfieffer (35) during the 1892 influenza pandemic; because of its frequent presence in postmortem cultures from patients dying of influenza, it was assumed to be the causative agent of this disease and was originally called the "influenza bacillus" (35). However, subsequent studies demonstrated the fallacy of this conclusion, and the organism was ultimately given the name *Haemophilus*, in recognition of the fact that hemin is required for growth. Nevertheless, *H. influenzae* has been clearly implicated as the cause of a number of human diseases, including both localized and systemic infections.

Isolates of *H. influenzae* are classified according to their polysaccharide capsule (37). There are a total of six structurally and antigenically distinct capsular types (serotypes), designated a through f. In addition, strains may be nonencapsulated; these strains are defined on the basis of their failure to agglutinate with typing antisera against capsular serotypes a through f and are considered nontypeable. The relationship between typeable and nontypeable strains remains poorly defined. In the past, nontypeable isolates of *H. influenzae* were considered by some to be phenotypic variants of serotype b strains (25, 41). However, studies by multilocus enzyme electrophoresis indicate that most nontypeable strains are distinguished from common type b cell lines (34, 38).

Historically, *H. influenzae* type b has been a leading cause of bacterial meningitis and other systemic diseases in infants and young children (12). However, in recent years efforts to develop a vaccine based on the type b capsule have come to fruition (11), and all infants residing in the United States should now receive a series of immunizations against *H. influenzae* type b (2, 11). Recent studies indicate that, as a consequence of immunization, the incidence of systemic type b disease has declined by more than 90% (1, 10, 33). On the other hand, the existing *H. influenzae* vaccines fail to protect against non-type b strains. Nontypeable *H. influenzae* represents a frequent etiology of localized respiratory tract disease,

Joseph W. St. Geme III • Department of Molecular Microbiology, Washington University School of Medicine, 660 South Euclid Avenue, Box 8230, St. Louis, Missouri 63110.

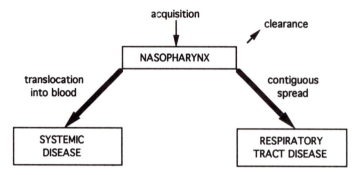

Figure 1. Pathogenic sequence for disease caused by *H. influenzae*.

including otitis media, sinusitis, conjunctivitis, bronchitis, and pneumonia (57). In addition, nontypeable strains are an occasional cause of serious systemic disease, including endocarditis, meningitis, and a recently described fulminant sepsis syndrome called Brazilian purpuric fever (7, 8, 47).

Disease caused by *H. influenzae* is believed to begin with colonization of the upper respiratory mucosa (Fig. 1). In most cases colonization persists in the absence of symptoms (45). On occasion bacteria penetrate the nasopharyngeal epithelial barrier and enter the bloodstream. If these organisms are able to survive and replicate within the intravascular space, bacteremia may be followed by dissemination to distant sites. Alternatively, in certain circumstances, colonization is followed by contiguous spread within the respiratory tract. Localization in the middle ear, the sinuses, the conjunctiva, or the lower respiratory tract results in disease at these sites.

Successful pharyngeal colonization requires that an organism overcome the mucociliary escalator of the respiratory tract. One strategy employed by *H. influenzae* to surmount mucociliary clearance involves attachment to the respiratory epithelium. Both pilus and nonpilus attachment factors appear to facilitate this process. In this chapter, I review the current understanding of *H. influenzae* hemagglutinating pili and the nontypeable *Haemophilus* high-molecular-weight adhesive proteins.

H. INFLUENZAE HEMAGGLUTINATING PILI

Bacterial pili are hair-like polymeric structures that protrude from the surface of the cell and are expressed by most gram-negative organisms (28). These surface organelles generally have adhesive properties and recognize specific host cell receptors (28).

In 1982, Guerina et al. (23) and Pichichero et al. (36) independently reported the presence of long peritrichous pili in selected isolates of *H. influenzae* type b. Both groups of investigators noted a correlation between piliation and increased levels of attachment to human oropharyngeal epithelial cells. In addition, piliation was associated with a capacity to agglutinate human erythrocytes. Consistent with the possibility that *Haemophilus* pili function as adhesins, several investigators have found that antipilus antibodies block in vitro adherence (18, 61). More recently, Weber et al. (65) demonstrated that in 1-year-old monkeys, the density of nasopharyngeal colonization was greater for piliated *H. influenzae* type b than for isogenic nonpiliated organisms containing a mutation in the structural pilin gene.

Prevalence of Hemagglutinating Pili

Although many *H. influenzae* type b isolates are devoid of hemagglutinating pili at the time of clinical isolation, in the vast majority of cases, piliated forms can be recovered following appropriate enrichment techniques (54). Studies of nontypeable *H. influenzae* indicate that these strains are also capable of expressing hemagglutinating pili. Brinton et al. (9) examined a series of nasopharyngeal isolates of nontypeable *H. influenzae* and found that 35% expressed hemagglutinating pili. Moreover, Southern blotting revealed that 95% of these strains contained at least some of the genetic material required for piliation (9), suggesting that even nonpiliated isolates may have the capacity for pilus expression. We examined a set of diverse *H. influenzae* biogroup aegyptius strains, including the Brazilian purpuric fever case clone (53). Initial analysis including electron microscopy and assessment of hemagglutinating ability demonstrated that 50% of these strains were piliated. Similar to the observations by Brinton and coworkers (9), Southern analysis confirmed the presence of a structural pilin gene in all strains in our collection. Consistent with this observation, piliated variants could be selected from each of the nonpiliated strains.

Structure

H. influenzae hemagglutinating pili are helical structures approximately 5 nm in width and up to 450 nm in length (Fig. 2). The major structural pilin subunit ranges in size between 22 and 27 kDa, depending on the strain. Whether the mature pilus contains minor subunits as well is unknown. At least 14 serological groups of hemagglutinating pili have been distinguished by enzyme-linked immunosorbent assay and whole-cell agglutination (9). While the majority of type b strains express pili belonging to the same serotype (serotype 4), nontypeable strains exhibit considerable diversity in this respect (9). Gilsdorf and colleagues (22) have characterized antisera directed against intact pili from two different type b strains, M43 and Eagan. Those workers found that rabbit antiserum specific for native pili from strain M43 bound to 18 of 21 unrelated piliated type b isolates in an immunodot assay (22). Sera prepared against purified pili from strain Eagan reacted with 11 of the 21 strains by the same assay (22). In contrast, neither the M43 nor the Eagan antiserum recognized any of 12 piliated nontypeable strains (20).

The Major Structural Pilin Subunit, HifA

The gene encoding the major structural pilin has now been cloned and sequenced from four different type b strains (M43p$^+$ [AO2], 770235f$^+$b^0, MinnA, and Eagan) and from three nontypeable strains (M37 and two BPF case clone

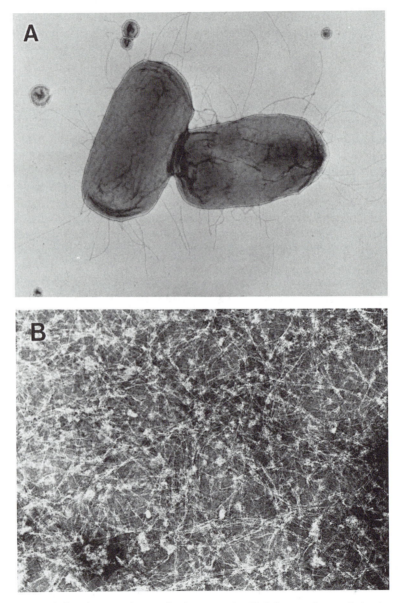

Figure 2. Transmission electron micrographs demonstrating *H. influenzae* hemagglutinating pili after negative staining with uranyl acetate. (A) Long, flexible, peritrichous hemagglutinating pili on the surface of *H. influenzae* type b Eagan. Magnification, ×27,000. (B) Purified hemagglutinating pili from strain Eagan. Magnification, ×65,000.

isolates designated F3031 and F3037) (13, 19, 21, 31, 51, 62, 66). In the type b strains there is evidence for a single gene, while in the nontypeable strains (M37, F3031, and F3037) Southern hybridization results suggest the possibility of two structural pilin genes. Whether both copies in the nontypeable isolates are complete and functional is unknown. Analysis of the nucleotide sequence of the pilin genes reveals that they are highly homologous to one another. The M43p$^+$, 770235f$^+$b^0, and MinnA genes are absolutely identical to one another and 80% identical to the genes from the other five strains. Analysis of the derived amino acid sequences demonstrates that the first 15 amino acids of the leader peptide are completely conserved among all of these strains and that the first 20 amino acids are the same for the four type b pilins and the two Brazilian purpuric fever clone pilins. Like pilin proteins from other gram-negative organisms, the *H. influenzae* pilins contain two cysteine residues and a penultimate tyrosine. Consistent with this finding, the mature pilins demonstrate significant sequence similarity with several *Escherichia coli* pilins, including PapA, F17, and type 1C. There is also significant homology with type 3 MR/K pili of *Klebsiella pneumoniae,* serotype 2 *Bordetella pertussis* pili, and *Serratia marcescens* US46 mannose-resistant pili.

Chaperone Protein Involved in Pilus Biogenesis

In general, the assembly and surface localization of pili requires multiple genes which are clustered together on the bacterial chromosome. For example, 11 genes clustered in an operon structure are involved in the biogenesis and expression of functional *E. coli* P pili (26). In the case of *H. influenzae* hemagglutinating pili, Smith and colleagues (44) have recently identified a gene designated *hifB* which is upstream of the pilin structural gene and is transcribed in the opposite direction. This gene shows significant homology to the *papD* gene of the P pilus operon. Extensive work by Hultgren and coworkers (27) has established that *papD* is a periplasmic chaperone that functions to stabilize pilin subunits in assembly-competent complexes prior to their ordered incorporation into the growing pilus. In support of the possibility that *hifB* is required for the assembly of *Haemophilus* pili, insertional inactivation of this gene eliminates the expression of pili (44, 64). In order to confirm a role for the *hifB* product as a chaperone protein, we prepared a series of constructs in *E. coli* ORN103, a strain that lacks any pilus chaperone (49). The results indicate that in the absence of *hifB* the structural pilin subunit is degraded in the periplasm. As predicted, HifB forms stable complexes with the structural pilin subunit, holding this subunit in a native conformation. Of note, PapD is known to consist of two immunoglobulin-like domains oriented in a boomerang shape such that a cleft is formed (24); the subunit binding site lies within this cleft and involves a critical solvent-exposed arginine at residue 8 (R8) of the mature protein (43). On the basis of molecular modeling, HifB appears to have a similar immunoglobulin-like structure and the same critical arginine residue within the cleft (27). Attempts to crystallize HifB are under way in order to confirm its three-dimensional structure.

Bacterial Usher Required for Pilus Assembly

Recently, Watson and colleagues (64) reported the identification of a third gene essential for the expression of hemagglutinating pili. Analysis of a series of mutants that produced structural pilin subunits but that were unable to express native pili revealed a 2.5-kb open reading frame upstream of the structural pilin gene and immediately downstream of the *hifB* gene. This locus has been designated *hifC* and encodes a protein with significant homology to pilus ushers, including PapC of the *E. coli* P pilus system (28%), FimD of the type 1 pilus system (29%), FimC of *B. pertussis* (30%), and MrkC of *K. pneumoniae* (30%). The derived amino acid sequence of the *hifC* gene includes two cysteine residues located 27 amino acid residues apart near the N terminus and two cysteines at the C terminus; this arrangement is common to all known pilus ushers. PapC represents the most thoroughly studied pilus usher and is believed to provide a structural platform for the assembly of pilins subunits into the growing pilus by facilitating interactions with the already assembled subunits (29). By analogy, HifC presumably plays a role in receiving structural pilin subunits from the HifB chaperone and directing their incorporation into hemagglutinating pili.

Phase Variation

Early on, investigators noted that expression of hemagglutinating pili was subject to reversible phase variation (14, 23, 36). During natural infection with *H. influenzae* type b, nasopharyngeal isolates are often piliated, while their isogenic counterparts from systemic sites are virtually always nonpiliated (32). At the same time, systemic isolates can be enriched in vitro for piliated variants, which themselves may revert to nonpiliated forms. On the basis of an analysis by Farley et al. (17), the frequency of variation is believed to be roughly 10^{-3} to 10^{-4} in both directions. An initial clue to the mechanism of phase variation came in a study by Langermann and Wright (31). These workers examined isogenic piliated and nonpiliated variants by Southern and Northern analyses and found that nonpiliated forms contained the structural pilin gene but failed to express this gene, indicating that expression of pili is controlled at the transcriptional level. More recently, van Ham et al. (63) reported that the promoter regions of the divergently transcribed structural pilin and *hifB* genes overlap; the area of overlap contains a variable number of TA repeats. In *H. influenzae* type b strains AM20 and AM30, nonpiliated variants contained 9 TA units in the reiterated sequence and failed to express either the structural pilin gene or *hifB;* in contrast, piliated variants had 10 TA repeats and expressed stable transcripts (63). Several lines of evidence suggest that the changes in the number of TA repeats in the structural gene-*hifB* promoter complex are caused by the process of slipped-strand mispairing (63).

Receptor Specificity and Tissue Tropism

Experiments by van Alphen and coworkers (59, 60) demonstrated that pilus-mediated hemagglutination correlates with erythrocyte expression of the AnWj antigen (formerly the Anton antigen) and is inhibited by anti-AnWj serum. On the other hand, human oropharyngeal epithelial cells lack this antigen, and pilus-dependent attachment to these cells is unaffected by anti-AnWj serum (59). However, the same antipilus monoclonal antibody is able to inhibit agglutination of erythrocytes as well as binding to oropharyngeal epithelial cells (61), suggesting the possibility that pili may recognize the same receptor structure on both erythrocytes and epithelial cells. Consistent with this possibility, recent work has demonstrated that compounds containing sialyllactosylceramide are able to inhibit hemagglutination and adherence to epithelial cells. Such experiments have been performed with piliated type b and nontypeable strains, including isolates of the Brazilian purpuric fever clone (49, 51, 58). Specifically, the gangliosides GM1, GM2, GM3, and GD1a in nanomolar concentrations are effective inhibitors. In contrast, the asialo derivative of GM1, sialyllactose, and sialoglycoproteins lack inhibitory activity. The relationship between the results of these inhibitor studies and the receptor epitope with which pili interact remains to be established. Similarly, the portion of the pilus involved in the adhesive interaction is unclear.

Recently, Sterk et al. (46) examined the capacity of a piliated *H. influenzae* type b strain to bind to a variety of fixed human tissues. As expected, these workers observed efficient binding to upper respiratory tract and pulmonary epithelia that was inhibited by an antipilus monoclonal antibody. In contrast, there was minimal binding to kidney and liver cells. These results suggest that hemagglutinating pili exhibit tissue specificity, presumably a reflection of the tissue distribution of the appropriate receptor structure.

NONTYPEABLE *H. INFLUENZAE* HIGH-MOLECULAR-WEIGHT ADHESINS

In the case of *H. influenzae* type b, several studies suggest the presence of adhesins distinct from hemagglutinating pili. Despite the growing evidence for a second adhesin among type b strains, the identity of this factor remains obscure.

Among nontypeable *H. influenzae* there is also evidence for adhesins separate from hemagglutinating pili. As mentioned earlier in this chapter, Brinton et al. (9) found that only 35% of nasopharyngeal isolates of nontypeable *H. influenzae* expressed hemagglutinating pili, an observation that implies the existence of other colonization factors. Bakaletz and coworkers (3) examined 60 clinical isolates of nontypeable *H. influenzae* and determined that all 60 expressed surface fibers of one or another morphology; however, there was no correlation between the expression of surface fibers and a capacity to bind to respiratory epithelium (3), suggesting the presence of nonfilamentous attachment factors. Using a model that employs human nasal turbinate tissue in organ culture, Read et al. (39) demonstrated that piliation was not required for attachment to the mucosal surface. We made similar observations using a tissue culture model (50).

Identification of High-Molecular-Weight Proteins

A number of studies indicate that serum bactericidal activity develops in the course of *H. influenzae* otitis media (15, 42). In a series of experiments aimed at defining the bacterial targets of this bactericidal activity, Barenkamp and Bodor (5) identified a group of nontypeable *H. influenzae* high-molecular-weight surface-exposed proteins. To further characterize these proteins, Barenkamp and Leininger (6) isolated and sequenced the genes encoding the two such proteins (HMW-1 and HMW-2) expressed by a prototypic nontypeable strain (strain 12) (6). The

HMW-1 protein is 125 kDa in size and is encoded by a 4.6-kb gene, while the HMW-2 protein has an apparent molecular mass of 120 kDa and is encoded by a 4.4-kb open reading frame. The first 1,259 bp of the two genes are identical; thereafter, the sequences diverge somewhat but are 80% identical overall, and the derived amino acid sequences show 70% identity. In order to determine the prevalence of antigenically related proteins among diverse *H. influenzae* strains, 125 epidemiologically distinct nontypeable isolates were examined by Western immunoblotting analysis by using a polyclonal antiserum raised against recombinant HMW-1 (6). This antiserum recognized both HMW-1 and HMW-2 and reacted with one or two high-molecular-mass (100 to 150 kDa) proteins in 75% of the isolates. Examination of a series of type b strains failed to reveal cross-reactive high-molecular-weight proteins (6, 49), suggesting that these proteins are unique to nontypeable strains.

Role in Adherence

On the basis of their derived amino acid sequences, both HMW-1 and HMW-2 show sequence similarity to filamentous hemagglutinin (6), a critical adherence factor of *B. pertussis* (30, 40). Immunoblot studies demonstrate that the HMW-1 and HMW-2 proteins are also antigenically related to filamentous hemagglutinin (6). To test the hypothesis that the high-molecular-weight proteins are functionally related to filamentous hemagglutinin and function as adhesins, initially we confirmed by immunogold electron microscopy that these proteins are surface exposed (Fig. 3). Subsequently, we constructed mutants of nontypeable *H. influen-*

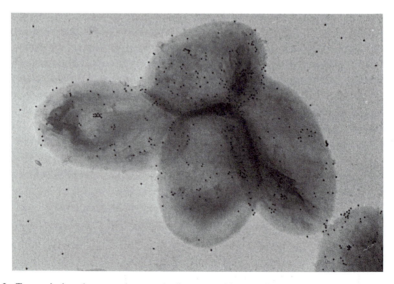

FIGURE 3. Transmission electron micrograph of nontypeable *H. influenzae* 12 after incubation of bacteria with a polyclonal antiserum reactive with HMW-1 and HMW-2 and then incubation with protein A-conjugated 10-nm-diameter gold particles and negative staining with uranyl acetate. Note the intense labeling, consistent with surface exposure of HMW-1 and HMW-2. Magnification, ×36,000.

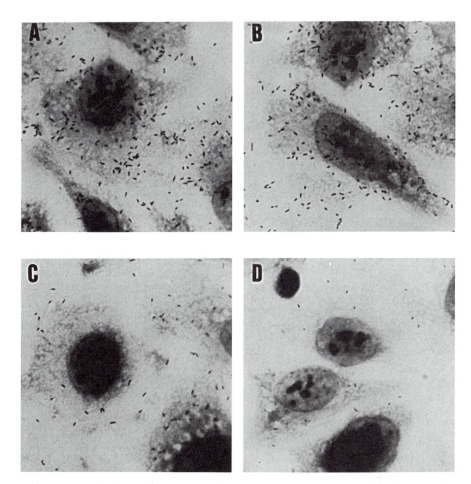

FIGURE 4. Light micrographs of nontypeable *H. influenzae* 12 derivatives incubated with Chang conjunctival epithelial cells and stained with Giemsa. (A) Parental strain 12; (B) HMW-2 mutant (HMW-1$^+$, HMW-2$^-$); (C) HMW-1 mutant (HMW-1$^-$, HMW-2$^+$); (D) HMW-1 and HMW-2 double mutant (HMW-1$^-$, HMW-2$^-$). Magnification, ×550. Reprinted with permission (52).

zae that are deficient in the expression of HMW-1 or HMW-2, or both, and examined these strains for their capacities to bind to cultured conjunctival epithelial cells (52). As shown in Fig. 4, adherence by the parent strain was quite efficient, with an average of more than 70 adherent organisms per cell. Adherence by the mutant expressing HMW-1 but not expressing HMW-2 (HMW-2$^-$) was also very efficient and comparable to that by the parent strain. In contrast, attachment by the strain expressing HMW-2 but not HMW-1 (HMW-1$^-$) was significantly decreased relative to that by the parent. Adherence by the double mutant (HMW-1$^-$, HMW-2$^-$) was decreased even further, roughly 50-fold relative to that by the parent, and nearly threefold compared with that by the HMW-1$^-$ mutant in quantitative assays.

To confirm an adherence function for both HMW-1 and HMW-2 and to evaluate the effects of these proteins independently of other *Haemophilus* surface structures, the *hmw-1* and *hmw-2* loci were introduced into *E. coli* DH5α (52), a normally nonadherent strain. DH5α expressing HMW-1 exhibited high-level attachment, and DH5α expressing HMW-2 was capable of an intermediate level of attachment (Fig. 5). In order to determine whether the high-molecular-weight proteins of other nontypeable strains also mediate attachment to cultured conjunctival cells, mutants were constructed in a second strain (52). Nontypeable *H. influenzae* 5 expresses two high-molecular-weight proteins that are antigenically related to HMW-1 and HMW-2 of strain 12 and to filamentous hemagglutinin. Preliminary nucleotide sequence analysis indicates that one of these proteins more

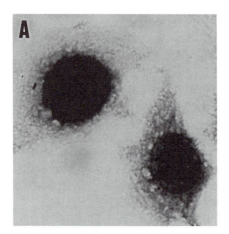
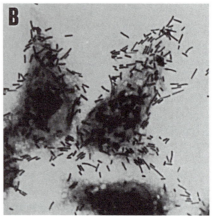
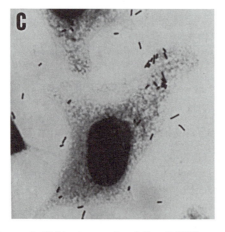

Figure 5. Light micrographs of *E. coli* DH5α expressing HMW-1 or HMW-2 incubated with Chang conjunctival epithelial cells and stained with Giemsa stain. (A) DH5α containing vector alone; (B) DH5α containing *hmw-1* locus; (C) DH5α containing *hmw-2* locus. Magnification, ×550. Reprinted with permission (52).

TABLE 1
Adherence to cultured cells by *E. coli* DH5α expressing the *H. influenzae* high-molecular-weight proteins HMW-1 or HMW-2

Cell line	Adherence[a]	
	DH5α (HMW-1)	DH5α (HMW-2)
Chang (human conjunctiva)	128.5 ± 17.9	15.8 ± 4.2
HEp-2 (human larynx)	72.3 ± 9.5	24.3 ± 3.8
KB (human oral cavity)	64.4 ± 2.9	2.6 ± 0.4
ME-180 (human cervix)	13.8 ± 1.7	6.0 ± 0.5
Hec-IB (human endometrium)	3.0 ± 0.4	11.0 ± 0.2

[a] Adherence values were determined in a 30-min assay by dividing the number of adherent bacteria by the number of inoculated bacteria. For the purposes of comparison, these values were normalized to adherence by DH5α lacking expression of HMW-1 or HMW-2, which was arbitrarily defined as 1.0. Numbers represent the means ± standard errors of the means of measurements made in triplicate from representative experiments. Reprinted with permission (26).

closely resembles HMW-1 (HMW-1-like), while the second is more similar to HMW-2 (HMW-2-like). Consistent with the findings with strain 12 and HMW-1 and HMW-2, loss of expression of the HMW-1-like protein resulted in a markedly decreased capacity to adhere, and the absence of expression of both the HMW-1-like and the HMW-2-like proteins was associated with a further decrease in attachment.

Cellular Specificity in Binding

To evaluate HMW-1 and HMW-2 cellular specificity, adherence by DH5α expressing either HMW-1 or HMW-2 was examined by using a variety of cultured epithelial cell lines (26). As shown in Table 1, HMW-1 mediated very efficient attachment to cells derived from human conjunctiva (Chang), larynx (HEp-2), and oral cavity (KB). In contrast, HMW-1 promoted only moderate-level adherence to cells derived from human cervix (ME-180) and virtually no adherence to cells from human endometrium (Hec-1B). Expression of HMW-2 was associated with moderate levels of attachment to conjunctival, laryngeal, and endometrial cells but low-level attachment to cervical cells and minimal binding to cells from the oral cavity. Comparison of HMW-1-mediated versus HMW-2-mediated attachment to each of these cell lines revealed that HMW-1 promoted more efficient binding than did HMW-2 to conjunctival, laryngeal, oral cavity, and cervical cells; the opposite was true with cells derived from human endometrium. These findings suggest that HMW-1 and HMW-2 interact with specific receptor molecules whose distribution varies from one cell type to another. In addition, these observations suggest that the HMW-1 and HMW-2 proteins interact with distinct eukaryotic cell receptors and perhaps function at different steps in the process of colonization.

HMW-1 Receptor

To begin to characterize the HMW-1 receptor present on cultured conjunctival cells, we examined the effect on HMW-1-dependent adherence of either *meta-*

periodate or proteinase K treatment of epithelial monolayers (48). *meta*-Periodate oxidation under mildly acidic conditions selectively hydrolyzes surface carbohydrates (16, 67), while proteinase K removes surface proteins (16). Both periodate pretreatment and proteinase K digestion resulted in a significant reduction in adherence, suggesting that HMW-1 interacts with a glycoprotein receptor. Consistent with this conclusion, treatment of epithelial cells with peptide: *N*-glycosidase F, an enzyme that specifically cleaves asparagine-linked carbohydrates (55), also caused a marked decrease in attachment. Further analysis suggested that sialic acid makes up at least part of the receptor recognized by HMW-1. In particular, desialylation of the epithelial cell surface by using either very gentle periodate oxidation or *Clostridium perfringens* sialidase reduced binding. Similarly, lectins specific for terminal α2,3-linked sialic acid were capable of inhibiting HMW-1-mediated attachment. Interestingly, despite the homology with filamentous hemagglutinin (which is known to mediate binding to galactose and lactosamine [56]), galactose-containing carbohydrates failed to inhibit interaction between HMW-1 and cultured conjunctival cells.

HMW-1 and HMW-2 Accessory Genes

Translation of the *hmw-1* and *hmw-2* open reading frames predicts the expression of proteins of 159 and 155 kDa in size, respectively (6). Instead, mature HMW-1 is 125 kDa, and mature HMW-2 has an apparent molecular mass of 120 kDa (6). Barenkamp and Leininger (6) have partially purified the native 125- and 120-kDa proteins and performed N-terminal amino acid sequence analysis; the sequences correspond to the derived amino acid sequences of the *hmw-1* and *hmw-2* genes, each beginning at residue 442. These findings suggest that generation of the mature proteins involves removal of lengthy segments from the amino termini of the *hmw-1* and *hmw-2* gene products. In this context, it interesting that the HMW-1 and HMW-2 preproteins lack typical signal sequences.

Further work by Barenkamp and myself (4) indicates that the *hmw-1* and *hmw-2* open reading frames are both flanked by two downstream genes, designated *b* and *c*, respectively (4). The derived amino acid sequences of the *hmw-1b* and *hmw-2b* genes are 99% identical and have significant similarity to the derived amino acid sequences of *shlB* of *S. marcescens* and *hpmB* of *P. mirabilis*; *shlB* and *hpmB* encode the outer membrane proteins required for activation, surface localization, and secretion of the *Serratia* and *Proteus* hemolysins, respectively. Additional analysis of the predicted amino acid sequences of HMW-1B and HMW-2B reveals the presence of a typical signal sequence and a conserved ATP-binding motif in these proteins. The derived amino acid sequences of the *hmw-1c* and *hmw-2c* open reading frames are 96% identical and lack homology to known sequences. Neither HMW-1C nor HMW-2C possesses a signal peptide.

In the absence of the B and C proteins, appropriate processing of the HMW-1 and HMW-2 proteins fails to occur and these adhesive proteins are not surface localized (4, 49). These observations suggest that the B and C proteins are required for protein processing and export. Whether processing is a prerequisite for surface localization and adhesive function is unknown. Experiments to define the specific functions of the B proteins and the C proteins are in progress.

CONCLUSION

In this chapter I reviewed the current understanding of *H. influenzae* hemagglutinating pili and a family of adhesive high-molecular-weight proteins expressed by nontypeable strains. The existing evidence supports a role for these factors in the process of *H. influenzae* colonization. An understanding of the interrelationship between these factors and other determinants of pathogenicity should provide important insights into the mechanism of *H. influenzae* disease and may suggest new approaches toward disease prevention.

ACKNOWLEDGMENTS. The work included in this chapter was supported by Public Health Service grants 00850 and HD29678 from the National Institutes of Health.

REFERENCES

1. **Adams, W. G., K. A. Deaver, S. L. Cochi, B. D. Plikaytis, E. R. Zell, C. V. Broome, and J. D. Wenger.** 1993. Decline of childhood Haemophilus influenzae type b (Hib) disease in the Hib vaccine era. *JAMA* **269**:221–226.
2. **American Academy of Pediatrics Committee on Infectious Diseases.** 1991. *Haemophilus influenzae* type b conjugate vaccines: recommendations for immunization of infants and children 2 months of age and older: update. *Pediatrics* **88**:169–172.
3. **Bakaletz, L. O., B. M. Tallan, T. Hoepf, T. F. DeMaria, H. G. Birck, and D. J. Lim.** 1988. Frequency of fimbriation of nontypable *Haemophilus influenzae* and its ability to adhere to chinchilla and human respiratory epithelium. *Infect. Immun.* **56**:331–335.
4. **Barenkamp, S. J., and J. W. St. Geme III.** Genes encoding high-molecular-weight adhesion proteins of nontypeable *Haemophilus influenzae* are part of gene clusters. *Infect. Immun.*, in press.
5. **Barenkamp, S. J., and F. F. Bodor.** 1990. Development of serum bactericidal activity following nontypable *Haemophilus influenzae* acute otitis media. *Pediatr. Infect. Dis. J.* **9**:333–339.
6. **Barenkamp, S. J., and E. Leininger.** 1992. Cloning, expression, and DNA sequence analysis of genes encoding nontypeable *Haemophilus influenzae* high-molecular-weight surface-exposed proteins related to filamentous hemagglutinin of *Bordetella pertussis*. *Infect. Immun.* **60**:1302–1313.
7. **Brazilian Purpuric Fever Study Group.** 1987. Brazilian purpuric fever: epidemic purpura fulminans associated with antecedent purulent conjunctivitis. *Lancet* **ii**:757–761.
8. **Brazilian Purpuric Fever Study Group.** 1987. *Haemophilus aegyptius* bacteremia in Brazilian purpuric fever. *Lancet* **ii**:761–763.
9. **Brinton, C. C., M. J. Carter, D. B. Derber, S. Kar, J. A. Kramarik, A. C.-C. To, S. C.-M. To, and S. W. Wood.** 1989. Design and development of pilus vaccines for *Haemophilus influenzae* diseases. *Pediatr. Infect. Dis. J.* **8**:S54–S61.
10. **Broadhurst, L. E., R. L. Erickson, and P. W. Kelley.** 1993. Decreases in invasive *Haemophilus influenzae* diseases in US Army children, 1984 through 1991. *JAMA* **269**:227–231.
11. **Centers for Disease Control and Prevention.** 1993. Recommendations for use of *Haemophilus b* conjugate vaccines and a combined diphtheria, tetanus, pertussis, and *Haemophilus b* vaccine. *Morbid. Mortal. Weekly Rep.* **42**:1–15.
12. **Cochi, S. L., and C. V. Broome.** 1986. Vaccine prevention of *Haemophilus influenzae* type b disease: past, present and future. *Pediatr. Infect. Dis.* **5**:12–19.
13. **Coleman, T., S. Grass, and R. Munson, Jr.** 1991. Molecular cloning, expression, and sequence of the pilin gene from nontypeable *Haemophilus influenzae* M37. *Infect. Immun.* **59**:1716–1722.
14. **Connor, E. M., and M. R. Loeb.** 1983. A hemadsorption method for detection of colonies of *Haemophilus influenzae* type b expressing fimbriae. *J. Infect. Dis.* **148**:855–860.

15. Faden, F., L. Brodsky, J. Bernstein, J. Stanievich, D. Krystofik, C. Shuff, J. J. Hong, and P. L. Ogra. 1989. Otitis media in children. I. The systemic immune response to nontypable *Haemophilus influenzae*. *J. Infect. Dis.* **160**:999–1004.

16. Falk, P., K. A. Roth, T. Boren, T. U. Westblom, J. I. Gordon, and S. Normark. 1993. An *in vitro* adherence assay reveals that *Helicobacter pylori* exhibits cell lineage-specific tropism in the human gastric epithelium. *Proc. Natl. Acad. Sci. USA* **90**:2035–2039.

17. Farley, M. M., D. S. Stephens, S. L. Kaplan, and E. O. Mason, JR. 1990. Pilus- and non-pilus-mediated interactions of *Haemophilus influenzae* type b with human erythrocytes and human nasopharyngeal mucosa. *J. Infect. Dis.* **161**:274–280.

18. Forney, L. J., J. R. Gilsdorf, and D. C. L. Wong. 1992. Effect of pili-specific antibodies on the adherence of *Haemophilus influenzae* type b to buccal epithelial cells. *J. Infect. Dis.* **165**:464–470.

19. Forney, L. J., C. F. Marrs, S. L. Bektesh, and J. R. Gilsdorf. 1991. Comparison and analysis of the nucleotide sequences of pilin genes from *Haemophilus influenzae* type b strains Eagan and M43. *Infect. Immun.* **59**:1991–1996.

20. Gilsdorf, J. R., H. Y. Chang, K. W. McCrea, and L. O. Bakaletz. 1992. Comparison of hemagglutinating pili of *Haemophilus influenzae* type b with similar structures of nontypeable *H. influenzae*. *Infect. Immun.* **60**:374–379.

21. Gilsdorf, J. R., C. F. Marrs, K. W. McCrea, and L. J. Forney. 1990. Cloning, expression, and sequence analysis of the *Haemophilus influenzae* type b strain M43p⁺ pilin gene. *Infect. Immun.* **58**:1065–1072.

22. Gilsdorf, J. R., K. McCrea, and L. Forney. 1990. Conserved and nonconserved epitopes among *Haemophilus influenzae* type b pili. *Infect. Immun.* **58**:2252–2257.

23. Guerina, N. G., S. Langermann, H. W. Clegg, H. W. Kessler, D. A. Goldmann, and J. R. Gilsdorf. 1982. Adherence of piliated *Haemophilus influenzae* type b to human oropharyngeal cells. *J. Infect. Dis.* **146**:564.

24. Holmgren, A., and C.-I. Branden. 1989. Crystal structure of chaperone protein PapD reveals an immunoglobulin fold. *Nature* (London) **342**:248–251.

25. Huber, P. S., and I. N. Egwu. 1985. Capsular variation in experimental strains of *Haemophilus influenzae*. Med. Microbiol. Immunol. **173**:345–353.

26. Hultgren, S. J., S. Abraham, M. Caparon, P. Falk, J. W. St. Geme III, and S. Normark. 1993. Pilus and nonpilus bacterial adhesins: assembly and function in cell recognition. *Cell* **73**:887–901.

27. Hultgren, S. J., F. Jacob-Dubuisson, C. H. Jones, and C.-I. Branden. 1993. PapD and superfamily of periplasmic immunoglobulin-like pilus chaperones. *Adv. Protein Chem.* **44**:99–123.

28. Hultgren, S. J., and S. Normark. 1991. Biogenesis of the bacterial pilus. *Curr. Opin. Genet. Dev.* **1**:313–318.

29. Jones, C. H., F. Jacob-Dubuisson, K. Dodson, M. Kuehn, L. Slonim, R. Striker, and S. J. Hultgren. 1992. Adhesin presentation in bacteria requires molecular chaperones and ushers. *Infect. Immun.* **60**:4445–4451.

30. Kimura, A., K. T. Mountzouros, D. A. Relman, S. Falkow, and J. L. Cowell. 1990. *Bordetella pertussis* filamentous hemagglutinin: evaluation as a protective antigen and colonization factor in a mouse respiratory infection model. *Infect. Immun.* **58**:7–16.

31. Langermann, S., and A. Wright. 1990. Molecular analysis of the *Haemophilus influenzae* type b pilin gene. *Mol. Microbiol.* **4**:221–230.

32. Mason, E. O., S. L. Kaplan, B. L. Wiedermann, E. Pinna Norrod, and W. A. Stenback. 1985. Frequency and properties of naturally occurring adherent piliated strains of *Haemophilus influenzae* type b. *Infect. Immun.* **49**:98–103.

33. Murphy, T. V., K. E. White, P. Pastor, L. Gabriel, F. Medley, D. M. Granoff, and M. T. Osterholm. 1993. Declining incidence of *Haemophilus influenzae* type b disease since introduction of vaccination. *JAMA* **269**:246–248.

34. Musser, J. M., S. J. Barenkamp, D. M. Granoff, and R. K. Selander. 1986. Genetic relationships of serologically nontypable and serotype b strains of *Haemophilus influenzae*. *Infect. Immun.* **52**:183–191.

35. Pfeiffer, R. 1893. Die Aetiologie der Influenzaa. *Z. Hyg. Infect.* **13**:357–386.

36. Pichichero, M. E., P. Anderson, M. Loeb, and D. H. Smith. 1982. Do pili play a role in pathogenicity of *Haemophilus influenzae* type b? *Lancet* **ii**:960–962.

37. **Pittman, M.** 1931. Variation and type specificity in the bacterial species *Haemophilus influenzae*. *J. Exp. Med.* **53:**471–493.

38. **Porras, O., D. A. Caugant, B. Gray, T. Lagergard, B. R. Levin, and C. Svanborg-Eden.** 1986. Difference in structure between type b and nontypable *Haemophilus influenzae* populations. *Infect. Immun.* **53:**79–89.

39. **Read, R. C., R. Wilson, A. Rutman, V. Lund, H. C. Todd, A. P. R. Brain, P. K. Jeffery, and P. J. Cole.** 1991. Interaction of nontypable *Haemophilus influenzae* with human respiratory mucosa in vitro. *J. Infect. Dis.* **163:**549–558.

40. **Relman, D. A., M. Domenighini, E. Tuomanen, R. Rappuoli, and S. Falkow.** 1989. Filamentous hemagglutinin of *Bordetella pertussis:* nucleotide sequence and crucial role in adherence. *Proc. Natl. Acad. Sci. USA* **86:**2637–2641.

41. **Sell, S. H., and D. T. Karzon.** 1973. *Haemophilus influenzae.* Vanderbilt University Press, Nashville, Tenn.

42. **Shurin, P. A., S. I. Pelton, I. B. Tager, and D. L. Kasper.** 1980. Bactericidal antibody and susceptibility to otitis media caused by nontypable strains of *Haemophilus influenzae. J. Pediatr.* **97:**364–369.

43. **Slonim, L. N., J. S. Pinkner, C.-I. Branden, and S. J. Hultgren.** 1992. Interactive surface in the PapD chaperone cleft is conserved in pilus chaperone superfamily and essential in subunit recognition and assembly. *EMBO J.* **11:**4747–4756.

44. **Smith, A. L., L. Slonim, L. Forney, M. Chanyangam, S. Lohrke, S. Moseley, S. Hultgren, and J. Haas.** Unpublished data.

45. **Spinola, S. M., J. Peacock, F. W. Denny, D. L. Smith, and J. G. Cannon.** 1986. Epidemiology of colonization by nontypable *Haemophilus influenzae* in children: a longitudinal study. *J. Infect. Dis.* **154:**100–109.

46. **Sterk, L. M. T., L. van Alphen, L. Geelen-van den Broek, H. J. Houthoff, and J. Dankert.** 1991. Differential binding of *Haemophilus influenzae* to human tissues by fimbriae. *J. Med. Microbiol.* **35:**129–138.

47. **St. Geme, J. W., III.** 1993. Nontypeable *Haemophilus influenzae* disease: epidemiology, pathogenesis, and prospects for prevention. *Infect. Agents Dis.* **2:**1–16.

48. **St. Geme, J. W., III.** Submitted for publication.

49. **St. Geme, J. W., III.** Unpublished data.

50. **St. Geme, J. W., III, and S. Falkow.** 1990. *Haemophilus influenzae* adheres to and enters cultured human epithelial cells. *Infect. Immun.* **58:**4036–4044.

51. **St. Geme, J. W., III, and S. Falkow.** 1993. Isolation, expression, and nucleotide sequencing of the pilin structural gene of the Brazilian purpuric fever clone of *Haemophilus influenzae* biogroup aegyptius. *Infect. Immun.* **61:**2233–2237.

52. **St. Geme, J. W., III, S. Falkow, and S. J. Barenkamp.** 1993. High-molecular-weight proteins of nontypable *Haemophilus influenzae* mediate attachment to human epithelial cells. *Proc. Natl. Acad. Sci. USA* **90:**2875–2879.

53. **St. Geme, J. W., III, J. R. Gilsdorf, and S. Falkow.** 1991. Surface structures and adherence properties of diverse strains of *Haemophilus influenzae* biogroup aegyptius. *Infect. Immun.* **59:**3366–3371.

54. **Stull, T. L., P. M. Mendelman, J. E. Haas, M. A. Schoenborn, K. D. Mack, and A. L. Smith.** 1984. Characterization of *Haemophilus influenzae* type b pili. *Infect. Immun.* **46:**787–796.

55. **Tarentino, A. L., C. M. Gomez, and T. H. Plummer, Jr.** 1985. Deglycosylation of asparagine-linked glycans by peptide:*N*-glycosidase F. *Biochemistry* **24:**4665–4671.

56. **Tuomanen, E., H. Towbin, G. Rosenfelder, D. Braun, G. Hanson, G. Larson, and R. Hill.** 1988. Receptor analogs and monoclonal antibodies which inhibit adherence of *Bordetella pertussis* to human ciliated respiratory epithelial cells. *J. Exp. Med.* **168:**267–277.

57. **Turk, D. C.** 1984. The pathogenicity of *Haemophilus influenzae. J. Med. Microbiol.* **18:**1–16.

58. **van Alphen, L., L. Geelen van den Broek, L. Blaas, M. van Ham, and J. Dankert.** 1991. Blocking of fimbria-mediated adherence of *Haemophilus influenzae* by sialyl gangliosides. *Infect. Immun.* **59:**4473–4477.

59. **van Alphen, L., J. Poole, L. Geelen, and H. C. Zanen.** 1987. The erythrocyte and epithelial cell receptors for *Haemophilus influenzae* are expressed independently. *Infect. Immun.* **55:**2355–2358.

60. **van Alphen, L., J. Poole, and M. Overbeeke.** 1986. The Anton blood group antigen is the erythrocyte receptor for *Haemophilus influenzae*. *FEMS Microbiol. Lett.* **37**:69–71.

61. **van Alphen, L., N. van den Berghe, and L. Geelen van den Broek.** 1988. Interaction of *Haemophilus influenzae* with human erythrocytes and oropharyngeal epithelial cells is mediated by a common fimbrial epitope. *Infect. Immun.* **56**:1800–1806.

62. **van Ham, S. M., F. R. Mooi, M. G. Sindhunata, W. R. Maris, and L. van Alphen.** 1989. Cloning and expression in *Escherichia coli* of *Haemophilus influenzae* fimbrial genes establishes adherence to oropharyngeal epithelial cells. *EMBO J.* **8**:3535–3540.

63. **van Ham, S. M., L. van Alphen, F. R. Mooi, and J. P. M. van Putten.** 1993. Phase variation of H. influenzae fimbriae: transcriptional control of two divergent genes through a variable combined promoter region. *Cell* **73**:1187–1196.

64. **Watson, W. J., J. R. Gilsdorf, M. A. Tucci, K. W. McCrea, L. J. Forney, and C. F. Marrs.** 1994. Identification of a gene essential for piliation in *Haemophilus influenzae* type b with homology to the pilus assembly platform genes of gram-negative bacteria. *Infect. Immun.* **62**:468–475.

65. **Weber, A., K. Harris, S. Lohrke, L. Forney, and A. L. Smith.** 1991. Inability to express fimbriae results in impaired ability of *Haemophilus influenzae* b to colonize the nasopharynx. *Infect. Immun.* **59**:4724–4728.

66. **Whitney, A. M., and M. M. Farley.** 1993. Cloning and sequence analysis of the structural pilin gene of Brazilian purpuric fever-associated *Haemophilus influenzae* biogroup aegyptius. *Infect. Immun.* **61**:1559–1562.

67. **Woodward, M. P., W. W. Young, Jr., and R. A. Bloodgood.** 1985. Detection of monoclonal antibodies specific for carbohydrate epitopes using periodate oxidation. *J. Immunol. Methods* **78**:143–153.

Molecular Genetics of Bacterial Pathogenesis
Edited by V. L. Miller, J. B. Kaper, D. A. Portnoy, and R. R. Isberg
© 1994 American Society for Microbiology, Washington, DC 20005

Chapter 12

Molecular Pathogenesis of Enteropathogenic *Escherichia coli*

James B. Kaper

INTRODUCTION

Escherichia coli can cause gastrointestinal disease by a variety of mechanisms (68). This chapter focuses on the molecular pathogenesis of one specific category of diarrheagenic *E. coli,* the enteropathogenic *E. coli* (EPEC). The term "enteropathogenic *E. coli*" has sometimes been used in a nonspecific manner to denote any *E. coli* strain capable of causing enteric disease. However, most investigators in this field now use the term in a very specific manner to describe one particular pathogen. Because of the plethora of recent information concerning the pathogenesis of the various diarrheagenic *E. coli,* it may be useful to first briefly review the different categories of *E. coli* before focusing on EPEC.

Six Classes of Diarrheagenic *E. coli*

There are currently six groups of *E. coli* that have been found to be associated with gastrointestinal disease by case-control epidemiological studies. EPEC is a common cause of infant diarrhea and produces a characteristic attaching and effacing histopathology in intestinal epithelial cells (see below). Enterotoxigenic *E. coli* (ETEC) is a common cause of dehydrating diarrhea in children in developing countries and of traveler's diarrhea in adults from industrialized countries traveling to developing countries. ETEC produces heat-labile enterotoxin (LT) and/or heat-stable enterotoxin (ST) and one or more intestinal colonization factors, such as CFA/I and CS1 to CS6; at least 12 such putative colonization factors have been described. Enteroinvasive *E. coli* (EIEC) is very similar to *Shigella* spp. and shares essentially all of the genes involved in epithelial cell entry, escape from the phagocytic vacuole, and spread into adjacent cells (54). Unlike *Shigella dysenteriae,* EIEC does not produce the powerful cytotoxin called Shiga toxin, Shiga-like toxin, or verotoxin (see chapter 26). Enterohemorrhagic *E. coli* (EHEC) has received much attention in the United States recently as a cause of bloody and nonbloody

James B. Kaper • Center for Vaccine Development, University of Maryland School of Medicine, 10 South Pine Street, Baltimore, Maryland 21201.

diarrhea as well as hemolytic uremic syndrome. EHEC produces Shiga or Shiga-like toxins and also produces the attaching and effacing lesion seen with EPEC. Enteroaggregative *E. coli* (EAggEC) has been found to be associated with persistent diarrhea (i.e., more than 2 weeks in duration) in a number of countries. EAggEC was so named because these strains adhere to HEp-2 or HeLa cells in an aggregative pattern (80). EAggEC produces an ST-like toxin which increases intracellular cyclic GMP levels (91), an LT toxin of ca. 120 kDa which is reported to increase intracellular calcium (8), and a fimbrial colonization factor called AAF/I (79). Diffusely adherent *E. coli* (DAEC) has also been found to be associated with diarrhea in several studies. These strains adhere to HEp-2 or HeLa cells in a diffuse pattern (80), but few details about their pathogenic mechanisms are known. Two different adhesions have been described: the F18454 adhesin (12), which is related to the afimbrial adhesin AFA-I of uropathogenic *E. coli*, and an adhesin called AIDA-1, which shares homology with the VirG (IcsA) protein of *Shigella flexneri* (11). In addition to these six classes of diarrheagenic *E. coli*, there are other potential classes of diarrheagenic *E. coli* that produce cytolethal distending toxin (61) or cytotoxic necrotizing factor toxin (21). However, as yet, there are no case-control epidemiological data linking those *E. coli* classes with diarrhea.

History

EPEC was the first *E. coli* to be associated with diarrhea. Major credit goes to Bray (15) in England, who in 1945 reported that antigenically homogeneous *E. coli* was associated with summer diarrhea in infants, whereas *E. coli* which did not share these surface antigens was not asociated with diarrhea. For the next 30 to 40 years, the only way by which EPEC could be distinguished from nonpathogenic *E. coli* was by agglutination reactions with antisera prepared against the several O:H serotypes associated with diarrhea. Indeed, as new information appeared from the laboratories of Stan Falkow and other investigators showing that the LT and ST toxins of ETEC were encoded on plasmids, many investigators argued that EPEC strains were not actually pathogens in their own right but were ETEC strains which had lost their LT and ST plasmids. Other investigators held that EPEC strains were true pathogens that possessed virulence factors that had yet to be described (reviewed by Robins-Browne [87]). This issue was resolved by volunteer experiments conducted by Levine et al. (69), who showed that EPEC could cause diarrhea in adult volunteers if it was fed in high doses after ingestion of sodium bicarbonate to neutralize stomach acid.

Epidemiology

EPEC is a major cause of infant diarrhea worldwide. There is a strong age distribution, with the peak incidence occurring at about 6 months of age, and disease rarely occurs in children over 1 year of age (70). EPEC consistently ranks among the most important enteric pathogens among individuals in this age group, and in some studies EPEC is the most important bacterial or viral enteric pathogen (47, 48, 86). In some studies, mortality caused by EPEC infection has exceeded

30% (95). Large outbreaks of acute nosocomial diarrhea in developed countries are uncommon now, but there are reports of chronic diarrhea caused by EPEC in the United States, including some child-care center outbreaks (14, 18, 82, 90).

HISTOPATHOLOGY

Histopathological examination of intestinal epithelial cells infected with EPEC reveals a striking pattern characterized by effacement of microvilli and intimate adherence between the bacterium and the epithelial cell membrane. Directly beneath the adherent bacterium, marked cytoskeletal changes are seen, including the accumulation of polymerized actin, and the bacteria sometimes sit upon a pedestal-like structure. This pattern was termed "attaching and effacing" by Moon et al. (77) (Fig. 1) and has been seen in tissue culture cells (3, 60, 63), animal models (77, 84, 96, 99), and humans infected with EPEC (90, 97, 101). This histopathology is quite different from that seen with ETEC and *Vibrio cholerae*, in which the organisms adhere in a nonintimate fashion without causing microvillus

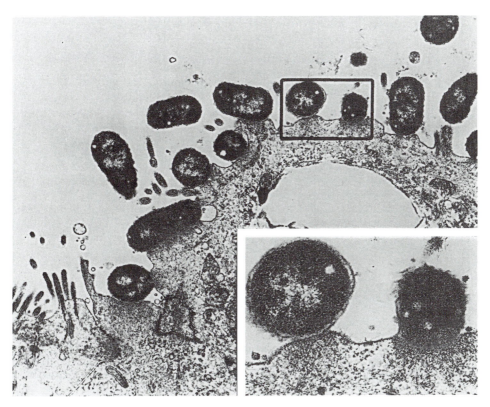

Figure 1. Attaching and effacing of intestinal epithelial cells from a gnotobiotic pig infected with EPEC. The inset is an enlargement of the box in the main part of the figure. Taken from Moon et al. (77).

effacement or actin polymerization. Similar attaching and effacing lesions are seen in animal models of EHEC (40, 100), children with diarrhea caused by *Hafnia alvei* (1), and mice with murine colonic hyperplasia caused by *Citrobacter freundii* (see chapter 13). Attaching and effacing activity in tissue culture cells can be detected by the fluorescent actin staining (FAS) test developed by Knutton et al. (64). In this test, fluorescein isothiocyanate-labeled phalloidin binds specifically to filamentous actin in the epithelial cells directly beneath the adherent bacteria.

Multiple steps are involved in producing this characteristic histopathology. Donnenberg and Kaper (27) have proposed a three-stage model of EPEC pathogenesis consisting of (i) localized adherence, (ii) signal transduction, and (iii) intimate adherence (Fig. 2). At least two separate bacterial adhesins are involved in the localized and intimate adherence. The identity of the bacterial factor(s) responsible

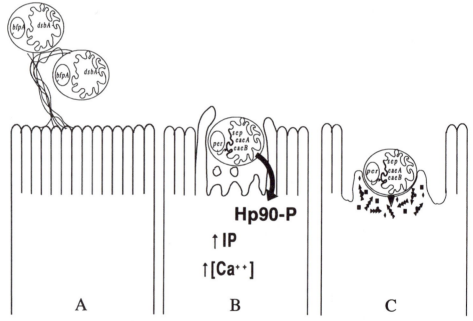

FIGURE 2. A three-stage model of EPEC pathogenesis (27). In the first stage (A), nonintimate adherence between the bacterium and the epithelial cells is mediated by the bundle-forming pilus (BFP) and probably other fimbriae. The *bfpA* structural gene is encoded on a large plasmid, and at least one additional gene required for formation of an active bundle-forming pilus is encoded on the chromosome (*dsbA*). In the second stage (B), a signal transduction event results in increased intracellular calcium levels, release of inositol phosphates, tyrosine phosphorylation of a 90-kDa epithelial cell protein (Hp90), and effacement of microvilli. Chromosomal *sep* genes encode a specialized protein secretion pathway which presumably allows presentation of the bacterial products triggering the signal transduction event. A plasmid-encoded regulator encoded by the *per* locus activates transcription of *eaeA*, *eaeB*, and other genes in the *per* regulon. In the third stage (C), intimate adherence of the bacterium to the epithelium is mediated by intimin (solid triangle), a 94-kDa outer membrane protein encoded by the *eaeA* locus. The intimate adherence of the bacterium amplifies the accumulation of filamentous actin and other cytoskeletal proteins (other geometric shapes) within the epithelial cell.

for the signal transduction triggering the dramatic cytoskeletal changes is not yet known. In this chapter, I discuss the various bacterial factors and genes involved in EPEC pathogenesis as well as the specific changes that occur in the host cell in response to EPEC infection. At the end of the chapter, I will return to the three-stage model and discuss it further in the overall context of EPEC pathogenesis.

LOCALIZED ADHERENCE

For several decades, the only way in which EPEC could even be presumptively identified was by serotyping. In 1979, Cravioto et al. (19) showed that EPEC, as defined by serotyping, could adhere to HEp-2 cells in culture, while most non-EPEC *E. coli* could not adhere to HEp-2 cells. EPEC adheres in a specific pattern whereby distinct microcolonies of bacteria rather than a uniform distribution of bacteria are seen on the HEp-2 cell. This pattern is called "localized adherence" (92) and is distinct from the aggregative or diffuse adherence patterns seen with other classes of diarrheagenic *E. coli* (80).

Baldini et al. (5) showed that the ability of EPEC E2348/69 (O127:H6) to adhere in a localized adherence pattern was dependent upon the presence of a 60-MDa plasmid designated pMAR2. Loss of this plasmid led to the loss of the localized adherence phenotype, and transfer of this plasmid to nonadherent *E. coli* HB101 allowed this strain to adhere to HEp-2 cells (although adherence was not as strong as that seen with the EPEC strain). An insertion mutation which inactivated localized adherence identified a plasmid region that was necessary for this phenotype and that led to the development of a diagnostic DNA probe (6, 78). This probe has been used in numerous epidemiological studies throughout the world to identify EPEC possessing 55- to 70-MDa plasmids encoding localized adherence, the so-called EPEC adherence factor (EAF) plasmids (32, 62, 72, 78). The EAF plasmids are fairly well conserved across different serotypes (81). The importance of this plasmid in human disease was shown by Levine et al. (71), who fed strain E2348/69 possessing the EAF plasmid and a derivative of this strain which had lost the EAF plasmid to adult volunteers. With the wild-type strain, diarrhea occurred in 9 of 10 volunteers (mean diarrheal stool volume, 1,178 ml), but diarrhea occurred in only 2 of 9 volunteers who ingested the cured derivative (mean stool volume, 433 ml; $P < 0.006$). Interestingly, although this plasmid is highly stable in vitro (<1% spontaneous plasmid cure rate), 67% of the challenge strain isolates recovered from volunteer stool specimens had lost this plasmid (71).

Bundle-Forming Pilus

The actual adhesin that mediates localized adherence was sought by several investigators for many years. In 1991, Girón et al. (45) described inducible fimbriae that were produced by EPEC and that tended to aggregate and form bundles, hence the name bundle-forming pilus (BFP). Antiserum prepared against purified BFP significantly but not completely reduced the localized adherence of EPEC B171 to HEp-2 cells. The N-terminal sequence of the purified pili showed homology to the toxin-coregulated pili of *V. cholerae*, a member of the type IV fimbrial family. Donnenberg et al. (25) identified the structural gene encoding the BFP by using a Tn*phoA* mutant of strain E2348/69 which no longer gave localized adherence. DNA sequence analysis of the *bfpA* gene revealed that the *V. cholerae tcpA* gene and the EPEC *bfpA* gene share 41% identity. In addition to a large stretch of plasmid DNA necessary to complement the Tn*phoA* insertion (between 5 and 15 kb), a chromosomal locus (*dsbA*) is also necessary for the expression of BFP. The *dsbA* gene encodes disulfide isomerase, which catalyzes disulfide bonds; inactivation of *dsbA* leads to the loss of localized adherence (107). The *bfpA* gene was used as a DNA probe to detect EPEC producing BFP, and a high correlation was seen between the presence of this gene and the expression of BFP (44). The results obtained with the *bfpA* probe and the EAF probe were nearly congruent, even though the *bfpA* and EAF probe sequences (the functions of which are still unknown) are separated by >20 kb on the pMAR2 plasmid.

Some EPEC isolates which cause the attaching and effacing lesion do not hybridize with the EAF or the *bfpA* probe and/or do not give localized adherence on HEp-2 or HeLa cells (44, 93, 94). Such strains are usually isolated from cases of sporadic diarrhea rather than from outbreaks. The percentage of EPEC strains which fall into this category varies greatly among different studies. In our own studies, my colleagues and I have found that ca. 90% of strains of EPEC serogroups which give a positive FAS test result hybridize to the EAF probe and exhibit localized adherence (58). In the United Kingdom, Scotland et al. (93) have suggested that localized adherence-positive, EAF probe-negative EPEC strains may be of greater importance than localized adherence-positive, EAF probe-positive strains. It is clear that an EPEC strain that lacks the EAF plasmid and localized adherence can still cause mild diarrhea in a minority of individuals, as shown in the adult volunteer studies by Levine et al. (71). However, case-control studies in Chile (72), Brazil (48, 49), and Thailand (32) found that only EAF-positive and/or localized adherence-positive strains of EPEC serogroups were associated with diarrhea. EPEC strains that did not react with the EAF probe, regardless of the FAS test result (32), were isolated at equal frequencies from healthy controls and infants with diarrhea. Strains not belonging to the EPEC serogroups but still reacting with the EAF probe and causing attaching and effacing were also not significantly associated with diarrhea. Thus, the available epidemiological and volunteer data indicate that to be fully virulent EPEC strains should (i) cause attaching and effacing, (ii) possess the EAF plasmid encoding the BFP, and (iii) belong to a recognized EPEC serogroup.

Other EPEC Fimbriae

There have been numerous reports of additional fimbrial structures produced by EPEC, some of which were subsequently shown to be type 1 fimbriae (for a review, see reference 66). Girón et al. (46) recently reported an extensive characterization of fimbriae produced by EPEC B171. In addition to BFP, this strain also produced rod-like fimbriae and fibrillae with subunit sizes of 16.5, 15.5, and 14.7 kDa. Their N-terminal amino acid sequences showed homology with the F9 and F7$_2$ fimbriae of uropathogenic *E. coli* (both are P fimbriae) and F1845 of diffusely adhering *E. coli,* respectively. Antisera prepared against a mixture of all three fimbriae (called FB171) reduced the adherence of strain B171 to HEp-2 cells by ca. 75% (46). By combining anti-FB171 and anti-BFP sera, localized adherence was inhibited by ca. 100%.

The need for multiple antisera to completely inhibit localized adherence and the results of ultrastructural studies of the "microcolonies" with the localized adherence phenotype indicate that this phenomenon is multifactorial (46). Scanning electron micrographs of the microcolonies reveal that multiple bacterium–HEp-2 cell and bacterium-bacterium interactions are involved. Thin fibers resembling fimbriae appear to link the bacterial cells and epithelial cells, while rope-like structures resembling BFP may primarily be involved in the bacterium-bacterium interactions. Construction of isogenic mutants specifically altered in genes encoding the various fimbrial structures of EPEC will be necessary before the localized adherence phenotype is completely understood.

INTIMATE ADHERENCE

EPEC *eaeA* Gene

Although localized adherence is associated with the presence of the 60-MDa pMAR2 plasmid, transfer of this plasmid into *E. coli* HB101 did not confer upon the recipient strain the ability to attach and efface (63). Moreover, EPEC cured of this plasmid could still produce attaching and effacing lesions on epithelial cells at a much lower frequency than that seen with the wild type (65). The development of the FAS test by Knutton et al. (64) provided an assay with which to begin genetic studies into the attaching and effacing phenotype. Jerse et al. (60) used Tn*phoA* to study this phenotype and isolated mutants which were deficient in the FAS test. When these mutants were added to the polarized intestinal epithelial cell line Caco-2, the intimate adherence of the bacteria to the eukaryotic membrane was not seen. These mutants still adhered to epithelial cells because of the presence of the plasmid-encoded adhesin on the EAF plasmid, but in the nonintimate pattern seen with *E. coli* HB101 containing the EAF plasmid (Fig. 3). The locus into which Tn*phoA* had inserted (named *eaeA* for *E. coli* attaching and effacing) was sequenced (60), and the predicted 102-kDa protein product was found to share 31% identity and 50% similarity with the invasin protein of *Yersinia pseudotuberculosis* (see chapter 15). Given the similarity of the *eaeA* gene product to invasin and the role of this factor in intimate adherence to epithelial cells, this protein was called intimin (Int). Alignment of intimin and invasin showed that the greatest homology is found within the N-terminal halves of the proteins, whereas the C-terminal portions of the proteins exhibited the least homology (Fig. 4) (106). Leong and colleagues (55, 67) have shown very elegantly that the C-terminal 192 amino acid residues of invasin bind to β$_1$ integrins. Recently, Frankel et al. (41) found that the C-terminal 280 residues of intimin contain receptor-binding activity. Furthermore, the binding of epithelial cells to invasin could be blocked by the addition of invasin but not by the addition of intimin, suggesting that intimin and invasin

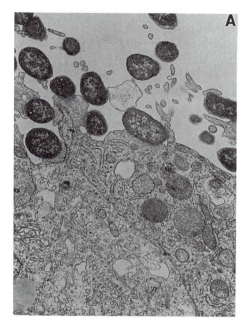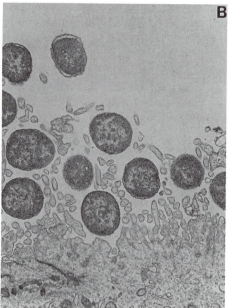

FIGURE 3. Loss of intimate adherence owing to mutation of the *eaeA* gene. (A) Intimate adherence of parent strain E2348/69 to Caco-2 cells. (B) Nonintimate adherence of isogenic *eaeA* deletion mutant CVD206. Taken from Donnenberg and Kaper (26).

do indeed bind to different receptors. A difference in receptor specificity would be consistent with the striking divergence at the C-terminal ends of intimin and invasin, but the available data do not definitively rule out or rule in the possibility that β_1 integrins or any other host cell molecule is the intimin receptor.

EHEC *eae* Gene

The *eaeA* gene from strain E2348/69 was used to probe a variety of enteric bacteria to detect the presence of homologous sequences. Of 99 strains of EPEC serogroups giving a positive FAS test result, 100% hybridized to the *eaeA* gene probe (58). The *eaeA* gene probe did not hybridize under low or high stringency to *Salmonella, Shigella,* or *Yersinia* spp., ETEC, EIEC, EAggEC, or DAEC, or normal-flora *E. coli* (60). The probe did hybridize to strains of EHEC, a cause of bloody and nonbloody diarrhea and hemolytic uremic syndrome. EHEC isolates produce an attaching and effacing lesion in newborn piglets that is very similar if not identical to that produced by EPEC isolates (98, 100). The sequence of the *eaeA* gene from EHEC EDL933 (O157:H7) shares 86 and 83% identities with the EPEC *eaeA* gene at the nucleotide and predicted amino acid sequence levels, respectively (106). Interestingly, the greatest divergence between the predicted protein products of these two genes is at their C-terminal ends (Fig. 5). Starting at the N terminus, the first 704 amino acid residues (making up 75% of the total)

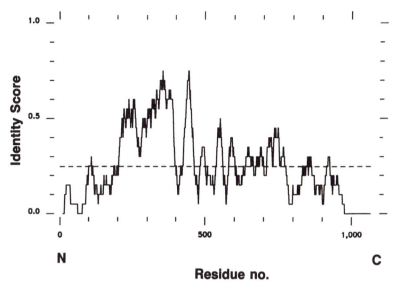

FIGURE 4. Alignment of predicted sequences of EPEC intimin and *Y. pseudotuberculosis* invasin (106). Sequences are displayed from the N terminus (left) to the C terminus by using the program PlotSimilarity from the Genetics Computer Group, Inc., Madison, Wis. The identity score is shown on the left axis, and the dashed line across the entire length of the proteins denotes the average identity between the proteins.

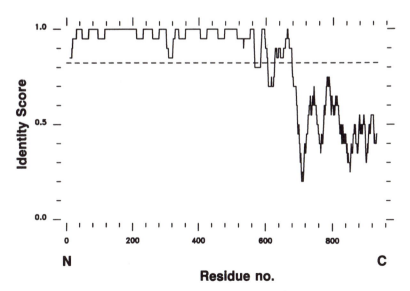

FIGURE 5. Alignment of predicted sequences of EPEC and EHEC intimins (106) arranged as described in the legend to Fig. 4.

share 94% identity, while the remaining 25% of the residues share only 49% identity. This difference in the putative receptor-binding region is intriguing, since EPEC is found primarily in the small intestine while EHEC is found primarily in the colon and distal ileum (98, 100). The C-terminal 280 residues of the intimin gene of EHEC O157 strains possess receptor-binding activity (41), and mutation of the *eaeA* gene in two different O157:H7 strains abolishes the ability to produce a positive FAS test result on HEp-2 cells (29, 73). Even within the EHEC category, divergence in the C-terminal end of intimin is seen. Hii et al. (53) sequenced the last 800 bp at the 3' ends of the *eaeA* genes in different EHEC strains and found 66% nucleotide similarity between O26:H11 and O157:H7 EHEC strains and 82% similarity between O111:H8 and O157:H7 EHEC strains.

eaeA in Other Species

Homologs of the EPEC *eaeA* gene have been found in other enteric pathogens that produce an attaching and effacing lesion. Elsewhere in this volume (chapter 13), David Schauer describes the role of *eaeA* in the pathogenesis of *C. freundii* biotype 4820, a cause of murine colonic hyperplasia. An *eaeA* homolog has also been found in strains of *H. alvei* associated with diarrhea in children in Bangladesh (1). The *eaeA* gene is also present in strains of *E. coli* which cause diarrhea in weanling rabbits, as exemplified by the strain RDEC-1 (60, 108). As with the intimin sequence of EPEC and EHEC strains, significant divergence in the C-terminal portions of the intimin sequences from these species is found.

eaeB

A second chromosomal locus is also necessary for the intimate attachment of EPEC to epithelial cells. This locus, *eaeB*, is located ca. 4.5 kb downstream from the end of the *eaeA* gene (30). An isogenic *eaeB* mutant of E2348/69 still produces the 94-kDa intimin protein but does not intimately attach to epithelial cells and is defective in the FAS test. Foubister et al. (37) have recently shown that the *eaeB* mutant fails to induce tyrosine phosphorylation or trigger the release of inositol phosphates (see below) from infected HeLa cells. The predicted 33-kDa *eaeB* gene product contains a pyridoxal-phosphate-binding site which is highly conserved in several aminotransferase enzymes involved in a variety of biosynthetic pathways. The exact function of the EaeB protein is not known, but the fact that the mutant is defective in the FAS test and in the induction of tyrosine phosphorylation and inositol phosphate release suggests that it might be directly or indirectly involved in signaling the dramatic cytoskeletal changes typical of the attaching and effacing lesion.

Role of *eaeA* in Disease

The role of *eaeA* in disease was tested by constructing an in-frame *eae* deletion mutation in EPEC E2348/69 and feeding the isogenic mutant to volunteers (28). Diarrhea was seen in 11 of 11 volunteers who ingested the wild-type E2348/69, whereas it occurred in 4 of 11 volunteers who ingested the isogenic mutant ($P =$

0.002). These results indicate that the *eaeA* gene is essential for the full virulence of EPEC E2348/69 but that additional virulence factors are clearly required for disease. Prior to the discovery of the *eaeA* gene, Levine et al. (71) reported that a 94-kDa outer membrane protein (OMP) engendered a strong antibody response in volunteers experimentally infected with EPEC. Subsequent studies by Jerse and Kaper (59) showed that this immunogenic 94-kDa OMP is intimin, the product of the *eaeA* gene. Interestingly, in the volunteer studies conducted by Levine et al. (71), the 9 of 10 volunteers who became ill upon challenge had no preexisting antibodies to the 94-kDa OMP. In the 10th volunteer, who did not get sick, antibodies to intimin were present in the volunteer's sera collected prior to challenge. This result hints that intimin may play a role in protective immunity to disease caused by EPEC. Secretory immunoglobulin A to a 94-kDa OMF of E2348/69 was also found in breast milk from women in a rural Mexican village (20).

The role of the EHEC *eae* gene in an animal model of EHEC infection was also investigated. An *eae* deletion-insertion mutation in wild-type EHEC O157: H7 86-24 was constructed and tested in newborn piglets by Donnenberg et al. (29). In this model, diarrhea was not observed because the animals were sacrificed at 48 h in order to maximize the probability of observing attaching and effacing lesions. The EHEC *eae* mutant was no longer capable of adhering intimately to intestinal epithelial cells and did not cause attaching and effacing lesions. Molecular Koch's postulates, as defined by Stan Falkow, were fulfilled by reintroducing the EHEC *eae* gene on a plasmid and restoring intimate adherence. Interestingly, complementation with the EPEC *eae* gene also restored intimate adherence in newborn piglets, indicating that, at least in that model, the *eae* genes of EHEC and EPEC are functionally homologous (29).

INVASION

EPEC has classically been considered to be noninvasive, principally because it does not cause keratoconjunctivitis in guinea pigs (Sereny reaction), a property associated with invasion by EIEC and *Shigella* and *Yersinia* spp. However, many published photographs of animal and human EPEC infections show apparently intracellular bacteria (77, 84, 96, 99, 101). In recent years, several investigators have shown that EPEC is capable of entering a variety of epithelial cell lines (3, 23, 39, 76). As with EIEC and *Shigella* and *Yersinia* spp., EPEC cell entry can be inhibited by cytochalasins, which block cell microfilaments, but unlike these organisms, cell entry can also be blocked by inhibitors of cellular microtubules (24, 39). Unlike true intracellular pathogens, EPEC does not multiply intracellularly or escape from a phagocytic vacuole. Thus, EPEC does not appear to be specifically adapted for intracellular survival. EPEC strains do not cause dysentery or a typhoid-like syndrome, and so the clinical significance of cell entry in the pathogenesis of disease caused by EPEC is not clear. In certain individuals, dissemination of EPEC has been noted late in infection (31, 43), and even a rare case of sepsis has been reported (14). Fecal leukocytes, a classic hallmark of invasive disease, are not observed in patients with EPEC infection, but local inflammatory responses have been observed with biopsy specimens (101).

Some studies (23, 24) report that the ability of EPEC to enter cell cultures exceeds that of EIEC, but other investigators (88) argue that the methodology of some invasion assays gives artificially elevated invasion levels for strongly adherent bacteria such as EPEC. Rather than expressing invasion as the percentage of the initial inoculum which resists killing by gentamicin (23, 24), those investigators suggest that invasion should be expressed as the proportion of cell-associated bacteria (at the time of gentamicin addition) that persisted after exposure to the antibiotic (88). This method could distinguish bacteria that adhere strongly but enter inefficiently from those that adhere poorly but enter efficiently once they have adhered. By this method, the strongly adherent EPEC isolates are less invasive than EIEC isolates, although the total number of EPEC isolates entering cells equals or exceeds the total number of EIEC isolates entering cells. By whatever method is used, it is clear that EPEC isolates do have the capacity to enter epithelial cells.

Donnenberg et al. (22) have used Tn*phoA* and the gentamicin protection assay to isolate mutants deficient in cell entry. One category of noninvasive mutants was found to have a mutation located in the *bfpA* or *dsbA* locus, both of which are required to produce functional BFP fimbriae (25, 107). A second category of mutants was deficient in intimate adherence, and the inserts that resulted in these mutants were in the *eaeA* and *eaeB* genes (22, 30). Another category of mutants (category 4 mutants, or *cfm*) is completely deficient in the FAS test and in the ability to induce a tyrosine kinase activity in the host cell (see below). These mutants contain mutations in a region of the genes whose predicted protein products share homology with proteins for a type III protein secretion system (see below). These results indicate that there is significant overlap between the genes responsible for the invasion process and the genes involved in producing attaching and effacing lesions. Thus, although the pathogenic significance of invasion by EPEC is unknown, the use of in vitro invasion assays has yielded significant information about the genetics of EPEC virulence factors.

REGULATION OF VIRULENCE

As for many other bacterial pathogens, expression of EPEC virulence factors is regulated by a *trans*-acting regulator. In the initial report of an immunogenic 94-kDa OMP, the EPEC derivative cured of the 60-MDa pMAR2 plasmid did not appear to express the 94-kDa OMP, suggesting that this protein was encoded on the plasmid (71). The discovery that the chromosomal *eaeA* gene encoded the 94-kDa OMP appeared to be inconsistent with the results of the earlier study (60). This apparent discrepancy was resolved by Jerse and Kaper (59), who showed that the 60-MDa pMAR2 plasmid encoded a positive regulatory factor which increased the level of expression of the chromosomal *eaeA* gene. In the presence of the pMAR2 plasmid, the alkaline phosphatase activity of a *eaeA*::Tn*phoA* gene fusion is increased up to 31-fold relative to that of the cured derivative.

By screening cloned fragments of pMAR2 for the ability to increase the alkaline phosphatase activity of the *eaeA*::Tn*phoA* gene fusion, Gomez and Kaper (52) cloned genes involved in this regulation. One gene, designated *perA* (plasmid-encoded regulator), consists of a 615-bp open reading frame which could potentially encode a 24-kDa protein. The predicted protein sequence of the *perA* gene shows homology to the AraC family of bacterial regulators, including regulators of virulence genes of *Shigella* (VirF) and ETEC (Rns and CfaD/CfaR). The PerA sequence shares 51% similarity and 31% identity with the predicted sequence of VirF and 51% similarity and 24% identity with the sequences of Rns and CfaD/CfaR. Like Rns, which is a plasmid-encoded regulator of chromosomal fimbrial genes in ETEC (17), the predicted pI of PerA is high (9.8) and the $G+C$ content of the *perA* gene is low (27%). An additional gene, *perC*, encodes a 10.6-kDa protein with a predicted pI of 10.6 and no striking homology to other bacterial regulators (52). The effect of Per on *eaeA* gene expression is at the level of transcription since the addition of *per* sequences greatly increases the level of production of *eaeA* mRNA (52).

The cloned *per* genes also regulate expression of the chromosomal *eaeB* gene (50) and the plasmid-borne *bfpA* gene (51). *eaeA* and *eaeB* are separated by ca. 4.5 kb, and a separate transcript was found for each gene (50). Additional genes may be under the control of Per since increased amounts of 50- and 33-kDa OMPs as well as decreased amounts of a 20-kDa OMP are also seen in the presence of Per. Thus, there appears to be a global regulatory system for the regulation of virulence in EPEC.

SECRETION OF EPEC PROTEINS

Approximately 9 kb upstream from the *eaeA* gene lies a region whose predicted protein products show striking homology to a recently recognized family of proteins that are involved in the secretion and translocation of virulence determinants of bacterial animal and plant pathogens (42, 102). This family includes the products of the *Salmonella invA* (42), *Yersinia lcrD* (83), and *S. flexneri mxiA* (4) genes. The LcrD protein has been shown to be involved in the export of *Yersinia* Yop proteins, and MxiA is involved in export of *Shigella* Ipa proteins. The role of InvA in the export of *Salmonella* virulence factors has not yet been reported, but *invA* mutants do not invade and do not alter the microvilli of polarized epithelial cells (42). Jarvis et al. (56) have found that the predicted protein product of one EPEC open reading frame, tentatively called *sepA*, shares 54% identity over 345 residues with LcrD, 35% identity over 639 residues with MxiA, and 38% identity over 633 residues with InvA. A second gene, *sepB*, encodes a protein with significant homology to the $F_1\beta$ subunit of the F_1F_0 proton-translocating ATPase found in *E. coli*, mitochondria, and chloroplasts as well as to Spa-47 of *S. flexneri* (33% identity in 281 residues), FliI of *Salmonella* spp. (37% identity in 284 residues), and FlaA ORF4 of *B. subtilis* (44% identity in 258 residues). The *spa-47* gene product is involved in export of the *Shigella* Ipa proteins (103), the *fliI* gene product is involved in the export of *Salmonella* flagellar proteins (104), and *flaA orf-4* is part of a large operon encoding flagella in *B. subtilis* (2). The exact mechanisms involved in this secretion pathway, called type III secretion in one recent review (102), are poorly characterized. The *Shigella* Ipa proteins are immunogenic in disease and are essential for the invasion of *Shigella* spp. They are present on the external surface of the bacterium but are also found in the extracellular medium. The proper translocation of these proteins depends upon the products of the *mxi* and *spa* genes. The Yop proteins are virulence factors for *Yersinia* spp. and include the tyrosine phosphatase YopH and the cytotoxic YopE. These proteins are exported into the external medium by the products of the *vir*, *lcr*, *ysc*, or *syc* genes (depending on the authors). Wattiau and Cornelis (105) recently reported homology between the predicted products of the *Yersinia sycH* gene and the open reading frame immediately upstream of *eaeA* called OrfU (106). SycH is involved in the translocation of YopH outside of the cell, and Wattiau and Cornelis (105) suggest that the EPEC OrfU gene product may be involved in the translocation of EPEC virulence factors.

The category 4 mutants (*cfm* mutants) of the noninvasive Tn*phoA* mutants described by Donnenberg et al. (22) are completely deficient in the FAS test and in the ability to induce a tyrosine kinase activity in the host cell (see below). A Tn*phoA* insertion from the *cfm* mutant 27-3-2(1) is located within the *sepA* gene (56). Jarvis et al. (57) have constructed a nonpolar mutation in *sepB* and examined the proteins present in culture supernatants of the parent strain E2348/69 and the *sepB* isogenic mutant. Prominent proteins of 38.5, 28, 25.5, and 24 kDa were seen in the culture supernatants of the parent strain but were absent from the supernatants of the *spa-2* mutant. These extracellular proteins may be the EPEC counterparts to the *Shigella* Ipa proteins and *Yersinia* Yop proteins and are likely candidates for EPEC proteins that induce the cytoskeletal changes leading to the attaching and effacing lesion.

LARGE CHROMOSOMAL REGION ENCODING EPEC VIRULENCE FACTORS

McDaniel et al. (75) have shown that the *eaeA, eaeB,* and *sep* (*cfm*) genes are all located within a ca. 35-kb region which is present in EPEC and in EHEC. This region, called LEE for locus of effacing *E. coli,* is not present in normal-flora *E. coli, E. coli* K-12, or ETEC. The LEE region is inserted into the *E. coli* K-12 chromosome at ca. 82 min, where the tRNA for selenocysteine (*selC*) is located. Interestingly, this location is also the site of insertion for the retronphage φR73 and a large (70-kb) insert (PAI) of uropathogenic *E. coli* containing genes for hemolysin (*hly*) and P-related fimbriae (*prf*) (13). The large insert for uropathogenic *E. coli* has been termed a "pathogenicity island" (13), and the insertion of the EPEC LEE at the same site suggests that this region of the *E. coli* chromosome is a hot spot for the insertion of virulence factor genes.

EPITHELIAL CELL RESPONSE TO EPEC INFECTION

Composition of Attaching and Effacing Lesions

Knutton et al. (64) first showed that the distinctive attaching and effacing lesions induced by EPEC contain high concentrations of filamentous actin. As noted above, this observation led to the development of the FAS test for the diagnosis of attaching and effacing EPEC and EHEC. Using immunofluorescence microscopy and antibodies specific for various cytoskeletal components, Finlay et al. (35) examined the cytoskeletal composition of the attaching and effacing lesions associated with adherence of EPEC to HeLa cells. The microfilament-associated proteins actin, α-actinin, talin, and ezrin were localized with adherent EPEC, whereas tropomyosin, keratin and vimentin (intermediate filaments), tubulin (microtubules), and vinculin were not localized with adherent EPEC. These cytoskeletal structures differed significantly from those associated with *Salmonella typhimurium* infection (35). The lack of tropomyosin accumulation with EPEC compared with the accumulation with *S. typhimurium* suggests that the cytoskeletal rearrangement caused by EPEC forms a relatively inert, stable cytoskeletal structure that the bacterium rests on but does not promote an actin-myosin-tropomyosin-mediated event involved in mechanical bacterial uptake. *S. typhimurium* invades HeLa cells quickly (within 10 min) and efficiently, whereas EPEC internalization occurs ca. 3 h after infection. Thus, it appears that the cytoskeletal rearrangements triggered by *S. typhimurium* are designed to efficiently internalize the bacterium, with a subsequent return to the normal cytoskeletal distribution, whereas those rearrangements caused by EPEC form a stable structure that remains for several hours at the site of localized bacterial adherence.

Increased Intracellular Calcium Concentrations

Baldwin and colleagues (10) have demonstrated that EPEC increases intracellular calcium concentrations ($[Ca^{2+}_i]$) in the HEp-2 cells to which they are attached. The increase in $[Ca^{2+}_i]$ is inhibited by dantrolene, an inhibitor of calcium mobilization from intracellular stores. Buffering of the intracellular $[Ca^{2+}_i]$ greatly reduced the polymerization of actin in the FAS test (9). The actin accumulation in the attaching and effacing lesion is dependent on calcium-calmodulin; prior treatment of HEp-2 cells with the calmodulin inhibitor compound 48/80 completely prevented actin accretion (9). The increase in $[Ca^{2+}_i]$ has been hypothesized to produce the cytoskeletal changes induced by EPEC by activation of a calcium-dependent, actin-severing protein which could break down actin in the microvillus core (10). Furthermore, since increases in $[Ca^{2+}_i]$ can inhibit Na^+ and Cl^- absorption and stimulate chloride secretion in enterocytes (33, 34), these data also suggest that changes in $[Ca^{2+}_i]$ may mediate the intestinal secretory response to EPEC.

Protein Kinase C

The involvement of protein kinase C in the pathogenesis of EPEC was first suggested by Baldwin et al. (7), who found that infection of HEp-2 monolayers with EPEC stimulated phosphorylation of several target cell proteins. The protein phosphorylation was not seen with the addition of culture supernatants but was seen only with the attachment of viable organisms. The most prominent of the phosphorylated proteins had molecular masses of 21 and 29 kDa, and the small protein was subsequently shown to be myosin light chain (74). Protein kinase was implicated in this phosphorylation when similar phosphorylation patterns were observed after the addition of phorbol esters to HEp-2 cells (7). Activation of protein kinase C induces rapid changes in intestinal water and electrolyte secretion in vivo and in vitro (85), thereby suggesting another possible intracellular mediator of the secretory response to EPEC infection.

Tyrosine Kinase

Rosenshine et al. (89) have found that binding of EPEC to HeLa cells induces tyrosine phosphorylation of three eukaryotic proteins. The major phosphorylation substrate is a 90-kDa protein (Hp90), the identity of which is unknown. The cytoskeletal lesions induced by EPEC also contain proteins phosphorylated on tyrosine residues. Using tyrosine kinase inhibitors such as staurosporine and genistein, those investigators showed that induction of Hp90 phosphorylation is involved in the initiation of the cytoskeletal structure assembly and in bacterial uptake by the HeLa cells. The addition of EPEC culture supernatants to the HeLa cells did not induce tyrosine phosphorylation of HeLa cell proteins, suggesting that binding of EPEC to the epithelial cells induces an endogenous HeLa cell tyrosine kinase. Interestingly, EPEC isolates grown in the absence of HeLa cells were also found to contain tyrosine-phosphorylated proteins of 30, 51, and 85 kDa. This activity was not found in *E. coli* K-12 strains and is in contrast to previous studies showing no specific tyrosine phosphorylation activity in *E. coli* (36).

Several EPEC mutants were examined for their abilities to induce phosphorylation of Hp90. A strain cured of the 60-MDa plasmid encoding BFP could still phosphorylate Hp90. Strains with mutations in the *eaeA* gene and which still contained the EAF plasmid were still capable of inducing Hp90 phosphorylation, although they were deficient in their invasion abilities. Mutants in category 4 (*cfm* mutants) were completely negative for phosphorylation of Hp90 and invaded HeLa cells at <1% of the rate of wild-type EPEC. As noted above, these mutants mapped to the *sep* genes which appear to be involved in the transport of virulence factors. Interestingly, a mixed infection with an *eaeA* mutant and a *cfm* mutant resulted in the internalization of the *cfm* mutant at nearly 50% of the efficiency of internalization of the wild-type EPEC strains. This result indicates that the *eaeA* mutant but not the *cfm* mutant is still capable of transducing a signal to the eukaryotic cell. Upon receipt of this signal, the eukaryotic cell then takes up those bacteria that are in intimate contact, i.e., those with a functional *eaeA* gene product, as is found in the *cfm* mutants. Foubister et al. (37) have recently shown that a mutation in the *eaeB* gene also abrogates the tyrosine phosphorylation of Hp90. Furthermore, coinfection with a noninvasive *eaeA* mutant and the noninvasive *eaeB* mutant allowed the entry of the *eaeB* mutant into HeLa cells. In each of the two coinfection experiments (*eaeA* + *cfm* and *eaeA* + *eaeB*), a mutant capable of producing intimin but deficient in signal transduction could be complemented by an *eaeA* mutant which is competent for signal transduction but is unable to produce intimin (37).

Inositol Phosphates

The inhibition of intracellular calcium increases by dantrolene suggested that the calcium was released from 1,4,5-inositol trisphosphate (IP_3)-sensitive stores (10). Foubister et al. (38) have recently shown that binding of EPEC to cultured epithelial cells triggers the release of inositol phosphates in infected cells. The most prominent phosphorylated inositol species released was inositol phosphate (IP); only small increases in IP_3 and IP_4 are seen. As with tyrosine phosphorylation, mutation of the *eaeB* gene abolished IP release (37). EPEC induced the release of IPs only after several hours of infection, in contrast to the rapid formation of IPs that occurs within seconds or minutes in response to many growth factors and hormones (38). This delayed onset is similar to that seen with other EPEC-induced changes in intracellular signals such as $[Ca^{2+}_i]$ and tyrosine phosphorylation.

Transepithelial Resistance

When EPEC isolates interact with polarized epithelial cells such as Caco-2 cells, the transepithelial resistance of the monolayers decreases (16). This occurs approximately 6 to 10 h after the addition of bacteria and is reversible if the monolayers are treated with tetracycline or gentamicin to kill the bacteria. The restoration of resistance seen by adding gentamicin suggests that intracellular bacteria do not play an important role in decreasing resistance. When various EPEC mutants were tested in this system, only those strains capable of causing the attaching and effacing lesion were able to decrease the monolayer resistance. This drop in tissue resistance did not appear to be due to an alteration in the paracellular pathway since actin filaments supporting tight junctions were not noticeably affected in the epithelial cells, nor was the distribution of ZO-1, a tight-junction protein, affected.

THREE-STAGE MODEL OF EPEC PATHOGENESIS

The rapidly accumulating data on EPEC pathogenesis led Donnenberg and Kaper (27) to propose a three-stage model of EPEC pathogenesis, which is shown in Fig. 2. In the first stage, a nonintimate adherence results from the interaction between the bacterium and the epithelial cell. This initial adherence appears to involve BFP and one or more additional fimbrial structures. In the second stage, a signal transduction event results in tyrosine phosphorylation of a 90-kDa eukaryotic protein (Hp90) and increased intracellular levels of calcium and the release of inositol phosphate, which leads to the effacement of microvilli. This signal transduction is perhaps mediated by the product of the *eaeB* gene and other EPEC gene products which require the products of the *sep* locus for surface presentation.

Simultaneously, the *eaeA* gene is transcriptionally activated by the product of the plasmid-encoded *per* locus. The third stage of infection results when intimin, the product of the *eaeA* locus, and possibly other products of the *eae* gene cluster mediate close attachment to the epithelial cell. With this intimate adherence, the effects on the epithelial cell are amplified, with the accumulation of filamentous actin and other cytoskeletal proteins. The second and third stages probably occur concurrently, but studies with EPEC mutants indicate that the second stage of signal transduction can occur in the absence of intimate adherence.

The details of this model of EPEC pathogenesis will undoubtedly change as more data accumulate, but the three stages provide a broad conceptual framework for further investigations. How these changes in epithelial cells result in diarrhea is not known. The dramatic loss of the absorptive microvilli could lead to diarrhea via malabsorption. However, the incubation period in adult volunteers can be very short: as little as 2.9 h between ingestion of the organisms and the onset of diarrhea (28). This rapidity suggests that a more active secretory mechanism is involved in diarrhea caused by EPEC, and a variety of intracellular mediators of intestinal ion transport, such as calcium, protein kinase C, inositol phosphates, and tyrosine kinase, are affected by EPEC infection. There is clearly much more to be learned about EPEC, but given that the study of EPEC pathogenesis languished for many years until the volunteer studies of Levine et al. in 1978 (69), remarkable progress in the understanding of this fascinating pathogen has recently been made.

REFERENCES

1. **Albert, M. J., S. M. Faruque, M. Ansaruzzaman, M. M. Islam, K. Haider, K. Alam, I. Kabir, and R. Robins-Browne.** 1992. Sharing of virulence-associated properties at the phenotypic and genetic levels between enteropathogenic *Escherichia coli* and *Hafnia alvei*. *J. Med. Microbiol.* **37:**310–314.

2. **Albertini, A. M., T. Caramori, W. D. Crabb, F. Scoffone, and A. Galizzi.** 1991. The *flaA* locus of *Bacillus subtilis* is part of a large operon coding for flagellar structures, motility functions, and an ATPase-like polypeptide. *J. Bacteriol.* **173:**3573–3579.

3. **Andrade, J. R., V. F. Da Veiga, M. R. De Santa Rosa, and I. Suassuna.** 1989. An endocytic process in HEp-2 cells induced by enteropathogenic *Escherichia coli*. *J. Med. Microbiol.* **28:**49–57.

4. **Andrews, G. P., and A. T. Maurelli.** 1992. *mxiA* of *Shigella flexneri* 2a, which facilitates export of invasion plasmid antigens, encodes a homolog of the low-calcium-response protein, LcrD, of *Yersinia pestis*. *Infect. Immun.* **60:**3287–3295.

5. **Baldini, M. M., J. B. Kaper, M. M. Levine, D. C. Candy, and H. W. Moon.** 1983. Plasmid-mediated adhesion in enteropathogenic *Escherichia coli*. *J. Pediatr. Gastroentrol. Nutr.* **2:**534–538.

6. **Baldini, M. M., J. P. Nataro, and J. B. Kaper.** 1986. Localization of a determinant for HEp-2 adherence by enteropathogenic *Escherichia coli*. *Infect. Immun.* **52:**334–336.

7. **Baldwin, T. J., S. F. Brooks, S. Knutton, H. A. Manjarrez Hernandez, A. Aitken, and P. H. Williams.** 1990. Protein phosphorylation by protein kinase C in HEp-2 cells infected with enteropathogenic *Escherichia coli*. *Infect. Immun.* **58:**761–765. (Erratum, **58:**2024.)

8. **Baldwin, T. J., S. Knutton, L. Sellers, H. A. M. Hernandez, A. Aitken, and P. H. Williams.** 1992. Enteroaggregative *Escherichia coli* strains secrete a heat-labile toxin antigenically related to *E. coli* hemolysin. *Infect. Immun.* **60:**2092–2095.

9. **Baldwin, T. J., M. B. Lee-Delaunay, S. Knutton, and P. H. Williams.** 1993. Calcium-calmodulin

dependence of actin accretion and lethality in cultured HEp-2 cells infected with enteropathogenic *Escherichia coli. Infect. Immun.* **61**:760–763.

10. **Baldwin, T. J., W. Ward, A. Aitken, S. Knutton, and P. H. Williams.** 1991. Elevation of intracellular free calcium levels in HEp-2 cells infected with enteropathogenic *Escherichia coli. Infect. Immun.* **59**:1599–1604.

11. **Benz, I., and M. A. Schmidt.** 1992. AIDA-I, the adhesin involved in diffuse adherence of the diarrhoeagenic *Escherichia coli* strain 2787 (O126:H27), is synthesized via a precursor molecule. *Mol. Microbiol.* **6**:1539–1546.

12. **Bilge, S. S., C. R. Clausen, W. Lau, and S. L. Moseley.** 1989. Molecular characterization of a fimbrial adhesin, F1845, mediating diffuse adherence of diarrhea-associated *Escherichia coli* to HEp-2 cells. *J. Bacteriol.* **171**:4281–4289.

13. **Blum, G., M. Ott, A. Lischewski, A. Ritter, H. Imrich, H. Tschäpe, and J. Hacker.** 1994. Excision of large DNA regions termed pathogenicity islands from tRNA-specific loci in the chromosome of an *Escherichia coli* wild-type pathogen. *Infect. Immun.* **62**:606–614.

14. **Bower, J. R., B. L. Congeni, T. G. Cleary, R. T. Stone, A. Wanger, B. E. Murray, J. J. Mathewson, and L. K. Pickering.** 1989. *Escherichia coli* O114:nonmotile as a pathogen in an outbreak of severe diarrhea associated with a day care center. *J. Infect. Dis.* **160**:243–247.

15. **Bray, J.** 1945. Isolation of antigenically homogeneous strains of *Bact. coli neapolitanum* from summer diarrhoea of infants. *J. Pathol. Bacteriol.* **57**:239–247.

16. **Canil, C., I. Rosenshine, S. Ruschkowski, M. S. Donnenberg, J. B. Kaper, and B. B. Finlay.** 1993. Enteropathogenic *Escherichia coli* decreases the transepithelial electrical resistance of polarized epithelial monolayers. *Infect. Immun.* **61**:2755–2762.

17. **Caron, J., L. M. Coffield, and J. R. Scott.** 1989. A plasmid-encoded regulatory gene, rns, required for expression of the CS1 and CS2 adhesins of enterotoxigenic *Escherichia coli. Proc. Natl. Acad. Sci. USA* **86**:963–967.

18. **Clausen, C. R., and D. L. Christie.** 1982. Chronic diarrhea in infants caused by adherent enteropathogenic *Escherichia coli. J. Pediatr.* **100**:358–361.

19. **Cravioto, A., R. J. Gross, S. M. Scotland, and B. Rowe.** 1979. An adhesive factor found in strains of *Escherichia coli* belonging to the traditional infantile enteropathogenic serotypes. *Curr. Microbiol.* **3**:95–99.

20. **Cravioto, A., A. Tello, H. Villafán, J. Ruiz, S. Del Vedovo, and J.-R. Neeser.** 1991. Inhibition of localized adhesion of enteropathogenic *Escherichia coli* to HEp-2 cells by immunoglobulin and oligosaccharide fractions of human colostrum and breast milk. *J. Infect. Dis.* **163**:1247–1255.

21. **De Rycke, J., E. A. Gonzalez, J. Blanco, E. Oswald, M. Blanco, and R. Boivin.** 1990. Evidence for two types of cytotoxic necrotizing factor in human and animal clinical isolates of *Escherichia coli. J. Clin. Microbiol.* **28**:694–699.

22. **Donnenberg, M. S., S. B. Calderwood, A. Donohue-Rolfe, G. T. Keusch, and J. B. Kaper.** 1990. Construction and analysis of Tn*phoA* mutants of enteropathogenic *Escherichia coli* unable to invade HEp-2 cells. *Infect. Immun.* **58**:1565–1571.

23. **Donnenberg, M. S., A. Donohue-Rolfe, and G. T. Keusch.** 1989. Epithelial cell invasion: an overlooked property of enteropathogenic *Escherichia coli* (EPEC) associated with the EPEC adherence factor. *J. Infect. Dis.* **160**:452–459.

24. **Donnenberg, M. S., A. Donohue-Rolfe, and G. T. Keusch.** 1990. A comparison of HEp-2 cell invasion by enteropathogenic and enteroinvasive *Escherichia coli. FEMS Microbiol. Lett.* **57**:83–86.

25. **Donnenberg, M. S., J. A. Girón, J. P. Nataro, and J. B. Kaper.** 1992. A plasmid-encoded type IV fimbrial gene of enteropathogenic *Escherichia coli* associated with localized adherence. *Mol. Microbiol.* **6**:3427–3437.

26. **Donnenberg, M. S., and J. B. Kaper.** 1991. Construction of an *eae* deletion mutant of enteropathogenic *Escherichia coli* by using a positive-selection suicide vector. *Infect. Immun.* **59**:4310–4317.

27. **Donnenberg, M. S., and J. B. Kaper.** 1992. Enteropathogenic *Escherichia coli. Infect. Immun.* **60**:3953–3961.

28. **Donnenberg, M. S., C. O. Tacket, S. P. James, G. Losonsky, J. P. Nataro, S. S. Wasserman, J. B. Kaper, and M. M. Levine.** 1993. Role of the *eaeA* gene in experimental enteropathogenic *Escherichia coli* infection. *J. Clin. Invest.* **92**:1412–1417.

29. **Donnenberg, M. S., S. Tzipori, M. L. McKee, A. D. O'Brien, J. Alroy, and J. B. Kaper.** 1993. The role of the *eae* gene of enterohemorrhagic *Escherichia coli* in intimate attachment in vitro and in a porcine model. *J. Clin. Invest.* **92:**1418–1424.

30. **Donnenberg, M. S., J. Yu, and J. B. Kaper.** 1993. A second chromosomal gene necessary for intimate attachment of enteropathogenic *Escherichia coli* to epithelial cells. *J. Bacteriol.* **175:** 4670–4680.

31. **Drucker, M. M., A. Polliack, R. Yeivin, and T. G. Sacks.** 1970. Immunofluorescent demonstration of enteropathogenic *Escherichia coli* in tissues of infants dying with enteritis. *Pediatrics* **46:** 855–864.

32. **Echeverria, P., F. Orskov, I. Orskov, S. Knutton, F. Scheutz, J. E. Brown, and U. Lexomboon.** 1991. Attaching and effacing enteropathogenic *Escherichia coli* as a cause of infantile diarrhea in Bangkok. *J. Infect. Dis.* **164:**550–554.

33. **Field, M., M. C. Rao, and E. B. Chang.** 1989. Intestinal electrolyte transport and diarrheal disease—part 1. *N. Engl. J. Med.* **321:**800–806.

34. **Field, M., M. C. Rao, and E. B. Chang.** 1989. Intestinal electrolyte transport and diarrheal disease—part 2. *N. Engl. J. Med.* **321:**879–883.

35. **Finlay, B. B., I. Rosenshine, M. S. Donnenberg, and J. B. Kaper.** 1992. Cytoskeletal composition of attaching and effacing lesions associated with enteropathogenic *Escherichia coli* adherence to HeLa cells. *Infect. Immun.* **60:**2541–2543.

36. **Foster, R., J. Thorner, and G. S. Martin.** 1989. Nucleotidylation, not phosphorylation, is the major source of the phosphotyrosine detected in enteric bacteria. *J. Bacteriol.* **171:**272–279.

37. **Foubister, V., I. Rosenshine, M. S. Donnenberg, and B. B. Finlay.** Submitted for publication.

38. **Foubister, V., I. Rosenshine, and B. B. Finlay.** 1994. A diarrheal pathogen, enteropathogenic *Escherichia coli* (EPEC) triggers a flux of inositol phosphates in infected epithelial cells. *J. Exp. Med.* **179:**993–998.

39. **Francis, C. L., A. E. Jerse, J. B. Kaper, and S. Falkow.** 1991. Characterization of interactions of enteropathogenic *Escherichia coli* O127:H6 with mammalian cells in vitro. *J. Infect. Dis.* **164:** 693–703.

40. **Francis, D. H., J. E. Collins, and J. R. Duimstra.** 1986. Infection of gnotobiotic pigs with an *Escherichia coli* O157:H7 strain associated with an outbreak of hemorrhagic colitis. *Infect. Immun.* **51:**953–956.

41. **Frankel, G., D. C. A. Candy, P. Everest, and G. Dougan.** 1994. Characterization of the C-terminal domains of intimin-like proteins of enteropathogenic and enterohemorrhagic *Escherichia coli*, *Citrobacter freundii*, and *Hafnia alvei*. *Infect. Immun.* **62:**1835–1842.

42. **Galan, J. E., C. Ginocchio, and P. Costeas.** 1992. Molecular and functional characterization of the *Salmonella* invasion gene *invA:* homology of InvA to members of a new protein family. *J. Bacteriol.* **174:**4338–4349.

43. **Giles, C., G. Sangster, and J. Smith.** 1949. Epidemic gastroenteritis of infants in Aberdeen during 1947. *Arch. Dis. Child.* **24:**45–53.

44. **Girón, J. A., M. S. Donnenberg, W. C. Martin, K. G. Jarvis, and J. B. Kaper.** 1993. Distribution of the bundle-forming pilus structural gene (*bfpA*) among enteropathogenic *Escherichia coli*. *J. Infect. Dis.* **168:**1037–1041.

45. **Girón, J. A., A. S. Y. Ho, and G. K. Schoolnik.** 1991. An inducible bundle-forming pilus of enteropathogenic *Escherichia coli*. *Science* **254:**710–713.

46. **Girón, J. A., A. S. Y. Ho, and G. K. Schoolnik.** 1993. Characterization of fimbriae produced by enteropathogenic *Escherichia coli*. *J. Bacteriol.* **175:**7391–7403.

47. **Gomes, T. A. T., P. A. Blake, and L. R. Trabulsi.** 1989. Prevalence of *Escherichia coli* strains with localized, diffuse, and aggregative adherence to HeLa cells in infants with diarrhea and matched controls. *J. Clin. Microbiol.* **27:**266–269.

48. **Gomes, T. A. T., V. Rassi, K. L. Macdonald, S. R. T. S. Ramos, L. R. Trabulsi, M. A. M. Vieira, B. E. C. Guth, J. A. N. Candeias, C. Ivey, M. R. F. Toledo, and P. A. Blake.** 1991. Enteropathogens associated with acute diarrheal disease in urban infants in Sao Paulo, Brazil. *J. Infect. Dis.* **164:** 331–337.

49. **Gomes, T. A. T., M. A. M. Vieira, I. K. Wachsmuth, P. A. Blake, and L. R. Trabulsi.** 1989.

Serotype-specific prevalence of *Escherichia coli* strains with EPEC adherence factor genes in infants with and without diarrhea in Sao Paulo, Brazil. *J. Infect. Dis.* **160**:131–135.

50. **Gomez, O., M. Donnenberg, and J. B. Kaper.** 1993. Plasmid encoded regulator A (PerA) is a transcriptional activator of *eaeA* and *eaeB* of enteropathogenic *Escherichia coli*, abstr. B-310, p. 81. *Abstr. 93rd Gen. Meet. Am. Soc. Microbiol. 1993.* American Society for Microbiology, Washington, D.C.

51. **Gomez, O., M. S. Donnenberg, and J. B. Kaper.** Unpublished data.

52. **Gomez, O., and J. B. Kaper.** Submitted for publication.

53. **Hii, J. H., J. De Azavedo, M. Louie, and J. L. Brunton.** 1993. The nucleotide sequence of the three prime end of the *eae* gene homologue of verotoxin-producing *E. coli* (VTEC) serotype O26: H11 differs from VTEC serotype O157:H7, abstr. D-197, p. 130. *Abstr. 93rd Gen. Meet. Am. Soc. Microbiol. 1993.* American Society for Microbiology, Washington, D.C.

54. **Hsia, R., P. L. C. Small, and P. M. Bavoil.** 1993. Characterization of virulence genes of enteroinvasive *Escherichia coli* by Tn*phoA* mutagenesis: identification of *invX*, a gene required for entry into HEp-2 cells. *J. Bacteriol.* **175**:4817–4823.

55. **Isberg, R. R., and J. M. Leong.** 1990. Multiple β-1 chain integrins are receptors for invasin, a protein that promotes bacterial penetration into mammalian cells. *Cell* **60**:861–871.

56. **Jarvis, K., T. McDaniel, M. Donnenberg, and J. Kaper.** 1994. Sequence analysis of a putative type III secretory pathway in enteropathogenic *E. coli* (EPEC), abstr. B295, p. 81. *Abstr. 94th Gen. Meet. Am. Soc. Microbiol. 1994.* American Society for Microbiology, Washington, D.C.

57. **Jarvis, K. G., T. K. McDaniel, J. A. Girón, and J. B. Kaper.** Unpublished data.

58. **Jerse, A. E., K. G. Gicquelais, and J. B. Kaper.** 1991. Plasmid and chromosomal elements involved in the pathogenesis of attaching and effacing *Escherichia coli*. *Infect. Immun.* **59**:3869–3875.

59. **Jerse, A. E., and J. B. Kaper.** 1991. The *eae* gene of enteropathogenic *Escherichia coli* encodes a 94-kilodalton membrane protein, the expression of which is influenced by the EAF plasmid. *Infect. Immun.* **59**:4302–4309.

60. **Jerse, A. E., J. Yu, B. D. Tall, and J. B. Kaper.** 1990. A genetic locus of enteropathogenic *Escherichia coli* necessary for the production of attaching and effacing lesions on tissue culture cells. *Proc. Natl. Acad. Sci. USA* **87**:7839–7843.

61. **Johnson, W. M., and H. Lior.** 1988. A new heat-labile cytolethal distending toxin (CLDT) produced by *Escherichia coli* isolates from clinical material. *Microb. Pathog.* **4**:103–113.

62. **Kain, K. C., R. L. Barteluk, M. T. Kelly, H. Xin, G. D. Hua, G. Yuan, E. M. Proctor, S. Byrne, and H. G. Stiver.** 1991. Etiology of childhood diarrhea in Beijing, China. *J. Clin. Microbiol.* **29**: 90–95.

63. **Knutton, S., M. M. Baldini, J. B. Kaper, and A. S. McNeish.** 1987. Role of plasmid-encoded adherence factors in adhesion of enteropathogenic *Escherichia coli* to HEp-2 cells. *Infect. Immun.* **55**:78–85.

64. **Knutton, S., T. Baldwin, P. H. Williams, and A. S. McNeish.** 1989. Actin accumulation at sites of bacterial adhesion to tissue culture cells: basis of a new diagnostic test for enteropathogenic and enterohemorrhagic *Escherichia coli*. *Infect. Immun.* **57**:1290–1298.

65. **Knutton, S., D. R. Lloyd, and A. S. McNeish.** 1987. Adhesion of enteropathogenic *Escherichia coli* to human intestinal enterocytes and cultured human intestinal mucosa. *Infect. Immun.* **55**: 69–77.

66. **Law, D.** 1994. Adhesion and its role in the virulence of enteropathogenic *Escherichia coli*. *Clin. Microbiol. Rev.* **7**:152–173.

67. **Leong, J. M., R. S. Fournier, and R. R. Isberg.** 1990. Identification of the integrin binding domain of the *Yersinia pseudotuberculosis* invasin protein. *EMBO J.* **9**:1979–1989.

68. **Levine, M. M.** 1987. *Escherichia coli* that cause diarrhea: enterotoxigenic, enteropathogenic, enteroinvasive, enterohemorrhagic, and enteroadherent. *J. Infect. Dis.* **155**:377–389.

69. **Levine, M. M., E. J. Bergquist, D. R. Nalin, D. H. Waterman, R. B. Hornick, C. R. Young, S. Sotman, and B. Rowe.** 1978. *Escherichia coli* strains that cause diarrhoea but do not produce heat-labile or heat-stable enterotoxins and are non-invasive. *Lancet* **i**:1119–1122.

70. **Levine, M. M., and R. Edelman.** 1984. Enteropathogenic *Escherichia coli* of classic serotypes associated with infant diarrhea: epidemiology and pathogenesis. *Epidemiol. Rev.* **6**:31–51.

71. **Levine, M. M., J. P. Nataro, H. Karch, M. M. Baldini, J. B. Kaper, R. E. Black, M. L. Clements,**

and A. D. O'Brien. 1985. The diarrheal response of humans to some classic serotypes of entero-pathogenic *Escherichia coli* is dependent on a plasmid encoding an enteroadhesiveness factor. *J. Infect. Dis.* **152**:550–559.

72. **Levine, M. M., V. Prado, R. Robins-Browne, H. Lior, J. B. Kaper, S. L. Moseley, K. Gicquelais, J. P. Nataro, P. Vial, and B. Tall.** 1988. Use of DNA probes and HEp-2 cell adherence assay to detect diarrheagenic *Escherichia coli. J. Infect. Dis.* **158**:224–228.

73. **Louie, M., J. C. S. De Azavedo, M. Y. C. Handelsman, C. G. Clark, B. Ally, M. Dytoc, P. Sherman, and J. Brunton.** 1993. Expression and characterization of the *eaeA* gene product of *Escherichia coli* serotype O157:H7. *Infect. Immun.* **61**:4085–4092.

74. **Manjarrez-Hernandez, H. A., B. Amess, L. Sellers, T. J. Baldwin, S. Knutton, P. H. Williams, and A. Aitken.** 1991. Purification of a 20-kDa phosphoprotein from epithelial cells and identification as a myosin light chain—phosphorylation induced by enteropathogenic *Escherichia coli* and phorbol ester. *FEBS Lett.* **292**:121–127.

75. **McDaniel, T. K., K. G. Jarvis, M. S. Donnenberg, and J. B. Kaper.** 1994. A large chromosomal gene cluster is present in enteropathogenic *Escherichia coli* (EPEC) but absent from non-patho-genic *E. coli*, abstr. B286, p. 80. *Abstr. 94th Gen. Meet. Am. Soc. Microbiol. 1994.* American Society for Microbiology, Washington, D.C.

76. **Miliotis, M. D., H. J. Koornhof, and J. I. Phillips.** 1989. Invasive potential of noncytotoxic entero-pathogenic *Escherichia coli* in an in vitro Henle 407 cell model. *Infect. Immun.* **57**:1928–1935.

77. **Moon, H. W., S. C. Whipp, R. A. Argenzio, M. M. Levine, and R. A. Giannella.** 1983. Attaching and effacing activities of rabbit and human enteropathogenic *Escherichia coli* in pig and rabbit intestines. *Infect. Immun.* **41**:1340–1351.

78. **Nataro, J. P., M. M. Baldini, J. B. Kaper, R. E. Black, N. Bravo, and M. M. Levine.** 1985. Detection of an adherence factor of enteropathogenic *Escherichia coli* with a DNA probe. *J. Infect. Dis.* **152**:560–565.

79. **Nataro, J. P., Y. Deng, D. R. Maneval, A. L. German, W. C. Martin, and M. M. Levine.** 1992. Aggregative adherence fimbriae I of enteroaggregative *Escherichia coli* mediate adherence to HEp-2 cells and hemagglutination of human erythrocytes. *Infect. Immun.* **60**:2297–2304.

80. **Nataro, J. P., J. B. Kaper, R. Robins-Browne, V. Prado, P. Vial, and M. M. Levine.** 1987. Patterns of adherence of diarrheagenic *Escherichia coli* to HEp-2 cells. *Pediatr. Infect. Dis. J.* **6**:829–831.

81. **Nataro, J. P., K. O. Maher, P. Mackie, and J. B. Kaper.** 1987. Characterization of plasmids encoding the adherence factor of enteropathogenic *Escherichia coli. Infect. Immun.* **55**: 2370–2377.

82. **Paulozzi, L. J., K. E. Johnson, L. M. Kamahele, C. R. Clausen, L. W. Riley, and S. D. Helgerson.** 1986. Diarrhea associated with adherent enteropathogenic *Escherichia coli* in an infant and toddler center, Seattle, Washington. *Pediatrics* **77**:296–300.

83. **Plano, G. V., S. S. Barve, and S. C. Straley.** 1991. LcrD, a membrane-bound regulator of the *Yersinia pestis* low-calcium response. *J. Bacteriol.* **173**:7293–7303.

84. **Polotsky, Y. E., E. M. Dragunskaya, V. G. Seliverstova, T. A. Avdeeva, M. G. Chakhutinskaya, I. Kétyi, A. Vertényi, B. Ralovich, L. Emödy, I. Málovics, N. V. Safonova, E. S. Snigirevskaya, and E. I. Karyagina.** 1977. Pathogenic effect of enterotoxigenic *Escherichia coli* and *Escherichia coli* causing infantile diarrhea. *Acta Microbiol. Acad. Sci. Hung.* **24**:221–236.

85. **Rao, M. C., and H. R. de Jonge.** 1990. Ca and phospholipid-dependent protein kinases, p. 209–232. *In* E. Lebenthal and M. Duffey (ed.), *Textbook of Secretory Diarrhea.* Raven Press, Ltd., New York.

86. **Robins-Browne, R., C. S. Still, M. D. Miliotis, N. J. Richardson, H. J. Koornhof, I. Frieman, B. D. Schoub, G. Lecatsas, and E. Hartman.** 1980. Summer diarrhoea in African infants and children. *Arch. Dis. Child.* **55**:923–928.

87. **Robins-Browne, R. M.** 1987. Traditional enteropathogenic *Escherichia coli* of infantile diarrhea. *Rev. Infect. Dis.* **9**:28–53.

88. **Robins-Browne, R. M., and V. Bennett-Wood.** 1992. Quantitative assessment of the ability of *Escherichia coli* to invade cultured animal cells. *Microb. Pathog.* **12**:159–164.

89. **Rosenshine, I., M. S. Donnenberg, J. B. Kaper, and B. B. Finlay.** 1992. Signal transduction be-tween enteropathogenic *Escherichia coli* (EPEC) and epithelial cells: EPEC induces tyrosine

phosphorylation of host cell proteins to initiate cytoskeletal rearrangement and bacterial uptake. *EMBO J.* **11:**3551–3560.

90. **Rothbaum, R., A. J. McAdams, R. Gianella, and J. C. Partin.** 1982. A clinicopathological study of enterocyte-adherent *Escherichia coli:* a cause of protracted diarrhea in infants. *Gastroenterology* **83:**441–454.

91. **Savarino, S. J., A. Fasano, J. Watson, B. M. Martin, M. M. Levine, S. Guandalini, and P. Guerry.** 1993. Enteroaggregative *Escherichia coli* heat-stable enterotoxin 1 represents another subfamily of *E. coli* heat-stable toxin. *Proc. Natl. Acad. Sci. USA* **90:**3093–3097.

92. **Scaletsky, I. C. A., M. L. M. Silva, and L. R. Trabulsi.** 1984. Distinctive patterns of adherence of enteropathogenic *Escherichia coli* to HeLa cells. *Infect. Immun.* **45:**534–536.

93. **Scotland, S. M., H. R. Smith, B. Said, G. A. Willshaw, T. Cheasty, and B. Rowe.** 1991. Identification of enteropathogenic *Escherichia coli* (EPEC) isolated in Britain as enteroaggregative or as members of a subclass of attaching-effacing *E. coli* not hybridising with the EPEC adherence factor probe. *J. Med. Microbiol.* **35:**278–283.

94. **Scotland, S. M., G. A. Willshaw, H. R. Smith, B. Said, N. Stokes, and B. Rowe.** 1993. Virulence properties of *Escherichia coli* belonging to serogroups O26, O55, O111 and O128 isolated in the United Kingdom in 1991 from patients with diarrhoea. *Epidemiol. Infect.* **111:**429–438.

95. **Senerwa, D., Ø. Olsvik, L. N. Mutanda, K. J. Lindqvist, J. M. Gathuma, K. Fossum, and K. Wachsmuth.** 1989. Enteropathogenic *Escherichia coli* serotype O111:HNT isolated from preterm neonates in Nairobi, Kenya. *J. Clin. Microbiol.* **27:**1307–1311.

96. **Staley, T. E., E. W. Jones, and L. D. Corley.** 1969. Attachment and penetration of *Escherichia coli* into intestinal epithelium of the ileum in newborn pigs. *Am. J. Pathol.* **56:**371–392.

97. **Taylor, C. J., A. Hart, R. M. Batt, C. McDougall, and L. McLean.** 1986. Ultrastructural and biochemical changes in human jejunal mucosa associated with enteropathogenic *Escherichia coli* (O111) infection. *J. Pediatr. Gastroenterol. Nutr.* **5:**70–73.

98. **Tzipori, S., R. Gibson, and J. Montanaro.** 1989. Nature and distribution of mucosal lesions associated with enteropathogenic and enterohemorrhagic *Escherichia coli* in piglets and the role of plasmid-mediated factors. *Infect. Immun.* **57:**1142–1150.

99. **Tzipori, S., R. M. Robins-Browne, G. Gonis, J. Hayes, M. Withers, and E. McCartney.** 1985. Enteropathogenic *Escherichia coli* enteritis: evaluation of the gnotobiotic piglet as a model of human infection. *Gut* **26:**570–578.

100. **Tzipori, S., I. K. Wachsmuth, C. Chapman, R. Birner, J. Brittingham, C. Jackson, and J. Hogg.** 1986. The pathogenesis of hemorrhagic colitis caused by *Escherichia coli* O157:H7 in gnotobiotic piglets. *J. Infect. Dis.* **154:**712–716.

101. **Ulshen, M. H., and J. L. Rollo.** 1980. Pathogenesis of *Escherichia coli* gastroenteritis in man—another mechanism. *N. Engl. J. Med.* **302:**99–101.

102. **Van Gijsegem, F., S. Genin, and C. Boucher.** 1993. Conservation of secretion pathways for pathogenicity determinants of plant and animal bacteria. *Trends Microbiol.* **1:**161–163.

103. **Venkatesan, M. M., J. M. Buysse, and E. V. Oaks.** 1992. Surface presentation of *Shigella flexneri* invasion plasmid antigens requires the products of the *spa* locus. *J. Bacteriol.* **174:**1990–2001.

104. **Volgler, A. P., M. Homma, V. M. Irikura, and R. M. Macnab.** 1991. *Salmonella typhimurium* mutants defective in flagellar filament regrowth and sequence similarity of FliI to F_0F_1, vacuolar, and archaebacterial ATPase subunits. *J. Bacteriol.* **173:**3564–3572.

105. **Wattiau, P., and G. R. Cornelis.** 1993. Response to 'yet another chaperone?'. *Trends Microbiol.* **1:**162–163.

106. **Yu, J., and J. B. Kaper.** 1992. Cloning and characterization of the *eae* gene of enterohaemorrhagic *Escherichia coli* O157:H7. *Mol. Microbiol.* **6:**411–417.

107. **Zhang, H.-Z., and M. S. Donnenberg.** 1993. Localized adherence by enteropathogenic *Escherichia coli* (EPEC) requires the chromosomal *dsbA* locus, abstr. B-312, p. 81. *Abstr. 93rd Gen. Meet. Am. Soc. Microbiol. 1993.* American Society for Microbiology, Washington, D.C.

108. **Zhang, W., M. Wolf, M. Bratoeva, R. Cantey, and E. Boedeker.** 1993. Cloning and characterization of the *eae*-like gene of rabbit enteropathogenic *Escherichia coli* RDEC-1, abstr. D-194, p. 129. *Abstr. 93rd Gen. Meet. Am. Soc. Microbiol. 1993.* American Society for Microbiology, Washington, D.C.

Molecular Genetics of Bacterial Pathogenesis
Edited by V. L. Miller, J. B. Kaper, D. A. Portnoy, and R. R. Isberg
© 1994 American Society for Microbiology, Washington, DC 20005

Chapter 13

Murine Colonic Hyperplasia

David B. Schauer

Transmissible murine colonic hyperplasia is a naturally occurring disease of laboratory mice caused by one particular biotype of *Citrobacter freundii*. The disease is not as clinically significant as it once was in laboratory animal facilities. This is probably due to the fact that the causative agent is widely recognized as an important pathogen of the gastrointestinal tract, and monitoring for the presence of the organism is routinely performed by commercial suppliers of laboratory mice. The disease is rarely seen in barrier-maintained mouse populations today. Despite the low incidence of clinical cases, transmissible murine colonic hyperplasia is of interest to microbiologists and gastroenterologists alike because of the unusual pathogenesis of the disease. Infection of mice with the pathogenic biotype of *C. freundii* results in an increased rate of epithelial cell proliferation in the distal large bowel. The increased rate of proliferation leads to mucosal hyperplasia, which is characterized by cytokinetic changes similar to those seen in humans suffering from proliferative bowel disorders. The hyperplastic state serves to promote the onset of chemically induced colon cancer in these mice, which may be analogous to the increased risk for colorectal cancer that ulcerative colitis patients are known to possess. In addition, the bacteria produce histopathological lesions in the large bowel which are indistinguishable from those associated with attaching and effacing *Escherichia coli* (AEEC) infections in humans. As a naturally occurring bacterial disease of laboratory mice, transmissible murine colonic hyperplasia provides a manageable system in which to better understand the pathogenesis of human AEEC disease.

The genus *Citrobacter* comprises three species: *C. freundii, C. diversus,* and *C. amalonaticus.* They occupy a distinct taxonomic position, intermediate between *E. coli* and the *Salmonella* species, in the family *Enterobacteriaceae. Citrobacter* species exhibit 40 to 50% DNA relatedness to both *E. coli* and *Salmonella* species (10). Isolation of *C. freundii* from the feces of humans and animals and from water, soil, and food is not uncommon (26). Most isolates are considered to be typical opportunistic pathogens, producing infection at a variety of extraintestinal sites, the most common of which is the urinary tract (33). On the other hand,

David B. Schauer • Division of Toxicology and Division of Comparative Medicine, Massachusetts Institute of Technology, 77 Massachusetts Avenue, Room E18-564, Cambridge, Massachusetts 02139.

isolates of *C. freundii* have occasionally been associated with diarrheal disease in humans (18). While a causal relationship between these *C. freundii* isolates and disease has not been established, some of these strains have been shown to express known virulence determinants. Strains producing heat-stable enterotoxin have been identified (19), while others, isolated from the stools of humans with diarrhea and from beef samples, have been shown to produce a Shiga-like toxin (30). Although the true significance of *C. freundii* in human diarrheal disease is not yet clear, it would appear that some strains have the ability to cause primary gastrointestinal disease in otherwise healthy individuals. This is in contrast to the opportunistic infections which are typically associated with the organism. *C. freundii* isolates which cause transmissible murine colonic hyperplasia also produce a primary gastrointestinal disease in otherwise healthy mice.

Sporadic outbreaks of rectal prolapse and diarrhea associated with moderate mortality were reported in several mouse colonies in the late 1960s and early 1970s (4, 9, 11, 15). Brennan et al. (9), at the Argonne National Laboratory, and Ediger et al. (15), at the Frederick Cancer Research Center, were each able to isolate a *C. freundii* strain from affected mice and use the isolates to fulfill Koch's postulates. The isolate of Brennan et al. (9) was identified as a biochemically atypical *C. freundii* strain, differing from typical strains in that it did not utilize citrate as a sole carbon source, it did not produce H_2S in the butt of Klingler iron agar, and it produced only slight H_2S on lead acetate paper. Also unlike typical *C. freundii*, this strain was not motile (9). Ediger et al. (15) reported that their strain was identical to that of Brennan et al. (9), except for the fact that it did not produce H_2S on lead acetate paper (15). A third outbreak was investigated by Barthold et al. (4), who characterized their *C. freundii* isolate more extensively. This atypical *C. freundii* strain was designated biotype 4280. Like the previous isolates, biotype 4280 did not utilize citrate as a sole carbon source after 24 h, although it did become positive for this activity after 48 to 72 h. This strain was also negative for H_2S production in the butt of triple sugar iron agar, but it did produce H_2S on lead acetate paper. In addition, biotype 4280 was found to decarboxylate ornithine and to be nonmotile (4). In contrast, 80% of *C. freundii* isolates produce H_2S in the butt of triple sugar iron agar, 99% readily utilize citrate as a sole carbon source, fewer than 20% decarboxylate ornithine, and 95% are motile (17). Koch's postulates have also been fulfilled with *C. freundii* biotype 4280, which produces transmissible murine colonic hyperplasia when pure cultures are inoculated into either conventional or germfree mice (4).

DISEASE

Transmissible murine colonic hyperplasia is primarily a disease of young mice. Affected animals exhibit nonspecific clinical signs, including listlessness, ruffled pelage, hunched posture, and retarded growth. The animals may develop accumulations of pasty feces about the anus, the base of the tail, and the perineum. Some mice also exhibit rectal prolapse. Adult mice are susceptible to infection, but rarely exhibit clinical signs of disease. Upon postmortem examination, both

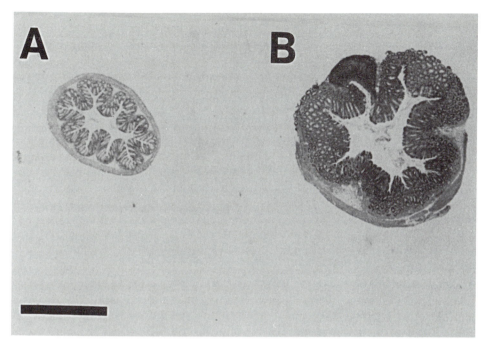

FIGURE 1. Low-power light micrograph of a transverse section of the distal colon from a normal mouse (A) and a mouse with transmissible murine colonic hyperplasia (B). Bar, 1 mm.

young animals and adults are found to have profound thickening and rigidity of the distal large bowel (Fig. 1). Occasionally, the entire colon, and even less frequently, the cecum, may be affected as well. Typically, the colon does not contain any formed feces, and the cecum is often empty and contracted. The progression of disease in experimentally infected mice has been documented (5, 21).

In suckling animals, a 7-day latent period after oral inoculation with *C. freundii* biotype 4280 is followed by the onset of clinical signs. These signs are most severe, and mortality is greatest 2 to 3 weeks after infection. Likewise, grossly detectable colonic thickening is first observed 1 week following infection and is most apparent at 2 to 3 weeks following infection, thereafter decreasing in severity. Mucosal hyperplasia, evidenced by elongation of the colonic crypts, is detectable microscopically by the fourth day postinfection. At this time, the cells in the lower half of the crypt appear crowded, a relative decrease in the number of goblet cells is evident, and mitotic activity is increased. By the end of the first week, the crypts are some 40 cells in height, compared to the normal crypt height of approximately 25 cells. Goblet cells are rarely seen, and mitotic figures are frequent. At the peak of mucosal hyperplasia, 2 to 3 weeks after infection, the crypts are crowded throughout, often appearing pseudostratified, and may contain three to four times the normal number of cells. Maximal hyperplasia is followed by regression of the lesions. Over a period of another 2 to 4 weeks the mucosa returns to normal. A rebound increase in the number of goblet cells often occurs, but by 8 weeks after infection the lesions are largely resolved. There is no significant difference in the onset and progression of hyperplasia between adult and suckling mice. However, superimposed on the hyperplasia lesions in suckling mice are marked inflammatory changes. If adult mice develop inflammation at all, it is less dramatic. While there appears to be no correlation between the presence of inflammation and the severity of hyperplasia, inflammation does correlate with morbidity and mortality (5).

Inflammatory changes are first seen in suckling mice 10 to 12 days after infection. Mucosal necrosis and erosions develop, as do infiltrates of neutrophils in the lamina propria and submucosa. With time, the inflammatory cells change from neutrophils to mononuclear cells, and eventually, lymphocytes and plasma cells predominate. In addition to age-related differences in the degree of inflammation, the genetic background of the mouse also seems to influence the response to infection (7). Mortality and inflammation are greater in C3H/HeJ mice than in DBA/2J, NIH Swiss, or C57BL/6J mice. In fact, the lesions in adult C3H/HeJ mice are similar to those in suckling NIH Swiss mice. Conversely, hyperplasia is no greater in C3H/HeJ mice than in either DBA/2J or C57BL/6J mice, while NIH Swiss mice get the most extensive hyperplasia. The bases for the genetic and age-related differences in susceptibility to disease and in the severity of hyperplasia are not clear. It is interesting that Swiss mice are also more susceptible to chemically induced colonic neoplasia than mice of the DBA/2 and C57BL/Ha strains (16).

COCARCINOGENESIS

The transient hyperplastic state which is induced by *C. freundii* biotype 4280 has a profound effect on the cytokinetics of the mucosal epithelial cells in the large bowel. These changes also serve to increase the susceptibility of the colonic epithelial cells to the carcinogenic effect of 1,2-dimethylhydrazine (DMH). In healthy mice, weekly administration of DMH leads to the development of neoplastic lesions in the large bowel. The earliest neoplastic change in the colon is focal epithelial cell proliferation in the superficial mucosa. The proliferation spreads into neighboring crypts, and eventually, adjacent foci coalesce to produce colonic tumors. These tumors are more frequent in the distal colon than in the proximal colon and do not involve the cecum. Colon cancer develops with a 70% incidence in DMH-treated mice after 3 months' time. In mice infected with *C. freundii* biotype 4280 early in the course of DMH administration, the latent period for carcinogenesis is reduced. In these animals a 70% incidence of colon cancer is apparent after 1 month of DMH administration (6). Transmissible murine colonic hyperplasia has no effect on established DMH-induced tumors (6). Even a single dose of DMH is sufficient to induce neoplastic lesions if it is given concurrently with or within 1 week of *C. freundii* infection. Single-dose administration of DMH 2 weeks after infection results in a low incidence of mild lesions, and if given 3 weeks after infection, neoplastic lesions do not develop at all (3, 6). Since DMH is metabolized and excreted within hours of administration, the temporal relationship between

DMH administration and *C. freundii* infection supports a model in which transmissible murine colonic hyperplasia promotes carcinogenesis followed by initiation by DMH. There is no evidence that infection with any single bacterial pathogen promotes colorectal cancer in humans, but the aberrant mucosal cytokinetic changes in people suffering from proliferative bowel disorders may be quite similar to those in mice with transmissible murine colonic hyperplasia.

The cytokinetics of the hyperplastic mucosa in transmissible murine colonic hyperplasia have been characterized by autoradiography of histological sections labeled with [³H]thymidine (1). In this manner, *C. freundii* biotype 4280 has been shown to cause a twofold increase in the labeling index of colonic epithelial cells. The proliferative zone, normally restricted to the basal one-third of the crypt, is also expanded, such that it encompasses the entire crypt column, including the surface epithelium. Increased labeling index, expansion of the proliferative zone, and surface labeling have also been demonstrated in ulcerative colitis patients (8). People with this disease are known to suffer an increased risk and an early onset of colorectal cancer. It may be that transmissible murine colonic hyperplasia is an appropriate model for some aspects of ulcerative colitis. However, intestinal cell renewal, like large bowel cancer, is influenced by many factors, including diet, intestinal microflora, and genotype. Only with a better understanding of their pathogenesis will it be possible to appreciate the relevance of transmissible murine colonic hyperplasia to human proliferative bowel disorders. The mechanism by which *C. freundii* biotype 4280 induces epithelial cell proliferation and promotes colonic neoplasia is poorly understood.

MICROBIAL PATHOGENESIS

Transient colonization of the colonic mucosa by *C. freundii* biotype 4280 precedes the onset of epithelial cell hyperplasia in infected mice. The presence of these bacteria in the large bowel is first detectable 3 to 4 days after oral inoculation into both suckling mice and adults. The number of organisms which can be recovered from the colon increases until day 10 postinfection, at which time the population of *C. freundii* may reach 10^{10} CFU/g of colon (2). When the number of organisms reaches maximal density in the colon, it appears to be at the expense of the resident flora. *C. freundii* biotype 4280 constitutes greater than 90% of the aerobic microflora in the colon 7 to 10 days after oral inoculation (9). Peak colonization is followed by a spontaneous decrease in the number of bacterial organisms. By 16 days postinfection, at a time when maximal hyperplasia is evident, the number of *C. freundii* present in the colon is already reduced. At 4 weeks following infection it is rare to recover the organism from the gastrointestinal tract at all (2). Elimination of the organism from the gastrointestinal tract of mice by administering neomycin in the drinking water demonstrates that the presence of *C. freundii* for as short a period as 2 days is sufficient to induce a proliferative response. The longer the bacteria are present, through 10 days after infection, the greater the degree of hyperplasia (2).

When *C. freundii* biotype 4280 colonizes the large bowel of mice, it adheres to the mucosal epithelial surface in a characteristic manner. When viewed by scanning electron microscopy the bacteria appear to be partially embedded in the apical plasma membrane of intestinal epithelial cells. Occasional vacant cavities can be observed; these vacant cavities appear to represent sites at which once-adherent bacteria are no longer attached. The bacteria themselves appear uniformly smooth and devoid of any obvious surface structures (21). Viewed by transmission electron microscopy, the outer membrane of the embedded portion of the bacteria is closely opposed with the enterocyte plasma membrane. A uniform space of some 25 nm separates the two membranes (Fig. 2). This intimate adherence gives the impression that the plasma membrane is forming pedestals or cups in which the bacteria rest. There is dissolution of the brush border at the site of bacterial attachment and destruction of the terminal web (21, 28). These histopathological changes are also characteristic of mucosal adherence by several

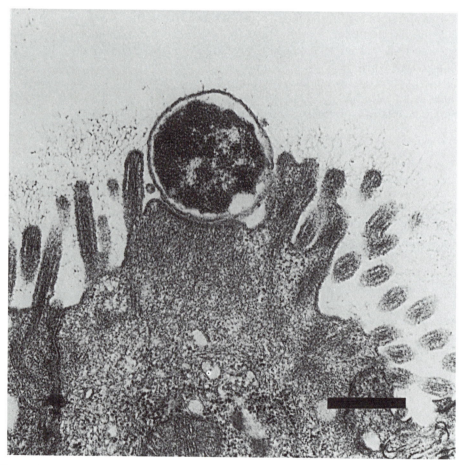

FIGURE 2. Transmission electron micrograph of *C. freundii* biotype 4280 producing a typical attaching and effacing lesion on an epithelial cell in the colon of a mouse. Bar, 0.5 μm.

pathogenic groups of *E. coli*. First observed by infecting neonatal pigs with strains from cases of human diarrhea (31) and subsequently termed attaching and effacing (AE) lesions (24), isolates that exhibit this activity have been referred to as AEEC, for attaching and effacing *E. coli*.

The AEEC can be broadly divided into two groups. Strains which cause small bowel diarrhea and which produce AE lesions primarily in the small intestine are termed enteropathogenic *E. coli* (EPEC) (see chapter 12). Strains which elaborate one or more Shiga-like toxins and which produce AE lesions primarily in the large bowel are termed enterohemorrhagic *E. coli* (EHEC). EHEC strains are associated with hemorrhagic colitis and hemolytic uremic syndrome and include strains of the O157:H7 serotype. While the role of AE lesion formation in AEEC infection is not completely understood, it is clearly involved in the pathogenesis of disease (14). To study the formation of AE lesions, tissue culture systems have been developed; in these systems AEEC strains produce typical histopathological changes in monolayers of cultured epithelial cells. Such a system has led to the development of a rapid assay for AE lesion formation. This assay exploits the fact that dense aggregations of filamentous actin accumulate within epithelial cells, in conjunction with the destruction of the terminal web, at the site of AEEC adherence. Fluorescein-conjugated phallotoxin, which specifically binds filamentous actin, is used to detect the actin accumulation within the cultured epithelial cells (22). The absence of actin staining activity by Tn*phoA* mutants of EPEC was used to identify a genetic locus necessary for the formation of AE lesions (20). This gene, now designated *eaeA*, for *E. coli* attaching and effacing, has been shown to encode an outer membrane protein termed intimin. Intimin is necessary, but not sufficient, for intimate adherence to cultured epithelial cells and has been shown to be a virulence determinant in a human volunteer study (14).

When a 1-kb restriction fragment of DNA derived from the central portion of the *eaeA* gene is used to probe DNA from *C. freundii* biotype 4280, it is found to hybridize. The *eaeA* gene probe does not hybridize with the DNAs of other biotypes of *C. freundii* (28). Since other biotypes of *C. freundii* do not effectively colonize the large bowels of laboratory mice and do not cause transmissible murine colonic hyperplasia (7), the presence of *eaeA* DNA homology is associated with the ability to infect a mouse and cause disease. On the basis of hybridization, a fragment of DNA containing the *eaeA* homology was identified from a genomic library of *C. freundii* biotype 4280 (28). The predicted amino acid sequence of the *C. freundii* gene product has 82% similarity to that of intimin. Interestingly, in the same reading frame, immediately 5' of the *C. freundii* coding sequence, there is a smaller open reading frame which is also highly conserved between *C. freundii* and EPEC. The predicted amino acid sequences of these two putative 156-amino-acid gene products share 96% similarity. It is not clear what the functions of these loci are or, indeed, if they are expressed at all.

The gene product of the *C. freundii* locus with homology to *eaeA* is predicted to be 936 amino acids in length, with a molecular mass of 102.1 kDa. This is close to the predicted size of 939 amino acids and a molecular mass of 102.4 kDa for intimin. There is 92% similarity between the N-terminal 702 residues of the predicted *C. freundii* gene product and the N-terminal 705 residues of intimin. The remaining 234 amino acids at the C terminus of each protein have only 68% similarity. Unlike the extensive homology 5' of the two genes, homology with *eaeA* ends abruptly 3' of the *C. freundii* open reading frame. In fact, the 15th nucleotide beyond the stop codon of the *C. freundii* gene begins a series of unique heptameric repeats with the sequence CCCGGCA. Approximately 40 directly repeating copies are present, and the function of the repeats has not been established. The *C. freundii* DNA 3' of the repeats exhibits interesting sequence homology. Over a region of approximately 200 bp there is greater than 80% nucleotide identity between the *C. freundii* sequence and the noncoding sequence 3' from the *yopH* gene of the *Yersinia pseudotuberculosis* virulence plasmid (27). Perhaps even more interesting is the homology between the complementary sequence of *C. freundii* DNA and the putative transposase from the insertion sequence IS*Rm3* of *Rhizobium meliloti*. This homology spans some 200 nucleotides, exhibits greater than 60% identity at the DNA level, and includes a partial open reading frame with 30% amino acid identity and 57% similarity to a portion of the putative transposase (27). While the presence of the repeats and the DNA homology with a putative transposase represent circumstantial evidence at best, they may indicate that horizontal transfer of *eaeA* and related genes has occurred.

The role of *eaeA* in the pathogenesis of transmissible murine colonic hyperplasia has been examined by the construction of an isogenic mutant of *C. freundii* biotype 4280 (29). When orally inoculated into susceptible mice, the *eaeA* mutant does not colonize the large bowel and fails to cause disease. In contrast, *eaeA* mutants of EPEC adhere to cultured epithelial cells and cause diarrhea in human volunteers, albeit with a reduced incidence (14). On the other hand, in the absence of plasmid-encoded pili, which mediate the initial bacterial interaction with epithelial cells, *eaeA* mutants of EPEC are not adherent (19). The similar phenotypes of *eaeA* mutants of *C. freundii* biotype 4280 and plasmid-cured EPEC strains may be due to the fact that *C. freundii* has no initial adhesin. Alternatively, it may be that, despite the presence of a putative initial adhesin, the *eaeA* gene product is necessary for persistent colonization by *C. freundii* in vivo.

The EPEC strains of humans and rabbits express plasmid-encoded adhesins, which appear to be important in the initial interaction of bacteria with epithelial cells. The genes for these fimbrial adhesins, or pili, are carried on high-molecular-weight virulence plasmids. *C. freundii* biotype 4280 contains an approximately 65-kb plasmid which fails to exhibit hybridization with the EPEC virulence plasmids in Southern blot analysis. Since a strain of *C. freundii* cured of this plasmid colonizes mice and produces disease in a manner which is indistinguishable from that by the plasmid-containing parent (29), it is probably appropriate to state that *C. freundii* biotype 4280 possesses a cryptic plasmid and not a virulence plasmid. However, the experiments which might demonstrate the absence of any role for the plasmid in pathogenesis, such as determination of median infectious doses of plasmid-containing and plasmid-cured strains, have not been performed.

The molecular basis of transmissible murine colonic hyperplasia remains enigmatic. Characteristic histopathological lesions are produced by *C. freundii* biotype 4280 in the mucosa of the large bowel of laboratory mice prior to the onset of gross hyperplasia. The relationship between the formation of these lesions and the induction of mucosal hyperplasia is not clear. A gene with significant homology to the *eaeA* gene of EPEC is present in *C. freundii* biotype 4280 but not in other biotypes of *C. freundii*. The *C. freundii eaeA* gene product exhibits C-terminal sequence divergence from intimin. This locus is necessary for both colonic colonization and disease in laboratory mice. The formation of AE lesions by human EPEC strains has been shown to involve the complex interaction of many gene products (13), and undoubtedly, many virulence determinants of *C. freundii* biotype 4280 remain to be identified.

SUMMARY

C. freundii biotype 4280 probably represents a clonal population of bacteria which asymptomatically infect adult laboratory mice. Spontaneous epizootic outbreaks have been documented, but given the nature of the disease, it is likely that many more cases of enzootic infection have existed and have gone unreported. Like many other naturally occurring diseases of laboratory animals, transmissible murine colonic hyperplasia is a casualty of modern research. Commercial vendors and scientists alike simply cannot afford to allow research animals to become infected with this agent. As a result, the derivation of animals by hysterectomy and their maintenance in barrier facilities has dramatically reduced the incidence of disease. This should not lessen interest in the pathogenesis of transmissible murine colonic hyperplasia, particularly given the remarkable pathophysiology of the disease, and the extent to which host factors may be characterized in naturally occurring infections of laboratory mice.

C. freundii biotype 4280 appears to be unusually well adapted for colonization of the murine large bowel. Experimental infection of rats and hamsters with this bacterium does not result in either colonization or disease (4). Mice, conversely, undergo a rapid increase in the number of bacteria in the large bowel, such that *C. freundii* constitutes almost the entire aerobic microflora in the colon; this is followed by a clearing of the organisms which is almost as rapid. Mice appear to be refractory to a second challenge, following resolution of the primary infection (2). Despite the fact that inflammation is not a primary component of the disease, humoral immunity to the bacteria may account for this subsequent protection. On the other hand, the proliferative response to infection may itself provide a mechanism for bacterial clearance. Newly arising enterocytes do not seem to become colonized by bacteria as they replace the surface epithelial cells to which the bacteria adhere (21). It is tempting to speculate that the terminally differentiated cells express a receptor which is absent or less abundant on the newly recruited cells. However, in those cases in which bacteria descend into the colonic crypts early in infection, they do adhere to the incompletely differentiated enterocytes (21). Perhaps some soluble factor in the gut or a regulatory change in the bacteria themselves accounts for this phenomenon.

Murine colonic hyperplasia is not pathognomonic for *C. freundii* infection.

Several targeted gene mutations in mice have recently been reported; these mutations lead to spontaneous hyperplastic colitis in the large bowel (23, 25). It has been speculated that the absence of T-cell regulation in these mice leads to an inappropriate B-cell response, which results in inflammatory bowel disease (32). These animals appear to be free of currently recognized pathogenic agents. Remarkably, when these animals are housed under germfree conditions, they do not develop disease (25). If the resident microflora is able to induce a condition in these mice with mutant genes that is similar to transmissible murine colonic hyperplasia, then one may well ask if there are common features of mucosal adherence by normal flora and *C. freundii*. At least one common, indigenous organism (12) has the ability to cause histopathological changes in enterocytes at the site of bacterial adherence which are reminiscent of AE lesions (Fig. 3). It seems likely that a better understanding of the molecular pathogenesis of transmissible murine

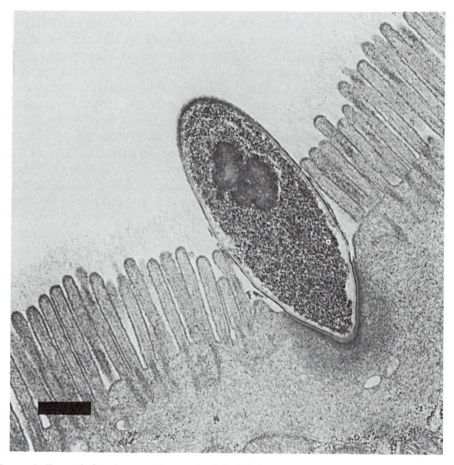

FIGURE 3. Transmission electron micrograph of the holdfast segment of an indigenous, filamentous organism (12) attached to an epithelial cell in the ileum of a mouse. Bar, 0.5 μm.

colonic hyperplasia will also bring a better understanding of the role of bacteria in proliferative bowel disease.

ACKNOWLEDGMENTS. I thank Stephen Barthold and Jim Kaper for generously providing strains, Nafisa Ghori for assistance with microscopy, and Stanley Falkow for making it all possible.

REFERENCES

1. **Barthold, S. W.** 1979. Autoradiographic cytokinetics of colonic mucosal hyperplasia in mice. *Cancer Res.* **39**:24–29.
2. **Barthold, S. W.** 1980. The microbiology of transmissible murine colonic hyperplasia. *Lab. Anim. Sci.* **30**:167–173.
3. **Barthold, S. W., and D. Beck.** 1980. Modification of early dimethylhydrazine carcinogenesis by colonic mucosal hyperplasia. *Cancer Res.* **40**:4451–4455.
4. **Barthold, S. W., G. L. Coleman, P. N. Bhatt, G. W. Osbaldiston, and A. M. Jonas.** 1976. The etiology of transmissible murine colonic hyperplasia. *Lab. Anim. Sci.* **26**:889–894.
5. **Barthold, S. W., G. L. Coleman, R. O. Jacoby, E. M. Livstone, and A. M. Jonas.** 1978. Transmissible murine colonic hyperplasia. *Vet. Pathol.* **15**:223–236.
6. **Barthold, S. W., and A. M. Jonas.** 1977. Morphogenesis of early 1,2-dimethylhydrazine-induced lesions and latent period reduction of colon carcinogenesis in mice by a variant of *Citrobacter freundii*. *Cancer Res.* **37**:4352–4360.
7. **Barthold, S. W., G. W. Osbaldiston, and A. M. Jonas.** 1977. Dietary, bacterial, and host genetic interactions in the pathogenesis of transmissible murine colonic hyperplasia. *Lab. Anim. Sci.* **27**:938–945.
8. **Bleiberg, H., P. Mainguet, P. Galand, J. Chretien, and N. Dupont-Mairesse.** 1970. Cell renewal in the human rectum. In vitro autoradiographic study on active ulcerative colitis. *Gastroenterology* **58**:851–855.
9. **Brennan, P. C., T. E. Fritz, R. J. Flynn, and C. M. Poole.** 1965. *Citrobacter freundii* associated with diarrhea in laboratory mice. *Lab. Anim. Care* **15**:266–275.
10. **Brenner, D. J.** 1984. Family I. *Enterobacteriaceae* Rahn 1937, Nom. fam. cons. Opin. 15, jud. Comm. 1958, 73; Ewing, Farmer, and Brenner 1980, 674; Judicial Commission 1981, 104, p. 408–420. *In* N. R. Krieg and J. G. Holt (ed.), *Bergey's Manual of Systematic bacteriology,* vol. 1. The Williams & Wilkins Co., Baltimore.
11. **Brynjolfsson, G., and L. S. Lombard.** 1969. Colitis cystica in mice. *Cancer* **23**:225–229.
12. **Chase, D. G., and S. L. Erlandsen.** 1976. Evidence for a complex life cycle and endospore formation in the attached, filamentous, segmented bacterium from murine ileum. *J. Bacteriol.* **127**:572–583.
13. **Donnenberg, M. S., and J. B. Kaper.** 1992. Enteropathogenic *Escherichia coli*. *Infect. Immun.* **60**:3953–3961.
14. **Donnenberg, M. S., C. O. Tacket, S. P. James, G. Losonsky, J. P. Nataro, S. S. Wasserman, J. B. Kaper, and M. M. Levine.** 1993. Role of the *eaeA* gene in experimental enteropathogenic *Escherichia coli* infection. *J. Clin. Invest.* **92**:1412–1417.
15. **Ediger, R. D., R. M. Kovatch, and M. M. Rabstein.** 1974. Colitis in mice with a high incidence of rectal prolapse. *Lab. Anim. Sci.* **24**:488–494.
16. **Evans, J. T., T. B. Shows, E. E. Sproul, N. S. Paolini, A. Mittelman, and T. S. Hauschka.** 1977. Genetics of colon carcinogenesis in mice treated with 1,2-dimethylhydrazine. *Cancer Res.* **37**:134–136.
17. **Farmer, J. J., III, B. R. Davis, F. W. Hickman-Brenner, A. McWhorter, G. P. Huntley-Carter, M. A. Asbury, C. Riddle, H. G. Wathen-Grady, C. Elias, G. R. Fanning, A. G. Steigerwalt, C. M. O'Hara, G. K. Morris, P. B. Smith, and D. J. Brenner.** 1985. Biochemical identification of new species and biogroups of *Enterobacteriaceae* isolated from clinical specimens. *J. Clin. Microbiol.* **21**:46–76.
18. **Guarino, A., G. Capano, B. Malamisura, M. Alessio, S. Guandalini, and A. Rubino.** 1987. Production of *Escherichia coli* STa-like heat-stable enterotoxin by *Citrobacter freundii* isolated from humans. *J. Clin. Microbiol.* **25**:110–114.

19. **Guarino, A., R. Giannella, and M. R. Thompson.** 1989. *Citrobacter freundii* produces an 18-amino-acid heat-stable enterotoxin identical to the 18-amino-acid *Escherichia coli* heat-stable enterotoxin (STIa). *Infect. Immun.* **57:**649–652.

20. **Jerse, A. E., J. Yu, B. D. Tall, and J. Kaper.** 1990. A genetic locus of enteropathogenic *Escherichia coli* necessary for the production of attaching and effacing lesions on tissue culture cells. *Proc. Natl. Acad. Sci. USA* **87:**7839–7843.

21. **Johnson, E., and S. W. Barthold.** 1979. The ultrastructure of transmissible murine colonic hyperplasia. *Am. J. Pathol.* **97:**291–301.

22. **Knutton, S., T. Baldwin, P. H. Williams, and A. S. McNeish.** 1989. Actin accumulation at sites of bacterial adhesion to tissue culture cells: basis of a new diagnostic test for enteropathogenic and enterohemorrhagic *Escherichia coli. Infect. Immun.* **57:**1290–1298.

23. **Mombaerts, P., E. Mizoguchi, M. J. Grusby, L. H. Glimcher, A. K. Bahn, and S. Tonegawa.** 1993. Spontaneous development of inflammatory bowel disease in T cell receptor mutant mice. *Cell* **75:** 275–282.

24. **Moon, H. W., S. C. Whipp, R. A. Argenzio, M. M. Levine, and R. A. Giannella.** 1983. Attaching and effacing activities of rabbit and human enteropathogenic *Escherichia coli* in pig and rabbit intestines. *Infect. Immun.* **41:**1340–1351.

25. **Sadlack, B., H. Merz, H. Schorle, A. Schimpl, A. C. Feller, and I. Horak.** 1993. Ulcerative colitis-like disease in mice with a disrupted interleukin-2 gene. *Cell* **75:**253–261.

26. **Sakazaki, R.** 1984. Genus IV. *Citrobacter* Werkman and Gillen 1932, 173^{AL}, p. 458–461. *In* N. R. Krieg and J. G. Holt (ed.), *Bergey's Manual of Systematic Bacteriology,* vol. 1. The Williams & Wilkins Co., Baltimore.

27. **Schauer, D. B.** Unpublished data.

28. **Schauer, D. B., and S. Falkow.** 1993. Attaching and effacing locus of a *Citrobacter freundii* biotype that causes transmissible murine colonic hyperplasia. *Infect. Immun.* **61:**2486–2492.

29. **Schauer, D. B., and S. Falkow.** 1993. The *eae* gene of *Citrobacter freundii* biotype 4280 is necessary for colonization in transmissible murine colonic hyperplasia. *Infect. Immun.* **61:**4654–4661.

30. **Schmidt, H., M. Montag, J. Bockemühl, J. Heesemann, and H. Karch.** 1993. Shiga-like toxin II-related cytotoxins in *Citrobacter freundii* strains from humans and beef samples. *Infect. Immun.* **61:**534–543.

31. **Staley, T. E., W. Jones, and L. D. Corley.** 1969. Attachment and penetration of *Escherichia coli* into intestinal epithelium of the ileum in newborn pigs. *Am. J. Pathol.* **56:**371–392.

32. **Strober, W., and R. O. Ehrhardt.** 1993. Chronic intestinal inflammation: an unexpected outcome in cytokine of T cell receptor mutant mice. *Cell* **75:**203–205.

33. **Zwadyk, P.** 1992. Opportunistic Enterobacteriaceae, p. 544–555. *In* W. K. Joklik, H. P. Willett, D. B. Amos, and C. M. Wilfert (ed.), *Zinsser Microbiology.* Appleton & Lange, Norwalk, Conn.

Part 3. Molecular and Cellular Biology of Intracellular Bacteria

A PERSONAL RETROSPECTIVE

I entered Stanley Falkow's laboratory in 1978 as a graduate student at the University of Washington. At that time, the emphasis of the laboratory was plasmids, transposons, and the molecular biology of enterotoxigenic and uropathogenic *Escherichia coli*. As for pathogenesis, toxins and pili were the central focus. There was no active interest in intracellular pathogens. Other graduate students in the lab included Steve Moseley and Ron Gill, while Walt Dallas and Marilyn Roberts both defended their theses that fall. Gordon Dougan had just returned to England, and Mark Achtman was on sabbatical from the Max Planck Institute in Berlin. Jorge Crosa, a previous postdoctoral fellow, was still there, but he had his own laboratory. Postdoctoral fellows in the lab included Richard and Sheila Hull, Bob Quackenbush, Mike Lovett, and Dennis Schaberg. Arriving the following year were two new students, Alison Weiss and Mike Koomey, and new post-doctoral fellows, Jim Kaper, Rod Welch, and Mitch Cohen; Pat Totten arrived as a technician. We represented the end of an era, because in 1981, Stan accepted the chair of microbiology at Stanford. Mike Koomey, Alison Weiss, Steve Moseley, Rod Welch, and I moved with Stan to California.

These were very exciting times as the laboratory moved into bacterial pathogenesis with a vengeance. The era of Tn*3* and plasmid biology came to an end, while studies of the pathogenesis of *Yersinia* spp., *Neisseria gonorrhoeae,* and *Bordetella pertussis* started in the lab and enterotoxigenic and uropathogenic *E. coli* remained areas under intense investigation. It still amazes me how Stan was able to have command of the intricacies of five different pathogens. He was the only person who knew every detail of each system. As a graduate student, this was an amazing environment in which to learn about bacterial pathogenesis.

In 1978, the lab used mammalian cells only to assay for toxic activity or to evaluate adherence. There was no active interest in intracellular pathogens or even bacterial entry for that matter. However, Steve Moseley (a second-year graduate student at the time) read an abstract at the 1977 Annual Meeting of the American Society for Microbiology authored by Don Zink; the abstract indicated that a plasmid was associated with a positive Sereny test in *Yersinia enterocolitica*. Since a positive Sereny test was believed to correlate with invasiveness in *Shigella* spp., Steve Moseley thought that the *Yersinia* plasmid might encode determinants required for invasion. Accordingly, he collected a number of natural isolates of *Y. enterocolitica* and ordered HEp-2 cells to begin invasion assays which he had done previously as a technician in Peter Gemski's lab in Sam Formal's group at the Walter Reed Army Institute of Research.

As a rotating graduate student in January of 1979, Stan assigned me to work with Steve Moseley and examine the relationship of plasmids and invasiveness in *Y. enterocolitica*. This rotation project became the basis of my doctoral research. In summary, we and others showed that all of the pathogenic yersiniae harbored a related plasmid which was essential for pathogenesis (6, 10, 11, 21, 22, 28). The plasmid was not necessary for invasion of cells but, rather, was associated with a cytotoxic activity (monolayer detachment assay) (22). The plasmid encoded a number of major outer membrane proteins (named Yops by Hans Wolf-Watz and later shown by Heeseman and colleagues to be secreted by yersiniae) which were expressed only at 37°C and during growth in medium with low calcium concentrations. Stan proposed that the plasmid was an "intracellular thermostat which senses whether the micro-organism should adopt a free-living or a parasitic mode of growth" (5). We believed that the plasmid acted subsequent to internalization. It is now thought that the plasmid-encoded products may act from outside of cells and prevent uptake of the bacteria by macrophages (see chapter 24). It is now known that among the plasmid-encoded proteins there is a tyrosine phosphatase, a serine or threonine kinase, and a cytotoxin (25). Nevertheless, in 1983 it was clear to both Stan and myself that the answers to many of the important questions regarding the interaction of *Yersinia* spp. with mammalian cells would require an appreciation for the cell biology of host cells. When I left in 1983 to do a postdoctoral fellowship with Jay Unkeless in the Zanvil Cohn Laboratory of Cellular Physiology and Immunology at The Rockefeller University, I fully anticipated returning to *Yersinia* spp. I thought the topic of invasion would be waiting. However, within months of my departure, Stan and a new postdoctoral fellow in Stan's lab, Ralph Isberg, made a phenomenal finding; that is, cloning of a single gene from *Yersinia* spp. into *E. coli* K-12 facilitated invasion of mammalian cells (16). This stunning observation ushered in a new era in the Falkow laboratory. As it turns out, I never returned to work with yersiniae but later chose to work on *Listeria monocytogenes*; I discuss much of that work in chapter 18.

THE GLOBAL IMPORTANCE OF INTRACELLULAR PATHOGENS

Intracellular pathogens are responsible for a devastating number of infectious diseases worldwide. For example, *Mycobacterium tuberculosis* remains one of the leading causes of death worldwide, and its prevalence is also increasing in the United States. *Chlamydia trachomatis*, an obligate intracellular pathogen, is one of the leading causes of blindness worldwide and is a major cause of sexually transmitted disease in the United States (see chapter 19 by Patrik Bavoil). The protozoan intracellular pathogens *Trypanosoma cruzi*, *Leishmania* spp., and *Plasmodium* spp. together infect more than 100 million people. Clearly, intracellular pathogens represent an enormous challenge to the scientific and medical communities, yet basic understanding of these pathogens is still in its infancy. This is due in part to the difficulty in cultivating many of these pathogens as well as the lack of sophisticated techniques of genetic manipulation.

Stan Falkow and most of his students have chosen to focus on pathogens which are amenable to genetic analysis. Hence, with the exception of the chapter on *C. trachomatis* (chapter 19), all of the organisms discussed in this part of the book are facultative intracellular pathogens, are easily manipulated, and serve as model systems for pathogens which are more difficult to study. It is a tribute to Stan's influence that, with very few exceptions, his students have chosen to remain within the field of bacterial pathogenesis rather than work on viral or protozoan pathogens.

A BRIEF INTRODUCTION TO THE CELLULAR BIOLOGY OF INTRACELLULAR PATHOGENS

For the purpose of this part of the volume, intracellular pathogens are defined as those microorganisms which enter mammalian cells at some point in the infection cycle. Thus, it is debatable whether the yersiniae actually replicate within mammalian cells, but all enter mammalian cells at some point. Likewise, the importance of the intracellular growth of *Salmonella* spp. is still debated (15). In contrast, *C. trachomatis* is obligately intracellular, while *Legionella* spp. and *Listeria monocytogenes* are probably intracellular for much of the infection process. Here, I will briefly introduce the potential pathogenic mechanisms used by intracellular pathogens, including entry into the cell, life in a vacuole, and entry into and growth within the cytoplasm.

The process of microbial entry is either host-specified endocytosis, as is observed in professional phagocytic cells such as macrophages, or "parasite-specified endocytosis," a term coined by Moulder (18). Although the term "invasion" is commonly used, it has a somewhat misleading connotation which does not reflect the active participation of the host cell. Terms like "entry" and "internalization" are more descriptive. In the case of ingestion by macrophages, pathogens are often recognized by receptors that may not trigger a detrimental host response (4). Thus, a number of pathogens enter macrophages via integrin receptors which do not trigger the oxidative burst. In the case of nonprofessional phagocytic cells, bacteria may bind to receptors not normally associated with phagocytosis and induce uptake by binding with an unusually high affinity (see chapter 15 by Dorothy Pierson on *Yersinia* entry). Alternatively, *Salmonella typhimurium* enters cells by stimulating host cell ruffling, which, surprisingly, results in the uptake of bystander bacteria (7, 8, 20). This process is reminiscent of macropinocytosis, which is induced in the presence of growth factors in both epithelial cells and macrophages (13, 23). A feature common to the entry processes of both *Yersinia* and *Salmonella* spp. is that bacterial entry is blocked by cytochalasin D, which disrupts actin-based processes. However, all cases of bacterial entry are not blocked by cytochalasin D (19). Also, as an aside, there is one example of entry by a protozoan pathogen which is actually enhanced by cytochalasin D (27).

Subsequent to entry, the majority of intracellular pathogens reside within a host membrane-bound compartment or vacuole. Two problems need to be dealt with by these pathogens: (i) avoidance of host bactericidal mechanisms and (ii) acquisition of nutrients. Not surprisingly, a number of different strategies have evolved. At one extreme, *Coxiella burnetii,* an obligate intracellular pathogen, has evolved to grow in an acidified compartment (12). In contrast, *Legionella pneumophila,* like a number of other intracellular pathogens, prevents the acidification of the host vacuole in which it resides (14). *S. typhimurium* delays acidification in macrophages and exists in spacious phagosomes (1, 2). In epithelial cells, *Salmonella* spp. have been observed in tubular lysosomes (9), although the significance of either of these morphological observations is not yet appreciated. Recently, it was reported that intracellular *Mycobacterium avium* was localized in vacuoles lacking components of the proton pump responsible for acidification (26). Although some characteristics of the vacuole have been described in a number of systems, bacterial determinants which govern the fate of the internalized microorganism have not been identified. Currently, a number of groups are taking a genetic approach to identifying the elusive bacterial factors (see chapters 14 and 17 on *Salmonella* spp. and *L. pneumophila,* respectively).

As an alternative to life in a vacuole, a number of microorganisms have evolved to lyse the vacuole and grow directly in the cytoplasm. Among the prokaryotes, this includes the shigellae, most rickettsiae, and pathogenic listeriae. There is a substantial amount of evidence that the cytoplasm of mammalian cells is a highly conducive environment for bacterial growth. First, *E. coli* harboring the *Shigella* virulence plasmid can grow in the cytoplasm of HeLa cells (24). Second, *Bacillus subtilis* strains expressing the *L. monocytogenes* hemolysin can escape from a vacuole in J774 cells and grow directly in the cytoplasm (3). Third, numerous auxotrophs of *L. monocytogenes* grow normally in cells (17). It is somewhat surprising that there are not more examples of intracytoplasmic bacterial pathogens.

The chapters in this part of the volume review the pathogenesis of *Salmonella, Legionella, Yersinia,* and *Chlamydia* spp. Catherine Lee writes on the use of genetics to study *Salmonella* pathogenesis (chapter 14), while B. Brett Finlay focuses on the entry and subsequent intracellular fate of *Salmonella* spp. (chapter 16). Dorothy E. Pierson reviews the entry of *Yersinia* spp. into mammalian cells (chapter 15). Ralph R. Isberg discusses intracellular trafficking of *Legionella* spp. (chapter 17). Patrik M. Bavoil examines determinants of chlamydial pathogenesis and immunity (chapter 19). Finally, I review the cellular biology of *L. monocytogenes* infection in chapter 18.

Daniel A. Portnoy

REFERENCES

1. **Alpuche-Aranda, C. M., E. L. Racoosin, J. A. Swanson, and S. I. Miller.** 1994. *Salmonella* stimulate macrophage macropinocytosis and persist within spacious phagosomes. *J. Exp. Med.* **179:**601–608.
2. **Alpuche-Aranda, C. M., J. A. Swanson, W. P. Loomis, and S. I. Miller.** 1992. *Salmonella typhimurium* activates virulence gene transcription within acidified macrophage phagosomes. *Proc. Natl. Acad. Sci. USA* **89:**10079–10083.

3. **Bielecki, J., P. Youngman, P. Connelly, and D. A. Portnoy.** 1990. *Bacillus subtilis* expressing a haemolysin gene from *Listeria monocytogenes* can grow in mammalian cells. *Nature* (London) **345:**175–176.

4. **Falkow, S., R. R. Isberg, and D. A. Portnoy.** 1992. The interaction of bacteria with mammalian cells. *Annu. Rev. Cell Biol.* **8:**333–363.

5. **Falkow, S., and D. A. Portnoy.** 1983. Bacterial plasmids—an overview. *Can. J. Clin. Invest.* **6:** 207–212.

6. **Ferber, D. M., and R. R. Brubaker.** 1981. Plasmids in *Yersinia pestis*. *Infect. Immun.* **31:**839–841.

7. **Francis, C. L., T. A. Ryan, B. D. Jones, S. J. Smith, and S. Falkow.** 1993. Ruffles induced by *Salmonella* and other stimuli direct macropinocytosis of bacteria. *Nature* (London) **364:**639–642.

8. **Galan, J. E., J. Pace, and M. J. Hayman.** 1992. Involvement of the epidermal growth factor receptor in the invasion of cultured mammalian cells by *Salmonella typhimurium*. *Nature* (London) **357:** 588–589.

9. **Garcia-del Portillo, F., M. B. Zwick, K. Y. Keung, and B. B. Finlay.** 1993. *Salmonella* induces the formation of filamentous structures containing lysosomal membrane glycoproteins in epithelial cells. *Proc. Natl. Acad. Sci. USA* **90:**10544–10548.

10. **Gemski, P., J. R. Lazere, and T. Casey.** 1980. Plasmid associated with pathogenicity and calcium dependency of *Yersinia enterocolitica*. *Infect. Immun.* **27:**682–685.

11. **Gemski, P., J. R. Lazere, T. Casey, and J. A. Wohlhieter.** 1980. Presence of a virulence-associated plasmid in *Yersinia pseudotuberculosis*. *Infect. Immun.* **28:**1044–1047.

12. **Hackstadt, T., and J. C. Williams.** 1981. Biochemical stratagem for obligate parasitism of eukaryotic cells by *Coxiella burnetii*. *Proc. Natl. Acad. Sci. USA* **78:**3240–3244.

13. **Haigler, H. T., J. A. McKanna, and S. Cohen.** 1979. Rapid stimulation of pinocytosis in human carcinoma cells A-431. *J. Cell Biol.* **83:**82.

14. **Horwitz, M. A., and F. R. Maxfield.** 1984. *Legionella pneumophila* inhibits acidification of its phagosome in human monocytes. *J. Cell Biol.* **99:**1936–1943.

15. **Hsu, H. S.** 1989. Pathogenesis and immunity in murine salmonellosis. *Microbiol. Rev.* **53:**390–409.

16. **Isberg, R. R., and S. Falkow.** 1985. A single genetic locus encoded by *Yersinia pseudotuberculosis* permits invasion of cultured animal cells by *Escherichia coli* K-12. *Nature* (London) **317:**262–264.

17. **Marquis, H., H. G. A. Bouwer, D. J. Hinrichs, and D. A. Portnoy.** 1993. Intracytoplasmic growth and virulence of *Listeria monocytogenes* auxotrophic mutants. *Infect. Immun.* **61:**3756–3760.

18. **Moulder, J. M.** 1985. Comparative biology of intracellular parasitism. *Microbiol. Rev.* **49:**298–337.

19. **Oelschlaeger, T. A., P. Guerry, and D. J. Kopecko.** 1993. Unusual microtubule-dependent endocytosis mechanisms triggered by *Campylobacter jejuni* and *Citrobacter freundii*. *Proc. Natl. Acad. Sci. USA* **90:**6884–6888.

20. **Pace, J., M. J. Hayman, and J. E. Galan.** 1993. Signal transduction and invasion of epithelial cells by *S. typhimurium*. *Cell* **72:**505–514.

21. **Portnoy, D. A., and S. Falkow.** 1981. Virulence-associated plasmids from *Yersinia enterocolitica* and *Yersinia pestis*. *J. Bacteriol.* **148:**877–883.

22. **Portnoy, D. A., S. L. Moseley, and S. Falkow.** 1981. Characterization of plasmid-associated determinants of *Yersinia enterocolitica* pathogenesis. *Infect. Immun.* **31:**775–782.

23. **Racoosin, E. L., and J. A. Swanson.** 1989. Macrophage colony-stimulating factor (rM-CSF) stimulates pinocytosis in bone marrow-derived macrophages. *J. Exp. Med.* **170:**1635–1648.

24. **Sansonetti, P. J., A. Ryter, P. Clerc, A. T. Maurelli, and J. Mounier.** 1986. Multiplication of *Shigella flexneri* within HeLa cells: lysis of the phagocytic vacuole and plasmid-mediated contact hemolysis. *Infect. Immun.* **51:**461–469.

25. **Straley, S. C., E. Skrzypek, G. V. Plano, and J. B. Bliska.** 1993. Yops of *Yersinia* spp. pathogenic for humans. *Infect. Immun.* **61:**3105–3110.

26. **Sturgill-Kosaycki, S., P. H. Schlesinger, P. Chakraborty, P. L. Haddix, H. L. Collins, A. K. Fok, R. D. Allen, S. L. Gluck, J. Heuser, and D. G. Russell.** 1994. Lack of acidification in *Mycobacterium* phagosomes produced by exclusion of the vesicular proton-ATPase. *Science* **263:**678–681.

27. **Tardieux, I., P. Webster, J. Ravesloot, W. Boron, J. A. Lunn, J. E. Heuser, and N. W. Andrews.** 1992. Lysosome recruitment and fusion are early events required for *Trypanosome* invasion of mammalian cells. *Cell* **71:**1117–1130.

28. **Zink, D. L., J. C. Feeley, J. G. Wells, C. Vanderzant, J. C. Vickery, W. D. Roof, and G. A. O'Donovan.** 1980. Plasmid-mediated tissue invasiveness in *Yersinia enterocolitica*. *Nature* (London) **283:**224–226.

Molecular Genetics of Bacterial Pathogenesis
Edited by V. L. Miller, J. B. Kaper, D. A. Portnoy, and R. R. Isberg
© 1994 American Society for Microbiology, Washington, DC 20005

Chapter 14

Genetic Approaches to Understanding *Salmonella* Pathogenicity

Catherine A. Lee

Stanley Falkow has put forth a challenge: study microbial pathogenesis by understanding the role of individual microbial genes (16). Researchers of *Salmonella* species have met this challenge head-on, using genetics to identify *Salmonella* virulence factors as well as to test the contributions of bacterial genes in pathogenesis. *Salmonella typhimurium,* in particular, is ideal for genetic studies since both classical transduction and modern molecular biology techniques are simple to do with this organism. It is possible to construct a defined *S. typhimurium* mutant and then examine the effect of a single genetic change on virulence.

This chapter presents selected examples of how genetic approaches have been used to identify *Salmonella* virulence factors. The genetic approaches are based on five distinct concepts and assumptions about what might characterize *Salmonella* virulence genes.

SCREENING MUTANTS FOR VIRULENCE

Insertion Mutants

The basic rationale behind screening mutants for virulence is that avirulent mutants must contain mutations in the genes required for virulence. Therefore, the most systematic way to identify *Salmonella* virulence genes should be to test tens of thousands of different bacterial mutants for their animal infectivities. This brute force approach is probably not realistic, but Groisman and Heffron (24) conducted a pilot study in which 400 random transposon mutants were individually screened for their virulences in mice. Interestingly, about 2% of the insertion mutations increased the intraperitoneal 50% lethal dose (LD_{50}) by $\geq$10,000-fold, whereas about 6% of the mutations increased the oral LD_{50}. If it is assumed that *S. typhimurium* contains 3,000 genes, results of the pilot study would indicate that approximately 60 to 180 genes are involved in pathogenesis.

However, the true meaning of these conclusions is unclear when one consid-

---

Catherine A. Lee • Department of Microbiology and Molecular Genetics, Harvard Medical School, 200 Longwood Avenue, Boston, Massachusetts 02115.

ers the definition of virulence used in that study as well as what defects are found among avirulent mutants. For technical reasons, it was difficult to identify mutants with weak effects on the LD_{50}, and so only mutants that dramatically increased the LD_{50} were scored as avirulent. This strict definition of avirulence may not be ideal, since *Salmonella* pathogenesis is not an all-or-none phenomenon. As described below, many different bacterial properties contribute to the success of *Salmonella* infection. Certain bacterial virulence factors affect the LD_{50} only <100-fold in the murine infection model, while other bacterial properties, like motility, may have no effect on the LD_{50} but are found to be important in alternate infection models. Therefore, because of the particular definition of virulence and the use of one infection model, the pilot mutagenesis study of Groisman and Heffron (24) would seem to underestimate the number of *S. typhimurium* genes involved in virulence. So, although a brute force analysis of mutants would seem to be the most comprehensive way to study pathogenesis, there are obvious limitations, both technical and interpretive.

Analysis of those mutations that significantly affect *S. typhimurium* virulence raises another issue. For example, *recA* mutants are avirulent. Does this mean that RecA is a *Salmonella* virulence factor? Well, in one sense, yes; in fact, recent studies indicate that RecA serves an important role during bacterial interaction with macrophages (9). Yet, in another sense, RecA is found in nonpathogenic microorganisms, obviously serving a basic function required by all bacteria. At issue is how one views *Salmonella* pathogenicity and virulence genes. Maybe one sees only those properties that are unique to a pathogen as conferring pathogenicity? If so, this approach is not appropriate; the comprehensive analysis of avirulent mutants means that a large number of genes involved in bacterial physiology will be identified. Or, is every aspect of bacterial physiology and adaptation that occurs during infection seen to be a contributor to pathogenicity? If this is the case, classification of avirulent mutants and their relevant phenotypes is worthwhile because it will reveal many essential features of *Salmonella*'s in vivo lifestyle.

Tn*phoA* Mutants

A slight modification of the approach described above has been to analyze Pho$^+$ Tn*phoA* insertion mutants for defects in virulence. Pho$^+$ Tn*phoA* mutations have been shown to affect genes encoding cell envelope proteins. In one study, Tn*phoA* mutants of *S. typhimurium*, identified as Pho$^+$ on L-agar plates, were tested for their virulences by oral infection of BALB/c mice. Interestingly, 15 of 150 Pho$^+$ mutants (10%) were avirulent; subsequent analysis showed that 9 of the 15 mutants were defective in lipopolysaccharide biosynthesis, and none had insertions in the virulence plasmid (40).

Although the use of this approach is subject to some of the same caveats discussed above, one advantage is that the unwieldy nature of random insertion mutant analysis is reduced by limiting the analysis of mutants to those defective in cell envelope factors. Unfortunately, the Pho$^+$ screen has one major disadvantage: it also limits the analysis to genes that are expressed in *S. typhimurium* on L agar. Thus, many virulence genes that are expressed at low levels or that are repressed on L-agar plates would be missed by this approach. The requirement for gene expression may explain why 9 of the 15 Pho$^+$ avirulent mutant strains contained insertions in lipopolysaccharide biosynthetic genes rather than insertions distributed over a variety of virulence loci. So, while the use of Tn*phoA* is a convenient way to restrict the scope of an analysis of random mutations, it should not be thought of as a comprehensive means of analyzing mutant strains with mutations in all cell envelope proteins. Another potential complication is that, in addition to disrupting a gene, the Pho$^+$ Tn*phoA* mutations produce alkaline phosphatase hybrid proteins. In one case, the phenotype of a Tn*phoA* mutant was not due to the loss of a gene but was solely attributable to the presence of a deleterious hybrid protein (44). Thus, other genetic analyses must be conducted on avirulent Pho$^+$ Tn*phoA* mutants before concluding that the disrupted gene is involved in virulence.

IDENTIFICATION OF *SALMONELLA*-SPECIFIC GENES

The identification of *Salmonella*-specific genes is based on the potential relationship between a pathogen-specific gene and the pathogenic phenotype. For example, if comparisons reveal that a pathogenic strain contains genetic material that is absent from a nonpathogenic strain or closely related nonpathogenic species, it is possible that this unique region of the genome encodes genes that confer pathogenicity.

Virulence Plasmid Genes

The advent and application of molecular biology techniques revealed that many *Salmonella* species carry a large plasmid (55 to 90 kb). Initially, this plasmid was termed "cryptic," but in the early 1980s, analysis of isogenic plasmid-carrying and plasmid-cured strains showed that the plasmid is essential for virulence. It is now clear that essential virulence genes, termed *spv* (*Salmonella* plasmid virulence), are located in an 8-kb region of the virulence plasmid (for a recent review, see reference 28). In fact, subclones carrying the minimal region restore virulence to plasmid-cured strains. In animal infection models, the *spv* genes are required for many *Salmonella* species to cause systemic infection of reticuloendothelial tissues. Although plasmid-cured strains are still able to colonize and infect the intestinal mucosa, their ability to proliferate in lymph node and spleen tissues is impaired (29). The biochemical and molecular functions of the *spv* gene products have remained elusive, but current speculations suggest that they may be involved in bacterial modulation of host immune cells. Interestingly, the *spvABCD* genes are regulated by *spvR,* a member of the LysR regulator family, and RpoS, the stationary-phase σ factor. During infection, *spv* gene expression may be induced in response to a specific growth-inhibiting environment in vivo, such as that within the phagolysosome. In this case, bacterial entry into phagocytes of the reticuloendothelial system may initiate the *spv*-dependent phase of pathogenesis.

S. typhimurium-Specific Sequences

Although the chromosomal maps of *S. typhimurium* and *Escherichia coli* K-12 are strikingly similar, *S. typhimurium* has some DNA sequences not present in *E. coli*. Since *S. typhimurium* is a pathogen and *E. coli* K-12 is not, certain of these *S. typhimurium*-specific regions might encode genes responsible for this pathogenic difference. As a first step in the analysis of *S. typhimurium*-specific genes, recombinant plasmids which contain *S. typhimurium*-specific sequences were identified on filter blots as those that were not able to hybridize to a probe of the total *E. coli* chromosome (20). Subsequent studies showed that one of these *S. typhimurium*-specific sequences maps on the chromosome at minute 7 and contains an open reading frame homologous to the LysR regulator family. Unfortunately, deletion of the gene had no apparent effect on bacterial growth or virulence properties, so the function of *sinR* (*Salmonella* insert regulator) remains a mystery (27). Analysis of *sinR*-regulated genes and identification of *sinR* mutant phenotypes may eventually reveal the function of *sinR*. Interestingly, the sequence hybridizes to DNA from *Klebsiella pneumoniae*.

Analysis of a second *S. typhimurium*-specific sequence was more fruitful. The 6.4-kb region maps to minute 60 of the chromosome, and deletions abolish the ability of *S. typhimurium* to enter epithelial cells (25). Interestingly, previous studies identified seven *Salmonella inv* (invasion) genes in this region which are also required for epithelial cell entry (2). Sequence analysis of the 6.4-kb region revealed an additional eight open reading frames, termed *spa*, downstream of the *inv* genes. Sequence comparisons showed that the gene order and the nucleotide and protein sequences of the *S. typhimurium inv* and *spa* genes are similar to those of the *Shigella mxi* and *spa* genes on the *Shigella* virulence plasmid. Consistent with this result, the 6.4-kb region hybridizes to the *Shigella* virulence plasmid under low-stringency conditions. The *mxi* and *spa* genes are required for *Shigella* species to have the ability to enter epithelial cells; the gene products facilitate export of the *Shigella* surface proteins that trigger the invasion process (*mxi*, for membrane expression of invasion protein antigens; *spa*, for surface presentation of antigen). These *Shigella* and *Salmonella* invasion genes are also homologous to the *Salmonella fli* and *flh* flagellar assembly and export genes. It appears that *Shigella* and *Salmonella* species have adapted components of the flagellar assembly and export mechanism to express invasion factors on their cell surfaces.

The *Salmonella inv* and *spa* genes are not the only *S. typhimurium*-specific virulence genes at this chromosomal region. Detailed analysis of the 60-min chromosomal region reveals that a contiguous 40 kb of *S. typhimurium* DNA that includes the 15 kb encoding the *inv* and *spa* genes is absent from *E. coli* K-12 (45). Interestingly, the remaining 25 kb also encodes invasion genes. One explanation for this finding is that this region of the *S. typhimurium* chromosome was acquired intact from an invasive ancestor. The 40 kb may encode *Salmonella* factors that specify the expression, export, assembly, and composition of surface components that interact with epithelial cells and trigger the invasion process.

In summary, the comparison of *Salmonella* strains with nonpathogenic strains or species has been a very useful approach to the identification of virulence genes.

The identification of the cryptic plasmid as being essential for virulence was an early success. More recently, comparisons with *E. coli* K-12 have identified interesting *S. typhimurium*-specific genes involved in virulence. However, because of evolutionary distance, *E. coli* is likely different from *S. typhimurium* in many ways that are not relevant to virulence. Some *S. typhimurium*-specific genes, such as *sinR*, may function for nonpathogenic aspects of the distinct *Salmonella* life cycle. Thus, the relatedness of the nonpathogenic strain or species chosen for comparison will greatly affect the results of this type of approach.

Use of this comparative approach has not been fully exploited. The comparison of *S. typhimurium* and *E. coli,* described above, has just begun. In addition, this approach might be adapted to the study of other aspects of *Salmonella* pathogenesis, such as host specificity and tissue tropism. For example, *Salmonella typhi* is a human-specific pathogen that lacks a virulence plasmid and *spv* genes, yet it causes systemic disease. In comparison, *Salmonella enteritidis* has a broad host range, and in humans it mainly causes localized infections. Thus, it might be informative to identify *S. typhi*-specific genes in comparison with *S. enteritidis* and vice versa. Subtractive hybridization and cloning of unique genomic sequences (53) from these *Salmonella* species may reveal whether unique genes determine the distinct pathogenic characteristics of *S. typhi* and *S. enteritidis*.

ANALYSIS OF *SALMONELLA* PHENOTYPES THOUGHT TO BE INVOLVED IN VIRULENCE

Researchers have engaged in "intelligent guessing" about what bacterial phenotypes might be necessary for virulence. In many cases, the subsequent identification and mutagenesis of bacterial genes involved in putative *Salmonella* virulence phenotypes have led to the identification of virulence genes.

Macrophage Survival Phenotype

Macrophages are an important primary host defense against microorganisms. The dogma of medical microbiology states that *Salmonella* species are able to avoid elimination by macrophages and are actually able to survive and replicate within these normally antimicrobial host cells. Early support for this pathogenic mechanism came from observations that bacteria are able to multiply in the liver, spleen, and peritoneal cavity, even when they are opsonized prior to inoculation into mice (11). There are numerous macrophage-mediated assaults that *Salmonella* species may encounter and resist during infection: oxygen derivatives, phagosome acidification, phagolysosome fusion, antimicrobial peptides, and nutrient limitation. The genetic basis and role in virulence of each of these individual bacterial resistance properties are being examined.

An alternative and more general approach has been to study in vitro systems in which *Salmonella* species are able to survive interaction with primary or cultured macrophages. For example, one approach to understanding the genetic basis and relevance of macrophage survival is to identify *S. typhimurium* mutants that are macrophage susceptible in vitro. Presumably, such mutants would no longer be resistant to certain macrophage-specific defense mechanisms. In one study, 9,516 individual transposon mutants were screened for their ability to survive macrophage interactions in vitro. Eighty-three of the mutants identified as being killed by macrophages in vitro were also found to be less virulent in vivo, providing further evidence that survival within macrophages is an essential aspect of *Salmonella* pathogenesis (18).

Further characterization of the mutant phenotypes showed that 12 mutants were auxotrophs, suggesting that certain nutrients are absent from the growth environment within macrophages. In addition to auxotrophs, one might expect the existence of mutants unable to resist other macrophage assaults. In fact, one class of mutants was found to be more susceptible to defensins, which are cationic peptides, from phagocyte granules. Subsequent studies have shown that the defensin-susceptible mutants are defective in *phoP* and *phoQ*, genes encoding sensor-regulator components which activate the expression of *pag* genes (*phoP*-activated genes) (17, 23, 41).

Since *phoP* null mutants are defensin susceptible and *phoP* constitutive mutants are as resistant as the wild type, one might predict that one or more *pag* genes specifies bacterial resistance to cationic peptides (43). Surprisingly, none of the *pag* genes characterized thus far appears to be essential for defensin resistance, and only one *pag* mutant, *pagC,* is defective in macrophage survival and virulence (41). One explanation for the inability to find a *pag* gene involved in resistance to cationic peptides is that, so far, *pag* mutants have been identified as Tn*phoA* mutants which demonstrate *phoP*-dependent expression of the Pho$^+$ phenotype. It is possible that if the *pag* genes, which are essential for cationic peptide resistance, are expressed at low levels, are polar on an essential bacterial gene, or do not encode cell envelope proteins, they would have been missed by the Tn*phoA* analysis. Interestingly, identification of the *S. typhimurium* genes that are required for bacterial resistance to cationic peptides in vitro have revealed none that are activated by *phoP* (26). Thus, it is possible that more than one *pag* gene is sufficient to specify the *phoP*-mediated defensin resistance mechanism. In this case, a mutant lacking a single *pag* gene would remain defensin resistant, while the *phoP* mutant which is defective in expression of all *pag* genes would be defensin susceptible. Further study of defensin resistance, the ability to survive against macrophages, and members of the *phoP* and *phoQ* regulon will eventually elucidate the relationship been these phenotypes and genes.

In summary, analysis of the genes required for *Salmonella* resistance to macrophages in vitro has demonstrated the importance and complexity of this bacterial virulence property. Characterization of macrophage-susceptible mutants shows that bacterial growth and resistance to cationic peptides are required for *Salmonella* virulence and survival within macrophages. Further characterization of the other macrophage-susceptible mutants will likely reveal additional *Salmonella* properties needed by the bacteria to resist the antimicrobial activities of macrophages in vitro and in vivo.

Invasion Phenotype

After oral inoculation of mice with *S. typhimurium*, the bacteria apparently escape the lumen of the small intestine and gain access to mesenteric lymph nodes (10). Researchers have been working to understand how salmonellae are able to cross the intestinal mucosa. One hypothesis is that the bacteria are able to penetrate intestinal epithelial cells. In fact, over 25 years ago, it was observed that *S. typhimurium* enters enterocytes during experimental infection of animals. It appears that contact between the bacteria and the microvilli of enterocytes results in disorganization of the epithelial brush border and engulfment of the bacteria in membrane-bounded vacuoles within the host cells. Thus, it has been suggested that transcytosis of *S. typhimurium* across enterocytes may provide bacterial access to the underlying lymphatics (54). More recently, by using ligated intestinal loop infection models, *S. typhimurium* has been observed to enter M cells of the intestinal epithelium (8). It is now generally accepted that M cells provide an important portal for *Salmonella* passage across the intestinal epithelium. Consequently, studies are under way to understand how *S. typhimurium* is able to enter intestinal cells and to determine the relevance of bacterial entry for virulence.

Fortunately, model systems have been developed in which *S. typhimurium* enters nonphagocytic cells in vitro. These in vitro systems with cultured mammalian cells have allowed detailed characterization of host cell processes (46) as well as identification of the bacterial genes involved in the entry process. The most common genetic approach to the identification of *Salmonella* invasion genes has been to study noninvasive mutants, since such mutants must be defective in the genes required for entry. Accordingly, mutant laboratory strains, traditional transposon mutants, and Pho⁺ Tn*phoA* mutants have been individually screened for their abilities to enter cultured cells in vitro (7, 19, 21, 52). Interestingly, some but not all of the noninvasive mutants were found to be defective in virulence. Further characterization of the noninvasive mutants is required to fully understand why some are unable to enter cells in vitro yet are apparently able to cross the intestinal epithelium in vivo. One explanation is that certain aspects of the in vitro entry systems may be irrelevant in the in vivo situation. For example, mutants defective in motility or lipopolysaccharide biosynthesis were found to be noninvasive in vitro (7, 19). The in vitro invasion defect of such mutants may be attributable to changes in contact efficiency between *S. typhimurium* and the cultured mammalian cells. So, although alteration of bacterial motility and chemotaxis have some effect on invasion of Peyer's patches in ligated intestinal loops (31), it is likely that the in vitro effect of bacterial motility is mainly an artifact caused by the position of the mammalian cells and the bacteria relative to gravitational forces. Apparently, motile bacteria swim downward, which increases their contact with the cell monolayer; consistent with this idea, the in vitro invasion defect of nonmotile bacteria can be suppressed by centrifuging the bacteria onto the cells.

A high percentage of transposon mutants tested were noninvasive, suggesting that a large number of bacterial genes are involved in the entry process. So far, most of the noninvasive mutants have not been fully characterized, but one essential invasion locus (*invHFGEABC*) was identified from both Tn*phoA* analysis and screening of mutant laboratory strains (2). The *S. typhimurium inv* genes are homologous to *Shigella* invasion genes as well as to bacterial flagellar export and assembly genes (see the section *S. typhimurium*-Specific Sequences above). Apparently, certain of the *inv* invasion genes are involved in facilitating the export and assembly of essential invasion components on the surface of *S. typhimurium*. It has been speculated that the *inv*-exported bacterial factors directly contact mammalian receptors, which induces localized membrane ruffling and engulfment of the bacteria.

Examination of certain noninvasive mutants in animal infection models provides strong support for the importance of bacterial entry into intestinal cells (4, 21). First, such mutants are less virulent after oral inoculation but are fully virulent after intraperitoneal inoculation. Second, after mixed oral inoculations of Inv⁻ and Inv⁺ strains, the noninvasive mutants have a significant disadvantage in colonization of the intestinal mucosa and spleen, whereas after mixed intraperitoneal inoculations, Inv⁻ strains infect the spleen as efficiently as wild-type strains. Third, studies of the kinetics of bacterial access to systemic sites show that Inv⁻ mutants are significantly delayed in their ability to reach the bloodstream and spleen after oral inoculation. Further study is required to determine whether enterocytes, M cells, or other sites are relevant portals for bacterial passage through the intestinal epithelium.

Another approach to identifying *Salmonella* invasion factors was to clone *Salmonella* genes into *E. coli* K-12 and select for recombinants that were able to enter cultured mammalian cells. In retrospect, it is not surprising that this approach was generally not successful. It is now known that one 40-kb region of the *S. typhimurium* genome encodes invasion genes and that other unlinked chromosomal loci are also required for invasion. Thus, it may not be possible to construct a single plasmid that confers all of the *Salmonella* invasion functions to *E. coli*. In addition, *Salmonella* invasion is regulated by bacterial growth conditions so that even if the invasion genes are transferred to *E. coli*, appropriate regulatory factors may be required for the expression and function of *Salmonella* invasion factors in *E. coli*.

Nevertheless, the cloning approach was successful in one situation; cosmids containing *S. typhi* genes were found to confer upon *E. coli* HB101 the ability to enter Henle 407 cells. Four genes, termed *invABCD*, on the cloned *S. typhi* DNA that are each required for the invasion phenotype in *E. coli* were identified (15). Interestingly, the homologous *S. typhimurium invABCD* genes mapped to the 60-min region of the chromosome. These genes should not be confused with the *invHFGEABC* genes in this region; instead, the *invABCD* genes were found to be interspersed with *recA* and *srl*, which are common housekeeping genes. As discussed above in the section *S. typhimurium*-Specific Sequences, the 60-min region of the *S. typhimurium* chromosome contains a 40-kb span that encodes many invasion genes. Surprisingly, the *invABCD* genes do not lie within this 40-kb region but lie to the left, toward minute 59 (45).

It seems odd that the *invABCD* genes are closely interspersed with genes that are present in nonpathogenic bacteria such as *E. coli* K-12. One explanation is that these *S. typhi* genes might not be involved in *Salmonella* invasion but, instead, indirectly alter the ability of *E. coli* to enter cells in vitro. In fact, it has been found that *E. coli* K-12 has an inherent, albeit weak, capacity to enter cultured cells in vitro, and interestingly, certain *E. coli* mutations significantly enhance this activity (50). Therefore, it is possible that the cloned *S. typhi* genes alter the physiology of HB101 in a manner analogous to the *E. coli* mutations and, thus, indirectly increases the ability of the *E. coli* strains to invade cells. However, contrary to this idea, mutations in the homologous *invABCD* genes in *S. typhimurium* have been found to disrupt the ability of *S. typhimurium* to enter cells in vitro (14). So, it seems likely that the *invABCD* genes truly play a direct role in *Salmonella* entry, but that they function at a step that is distinct from the membrane-ruffling process specified by the other invasion genes. In this case, the invasion phenotype conferred to *E. coli* by the *S. typhi invABCD* gene products may reflect a relevant, independent *Salmonella* invasion function.

Other genetic approaches to understanding *Salmonella* invasion have been based on the observation that bacterial entry into cultured cells is regulated by bacterial growth conditions. These approaches, aimed at identifying the regulated invasion factors, are discussed below in the section Oxygen-Regulated Invasion Genes.

In summary, intelligent guessing about *Salmonella* virulence phenotypes is a valuable approach to understanding *Salmonella* pathogenicity. Since, in most cases, there is an in vitro assay for the putative virulence property, analysis is amenable to both genetic and biochemical approaches. Unfortunately, there may be times when intelligent guessing is misleading. For example, even if a correlation is found between an in vitro property and in vivo virulence, it is possible that the effect on virulence may reflect the alteration of a totally different bacterial property. Take the case of *Salmonella cholerae-suis* passage through epithelial cells in vitro. The identification of mutants defective in this in vitro phenotype was undertaken to understand bacterial passage through the intestinal epithelium. However, one class of noninvasive mutants, found to be avirulent, was defective in lipopolysaccharide biosynthesis (19). Although it is possible that lipopolysaccharide is involved in bacterial passage through the intestinal epithelium, it is also known that such mutants are more susceptible to serum. Thus, the correlation with virulence may be attributable to the change in serum resistance, not the invasion defect. In this case, further study is required to determine if the influence of bacterial lipopolysaccharide in the in vitro interaction with epithelial cells is an artifact or if *S. cholerae-suis* lipopolysaccharide actually plays an important role in the in vivo interaction with epithelial cells.

ANALYSIS OF KNOWN *SALMONELLA* COMPONENTS AND SYSTEMS FOR THEIR INVOLVEMENT IN VIRULENCE

Much is known about the biology of *S. typhimurium*. This has allowed researchers to engage in informed guessing about what bacterial factors might be necessary for virulence. In this way, informed guessing has prompted analysis of known bacterial components and systems for their involvement in virulence.

Enterochelin

More than a century ago, it was observed that iron supplementation or disease states that release iron stores increase the host's susceptibility to pathogens. It is now known that host systems, such as transferrin and lactoferrin, normally chelate and reduce the levels of free iron available in serum and tissue. Increases in free iron levels might augment infections by increasing microbial growth at certain sites within the host. Clearly, microbes sense a low-iron environment within the host; in fact, certain pathogens respond to this environmental cue by inducing virulence factors (34). Thus, one question has been, how do pathogens grow in host fluids depleted of free iron? One answer may be that many bacteria possess high-affinity iron-binding systems which can capture iron from the host iron chelators. For example, *S. typhimurium* secretes enterochelin, a cyclic trimeric ester of 2,3-dihydroxylbenzoyl serine, which has a very high affinity for ferric iron. Bacterial binding and utilization of the enterochelin-iron complex is facilitated by Chr-TonB, an outer membrane receptor.

Since *S. typhimurium ent* mutants cannot grow in mouse serum or in minimal medium containing iron chelators without the addition of exogenous iron, enterochelin was considered a likely candidate as a virulence factor. Presumably, enterochelin would be necessary for bacterial iron acquisition during infection. Surprisingly, *ent* mutant strains were found to be fully virulent in mice, indicating that the bacteria are able to acquire iron and grow in vivo without enterochelin (5). Consistent with this result, an *S. typhimurium chr* mutation caused minimal, if any, alteration of virulence (51). One possible explanation for these results is that *S. typhimurium* may replicate at a site within the host that is not limiting for free iron, for example, within host cells. Alternatively, *S. typhimurium* may possess other iron acquisition systems that are expressed in vivo but not in vitro. This latter possibility might be examined by identifying *S. typhimurium* mutants that express other iron acquisition systems in vitro as suppressors of *ent* mutants.

Motility and Chemotaxis

Bacterial motility and chemotaxis require 50 different genes which encode chemical sensors, physical motors, directional switches, and components of the flagellar structure, as well as specific factors which export and assemble the flagellar structure. Maintenance and operation of this system appear to significantly tax the metabolic capacity of bacteria so that Mot$^-$ mutants have a selective advantage during growth in the laboratory (37). Yet, natural isolates of most *Salmonella* species have retained motility and chemotaxis properties in the face of this apparent growth disadvantage. Thus, it would seem that bacterial motility and chemotaxis have been conserved and must play an important role for the natural life of *Salmonella* species.

In fact, it has been speculated that *Salmonella* motility and chemotaxis may be important for virulence. For example, bacteria might need to sense chemical signals and move toward appropriate host sites for attachment or growth, or both. In addition, flagella have been hypothesized to serve directly as adherence and invasion factors during infection of intestinal epithelial cells or macrophages, or both. Consequently, many studies have been conducted to examine the contribution of *Salmonella* flagella, motility, and chemotaxis in animal infection models.

One such study yielded very disappointing results. As judged by the LD_{50} in mice infected orally or intraperitoneally, the virulences of nonmotile, flagellated (*mot*) as well as nonflagellated (*fli* and *flh*) mutants of *S. typhimurium* were found to be no different from those of wild-type bacteria (35). These results indicate that bacterial motility and chemotaxis have no effect on the murine typhoid fever model of infection. Interestingly, in the same oral infection model, *fim fli* double mutants which lack both type 1 fimbriae and flagella were found to be significantly attenuated for virulence, whereas the *fim* or *fli* mutation alone did not attenuate virulence (36). One explanation for these results is that fimbriae and flagella mediate a common pathogenic mechanism such as colonization. In this case, loss of a single factor would have little effect on colonization caused by the presence of the second factor, but loss of both factors would cause a severe colonization defect and, thus, attenuate infection. Thus, in this infection model, the importance of bacterial motility and chemotaxis for virulence appears to be influenced by the background of the *Salmonella* strain used.

The importance of bacterial motility for virulence is also influenced by the infection model used. In an experimental urinary tract infection model in which *S. typhimurium* colonizes the bladder, nonmotile strains were 100- to 1,000-fold less able than motile derivatives to colonize the bladder (49). Possibly, motility is essential for bacterial access and contact with endothelial cells of the urinary bladder. Thus, in this alternate infection model, bacterial motility was found to be a very important bacterial virulence mechanism.

Systemic infection of BALB/c mice is one of the best-characterized models for the study of *Salmonella* virulence. Therefore, evaluation of *S. typhimurium* virulence is most often conducted by measuring LD_{50}s in BALB/c mice after oral, intravenous, or intraperitoneal inoculation. Unfortunately, it appears that this virulence model has particular quirks. For example, BALB/c mice are homozygous for the *Ity*^s mutation, which results in impaired macrophage function (55). Thus, in *Ity*^s mice, the most critical stage of infection might be whether phagocytes are able to kill the infecting microorganism, whereas in *Ity*^r mice, phagocytes are not impaired and another stage of infection may be the limiting step. In fact, the *S. typhimurium mviA* (mouse virulence) allele has a dramatic influence on infection of *Ity*^s mice but has a minimal effect on infection of *Ity*^r mice (6). So, the distinction between bacterial properties that are important for virulence versus those that are not might depend on the most susceptible step of pathogenesis inherent in a particular infection model.

In summary, analysis of known *Salmonella* components and systems for their involvement in virulence has had mixed results. Several successes came from examination of mutants affecting global regulatory systems, such as *crp-cya*, *ompR-envZ*, and *phoP-phoQ* (12, 13, 41). Such mutants are severely attenuated in most infection models. In these cases, it is likely that the deregulation of more than one bacterial factor specifies the loss of virulence. Unfortunately, there are also many instances of studies showing that *Salmonella* factors play no role in virulence, such as the *ent-chr* iron acquisition system or the *btuB-cob* genes required for vitamin B_{12} uptake and synthesis (48). In other cases, the requirement of a *Salmonella* factor for virulence is found to be dependent on the infection model and the bacterial strain used, such as for the bacterial motility system or for fimbrial adherence factors.

Results showing that mutants are fully virulent may indicate that a bacterial factor is conserved and is important for another aspect of the life cycle of *Salmonella* species. For example, the enterochelin system may function for bacterial iron acquisition when *S. typhimurium* is outside of the host. In fact, nonpathogenic bacteria possess homologous iron acquisition factors. While the obvious advantage of this approach is that the potential virulence factors are already known and mutants may already exist, many times this approach yields negative or ambiguous results. This may be due to misconceptions of what occurs during infection as well as to dependence on potentially quirky infection models that may not reflect all of the events that influence the outcomes of natural infections.

ANALYSIS OF *SALMONELLA* GENES ON THE BASIS OF THEIR REGULATION

It is apparent that gene regulation by environmental signals during infection is important for bacterial pathogenesis (39). Although this section discusses three cases in which *Salmonella* virulence genes were identified on the basis of their regulation, each example has a distinct starting point. For example, the first case is based on the knowledge that a sensor-regulator system is important for virulence. Consequently, studies of genes regulated by the sensor-regulator were conducted to identify virulence genes. In the second example, regulated virulence genes were identified without even knowing the regulator involved. In this case, environmental conditions known to modulate a virulence property were used to identify the regulated virulence factors. The third example is a more general approach designed to identify regulated genes that are expressed specifically during infection. Presumably, their selective expression indicates that such genes act during infection and are involved in virulence. Although initiated quite recently, this last approach shows great potential for helping investigators understand the complexities of *Salmonella* pathogenicity.

PhoPQ-Regulated Genes

phoP was originally identified as a regulator of the *S. typhimurium* nonspecific acid phosphatase. It is now known that *phoP* and *phoQ* comprise a two-component regulatory system in which PhoP serves as a transcriptional activator and PhoQ functions as a transmembrane sensor and protein kinase. Together, these proteins modulate the expression of many genes in response to several environmental conditions: phosphate, carbon, sulfur, or nitrogen starvation as well as acidity. Two genetic approaches—analysis of macrophage-susceptible mutants and analysis of known bacterial regulatory systems—revealed that *phoP* and *phoQ* are essential virulence factors in *S. typhimurium; phoP* null mutants are severely attenuated in the murine infection model after oral or intraperitoneal administration (23, 41).

One explanation for the avirulence of *phoP* null mutants is that *phoP*-activated genes (*pag* genes) encode essential virulence factors. In order to identify any *phoP*-activated virulence factors, Tn*phoA* mutants were screened for those that exhibit *phoP*-dependent expression of the Pho⁺ phenotype. So far, only one *pag* gene, *pagC,* has been found to be important for virulence. Consistent with the finding that *phoP* null mutants are also susceptible to killing by macrophages, one *pag* gene, *pagC,* was found to be required for macrophage survival in vitro (41). *pagC* encodes an outer membrane protein that is homologous to a family of virulence factors (47).

As discussed above in the section Macrophage Survival Phenotype, it is likely that many *pag* genes have yet to be discovered. In fact, analysis of two-dimensional gels shows that the expression of at least 10 proteins is dependent on *phoP*. Thus far, six *phoP*-activated genes have been identified: *phoN, psiD, pagA, pagB, pagC,* and *pagD* (23, 41). Interestingly, expression of *pag* genes is induced after bacterial entry into macrophages. It appears that the acidic environment of the phagosomal compartment serves to induce gene expression, since treatment of macrophages with chloroquine or NH₄Cl, which raises the pHs of intracellular compartments, suppresses *pag* gene expression (1).

The *pho-24* allele was isolated and was found to cause constitutive expression of the nonspecific acid phosphatase; basically, the *pho-24* mutation results in expression of the *phoPQ* regulon independent of environmental conditions. Interestingly, when tested in infection models, the *pho-24* mutant was avirulent (42). Possible explanations for this finding are that inappropriate expression of *pag* genes during infection might reduce virulence, or there may be *phoP*-repressed genes (*prg* genes) that encode virulence factors. In order to identify *phoP*-repressed virulence genes, Tn*phoA* mutants were screened for those that are Pho⁺ in a *phoP* null strain and Pho⁻ in a *pho-24* strain; five such *prg* genes were identified. Consistent with the hypothesis that *prg* genes might contribute to virulence, one *prg* mutation, the *prgH* mutation, was found to decrease virulence. Interestingly, *prgH* was found to map on the chromosome at minute 60, within the 40-kb span that contains *Salmonella* invasion genes. In fact, examination of the *prgH*::Tn*phoA* mutant in in vitro invasion assays suggests that *prgH* is an essential invasion gene (4).

It is interesting that an invasion gene is repressed by *phoPQ*, whereas a macrophage survival gene is activated by *phoPQ; prg* virulence factors would be expressed in environments that repress *pag* gene expression and vice versa. Presumably, this means that the stage of infection during which invasion factors are expressed is exclusive and distinct from the stage during which macrophage survival factors are expressed. The opposing regulation of these virulence phenotypes may be important for the successful course of pathogenesis.

Oxygen-Regulated Invasion Genes

Study of in vitro invasion systems revealed that the ability of *Salmonella* species to enter cultured cells is affected by bacterial growth conditions; growth phase, oxygen concentration and osmolarity were each found to modulate invasiveness (22, 32). The relevant environment during infection in which the bacteria are growing in high-osmolarity and low-oxygen conditions might be within the intestinal lumen. In this case, the regulated invasion factors might be those that allow *Salmonella* species to cross the intestinal mucosa and access systemic sites.

In order to identify oxygen-regulated invasion factors, a promoterless *lacZ* gene was inserted randomly in the *S. typhimurium* genome, and mutants were screened to identify the genes that are induced (Lac$^+$) by low-oxygen conditions. Since insertion of *lacZ* into genes disrupts their function, oxygen-regulated *lacZ* mutants were subsequently examined in in vitro invasion assays to see if any were disrupted in an essential invasion gene. In this way, one gene, *orgA* (oxygen-regulated gene), was identified; this gene is repressed by oxygen and is essential for entry into cultured cells. In addition, *orgA* mutants were found to be defective in their abilities to enter murine M cells in ligated intestinal loop studies (30). Interestingly, *orgA* maps to the 60-min region of the *S. typhimurium* chromosome and lies downstream of *prgH*. So, *orgA* defines yet another essential invasion gene in the *Salmonella*-specific invasion region.

Another approach to identifying the regulated invasion factors was to isolate hyperinvasive mutants which can enter cultured cells even when they are grown under repressing aerobic conditions. It was hoped that the phenotypes of hyperinvasive mutants might be due to the constitutive expression of the normally regulated invasion factors. Interestingly, one class of hyperinvasive mutants, termed *hil* (hyperinvasion locus), was obtained by insertion of the constitutive *neo* promoter in the *S. typhimurium* genome (33). Genetic analysis revealed that *hil* encodes essential invasion genes and also maps to the invasion-specific region of the *S. typhimurium* chromosome at minute 60 (45). As hoped, isolation of hyperinvasive mutants led to the identification of invasion genes. However, these results do not indicate whether the *hil* genes are normally involved in the oxygen-regulated invasion mechanism, nor do they explain the mechanism of the hyperinvasion phenotype.

Since the two *hil* hyperinvasion mutations, *hil-378* and *hil-380,* were obtained by insertion of the *neo* promoter in the chromosome, it is possible that the mutations increase the organism's invasion ability by directly increasing the level of expression of a rate-limiting invasion component from the inserted *neo* promoter. Such a rate-limiting factor might be an invasion protein that interacts with mammalian receptors and triggers membrane ruffling, or it might be an activator of invasion gene expression which, when overproduced, increases the level of production of many invasion factors to cause the hyperinvasion phenotype. If this is the case, the *neo* promoter would be essential for the *hil* hyperinvasion phenotype. In addition, expression of genes downstream of the *neo* promoter, in merodiploid, would confer the hyperinvasion phenotype to wild-type *S. typhimurium* strains. In fact, this appears to be the case for the *hil-378* hyperinvasive mutant. Apparently, constitutive expression of a protein, HilA, from the *neo* promoter increases the ability of *S. typhimurium* to enter cultured cells (3). It is not known whether the *hilA* gene is normally regulated by growth conditions, and further study is required to determine the function of the HilA protein in normal invasion and in hyperinvasion.

Although the *hil* hyperinvasive mutants were obtained only by insertion of the *neo* promoter in the chromosome, it appears that the *hil-380* insertion mutation increases the organism's invasion ability by disrupting a negative regulatory element, *hilR* (3). In fact, the *hilR* mutation increases the level of expression of an invasion gene, *prgH*. Furthermore, since *prgH* lies upstream of *hilR,* the HilR regulator appears to act in *trans* on gene expression (4). So, the hyperinvasion phenotype of the *hil-380* mutation can be accounted for by disruption of a repressor of invasion gene expression. Further study is required to see whether other invasion genes are under the control of *hilR* and whether *hilR* mediates the regulation of invasion genes in response to environmental conditions.

In a direct search for oxygen-regulated invasion factors, one analysis (30) directly sought invasion genes regulated by growth conditions and identified *orgA*. In contrast, the analysis of hyperinvasive mutants was an indirect approach to identifying the regulated invasion genes. Basically, the hope was that hyperinvasion mutations increase the invasion abilities of aerobically grown bacteria and might increase the level of expression of the normally oxygen-regulated invasion factors. However, it is possible that the hyperinvasion phenotypes are due to changes which increase other rate-limiting aspects of invasion. Because of this alternative possibility, further study of the *hilA* and *hilR* genes is required to understand their normal roles in invasion and invasion gene expression before concluding that they are regulated by oxygen or that they mediate regulation of invasion genes in response to oxygen.

In Vivo-Induced Genes

The course of *Salmonella* pathogenesis appears to resemble a bacterial *Fantastic Voyage* during which the bacteria interact with different host cells and encounter a series of environments in vivo. Throughout infection, different cues apparently inform the bacteria where they are in the host and serve to coordinate the appropriate bacterial response for each situation. A recent approach to identifying virulence factors has been to isolate genes that are specifically expressed during infection. In contrast to the previous examples, this approach does not depend on knowledge of regulatory factors or specific environmental signals.

In vivo expression technology (IVET) was designed to identify bacterial genes that are expressed in vivo but that are not expressed in vitro (38). Presumably, their selective in vivo expression indicates that the corresponding gene products function during infection and are likely to be involved in virulence. The first application of IVET relied on the fact that *S. typhimurium purA* mutants are avirulent. Basically, random DNA segments were positioned upstream of a promoterless *purA* gene and the fusions were introduced into an *S. typhimurium* strain that could express all virulence functions except *purA*. In order to infect an animal, the *purA*-attenuated bacteria must express the *purA* reporter gene from a promoter on the DNA fragment, thus providing a means of selecting for those random genetic segments that correspond to genes that are normally expressed in vivo.

It was simple to select such infective fusion strains from random pools by inoculating mice and then recovering those bacteria that could survive and replicate within the host. Of course, most of the recovered strains are constitutively Pur$^+$, expressing the *purA* fusion when in the laboratory as well as when within the animal. Fortunately, such in vitro-expressed fusions are easily detected and ignored since the IVET system generates fusions to both the *purA* and *lacZY* genes. Simply by plating for CFU samples from the infected animal onto lactose indicator medium, strains that contain fusions expressed both in vivo and in vitro are seen as Lac$^+$. In contrast, the desired fusion strains that specifically express the fusion in vivo are recovered from animals as rare Lac$^-$ isolates. In this way, the design of IVET makes it simple to identify strains that contain in vivo-specific fusions. Analysis of *ivi* (in vivo-induced) genes has revealed many distinct loci and, at this point, is far from saturating. So far, of three *ivi* genes tested, all three are required for virulence. Further study is required to test the role of each *ivi* gene in virulence and to determine where each is expressed in vivo and what signals regulate their expression.

Many modifications of the IVET system are possible; one could use different reporter genes, different infection models, as well as different in vitro growth conditions. Interestingly, each modification may yield different sets of virulence genes. For example, in the case of the *purA* reporter system, if the *purA* gene product is required during bacterial interaction with macrophages, an *ivi-purA* fusion will be selected for as long as the fusion is expressed at those stages of infection prior to phagocytosis. However, if the *purA* gene is required at several stages during infection, e.g., during intestinal colonization, in the bloodstream, in host tissues, or in different host cell types, an *ivi-purA* fusion may have to be expressed throughout infection to answer the selection. In either case, the selection may miss an *ivi* gene that is expressed exclusively at a single site in the host. So, the use of different reporter gene and attentuating mutation combinations might yield different sets and subsets of *ivi* genes.

Another advantage of using different reporter systems is that the sensitivity of the in vivo selection could be changed. For example, the in vitro versus in vivo expression of one *ivi* gene might be 0.01 versus 1 unit, respectively, whereas that of another might be 10 versus 1,000 units, respectively. In the former case, the properties of the reporter gene and the attenuating mutation might not be able to select for such an *ivi* fusion if more than 1 unit of expression is required for in vivo complementation. Alternatively, a reporter and mutation combination might be too sensitive to identify the latter *ivi* fusion if 10 units of expression in vitro is scored as "on."

The first IVET selection was conducted in BALB/c mice by intraperitoneal inoculation of *S. typhimurium* fusion strains. It is likely that infection via different modes of inoculation or even of different hosts is characterized by unique bacterium-host cell interactions and distinct challenges to bacterial survival. So, the use of different infection models might alter the in vivo selection conditions and, thus, affect what genes are identified by IVET as being induced in vivo. Another variable in the identification of in vivo-induced genes is the choice of the in vitro growth condition. If some aspect of the in vitro growth condition mimics an important aspect of the in vivo environment, certain regulated virulence genes might be discarded as not being in vivo specific. For example, it is likely that simple environmental cues such as host body temperature, high osmolarity, or nutrient conditions may play important regulatory roles in vivo. For this reason, a number of *ivi* loci may have been missed by the initial IVET analysis in which MacConkey lactose medium at 37°C was chosen for the in vitro growth condition.

In summary, the IVET system provides a novel approach to the study of genes involved in infection by identifying genes whose expression is regulated in vivo. One feature of IVET selection is that simple alterations in the design of the system could change what subset of *ivi* genes is identified. Further application of IVET approaches to *Salmonella* species as well as to other systems will undoubtedly aid in understanding the complexities of bacterial adaptation during pathogenesis.

SUMMARY

This chapter has described five different genetic approaches used to identify virulence genes in *Salmonella* species. Each approach has its advantages and disadvantages, but each has been valuable for understanding *Salmonella* pathogenicity. It is gratifying that, in several cases, the application of different approaches has identified the same virulence factors. Although a lot has been learned about *Salmonella* virulence, much remains to be understood. For example, extensive molecular genetic analysis of *Salmonella* invasion has yet to identify the bacterial invasion factors that directly interact with host cells. Although a large number of genes have been identified to be required for invasion, determination of each of their functions will be time-consuming. In retrospect, it may turn out that biochemical characterization will have been a more direct approach to the identification of the interactive invasion components. However, the use of genetics will continue to serve as an essential tool in future studies of *Salmonella* pathogenesis, both to identify virulence genes and to elucidate their roles in pathogenesis. As researchers develop genetic systems in less tractable microbes, this variety of genetic approaches that has been so useful for understanding *Salmonella* species might help in studies of other pathogens.

ACKNOWLEDGMENTS. I thank Stanley Falkow for giving me the opportunity to be a postdoctoral fellow in his lab (January 1988 to December 1991). Stanley continues to inspire and encourage my research. I dedicate this chapter to him. I also thank my many colleagues in the *Salmonella* invasion field who have freely shared information and materials prior to publication.

C. Lee is a recipient of an American Cancer Society Junior Faculty Research Award and a Charles H. Hood Foundation grant.

REFERENCES

1. **Alpuche Aranda, C. M., J. A. Swanson, W. P. Loomis, and S. I. Miller.** 1992. *Salmonella typhimurium* activates virulence gene transcription within acidified macrophage phagosomes. *Proc. Natl. Acad. Sci. USA* **89:**10079–10083.
2. **Altmeyer, R. M., J. K. McNern, J. C. Bossio, I. Rosenshine, B. B. Finlay, and J. E. Galán.** 1993. Cloning and molecular characterization of a gene involved in *Salmonella* adherence and invasion of cultured epithelial cells. *Mol. Microbiol.* **7:**89–98.
3. **Bajaj, V., and C. A. Lee.** Unpublished data.

4. **Behlau, I., and S. I. Miller.** 1993. A PhoP-repressed gene promotes *Salmonella typhimurium* invasion of epithelial cells. *J. Bacteriol.* **175:**4475–4484.

5. **Benjamin, W. H., Jr., C. L. Turnbough, Jr., B. S. Posey, and D. E. Briles.** 1985. The ability of *Salmonella typhimurium* to produce the siderophore enterobactin is not a virulence factor in mouse typhoid. *Infect. Immun.* **50:**392–397.

6. **Benjamin, W. H., Jr., J. Yother, P. Hall, and D. E. Briles.** 1991. The *Salmonella typhimurium* locus *mviA* regulates virulence in Ity^s but not Ity^r mice: functional *mviA* results in avirulence; mutation (nonfunctional) *mviA* results in virulence. *J. Exp. Med.* **174:**1073–1083.

7. **Betts, J., and B. B. Finlay.** 1992. Identification of *Salmonella typhimurium* invasiveness loci. *Can. J. Microbiol.* **38:**852–857.

8. **Bliska, J. B., J. E. Galán, and S. Falkow.** 1993. Signal transduction in the mammalian cell during bacterial attachment and entry. *Cell* **73:**903–920.

9. **Buchmeier, N. A., C. J. Lipps, M. Y. So, and F. Heffron.** 1993. Recombination-deficient mutants of *Salmonella typhimurium* are avirulent and sensitive to the oxidative burst of macrophages. *Mol. Microbiol.* **7:**933–936.

10. **Carter, P. B., and F. M. Collins.** 1974. The route of enteric infection in normal mice. *J. Exp. Med.* **139:**1189–1203.

11. **Collins, F. M.** 1969. Effect of specific immune mouse serum on the growth of *Salmonella enteritidis* in nonvaccinated mice challenged by various routes. *J. Bacteriol.* **97:**667–675.

12. **Curtiss, R., III, and S. M. Kelly.** 1987. *Salmonella typhimurium* deletion mutants lacking adenylate cyclase and cyclic AMP receptor protein are avirulent and immunogenic. *Infect. Immun.* **55:** 3035–3043.

13. **Dorman, C. J., S. Chatfield, C. F. Higgins, C. Hayward, and G. Dougan.** 1989. Characterization of porin and *ompR* mutants of a virulent strain of *Salmonella typhimurium: ompR* mutants are attenuated in vivo. *Infect. Immun.* **57:**2136–2140.

14. **Elsinghorst, E. A.** Personal communication.

15. **Elsinghorst, E. A., L. S. Baron, and D. J. Kopecko.** 1989. Penetration of human intestinal epithelial cells by *Salmonella:* molecular cloning and expression of *Salmonella typhi* invasion determinants in *Escherichia coli. Proc. Natl. Acad. Sci. USA* **86:**5173–5177.

16. **Falkow, S.** 1988. Molecular Koch's postulates applied to microbial pathogenicity. *Rev. Infect. Dis.* **10:**S274–S276.

17. **Fields, P. I., E. A. Groisman, and F. Heffron.** 1989. A *Salmonella* locus that controls resistance to microbicidal proteins from phagocytic cells. *Science* **243:**1059–1062.

18. **Fields, P. I., R. V. Swanson, C. G. Haidaris, and F. Heffron.** 1986. Mutants of *Salmonella typhimurium* that cannot survive within the macrophage are avirulent. *Proc. Natl. Acad. Sci. USA* **83:** 5189–5193.

19. **Finlay, B. B., M. N. Starnbach, C. L. Francis, B. A. Stocker, S. Chatfield, G. Dougan, and S. Falkow.** 1988. Identification and characterization of Tn*phoA* mutants of *Salmonella* that are unable to pass through a polarized MDCK epithelial cell monolayer. *Mol. Microbiol.* **2:**757–766.

20. **Fitts, R.** 1985. Development of a DNA-DNA hybridization test for the presence of *Salmonella* in foods. *Food Technol.* **39:**95–102.

21. **Galán, J. E., and R. Curtiss III.** 1989. Cloning and molecular characterization of genes whose products allow *Salmonella typhimurium* to penetrate tissue culture cells. *Proc. Natl. Acad. Sci. USA* **86:**6383–6387.

22. **Galán, J. E., and R. Curtiss III.** 1990. Expression of *Salmonella typhimurium* genes required for invasion is regulated by changes in DNA supercoiling. *Infect. Immun.* **58:**1879–1885.

23. **Groisman, E. A., E. Chiao, C. J. Lipps, and F. Heffron.** 1989. *Salmonella typhimurium phoP* virulence gene is a transcriptional regulator. *Proc. Natl. Acad. Sci. USA* **86:**7077–7081.

24. **Groisman, E. A., and F. Heffron.** Personal communication.

25. **Groisman, E. A., and H. Ochman.** 1993. Cognate gene clusters govern invasion of host epithelial cells by *Salmonella typhimurium* and *Shigella flexneri. EMBO J.* **12:**3779–3787.

26. **Groisman, E. A., C. Parra-Lopez, M. Salcedo, C. J. Lipps, and F. Heffron.** 1992. Resistance to host antimicrobial peptides is necessary for *Salmonella* virulence. *Proc. Natl. Acad. Sci. USA* **89:**11939–11943.

27. **Groisman, E. A., M. A. Sturmoski, F. R. Solomon, R. Lin, and H. Ochman.** 1993. Molecular,

functional, and evolutionary analysis of sequences specific to *Salmonella*. *Proc. Natl. Acad. Sci. USA* **90**:1033–1037.

28. **Gulig, P. A., H. Danbara, D. G. Guiney, A. J. Lax, F. Norel, and M. Rhen.** 1993. Molecular analysis of *spv* virulence genes of the salmonella virulence plasmids. *Mol. Microbiol.* **7**:825–830.

29. **Gulig, P. A., and T. J. Doyle.** 1993. The *Salmonella typhimurium* virulence plamid increases the growth rate of salmonellae in mice. *Infect. Immun.* **61**:504–511.

30. **Jones, B. D., and S. Falkow.** Personal communication.

31. **Jones, B. D., C. A. Lee, and S. Falkow.** 1992. Invasion of *Salmonella typhimurium* is affected by the direction of flagellar rotation. *Infect. Immun.* **60**:2475–2480.

32. **Lee, C. A., and S. Falkow.** 1990. The ability of *Salmonella* to enter mammalian cells is affected by bacterial growth state. *Proc. Natl. Acad. Sci. USA* **87**:4304–4308.

33. **Lee, C. A., B. D. Jones, and S. Falkow.** 1992. Identification of a *Salmonella typhimurium* invasion locus by selection for hyperinvasive mutants. *Proc. Natl. Acad. Sci. USA* **89**:1847–1851.

34. **Litwin, C. M., and S. B. Calderwood.** 1993. Role of iron in regulation of virulence genes. *Clin. Microbiol. Rev.* **6**:137–149.

35. **Lockman, H. A., and R. Curtiss III.** 1990. *Salmonella typhimurium* mutants lacking flagella or motility remain virulent in BALB/c mice. *Infect. Immun.* **58**:137–143.

36. **Lockman, H. A., and R. Curtiss III.** 1992. Virulence of non-type 1-fimbriated and nonfimbriated nonflagellated *Salmonella typhimurium* mutants in murine typhoid fever. *Infect. Immun.* **60**:491–496.

37. **Macnab, R. M.** 1992. Genetics and biogenesis of bacterial flagella. *Annu. Rev. Genet.* **26**:131–158.

38. **Mahan, M. J., J. M. Slauch, and J. J. Mekalanos.** 1993. Selection of bacterial virulence genes that are specifically induced in host tissues. *Science* **259**:686–688.

39. **Mekalanos, J. J.** 1992. Environmental signals controlling the expression of virulence determinants in bacteria. *J. Bacteriol.* **174**:1–7.

40. **Miller, I., D. Maskell, C. Hormaeche, K. Johnson, D. Pickard, and G. Dougan.** 1989. Isolation of orally attenuated *Salmonella typhimurium* following Tn*phoA* mutagenesis. *Infect. Immun.* **57**:2758–2763.

41. **Miller, S. I., A. M. Kukral, and J. J. Mekalanos.** 1989. A two component regulatory system (*phoP* and *phoQ*) controls *Salmonella typhimurium* virulence. *Proc. Natl. Acad. Sci. USA* **86**:5054–5058.

42. **Miller, S. I., and J. J. Mekalanos.** 1990. Constitutive expression of the *PhoP* regulon attenuates *Salmonella* virulence and survival within macrophages. *J. Bacteriol.* **172**:2485–2490.

43. **Miller, S. I., W. S. Pulkkinen, M. E. Selsted, and J. J. Mekalanos.** 1990. Characterization of defensin resistance phenotypes associated with mutations in the *phoP* virulence regulon of *Salmonella typhimurium*. *Infect. Immun.* **58**:3706–3710.

44. **Miller, V. M., K. B. Beer, W. P. Loomis, J. A. Olson, and S. I. Miller.** 1992. An unusual *pagC*::Tn-*phoA* mutation leads to an invasion- and virulence-defective phenotype in *Salmonellae*. *Infect. Immun.* **60**:3763–3770.

45. **Mills, D. M., and C. A. Lee.** Unpublished data.

46. **Pace, J., M. J. Hayman, and J. E. Galán.** 1993. Signal transduction and invasion of epithelial cells by *S. typhimurium*. *Cell* **72**:505–514.

47. **Pulkkinen, W. S., and S. I. Miller.** 1991. A *Salmonella typhimurium* virulence protein is similar to a *Yersinia enterocolitica* invasion protein and a bacteriophage lambda outer membrane protein. *J. Bacteriol.* **173**:86–93.

48. **Sampson, B. A., and E. C. Gotschlich.** 1992. Elimination of the vitamin B_{12} uptake or synthesis pathway does not diminish the virulence of *Escherichia coli* K1 or *Salmonella typhimurium* in three model systems. *Infect. Immun.* **60**:3518–3522.

49. **Siitonen, A., and M. Nurminen.** 1992. Bacterial motility is a colonization factor in experimental urinary tract infection. *Infect. Immun.* **60**:3918–3920.

50. **Sinai, A. P., and P. M. Bavoil.** 1993. Hyper-invasive mutants define a novel *pho*-regulated invasion pathway in *Escherichia coli*. *Mol. Microbiol.* **10**:1125–1137.

51. **Stocker, B. A. D., and P. H. Mäkëla.** 1986. Genetic determination of bacterial virulence, with special reference to *Salmonella*. *Curr. Top. Microbiol. Immunol.* **124**:149–172.

52. **Stone, B. J., C. M. Garcia, J. L. Badger, T. Hassett, R. I. F. Smith, and V. L. Miller.** 1992.

Identification of novel loci affecting entry of *Salmonella enteritidis* into eukaryotic cells. *J. Bacteriol.* **174:**3945–3952.

53. **Straus, D., and F. M. Ausubel.** 1990. Genomic substraction for cloning DNA corresponding to deletion mutations. *Proc. Natl. Acad. Sci. USA* **87:**1889–1893.

54. **Takeuchi, A.** 1967. Electron microscope studies of experimental *Salmonella* infection. I. Penetration into the intestinal epithelium by *Salmonella typhimurium. Am. J. Pathol.* **50:**109–136.

55. **Vidal, S. M., D. Malo, K. Vogan, E. Skamene, and P. Gros.** 1993. Natural resistance to infection with intracellular parasites: isolation of a candidate for Bcg. *Cell* **73:**469–485.

Molecular Genetics of Bacterial Pathogenesis
Edited by V. L. Miller, J. B. Kaper, D. A. Portnoy, and R. R. Isberg
© 1994 American Society for Microbiology, Washington, DC 20005

Chapter 15

Mechanisms of *Yersinia* Entry into Mammalian Cells

Dorothy E. Pierson

The pathogenic yersiniae, *Yersinia pestis, Yersinia pseudotuberculosis,* and *Yersinia enterocolitica,* share tissue tropism in the host, all eventually ending up in lymphatic tissue. The means of initiating infection differ, and the ultimate disease outcomes of infection with these three organisms differ. This variation in disease outcomes is presumably a result of the alternative sites of initial infection. *Y. pestis* enters the host via the bite of an infected flea. Thus, this organism is deposited across the epithelial cell boundary of the skin in close proximity to the lymphatic system below. *Y. pseudotuberculosis* and *Y. enterocolitica* enter the gastrointestinal tract of the host via contaminated food or water. To ultimately establish infection, these two organisms must cross the epithelial cell boundary in the intestinal tract, specifically, the epithelial cell layer in the terminal ileum. There is evidence that these bacteria cross this cell layer through the specialized absorptive epithelial cells overlying the lymphatic system, that is, M cells (16).

To begin to examine the process whereby the enteric pathogenic yersiniae enter the host, in 1981 Devenish and Schiemann (13) developed the now famous invasion assay, an in vitro model for the entry of *Y. enterocolitica* into mammalian cells. In this model, bacteria are added to a monolayer of tissue culture cells and are incubated for various periods of time to allow interaction between the bacteria and mammalian cells. Nonadherent bacteria are washed away and medium containing the antibiotic gentamicin, an antibiotic which only very poorly crosses the eukaryotic cell membrane, is added back to the tissue culture. After incubation to allow the gentamicin to kill the extracellular bacteria, the medium is removed and the tissue culture cells are lysed to release bacteria, which are plated to determine viable counts. By using this type of assay, it was shown that *Y. pseudotuberculosis* and *Y. enterocolitica* enter a number of cell types that are epithelial in nature as well as Chinese hamster ovary (CHO) cells, a fibroblast-derived cell line (9, 13, 25, 32, 40). In contrast, *Y. pestis* does not enter these cell types in this assay (48). The entry process did not seem to be dependent on the large plasmid

---

Dorothy E. Pierson • Department of Microbiology, University of Colorado Health Sciences Center, Denver, Colorado 80262.

found in the pathogenic *Yersinia* species that had been shown to be required in animal models for virulence (see chapter 24). That is, bacteria cured of this plasmid could enter cells at a frequency similar to that of their plasmid-containing parental strains (40, 46).

IDENTIFICATION OF GENES REQUIRED FOR ENTRY OF ENTEROPATHOGENIC YERSINIAE INTO EUKARYOTIC CELLS

Analysis of the genes involved in the entry of *Yersinia* spp. into eukaryotic cells in vitro was initiated by Isberg and Falkow (20), who took the then novel (but now familiar) tactic of trying to clone the gene(s) that allowed *Y. pseudotuber-culosis* to enter HEp-2 cells, an epithelial cell line. A library of *Y. pseudotuberculo-sis* chromosomal DNA was introduced into a strain of *Escherichia coli* K-12 that does not normally interact with these cells, and the bacteria that entered the cells were selected (20). This method required that the cell entry phenotype be fairly simple in *Y. pseudotuberculosis,* being encoded by a relatively small number of linked genes that could be cloned into a cosmid. As luck would have it, a single gene which conferred upon *E. coli* the ability to enter cultured cells was isolated. This gene was named *inv,* for the invasion phenotype that it specified (Table 1).

Using a similar protocol, Miller and Falkow (32) isolated two genes from *Y. enterocolitica* that independently acted to promote *E. coli* entry into tissue culture cells. One of these genes turned out to be homologous to the *inv* gene of *Y. pseudotuberculosis* and was therefore named *inv.* The second gene identified was named *ail,* for attachment invasion locus, reflecting the fact that this locus pro-motes high levels of bacterial attachment to cells (Table 2), in addition to the invasion phenotype.

A final *Yersinia* gene identified as being involved in bacterial entry into cells is one that is located on the virulence-associated plasmid. Although plasmidless bacteria will enter cultured cells, the plasmid has been implicated in the entry process, because *inv* mutant *Y. pseudotuberculosis* exhibits a low level of entry that is dependent upon the presence of the virulence plasmid. By subcloning por-tions of the virulence plasmid and using plasmids with insertions in known loci, Yang and Isberg (56) and Bliska et al. (6) determined that the plasmid locus neces-sary for the low level of entry by *inv* mutant *Y. pseudotuberculosis* is the *yadA* gene (Table 3).

The *inv, ail,* and *yadA* genes were identified as having independent roles in the entry of yersiniae into mammalian cells. The fact that these three genes could be identified as being involved in cell entry by isolating clones that exhibited increased levels of entry implies that the *Yersinia* cell entry process is fairly simple. This simplicity in entry systems has not been seen for other invasive bacterial pathogens. For example, in the case of *Salmonella* spp., it has not been possible to identify cell entry factors by cloning genes encoding these factors into *E. coli,* suggesting that entry by *Salmonella* spp. is more complex (for further discussions of *Salmonella* entry mechanisms, see chapters 14 and 16).

Table 1

Enrichment procedure yields *E. coli* strains that invade cell culture monolayers[a]

Strain	% Invasion[b]
YPIII(p⁻)[c]	9.0
HB101[d]	0.005
HB101(pINVA2)[e]	8.3
HB101(pINVA7)	7.9
HB101(pINVG10)	8.7
HB101(pRI203)	9.2

[a] From Isberg, R. R., and S. Falkow. 1985. *Nature* (London) **317**:262–264 (20). Reprinted with permission from *Nature* (London) **317**:262–264. Copyright 1985 Macmillan Magazines Limited. Saturated cultures of bacteria, grown at 28°C, were washed twice in phosphate-buffered saline (PBS) and resuspended to a concentration of 3×10^8 bacteria ml^{-1}. Aliquots of 50 μl of each strain were added to monolayer cultures of HEp-2 cells seeded at a concentration of 2×10^5 animal cells per microtiter well (24-well Falcon 3047 microtiter dishes) in RPMI 1640 medium. Bacteria were centrifuged onto the monolayer at $600 \times g$ as described previously (13), and the infected cultures were incubated at 36°C for 3 h in a 5% CO_2 atmosphere, to allow binding and invasion of bacteria. Nonadherent bacteria were removed from the monolayer by washing three times with sterile PBS, and RPMI 1640 medium containing 40 μg of gentamicin (Sigma) ml^{-1} was added to each microtiter well. The incubation was continued at 36°C for 2 h in the presence of the antibiotic before washing of the monolayers in PBS twice more. Internalized bacteria were then released from the monolayers by the addition of 1% Triton X-100, and titers were determined on L-agar plates (28).

[b] Percentage of bacteria added to HEp-2 monolayers that resist treatment by gentamicin.

[c] *Y. pseudotuberculosis* strain (8).

[d] *E. coli* K-12 strain HB101.

[e] HB101 harboring cosmids that are denoted in parentheses.

Table 2

Relative entry of *Y. enterocolitica* clones into HEp-2 cells[a]

Infecting strain	% Invasion
Y. enterocolitica 8081c	26.8 ± 3.6
E. coli HB101	0.0075 ± 0.0035
HB101(pBR322-8081c) (clone 3 = pVM103) (*ail*)	0.82 ± 0.24
HB101(pBR322-8081c) (clone 7 = pVM102) (*ail*)	0.37 ± 0.19
HB101(pBR322-8081c) (clone 6 = pVM101) (*inv*)	6.2 ± 2.1
HB101(pBR322-8081c) (clone 8) (*inv*)	10.8 ± 0.1
HB101(pBR322-8081c) (clone 9) (*inv*)	6.9 ± 0.32

[a] Strains were used to infect a monolayer of the human larynx epithelial cell line HEp-2. Percent invasion is as described in footnote b of Table 1. Adapted from Miller, V. L., and S. Falkow. 1988. *Infect. Immun.* **56**:1242–1248 (32).

TABLE 3
YadA-mediated entry of *Y. pseudotuberculosis* into HEp-2 cells[a]

Strain	Description	% Invasion[a]
YPIII (P$^+$)	*inv*$^+$, P$^+$	1.6 ± 0.2
YPIII (P$^-$)	*inv*$^+$, P$^-$	7.2 ± 1.5
YP306	*inv*$^+$, P$^-$, pACYC184 *yadA*	9.7 ± 2.0
YP332	*inv*::*tet*, P$^-$	0.022 ± 0.008
YP336	*inv*::*tet*, P$^-$, pACYC184 *yadA*	2.7 ± 0.87

[a] Percent invasion was determined as described in footnote *b* of Table 1. Adapted from Yang, Y., and R. Isberg. 1993. *Infect. Immun.* **61**:3907–3913 (56).

Further characterization of each gene is described below. The case of these genes in *Y. pestis* is discussed at the end of this chapter.

ANALYSIS OF *inv*

The *inv* gene was originally identified as having a role in bacterial entry into eukaryotic cells by demonstrating that *E. coli* containing the cloned *inv* gene entered cells in vitro (20, 32). The fact that this gene has a role in this process in *Yersinia* spp. was confirmed by the demonstration that *inv* mutant *Y. pseudotuberculosis* and *Y. enterocolitica* are greatly reduced in their abilities to enter cells compared with the abilities of their wild-type parents (23, 36, 45). When the wild-type *inv* gene is reintroduced into the *inv* mutants, the bacteria are again able to enter cells.

The invasin protein encoded by the *inv* gene from *Y. pseudotuberculosis* is predicted to be a 103-kDa membrane protein (23), whereas the invasin protein from *Y. enterocolitica* is predicted to be a 91-kDa membrane protein (58). There is a high degree of homology between these two proteins, with 73% identity in the amino acid sequence (58). The main difference between the two proteins is that the *Y. enterocolitica* invasin is missing a stretch of 99 amino acids in the middle of the protein that is present in the invasin from *Y. pseudotuberculosis*. Additionally, there are two short stretches present in the amino-terminal portion of the *Y. pseudotuberculosis* invasin that are absent from the *Y. enterocolitica* invasin. These two proteins have been demonstrated to be cell surface proteins by cell fractionation and by showing that the proteins are accessible to trypsin cleavage in whole cells (23, 58).

The receptor on the mammalian cell surface that the *Y. pseudotuberculosis* invasin protein binds to was identified by a number of elegant studies from the Isberg lab. Proteins that bind invasin were purified by affinity chromatography of extracts of various human cell lines over columns containing invasin covalently linked to agarose beads (21). The invasin receptors identified in this manner are members of the β$_1$ family of integrins. Integrins are heterodimeric cell surface molecules with α and β subunits which are involved in the interaction with both the extracellular matrix and the cytoskeletal apparatus (18). A number of α and β subunits have been identified, and integrins are grouped into classes on the basis of the type of subunit that they possess. The β_1 family of integrins is distributed over a wide variety of cell types, including, but not limited to, epithelial cells, monocytes, and T cells (19). Antibodies directed against the β_1 integrin subunit or against members of the β_1 family block entry into the eukaryotic cell mediated by either the *Y. pseudotuberculosis* or the *Y. enterocolitica* invasin protein (21, 57).

Invasin-mediated entry does not require viable bacteria; latex beads coated with invasin from either organism are taken up by mammalian cells. Using truncated proteins and protein fusions to maltose-binding protein and by monoclonal antibody mapping, the portion of the *Y. pseudotuberculosis* invasin that binds to the β_1 integrin molecule has been narrowed down to the carboxyl-terminal 192 amino acids of the protein (26). The integrin-binding domain of invasin has since been narrowed down further to a 76-amino-acid disulfide loop at the carboxyl terminus of the protein by identification of mutants that alter the formation of the disulfide bond and reduce integrin binding (27). There is a high degree of homology between the *Y. enterocolitica* and *Y. pseudotuberculosis* invasins in this region, suggesting that this portion of the *Y. enterocolitica* protein is involved in receptor binding as well (58).

Uptake of invasin-coated substances, be they bacteria or beads, appears to be dependent upon the affinity of the invasin protein for its integrin receptor molecule. Entry is hypothesized to occur by zippering of the host cell surface around the invasin-coated bacterium because of the strong interactions between invasin and its receptor (19). Invasin and fibronectin both bind to the same integrins, probably at the same site (53), yet fibronectin-coated beads or bacteria are not taken up by cells (42). Invasin binds to the integrin $\alpha_5\beta_1$ over 100 times better than fibronectin, leading to the suggestion that it is the difference in affinity that determines whether uptake occurs (53). The role of affinity in uptake has been confirmed by Tran Van Nhieu and Isberg (54). Those investigators examined the interaction of host cells with monoclonal antibodies directed against β_1 chain integrins with various affinities. Binding to or entry into cells by *Staphylococcus aureus* coated with different anti-β_1 monoclonal antibodies is dependent upon the affinity of the particular antibody for its ligand. *S. aureus* coated with high-affinity antibodies enters tissue culture cells, whereas low-affinity antibody-coated *S. aureus* binds to but does not enter cells.

A number of groups have examined the regulation of *inv* gene expression by environmental factors. In both *Y. pseudotuberculosis* and *Y. enterocolitica,* the *inv* gene is expressed well at 30°C and at lower temperatures, but not at the host temperature of 37°C (22, 38). The regulator of *inv* gene expression in response to temperature has been identified as one that regulates the expression of other putative virulence genes in *Yersinia* spp., including the enterotoxin gene *yst* and the *rfb* genes for lipopolysaccharide O polysaccharide side chain biosynthesis (50). The fact that the *inv* gene is not expressed at the host temperature, at least not in bacteria growing in rich medium in the laboratory, suggests that if invasin is important in virulence it must be required early, before the bacteria have grown at the host temperature and turned off *inv* gene expression. Alternatively, the actual in vivo conditions that induce *inv* gene expression at 37°C may not be the same as those conditions found in the laboratory, and therefore, they have yet to be identified. This hypothesis is supported by recent data suggesting that the *inv* gene is expressed in bacteria found within the Peyer's patches during experimental infection. In addition, growth in low-pH medium in vitro has been shown to induce *inv* gene expression at 37°C (35). Whether low pH is the in vivo signal for *inv* gene expression in the Peyer's patches or whether another in vivo signal is responsible for the activation of *inv* gene expression is not known.

Analysis of the distribution of *inv*-homologous sequences in pathogenic and nonpathogenic yersiniae revealed that *inv*-homologous sequences are present in all *Yersinia* spp., pathogenic or not (33). However, in at least four nonpathogenic *Y. enterocolitica* isolates, the *inv*-homologous sequences are not expressed (38). Furthermore, introduction of the *inv* gene from a pathogenic strain into three of these strains is sufficient to confer upon them the ability to enter cultured cells (38). Thus, the presence of *inv*-homologous sequences in nonpathogenic yersiniae does not preclude a role for this protein in virulence.

A direct analysis of the role of invasin in virulence was performed by Rosqvist et al. (45), who examined the role of the *inv* gene in infection of mice by *Y. pseudotuberculosis*. An *inv* mutant derivative of *Y. pseudotuberculosis* YPIII, the strain from which the *inv* gene was originally isolated, had a 50% lethal dose (LD_{50}) similar to that of its wild-type parent when the strains were administered orally or intraperitoneally into mice (45). However, those investigators did note that the time course of infection by the mutant was delayed in comparison with that of the wild type. Subsequent studies on an *inv* mutant of *Y. enterocolitica* by Pepe and Miller (36) confirmed these observations. However, although ultimately the outcomes for the wild type and the *inv* mutant were similar, the initial courses of infection differed. The numbers of *inv* mutant bacteria colonizing the Peyer's patches were drastically reduced in comparison with the numbers of wild-type bacteria that did so. Presumably, this lower level of colonization is responsible for the delay in the infection process by the *inv* mutant. These data suggest that invasin is required early in infection, as had been suggested previously by the temperature regulation and Peyer's patch expression data. Alternative cell entry factors (perhaps Ail or YadA) may substitute for invasin, eventually allowing the mutants to progress from the gastrointestinal tract, leading to systemic infection. Once the organism has successfully crossed the epithelial cell boundary, invasin is apparently not required for subsequent steps in the infection. The careful studies of Pepe and Miller (36) allow examination of the role of invasin in *Y. enterocolitica* virulence in more depth than can be done by using the crude measurement of LD_{50}. As Pepe and Miller suggest (36), LD_{50} measurements are not the best way to study the virulence factors of enteric pathogens such as *Y. enterocolitica*, because death is not a normal outcome of infection with these organisms.

ANALYSIS OF *ail*

The Ail protein is predicted to be a 17-kDa membrane protein with eight membrane-spanning domains (3, 31). Ail has been localized to the outer membrane of *Y. enterocolitica* by cellular fractionation (31).

E. coli containing the cloned *ail* gene does not enter cells at nearly the efficiency that *E. coli* containing the *inv* gene does (32). Further analysis showed that although *ail*-containing *E. coli* can attach to a wide variety of cell types, significant levels of entry are seen into CHO cells only (Table 2). Thus, in the case of *ail*, adherence and entry are separable phenomena. *ail* mutants of *Y. enterocolitica* do not adhere to or enter into cultured cells when they are grown under conditions in which the *inv* gene is not expressed (39). An *inv ail* double mutant does not enter cells to any measurable degree (37). Thus, both *inv* and *ail* are involved in *Y. enterocolitica* entry into cultured cells. Although there are *ail*-homologous sequences in *Y. pseudotuberculosis* (33), these have not been analyzed, and their roles in bacterial uptake by eukaryotic cells are unknown.

As had been seen with the *inv* gene, *ail* gene expression is regulated by temperature (39). However, unlike the *inv* gene, the *ail* gene is expressed at the host temperature, 37°C. The *ail* gene is expressed in bacteria growing logarithmically at 20, 28, and 37°C. In stationary-phase bacteria, however, *ail* transcripts are detected only at 37°C. The expression of the *ail* gene at the host temperature is consistent with a role for this gene in the host.

Further support for *ail* as a virulence factor comes from examination of pathogenic and nonpathogenic yersiniae for *ail*-homologous sequences. Only pathogenic *Yersinia* spp. and pathogenic isolates of *Y. enterocolitica* have sequences homologous to the *ail* gene (33). This information suggests that *ail* may be important for pathogenesis. Curiously, though, when *ail* is introduced into these nonpathogenic strains, they do not enter cells any better than their parents that lack the *ail* gene (39). The introduced *ail* sequences are expressed, and Ail protein can be detected on the bacterial cell surface. One explanation for the absence of Ail-mediated entry in these nonpathogenic strains containing the *ail* gene is that other factors, in addition to Ail, that affect yersinia entry are required.

Y. enterocolitica isolates have been classified into American and non-American strains on the bases of their O serotypes and virulence characteristics. The American strains cause a more severe disease than that caused by non-American strains (10). Besides a difference in O serotype, the *ail* sequences differ in these strains. A sequence homologous to IS*3* is found downstream of the *ail* gene and in multiple other locations in the genomes of the American strains but is absent from the genomes of non-American strains (3, 33). In addition, the sequence of the *ail* gene itself differs in the two strain types, with the majority of the differences being found in sequences in portions of the Ail protein predicted to be located on the cell surface (3). *E. coli* containing the *ail* gene cloned from four different American strains enters cells 5- to 10-fold better than *E. coli* containing the *ail* gene from four non-American strains (3). Whether these differences help to account for the difference in virulence between the strains is not known. The definitive proof for Ail as a virulence factor awaits animal studies with *ail* mutants and *ail inv* double mutants.

Sequences homologous to the Ail protein have been found in a number of other members of the family *Enterobacteriaceae* (17, 41). The OmpX protein, an outer membrane protein of unknown function, from *Enterobacter cloacae* has 43.5% amino acid identity to Ail. An outer membrane protein, also of unknown function, encoded by the *lom* gene of bacteriophage λ, which is expressed in λ lysogens (2), also has homology to Ail. Finally, two different virulence factors of *Salmonella typhimurium* have been identified as being homologous to Ail. One of these, PagC, is required for the survival of this organism in macrophages and virulence in animal models (29). *pagC* mutants of *S. typhimurium* are unaffected in their ability to enter eukaryotic cells, suggesting that PagC does not have the same function in *S. typhimurium* that the *ail* gene product does in *Y. enterocolitica* (30). The other *S. typhimurium* virulence factor with homology to Ail is the product of the *rck* gene, a plasmid gene involved in the survival of *S. typhimurium* in human serum (17). The homologous regions of these different proteins lie primarily in domains predicted to span the membrane, suggesting that the homology is due more to structural considerations rather than to any functional considerations. There is evidence that the Ail protein, in addition to functioning as a cell entry factor, has a role in *Y. enterocolitica* survival in serum. *ail* mutant derivatives of *Y. enterocolitica* are over 10^5 times more susceptible to serum than their wild-type parents (5, 39). Nonpathogenic *Y. enterocolitica* strains that contain the cloned *ail* gene are more resistant to serum killing than their wild-type parents as well (39). Thus, *ail* appears to have more than one function that might be involved in the survival of *Y. enterocolitica* in the host.

ANALYSIS OF *yadA*

The YadA protein is a 45-kDa plasmid-encoded protein that has been demonstrated to form fibrillar structures on the bacterial cell surface (24). Similar to what has been seen in the case of the Ail protein, a number of other properties in addition to cell entry have been identified as being mediated by YadA. Two of these properties, adherence to mammalian cells and increased resistance to serum killing, are mediated independently by both Ail and YadA (1, 12). The YadA protein has also been shown to bind to extracellular matrix proteins such as collagen and fibronectin (14, 47, 52). YadA has also been implicated in bacterial binding to mucus (34); whether this binding is beneficial to the bacterium or is a mechanism that the host uses to clear the organism is not known.

As is seen for the regulation of *inv* and *ail* gene expression, *yadA* gene expression is regulated by temperature (7). The *yadA* gene is expressed at 37°C, the host temperature, under the control of another plasmid gene, *virF* (11, 49). This expression of *yadA* under conditions that exist in the host supports a role for this gene in *Yersinia* infection.

A role for YadA in colonization has been confirmed by animal studies with *yadA* mutant derivatives of *Y. enterocolitica*. *yadA* mutants of *Y. enterocolitica* were rapidly cleared from the intestinal lumen, whereas their *yadA*⁺ parents were not (24). However, *yadA* mutants of *Y. pseudotuberculosis* did not have reduced levels of colonization of the intestinal lumen (24), nor were they reduced in their virulence compared with their parental strains (8). This difference between *yadA* mutants of different *Yersinia* spp. raises questions about which, if any, of the in vitro properties ascribed to *yadA* is important in virulence.

ROLES OF THE GENES IN THE INFECTION PROCESS

Analysis of *inv* and *yadA* mutants supports a role for these two genes in infection. In addition, correlation of the presence of *ail* sequences with pathogenicity supports a role for this gene in infection. All three genes described above have been identified as being able to promote bacterial entry into mammalian cells in vitro. However, there is no evidence that this in vitro property has any relevance in a natural infection. That is, it is not clear that bacteria are found within host cells during the infection process, and if they are, it is not clear that this is important for the infection process. One problem with the experiments performed in the in vitro system is that the strains used were primarily those that lack the virulence plasmid. Plasmidless bacteria were used specifically to make the assays easier. One reason that plasmidless bacteria were used is that the plasmid encodes a cytotoxin(s) that causes detachment of the tissue culture cell monolayer within the time that it takes to perform an assay (15, 40). In addition, since plasmidless bacteria enter cells, it seemed likely that the cell entry factors were chromosomally encoded. This information aided in the characterization of cell entry in the initial stages. Unfortunately, by ignoring the plasmid, which has been shown to be essential for infection, it is possible that important plasmid-encoded functions that alter the ability of the bacteria to enter cells may have been missed. Evidence from Rosqvist et al. (43, 44) and Bliska et al. (6) suggests that this may be the case. That is, two plasmid-encoded proteins, YopE and YopH, may prevent the entry of *Yersinia* spp. into cells during the natural infection. Until more in vivo studies are performed and the exact location of the bacteria within the host is known (i.e., intracellular versus extracellular), it is not possible to resolve this dilemma.

THE CASE OF *Y. PESTIS*

Y. pestis is over 88% homologous to *Y. pseudotuberculosis* at the DNA level (4); therefore, it is not surprising that sequences homologous to *inv, ail,* and *yadA* are present in *Y. pestis* (33, 55). *Y. pestis* isolates do not enter mammalian cells in vitro, however (48). The *yadA* gene found on the *Y. pestis* virulence plasmid has a deletion of a single base, resulting in the formation of a nonfunctional truncated YadA protein (51). As yet, it is not known if either the *inv* or the *ail* gene is expressed in *Y. pestis*. However, it is known that when the *inv* gene from *Y. pseudotuberculosis* is introduced into *Y. pestis,* the resulting strain is able to enter mammalian cells (44). That result suggests that the *inv* gene of *Y. pestis* does not produce a functional invasin protein.

One explanation for the lack of functioning cell entry factors in *Y. pestis* is that the route of infection is such that the organism does not need to enter epithelial cells. That is, since the bacterium is directly deposited across the epidermal epithelium, factors that promote entry into epithelial cells are not necessary for the survival of this organism in the host. If it were true that *Y. pestis* lacks cell entry factors because of the alternate route of infection for this organism in comparison with those for the other pathogenic yersiniae, that would certainly support the idea that these factors are important in infection by the other pathogenic yersiniae. That is, in the enteropathogenic *Yersinia* spp., the role of these factors is to allow the organism to cross the epithelium to access the lymphatic tissue.

As an alternative, it could be that the lack of cell entry factors has an effect later in the *Y. pestis* infection process (i.e., after entry into the host) and could account for the differences in the infection course for this organism compared with that for the other pathogenic yersiniae. *Y. pestis* causes infections that are more fulminant than infections caused by *Y. pseudotuberculosis* or *Y. enterocolitica*, with bacteria spreading throughout the host. Perhaps the spread of *Y. pestis* throughout the host is due to the inability of this organism to enter host cells during the infection process. One provocative piece of evidence in support of the idea that a decrease in entry efficiency results in an increase in virulence comes from studies of Rosqvist et al. (45), who used a *Y. pseudotuberculosis inv* and *yadA* double mutant. The LD_{50} for this mutant was 5×10^2 to 1×10^3 lower than those for wild-type strains or strains with mutations in just one or the other gene (45). How cell entry would result in a decrease in spreading of the organism within the host is unclear. One explanation is that bacterial entry into host cells could signal the host to limit the infection.

SUMMARY

Invasin, Ail, and YadA proteins have been identified as having a role in promoting the uptake of *Yersinia* spp. into mammalian cells in vitro. These three factors can act independently in vitro to stimulate bacterial entry into cells. Mutations resulting in the loss of either of two of these cell entry factors, YadA and invasin, result in a decrease in the level of virulence in animal models, confirming a role for both proteins in virulence. Both Ail and YadA have been shown to mediate other phenotypes in *Yersinia* spp., including resistance to serum killing. This apparent redundancy in function may yet turn out to be an in vitro artifact or may instead have evolved to allow the bacterium to survive in the myriad of environments it encounters during the normal course of infection.

REFERENCES

1. **Balligand, G., Y. Laroche, and G. Cornelis.** 1985. Genetic analysis of virulence plasmid from a serogroup 9 *Yersinia enterocolitica* strain: role of outer membrane protein P1 in resistance to human serum and autoagglutination. *Infect. Immun.* **48:**782–786.
2. **Barondess, J. J., and J. Beckwith.** 1990. A bacterial virulence determinant encoded by lysogenic coliphage λ. *Nature* (London) **346:**871–874.
3. **Beer, K. B., and V. L. Miller.** 1992. Amino acid substitutions in naturally occurring variants of Ail result in altered invasion activity. *J. Bacteriol.* **174:**1360–1369.
4. **Bercovier, H., and H. H. Mollaret.** 1984. Genus XIV. *Yersinia. In* N. R. Krieg and J. G. Holt (ed.), *Bergey's Manual of Systematic Bacteriology.* The Williams & Wilkins Co., Baltimore.
5. **Bliska, J., and S. Falkow.** 1992. Bacterial resistance to complement killing mediated by the Ail protein of *Yersinia enterocolitica. Proc. Natl. Acad. Sci. USA* **89:**3561–3565.
6. **Bliska, J. B., M. C. Copass, and S. Falkow.** 1993. The *Yersinia pseudotuberculosis* adhesin *yadA* mediates intimate bacterial attachment to and entry into HEp-2 cells. *Infect. Immun.* **61:**3914–3921.
7. **Bolin, I., L. Norlander, and H. Wolf-Watz.** 1982. Temperature-inducible outer membrane protein

of *Yersinia pseudotuberculosis* and *Yersinia enterocolitica* is associated with the virulence plasmid. *Infect. Immun.* 37:506–512.

8. **Bolin, I., and H. Wolf-Watz.** 1984. Molecular cloning of the temperature-inducible outer membrane protein 1 of *Yersinia pseudotuberculosis*. *Infect. Immun.* 43:72–78.

9. **Bovallius, A., and G. Nilsson.** 1975. Ingestion and survival of *Y. pseudotuberculosis* in HeLa cells. *Can. J. Microbiol.* 21:1997–2007.

10. **Cornelis, G., Y. Larouche, G. Balligand, M.-P. Sory, and G. Wauters.** 1987. *Yersinia enterocolitica,* a primary model for bacterial invasiveness. *Rev. Infect. Dis.* 9:64–87.

11. **Cornelis, G., C. Sluiters, D. Lambert de Rouvroit, and T. Michiels.** 1989. Homology between VirF, the transcriptional activator of the *Yersinia* virulence regulon, and AraC, the *Escherichia coli* arabinose operon regulator. *J. Bacteriol.* 171:254–262.

12. **Cornelis, G. R., T. Biot, C. Lambert de Rouvroit, T. Michiels, B. Mulder, C. Sluiters, M. Sory, M. Van Bouchaute, and J. Vanooteghem.** 1989. The *Yersinia yop* regulon. *Mol. Microbiol.* 3: 1455–1459.

13. **Devenish, J. A., and D. A. Schiemann.** 1981. HeLa cell infection by *Yersinia enterocolitica:* evidence for lack of intracellular multiplication and development of a new procedure for quantitative expression of infectivity. *Infect. Immun.* 32:48–55.

14. **Emody, L., J. Heesemann, H. Wolf-Watz, M. Skurnik, G. Kapperud, P. O'Toole, and T. Wadstrom.** 1989. Binding to collagen by *Yersinia enterocolitica* and *Yersinia pseudotuberculosis:* evidence for *yopA*-mediated and chromosomally encoded mechanisms. *J. Bacteriol.* 171:6674–6679.

15. **Goguen, J. D., W. S. Walker, T. P. Hatch, and J. Yother.** 1986. Plasmid-determined cytotoxicity in *Yersinia pestis* and *Yersinia pseudotuberculosis*. *Infect. Immun.* 51:788–794.

16. **Grutzkau, S., C. Hanski, H. Hahn, and E. O. Riecken.** 1990. Involvement of M cells in the bacterial invasion of Peyer's patches: a common mechanism shared by *Yersinia enterocolitica* and other enteroinvasive bacteria. *Gut* 31:1011–1015.

17. **Heffernan, E. J., J. Harwood, J. Fierer, and D. Guiney.** 1992. The *Salmonella typhimurium* virulence plasmid complement resistance gene *rck* is homologous to a family of virulence-related outer membrane protein genes, including *pagC* and *ail*. *J. Bacteriol.* 174:84–91.

18. **Hynes, R. O.** 1987. Integrins: a family of cell surface receptors. *Cell* 48:549–554.

19. **Isberg, R. R.** 1991. Discrimination between intracellular uptake and surface adhesion of bacterial pathogens. *Science* 252:934–938.

20. **Isberg, R. R., and S. Falkow.** 1985. A single genetic locus encoded by *Yersinia pseudotuberculosis* permits invasion of cultured animal cells by *Escherichia coli* K-12. *Nature* (London) 317:262–264.

21. **Isberg, R. R., and J. M. Leong.** 1990. Multiple β1 chain integrins are receptors for invasin, a protein that promotes bacterial penetration into mammalian cells. *Cell* 60:861–871.

22. **Isberg, R. R., A. Swain, and S. Falkow.** 1988. Analysis of expression and thermoregulation of the *Yersinia pseudotuberculosis inv* gene with hybrid proteins. *Infect. Immun.* 56:2133–2138.

23. **Isberg, R. R., D. L. Voorhis, and S. Falkow.** 1987. Identification of invasin: a protein that allows enteric bacteria to penetrate cultured mammalian cells. *Cell* 50:769–778.

24. **Kapperud, G., E. Namork, M. Skurnik, and T. Nesbakken.** 1987. Plasmid-mediated surface fibrillae of *Yersinia pseudotuberculosis* and *Yersinia enterocolitica:* relationship to the outer membrane protein YOP1 and possible importance for pathogenesis. *Infect. Immun.* 55:2247–2254.

25. **Lee, W. H., P. P. McGrath, P. H. Carter, and E. L. Eide.** 1977. The ability of some *Yersinia enterocolitica* strains to invade HeLa cells. *Can. J. Microbiol.* 23:1714–1722.

26. **Leong, J. M., R. S. Fournier, and R. R. Isberg.** 1990. Identification of the integrin-binding domain of the *Yersinia pseudotuberculosis* invasin protein. *EMBO J.* 9:1979–1989.

27. **Leong, J. M., P. E. Morrissey, and R. R. Isberg.** 1993. A 76-amino-acid disulfide loop in the *Yersinia pseudotuberculosis* invasin protein is required for integrin receptor recognition. *J. Biol. Chem.* 268:20524–20532.

28. **Miller, J. H.** 1972. *Experiments in Molecular Genetics*. Cold Spring Harbor Laboratory, Cold Spring Harbor, N.Y.

29. **Miller, S. I., A. M. Kukral, and J. J. Mekalanos.** 1989. A two-component regulatory system (*phoP phoQ*) controls *Salmonella typhimurium* virulence. *Proc. Natl. Acad. Sci. USA* 86:5054–5058.

30. **Miller, V. L., K. B. Beer, W. P. Loomis, J. A. Olson, and S. I. Miller.** 1992. An unusual *pagC*::Tn-

phoA mutation leads to an invasion- and virulence-defective phenotype in salmonellae. *Infect. Immun.* **60:**3763–3770.

31. **Miller, V. L., J. B. Bliska, and S. Falkow.** 1990. Nucleotide sequence of the *Yersinia enterocolitica ail* gene and characterization of the Ail protein product. *J. Bacteriol.* **172:**1062–1069.

32. **Miller, V. L., and S. Falkow.** 1988. Evidence for two genetic loci in *Yersinia enterocolitica* that can promote invasion of epithelial cells. *Infect. Immun.* **56:**1242–1248.

33. **Miller, V. L., J. J. Farmer III, W. E. Hill, and S. Falkow.** 1989. The *ail* locus is found uniquely in *Yersinia enterocolitica* serotypes commonly associated with disease. *Infect. Immun.* **57:**121–131.

34. **Paerregaard, A., F. Espersen, O. M. Jensen, and M. Skurnik.** 1991. Interactions between *Yersinia enterocolitica* and rabbit ileal mucus: growth, adhesion, penetration, and subsequent changes in surface hydrophobicity and ability to adhere to ileal brush border membrane vesicles. *Infect. Immun.* **59:**253–260.

35. **Pepe, J. C., J. L. Badger, and V. L. Miller.** 1994. Growth phase and low pH affect the thermal regulation of the *Yersinia enterocolitica inv* gene. *Mol. Microbiol.* **11:**123–135.

36. **Pepe, J. C., and V. L. Miller.** 1993. *Yersinia enterocolitica* invasin: a primary role in the initiation of infection. *Proc. Natl. Acad. Sci. USA* **90:**6473–6477.

37. **Pierson, D. E.** 1994. Mutations affecting lipopolysaccharide enhance Ail-mediated entry of *Yersinia enterocolitica* into mammalian cells. *J. Bacteriol.* **176:**4043–4051.

38. **Pierson, D. E., and S. Falkow.** 1990. Nonpathogenic isolates of *Yersinia enterocolitica* do not contain functional *inv*-homologous sequences. *Infect. Immun.* **58:**1059–1064.

39. **Pierson, D. E., and S. Falkow.** 1993. The *ail* gene of *Yersinia enterocolitica* has a role in the ability of this organism to survive serum killing. *Infect. Immun.* **61:**1846–1852.

40. **Portnoy, D. A., S. L. Moseley, and S. Falkow.** 1981. Characterization of plasmids and plasmid-associated determinants of *Yersinia enterocolitica* pathogenesis. *Infect. Immun.* **31:**775–782.

41. **Pulkkinen, W. S., and S. I. Miller.** 1991. A *Salmonella typhimurium* virulence protein is similar to a *Yersinia enterocolitica* invasion protein and a bacteriophage lambda outer membrane protein. *J. Bacteriol.* **173:**86–93.

42. **Rankin, S., R. R. Isberg, and J. M. Leong.** 1992. The integrin-binding domain of invasin is sufficient to allow bacterial entry into mammalian cells. *Infect. Immun.* **60:**3909–3912.

43. **Rosqvist, R., I. Bolin, and H. Wolf-Watz.** 1988. Inhibition of phagocytosis in *Yersinia pseudotuberculosis:* a virulence plasmid-encoded ability involving the Yop2b protein. *Infect. Immun.* **56:**2139–2143.

44. **Rosqvist, R., A. Forsberg, M. Rimpilainen, T. Bergman, and H. Wolf-Watz.** 1990. The cytotoxic protein YopE of *Yersinia* obstructs the primary host defence. *Mol. Microbiol.* **4:**657–667.

45. **Rosqvist, R., M. Skurnik, and H. Wolf-Watz.** 1988. Increased virulence of *Yersinia pseudotuberculosis* by two independent mutations. *Nature* (London) **334:**522–525.

46. **Schiemann, D. A., and J. A. Devenish.** 1982. Relationship of HeLa cell infectivity to biochemical, serological, and virulence characteristics of *Yersinia enterocolitica. Infect. Immun.* **35:**497–506.

47. **Schulze-Koops, H., H. Burkhardt, J. Heesemann, K. von der Mark, and F. Emmrich.** 1992. Plasmid-encoded outer membrane protein YadA mediates specific binding of enteropathogenic yersiniae to various types of collagen. *Infect. Immun.* **60:**2153–2159.

48. **Sikkema, D. J., and R. R. Brubaker.** 1987. Resistance to pesticin, storage of iron, and invasion of HeLa cells by yersiniae. *Infect. Immun.* **55:**572–578.

49. **Skurnik, M., and P. Toivanen.** 1992. LcrF is the temperature-regulated activator of the *yadA* gene of *Yersinia enterocolitica* and *Yersinia pseudotuberculosis. J. Bacteriol.* **174:**2047–2051.

50. **Skurnik, M., and P. Toivanen.** 1993. *Yersinia enterocolitica* lipopolysaccharide: genetics and virulence. *Trends Microbiol.* **1:**148–152.

51. **Skurnik, M., and H. Wolf-Watz.** 1989. Analysis of the *yopA* gene encoding the Yop1 virulence determinants of *Yersinia* spp. *Mol. Microbiol.* **3:**517–529.

52. **Tertti, R., M. Skurnik, T. Vartio, and P. Kuusela.** 1992. Adhesion protein YadA of *Yersinia* species mediates binding of bacteria to fibronectin. *Infect. Immun.* **60:**3021–3024.

53. **Tran Van Nhieu, G., and R. R. Isberg.** 1991. The *Yersinia pseudotuberculosis* invasin protein and human fibronectin bind to mutually exclusive sites on the α5β1 integrin receptor. *J. Biol. Chem.* **266:**24367–24375.

54. **Tran Van Nhieu, G., and R. R. Isberg.** 1993. Bacterial internalization mediated by β1 chain integrins is determined by ligand affinity and receptor density. *EMBO J.* **12:**1887–1895.

55. **Wolf-Watz, H., D. A. Portnoy, I. Bolin, and S. Falkow.** 1985. Transfer of the virulence plasmid of *Yersinia pestis* to *Yersinia pseudotuberculosis*. *Infect. Immun.* **48:**241–243.

56. **Yang, Y., and R. Isberg.** 1993. Cellular internalization in the absence of invasin expression is promoted by the *Yersinia pseudotuberculosis yadA* product. *Infect. Immun.* **61:**3907–3913.

57. **Young, V. B., S. Falkow, and G. K. Schoolnik.** 1992. The invasin protein of *Yersinia enterocolitica:* internalization of invasin-bearing bacteria by eukaryotic cells is associated with reorganization of the cytoskeleton. *J. Cell Biol.* **116:**197–207.

58. **Young, V. B., V. L. Miller, S. Falkow, and G. K. Schoolnik.** 1990. Sequence, localization and function of the invasin protein of *Yersinia enterocolitica*. *Mol. Microbiol.* **4:**1119–1128.

Molecular Genetics of Bacterial Pathogenesis
Edited by V. L. Miller, J. B. Kaper, D. A. Portnoy, and R. R. Isberg
© 1994 American Society for Microbiology, Washington, DC 20005

Chapter 16

Cell Biology of *Salmonella* Pathogenesis

B. Brett Finlay

Bacteria are excellent cell biologists.

—Stanley Falkow

Salmonella species are masters at subverting various host processes for their own use. The use of these bacteria as models for studying the interactions that occur between host cells and salmonellae has been facilitated by the ease with which their growth can be accomplished, the ease of performing molecular genetic techniques, and the availability of appropriate tissue culture and animal models. The interactions that occur between *Salmonella* species and host cells are complex. They have been studied more extensively with nonphagocytic cells such as epithelial cells than with macrophages, although some of the processes are common between these two cell types.

For the review in this chapter, I have chosen to concentrate on the events that occur in the host cell rather than to focus on the bacterial genes that mediate these events. Study of the cell biology of these interactions has revealed several interesting processes and provided new tools for the study of eukaryotic cell function. For example, invasion into epithelial cells is a spectacular event that invokes the host cell cytoskeleton, cell surface rearrangement, and numerous signaling processes. Once inside, the bacterium remains inside a vacuole, yet this vacuole undergoes morphogenesis and is unlike the endocytic or phagocytic vacuoles normally found within cells. Bacterial intracellular replication is closely linked to the formation of a unique and striking host cell structure, a tubular lysosome. Formation of this filamentous structure is mediated by the intracellular bacteria. These types of examples highlight the complexity of the interactions between *Salmonella* species and their host cells. They have also provided alternate methods of studying cell biology. In addition, these studies have provided much information about the molecular mechanisms used by these pathogens to mediate disease.

INVASION

In 1967, Takeuchi (46) provided a detailed morphological description of *Salmonella typhimurium* interacting with and invading guinea pig ileal intestinal epi-

B. Brett Finlay • Biotechnology Laboratory and Departments of Biochemistry and Microbiology, University of British Columbia, Vancouver, British Columbia, Canada V6T 1Z3.

thelial cells. This informative description is representative of *Salmonella* interactions with most nonphagocytic cells, has been documented by other workers with other model systems and cell lines, and can be summarized as follows (for examples, see references 15, 21, 35, and 41). Prior to initial bacterial contact with the intestinal epithelium, the brush border remains intact. However, when bacteria come close to the epithelial surface, the microvilli in the immediate vicinity begin to degenerate through elongation, swelling, and budding (a process called "ruffling"). There are often long fibrous structures linking the organism with the apical surface, although the organism always retains space between the bacterial and host surfaces. As this process progresses, the apical cytoplasm close to the organism begins to bleb and swell, distorting outward. As the cell surface is distorted, the organism is internalized within a membrane-bound vesicle and is often surrounded by the cytoplasmic extrusion (Fig. 1). Accompanying this extrusion is a marked increase in localized endocytic activity, resulting in internalization of many vesicles. Initially, each invading organism is internalized within an individual

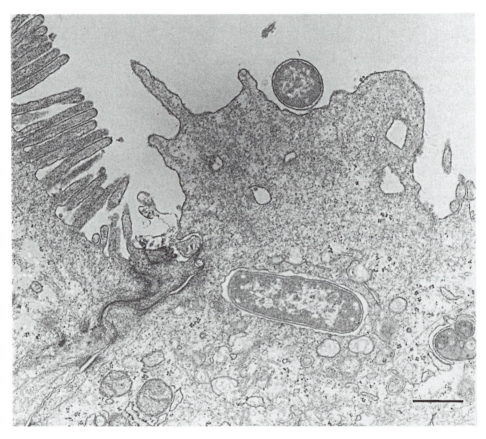

FIGURE 1. Transmission electron micrograph of *S. typhimurium* entering polarized Caco-2 human intestinal epithelial cells. Note the microvillus distortions and the bacterium localized within membrane-bound inclusions. Bar, 1 μm.

vacuole, although at later times these vacuoles may fuse. Although environmental factors regulate *Salmonella* invasion, once the bacteria are committed to invade the cell, the entire invasion process occurs within minutes (21).

The invasion process in macrophages has not been extensively characterized, although salmonellae are rapidly taken up by the macrophages underlying the intestinal mucosa, presumably following penetration of the intestinal epithelial barrier. It has recently been shown that *Salmonella* species trigger extensive membrane ruffling and macropinocytosis in macrophages and enter into a spacious phagosome (2). Bacterial invasins appear to significantly enhance uptake into phagocytic cells such as macrophages, since noninvasive *S. typhimurium* mutants have decreased levels of invasion into cultured macrophages (4, 22). Once inside the macrophages, the salmonellae persist within spacious phagosomes (2).

Several invasiveness loci have been identified in various *Salmonella* species by several groups. Unfortunately, of those characterized, nearly all loci are involved in regulation of invasion, secretion of products to the bacterial surface, lipopolysaccharide (LPS), or motility (13). Perhaps the best-characterized invasion locus from *S. typhimurium* (*inv*) has been cloned and characterized by Galan and Curtiss (23). There appear to be several genes in this locus that are required for adherence and/or invasion. These include *invABCDE* (23–25). Recent results indicate that several members of the *inv* locus are homologous to virulence factors in other pathogens and suggest that these products are involved in transporting and secreting proteins to the bacterial surface (24, 30, 47). The actual *Salmonella* invasins that are being exported by this machinery remain uncharacterized.

Uptake of *Salmonella* species into epithelial cells requires host cell metabolism and energy (34). This observation suggests that *Salmonella* uptake into nonphagocytic cells is an active process and, given the morphological alterations that occur, that the bacteria are capable of transmitting a localized signal at the host cell surface which mediates bacterial uptake. Recent evidence suggests that such a process occurs.

It is clear that host actin-containing microfilaments are required for *Salmonella* uptake. Treatment of cultured cells with cytochalasins, which disrupt actin filaments, blocks *Salmonella* uptake in several systems (for example, see references 14 and 34), although inhibitors of microtubules do not affect bacterial invasion. Additionally, it has been shown that *S. typhimurium* triggers localized rearrangement of polymerized actin and other microfilament-related proteins including α-actinin, tropomyosin, talin, and ezrin (18). This rearrangement consists of loose "strings" of actin filaments that accumulate in the vicinity of the invading organism. This rearrangement is also closely correlated with invasion, and once the bacterium is internalized, the cytoskeleton returns to its normal distribution (18). This rearrangement again suggests that signals are being transmitted through the host membrane to mediate cytoskeletal rearrangement.

In addition to the cytoskeltal rearrangements triggered by *S. typhimurium* in epithelial cells, this bacterium triggers a marked capping of host proteins that are associated with the cytoskeleton (Fig. 2) (27). Furthermore, the addition of cytochalasin D blocks membrane protein aggregation. An *S. typhimurium* invasion mutant (*invA*) and *Yersinia enterocolitica* do not trigger capping. Thus, *S. typhi-*

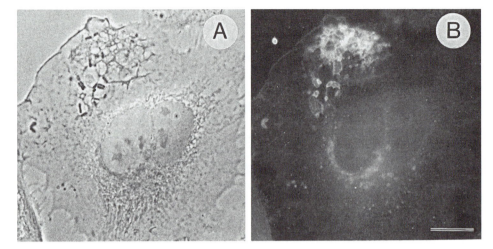

Figure 2. *S. typhimurium* triggers capping of host membrane proteins in HeLa cells. (A) Phase-contrast micrograph of *S. typhimurium* invading HeLa cells after 30 min. (B) Corresponding fluorescent micrograph after staining with antibodies to the cell surface-associated marker, the class I major histocompatibility complex. Note the capped localization of this marker corresponding to the area of bacterial invasion. Bar, 10 μm.

murium induces selective aggregation of host cell surface proteins, and this aggregation is linked to the cytoskeletal rearrangements and membrane ruffling.

As mentioned above, *Salmonella* species cause extensive ruffling in the host cell surface (2, 15, 20, 33). These membrane extrusions are localized to the vicinity of the invading bacteria and cause marked extrusions in the host cell cytoplasm and increased endocytic activity. These ruffles are morphologically similar to the membrane ruffles induced by growth factors, mitogens, and oncogene expression. In a recent study (20), it was shown that cytochalasin D, which blocks actin rearrangement and bacterial invasion, also blocks the membrane ruffling induced by *S. typhimurium*. This suggests that cytoskeletal rearrangement precedes membrane ruffling. It was also found (20) that *S. typhimurium*-induced ruffling also triggers uptake of noninvasive adherent bacteria and latex beads. Furthermore, the addition of epidermal growth factor (EGF), which also induces ruffling, facilitates uptake of adherent noninvasive *S. typhimurium* and other noninvasive bacteria. These results suggest that *Salmonella* invasion could be mediated simply by activating a host cell signal transduction cascade that mediates membrane ruffling in the host cell, which then causes nonspecific internalization of adherent organisms. Signaling of the bacterium-induced ruffling event appears to involve a different or distal pathway than that involving EGF, since inhibition of the small GTPases, Rac and Rho, had no effect on *S. typhimurium*-induced ruffling yet blocked EGF-induced events (33).

As with most invasive enteric organisms, there appear to be several signals that are transduced in the host cell that are involved in *Salmonella* uptake (for reviews, see references 5 and 43). Given the marked cytoskeletal rearrangement triggered by *S. typhimurium* and the role that intracellular Ca^{2+} plays in cytoskeletal rearrangements, it was not unexpected to find that *S. typhimurium* triggers a Ca^{2+} flux in cultured epithelial cells (29). Mutations in *invE* (which encodes a putative secretory machinery product) were unable to trigger an intracellular Ca^{2+} flux or actin rearrangement, although these mutations could be rescued in *trans* by adding the parental strain. Additionally, chelators of intracellular Ca^{2+}, but not extracellular Ca^{2+}, block *S. typhimurium* entry into cultured epithelial cells (44). Release of intracellular Ca^{2+} is often mediated by fluxes in the inositol phosphate IP_3, and it has been shown that *S. typhimurium* also triggers fluxes in inositol phosphates and that this flux correlates with invasion (44).

Further information about the signal(s) used by *S. typhimurium* to enter cultured cells came from work that described the activation of the EGF receptor (EGFR) by invading *S. typhimurium* in Henle-407 cells (25). Galan and coworkers (25) showed that *S. typhimurium* triggered tyrosine phosphorylation of the EGFR, and mutants with the *invA* mutation (which encodes another putative secretory machinery product) were unable to induce such activation. The addition of EGF to cells increased the invasiveness of mutants with the *invA* mutation. These workers then extended their findings and proposed a complex sequence of events that mediate *S. typhimurium* invasion (40). The order of this cascade is activation of the EGFR, activation of MAP kinase by EGFR, activation of phospholipase A_2 (PLA_2) by MAP kinase, generation of arachidonic acid by PLA_2, conversion of arachidonic acid into leukotriene LTD_4 by 5-lipoxygenase, opening of Ca^{2+} channels by LTD_4, and membrane ruffling, cytoskeletal rearrangements, and bacterial uptake, which are caused by the influx of extracellular Ca^{2+}. Consistent with this model, LTD_4 alone can complement *invA* mutants for the ability to invade.

This model has several appealing features. EGF activation is known to trigger membrane ruffles morphologically similar to those seen with *S. typhimurium*. It also causes Ca^{2+} fluxes and several other signals. However, despite the appeal of this pathway, there are several unexplained and contradictory results. For example, this pathway does not invoke any role for intracellular Ca^{2+} (although it does predict a role for extracellular Ca^{2+}), nor is there any involvement of phospholipase C, which presumably generates the inositol phosphates seen during *S. typhimurium* invasion (44). Additionally, cells which are lacking the EGFR are still invaded efficiently (20, 25, 33), and cells treated with potent inhibitors of the EGFR tyrosine kinase do not affect *S. typhimurium* invasion (42), although they block EGF-mediated signal transduction. Further supporting evidence that the EGFR does not participate in *S. typhimurium* invasion comes from two other studies. The first, as discussed above, demonstrated that inhibition of Rac and Rho, two host proteins that are essential in EGF-mediated cytoskeletal ruffling, are not involved in *S. typhimurium*-mediated ruffling (33). Jones and coworkers (33) also found that complete downregulation of the EGFR in Henle-407 cells had no effect on the ability of *S. typhimurium* to invade host cells. In another report (20), this group found that *S. typhimurium* promoted ruffling and bacterial uptake via an EGF-independent mechanism and that *S. typhimurium* could trigger ruffling in cell lines that do not possess the EGFR. It was concluded that *S. typhimurium* invades via an EGFR-independent pathway, although this conclusion does not rule out coincidental activation of the EGFR and its signals by invading *S. typhimurium*.

PENETRATION THROUGH POLARIZED EPITHELIAL MONOLAYERS

Although the study of cultured epithelial cells has been useful in defining some of the events that occur during *Salmonella* infection, additional knowledge has come from the study of infected polarized epithelial monolayers. These monolayers have defined apical and basolateral surfaces and have well-developed brush borders and functional tight junctions. Thus, they represent a model that more closely resembles the intestinal barrier than cultured nonpolarized cells do, and since they are grown on permeable filters, bacterial penetration through the monolayer can be studied. Apical infection of polarized MDCK (dog kidney) and Caco-2 (human intestinal) cell monolayers with either *S. typhimurium* or *Salmonella cholerae-suis* triggers events remarkably similar to those that occur in vivo, as described by Takeuchi (46). As with nonpolarized cells, membrane ruffling occurs, as does cytoskeletal rearrangement (15). The microvilli degenerate, and there is a marked extrusion of host cell cytoplasm, which may be mediated by the cytoskeletal rearrangements. It appears that invasion into these cells may occur in a manner analogous to that described for nonpolarized cells.

Tight junctions are specialized structures that link adjacent epithelial cells, thereby forming a barrier that is impermeable to most molecules, including ions. These structures form the basis of epithelial barriers. However, *Salmonella* infection of polarized monolayers causes a rapid loss in electrical resistance (15, 17), which reflects tight-junction integrity. This loss is correlated with invasiveness, since noninvasive mutants are unable to cause the loss in electrical resistance (19). In addition to the decrease in electrical resistance, salmonellae also cause depolarization of apical and basolateral proteins, which is presumably a direct result of tight-junction disruption (17). Although the mechanisms used by salmonellae to disrupt tight junctions are not well understood, this loss may be due to salmonella-induced actin filament disruption, since actin filaments are responsible for maintaining tight junctions and cell polarity.

Since polarized monolayers are grown on permeable filters, it has been possible to examine the capacity of salmonellae to infect both apical and basolateral surfaces. It was shown that salmonellae preferentially and rapidly infect the apical surfaces of MDCK and Caco-2 cell monolayers (15, 17). However, basolateral infection occurred much more slowly and did not result in tight-junction disruption, although this difference may have been due to steric hindrance by the filter.

Perhaps the most significant benefit obtained from using polarized monolayers is that bacterial penetration through a polarized epithelial monolayer can be measured. It was found that salmonellae penetrate polarized monolayers within 2 to 4 h when they are added to the apical surface (15, 17). However, noninvasive *Escherichia coli* and noninvasive *Salmonella* mutants were unable to penetrate monolayers, even when added to those infected with virulent *Salmonella* species. Despite the translocation of salmonellae through polarized monolayers, this process is quite different from transcytosis, a process that transports vesicles across epithelial barriers for two reasons. First, salmonellae depolarize monolayers, and thus their penetration is not through a true "polarized" cell. Second, most (90%) salmonellae that invade polarized monolayers remain within the monolayer, while approximately 9% return to the apical medium and only 1% appear in the basolateral medium (16). Thus, penetration through this monolayer is not efficient, and in vivo, the small number that presumably penetrate the epithelial barrier would be sufficient to cause disease. Alternatively, *Salmonella* species may penetrate through the M cells in Peyer's patches (35), which preferentially transport materials across the intestinal barrier to underlying lymphoid cells.

THE INTRACELLULAR ENVIRONMENT

There is general agreement that *Salmonella* species reside within a membrane-bound vacuole within both phagocytic and nonphagocytic cells. More recently, the trafficking of this vacuole has been examined. In cultured epithelial cells, as discussed above, *S. typhimurium* triggers the capping of several host cell surface proteins. However, only one of several surface proteins is present at early times postinfection in vacuoles containing this bacterium (27). Other host cell surface markers are internalized into the host cell but do not colocalize with bacteria. This finding implies that there is a sorting mechanism at the host cell surface that excludes some cytoskeleton-associated proteins from entering the bacteria-containing vacuole or fast recycling to plasma membranes. No apparent selectivity in host surface markers is observed in vacuoles containing bacteria which invade cells by using the *Yersinia* invasin pathway.

Once the *Salmonella* vacuole is formed, its composition changes. By 90 min after bacterial entry, other markers begin to appear, and the vacuole containing these bacteria becomes stained with the lysosomal marker lysosomal glycoprotein (lgp) and other lysosomal markers (28). Collectively, these data suggest that *S. typhimurium* initially enters into a specialized vacuole which then proceeds to a lysosomal-like compartment that is isolated from other endocytic traffic.

The intracellular environment that *S. typhimurium* resides within inside epithelial cells has been characterized indirectly by using bacterial reporter genes (26). Measurement of the β-galactosidase activities of various *lacZ* fusions by using a fluorescent substrate led to the conclusion that the concentrations of free Fe^{2+} and Mg^{2+} in the vacuoles of epithelial cells are low, that the vacuole has a mild acidic pH, and that lysine and oxygen are present within the intracellular environment. That work demonstrates the utility of using bacterial gene fusions to measure genes that are expressed intracellularly and gives a glimpse of the intracellular environment in which salmonellae reside within epithelial cells.

Several investigators have begun to characterize the intracellular targeting and environment of *S. typhimurium* inside macrophages. One report (9) indicated that *S. typhimurium* resides within phagosomes that have fused with lysosomes. Other workers (7, 31) concluded that *S. typhimurium* inhibited phagosome-lysosome fusion within several types of mouse-derived macrophages. Perhaps these discrepancies are due to incomplete inhibition of phagosome-lysosome fusion or are a result of the fact that only part of the intracellular population is capable of blocking this event. It was also suggested that viable intracellular bacteria are needed for this inhibition (7). It has been reported (3) that phagosomes containing *S. typhimurium* are acidified slowly and that it takes 4 to 5 h before the pH drops below 5.0. In contrast, vacuoles containing killed organisms were rapidly acidified (pH < 4.5 within 1 h). These data suggest that viable organisms either are needed for the inhibition of acidification or need to invade cells via a bacterium-mediated pathway that delivers the organism to an intracellular location which is acidified more slowly than if it had been delivered by a phagocytic pathway. Alpuche-Aranda and coworkers (3) also found that fluid-phase markers fused with internalized bacteria, which is different from that seen with epithelial cells. They concluded that *S. typhimurium* resides within a lysosome, yet is capable of blocking endosome acidification. It is also possible that the organisms reside within an intracellular environment which contains some of the lysosomal markers, yet this environment is not a normal phagolysosome.

The past few years have seen the identification of several bacterial factors that enhance the ability of *S. typhimurium* to survive within macrophages (12). These survival factors are macrophage specific, since these mutants survive equally well within nonphagocytic cells (22). The PhoP-PhoQ system is a two-component regulatory system that activates at least five bacterial products (*pag*) and represses others (*prg*) (for a review, see reference 38). One of the phenotypes that the PhoP-PhoQ system regulates is the capacity to survive bactericidal cationic peptides, which are thought to be involved in killing intracellular bacteria (11, 39). It has been demonstrated directly (by measuring β-galactosidase fusions) that the PhoP-PhoQ system is induced by low pH within macrophages and that inhibition of endosome acidification blocks activation of the PhoP-PhoQ system (3). A cytolysin that is required for survival within macrophages and virulence has recently been identified in *Salmonella* species (37).

Other bacterial products may also contribute to intracellular survival within phagocytic cells. For example, a 59-kDa outer membrane protein from *S. typhimurium* has been reported to provide protection from oxidative killing within polymorphonuclear leukocytes (45). Additionally, mutations in *recA* and *recBC* in *S. typhimurium* are avirulent and are susceptible to the oxidative burst of macrophages, indicating that the ability to repair DNA damage is essential for survival within macrophages and virulence (8). A Tn5 mutant of *S. typhimurium* that lacks the ability to block phagosome-lysosome fusion has been described (32). Interestingly, although this mutant was susceptible to intracellular killing, it was still virulent in mice.

INTRACELLULAR REPLICATION

There are conflicting data on the growth of *Salmonella* species within macrophages (6, 9). The growth of mutant strains that are unable to grow within epithelial cells is unaffected within macrophages (36). However, it has recently been proposed that two populations of *S. typhimurium* exist within macrophages: one which is static and the other which is rapidly growing (1). The existence of these two pools may perhaps explain the conflicting data regarding lysosome fusion and intracellular growth within phagocytic cells.

Salmonella species have the capacity to multiply within vacuoles in nonphagocytic cells, after an initial lag of approximately 4 h (10, 14, 22, 48). The lag period seen in nonphagocytic cells that precedes the initiation of bacterial replication indicates that processes occur prior to bacterial replication and that specific bacterial genes may be required for replication in this unique niche. (Nonvirulent *E. coli* does not replicate within vacuoles in epithelial cells.) Although vacuole acidification is a signal used by some intracellular pathogens to initiate replication, *Salmonella* species probably do not use this signal, because blockage of endosome acidification with inhibitors and defective cell lines does not block intracellular replication (14).

Mutants of *S. typhimurium* that are defective for intracellular replication in epithelial cells have been identified (36). Both auxotrophic and prototrophic mutants were found, and the auxotrophic mutants could be complemented by the addition of the appropriate nutrients to the tissue culture medium. The three prototrophic mutants identified were highly attenuated for virulence in mice, yet they persisted within livers and spleens for at least 3 weeks. The identification of these mutants suggests that there are bacterial genes specific for intracellular replication and that these participate in virulence.

A possible function of these bacterial genes has been reported recently (28). As discussed above, *S. typhimurium* is localized to vacuoles that contain lysosomal glycoproteins (lgps) in epithelial cells. However, 4 to 6 h after invasion, intracellular *S. typhimurium* induces the formation of stable filamentous structures that contain lgps that are connected to the vacuoles containing bacteria (Fig. 3) (28). The kinetics of formation of these lgp-rich filamentous structures parallels the rate of intracellular replication, including the initial lag period. Filament formation requires viable intracellular bacteria, since the addition of antibiotics blocks the formation of these novel structures. Endosome acidification inhibitors or microtubule-disrupting agents also inhibit tubule formation. These unique structures are never observed in uninfected cells or in those infected with *Yersinia* species, although all *Salmonella* species tested trigger their formation. The prototrophic *Salmonella* mutants that are unable to multiply inside epithelial cells are also completely defective at triggering the formation of these filamentous structures.

Molecular characterization of one locus that mediates tubule formation indicates that it is a unique *Salmonella* sequence that is inserted between housekeeping genes and that it has no homology to other sequences within the gene banks. Thus, it appears that *Salmonella* species have specific loci which are responsible for triggering filament formation from within the host cell, and filament formation is correlated with intracellular replication. Whether these filaments mediate intracellular replication or how the bacteria direct the formation of these novel host processes remains a mystery. One potential function for these structures is that they provide access to nutrients for the intracellular bacteria, possibly by intersecting with endocytic or exocytic vesicular transport pathways.

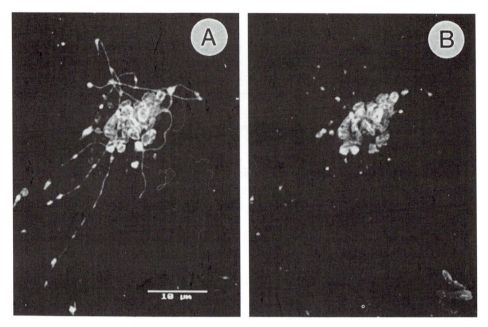

FIGURE 3. Fluorescent confocal micrograph of filamentous structures in HeLa epithelial cells infected with *S. typhimurium*. Cells were infected for 6 h and then fixed and labeled with antibodies to a lysosomal glycoprotein (lgp) (A) or anti-*S. typhimurium* lipopolysaccharide (B). Note the blebs associated with the filamentous structures distal to intracellular bacteria. Bar, 10 μm.

CONCLUSIONS

The interactions between *Salmonella* species and host cells are intimate and complex. It is apparent that these bacteria are skilled in exploiting preexisting host cell functions and subverting these functions for their own benefit. This is seen during invasion, when salmonellae exploit host signal transduction pathways, which affect cytoskeletal rearrangement, membrane protein capping, membrane ruffling, and ultimately, bacterial invasion. It also occurs when the salmonellae are within a membrane-bound vacuole in both epithelial cells and macrophages. The targeting of this vacuole bypasses normal endocytic pathways and the vacuole fuses with lysosomal markers, although the vacuole does not have all the characteristics of a true lysosome. Finally, from within the vacuole inside epithelial cells, these bacteria direct the formation of a novel host structure, the tubular lysosome. It appears that the bacteria have specific genes that mediate filamentous lysosome formation and that this process is linked to intracellular replication. One recurring theme about all of these exploitations of host cell function is that the bacteria achieve the desired effect by "nonconventional" mechanisms, as viewed by cell biologists. These alternate mechanisms have presumably arisen from evolutionary selection for mechanisms that work, as viewed from an organism in intimate contact with a host cell.

In the future, the details of several aspects of *Salmonella* interaction with host cells will be uncovered. The advancement of this field is heavily dependent on other fields, especially cell biology. Additional elucidation of basic host cell processes such as signal transduction, membrane ruffling, cytoskeletal rearrangement, vesicular targeting, and endocytic trafficking will assist in further defining the mechanisms used by *Salmonella* species. In addition, further characterization of the bacterial products that are involved in the various stages of infection will enhance the cell biology studies. Together, this knowledge will provide unexpected findings concerning mechanisms in both the bacteria and the host cell.

Despite the knowledge gained from detailed studies of the interaction of *Salmonella* species with the host cell, much remains to be discovered. The weakest area of this field is extrapolating the knowledge that has been gained from in vitro studies to in vivo work and defining the role of each of these mechanisms in pathogenesis. This is the biggest challenge to those of us working in this field and will provide fertile results in the years to come.

ACKNOWLEDGMENTS. I thank the members of my laboratory who have contributed much to this field, including F. Garcia del Portillo, K. Leung, I. Rosenshine, S. Ruschkowski, and M. Stein.

Work in my laboratory is supported by operating grants from the British Columbia Health Care Research Foundation, the Medical Research Council, the Canadian Bacterial Disease Centre of Excellence, and a Howard Hughes International Research Scholar award.

REFERENCES

1. **Abshire, K. Z., and F. C. Neidhardt.** 1993. Growth rate paradox of *Salmonella typhimurium* within host macrophages. *J. Bacteriol.* **175:**3744–3748.

2. **Alpuche-Aranda, C. M., E. L. Racoonsin, J. A. Swanson, and S. I. Miller.** 1994. *Salmonella* stimulate macrophage macropinocytosis and persist within spacious phagosomes. *J. Exp. Med.* **179:**601–608.

3. **Alpuche-Aranda, C. M., J. A. Swanson, W. P. Loomis, and S. I. Miller.** 1992. *Salmonella typhimurium* activates virulence gene transcription within acidified macrophage phagosomes. *Proc. Natl. Acad. Sci. USA* **89:**10079–10083.

4. **Betts, J., and B. B. Finlay.** 1992. Identification of *Salmonella typhimurium* invasiveness loci. *Can. J. Microbiol.* **38:**852–857.

5. **Bliska, J. B., J. E. Galan, and S. Falkow.** 1993. Signal transduction in the mammalian cell during bacterial attachment and entry. *Cell* **73:**903–920.

6. **Buchmeier, N. A., and F. Heffron.** 1989. Intracellular survival of wild-type *Salmonella typhimurium* and macrophage-sensitive mutants in diverse populations of macrophages. *Infect. Immun.* **57:**1–7.

7. **Buchmeier, N. A., and F. Heffron.** 1991. Inhibition of macrophage phagosome-lysosome fusion by *Salmonella typhimurium*. *Infect. Immun.* **59:**2232–2238.

8. **Buchmeier, N. A., C. J. Lipps, M. Y. So, and F. Heffron.** 1993. Recombination-deficient mutants of *Salmonella typhimurium* are avirulent and sensitive to the oxidative burst of macrophages. *Mol. Microbiol.* **7:**933–936.

9. **Carrol, M. E., P. S. Jackett, V. R. Aber, and D. B. Lowrie.** 1979. Phagolysosome formation, cyclic adenosine 3′,5′-monophosphate and the fate of *Salmonella typhimurium* within mouse peritoneal macrophages. *J. Gen. Microbiol.* **110:**421–429.

10. **Conlan, J. W., and R. J. North.** 1992. Early pathogenesis of infection in the liver with the facultative intracellular bacteria *Listeria monocytogenes, Francisella tularensis*, and *Salmonella typhimurium* involves lysis of infected hepatocytes by leukocyte. *Infect. Immun.* **60:**5164–5171.

11. **Fields, P. I., E. A. Groisman, and F. Heffron.** 1989. A *Salmonella* locus that controls resistance to microbicidal proteins from phagocytic cells. *Science* **243:**1059–1062.

12. **Fields, P. I., R. V. Swanson, C. G. Haidaris, and F. Heffron.** 1986. Mutants of *Salmonella typhimurium* that cannot survive within the macrophage are avirulent. *Proc. Natl. Acad. Sci. USA* **83:**5189–5193.

13. **Finlay, B. B.** Molecular and cellular mechanisms of *Salmonella pathogenesis. Curr. Top. Microbiol.*, in press.

14. **Finlay, B. B., and S. Falkow.** 1988. Comparison of the invasion strategies used by *Salmonella cholerae-suis, Shigella flexneri,* and *Yersinia enterocolitica* to enter cultured animal cells: endosome acidification is not required for bacterial invasion or intracellular replication. *Biochimie* **70:** 1089–1099.

15. **Finlay, B. B., and S. Falkow.** 1990. *Salmonella* interactions with polarized human intestinal Caco-2 epithelial cells. *J. Infect. Dis.* **162:**1096–1106.

16. **Finlay, B. B., J. Fry, E. P. Rock, and S. Falkow.** 1989. Passage of *Salmonella* through polarized epithelial cells: role of the host and bacterium. *J. Cell Sci.* **11(Suppl.):**99–107.

17. **Finlay, B. B., B. Gumbiner, and S. Falkow.** 1988. Penetration of *Salmonella* through a polarized Madin-Darby canine kidney epithelial cell monolayer. *J. Cell Biol.* **107:**221–230.

18. **Finlay, B. B., S. Ruschkowski, and S. Dedhar.** 1991. Cytoskeletal rearrangements accompanying *Salmonella* entry into epithelial cells. *J. Cell Sci.* **99:**283–296.

19. **Finlay, B. B., M. N. Starnbach, C. L. Francis, B. A. D. Stocker, S. Chatfield, G. Dougan, and S. Falkow.** 1988. Identification and characterization of Tn*phoA* mutants of *Salmonella* that are unable to pass through a polarized MDCK epithelial cell monolayer. *Mol. Microbiol.* **2:**757–766.

20. **Francis, C. L., T. A. Ryan, B. D. Jones, S. J. Smith, and S. Falkow.** 1993. Ruffles induced by *Salmonella* and other stimuli direct macropinocytosis of bacteria. *Nature* (London) **364:**639–642.

21. **Francis, C. L., M. N. Starnbach, and S. Falkow.** 1992. Morphological and cytoskeletal changes in epithelial cells occur immediately upon interaction with *Salmonella typhimurium* grown under low-oxygen conditions. *Mol. Microbiol.* **6:**3077–3087.

22. **Gahring, L. C., F. Heffron, B. B. Finlay, and S. Falkow.** 1990. Invasion and replication of *Salmonella typhimurium* in animal cells. *Infect. Immun.* **58:**443–448.

23. **Galan, J. E., and R. Curtiss III.** 1989. Cloning and molecular characterization of genes whose products allow *Salmonella typhimurium* to penetrate tissue culture cells. *Proc. Natl. Acad. Sci. USA* **86:**6383–6387.

24. **Galan, J. E., C. Ginocchio, and P. Costeas.** 1992. Molecular and functional characterization of the *Salmonella* invasion gene *invA:* homology of InvA to members of a new protein family. *J. Bacteriol.* **174:**4338–4349.

25. **Galan, J. E., J. Pace, and M. J. Hayman.** 1992. Involvement of the epidermal growth factor receptor in the invasion of cultured mammalian cells by *Salmonella typhimurium. Nature* (London) **357:** 588–589.

26. **Garcia del Portillo, F., J. W. Foster, M. E. Maguire, and B. B. Finlay.** 1992. Characterization of the micro-environment of *Salmonella typhimurium*-containing vacuoles within MDCK epithelial cells. *Mol. Microbiol.* **6:**3289–3297.

27. **Garcia-del Portillo, F., M. G. Pucciarelli, W. A. Jefferies, and B. B. Finlay.** *Salmonella typhimurium* induces selective aggregation and internalization of host cell surface proteins during invasion of epithelial cells. *J. Cell Sci.,* in press.

28. **Garcia del Portillo, F., M. B. Zwick, K. Y. Leung, and B. B. Finlay.** 1993. *Salmonella* induces the formation of filamentous structures containing lysosomal membrane glycoproteins in epithelial cells. *Proc. Natl. Acad. Sci. USA* **90:**10544–10548.

29. **Ginocchio, C., J. Pace, and J. E. Galan.** 1992. Identification and molecular characterization of a *Salmonella typhimurium* gene involved in triggering the internalization of salmonellae into cultured epithelial cells. *Proc. Natl. Acad. Sci. USA* **89:**5976–5980.

30. **Groisman, E. A., and H. Ochman.** 1993. Cognate gene clusters govern invasion of host epithelial cells by *Salmonella typhimurium* and *Shigella flexneri. EMBO J.* **10:**3779–3787.

31. **Ishibashi, Y., and T. Arai.** 1990. Specific inhibition of phagosome-lysosome fusion in murine macrophages mediated by *Salmonella typhimurium* infection. *FEMS Microbiol. Immunol.* **2:** 35–43.

32. **Ishibashi, Y., K. Nobuta, and T. Arai.** 1992. Mutant of *Salmonella typhimurium* lacking the inhibitory function for phagosome-lysosome fusion in murine macrophages. *Microb. Pathog.* **13:** 317–323.

33. **Jones, B. D., H. F. Paterson, A. Hall, and S. Falkow.** 1993. *Salmonella typhimurium* induces

membrane ruffling by a growth factor-receptor-independent mechanism. *Proc. Natl. Acad. Sci. USA* **90:**10390–10394.

34. **Kihlstrom, E., and L. Nilsson.** 1977. Endocytosis of *Salmonella typhimurium* 395 MS and MR10 by HeLa cells. *Acta Pathol. Microbiol. Scand. Sect. B* **85:**322–328.

35. **Kohbata, S., H. Yokoyama, and E. Yabuuchi.** 1986. Cytopathogenic effect of *Salmonella typhi* GIFU 10007 on M cells of murine ileal Peyer's patches in ligated ileal loops: an ultrastructural study. *Microbiol. Immunol.* **30:**1225–1237.

36. **Leung, K. Y., and B. B. Finlay.** 1991. Intracellular replication is essential for the virulence of *Salmonella typhimurium. Proc. Natl. Acad. Sci. USA* **88:**11470–11474.

37. **Libby, S. J., W. Goebel, A. Ludwig, N. Buchmeier, F. Bowe, F. C. Fang, D. G. Guiney, J. G. Songer, and F. Heffron.** 1994. A cytolysin encoded by *Salmonella* is required for survival within macrophages. *Proc. Natl. Acad. Sci. USA* **91:**489–493.

38. **Miller, S. I.** 1991. PhoP/PhoQ: macrophage-specific modulators of *Salmonella* virulence? *Mol. Microbiol.* **5:**2073–2078.

39. **Miller, S. I., A. M. Kukral, and J. J. Mekalanos.** 1989. A two-component regulatory system (*phoP phoQ*) controls *Salmonella typhimurium* virulence. *Proc. Natl. Acad. Sci. USA* **86:**5054–5058.

40. **Pace, J., M. J. Hayman, and J. E. Galan.** 1993. Signal transduction and invasion of epithelial cells by *S. typhimurium. Cell* **72:**505–514.

41. **Popiel, I., and P. C. Turnbull.** 1985. Passage of *Salmonella enteritidis* and *Salmonella thompson* through chick ileocecal mucosa. *Infect. Immun.* **47:**786–792.

42. **Rosenshine, I., V. Duronio, and B. B. Finlay.** 1992. Tyrosine protein kinase inhibitors block invasin-promoted bacterial uptake by epithelial cells. *Infect. Immun.* **60:**2211–2217.

43. **Rosenshine, I., and B. B. Finlay.** 1993. Exploitation of host signal transduction pathways and cytoskeletal functions by invasive bacteria. *Bioessays* **15:**17–24.

44. **Ruschkowski, S., I. Rosenshine, and B. B. Finlay.** 1992. *Salmonella typhimurium* induces an inositol phosphate flux in infected epithelial cells. *FEMS Microbiol. Lett.* **95:**121–126.

45. **Stinavage, P. S., L. E. Martin, and J. K. Spitznagel.** 1990. A 59 kilodalton outer membrane protein of *Salmonella typhimurium* protects against oxidative intraleukocytic killing due to human neutrophils. *Mol. Microbiol.* **4:**283–293.

46. **Takeuchi, A.** 1967. Electron microscope studies of experimental Salmonella infection. I. Penetration into the intestinal epithelium by *Salmonella typhimurium. Am. J. Pathol.* **50:**109–136.

47. **Van Gijsegem, F., S. Genin, and C. Boucher.** 1993. Conservation of secretion pathways for pathogenicity determinants of plant and animal bacteria. *Trends Microbiol.* **1:**175–180.

48. **Yokoyama, H., M. Ikedo, S. Kohbata, T. Ezaki, and E. Yabuuchi.** 1987. An ultrastructural study of HeLa cell invasion with *Salmonella typhi* GIFU 10007. *Microbiol. Immunol.* **31:**1–11.

Molecular Genetics of Bacterial Pathogenesis
Edited by V. L. Miller, J. B. Kaper, D. A. Portnoy, and R. R. Isberg
© 1994 American Society for Microbiology, Washington, DC 20005

Chapter 17

Intracellular Trafficking of *Legionella pneumophila* within Phagocytic Cells

Ralph R. Isberg

A number of bacterial pathogens replicate within macrophages during host infections. Pathogens that grow within this niche must either express factors that allow survival in the presence of antimicrobial factors or be internalized into a compartment that bypasses exposure of the microorganism to phagocytic killing mechanisms. Bacteria that establish a novel intracellular route either destroy the nascent phagosomal membrane to gain access to the host cell cytosol (50, 63) or else are targeted to a special membrane-bound compartment that allows exclusion of antimicrobial factors, and presumably access to nutrients (33, 34, 42). This latter group of pathogens includes *Legionella pneumophila* (34), *Brucella abortus* (17), *Chlamydia* species (20), and the eukaryotic parasite *Toxoplasma gondii* (42). It is hoped that insight into the growth strategy of one of these microorganisms will give some clues regarding a general pattern of how the host cell is manipulated to suit the lifestyles of these pathogens.

PATHOGENESIS AND ECOLOGY OF *L. PNEUMOPHILA*

L. pneumophila is a gram-negative facultative intracellular bacterium that is the causative agent of Legionnaires' pneumonia as well as the self-limiting Pontiac fever (52, 57). A ubiquitous environmental microorganism, *Legionella* species often cause small-scale outbreaks as the result of contamination of water supplies or air-conditioning cooling towers, which allow infection of the susceptible host by aerosolization (24, 85). The fastidious nature of the microorganism is apparently inconsistent with the colonization of *Legionella* species in such nutrient-deprived sites. Replication in the environment probably occurs intracellularly within a variety of amoebal species and extracellularly within heterogeneous biofilms (27, 70).

After infection of humans, the bacteria replicate within alveolar macrophages (16). In some ways, replication in these cells resembles an opportunistic disease,

Ralph R. Isberg • Howard Hughes Medical Institute and Department of Molecular Biology and Microbiology, Tufts University School of Medicine, 136 Harrison Avenue, Boston, Massachusetts 02111.

mimicking colonization of amoeba. Lending support to this model, serious disease is usually associated with immunocompromised patients, such as alcoholics or those suffering from hairy cell leukemia (57). The microorganism is a common cause of nosocomial pneumonias (31), and hospitalized individuals most at risk are those having impaired respiratory or mucociliary function.

Intracellular replication of *L. pneumophila* can be studied in culture, and much progress has been made by using in vitro models. Since most media for host cells cannot support replication of *Legionella* species, growth is usually assayed either by demonstrating an increase in viable counts that requires the presence of cultured mammalian or amoebal cells (49, 55, 62) or by the use of a plaque assay that is similar in nature to that previously used to analyze a number of other intracellular bacteria (49, 58, 78). In the plaque assay, limiting numbers of bacteria are plated onto a lawn of host cells; this is followed by immobilization in agar. After allowing intracellular growth and extracellular spread of the bacteria, a small zone of infected mammalian host cells is revealed by the vital stain neutral red. Plaques arise in approximately 4 days, facilitating the screening of mutants defective for intracellular replication.

A wide variety of cell types act as hosts for *Legionella* species. Fully virulent bacteria are able to grow within a variety of amoebal species in culture, including *Acanthamoeba castellani* (75) and *Hartmannella vermiformis* (26). Growth within mammalian phagocytic cells is less promiscuous, because the cell and host ranges of intracellular growth are restricted. For instance, the bacteria can survive but not grow within neutrophils from any species. Human or guinea pig peripheral blood monocytes and macrophages can support replication of the bacterium (25), but *L. pneumophila* is unable to grow in phagocytic cells isolated from all mouse strains that have been tested except for the A/J strain (90, 91). This parallels results from animal infection models, in which experimental pneumonia can be induced in guinea pigs and A/J mice, whereas other mouse strains are resistant to disease (89). Human monocytoid cell lines terminally differentiated into adherent macrophages by phorbol esters can quite conveniently be used as host cells in culture. Both U937 and HL60 cells have been used for this purpose, because they grow quickly and can support high levels of bacterial replication (49, 62). The disadvantage of using these lines is that they spread poorly after adherence to solid substrates, making it difficult to analyze infected cells by fluorescence microscopy (82). For this reason, fluorescence assays are usually confined to macrophages directly isolated from humans or A/J mice (82).

UPTAKE OF *L. PNEUMOPHILA*

Much of the understanding of the events that occur after interaction of *Legionella* species with human monocytes is the result of studies performed during the 1980s by Horwitz and Silverstein (38) on *L. pneumophila* Philadelphia 1. Binding of the bacterium to the phagocyte is stimulated by opsonization with complement. This is apparently due to the fixation of complement components C3b and C3bi to the surface of the bacterium (60), because C3 derivatives can bind to the product of the *ompS* gene (3), the major outer membrane protein of the bacterium (32). Uptake of the microorganism into monocytes is reduced significantly in the presence of a mixture of antibodies directed against CR1 and CR3, receptors for C3b and C3bi, respectively (49, 60). Several lines of evidence, however, indicate that this choice of receptors is not necessary for intracellular growth. First, bacteria internalized in the presence of antireceptor antibodies are able to grow intracellularly. Second, bacteria coated with antibodies are internalized via Fc receptors, yet they are only partially impaired for growth (34, 56). Finally, *L. pneumophila* can grow in a variety of cell lines that lack CR1 and CR3, as well as in amoeba which clearly lack these receptors (26). The primary significance of this choice of receptors may be that they reduce the stress response of the microorganism during uptake, since internalization via complement receptors reduces the exposure of the bacterium to toxic oxygen compounds (88).

Internalization of the microorganism occurs via a unique process called "coiling phagocytosis," in which the phagosomal membrane wraps in a serpentine fashion about the bacterium during uptake (35). The coils rapidly dissipate after internalization, targeting the bacterium to a membrane-bound phagosome. A few other intracellular microorganisms, such as *Leishmania donovani*, are internalized via a similar coiling process, as are some microorganisms that show no intracellular growth, such as *Borrelia burgdorferi* (69). Formalin-killed *Legionella* species are internalized by coiling phagocytosis, yet their subsequent intracellular targeting is very different from the route taken by live bacteria (described below [34]). Furthermore, only the *L. pneumophila* Philadelphia 1 strain has clearly been shown to undergo coiling. Finally, there is no direct connection between binding of complement receptors and coiling, since a number of microorganisms bind these receptors without inducing coiling and liposomes coated with C3 components are internalized by conventional phagocytosis (3). Therefore, the relationship between coiling phagocytosis and subsequent intracellular growth is unclear.

The nascent phagosome formed by the coil has a distinct protein composition, an interesting observation likely to be important for subsequent events during intracellular growth. A number of cell surface proteins, including major histocompatibility complex class I molecules, are excluded from the phagosome, whereas CR3 is concentrated about the bacterium (15). These phenomena could help to explain the next events that occur during interaction with the phagocyte.

MATURATION OF THE *L. PNEUMOPHILA* PHAGOSOME

From a number of studies on the behavior of *L. pneumophila* within phagocytic cells, it is apparent that the phagosome bearing the microorganism is routed within the cell in a fashion distinct from that seen for phagosomes bearing nonpathogens. Unfortunately, very few studies that systematically investigate any default routing pathways that may exist for phagocytosis of nonpathogens have been performed. Some recent studies that used indirect immunofluorescence techniques with monoclonal antibodies directed against host cell components have shed some light on the process (71, 82) and indicate that phagocytic uptake has many similarities to the routing of endosomes within eukaryotic cells. In order to appreciate the distinctive nature of the *L. pneumophila* interaction with host cells, the following discussion presents what are thought to be the salient features of a default pathway.

During uptake of inert particles or nonpathogens into phagocytes, the phagosome matures or is directed toward a pathway in which new host cell proteins are acquired (79). The phagosomal membrane accumulates the vacuolar proton-pumping ATPase, which is apparently responsible for luminal acidification to a range of between pH 4.5 and 5.3 (48). Shortly after formation of the membrane-bound compartment, the late endosomal membrane protein LGP (or LAMP-1 [11]) and the small GTP-binding protein Rab-7 can be detected on the surface of the phagosome (12, 66, 67). As determined by immunofluorescence studies, the phagosome remains covered by LGP for the duration of its existence, whereas phagosomal membrane association with Rab-7 is transient, as might be expected for a protein believed to be involved in regulating membrane-targeting events (12, 71). Eventually, lysosomal contents, such as acid phosphatase, can be detected within the phagosome, and at this point it is said that phagosome-lysosome fusion has occurred (34). Detection of phagosome-lysosome fusion is usually facilitated by prelabeling of phagocytes with either the electron-dense heavy metal marker thorium dioxide (34) or a fluorescent fluid-phase marker, such as Texas red-ovalbumin. If the phagosome fuses with the secondary lysosomes formed by uptake of these markers, then phagosome-lysosome fusion can be detected by electron microscopy, which shows colocalization with the electron-dense marker, or by fluorescence microscopy, which detects phagosomes bearing Texas red-ovalbumin (82).

Two observations indicate that phagosomes bearing *L. pneumophila* deviate from this proposed default pathway during or very soon after uptake. By using fluorescent pH-sensitive probes combined with microscopic techniques, Horwitz and Maxfield (37) demonstrated that internalized virulent *L. pneumophila* is found in a compartment that is significantly less acidic than that of nonpathogens, with the pH of the average phagosome differing by at least 1 pH unit. Since Clemens and Horwitz (15) have clearly shown that the phagosome excludes a variety of membrane proteins, it is tempting to speculate that the lack of acidity may be due to exclusion of the vacuolar ATPase from the phagosomal membrane. Precedence for this strategy has recently been obtained by Russell and coworkers (77), who demonstrated that *Mycobacterium avium* inhibits acidification of the phagosome by becoming localized in a compartment devoid of the vacuolar ATPase. A second piece of evidence that the phagosome maturation pathway deviates very early from that of a nonpathogen comes from immunofluorescence studies on LAMP-1. Phagosomes bearing virulent *L. pneumophila* have much lower levels of this protein than those bearing nonpathogens, which acquire LAMP-1 within 10 min after the initiation of uptake (71).

The most striking series of observations regarding the nature of the *L. pneumophila* phagosome was made by Horwitz (34), who showed that this compartment is totally devoid of lysosomal acid phosphatase and resists fusion with thorium-labeled compartments, indicating that phagosome-lysosome fusion is bypassed or inhibited by virulent organisms. This phenomenon may be a key step in the intracellular growth of the microorganism, because it establishes a replicative niche protected from the antimicrobial components of the lysosome. A number of other intracellular microorganisms that grow within a phagosome also have this property, such as *Chlamydia* species, *B. abortus,* and *T. gondii* (17, 20, 74). In contrast, formalin-treated *L. pneumophila* is found in phagolysosomes, where it is presumably degraded.

After formation of the phagosome, localized changes in the host cell cytoplasm appear in response to the presence of the phagosome (33). Smooth vesicles accumulate around the compartment because of either budding from the phagosome or trafficking from another site in the host cell to the phagosomal surface. In addition, mitochondria line up around the phagosomal surface, initiating a process termed "organelle recruitment" (33). Within a few hours after uptake, the close association with the mitochondria dissipates, and in its place reticular membranes and what appear to be polysomes wrap around the cytoplasmic surface of the phagosome membrane (33). These latter structures strongly resemble rough endoplasmic reticulum (RER), and similar structures have been seen after *T. gondii* (41) and *B. abortus* (17) uptake into phagocytes. To demonstrate that the reticular structures are RER derived, primary macrophages infected with virulent *L. pneumophila* were analyzed by indirect immunofluorescence by using antibodies directed against the luminal RER protein BiP (Hsc78 [10]). Distinct rings of BiP localize about 60% of the phagosomes within 4.5 h after initial exposure of the phagocytes to the bacteria (82). This compartment is called the "replicative phagosome," because the appearance of these rings coincides with the first evidence of bacterial division.

The role that each of these events plays in intracellular growth is still unknown, although the studies with mutant strains described below contribute to understanding of these processes. A compelling argument can be made, that is, that prevention of acidification and avoidance of phagosome-lysosome fusion are critical for growth of the bacteria, because *L. pneumophila* is relatively susceptible to a variety of agents that induce stress in broth culture, and the microorganism is growth inhibited by even mildly acidic culture conditions. The recruitment of organelles and the sequestering of RER, however, are more obscure phenomena. Immunoelectron microscopy indicates that neither the bacterium nor the phagosome enters directly into the lumen of the RER but is surrounded by the cytoplasmic face of the RER (81). It seems unlikely, therefore, that the unique oxidizing environment of RER is being parasitized. On the other hand, as bacterial division occurs, the phagosome must grow and phospholipid must be targeted to the surface of the phagosome. Perhaps the RER becomes a ready source of newly synthesized phospholipid which is then transported to the nearby phagosome.

Another explanation for the distinctive trafficking properties of an infected cell is that it could be a reflection of the response of the phagocyte to changes that occur on the surface of the phagosome. The microorganism spends the early hours of the infection establishing a replicative niche, in which the phagosomal membrane is primed to transport the appropriate nutrients from the cytosol into the lumen of the phagosome and to coordinate membrane growth with bacterial cell division. As maturation occurs, the phagosomal membrane may appear abnormal to the host cell, and a sensing mechanism within the phagocyte may direct the endoplasmic reticulum to wall off this site from cytosolic components. By this formulation, recruitment plays no role in bacterial growth but is a host cell response to intracellular stress that occurs if the assembled phagosome can promote bacterial growth. Morphologically, the formation of the replicative phagosome resembles autophagy (80), a nonselective bulk process in which the endoplasmic reticulum of mammalian cells encircles the cytoplasm and organelles (18). Unlike the replicative phagosome, organelles cleared by autophagy are targeted for fusion with the lysosome (19). Presumably, because of the absence of appropriate surface factors, the replicative phagosome is not competent to fuse with the lysosome, even though it may be sequestered in a locale associated with this fusion event. Further studies are necessary to determine if sequestration by RER provides some advantage to the replicating *L. pneumophila*.

BACTERIAL FACTORS INVOLVED IN INTRACELLULAR GROWTH OF *L. PNEUMOPHILA*

Several approaches have been taken to identify the factors involved in the intracellular life cycle of *L. pneumophila*. The slow growth of the microorganism offers some challenges to workers, as do the normal technical problems that arise when studying a poorly characterized pathogen. After accommodations are made, such as finding appropriate plasmid and transposon vectors, it is definitely possible for the molecular biologist to adapt to the quirks of the microorganism. A variety of approaches have allowed the identification of a number of factors potentially involved in the intracellular growth of pathogenesis of *L. pneumophila*.

Analyses of animal infections originally implicated a role for an extracellular Zn^{2+} metalloprotease, the most abundant protein secreted by *L. pneumophila*, in pathogenesis. The protease has been shown to possess cytolytic and hemolytic activities and apparently plays some role in bypassing killing by neutrophils and other immune cells (64, 65), perhaps by cleaving key regulatory molecules (54). This protein originally sparked a great deal of interest when it was shown that intratracheal inoculation of guinea pigs with the protease resulted in histopathological lesions in the lung that mimicked Legionnaires' pneumonia (2, 86). Furthermore, immunization with the protease apparently protects animals from disease (8). Molecular studies on the protease suggest that its role is more subtle. Mutations that affect its production have been isolated, and they indicate that it is clearly dispensable for intracellular replication (83). This is despite the fact that intracellular organisms produce large amounts of the protease. Protease mutants also appear to have virulences equal to those of wild-type microorganisms, although this result should be interpreted cautiously since partial differences in virulence or histopathology between the mutant and wild type may be difficult to detect in the guinea pig model (9). At this point, however, it can be said that the protease is dispensable for the intracellular trafficking pathway of *L. pneumophila* described above.

A second protein that has been studied extensively was originally identified as a major 24-kDa cell membrane-associated antigen recognized by animals immunized with whole *L. pneumophila* (22, 23, 61). *Escherichia coli* strains expressing this protein were isolated by introducing an *L. pneumophila* gene bank into *E. coli;* this was followed by probing the colonies with antibody raised against whole *L. pneumophila*. The strains identified that expressed the 24-kDa species apparently properly localized the protein, facilitating its analysis and isolation of null mutations (21). Mutants containing one such mutation that crossed onto the *L. pneumophila* chromosome had a very distinct phenotype: the initiation of replication of mutants within phorbol ester-transformed U937 cells was delayed, but once replication was established, the bacteria grew with kinetics that were close to that of the wild type (14). Furthermore, there was no defect in uptake of the mutants by cultured cells. On the basis of these phenotypes, the gene expressing the 24-kDa protein was called *mip,* for macrophage infectivity protein (14). Mutants that failed to express *mip* required higher doses of bacteria than the wild type to cause a lethal pneumonia in guinea pigs, consistent with a role for the speedy establishment of intracellular replication in pathogenesis (13).

The phenotype of *mip* mutants suggests that they are delayed in establishing a replicative phagosome, and it would be important to determine if recruitment of RER to the phagosomal surface is delayed in *mip* mutants relative to that in wild-type strains. It seems particularly important to determine this in light of recent results regarding the probable biochemical activity of the Mip protein. The sequence of the open reading frame corresponding to *mip* shows high amino acid similarity to the large FKBP family of immunophilins, whose apparent normal role in the cell is to facilitate proper folding of proteins by virtue of a peptidyl *cis-trans* prolyl isomerase (PPIase) activity (30). In fact, Mip has been demonstrated to be a PPIase (29). Homologs of *mip* are found in a wide variety of eukaryotes and prokaryotes, indicating that it has a central role in cellular assembly processes (30, 46). Originally identified as the target of the immunosuppressive drug FK506 in T cells, these proteins apparently have an altered substrate specificity in the presence of the drug, such that they inhibit the proper nuclear localization of at least one transcription factor found in T cells and also act as inhibitors of calcineurin activity (44). Although this phenomenon presumably has no impact on *L. pneumophila* replication, it does indicate that binding of Mip to its target may be modulated by a third factor, so that the Mip protein may have unexpected activities that have little resemblance to prolyl isomerization.

There are two obvious models for the role that Mip plays in the infection process. First, the protein could directly affect the folding of a host cell protein or could inhibit the activity of such a protein. This model appears consistent with a preliminary report that indicates that Mip may interfere with the enzymatic activity of a host cell protein kinase C (40). Second, null mutations in *mip* could cause pleiotropic defects in the bacterium. A wide variety of bacterial proteins that require PPIase activity for rapid folding may mature or assemble slowly in the absence of Mip. By this formulation, mutations in *mip* may cause slowed assembly or export of some bacterium-encoded factors important for intracellular replication. This defect in macromolecular assembly would delay the ability of the microorganism to establish a productive replication cycle.

The most systematic search for mutants defective for intracellular replication was performed by Shuman and coworkers (72). By screening 4,564 independent Tn*903*dII*lacZ* insertions in the *L. pneumophila* chromosome, 55 mutants that were unable to kill phorbol ester-transformed HL60 cells were identified. On the basis of restriction fragment disruption, the mutants could be classified into at least nine groups that were absolutely defective for intracellular growth. Unlike the *mip* mutant, each isolate in these groups was unable to initiate replication even after 48 h of incubation with cultured cells. Nine of the 55 mutations were located in a region of the chromosome known to contain genes critical for intracellular replication (called class I [72]). The remaining mutations apparently defined novel loci involved in intracellular replication. Although the fates of the phagosomes bearing the mutants were not studied, the vast number of insertions obtained in that study should facilitate the identification of many important factors necessary for intracellular replication.

The region defined by the class I mutants described above is apparently a multigenic locus; mutations in this region have been previously shown to disrupt the phagosomal trafficking pathway taken by virulent *L. pneumophila*. For example, Horwitz (36) isolated spontaneous derivatives of *L. pneumophila* Philadelphia 1 that survived ten passages on plates containing brain heart infusion agar, a medium that is suboptimal for the growth of virulent *L. pneumophila*. The faster-growing derivatives were found not to grow in human monocyte cultures and were completely avirulent in the guinea pig pneumonia model (7). Most importantly, internalized mutants were found in a phagosome that was unable to avoid fusion with thorium-labeled compartments and that was unable to recruit organelles. The chromosomal region corresponding to the site of these mutations was identified recently by introducing into the intracellular replication-deficient derivative a cosmid bank of the wild-type *L. pneumophila* chromosome (48). Strains able to grow intracellularly were selected and analyzed. The chromosomal region identified contains at least four open reading frames that are potentially important for intracellular replication, because the nine class I insertion mutations defective for intracellular replication disrupt each of these open reading frames (73).

The mutations near the region of the *L. pneumophila* chromosome containing the lesions originally isolated by Horwitz were independently isolated by a third strategy. Thymine-requiring auxotrophs of *L. pneumophila* die as a result of a lack of thymine if they are internalized by cultured cells growing in standard tissue culture medium (53). This phenomenon depends on the attempted initiation of DNA replication by the microorganism. If the bacterium cannot grow intracellularly, then the failure to replicate DNA saves the thymine auxotroph from starvation-induced death. By passing *L. pneumophila thyA* strains through phorbol ester-transformed U937 cells, mutants that were defective for intracellular growth and that survived the intracellular environment were selected (4). Survivors from thymineless death either showed no growth intracellularly, showed slow growth, or were delayed for the initiation of replication. Those in the first group caused distinct defects in the normal intracellular routing of *L. pneumophila*. Some of the totally replication-deficient mutants were localized in phagosomes that were defective in association with RER, but they persisted in avoiding fusion with the thorium-labeled compartment (4). Most mutants, however, were found in phagosomes that aberrantly fused with the heavy metal compartment and were defective for RER association.

Molecular clones derived from the wild-type *L. pneumophila* chromosome were isolated; these clones could complement the mutants defective for intracellular replication (4). Surprisingly, a single gene called *dotA* (defect in organelle trafficking) could complement all of these mutants. Apparently, the different intracellular phenotypes of the replication-deficient mutants were due to less severe defects in some of the *dotA* mutants. Marker rescue experiments followed by DNA sequencing demonstrated that all the mutations complemented by this molecular clone caused premature translation termination of the *dotA* gene (6). It is not clear why some mutants caused a more defective phenotype than others, although it is presumed that some of the truncation products retain partial function. The gene encodes a protein of 1,048 amino acids that contains a hydrophobic region of over 100 amino acids lacking charged residues. Most of the mutations are predicted to be missing this large region, so it is tempting to postulate that this structure is involved in altering the phagosomal membrane in a fashion that may be important for the unique trafficking pathway of *L. pneumophila*.

The *dotA* gene product appears to perform its function during or shortly after phagocytic uptake of the bacterium. Wild-type *L. pneumophila* is internalized into phagosomes that are relatively deficient in the membrane protein LAMP-1. Mutations in *dotA*, on the other hand, cause the bacteria to be targeted into compartments that are rich in LAMP-1. This phenotypic distinction can be detected within 10 min after the initiation of phagocytosis: the absence of the *dotA* gene product has an immediate effect on intracellular targeting of the microorganism. Results of kinetic experiments on the regulated expression of *dotA* are consistent with the fact that its product plays an early role in proper targeting. By placing the *dotA* gene under the control of a highly attenuated *lac* promoter, expression can be regulated relative to the time of bacterial contact with the phagocyte. Bacteria that are exposed to the inducer isopropyl-β-D-thiogalactopyranoside prior to contact with phagocytes are able to grow intracellularly for at least eight generations, even in the absence of *dotA* expression during a 24-h period of intracellular incubation. This indicates that although the product of *dotA* is required for an early event during establishment of the replicative phagosome, the protein is dispensable once the bacterium is targeted to this site and replication has been initiated (5).

The model that arises from these mutant studies is that the product of the *dotA* gene is one of several factors that are expressed by *L. pneumophila* prior to contact with the host cell and that are required to target the microorganism into a replicative phagosome. Some of the consequences of this activity, however, cannot be seen for several hours after contact, since the coating of the phagosome by RER is not completed until at least 4 h after uptake. There must be a series of events that occur before this point that allows assembly of the phagosome in the absence of any morphologically obvious changes. Presumably, DotA plays some role in promoting a signaling event that causes an elimination of the proteins involved in targeting the phagosome to the lysosome or that informs the host cell to route this compartment in a novel fashion.

L. PNEUMOPHILA FACTORS INDUCED IN RESPONSE TO INTERNALIZATION BY PHAGOCYTES

A central property of *L. pneumophila* is its ability to grow both inside and outside of host cells. These two environments have distinctly different requirements for bacterial replication, so the microorganism presumably expresses different sets of factors in response to its surroundings. In addition, although the composition of the phagosome environment is poorly defined, presumably any extracellular site will lack some of the components found intracellularly. An important illustration of these points can be found in how the microorganism handles pH-induced stress. *L. pneumophila* grows rapidly within the phagosome at a pH of ≈6.2 (37), yet bacteriological medium at this acidity prevents extracellular replication. There must be unknown components in the phagosomal environment or new bacterial proteins expressed in the phagosome that facilitate replication under acidic conditions.

Evidence that different sets of proteins are expressed extracellularly and intracellularly was obtained by Engleberg and coworkers (1). In that study, the *L. pneumophila* internalized by phorbol ester-transformed U937 cells was radiolabeled biosynthetically with [^{35}S]methionine, and lysates were fractionated by two-dimensional gel electrophoresis. At least 35 proteins were synthesized at higher rates by intracellular bacteria than by bacteria grown in bacteriological medium. The expression of 13 of these proteins could be induced in extracellular medium by a variety of stress-inducing agents, such as high temperature and oxidizing agents. These results indicated that a variety of global stress response regulons are probably induced as a result of internalization within cultured cells. That these regulons express factors required specifically for intracellular growth remains to be determined.

In order to systematically identify genes that are regulated transcriptionally in response to an intracellular environment, a selection scheme that allows the isolation of promoters that are preferentially functional during intracellular growth of *L. pneumophila* was devised (68). This strategy takes advantage of the observation that *L. pneumophila thyA* strains are unable to grow intracellularly without thymine added to the tissue culture medium. Expression of thymidylate synthase located on a plasmid rescued this intracellular growth defect. By using a plasmid containing a promoterless phage T4 *td*Δi gene encoding thymidylate synthase, promoters were identified by placing random fragments in front of the gene, which selected for intracellular growth of *L. pneumophila* strains. In order to identify promoters that were expressed only intracellularly, bacterial strains also either were screened for the inability to grow extracellularly in the absence of thymine or were plated on trimethoprim, which selects for the lack of thymidylate synthase expression. A similar strategy has been described by Mekalanos and coworkers (47) for identifying *Salmonella typhimurium* genes induced in response to growth in an animal.

Among the fragments that repeatedly arose in this procedure was one that corresponded to a gene highly similar to *S. typhimurium ahpC,* whose product has been identified both as a subunit of alkyl hydrogen peroxide reductase and as a reductase that acts on aberrantly oxidized sulfhydryl groups on proteins (76, 84). In enteric bacteria, the product of *ahpC* is expressed after exposure to hydrogen peroxide, with its synthesis controlled by the OxyR protein (84). This protein therefore appears to protect against toxicity that results from oxidative stresses and is expressed under conditions in which the bacterial cytoplasm has become highly oxidized. These characteristics are consistent with those described in a previous report (1) that indicated that some of the major *L. pneumophila* proteins expressed in response to an intracellular environment are also induced by oxidative stresses extracellularly.

IMPLICATIONS FOR INTRACELLULAR GROWTH

Mutants of *L. pneumophila* defective for targeting to a replicative phagosome are unable to grow intracellularly. Several lines of evidence indicate that the defective phagosome is not the result of the absence of bacterial replication but, rather, that the failure of the phagosome to assume proper morphology is the cause of the replication defects. First, blockage of DNA replication by thymine starvation of *thyA* auxotrophs is not sufficient to prevent the bacteria from being targeted to a phagosome surrounded by RER (4). Second, inhibition of protein synthesis by erythromycin treatment at the initiation of bacterial contact with phagocytes has no effect on the ability of the microorganism to target the appropriate compartment, and withdrawal of this antibiotic from cell cultures results in the immediate initiation of bacterial replication within monocytes (39). Finally, *dotA* mutants are found in phagosomes that are morphologically distinct from *dotA* in the wild-type microorganism within 10 min after bacterial uptake. These results point to a model in which the microorganism expresses factors, which are translated prior to bacterial contact with the host cell, that operate during the initial moments of formation of the phagosome. It is interesting to note that these factors expressed on contact with the cells are not likely to be ligands for traditional macrophage receptors, because *L. pneumophila* can grow and can be targeted properly in a variety of cells that have very different membrane receptor compositions.

There appear to be linked events that occur during the binding of the bacterium to phagocytes and that may be key to promoting proper targeting of the bacteria. High-multiplicity infections of primary cultured macrophages, in the absence of bacterial replication, result in rapid destruction of the target cells soon after bacterial contact (43). This killing may be due to the combined actions of previously identified secreted proteins, such as metalloprotease and legiolysin (87), or it may be a consequence of the inhibition of host protein synthesis (51). It seems more likely, however, that target cell disruption is the consequence of an important event that occurs during the formation of the phagosome and that leads to localized membrane damage. Under low-multiplicity conditions, this damage may be repaired by the host cell in a fashion that causes modification of the phagosome so that it is unable to be intracellularly targeted via the default pathway. The repaired and modified phagosome, which is unable to participate in a variety of fusion events, is then sequestered by the host cell into a compartment encompassed by a variety of cytoplasmic organelles.

A likely source of cell damage would be the presence of a complex of proteins on the bacterial cell surface that is deposited within the macrophage, causing localized disruption of the target cell. Precedence for such a complex causing changes limited to the site of bacterial binding has been seen previously with *S. typhimurium*, which causes localized recruitment of both cytoskeletal components and members of the host cell signaling apparatus (28, 59). Perhaps *L. pneumophila*, by disrupting the phagosomal membrane, causes a series of similar localized changes, with the result that the microorganism is eventually targeted to its replicative niche. Future work will certainly be directed toward uncovering an activity involved in these events.

ACKNOWLEDGMENTS. I thank the members of my laboratory for contributing to the work described in this chapter. In particular, I thank Karen Berger for having the courage to initiate our first studies on *L. pneumophila*, Susannah Rankin for development of the selection for promoters expressed in response to an intracellular environment, and Michele Swanson for pointing out the similarity of autophagy to the replicative phagosome. Finally, I thank Stanley Falkow for encouraging me to work on challenging problems and guiding me into the world of intracellular microorganisms.

REFERENCES

1. **Abu Kwaik, Y., B. I. Eisenstein, and N. C. Engleberg.** 1993. Phenotypic modulation by *Legionella pneumophila* upon infection of macrophages. *Infect. Immun.* **61:**1320–1329.
2. **Baskerville, A., J. W. Conlan, L. A. Ashworth, and A. B. Dowsett.** 1986. Pulmonary damage caused by a protease from *Legionella pneumophila*. *Br. J. Exp. Pathol.* **67:**527–536.
3. **Bellinger-Kawahara, C., and M. A. Horwitz.** 1990. Complement component C3 fixes selectively to the major outer membrane protein (MOMP) of *Legionella pneumophila* and mediates phagocytosis of liposome-MOMP complexes by human monocytes. *J. Exp. Med.* **172:**1201–1210.
4. **Berger, K., and R. R. Isberg.** 1993. Two distinct defects in intracellular growth complemented by a single genetic locus in *Legionella pneumophila*. *Mol. Microbiol.* **7:**7–19.
5. **Berger, K., and R. R. Isberg.** Unpublished data.
6. **Berger, K. H., J. J. Merriam, and R. R. Isberg.** Submitted for publication.
7. **Blander, S. J., R. F. Breiman, and M. A. Horwitz.** 1989. A live avirulent mutant *Legionella pneumophila* vaccine induces protective immunity against lethal aerosol challenge. *J. Clin. Invest.* **83:**810–815.
8. **Blander, S. J., and M. A. Horwitz.** 1989. Vaccination with the major secretory protein of *Legionella pneumophila* induces cell-mediated and protective immunity in a guinea pig model of Legionnaires' disease. *J. Exp. Med.* **169:**691–705.
9. **Blander, S. J., L. Szeto, H. A. Shuman, and M. A. Horwitz.** 1990. An immunoprotective molecule, the major secretory protein of *Legionella pneumophila*, is not a virulence factor in a guinea pig model of Legionnaires' disease. *J. Clin. Invest.* **86:**817–824.
10. **Bole, D. G., R. Dowin, M. Doriaux, and J. D. Jamieson.** 1990. Immunocytochemical localization of BiP to the rough endoplasmic reticulum: evidence for protein sorting by selective retention. *J. Biol. Chem.* **265:**6879–6883.
11. **Cha, Y., S. M. Holland, and J. T. August.** 1990. The cDNA sequence of mouse LAMP-2. Evidence for two classes of lysosomal membrane glycoproteins. *J. Biol. Chem.* **265:**5008–5013.
12. **Chavrier, P., R. G. Parton, H. P. Hauri, K. Simons, and M. Zerial.** 1990. Localization of low molecular weight GTP binding proteins to exocytic and endocytic compartments. *Cell* **62:**317–329.
13. **Cianciotto, N. P., B. I. Eisenstein, C. H. Mody, and N. C. Engleberg.** 1990. A mutation in the mip gene results in an attenuation of *Legionella pneumophila* virulence. *J. Infect. Dis.* **162:**121–126.

14. **Cianciotto, N. P., B. I. Eisenstein, C. H. Mody, G. B. Toews, and N. C. Engleberg.** 1989. A *Legionella pneumophila* gene encoding a species-specific surface protein potentiates initiation of intracellular infection. *Infect. Immun.* **57:**1255–1262.

15. **Clemens, D. L., and M. A. Horwitz.** 1992. Membrane sorting during phagocytosis: selective exclusion of major histocompatibility complex molecules but not complement receptor CR3 during conventional and coiling phagocytosis. *J.Exp. Med.* **175:**1317–1326.

16. **Davis, G. S., W. C. Winn, Jr., D. W. Gump, and H. N. Beaty.** 1983. The kinetics of early inflammatory events during experimental pneumonia due to *Legionella pneumophila* in guinea pigs. *J. Infect. Dis.* **148:**823–835.

17. **Detilleux, P. G., B. L. Deyoe, and N. F. Cheville.** 1990. Penetration and intracellular growth of *Brucella abortus* in nonphagocytic cells in vitro. *Infect. Immun.* **58:**2320–2328.

18. **Dunn, W. A., Jr.** 1990. Studies on the mechanisms of autophagy: formation of the autophagic vacuole. *J. Cell Biol.* **110:**1935–1945.

19. **Dunn, W. A., Jr.** 1992. Studies on the mechanisms of autophagy: maturation of the autophagic vacuole. *J. Cell. Physiol.* **152:**458–466.

20. **Eissenberg, L. G., P. B. Wyrick, C. H. Davis, and J. W. Rumpp.** 1983. *Chlamydia psittaci* elementary body envelopes: ingestion and inhibition of phagolysosome fusion. *Infect. Immun.* **40:**741–751.

21. **Engleberg, N. C., C. Carter, D. R. Weber, N. P. Cianciotto, and B. I. Eisenstein.** 1989. DNA sequence of *mip*, a *Legionella pneumophila* gene associated with macrophage infectivity. *Infect. Immun.* **57:**1263–1270.

22. **Engleberg, N. C., E. Pearlman, D. Dixon, and B. I. Eisenstein.** 1986. Antibodies isolated by using cloned surface antigens recognize antigenically related components of *Legionella pneumophila* and other *Legionella* species. *J. Immunol.* **136:**1415–1417.

23. **Engleberg, N. C., E. Fearlman, and B. I. Eisenstein.** 1984. *Legionella pneumophila* surface antigens cloned and expressed in *Escherichia coli* are translocated to the host cell surface and interact with specific anti-*Legionella* antibodies. *J. Bacteriol.* **160:**199–203.

24. **Fallon, R. J., and T. J. Rowbotham.** 1990. Microbiological investigations into an outbreak of Pontiac fever due to *Legionella micdadei* associated with use of a whirlpool. *J. Clin. Pathol.* **43:**479–483.

25. **Fields, B. S., J. M. Barbaree, E. B. Shotts, J. C. Feeley, W. E. Morrill, G. N. Sanden, and M. J. Dykstra.** 1986. Comparison of guinea pig and protozoan models for determining virulence of *Legionella* species. *Infect. Immun.* **53:**553–559.

26. **Fields, B. S., S. R. Fields, J. N. Loy, E. H. White, W. L. Steffens, and E. B. Shotts.** 1993. Attachment and entry of *Legionella pneumophila* in *Hartmannella vermiformis. J. Infect. Dis.* **167:**1146–1150.

27. **Fields, B. S., E. B. Shotts, J. C. Feeley, G. W. Gorman, and W. T. Martin.** 1984. Proliferation of *Legionella pneumophila* as an intracellular parasite of the ciliated protozoan *Tetrahymena pyriformis. Appl. Environ. Microbiol.* **47:**467–471.

28. **Finlay, B. B., S. Ruschowski, and S. Dedhar.** 1991. Cytoskeletal rearrangements accompanying *Salmonella* entry into epithelial cells. *J. Cell Sci.* **99:**283–296.

29. **Fischer, G., H. Bang, B. Ludwig, K. Mann, and J. Hacker.** 1992. Mip protein of *Legionella pneumophila* exhibits peptidyl-prolyl-cis/trans isomerase (PPlase) activity. *Mol. Microbiol.* **6:**1375–1383.

30. **Hacker, J., and G. Fischer.** 1993. Immunophilins: structure-function relationship and possible role in microbial pathogenicity. *Mol. Microbiol.* **10:**445–456.

31. **Hart, C. A., and T. Makin.** 1991. *Legionella* in hospitals: a review. *J. Hosp. Infect.* **Suppl. A:**481–489.

32. **Hoffman, P. S., M. Ripley, and R. Weeratna.** 1992. Cloning and nucleotide sequence of a gene (*ompS*) encoding the major outer membrane protein of *Legionella pneumophila. J. Bacteriol.* **174:**914–920.

33. **Horwitz, M. A.** 1983. Formation of a novel phagosome by the Legionnaires' disease bacterium (*Legionella pneumophila*) in human monocytes. *J. Exp. Med.* **158:**1319–1331.

34. **Horwitz, M. A.** 1983. The Legionnaires' disease bacterium (*Legionella pneumophila*) inhibits phagosome-lysosome fusion in human monocytes. *J. Exp. Med.* **158:**2108–2126.

35. **Horwitz, M. A.** 1984. Phagocytosis of the Legionnaires' disease bacterium (*Legionella pneumophila*) occurs by a novel mechanism: engulfment within a pseudopod coil. *Cell* **36**:27–33.

36. **Horwitz, M. A.** 1987. Characterization of avirulent mutants of *Legionella pneumophila* that survive but do not multiply within human monocytes. *J. Exp. Med.* **166**:1310–1328.

37. **Horwitz, M. A., and F. R. Maxfield.** 1984. *Legionella pneumophila* inhibits acidification of its phagosome in human monocytes. *J. Cell Biol.* **99**:1936–1943.

38. **Horwitz, M. A., and S. C. Silverstein.** 1980. Legionnaires' disease bacterium (*Legionella pneumophila*) multiples intracellularly in human monocytes. *J. Clin. Invest.* **66**:441–450.

39. **Horwitz, M. A., and S. C. Silverstein** 1983. Intracellular multiplication of Legionnaires' disease bacteria (*Legionella pneumophila*) in human monocytes is reversibly inhibited by erythromycin and rifampin. *J. Clin. Invest.* **71**:15–26.

40. **Hurley, M. C., K. Balazovich, M. Albano, N. C. Engleberg, and B. I. Eisenstein.** 1993. *Legionella pneumophila* Mip inhibits protein kinase C, p. 69–70. *In* J. M. Barbaree, R. F. Breiman, and A. P. Dufour (ed.), *Legionella: Current Status and Emerging Perspectives*. American Society for Microbiology, Washington, D.C.

41. **Joiner, K.** Personal communication.

42. **Joiner, K. A., S. A. Fuhrman, H. M. Miettinen, L. H. Kasper, and I. Mellman.** 1990. *Toxoplasma gondii:* fusion competence of parasitophorous vacuoles in Fc receptor-transfected fibroblasts. *Science* **249**:641–646.

43. **Kishimoto, R. A., J. D. White, F. G. Shirey, V. G. McGann, R. F. Berendt, E. W. Larson, and K. W. Hedlund.** 1981. In vitro responses of guinea pig peritoneal macrophages to *Legionella pneumophila. Infect. Immun.* **31**:1209–1213.

44. **Liu, J., J. D. Farmer, Jr., W. S. Lane, J. Friedman, I. Weissman, and S. L. Schreiber.** 1991. Calcineurin is a common target of cyclophilin-cyclosporin A and FKBP-FK506 complexes. *Cell* **66**:807–815.

45. **Lukacs, G. L., O. D. Rotstein, and S. Grinstein.** 1990. Phagosomal acidification is mediated by a vacuolar-type H(+)-ATPase in murine macrophages. *J. Biol. Chem.* **265**:21099–21107.

46. **Lundemose, A. G., D. A. Rouch, S. Birkelund, G. Christiansen, and J. H. Pearce.** 1992. *Chlamydia trachomatis* Mip-like protein. *Mol. Microbiol.* **6**:2539–2548.

47. **Mahan, M. J., J. M. Slauch, and J. J. Mekalanos.** 1993. Selection of bacterial virulence genes that are specifically induced in host tissues. *Science* **259**:686–688.

48. **Marra, A., S. J. Blander, M. A. Horwitz, and H. A. Shuman.** 1992. Identification of a *Legionella pneumophila* locus required for intracellular multiplication in human macrophages. *Proc. Natl. Acad. Sci. USA* **89**:9607–9611.

49. **Marra, A., M. A. Horwitz, and H. A. Shuman.** 1990. The HL-60 model for the interaction of human macrophages with the Legionnaires' disease bacterium. *J. Immunol.* **144**:2738–2744.

50. **Maurelli, A. T., and P. J. Sansonetti.** 1988. Genetic determinants of *Shigella* pathogenicity. *Annu. Rev. Microbiol.* **42**:127–150.

51. **McCusker, K. T., B. A. Braaten, M. W. Cho, and D. A. Low.** 1991. *Legionella pneumophila* inhibits protein synthesis in Chinese hamster ovary cells. *Infect. Immun.* **59**:240–246.

52. **Meyer, R. D.** 1983. *Legionella* infections: a review of five years of research. *Rev. Infect. Dis.* **5**:258–278.

53. **Mintz, C. S., J. X. Chen, and H. A. Shuman.** 1988. Isolation and characterization of auxotrophic mutants of *Legionella pneumophila* that fail to multiply in human monocytes. *Infect. Immun.* **56**:1449–1455.

54. **Mintz, C. S., R. D. Miller, N. S. Gutgsell, and T. Malek.** 1993. *Legionella pneumophila* protease inactivates interleukin-2 and cleaves CD4 on human T cells. *Infect. Immun.* **61**:3416–3421.

55. **Moffat, J. F., and L. S. Tompkins.** 1992. A quantitative model of intracellular growth of *Legionella pneumophila* in *Acanthamoeba castellanii. Infect. Immun.* **60**:296–301.

56. **Nash, T. W., D. M. Libby, and M. A. Horwitz.** 1984. Interaction between the Legionnaires' disease bacterium (*Legionella pneumophila*) and human alveolar macrophages. Influence of antibody, lymphokines, and hydrocortisone. *J. Clin. Invest.* **74**:771–782.

57. **Nguyen, M. H., J. E. Stout, and V. L. Yu.** 1991. Legionellosis. *Infect. Dis. Clin. North Am.* **5**:561–584.

58. **Oaks, E. V., M. E. Wingfield, and S. B. Formal.** 1985. Plaque formation by virulent *Shigella flexneri*. *Infect. Immun.* **48:**124–129.

59. **Pace, J., M. J. Hayman, and J. E. Galán.** 1993. Signal transduction and invasion of epithelial cells by *S. typhimurium*. *Cell* **72:**505–514.

60. **Payne, N. R., and M. A. Horwitz.** 1987. Phagocytosis of *Legionella pneumophila* is mediated by human monocyte complement receptors. *J. Exp. Med.* **166:**1377–1389.

61. **Pearlman, E., N. C. Engleberg, and B. I. Eisenstein.** 1985. Identification of protein antigens of *Legionella pneumophila* serogroup 1. *Infect. Immun.* **47:**74–79.

62. **Pearlman, E., A. H. Jiwa, N. C. Engleberg, and B. I. Eisenstein.** 1988. Growth of *Legionella pneumophila* in a human macrophage-like (U937) cell line. *Microb. Pathog.* **5:**87–95.

63. **Portnoy, D. A., P. S. Jacks, and D. J. Hinrichs.** 1988. Role of hemolysin for the intracellular growth of *Listeria monocytogenes*. *J. Exp. Med.* **167:**1459–1471.

64. **Quinn, F. D., M. G. Keen, and L. S. Tompkins.** 1989. Genetic, immunological, and cytotoxic comparisons of *Legionella* proteolytic activities. *Infect. Immun.* **57:**2719–2725.

65. **Quinn, F. D., and L. S. Tompkins.** 1989. Analysis of a cloned sequence of *Legionella pneumophila* encoding a 38 kD metalloprotease possessing haemolytic and cytotoxic activities. *Mol. Microbiol.* **3:**797–805.

66. **Rabinowitz, S., H. Horstmann, S. Gordon, and G. Griffiths.** 1992. Immunocytochemical characterization of the endocytic and phagolysosomal compartments in peritoneal macrophages. *J. Cell Biol.* **116:**95–112.

67. **Racoosin, E. L., and J. A. Swanson.** 1993. Macropinosome maturation and fusion with tubular lysosomes in macrophages. *J. Cell Biol.* **121:**1011–1020.

68. **Rankin, S., and R. R. Isberg.** 1993. Identification of *Legionella pneumophila* loci expressed in response to an intracellular environment. *Infect. Agents Dis.*, **2:**269–271.

69. **Rittig, M. G., A. Krause, T. Haupl, U. E. Schaible, M. Modolell, M. D. Kramer, E. Lutjen-Drecoll, M. M. Simon, and G. R. Burmester.** 1992. Coiling phagocytosis is the preferential phagocytic mechanism for *Borrelia burgdorferi*. *Infect. Immun.* **60:**4205–4212.

70. **Rowbotham, T. J.** 1983. Isolation of Legionella pneumophila from clinical specimens via amoebae, and the interaction of those and other isolates with amoebae. *J. Clin. Pathol.* **36:**978–986.

71. **Roy, C., and R. Isberg.** Unpublished data.

72. **Sadosky, A. B., L. A. Wiater, and H. A. Shuman.** 1993. Identification of *Legionella pneumophila* genes required for growth within and killing of human macrophages. *Infect. Immun.* **61:**5361–5373.

73. **Shuman, H.** Personal communication.

74. **Sibley, L. D., E. Weidner, and J. L. Kraehenbuhl.** 1985. Phagosome acidification blocked by intracellular *Toxoplasma gondii*. *Infect. Immun.* **49:**760–764.

75. **Skinner, A. R., C. M. Anand, A. Malic, and J. B. Kurtz.** 1983. *Acanthamoebae* and environmental spread of *Legionella pneumophila*. *Lancet* **ii:**289–290.

76. **Storz, G., L. A. Tartaglia, S. B. Farr, and B. N. Ames.** 1990. Bacterial defenses against oxidative stress. *Trends Genet.* **6:**363–368.

77. **Sturgill-Koskycki, S., P. H. Schlesinger, P. Chakraborty, P. Haddix, H. L. Collins, S. Gluck, A. K. Fok, R. D. Allen, J. Heuser, and D. G. Russell.** 1994. *Mycobacteria* resist acidification of their phagosomes by selectively blocking incorporation of the vesicular proton-ATPase. *Science* **263:**678–681.

78. **Sun, A. N., A. Camilli, and D. A. Portnoy.** 1990. Isolation of *Listeria monocytogenes* small plaque mutants defective for intracellular growth and cell-to-cell spread. *Infect. Immun.* **58:**3770–3778.

79. **Swanson, J., E. Burke, and S. C. Silverstein.** 1987. Tubular lysosomes accompany stimulated pinocytosis in macrophages. *J. Cell Biol.* **104:**1217–1222.

80. **Swanson, M.** Unpublished data.

81. **Swanson, M., and R. Isberg.** Unpublished data.

82. **Swanson, M. S., and R. R. Isberg.** 1993. Formation of the *Legionella pneumophila* replicative phagosome. *Infect. Agents Dis.* **2:**224–226.

83. **Szeto, L., and H. A. Shuman.** 1990. The *Legionella pneumophila* major secretory protein, a protease, is not required for intracellular growth or cell killing. *Infect. Immun.* **58:**2585–2592.

84. **Tartaglia, L. A., G. Storz, and B. N. Ames.** 1989. Identification and molecular analysis of *oxyR*-

regulated promoters important for the bacterial adaptation to oxidative stress. *J. Mol. Biol.* **210:** 709–719.

85. **Wadowsky, R. M., L. J. Butler, M. K. Cook, S. M. Verma, M. A. Paul, B. S. Fields, G. Keleti, J. L. Sykora, and R. B. Yee.** 1988. Growth-supporting activity for *Legionella pneumophila* in tap water cultures and implication of *Hartmannella* amoebae as growth factors. *Appl. Environ. Microbiol.* **54:**2677–2682.

86. **Williams, A., A. Baskerville, A. B. Dowsett, and J. W. Conlan.** 1987. Immunocytochemical demonstration of the association between *Legionella pneumophila,* its tissue-destructive protease, and pulmonary lesions in experimental Legionnaires' disease. *J. Pathol.* **153:**257–264.

87. **Wintermeyer, E., U. Rdest, B. Ludwig, A. Debes, and J. Hacker.** 1991. Characterization of legiolysin (*lly*), responsible for haemolytic activity, colour production and fluorescence of *Legionella pneumophila. Mol. Microbiol.* **5:**1135–1143.

88. **Wright, S. D., and S. C. Silverstein.** 1983. Receptors for C3b and C3bi promote phagocytosis but not the release of toxic oxygen from human phagocytes. *J. Exp. Med.* **158:**2016–2023.

89. **Yamamoto, Y., T. W. Klein, and H. Friedman.** 1991. *Legionella pneumophila* growth in macrophages from susceptible mice is genetically controlled. *Proc. Soc. Exp. Biol. Med.* **196:**405–409.

90. **Yamamoto, Y., T. W. Klein, C. A. Newton, and H. Friedman.** 1988. Interaction of *Legionella pneumophila* with peritoneal macrophages from various mouse strains. *Adv. Exp. Med. Biol.* **239:** 89–98.

91. **Yoshida, S., Y. Goto, Y. Mizuguchi, K. Nomoto, and E. Skamene.** 1991. Genetic control of natural resistance in mouse macrophages regulating intracellular *Legionella pneumophila* multiplication in vitro. *Infect. Immun.* **59:**428–432.

Molecular Genetics of Bacterial Pathogenesis
Edited by V. L. Miller, J. B. Kaper, D. A. Portnoy, and R. R. Isberg
© 1994 American Society for Microbiology, Washington, DC 20005

Chapter 18

Cellular Biology of *Listeria monocytogenes* Infection

Daniel A. Portnoy

HISTORICAL PERSPECTIVE

It can be argued that more is known about the murine immune response to *Listeria monocytogenes* than to any other bacterial pathogen. This is because *L. monocytogenes* has been used by immunologists for decades as a model pathogen for the study of cell-mediated immunity in mice. Even in 1993 experimental listeriosis was often the model of choice for characterizing components of the host immune system in transgenic or gene-knockout mice (43, 44, 67, 80, 89, 92).

The classic work by Mackaness (58, 59) demonstrated that mice immunized with *L. monocytogenes* were resistant to subsequent challenge with *L. monocytogenes* and that resistance was transferable with cells, but not with serum, i.e., cellular immunity. The effector cells responsible for listericidal activity were activated macrophages (75). Resistance of this type was not specific, because mice immunized with *L. monocytogenes* were subsequently resistant to *L. monocytogenes* as well as *Bacillus abortus* or *Mycobacterium tuberculosis* (58, 59). This nonspecific resistance to infection lasted only a few weeks, while specific immunity was long-lived (3 months) (74). The discovery of T cells as the cellular component necessary for immunity was also discovered by using the *L. monocytogenes* model (8, 54, 73). On the basis of these observations, a simple model emerged to explain immunity to *L. monocytogenes,* a model which still dominates the textbooks and which can be summarized as follows. *L. monocytogenes* normally grows in resident macrophages. Infection results in the generation of specific T cells which secrete macrophage-activating factors, including gamma interferon (IFN-γ), which activate macrophages to kill *L. monocytogenes* either directly or through granuloma formation. This model is an oversimplification and does not deal with at least two important observations. First, there is compelling evidence that *L. monocytogenes* grows in cells other than macrophages in vivo (19, 72, 88, 91), and second, mice which lack T cells do surprisingly well in response to *L. monocytogenes* infection (2, 17, 28, 69, 70). It is now clear that immunity to *L.*

Daniel A. Portnoy • Department of Microbiology, University of Pennsylvania School of Medicine, Philadelphia, Pennsylvania 19104-6076.

monocytogenes is quite complex, involving components of innate immunity (82) and acquired immunity (77), which is necessary for sterilizing immunity and memory.

A contemporary view of the murine response to experimental *L. monocytogenes* infection in mice can be summarized as follows. Within 10 min after intravenous injection, approximately 90% of the inoculum can be recovered from the liver and 10% can be recovered from the spleen (18). In the liver, the bacteria are initially found in Kupffer cells, which are a dividing population of macrophages lining the sinusoids of the liver, or in immigrant macrophages. The majority of the bacteria initially in the liver are killed, although the extent of the initial kill may depend on the level of macrophage activation, which may be further dependent on the activation state of the infected animals. Infected macrophages probably initiate the host response by secreting chemokines and cytokines in response to the bacterial challenge. Thus, within minutes after infection, neutrophils begin to migrate toward the site of infection. A small percentage of the original inoculum will enter the Kupffer cell cytoplasm, grow, and spread into hepatocytes by the cell-to-cell spreading mechanism which will be described in detail later in this chapter. Within 12 h of infection, the bulk of replicating bacteria can be observed in hepatocytes (18). Bacteria grow logarithmically for 2 to 3 days, although the doubling time observed in vitro is far faster than that observed in vivo (3 to 4 h in vivo compared with 1 h in vitro), which may reflect host control of the infection to some extent during the first day of infection. It is clear that neutrophils are necessary for resolving the infection (18, 90, 91), and it has been suggested that neutrophils can recognize infected hepatocytes and dissolve them (18). Natural killer (NK) cells are also important, probably as early producers of IFN-γ (27). Direct lysis of infected cells by NK cells has yet to be demonstrated. The role of T cells has been the subject of hundreds of reports (48) and will not be dealt with in detail here. However, it is clear that CD8$^+$ T cells are capable of lysing infected target cells in vitro and can adoptively transfer immunity to naive mice (7, 9, 11, 22, 66, 78). Sterilizing immunity probably requires enhanced bactericidal capacities of phagocytic cells, granuloma formation, and cytotoxic T cells.

Although the immune response to *L. monocytogenes* has received an enormous amount of attention, it was not until the mid-1980s that bacteriologists began to appreciate the potential of *L. monocytogenes* as an excellent model system for use in dissecting the molecular determinants of pathogenicity (35, 47). Whereas immunologists had routinely administered *L. monocytogenes* by the intravenous route, it was clear to some that the oral route was the natural route of infection (57, 88, 112). Indeed, *L. monocytogenes* is a food-borne pathogen responsible for serious infections in immunocompromised individuals and pregnant women (29, 36). On the basis of the findings that some enteric pathogens invade nonprofessional phagocytes, it was not surprising that *L. monocytogenes* was found to enter fibroblasts and epithelial cells (34, 41). However, more important was the observation that intracellular *L. monocytogenes* grew rapidly in the cytoplasm of cells and spread cell to cell by using components of the host's cytoskeleton (34, 41, 68, 84, 105). In this chapter, the molecular and cellular biology of *L. monocyto-*

genes infection is reviewed and, at the end, a model which attempts to bridge the immunology with the pathogenesis is discussed.

TISSUE CULTURE MODELS OF INFECTION

Among the most important advances in *L. monocytogenes*-related research were the development of simple and quantitative methods for evaluating intracellular growth in vitro. A number of factors facilitate in vitro assays. First, and most importantly, *L. monocytogenes* actually grows intracellularly. The intracellular growth rate approximates that seen in rich medium. Second, gentamicin can be used to sterilize the extracellular medium without affecting the intracellular growth rate. This is critical, because *L. monocytogenes* grows in tissue culture medium containing serum. Third, the infection of host cells is relatively nontoxic, so that cells do not detach from the substrate and gentamicin does not enter the cell cytoplasm. Indeed, *L. monocytogenes* can multiply from 1 bacterium to greater than 200 bacteria per cell during an 8-h infection in the presence of 50 μg of gentamicin per ml, with a doubling time of 51 min (10). The intracellular doubling time can be easily calculated and used to compare mutants with wild-type strains (84, 97). *L. monocytogenes* forms large homogeneous plaques in fibroblast monolayers which reflect cell-to-cell spread (53, 97). The diameter of the plaque can be accurately measured as a correlate of virulence (15). The ability of the bacteria to spread by the direct cell-to-cell route compared with an extracellular route can be examined by using either low or high concentrations of gentamicin (10). Lastly, the stages in the infection can be examined and quantitated by electron microscopy (68, 85, 103, 108) and video microscopy (21, 101).

GENETIC TOOLS IN *L. MONOCYTOGENES*

The first major advance in the genetics of *L. monocytogenes* was the use of conjugative transposons to isolate hemolysin-negative mutants (35, 47). These seminal papers brought the study of *L. monocytogenes* into the modern era. This was followed by the development of vectors for the delivery of Tn*917* and Tn*917lac* (14, 20). A number of plasmid transformation systems have been developed, the best of which uses electroporation of penicillin-treated bacteria (79). Unfortunately, this method generates only approximately 1,000 transformants per μg of plasmid DNA in most strains. A protocol has been developed for insertional inactivation by site-specific inactivation of genes by homologous recombination using a thermal-sensitive plasmid (49, 110). However, insertion of the thermal-sensitive plasmid into random sites on the *L. monocytogenes* chromosome resulted in a defect in plaque size (15). Thus, the use of such vectors is limited. The most appropriate strategy for generating point mutations or in-frame deletions is allelic exchange (15, 65). Lastly, *L. monocytogenes* genes can be expressed in the gram-positive bacteria *Bacillus subtilis* (6, 32) and *Listeria innocua* (33). Unfortunately, it has not yet been possible to map or complement random or chemically induced

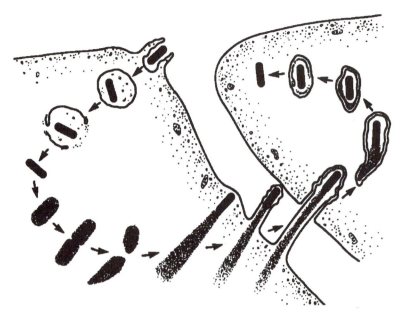

FIGURE 1. Model for the intracellular life cycle of *L. monocytogenes*. The stippled material represents polymerized actin. Adapted from Tilney and Portnoy (105), with copyright permission of the Rockefeller University Press.

point mutations, although a physical map of the *L. monocytogenes* chromosome has been reported (64).

CELL BIOLOGY

The morphological stages which occur during growth and the cell-to-cell spread of *L. monocytogenes* have been identified by electron microscopy (68, 105) and are depicted in Fig. 1. Many of the genes required to complete the cycle have also been identified and are shown in Fig. 2 (83). Rather than concentrating

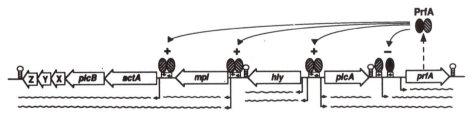

FIGURE 2. *prfA* regulon of *L. monocytogenes*. Genes are represented by large arrows, transcripts are represented by wavy lines, putative PrfA-binding sites are represented by small boxes containing short arrows, and putative rho-independent terminators are represented by a short post terminated by a circle. PrfA protein is shown as a dimer, although this has not been demonstrated. This model is based on information presented in previous publications (15, 16, 31, 32, 55, 56, 62, 63, 108).

on the separate role of each gene, each step in the intracellular cycle will be discussed.

Entry

Entry of *L. monocytogenes* into phagocytic cells occurs through the complement receptor CR3 as well as other unidentified receptors (25, 26). However, in addition to phagocytic cells, *L. monocytogenes* enters into all adherent cells so far evaluated, although the efficiency of entry ranges over 4 orders of magnitude. For example, the plaque-forming efficiency in the J774 macrophage-like cell line is 10,000 times greater than that in L2 fibroblasts. The biological significance of direct entry into nonprofessional phagocytes is unknown. However, the bacteria are thought to enter via Peyer's patches and, presumably, M cells after oral infection (57) and have been found in intestinal epithelial cells (88).

Gaillard et al. (33) screened 2,500 Tn*1545* insertions and identified a locus, *inlAB*, which is necessary for *L. monocytogenes* entry into epithelial cells. The product of *inlA*, internalin, is a repeat-containing surface protein with structural features similar to those of the M protein of *Streptococcus pyogenes*. This is the only example of such a protein in a bacterium other than gram-positive cocci (30). Interestingly, *inlA* may be a member of a gene family in *L. monocytogenes,* because a number of restriction fragments hybridize to an *inlA*-specific probe under conditions of low stringency (33). This raises the possibility that proteins related to internalin could bind to different host receptors or to analogous receptors in different animal species. This is consistent with the observation that *L. monocytogenes* has been isolated from over 50 different animal species (29).

Another protein, called Iap, has been associated with invasion in *L. monocytogenes* (51). Iap is a major secreted protein whose secretion is greatly diminished in spontaneous rough isolates of *L. monocytogenes*. Rough mutants tend to form long chains containing septa which divide upon the addition of Iap protein (51). Rough mutants show a decrease in their level of invasion of fibroblasts, but this is probably because of the difficulty of invasion by the long chains. Iap shows amino acid relatedness to a family of muramidases (46). Why *L. monocytogenes* secretes a large amount of a muramidase necessary for cell division is unknown.

Escape from a Vacuole (the Role of Listeriolysin O)

Subsequent to internalization, *L. monocytogenes* is found in a membrane-lined vacuole or phagosome. Almost nothing is known about the physical properties of the vacuole with regard to pH, the presence of lysosomal markers, etc. In primary macrophages, the majority of the bacteria fail to escape from the vacuole and presumably are killed (15, 85). In the J774 macrophage-like cell line, more than 50% of the bacteria are free in the cytoplasm within 2 h of infection (105). It is presumed that greater than 99% of the bacteria eventually escape from the J774 vacuole because methicillin added after 2 h kills 99.9% of wild-type *L. monocytogenes* but does not kill the hemolysin-negative mutants which fail to escape from the vacuole (13). Therefore, if live bacteria remained in the vacuole, they would be spared killing by methicillin. However, this argument assumes that wild-type *L. monocytogenes* does not grow in the vacuole, but this has not been for-mally proven. What is known, however, is that hemolysin-negative mutants fail to escape from the J774 vacuole and fail to grow but can survive in the vacuole for up to 24 h. In contrast, hemolysin-negative mutants continue to be killed in peritoneal macrophages (85). Therefore, it is clear that the vacuolar environment in primary macrophages is different from that in J774 cells.

Escape from a vacuole is largely dependent on the production of listeriolysin O (LLO), as LLO-negative (LLO⁻) mutants fail to escape from a vacuole or grow in most cells examined (34, 52, 84, 105). Consequently, LLO⁻ mutants are 5 log units less virulent for mice (35, 47, 84). Absolute proof of a role of LLO in pathogenicity was provided by the introduction of the cloned gene on a plasmid into a strain containing a structural gene mutation and then restoration of 4 log units of virulence (20). Further evidence supporting the role of LLO was obtained when LLO was expressed in *B. subtilis,* which then acquired the capacity to escape from a J774 vacuole and grow in the cytoplasm (6). In *L. monocytogenes,* LLO probably does not act alone, because mutants lacking phosphatidylinositol-specific phospholipase C (PI-PLC) are somewhat defective in escaping from the vacuole of bone marrow-derived macrophages (15).

LLO is a member of a family of sulfhydryl-activated, oxygen-labile cytolysins present in 19 diverse species of gram-positive bacteria (1, 95). Members of the family share approximately 50% amino acid identity and share an 11-amino-acid region which contains the conserved but nonessential cysteine residue (65, 81). On the basis of the available information, LLO is the only member of the family whose role is to lyse the host vacuole. In contrast, streptolysin O and perfringo-lysin O (PFO) are thought to block neutrophil chemotaxis at sublytic concentra-tions (96, 107). A direct comparison of the related cytolysins might shed some light on the features of LLO which are necessary for its unique role. The most obvious difference is that LLO has an acidic pH optimum (37), while the others are active at both neutral and acidic pHs (86). PFO has the same specific activity as LLO when assayed at pH 5.5 but is 20 times as active as LLO at pH 7.0 (86). *B. subtilis* expressing PFO can escape from the J774 vacuole and grow in the cytoplasm. However, the infected J774 cells appear to be physically damaged (86). These data indicate that PFO can functionally replace LLO in a vacuole, but suggest that PFO may be toxic to the host cell. It has been speculated that the low pH optimum of LLO reflects a mechanism that prevents its activity in the cytoplasm (86). However, there are other possible reasons that PFO may have damaged the host cell; e.g., PFO may have a longer half-life than LLO in the cytoplasm.

To directly evaluate the capacity of PFO to replace LLO, the structural gene encoding PFO was used to replace that encoding LLO on the *L. monocytogenes* chromosome. The construction retained the *L. monocytogenes* promoter and rib-osome-binding sites (45). The chimeric strain secreted active PFO but was com-pletely avirulent in mice. In J774 cells, the PFO-expressing strain was able to escape from the vacuole, commense growth, and then cause the permeabilization of the host cell, resulting in the influx of gentamicin, which killed the bacteria. Although differences in regulation cannot be ruled out at this time, the data are consistent with a model in which LLO is unique in that it does not adversely affect the host cell. This is probably due to a combination of factors including regulation, pH optimum, and half-life in the cytoplasm. As a good pathogen does not kill its host, a good intracellular pathogen does not kill its host cell. LLO has evolved to act within a vacuole, not in the cytoplasm.

It is clear that LLO is an essential determinant of pathogenicity whose role is to mediate lysis of the host vacuole. However, LLO⁻ mutants still escape from a vacuole and grow normally in human epithelial cell lines such as Henle 407 and HeLa (84). The significance of this observation is unknown. However, it raises the question as to the precise role of LLO in lysis of the vacuole. Perhaps LLO does not mediate lysis directly, but acts indirectly by facilitating the action of either the bacterial or the host phospholipases. By creating large pores, LLO could affect the fate of the vacuole, e.g., fusion with lysosomes and/or its pH. Vacuolar maturation may be altered in different cell types so that the phospholipases can act even in the absence of LLO under some conditions. In support of a role for the phospholipases C (PLCs) in vacuolar lysis, an LLO and *plcB* double mutant of *L. monocytogenes* cannot escape from a vacuole in human epithelial cells (61) and a *plcA* mutant is defective in its ability to escape from a vacuole in bone marrow-derived macrophages (15).

Growth in the Cytoplasm

There is a growing body of evidence that the mammalian cytoplasm is a nutritionally rich environment. First, wild-type *L. monocytogenes* is naturally auxotrophic for several amino acids and vitamins, and additional auxotrophic mutants requiring aromatic amino acids or purines were able to grow normally intracellularly (60). Interestingly, amino acid auxotrophs even grew in cells which were starved of the required amino acid, a result which was consistent with growth of the bacteria on host peptides (60). Second, no mutant which enters the cytoplasm but fails to grow has yet been isolated. Lastly, as mentioned above, a *B. subtilis* strain that expresses LLO was able to grow in the mammalian cell cytoplasm.

Actin-Based Movement

Within a short time after *L. monocytogenes* arrives in the cytoplasm, it becomes surrounded by a cloud of host actin filaments and actin-binding proteins (21, 68, 103, 105) (Fig. 1). Examination of the literature reveals that in 1972 *L. monocytogenes* was first observed within gut epithelial cells surrounded by a cloud of actin filaments, although it was not appreciated at the time (88). It is now known that three unrelated bacterial pathogens, *L. monocytogenes*, *Shigella flexneri*, and some species of *Rickettsia*, all utilize host actin filaments to mediate the spread of infection (5, 39, 42, 76, 99).

After a couple of divisions, *L. monocytogenes* begins to move at rates of up to 1.5 μm/s in J774 cells and 0.4 μm/s in Ptk2 cells, and after 4 h of infection, approximately 80% of the bacteria are moving throughout the cytoplasm (21, 93, 101). The addition of cytochalasin D results in the immediate cessation of bacterial movement (21). During actin-based movement, the bacteria can be observed with a tail of short actin filaments. The actin filaments are stationary in the cytoplasm and turn over with an average half-life of 33 s (101). New filament growth occurs at the bacterium-tail interface (93, 101, 104). The rate of bacterial movement is equal to the rate of actin polymerization, which is consistent with a model in which actin polymerization is providing the driving force for propulsion (93, 101).

The *L. monocytogenes* ActA protein is absolutely required for actin nucleation and intracellular motility (24, 49). An in-frame deletion within *actA* resulted in a 3-log-unit decrease in virulence for mice, supporting the contention that actin-based movement is essential for *L. monocytogenes* pathogenesis (10). In most strains of *L. monocytogenes,* ActA is minor surface protein after in vitro growth, whereas in host cells, it is the major detectable surface or secreted protein (10). ActA is localized to the bacterial surface and has a polar localization, being absent from the most recent pole that has undergone cell division (50, 71).

The *actA* gene predicts a protein of 610 amino acids, which contains a signal sequence, a region of four proline-rich repeats, and a hydrophobic stretch of amino acids at its C terminus which may serve the role of membrane anchor (24, 49). The ActA protein migrates upon sodium dodecyl sulfate-polyacrylamide gel electrophoresis (SDS-PAGE) at an apparent molecular mass of about 97 kDa, although the sequence predicts a mature protein of 67 kDa. The aberrant migration on gels may be due to its high proline content (9.2% overall).

The capacity of *L. monocytogenes* to use a host system of actin-based motility has made it a very attractive model for use in dissecting the host components required for this complex process. This is no simple task, because over 100 actin-binding proteins have been identified. *L. monocytogenes* is attractive because it can be studied in the absence of host membranes. Indeed, Theriot et al. (102) have developed a cell-free system that uses extracts from *Xenopus* eggs which faithfully supports the actin-based motility of *L. monocytogenes.*

Profilin is a host actin-binding protein which is known to bind polyproline in vitro (98), which led Theriot et al. (102) to hypothesize that the four proline-rich repeats within ActA might function to bind profilin. Accordingly, profilin was shown by immunofluorescence to localize at the interface of *L. monocytogenes* and its actin-based tail but not to *actA* mutants. In addition, depletion of profilin from *Xenopus* extracts ablated the ability of the extract to support bacterial movement (102). Although bacteria failed to move in the profilin-depleted extracts, the bacteria still had actin filaments on their surfaces. These data are consistent with a model in which ActA has at least two functions, the first of which is to nucleate or bind actin filaments and the second of which is to bind host profilin, which in turn mediates actin-based motility by a mechanism which is still controversial (100). Direct binding of profilin to *L. monocytogenes* has not been demonstrated in vitro, suggesting that another host component may be involved or that ActA must be modified by the host. Interestingly, ActA is phosphorylated in host cells, resulting in migratory shifts upon SDS-PAGE (10). Whether phosphorylation of ActA is necessary for its function or has an inhibitory effect as suggested for IcsA in *S. flexneri* (23) awaits further study.

Cell-to-Cell Spread

There is a reasonable amount of data to support the model that *L. monocytogenes* spreads from cell to cell without an extracellular phase as depicted in Fig. 1. Bacteria spread from cell to cell even in the presence of high concentrations of gentamicin. This can be observed by light microscopy or by the formation of macroscopic plaques. Infection of one bacterium per 50 cells results in an average of 10 cells infected after 8 h of intracellular growth (97). Mutants which cannot nucleate actin filaments have been found in one heavily infected cell (10, 24, 49, 105). Electron microscopy has shown that after 4 h of infection, bacteria are found in double-membrane vacuoles which are probably formed by the ingestion of *L. monocytogenes* on cellular projections (68, 105). These projections are very commonly seen by light, video, or electron microscopy. However, what one commonly sees by video microscopy is that bacteria move into pseudopod-like projections, but the vast majority are not ingested by neighboring cells. Thus, the efficiency of cell-to-cell spread may be quite low.

Whether cell-to-cell spread occurs by receptor-ligand interaction is not known. What is known, however, is that the *L. monocytogenes* PLCs are probably involved in the spread. Mutation of *plcB*, which encodes a broad-range PLC (38, 40), resulted in a mutant which was found in double-membrane vacuoles at a higher frequency than the wild type (108). However, a mutant with in-frame deletions in both *plcA*, which encodes a PI-PLC, and *plcB* resulted in a plaque size much smaller than that of either single mutant (94). Thus, it appears that both PLCs may be involved in cell-to-cell spread. Another level of complexity is added by the observation that PlcB is made as an inactive precursor which is activated in vitro by the product of the *mpl* gene which encodes a zinc metalloprotease (87).

There are at least two potential roles for the PLCs in cell-to-cell spread. The most obvious role is that the PLCs are responsible for hydrolysis of membrane phospholipids, resulting in the dissolution of the membranes. Alternatively, the action of PLCs will result in the formation of diacylglycerol, which could induce membrane fusion and/or activate protein kinase C (106). The latter could conceivably play a role in cell-to-cell spread or have global effects on host cell physiology. Lastly, the role of LLO in cell-to-cell spread has yet to be evaluated.

BRIDGING THE CELL BIOLOGY AND IMMUNOLOGY OF *L. MONOCYTOGENES* INFECTION

The following are the most important points to remember with regard to immunity to experimental listeriosis. (i) Immunity is cellular, and antibodies play no measurable role. (ii) Immunity is sterilizing within a couple of weeks. Activated macrophages and neutrophils are among the critical cells necessary to mediate resistance. (iii) Administration of a sublethal dose of live *L. monocytogenes* leads to long-lived immunity, while administration of dead *L. monocytogenes* does not lead to immunization (109). Importantly, live, LLO⁻ mutants also fail to lead to immunization (4). (iv) T cells play a significant but relatively minor role in resistance to the initial infection (82), although their role in long-lived immunity is critical. (v) The bulk of bacterial replication occurs in cells other than macrophages.

The model presented in Fig. 1 is consistent with the finding that antibodies play no role in resistance to *L. monocytogenes*. That is, once a bacterium enters a cell, it may never again confront the humoral immune response. However, examination of the model reveals a number of sites in which the host could abort the infection. First, bacteria can be killed upon phagocytosis, as is clearly the case with neutrophils and activated macrophages (12). Bacteria could be denied access to the host cytoplasm, as is seen in IFN-γ-treated cells (85). Infected cells can be lysed by cytotoxic T cells, as has been shown in vitro (9, 22). Lastly, the spread of the infection could be halted by granuloma formation, which might deny access of the bacteria to uninfected cells.

The current model also has serious implications for the nature of antigen presentation. Current models describe two distinct pathways of antigen presentation (111). The first involves the processing of antigens within an acidic vacuolar compartment and presentation to CD4$^+$ T cells in the context of major histocompatibility complex (MHC) class II antigens. The second involves presentation of cytosolic antigens and presentation to CD8$^+$ T cells in the context of MHC class I antigens. Thus, the first thing that occurs to an immunologist when confronted with the *L. monocytogenes* system is that secreted *L. monocytogenes* antigens will be introduced into the class I pathway of antigen presentation. Indeed, immunity to *L. monocytogenes* is largely restricted to class I, and CD8$^+$ T cells are thought to be largely responsible for immunity (77). Therefore, the observation that neither dead nor LLO$^-$ *L. monocytogenes* can immunize mice (4, 109) is consistent with the fact that neither type of organism enters the cytoplasm, and the organisms thereby avoid the class I pathway of antigen presentation. LLO$^-$ bacteria are not recognized by CD8$^+$ T cells in vitro (9, 11). The interpretation of these results was complicated by the provocative observation that LLO is itself a major antigen recognized by immune CD8$^+$ T cells (9, 78). Nevertheless, all mutants which enter the cytoplasm were shown to immunize mice to subsequent challenge (3). Thus, the data are consistent with the requirement that the organism must enter the cytoplasm for the induction of immunity. Whether entry into the cytoplasm is sufficient for long-lived immunity or whether cell-to-cell spread is also necessary remains to be determined.

REFERENCES

1. **Alouf, J. E., and C. Geoffroy.** 1991. The family of the antigenically-related, cholesterol-binding ('sulphydryl-activated') cytolytic toxins, p. 147–186. *In* J. E. Alouf and J. H. Freer (ed.), *Sourcebook of Bacterial Protein Toxins*. Academic Press, London.
2. **Bancroft, G. J., M. J. Bosma, G. C. Bosma, and E. R. Unanue.** 1986. Regulation of macrophage Ia expression in mice with severe immunodeficiency: induction of Ia expression by a T cell-independent mechanism. *J. Immunol.* **137:**4–9.
3. **Barry, R. A., H. G. A. Bouwer, D. A. Portnoy, and D. J. Hinrichs.** 1992. Pathogenicity and immunogenicity of *Listeria monocytogenes* small-plaque mutants defective for intracellular growth and cell-to-cell spread. *Infect. Immun.* **60:**1625–1632.

4. Berche, P., J. Gaillard, and P. J. Sansonetti. 1987. Intracellular growth of *Listeria monocytogenes* as a prerequisite for in vivo induction of T cell-mediated immunity. *J. Immunol.* **138:**2266–2271.

5. Bernardini, M. L., J. Mounier, H. d'Hauterville, M. Coquis-Rondon, and P. J. Sansonetti. 1989. *ics*A, a plasmid locus of *Shigella flexneri*, governs bacterial intra- and intercellular spread through interaction with F-actin. *Proc. Natl. Acad. Sci. USA* **86:**3867–3871.

6. Bielecki, J., P. Youngman, P. Connelly, and D. A. Portnoy. 1990. *Bacillus subtilis* expressing a haemolysin gene from *Listeria monocytogenes* can grow in mammlian cells. *Nature* (London) **345:**175–176.

7. Bishop, D. K., and D. J. Hinrichs. 1987. Adoptive transfer of immunity to *Listeria monocytogenes*: the influence of in vitro stimulation on lymphocyte subset requirements. *J. Immunol.* **139:** 2005–2009.

8. Blanden, R. V., and R. E. Langman. 1972. Cell-mediated immunity to bacterial infection in the mouse. Thymus-derived cells as effectors of acquired resistance to *Listeria monocytogenes*. *Scand. J. Immunol.* **1:**379–391.

9. Bouwer, H. G. A., C. S. Nelson, B. L. Gibbins, D. A. Portnoy, and D. J. Hinrichs. 1992. Listeriolysin O is a target of the immune response to *Listeria monocytogenes*. *J. Exp. Med.* **175:** 1467–1471.

10. Brundage, R. A., G. A. Smith, A. Camilli, J. A. Theriot, and D. A. Portnoy. 1993. Expression and phosphorylation of the *Listeria monocytogenes* ActA protein in mammalian cells. *Proc. Natl. Acad. Sci. USA* **90:**11890–11894.

11. Brunt, L. M., D. A. Portnoy, and E. R. Unanue. 1990. Presentation of *Listeria* by CD8$^+$ T cells requires secretion of hemolysin and intracellular growth. *J. Immunol.* **145:**3540–3546.

12. Cambell, P. A. 1986. Are inflammatory phagocytes responsible for resistance to facultative intracellular bacteria? *Immunol. Today* **7:**70–72.

13. Camilli, A., C. R. Paynton, and D. A. Portnoy. 1989. Intracellular methicillin selection of *Listeria monocytogenes* mutants unable to replicate in a macrophage cell line. *Proc. Natl. Acad. Sci. USA* **86:**5522–5526.

14. Camilli, A., D. A. Portnoy, and P. Youngman. 1990. Insertional mutagenesis of *Listeria monocytogenes* with a novel Tn*917* derivative that allows direct cloning of DNA flanking transposon insertions. *J. Bacteriol.* **172:**3738–3744.

15. Camilli, A., L. G. Tilney, and D. A. Portnoy. 1993. Dual roles of *plcA* in *Listeria monocytogenes* pathogenesis. *Mol. Microbiol.* **8:**143–157.

16. Chakraborty, T., M. Leimeister-Wachter, E. Domann, M. Hartl, W. Goebel, T. Nichterlein, and S. Notermans. 1992. Coordinate regulation of virulence genes in *Listeria monocytogenes* requires the product of the *prfA* gene. *J. Bacteriol.* **174:**568–574.

17. Cheers, C., and R. Waller. 1975. Activated macrophages in congenitally athymic "nude" mice and in lethally irradiated mice. *J. Immunol.* **115:**844–847.

18. Conlan, J. W., and R. J. North. 1991. Neutrophil-mediated dissolution of infected host cells as a defense strategy against a facultative intracellular bacterium. *J. Exp. Med.* **174:**741–744.

19. Conlan, J. W., and R. J. North. 1992. Early pathogenesis of infection in the liver with the facultative intracellular bacteria *Listeria monocytogenes*, *Francisella tularensis*, and *Salmonella typhimurium* involves lysis of infected hepatocytes by leukocytes. *Infect. Immun.* **60:**5164–5171.

20. Cossart, P., M. F. Vincente, J. Mengaud, F. Baquero, J. C. Perez-Diaz, and P. Berche. 1989. Listeriolysin O is essential for virulence of *Listeria monocytogenes*: direct evidence obtained by gene complementation. *Infect. Immun.* **57:**3629–3636.

21. Dabiri, G. A., J. M. Sanger, D. A. Portnoy, and F. S. Southwick. 1990. *Listeria monocytogenes* moves rapidly through the host cytoplasm by inducing directional actin assembly. *Proc. Natl. Acad. Sci. USA* **87:**6068–6072.

22. De Libero, G., and S. H. E. Kaufmann. 1986. Antigen-specific Lyt-2$^+$ lymphocytes from mice infected with the intracellular bacterium *Listeria monocytogenes*. *J. Immunol.* **137:**2688–2694.

23. d'Hauteville, H., and P. J. Sansonetti. 1992. Phosphorylation of IcsA by cAMP-dependent protein kinase and its effect on intracellular spread of *Shigella flexneri*. *Mol. Microbiol.* **6:**833–841.

24. Domann, E., J. Wehland, M. Rohde, S. Pistor, M. Hartl, W. Goebel, M. Leimeister-Wachter, M. Wuenscher, and T. Chakraborty. 1992. A novel bacterial virulence gene in *Listeria monocyto-*

genes required for host cell microfilament interaction with homology to the proline-rich region of vinculin. *EMBO J.* **11**:1981–1990.

25. **Drevets, D. A., and P. A. Campbell.** 1991. Roles of complement and complement receptor type 3 in phagocytosis of *Listeria monocytogenes* by inflammatory mouse peritoneal macrophages. *Infect. Immun.* **59**:2645–2652.

26. **Drevets, D. A., B. P. Canono, and P. A. Campbell.** 1992. Listericidal and nonlistericidal mouse macrophages differ in complement receptor type 3-mediated phagocytosis of *L. monocytogenes* and in preventing escape of the bacteria into the cytoplasm. *J. Leukocyte Biol.* **52**:70–79.

27. **Dunn, P. L., and R. J. North.** 1991. Early gamma interferon production by natural killer cells is important in defense against murine listeriosis. *Infect. Immun.* **59**:2892–2900.

28. **Emmerling, P., H. Finger, and H. Hof.** 1977. Cell-mediated resistance to infection with *Listeria monocytogenes* in nude mice. *Infect. Immun.* **15**:382–385.

29. **Farber, J. M., and P. I. Peterkin.** 1991. *Listeria monocytogenes,* a food-borne pathogen. *Microbiol. Rev.* **55**:476–511.

30. **Fischetti, V. A., V. Pancholi, and O. Schneewind.** 1990. Conservation of a hexapeptide sequence in the anchor region of surface proteins from gram-positive cocci. *Mol. Microbiol.* **4**:1603–1605.

31. **Freitag, N. E., L. Rong, and D. A. Portnoy.** 1993. Regulation of the *prfA* transcriptional activator of *Listeria monocytogenes:* multiple promoter elements contribute to intracellular growth and cell-to-cell spread. *Infect. Immun.* **61**:2537–2544.

32. **Freitag, N. E., P. Yongman, and D. A. Portnoy.** 1992. Transcriptional activation of the *Listeria monocytogenes* hemolysin gene in *Bacillus subtilis. J. Bacteriol.* **174**:1293–1298.

33. **Gaillard, J. L., P. Berche, C. Frehel, E. Gouin, and P. Cossart.** 1991. Entry of *L. monocytogenes* into cells is mediated by internalin, a repeat protein reminiscent of surface antigens from gram-positive cocci. *Cell* **65**:1127–1141.

34. **Gaillard, J. L., P. Berche, J. Mounier, S. Richard, and P. Sansonetti.** 1987. In vitro model of penetration and intracellular growth of *Listeria monocytogenes* in the human enterocyte-like cell line Caco-2. *Infect. Immun.* **55**:2822–2829.

35. **Gaillard, J. L., P. Berche, and P. Sansonetti.** 1986. Transposon mutagenesis as a tool to study the role of hemolysin in the virulence of *Listeria monocytogenes. Infect. Immun.* **52**:50–55.

36. **Gellin, B. G., and C. V. Broome.** 1989. Listeriosis. *JAMA* **261**:1313–1320.

37. **Geoffroy, C., J. L. Gaillard, J. E. Alouf, and P. Berche.** 1987. Purification, characterization, and toxicity of the sulfhydryl-activated hemolysin listeriolysin O from *Listeria monocytogenes. Infect. Immun.* **55**:1641–1646.

38. **Geoffroy, C., J. Raveneau, J. Beretti, A. Lecroisey, J.-A. Vazques-Boland, J. E. Alouf, and P. Berche.** 1991. Purification and characterization of an extracellular 29-kilodalton phospholipase C from *Listeria monocytogenes. Infect. Immun.* **59**:2382–2388.

39. **Goldberg, M. B., and P. J. Sansonetti.** 1993. *Shigella* subversion of the cellular cytoskeleton: a strategy for epithelial colonization. *Infect. Immun.* **61**:4941–4946.

40. **Goldfine, H., N. C. Johnston, and C. Knob.** 1993. The nonspecific phospholipase C of *Listeria monocytogenes:* activity on phospholipids in Triton X-100 mixed micelles and in biological membranes. *J. Bacteriol.* **175**:4298–4306.

41. **Havell, E. A.** 1986. Synthesis and secretion of interferon by murine fibroblasts in response to intracellular *Listeria monocytogenes. Infect. Immun.* **54**:787–792.

42. **Heinzen, R. A., S. F. Hayes, M. G. Peacock, and T. Hackstadt.** 1993. Directional actin polymerization associated with spotted fever group rickettsia infection of Vero cells. *Infect. Immun.* **61**: 1926–1935.

43. **Hsieh, C.-S., S. E. Macatonia, C. S. Tripp, S. F. Wolf, A. O'Garra, and K. M. Murphy.** 1993. Development of TH1 CD4[+] T cells through IL-12 produced by *Listeria*-induced macrophages. *Science* **260**:547–549.

44. **Huang, S., W. Hendriks, A. Althage, S. Hemmi, H. Bluethmann, R. Kamijo, J. Vilcek, R. M. Zinkernagel, and M. Aguet.** 1993. Immune responses in mice that lack the interferon-γ receptor. *Science* **259**:1742–1745.

45. **Jones, S., and D. A. Portnoy.** Unpublished data.

46. **Joris, B., S. Englebert, C.-P. Chu, R. Kariyama, L. Daneo-Moore, G. D. Shockman, and J.-M.**

Ghuysen. 1992. Modular design of the *Enterococcus hirae* muramidase-2 and *Streptococcus faecalis* autolysin. *FEMS Microbiol. Lett.* **91:**257–264.

47. **Kathariou, S., P. Metz, H. Hof, and W. Goebel.** 1987. Tn*916*-induced mutations in the hemolysin determinant affecting virulence of *Listeria monocytogenes. J. Bacteriol.* **169:**1291–1297.

48. **Kaufmann, S. H. E.** 1993. Immunity to intracellular bacteria. *Annu. Rev. Immunol.* **11:**129–163.

49. **Kocks, C., E. Gouin, M. Tabouret, P. Berche, H. Ohayon, and P. Cossart.** 1992. *L. monocytogenes*-induced actin assembly requires the *actA* gene product, a surface protein. *Cell* **68:**521–531.

50. **Kocks, C., R. Hellio, P. Gounon, H. Ohayon, and P. Cossart.** 1993. Polarized distribution of *Listeria monocytogenes* surface protein ActA at the site of directional actin assembly. *J. Cell Sci.* **105:**699–710.

51. **Kuhn, M., and W. Goebel.** 1989. Identification of an extracellular protein of *Listeria monocytogenes* possibly involved in intracellular uptake by mammalian cells. *Infect. Immun.* **57:**55–61.

52. **Kuhn, M., S. Kathariou, and W. Goebel.** 1988. Hemolysin supports survival but not entry of the intracellular bacterium *Listeria monocytogenes. Infect. Immun.* **56:**79–82.

53. **Kuhn, M., M. C. Prevost, J. Mounier, and P. J. Sansonetti.** 1990. A nonvirulent mutant of *Listeria monocytogenes* does not move intracellularly but still induces polymerization of actin. *Infect. Immun.* **58:**3477–3486.

54. **Lane, F. C., and E. R. Unanue.** 1972. Requirement of thymus (T) lymphocytes for resistance to listeriosis. *J. Exp. Med.* **135:**1104–1112.

55. **Leimeister-Wachter, M., E. Domann, and T. Chakraborty.** 1992. The expression of virulence in *Listeria monocytogenes* is thermoregulated. *J. Bacteriol.* **174:**947–952.

56. **Leimeister-Wachter, M., C. Haffner, E. Domann, W. Goebel, and T. Chakraborty.** 1990. Identification of a gene that positively regulates expression of listeriolysin, the major virulence factor of *Listeria monocytogenes. Proc. Natl. Acad. Sci. USA* **87:**8336–8340.

57. **MacDonald, T. T., and P. B. Carter.** 1980. Cell-mediated immunity to intestinal infection. *Infect. Immun.* **28:**516–523.

58. **Mackaness, G. B.** 1962. Cellular resistance to infection. *J. Exp. Med.* **116:**381–406.

59. **Mackaness, G. B.** 1964. The immunological basis of acquired cellular resistance. *J. Exp. Med.* **120:**105–120.

60. **Marquis, H., H. G. A. Bouwer, D. J. Hinrichs, and D. A. Portnoy.** 1993. Intracytoplasmic growth and virulence of *Listeria monocytogenes* auxotrophic mutants. *Infect. Immun.* **61:**3756–3760.

61. **Marquis, H., and D. A. Portnoy.** Unpublished data.

62. **Mengaud, J., S. Dramsi, E. Gouin, J. A. Vazquez-boland, G. Milon, and P. Cossart.** 1991. Pleiotropic control of *Listeria monocytogenes* virulence factors by a gene that is autoregulated. *Mol. Microbiol.* **5:**2273–2283.

63. **Mengaud, J., M. F. Vicente, and P. Cossart.** 1989. Transcriptional mapping and nucleotide sequence of the *Listeria monocytogenes hlyA* region reveal structural features that may be involved in regulation. *Infect. Immun.* **57:**3695–3701.

64. **Michel, E., and P. Cossart.** 1992. Physical map of the *Listeria monocytogenes* chromosome. *J. Bacteriol.* **174:**7098–7103.

65. **Michel, E., K. A. Reich, R. Favier, P. Berche, and P. Cossart.** 1990. Attenuated mutants of the intracellular bacterium *Listeria monocytogenes* obtained by single amino acid substitutions in listeriolysin O. *Mol. Microbiol.* **4:**2167–2178.

66. **Mielke, M. E. A., G. Niedobitek, H. Stein, and H. Hahn.** 1989. Acquired resistance to *Listeria monocytogenes* is mediated by Lyt-2[+] T cells independently of the influx of monocytes into granulomatous lesions. *J. Exp. Med.* **170:**589–594.

67. **Mombaerts, P., J. Arnoldi, F. Russ, S. Tonegawa, and S. H. E. Kaufmann.** 1993. Different roles of αβ and γδ T cells in immunity against an intracellular bacterial pathogen. *Nature* (London) **365:**53–56.

68. **Mounier, J., A. Ryter, M. Coquis-Rondon, and P. J. Sansonetti.** 1990. Intracellular and cell-to-cell spread of *Listeria monocytogenes* involves interaction with F-actin in the enterocytelike cell line Caco-2. *Infect. Immun.* **58:**1048–1058.

69. **Newborg, M. F., and R. J. North.** 1980. The mechanism of T cell independent anti-*Listeria* resistance in nude mice. *J. Immunol.* **124:**571–576.

70. **Nickol, A. D., and P. F. Bonventre.** 1977. Anomalous high native resistance of athymic mice to bacterial pathogens. *Infect. Immun.* **18:**636–645.

71. **Niebuhr, K., T. Chakraborty, M. Rohde, T. Gazlig, B. Jansen, P. Kollner, and J. Wehland.** 1993. Localization of the ActA polypeptide of *Listeria monocytogenes* in infected tissue culture cell lines: ActA is not associated with actin "comets." *Infect. Immun.* **61:**2793–2802.

72. **North, R. J.** 1970. The relative importance of blood monocytes and fixed macrophages to the expression of cell-mediated immunity to infection. *J. Exp. Med.* **132:**521–534.

73. **North, R. J.** 1973. Cellular mediators of anti-*Listeria* immunity as an enlarged population of short-lived replicating T cells. Kinetics of their production. *J. Exp. Med.* **138:**342–355.

74. **North, R. J.** 1975. Nature of "memory" in T-cell-mediated antibacterial immunity: anamnestic production of mediator T cells. *Infect. Immun.* **12:**754–760.

75. **North, R. J.** 1978. The concept of the activated macrophage. *J. Immunol.* **121:**806–809.

76. **Pal, T., J. W. Newland, B. D. Tall, S. B. Formal, and T. L. Hale.** 1989. Intracellular spread of *Shigella flexneri* associated with the *kcpA* locus and a 140-kilodalton protein. *Infect. Immun.* **57:** 477–486.

77. **Pamer, E. G.** 1993. Cellular immunity to intracellular bacteria. *Curr. Opin. Immunol.* **5:**492–496.

78. **Pamer, E. G., J. T. Harty, and M. J. Bevan.** 1991. Precise prediction of a dominant class I MHC-restricted epitope of *Listeria monocytogenes. Nature* (London) **353:**852–855.

79. **Park, S. F., and S. A. B. Stewart.** 1990. High-efficiency transformation of *Listeria monocytogenes* by electroporation of penicillin-treated cells. *Gene* **94:**129–132.

80. **Pfeffer, K., T. Matsuyama, T. M. Kundig, A. Wakeham, K. Kishihara, A. Shahinian, K. Wiegmann, P. S. Ohashi, M. Kronke, and T. W. Mak.** 1993. Mice deficient for the 55 kd tumor necrosis factor receptor are resistant to endotoxic shock yet succumb to L. monotogenes infection. *Cell* **73:**457–467.

81. **Pinkney, M., E. Beachey, and M. Kehoe.** 1989. The thiol-activated toxin streptolysin O does not require a thiol group for cytolytic activity. *Infect. Immun.* **57:**2553–2558.

82. **Portnoy, D. A.** 1992. Innate immunity to a facultative intracellular bacterial pathogen. *Curr. Opin. Immunol.* **4:**20–24.

83. **Portnoy, D. A., T. Chakraborty, W. Goebel, and P. Cossart.** 1992. Molecular determinants of *Listeria monocytogenes* pathogenesis. *Infect. Immun.* **60:**1263–1267.

84. **Portnoy, D. A., P. S. Jacks, and D. J. Hinrichs.** 1988. Role of hemolysin for the intracellular growth of *Listeria monocytogenes. J. Exp. Med.* **167:**1459–1471.

85. **Portnoy, D. A., R. D. Schreiber, P. Connelly, and L. G. Tilney.** 1989. Gamma interferon limits access of *Listeria monocytogenes* to the macrophage cytoplasm. *J. Exp. Med.* **170:**2141–2146.

86. **Portnoy, D. A., R. K. Tweten, M. Kehoe, and J. Bielecki.** 1992. Capacity of listeriolysin O, streptolysin O, and perfringolysin O to mediate growth of *Bacillus subtilis* within mammalian cells. *Infect. Immun.* **60:**2710–2717.

87. **Poyart, C., E. Abachin, I. Razafimanantsoa, and P. Berche.** 1993. The zinc metalloprotease of *Listeria monocytogenes* is required for maturation of phosphatidylcholine phospholipase C: direct evidence obtained by gene complementation. *Infect. Immun.* **61:**1576–1580.

88. **Racz, P., K. Tenner, and E. Mero.** 1972. Experimental *Listeria enteritis.* I. An electron microscopic study of the epithelial phase in experimental Listeria infection. *Lab. Invest.* **26:**694–700.

89. **Roberts, A. D., D. J. Ordway, and I. M. Orme.** 1993. *Listeria monocytogenes* infection in β2 microglobulin-deficient mice. *Infect. Immun.* **61:**1113–1116.

90. **Rogers, H. W., and E. R. Unanue.** 1993. Neutrophils are involved in acute, nonspecific resistance to *Listeria monocytogenes* in mice. *Infect. Immun.* **61:**5090–5096.

91. **Rosen, H., S. Gordon, and R. J. North.** 1989. Exacerbation of murine listeriosis by a monoclonal antibody specific for the type 3 complement receptor of myelomonocytic cells. Absence of monocytes at infective foci allows *Listeria* to multiply in nonphagocytic cells. *J. Exp. Med.* **170:**27–37.

92. **Rothe, J., W. Lesslauer, H. Lotscher, Y. Lang, P. Koebel, F. Kontgen, A. Althage, R. Zinkernagel, M. Steinmetz, and H. Bluethmann.** 1993. Mice lacking the tumour necrosis factor receptor 1 are resistant to TNF-mediated toxicity but highly susceptible to infection by *Listeria monocytogenes. Nature* (London) **364:**798–802.

93. **Sanger, J. M., J. W. Sanger, and F. S. Southwick.** 1992. Host cell actin assembly is necessary

and likely to provide the propulsive force for intracellular movement of *Listeria monocytogenes*. *Infect. Immun.* **60:**3609–3619.

94. **Smith, G. A., and D. A. Portnoy.** Unpublished data.

95. **Smyth, C. J., and J. L. Duncan.** 1978. Thiol-activated (oxygen-labile) cytolysins. *In* J. Jeljaszewicz and T. Wasstrom (ed.), *Bacterial Toxins and Cell Membranes*. Academic Press, Inc., New York.

96. **Stevens, D. L., J. Mitten, and C. Henry.** 1987. Effects of α and θ toxins from *Clostridium perfringens* on human polymorphonuclear leukocytes. *J. Infect. Dis.* **156:**324–333.

97. **Sun, A. N., A. Camilli, and D. A. Portnoy.** 1990. Isolation of *Listeria monocytogenes* small-plaque mutants defective for intracellular growth and cell-to-cell spread. *Infect. Immun.* **58:**3770–3778.

98. **Tanaka, M., and H. Shibata.** 1985. Poly (L-proline)-binding proteins from chick embryos are profilin and profilactin. *Eur. J. Biochem.* **151:**291–297.

99. **Teysseire, N., C. Chiche-Portiche, and D. Raoult.** 1992. Intracellular movements of *Rickettsia conorii* and *R. typhi* based on actin polymerization. *Res. Microbiol.* **143:**821–829.

100. **Theriot, J. A., and T. J. Mitchison.** 1993. The three faces of profilin. *Cell* **75:**835–838.

101. **Theriot, J. A., T. J. Mitchison, L. G. Tilney, and D. A. Portnoy.** 1992. The rate of actin-based motility of intracellular *Listeria monocytogenes* equals the rate of actin polymerization. *Nature* (London) **357:**257–260.

102. **Theriot, J. A., J. Rosenblatt, D. A. Portnoy, P. J. Goldschmidt-Clermont, and T. J. Mitchison.** 1994. Involvement of profilin in the actin-based motility of Listeria monocytogenes in cells and in cell-free extracts. *Cell* **74:**505–517.

103. **Tilney, L. G., P. S. Connelly, and D. A. Portnoy.** 1990. The nucleation of actin filaments by the bacterial intracellular pathogen, *Listeria monocytogenes*. *J. Cell Biol.* **111:**2979–2988.

104. **Tilney, L. G., D. J. DeRosier, A. Weber, and M. S. Tilney.** 1992. How *Listeria* exploits host cell actin to form its own cytoskeleton. II. Nucleation, actin filament polarity, filament assembly, and evidence for a pointed end capper. *J. Cell Biol.* **118:**83–93.

105. **Tilney, L. G., and D. A. Portnoy.** 1989. Actin filaments and the growth, movement, and spread of the intracellular bacterial parasite, *Listeria monocytogenes*. *J. Cell Biol.* **109:**1597–1608.

106. **Titball, R. W.** 1993. Bacterial phospholipases C. *Microbiol. Rev.* **57:**347–366.

107. **Van Epps, D. E., and B. R. Andersen.** 1974. Streptolysin O inhibition of neutrophil chemotaxis and mobility: nonimmune phenomenon with species specificity. *Infect. Immun.* **9:**27–33.

108. **Vazquez-Boland, J.-A., C. Kocks, S. Dramsi, H. Ohayon, C. Geoffrey, J. Mengaud, and P. Cossart.** 1992. Nucleotide sequence of the lecithinase operon of *Listeria monocytogenes* and possible role of lecithinase in cell-to-cell spread. *Infect. Immun.* **60:**219–230.

109. **von Koenig, C. H. W., H. Finger, and H. Hof.** 1982. Failure of killed *Listeria monocytogenes* vaccine to produce protective immunity. *Nature* (London) **297:**233–234.

110. **Wuenscher, M. D., S. Kohler, W. Goebel, and T. Chakraborty.** 1991. Gene disruption by plasmid integration in *Listeria monocytogenes:* insertional inactivation of the listeriolysin determinant *lisA*. *Mol. Gen. Genet.* **228:**177–182.

111. **Yewdell, J. W., and J. R. Bennink.** 1990. The binary logic of antigen processing and presentation to T cells. *Cell* **62:**203–206.

112. **Zachar, Z., and D. C. Savage.** 1979. Microbial interference and colonization of the murine gastrointestinal tract by *Listeria monocytogenes*. *Infect. Immun.* **23:**168–174.

Molecular Genetics of Bacterial Pathogenesis
Edited by V. L. Miller, J. B. Kaper, D. A. Portnoy, and R. R. Isberg
© 1994 American Society for Microbiology, Washington, DC 20005

Chapter 19

Determinants of Chlamydial Pathogenesis and Immunity

Patrik M. Bavoil

Incense-ground-in-Lizard's-dung, cow's Blood, ass's Blood, pig's Blood, dog's Blood, stag's Blood, collyrium, incense.
Crush, rub into one in the different kinds of Blood, and place on the part whence the Hair has been pulled so that it may not grow again.
— Ancient Egyptian remedy for trachoma, excerpt from *The Papyrus Ebers*

The spectrum of chlamydial disease in humans includes ocular, respiratory tract, and urogenital tract infections. It has been estimated that 500 million people worldwide are affected by the chronic ocular disease trachoma, often referred to as the leading cause of preventable blindness (11). In the United States alone, it is estimated that more than 4 million individuals contract chlamydial sexually transmitted diseases (STDs) annually (12), and while no estimates are available worldwide, a conservative guess is that several hundred millions of people are infected. Thus, the magnitude of the *Chlamydia* problem in any given year is close to an astounding 1 billion or so infected people worldwide (24). Moreover, *Chlamydia* infection is not new; as illustrated by the quotation given above, chlamydial diseases are at least as old as papyrus.

There are several possible causes for both the resilience and magnitude of chlamydial disease. One major aspect of disease is that the primary infection is often asymptomatic or a relatively minor irritant in men and women. Sequelae, such as blinding trachoma, tubal infertility, ectopic pregnancy, heart disease, or reactive arthritis, are usually observed long after infectious chlamydiae have all but vanished from the diseased host. Indeed, trachoma is a lifelong chronic disease, the end point of which, blindness, usually affects older individuals past their "productive" years. Moreover, trachoma occurs in developing countries that often have far more serious problems to cope with. Chlamydial STDs, while clinically resembling gonococcal STDs, are frequently milder and, for a variety of reasons, may go either undetected, unreported, or untreated. Thus, in either disease setting, while the primary infection is theoretically readily treatable, appro-

Patrik M. Bavoil • Department of Microbiology and Immunology, University of Rochester Medical Center, Box 672, Rochester, New York 14642.

priate therapeutic measures often are not applied, allowing chronic disease and associated immunopathologies to develop. A direct result of this neglect of the primary infection is that the chlamydial infectious potential, i.e., the global reservoir, is never affected significantly.

A second factor in the elusiveness of chlamydial disease lies in the biology of the organism itself. *Chlamydia*'s preferred niche is the intracellular compartments of the mucosal epithelial cells of the host. There the metabolically active form of the organism, the reticulate body (RB), can grow sheltered from the various effector cells and products of the immune system. About 40 h after the initial infection, a differentiated form of the organism, the elementary body (EB), is released in large numbers. While metabolically dormant, the EB is specialized for survival outside of the host cell and for infection of new cells, thus spreading the infection within the host and to new hosts. Infection of new cells initiates with the endocytic uptake of the infectious EB particle, which in turn differentiates into the growing RB, thereby closing the cycle (Fig. 1).

Overall, the combination of psycho-socio-economic factors surrounding human management of chlamydial diseases and the stealth qualities of the chlamydial life cycle have made *Chlamydia* species one of the most successful microbial pathogens of humans, uniquely both on human and biological scales.

A last, contemporary factor in the success of *Chlamydia* species at human expense is investigators' continuing inability to carry out genetic analysis of this organism. Thus, while many genes have now been cloned and characterized, it has not been possible to return mutant genes to *Chlamydia* organisms, therefore

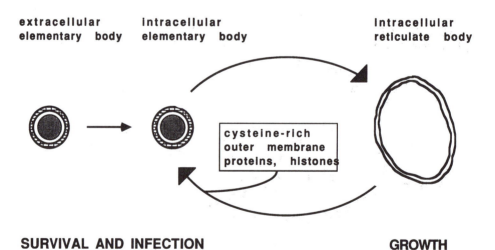

**extracellular   Intracellular   Intracellular
elementary body   elementary body   reticulate body**

cysteine-rich
outer membrane
proteins, histones

SURVIVAL AND INFECTION   **GROWTH**

FIGURE 1. A reductionist view of the chlamydial life cycle. At time zero, the infectious EB is internalized and subsequently differentiates into the metabolically active RB. At time 24 to 40+ h, RBs differentiate back to EBs. Two classes of proteins are central to the latter differentiation event: (i) cysteine-rich proteins, including the constitutively expressed MOMI and the late-expressed Omp-2 and Omp-3 proteins, which form disulfide bonds rendering the outer cell wall rigid and impermeable (8), and (ii) histone-like proteins which likely mediate chromosomal condensation (5).

preventing the fulfillment of the molecular Koch's postulates for chlamydial genes. Functional characterization of cloned chlamydial genes is usually performed the poor man's way, i.e., by analysis of sequence homology with genes from other, better-characterized systems.

The review in this chapter focuses on the molecular determinants of the initial step of chlamydial pathogenesis, adherence, and on the antigens responsible for immunopathogenesis and immunity to chlamydial infection. Each of these currently intense areas of *Chlamydia* research has reached a threshold of information and polemic, to the extent that the advent of chlamydial genetics would likely have a major beneficial impact. Since the following text reflects personal views and interests, the reader is also referred to several excellent reviews which address other aspects of *Chlamydia* research as well as other views (27, 30, 31, 50, 54, 64).

DETERMINANTS OF CHLAMYDIAL ADHERENCE TO EUKARYOTIC CELLS

Every chlamydial infection starts with the physical interaction of an infectious EB with a susceptible mucosal epithelial cell. The endocytic events which follow the initial interaction will not be considered here. A consensus based on indirect evidence is that chlamydial adherence to susceptible eukaryotic cells is a two-step mechanism involving a nonspecific electrostatic interaction followed by at least one, but probably several, high-affinity, specific interactions. The former is likely contributed to a degree by all charged, surface-exposed components including outer membrane proteins, lipopolysaccharide, and a recently described heparan sulfate-like glycosaminoglycan (GAG) (see below). The latter involves one or several surface-exposed chlamydial adhesins paired with one or several eukaryotic receptors. Wyrick and collaborators (19, 20) have provided the most compelling evidence for a receptor-mediated endocytic mechanism by demonstrating the presence of infecting EBs in clathrin-coated pits (Fig. 2) and clathrin-coated vesicles. In contrast to the clarity of this basic observation, the identification of the chlamydial determinant(s) of the initial interaction, the adhesins, has been far more problematic. A search through the scientific literature provides the reader with a wide selection of candidate adhesins or adherence factors with various degrees of underlying experimental foundation. While this plethora of candidate adhesins may actually reflect multiple mechanisms of adherence and coupled entry, as has been proposed previously (49), the absence of isogenic mutant pairs in each separate pathway makes direct verification of this proposition difficult. Moreover, a basic question which should precede the identification of a putative adhesin, the question of surface accessibility, is often unresolved or open to doubts. In this light, the following paragraphs are a brief review of the most recently proposed chlamydial adhesins. These include the chlamydial major outer membrane protein (MOMP) (60, 61), the chlamydial cytadhesin (CCA) (22), a heparan sulfate-like GAG (68), Hsp-70 or a Hsp-70–genetically linked product (48, 52), and the cysteine-rich outer membrane protein Omp-2.

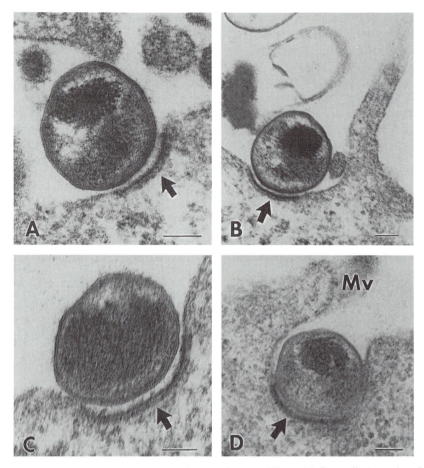

FIGURE 2. Receptor-mediated attachment of *C. trachomatis* EBs to McCoy cells (reproduced with permission from P. B. Wyrick and reference 19). Electron micrographs show EBs tightly attached at the base of microvilli (Mv) and in clathrin-coated pits (arrows). Bars, 0.1 μm.

MOMP

MOMP is the most abundant chlamydial protein and the major surface constituent of the infectious EB, representing approximately 60% of the total outer membrane proteins (10). A role for MOMP in adherence has been proposed on the basis of the inhibitory capacities of trypsin (61) and specific monoclonal antibodies (60). The latter studies by Su et al. (60) indicate that both surface-exposed antigenically variable domains (VDs) VD2 and VD4 of *Chlamydia trachomatis* MOMP may be involved in adherence. Most compelling was the finding that loss of EB adherence upon mild heat treatment correlated with a conformational change in a hydrophobic nonapeptide of VD4.

While it is almost predictable that MOMP will, by virtue of its abundance, at least indirectly affect adherence at some level, and since there is no doubt about

MOMP's surface accessibility, the key question left to ponder is whether MOMP actually functions as a high-affinity specific ligand, i.e., as the chlamydial adhesin. This question cannot, unfortunately, be answered by use of currently available methods since both inhibitory protease and antibody treatments may indirectly affect other surface components which themselves possess adhesive properties.

CAA

Joseph and Bose (22) have proposed that the 38-kDa CCA is a chlamydial adhesin on the basis of the fact that n-octyl β-D-glucopyranoside (OGP)-extracted CCA can selectively bind to glutaraldehyde-fixed HeLa cells. Furthermore, CCA-containing OGP extracts inhibited EB adherence, and mild heat treatment of the extracts abolished CCA binding.

Two separate observations were taken as suggestive of surface exposure. (i) CCA can be iodinated in situ by using Iodo-Beads, and (ii) trypsin treatment of EBs results in the complete loss of the CCA band. Both experiments may be prone to artifactual results, particularly those done with purified EB preparations, which often contain disrupted organisms. Nevertheless, the most important question concerning CCA that must be resolved is its identity. In view of the closeness of their molecular weights, it will be important to determine whether CCA and MOMP are structurally related.

GAG

Zhang and Stephens (68) have recently proposed a novel mechanism for adherence. The model proposes that a heparan sulfate-like chlamydial GAG is involved in a trimolecular mechanism of attachment of EBs to eukaryotic cells. This is essentially based on two observations: (i) while high concentrations of heparin or heparan sulfate inhibit EB adherence, low concentrations are stimulatory, and (ii) EB adherence is susceptible to heparitinase and the adherence of heparitinase-treated EBs can be restored by the addition of exogenous heparan sulfate. Furthermore, mild heat treatment provoked the release of exogenously added GAG in the reconstituted system, paralleling the effect of this treatment on EB adherence.

It can probably be safely assumed that GAG is a surface component since it is accessible to heparitinase (although GAG lyases are notoriously contaminated by proteases, which could affect surface accessibility). Nevertheless, while the data of Zhang and Stephens (68) leave little doubt of the existence of such a mechanism of adherence of EBs toward HeLa cells, important questions remain. Elucidation of the biochemical and genetic bases of the proposed mechanism may eventually provide the tools necessary to evaluate the in vivo significance of GAG-mediated infectivity.

Hsp-70 or a Genetically Linked Product

Using a recombinant approach, Schmiel et al. have isolated a 6.7-kb segment of chlamydial DNA which confers adherence properties upon host *Escherichia coli* bacteria (52). Three polypeptides with molecular masses of 18, 28, and 82 kDa were expressed from this DNA. It was later shown that the 82-kDa species corresponds to the chlamydial *dnaK* gene product, the Hsp-70 protein homolog of *Chlamydia* species (48). Other genetically linked products which may also play a role in the observed phenotype have not been described yet. Surprisingly, immunogold electron microscopy revealed that Hsp-70 is associated with outer membrane fractions, albeit mostly on the inner leaflet of the outer membrane (48). An earlier observation by Danilition et al. (14) also suggested a surface association of Hsp-70 on the basis of its immunoaccessibility and the neutralizing activities of anti-Hsp-70 antibodies.

Cumulatively, these observations pose a serious paradox, since it is well established that, lacking a leader sequence, Hsp-70 is "normally" a nonsecreted protein and is universally located within the prokaryotic or eukaryotic cytoplasmic compartment where it carries out its protein chaperoning function (25). Thus, the presence of Hsp-70 at the EB surface and a role for this protein in adherence should not be considered without extreme caution. Notwithstanding these reservations, the possibility remains that some normally unexposed proteins gain access to the EB surface by nonclassical secretory pathways. This may occur upon lysis of the osmotically fragile RB and of intermediate forms caused by hypo-osmotic shock upon rupture of the chlamydial inclusion membrane. If the released proteins are relatively resistant to host proteases and if, in addition, they possess some nonspecific affinity for the EB surface, they may then concentrate at the EB surface, thus providing an additional potential ligand for eukaryotic receptors. The heat shock proteins and potentially other normally nonsecreted chlamydial proteins may satisfy some of these criteria and, hence, should not be dismissed outright as effectors, direct or indirect, of the initial interaction.

Omp-2 Revisited

Studies by Batteiger et al. (6) revealed that the pIs of the cysteine-rich Omp-2 and Omp-3 proteins were significantly more basic in strains of the more virulent lymphogranuloma venereum biovar than in strains of the less virulent trachoma biovar of *C. trachomatis*. To my knowledge, this is still the only known significant structural property which differentiates strains of the two biovars. Since the 60-kDa Omp-2 and the 12-kDa Omp-3 proteins are present only in the infectious EB (17, 18, 35), possess a leader sequence (1, 3), and copurify with outer membrane fractions (6, 18), it was suggested that one or both of these proteins may play a role pertinent to chlamydial pathogenesis such as at the attachment stage (6). This hypothesis lost momentum when Omp-3- and Omp-2-specific monoclonal antibodies and a monospecific antiserum prepared against high-pressure liquid chromatography-purified Omp-2 protein failed to surface label EBs (13). Later, Everett and Hatch (15) suggested that Omp-2 is an inner membrane protein exposed into the periplasm on the basis of primary structure considerations. Notwithstanding these negative results and adverse structural considerations, no direct evidence exists of Omp-2's exposure, or lack thereof, at the EB surface. In fact, surface iodination by using lactoperoxidase was successful in labeling an EB polypeptide in the same molecular size range as Omp-2 (10). In my laboratory, a recombinant Omp-2-specific antiserum raised in a rabbit was shown to reproducibly surface label whole EBs in suspension at as low as a 1:10,000 dilution (unpublished data). In contrast, neither the same antiserum depleted of Omp-2-specific antibodies nor the preimmune serum surface labeled whole EBs. Since EB preparations contain a significant proportion of disrupted organisms and ghost-like particles, apparent, artifactual surface labeling may actually be observed. This can be resolved only by using immunogold electron microscopy, which has not yet been attempted with this serum. By using an assay system similar to that of Joseph and Bose (22), both MOMP and Omp-2 were shown to bind to glutaraldehyde-fixed HeLa cells (63). Inhibitory mild heat treatment abolished Omp-2 binding but not MOMP binding in this assay. Thus, the possibility that Omp-2 is surface exposed and functions in EB adherence to susceptible cells still exists, and renewed consideration of the earlier suggestion by Batteiger et al. (6) is warranted.

Regardless of the identity of the adhesin(s), none of the mechanisms so far proposed provides a satisfactory explanation for the observed differences in tissue tropism and virulence between strains and serovars. An attractive possibility is that an equilibrium between various mechanisms of adherence and entry exists and that it varies within the genus and species. Thus, in a given strain, dominance of one pathway over others may translate into preferential affinity toward receptors which are specific to a particular anatomical site. Alternatively, the step of adherence may be "blind" with respect to infection, and tissue selectivity may be imposed at a later step of the pathogenic process.

IMMUNITY, IMMUNOPATHOLOGY, AND THE CONTINUING SEARCH FOR A *CHLAMYDIA* VACCINE

A practical benefit of the identification of chlamydial adhesins will be the provision of appropriate target molecules for the development of a chlamydial subunit vaccine. Although great advances have been made in understanding the chlamydial antigenic structure and the immunology of chlamydial infection, a vaccine is not yet available, and one can still be only cautiously optimistic as to its eventual existence. Three reasons to be optimistic are that (i) there exists information to suggest that natural immunity to reinfection does occur (4, 23), (ii) a deleterious antigen thought to elicit a hallmark hypersensitivity reaction has been identified as the chlamydial Hsp-60 homolog (28, 29), and (iii) a demonstrated protective antigen, MOMP, is at hand (55, 56, 67). Three equally tangible reasons to be pessimistic are that (i) natural immunity is of short duration, (ii) there may be more, or less, to hypersensitivity than Hsp-60, and (iii) various synthetic or recombinant MOMP-based immunization regimens have so far failed to generate significant protection against either infection or pathology.

Protective Immunity

Humans mount strong humoral and cell-mediated immune responses upon chlamydial infection (24). The role of each arm of the immune response in resolving the infection is not directly demonstrated in humans but can be inferred from various animal models. Animal models can be grouped into three general categories: the primate models, various murine models, and a guinea pig model. Each presents advantages and inconveniences.

The animal models most closely related to humans involve nonhuman primates and the human pathogen *C. trachomatis*. Unfortunately, primates are prohibitively expensive, and results obtained with these systems are usually statistically unreliable. Various murine models also involve *C. trachomatis,* either the human biovars or the lone representative of the mouse biovar, strain MoPn. While the murine systems offer great flexibility for the detailed analysis of immunological parameters during infection and recovery from infection, they are lacking in that of all models, they least resemble human disease in many biological and immunological aspects. In particular, immune protection in mice appears to be dependent primarily on cell-mediated immunity (38, 39, 46), a clear distinction from the primate and guinea pig models (see below), in which serum and secretory antibodies may play a major role. A third model, which my colleagues and I use in our laboratory in collaboration with the laboratory of Roger Rank at the University of Arkansas for Medical Sciences, involves the guinea pig and *Chlamydia psittaci* strain guinea pig inclusion conjunctivitis (GPIC), a natural guinea pig pathogen. While this model does not involve a chlamydial pathogen of humans, it perhaps provides the best small animal model for chlamydial disease. Among parallels between the two systems are that (i) GPIC causes naturally transmissible ocular and genital infections (32, 34), (ii) a genital infection of the female guinea pig may ascend to the upper genital tract, where it may cause salpingitis and salpingitis-associated sequelae (tubal blockage and sterility) (45), (iii) both the humoral and cell-mediated arms of the immune response are required for resolution of the infection (40, 47), (iv) naturally acquired protective immunity postinfection is temporally limited (42), (v) sensitized animals display the hallmark hypersensitivity reaction thought to be mediated by the chlamydial Hsp-60 homolog (28, 29, 65), and (vi) immunization with preparations enriched for GPIC-MOMP yields partial immune protection (7).

In 1987, Zhang et al. (69) made the important observation that monoclonal antibodies specific for certain MOMP epitopes protected mice from toxic death after intravenous injection of *C. trachomatis* EBs and, likewise, neutralized EB infectivity in a monkey ocular model. Further investigation by the same group led to the identification in MOMPs of several linear epitopes, antibodies to which were capable of neutralizing infectivity toward cultured HaK cells (60). In other experiments, T-helper epitopes were characterized on MOMPs as well (2, 59). Immunization of mice or nonhuman primates with chimeric synthetic peptides which include both the relevant B- and T-cell epitopes elicited high levels of antibodies in serum (57, 58). However, while these antibodies were neutralizing in vitro, immunized animals failed to mount a significant immune protection at mucosal surfaces (57, 58). This apparent paradox is not explained by the presence of the mucosal epithelial barrier since, at least in the genital tract, serum immunoglobulin G (IgG) is able to transudate across the mucosa, and high levels of serum IgG are found in mucosal secretions (26). That serum IgG plays an essential role in immune clearance in the genital tract is also confirmed in the guinea pig model

(41). In this system, parenteral immunization with live EBs, UV-inactivated EBs, or MOMP-enriched extracts all elicited protective immunity, albeit to various degrees, in the female genital tract (7, 43).

A plausible explanation for the inability of synthetic peptides to elicit a protective immune response is that while high levels of MOMP-specific antibodies are generated, these antibodies target epitopes which either are irrelevant to neutralization or are not immunoaccessible. Support for this hypothesis comes from the work of Batteiger et al. (7), who showed that when an OGP extract highly enriched for MOMP was used to immunize guinea pigs, it elicited MOMP-specific serum antibodies capable of binding to whole EBs by enzyme-linked immunosorbent assay (ELISA) and significant immune protection was achieved. In contrast, when sodium dodecyl sulfate-purified MOMP (7) or a number of recombinant forms of MOMP, denatured or misfolded (9), were used, antibodies to MOMP were also found in high levels, but these did not bind to whole EBs by ELISA, nor was any measure of protection observed upon genital challenge. Furthermore, when sera from the protected and nonprotected animals were compared by immunoblot analysis using LamB-MOMP fusions, no significant difference in fine specificity was observed, in particular with respect to linear peptides contained within the MOMP VDs (7). The inference from these results is that, where MOMP antibodies are concerned, surface-exposed discontinuous (conformational) epitopes, but not linear epitopes, are essential for recognition by antibodies which may neutralize infectivity. Consistent with this hypothesis is the observation that a *C. trachomatis* MOMP VD1-derived peptide is more immunogenic when conformationally constrained, i.e., when integrated within a loop of the poliovirus capsid protein VP1, than as a synthetic peptide (33). Taken together, these results suggest that, to be effective as a subunit vaccine, MOMP must be at least partially properly folded, a task that has proven insurmountable so far with both synthetic peptides and recombinant products.

Hypersensitivity

Early human vaccine trials with whole inactivated or attenuated organisms were sometimes disastrous in that, instead of eliciting protection in immunized patients, they actually primed for a more intense inflammatory response than in control nonimmunized individuals (for a review, see reference 51). In 1986, Watkins et al. (65) demonstrated that a chlamydial Triton X-100 extract could elicit an ocular delayed-type hypersensitivity (DTH) reaction in sensitized animals. Morrison et al. (28, 29) later showed that the agent responsible for the DTH reaction was a protein of 57 kDa related to the Hsp-60 family of stress-induced antigens. Hypersensitivity was demonstrated by using immunoaffinity-purified chlamydial Hsp-60 as well as recombinant Hsp-60. While the results of those experiments remain provocative, questions linger about the actual role of chlamydial Hsp-60 in the observed hypersensitivity reaction. First, it is not clear that Hsp-60 is solely responsible for eliciting DTH; i.e., other deleterious chlamydial antigens may exist. Second, the original experiments performed by Watkins et al. (65) and Morrison et al. (28, 29) included Triton X-100 in the Hsp-60 preparations.

While the presence of the detergent may substitute for the invasive capability of the organism, it remains that the DTH reaction was not observed without it. Third, it is unclear what role other contaminants of the preparations may have played in the observed immunopathologies; in this regard, it is worrisome that the recombinant Hsp-60 used in the experiments of Morrison et al. (28, 29) appeared to be contaminated with significant quantities of endotoxin (66). Lastly, in reverse experiments in which guinea pigs were primed with sucrose gradient-purified Hsp-60 and challenged in the eye with live GPIC organisms, enhanced immunopathology was not observed (44). Furthermore, animals challenged with attenuated, but invasive, *Salmonella typhimurium aroA* expressing recombinant Hsp-60 did not exhibit enhanced immunopathology over controls. Whatever the role of Hsp-60 in eliciting a deleterious response, it needs to be confirmed by using more highly purified reagents. Moreover, and in light of the uncertainty, the search for other deleterious antigens should be resumed.

CONCLUDING REMARKS

The field of *Chlamydia* research has moved forward considerably in the last decade. Outstanding landmarks can retrospectively be noted: the molecular characterization of the chlamydial outer membrane (10, 17, 36), the molecular cloning of MOMP (55), and more recently, the discovery of the chlamydial histones (16). Yet, in great part because of the paucity of molecular tools for studying this organism, much remains to be learned. A next landmark of *Chlamydia* research should be the much awaited development of genetics for this organism. This would enable mutant analysis, which would potentially constitute a major advance in the study of chlamydial adherence. There, the existence of multiple mechanisms of adherence may be an asset to chlamydiologists since knockout mutations in any one pathway would not be "lethal," i.e., would not prevent culture of the mutant. Furthermore, the construction of mutants altered in their adherence or deleterious antigenic properties may lead to the generation of *Chlamydia* strains that are attenuated in vivo. Together with improved methods of growth now available (62), these may ultimately hold the key to the development of effective *Chlamydia* vaccines.

The development of methods of genetic transfer for *Chlamydia* species presents unique challenges. One is that the most reasonable target for transformation, the RB, is noninfectious; therefore, it must be transformed while inside the chlamydial inclusion, which itself resides inside the host cell. In other words, transforming DNA would have to go through four separate membranes: the plasma membrane of the host cell, the inclusion membrane, and the two membranes of the RB envelope. Transformation of the EB may, alternatively, be possible. However, rendering these small, spore-like particles competent for DNA uptake may also constitute a formidable task. A common theme has emerged from unpublished accounts of transformation attempts from various laboratories. DNA uptake appears to be successful to the extent that reporter gene expression from a chlamydial promoter usually inserted within the chlamydial cryptic plasmid (37) can be detected. However, expression is characteristically low and/or transient. Furthermore, sufficient replication of the putatively transformed DNA does not take place because the plasmid is rapidly diluted or lost upon serial passage. Notwithstanding the apparent lack of progress, one should be heartened that other "difficult" intracellular parasites have recently been transformed (21, 53). *Chlamydia* genetics should be attainable. They transform mitochondria, don't they?

ACKNOWLEDGMENTS. I am grateful to all of my collaborators and students for their hard work. Special thanks go to Ru-ching Hsia, Yvonne Pannekoek, Byron E. Batteiger, and Roger G. Rank for help in shaping the review in this chapter and to Ru-ching Hsia and Priscilla Wyrick for providing illustrations.

The author is a recipient of Public Health Service Research Career Development Award AI01057-01.

REFERENCES

1. **Allen, J. E., M. C. Cerrone, P. R. Beatty, and R. S. Stephens.** 1990. Cysteine-rich outer membrane proteins of *Chlamydia trachomatis* display compensatory sequence changes between biovariants. *Mol. Microbiol.* **4:**1543–1550.

2. **Allen, J. E., R. M. Locksley, and R. S. Stephens.** 1991. A single peptide from the major outer membrane protein of *Chlamydia trachomatis* elicits T cell help for the production of antibodies to protective determinants. *J. Immunol.* **147:**674–679.

3. **Allen, J. E., and R. S. Stephens.** 1989. Identification by sequence analysis of two-site posttranslational processing of the cysteine-rich outer membrane protein 2 of *Chlamydia trachomatis* serovar L2. *J. Bacteriol.* **171:**285–291.

4. **Barnes, R. C., B. P. Katz, R. T. Rolfs, B. E. Batteiger, V. Caine, and R. B. Jones.** 1990. Quantitative culture of endocervical *Chlamydia trachomatis. J. Clin. Microbiol.* **28:**774–780.

5. **Barry, C. E., III, S. F. Hayes, and T. Hackstadt.** 1992. Nucleoid condensation in *Escherichia coli* that express a chlamydial histone homolog. *Science* **256:**377–378.

6. **Batteiger, B. E., W. J. Newhall, and R. B. Jones.** 1985. Differences in outer membrane proteins of the lymphogranuloma venereum and trachoma biovars of *Chlamydia trachomatis. Infect. Immun.* **50:**488–494.

7. **Batteiger, B. E., R. G. Rank, P. M. Bavoil, and L. S. F. Soderberg.** 1993. Partial protection against genital reinfection by immunization of guinea pigs with isolated outer membrane proteins of the chlamydial agent of guinea pig inclusion conjunctivitis. *J. Gen. Microbiol.* **139:**2965–2972.

8. **Bavoil, P. M., A. Ohlin, and J. Schachter.** 1984. Role of disulfide bonding in outer membrane structure and permeability in *Chlamydia trachomatis. Infect. Immun.* **44:**479–485.

9. **Bavoil, P. M., and R. G. Rank.** Unpublished data.

10. **Caldwell, H. D., J. Kromhout, and J. Schachter.** 1981. Purification and partial characterization of the major outer membrane protein of *Chlamydia trachomatis. Infect. Immun.* **31:**1161–1176.

11. **Centers for Disease Control.** 1982. Prevention of blindness: trachoma control. *Morbid. Mortal. Weekly Rep.* **31:**561–562.

12. **Centers for Disease Control and Prevention.** 1993. Recommendations for the prevention and management of *Chlamydia trachomatis* infections. *Morbid. Mortal. Weekly Rep.* **42**(RR-12):1–39.

13. **Collett, B. A., W. J. Newhall, R. A. Jersild, Jr., and R. B. Jones.** 1989. Detection of surface-exposed epitopes on *Chlamydia trachomatis* by immune electron microscopy. *J. Gen. Microbiol.* **135:**85–94.

14. **Danilition, S. L., I. W. Maclean, R. Peeling, S. Winston, and R. C. Brunham.** 1990. The 75-kilodalton protein of *Chlamydia trachomatis:* a member of the heat shock protein 70 family. *Infect. Immun.* **58:**189–196.

15. **Everett, K. D., and T. P. Hatch.** 1991. Sequence analysis and lipid modification of the cysteine-rich envelope proteins of *Chlamydia psittaci* 6BC. *J. Bacteriol.* **173:**3821–3830.

16. **Hackstadt, T., W. Baehr, and Y. Ying.** 1991. *Chlamydia trachomatis* developmentally regulated protein is homologous to eukaryotic histone H1. *Proc. Natl. Acad. Sci. USA* **88:**3937–3941.

17. **Hatch, T. P., I. Allan, and J. H. Pearce.** 1984. Structural and polypeptide differences between envelopes of infective and reproductive life cycle forms of *Chlamydia* spp. *J. Bacteriol.* **157:**13–20.

18. **Hatch, T. P., M. Miceli, and J. E. Sublett.** 1986. Synthesis of disulfide-bonded outer membrane proteins during the developmental cycle of *Chlamydia psittaci* and *Chlamydia trachomatis. J. Bacteriol.* **165:**379–385.

19. **Hodinka, R. L., C. H. Davis, J. Choong, and P. B. Wyrick.** 1988. Ultrastructural study of endocytosis of *Chlamydia trachomatis* by McCoy cells. *Infect. Immun.* **56:**1456–1463.

20. **Hodinka, R. L., and P. B. Wyrick.** 1986. Ultrastructural study of mode of entry of *Chlamydia psittaci* into L-929 cells. *Infect. Immun.* **54:**855–863.

21. **Jacobs, W. R., Jr., G. V. Kalpana, J. D. Cirillo, L. Pascopella, S. B. Snapper, R. A. Udani, W. Jones, R. G. Barletta, and B. R. Bloom.** 1991. Genetic systems for mycobacteria. *Methods Enzymol.* **204:**537–555.

22. **Joseph, T. D., and S. K. Bose.** 1991. A heat-labile protein of *Chlamydia trachomatis* binds to HeLa cells and inhibits the adherence of chlamydiae. *Proc. Natl. Acad. Sci. USA* **88:**4054–4058.

23. **Katz, B. P., B. E. Batteiger, and R. B. Jones.** 1987. Effect of prior sexually transmitted disease on the isolation of *Chlamydia trachomatis. Sex. Transm. Dis.* **14:**160–164.

24. **Kunimoto, D., and R. C. Brunham.** 1985. Human immune response and *Chlamydia trachomatis* infection. *Rev. Infect. Dis.* **7:**665–673.

25. **Langer, T., C. Lu, H. Echols, J. Flanagan, M. K. Hayer, and F. U. Hartl.** 1992. Successive action of DnaK, DnaJ and GroEL along the pathway of chaperone-mediated protein folding. *Nature* (London) **356:**683–689.

26. **Masson, P. L., J. F. Heremans, and J. Ferin.** 1969. Clinical importance of the biochemical changes in the female genital tract. I. Studies on the proteins of the cervical mucus. *Int. J. Fertil.* **14:**1–7.

27. **Morrison, R. P.** 1991. Chlamydial hsp60 and the immunopathogenesis of chlamydial disease. *Semin. Immunol.* **3:**25–33.

28. **Morrison, R. P., R. J. Belland, K. Lyng, and H. D. Caldwell.** 1989. Chlamydial disease pathogenesis. The 57-kD chlamydial hypersensitivity antigen is a stress response protein. *J. Exp. Med.* **170:** 1271–1283.

29. **Morrison, R. P., K. Lyng, and H. D. Caldwell.** 1989. Chlamydial disease pathogenesis. Ocular hypersensitivity elicited by a genus-specific 57-kD protein. *J. Exp. Med.* **169:**663–675.

30. **Morrison, R. P., D. S. Manning, and H. D. Caldwell.** 1992. Immunology of *Chlamydia trachomatis* infections: immunoprotective and immunopathogenetic responses, p. 57–84. *In* T. C. Quinn (ed.), *Sexually Transmitted Diseases.* Raven Press Ltd., New York.

31. **Moulder, J. W.** 1991. Interaction of chlamydiae and host cells in vitro. *Microbiol. Rev.* **55:**143–190.

32. **Mount, D. T., P. E. Bigazzi, and A. L. Barron.** 1972. Infection of genital tract and transmission of ocular infection to newborns by the agent of guinea pig inclusion conjunctivitis. *Infect. Immun.* **5:**921–926.

33. **Murdin, A. D., H. Su, D. S. Manning, M. H. Klein, M. J. Parnell, and H. D. Caldwell.** 1993. A poliovirus hybrid expressing a neutralization epitope from the major outer membrane protein of *Chlamydia trachomatis* is highly immunogenic. *Infect. Immun.* **61:**4406–4414.

34. **Murray, E. S.** 1964. Guinea pig inclusion conjunctivitis. I. Isolation and identification as a member of the psittacosis-lymphogranuloma-trachoma group. *J. Infect. Dis.* **114:**1–12.

35. **Newhall, W. J.** 1987. Biosynthesis and disulfide cross-linking of outer membrane components during the growth cycle of *Chlamydia trachomatis. Infect. Immun.* **55:**162–168.

36. **Newhall, W. J., and R. B. Jones.** 1983. Disulfide linked oligomers of the major outer membrane protein of chlamydiae. *J. Bacteriol.* **154:**998–1001.

37. **Palmer, L., and S. Falkow.** 1986. A common plasmid of *Chlamydia trachomatis. Plasmid* **16:** 52–62.

38. **Ramsey, K. H., and R. G. Rank.** 1991. Resolution of chlamydial genital infection with antigen-specific T-lymphocyte lines. *Infect. Immun.* **59:**925–931.

39. **Ramsey, K. H., L. S. F. Soderberg, and R. G. Rank.** 1988. Resolution of chlamydial genital infection in B-cell-deficient mice and immunity to reinfection. *Infect. Immun.* **56:**1320–1325.

40. **Rank, R. G., and A. L. Barron.** 1983. Effect of antithymocyte serum on the course of chlamydial genital infection in female guinea pigs. *Infect. Immun.* **41:**876–879.

41. Rank, R. G., and B. E. Batteiger. 1989. Protective role of serum antibody in immunity to chlamydial genital infection. *Infect. Immun.* **57:**299–301.

42. Rank, R. G., B. E. Batteiger, and L. S. F. Soderberg. 1988. Susceptibility to reinfection after a primary chlamydial genital infection. *Infect. Immun.* **56:**2243–2249.

43. Rank, R. G., B. E. Batteiger, and L. S. F. Soderberg. 1990. Immunization against chlamydial genital infection in guinea pigs with UV-inactivated and viable chlamydiae administered by different routes. *Infect. Immun.* **58:**2599–2605.

44. Rank, R. G., and P. M. Bavoil. Unpublished data.

45. Rank, R. G., and M. M. Sanders. 1990. Ascending genital tract infection as a common consequence of vaginal inoculation with the guinea pig inclusion conjunctivitis agent in normal guinea pigs, p. 249–252. *In* W. R. Bowie, H. D. Caldwell, R. B. Jones, P.-A. Mardh, G. L. Ridgway, J. Schachter, W. E. Stamm, and M. E. Ward (ed.), *Chlamydial Infections.* Cambridge University Press, New York.

46. Rank, R. G., L. S. F. Soderberg, and A. L. Barron. 1985. Chronic chlamydial genital infection in congenitally athymic nude mice. *Infect. Immun.* **48:**847–849.

47. Rank, R. G., H. J. White, and A. L. Barron. 1979. Humoral immunity in the resolution of genital infection in female guinea pigs infected with the agent of guinea pig inclusion conjunctivitis. *Infect. Immun.* **26:**573–579.

48. Raulston, J. E., C. H. Davis, D. H. Schmiel, M. W. Morgan, and P. B. Wyrick. 1993. Molecular characterization and outer membrane association of a *Chlamydia trachomatis* protein related to the hsp70 family of proteins. *J. Biol. Chem.* **268:**23139–23147.

49. Reynolds, D. J., and J. H. Pearce. 1991. Endocytic mechanisms utilized by chlamydiae and their influence on induction of productive infection. *Infect. Immun.* **59:**3033–3039.

50. Schachter, J. 1989. Pathogenesis of chlamydial infections. *Pathol. Immunopathol. Res.* **8:**206–220.

51. Schachter, J., and C. R. Dawson. 1978. *Human Chlamydial Infections.* PSG Publishing Company, Inc., Littleton, Mass.

52. Schmiel, D. H., S. T. Knight, J. E. Raulston, J. Choong, C. H. Davis, and P. B. Wyrick. 1991. Recombinant *Escherichia coli* clones expressing *Chlamydia trachomatis* gene products attach to human endometrial epithelial cells. *Infect. Immun.* **59:**4001–4012.

53. Soldati, D., and J. C. Boothroyd. 1993. Transient transfection and expression in the obligate intracellular parasite *Toxoplasma gondii. Science* **260:**349–352.

54. Stephens, R. S. 1993. Challenge of *Chlamydia* research. *Infect. Agents Dis.* **1:**279–293.

55. Stephens, R. S., G. Mullenbach, R. Sanchez-Pescador, and N. Agabian. 1986. Sequence analysis of the major outer membrane protein gene from *Chlamydia trachomatis* serovar L2. *J. Bacteriol.* **168:**1277–1282.

56. Stephens, R. S., R. Shanchez-Pescador, E. A. Wagar, C. Inouye, and M. S. Urdea. 1987. Diversity of *Chlamydia trachomatis* major outer membrane protein genes. *J. Bacteriol.* **169:**3879–3885.

57. Su, H., and H. D. Caldwell. 1992. Immunogenicity of a chimeric peptide corresponding to T helper and B cell epitopes of the *Chlamydia trachomatis* major outer membrane protein. *J. Exp. Med.* **175:**227–235.

58. Su, H., and H. D. Caldwell. 1993. Immunogenicity of a synthetic oligopeptide corresponding to antigenically common T-helper and B-cell neutralizing epitopes of the major outer membrane protein of *Chlamydia trachomatis. Vaccine* **11:**1159–1166.

59. Su, H., R. P. Morrison, N. G. Watkins, and H. D. Caldwell. 1990. Identification and characterization of T helper cell epitopes of the major outer membrane protein of *Chlamydia trachomatis. J. Exp. Med:* **172:**203–212.

60. Su, H., N. G. Watkins, Y.-X. Zhang, and H. D. Caldwell. 1990. *Chlamydia trachomatis*-host cell interactions: role of the chlamydial major outer membrane protein as an adhesin. *Infect. Immun.* **58:**1017–1025.

61. Su, H., Y.-X. Zhang, O. Barrera, N. G. Watkins, and H. D. Caldwell. 1988. Differential effect of trypsin on infectivity of *Chlamydia trachomatis:* loss of infectivity requires cleavage of major outer membrane protein variable domains II and IV. *Infect. Immun.* **56:**2094–2100.

62. Tam, J. E., S. T. Knight, C. H. Davis, and P. B. Wyrick. 1992. Eukaryotic cells grown on microcarrier beads offer a cost-efficient way to propagate *Chlamydia trachomatis. BioTechniques* **13:**374–378.

63. **Ting, L.-M., and P. M. Bavoil.** Unpublished data.

64. **Ward, M. E.** 1992. Chlamydia vaccines—future trends. *J. Infect.* **25:**S11–S26.

65. **Watkins, N. G., W. J. Hadlow, A. B. Moos, and H. D. Caldwell.** 1986. Ocular delayed hypersensitivity: a pathogenetic mechanism of chlamydial conjunctivitis in guinea pigs. *Proc. Natl. Acad. Sci. USA* **83:**7480–7484.

66. **Witkin, S. S., J. Jeremias, M. Toth, and W. J. Ledger.** 1993. Cell-mediated immune response to the recombinant 57-kDa heat-shock protein of *Chlamydia trachomatis* in women with salpingitis. *J. Infect. Dis.* **167:**1379–1383.

67. **Yuan, Y., Y.-X. Zhang, N. G. Watkins, and H. D. Caldwell.** 1989. Nucleotide and deduced amino acid sequences for the four variable domains of the major outer membrane proteins of the 15 *Chlamydia trachomatis* serovars. *Infect. Immun.* **57:**1040–1049.

68. **Zhang, J. P., and R. S. Stephens.** 1992. Mechanism of *Chlamydia trachomatis* attachment to eukaryotic host cells. *Cell* **69:**861–869.

69. **Zhang, Y. X., S. Stewart, T. Joseph, H. R. Taylor, and H. D. Caldwell.** 1987. Protective monoclonal antibodies recognize epitopes located on the major outer membrane protein of *Chlamydia trachomatis. J. Immunol.* **138:**575–581.

Part 4. Extracellular Structures and Products

The export and proper assembly of extracellular complexes by pathogenic bacteria are central to the initiation and maintenance of infectious diseases. Multimolecular complexes associated with virulence either (i) interact directly with target host cells, (ii) promote proper secretion, folding, and assembly of virulence factors, or (iii) provide important colonization factors for pathogens within host tissues. Several model systems that promote these processes are presented in this part. One of the striking characteristics found in the studies summarized here is the high degree of sequence and functional conservation among factors involved in disparate virulence-associated processes. The molecular details of the assembly and secretion of these structures are largely unknown, and in the coming years, work dedicated to uncovering these details should be a source of much excitement. The following describes some of the common features of the biosynthesis and assembly of these virulence-associated structures.

DEDICATED FACTORS THAT ALLOW TRANSLOCATION ACROSS THE GRAM-NEGATIVE OUTER MEMBRANE

Virulence factors that are encoded by gram-negative bacteria and that are targeted at host cells must be translocated across the outer membrane. Movement of macromolecules across this membrane barrier raises unusual problems because of the unique composition of the outer membrane, its permeability to molecules involved in maintaining a proper energy charge, and its displacement from the cytoplasmic membrane. As a result, dedicated secretion and assembly machines are used to promote translocation out of the cytoplasm and across the periplasm and outer membrane. Richard Silver (chapter 20) describes a model for how a biosynthetic machine could coordinate the assembly of cytosolic capsule intermediates with translocation across the outer membrane. Alison Weiss (chapter 22) describes how secretion of pertussis toxin follows folding and assembly of the toxin complex into an enzymatically active form. Interestingly, a large array of gene products that are associated with translocation and assembly of this toxin are highly homologous to factors involved in the translocation of DNA in *Agrobacterium* species, the formation of pili in *Pseudomonas* species, and the secretion of factors expressed by the plant pathogens *Erwinia* and *Xanthomonas* species. A similar translocation machine is outlined by James Bliska (chapter 24) in his description of *Yersinia* deposition of the cytotoxic proteins YopE and YopH into target host cells.

TRANSLOCATION OF ATPases PERIPHERALLY ASSOCIATED WITH THE GRAM-NEGATIVE INNER MEMBRANE

Input of energy is required for both movement of structural intermediates across the cytoplasmic membrane and maintenance of these intermediates in a translocation-competent conformation. No direct studies are available to test whether the proton motive force is coupled to these processes, but sequence information indicates that most translocation machines involved in the extracellular localization of virulence factors have a dedicated cytoplasmic membrane-associated ATPase. Many of these ATPases are highly homologous to the eukaryotic multidrug resistance protein and apparently act as dimers bound to cytoplasmic membrane components of the translocation machine. Examples of these factors are HlyB protein, which is involved in hemolysin secretion and which is described by Rodney Welch (chapter 23), and KpsT, which is involved in translocation of K1 capsule intermediates and which is described by Richard Silver (chapter 20). Other ATPases not obviously in this family, but apparently similarly able to couple ATP hydrolysis with movement of macromolecules into the extracellular milieu, are PtlH, which is involved in *Pertussis* toxin secretion (described by Alison Weiss in chapter 22), as well as YscN, which is necessary for translocation of Yops in *Yersinia* species (chapter 24). It is not known how ATP hydrolysis is coupled to the movements of any of these molecules.

ENZYMATIC MODIFICATION THAT ACTIVATES VIRULENCE FACTORS

An enzymatic modification may be necessary to activate many virulence-associated factors. Several of the structures described in this part follow this paradigm. The *Escherichia coli* hemolysin described by Rodney Welch (chapter 23) must be acylated for activity, while sialylation of capsule intermediates in the type III capsule of group B streptococci (chapter 21) and the *E. coli* K1 capsule (chapter 20) appears to be an important determinant of bloodstream colonization by these microorganisms. Furthermore, an unknown posttranslational modification is involved in flagellar assembly by *Campylobacter* species (chapter 25). Again, it is not known why these modifications are required, and future work will certainly focus attention on the roles that modifications by small molecules play in the assembly and activities of these complexes.

DISRUPTION OF HOST CELL PHAGOCYTOSIS

Several of the secreted products described in this part can be argued to be associated with inactivation of host phagocytes or are involved in bypassing attack by phagocytes. *E. coli* hemolysin is assayed by its ability to lyse erythrocytes, but its importance in maintaining an extraintestinal and extracellular site of infection is reminiscent of the properties of the antiphagocytic factors expressed by other microorganisms. Perhaps the true targets of this protein are phagocytic cells. Similarly, the proteins YopE and YopH from *Yersinia* species that are deposited within host cells by specialized translocation machinery have been demonstrated to play a role in antagonizing the phagocytosis of bound *Yersinia* species. Finally, the most important role of capsular polysaccharide in group B streptococci and *E. coli* K1 strains is to allow colonization of the bloodstream prior to the onset of meningitis. Colonization at this site depends on interference of these capsules with phagocytosis.

The common themes described in this part regarding the export and assembly of extracellular complexes associated with the virulence of gram-negative bacteria are well covered in the following chapters. Because most of the critical components in these structures have been identified, the large challenges that lay ahead will involve dissection of their molecular mechanisms of action.

Ralph R. Isberg

Molecular Genetics of Bacterial Pathogenesis
Edited by V. L. Miller, J. B. Kaper, D. A. Portnoy, and R. R. Isberg
© 1994 American Society for Microbiology, Washington, DC 20005

Chapter 20

Genetic Analysis of the *Escherichia coli* K1 Capsule Gene Cluster

Richard P. Silver

Bacterial surface structures play a major role in the pathogenesis of infectious diseases (41, 62). The ability of the host to respond to microbial infection is, in large part, dependent on the chemical composition and physical characteristics of the cell surface. Capsular polysaccharides are the outermost structures on the bacterial cell and play a critical role in the interaction between a microorganism and its environment. They have long been recognized as important virulence determinants of bacteria able to invade the human host and survive in the bloodstream (9, 34). Capsules provide the bacterium with mechanisms to avoid an array of nonspecific host defenses that includes complement-mediated bacteriolysis and opsonophagocytosis (9, 22, 34, 44, 62). Encapsulated bacteria such as meningococci, pneumococci, *Haemophilus influenzae* type b, *Salmonella typhi,* group B streptococci, and *Escherichia coli* are of primary concern as pathogens that invade the bloodstream of otherwise healthy individuals. Indeed, many of the serious invasive infections in humans, including meningitis, septicemia, pneumonia, osteomyelitis, septic arthritis, and pyelonephritis, are caused by encapsulated organisms (34). Interest in the prevention and control of disease caused by these organisms has been stimulated by the continued emergence of antibiotic-resistant strains (48). The high degree of morbidity and mortality of diseases caused by encapsulated bacteria has also heightened interest in disease prevention and control (48).

THE POLYSIALIC CAPSULE OF *E. COLI* K1

E. coli is a common component of the normal enteric flora of most warm-blooded animals. Although most strains are harmless commensal organisms, certain isolates can be considered primary pathogens that possess an array of virulence determinants that allow them to evade host defenses, leading to infection and overt disease (67). Infection results in a variety of clinical syndromes, including several distinct forms of diarrheal disease as well as urinary tract infections

Richard P. Silver • Department of Microbiology and Immunology, University of Rochester Medical Center, Rochester, New York 14642.

(67). *E. coli* also continues to be the most common gram-negative organism that causes sepsis and meningitis in newborn infants (42). The prognosis for infants with *E. coli* meningitis is poor. Mortality remains high, and neurological sequelae are common among survivors (32).

In 1927, Smith and Bryant (64) showed that capsules are essential virulence determinants of *E. coli* that cause extraintestinal infections. They observed that naturally occurring unencapsulated mutants of *E. coli,* isolated from bacteremic calves, were less pathogenic in guinea pigs than the wild-type organism and were more susceptible to phagocytosis (64). It is now known that *E. coli* produces more than 70 serologically distinct capsular antigens, or K antigens (35). Indeed, a striking feature of capsules in both gram-negative and gram-positive bacteria is their diversity of structure (25, 48). The K antigens of *E. coli* have been classified into two distinct groups (24). Group I antigens are characterized by a high molecular mass (>100 kDa); the presence of glucuronic acid, galacturonic acid, and pyruvate; expression below 20°C; lipid A substitution at the reducing end of the polymer; coexpression with lipopolysaccharide (LPS) serotype O8, O9, or O20; and expression from the chromosomal *rfb* and *rfc* loci (24). In contrast, group II capsules have a lower molecular mass (~50 kDa); contain glucuronic acid, *N*-acetylneuraminic acid (NeuNAc), 2-keto-3-deoxyoctulosonic acid, *N*-acetylmannosamine, and phosphate; are not expressed below 20°C; are substituted with phosphatidic acid at the reducing end of the polymer; are coexpressed with a variety of LPS serotypes; and are encoded by the chromosomal *kps* locus (24). *E. coli* strains that produce group II capsules, but not group I capsules, are commonly associated with invasive disease (24, 35). For example, strains that express the group II K1, K2, K3, K5 and K12, and K13 capsular serotypes are commonly isolated from patients with urinary tract infections (27). Strains that produce the K1 capsule account for 80% of the isolates that cause *E. coli* neonatal meningitis and comprise the majority of isolates from neonates with septicemia without meningitis and acute pediatric pyelonephritis (27, 49). The K1 polysaccharide, an α2,8-linked linear polymer of sialic acid (NeuNAc), is identical to the capsule of *Neisseria meningitidis* group B (28). Sialic acids are essential constituents of many mammalian glycoconjugates, yet they are relatively rare among prokaryotes (53). However, sialic acids are frequent components of capsules associated with bacterial disease (25, 48).

The polysialic acid (PSA) capsule of *E. coli* K1 is essential for the virulence of the organism (60). Nonencapsulated derivatives of invasive K1 strains did not cause bacteremia in the infant rat model, although intestinal colonization was the same as that with encapsulated strains (45, 63). The passive administration of K1 antibodies prevented bacteremia and meningitis in infant rats fed *E. coli* K1 (49). Invasiveness is correlated with the ability of terminal sialic acid to inhibit complement activation by the alternative pathway (16, 26, 60). The K1 polysaccharide is also poorly immunogenic (78), a property attributed to the identity of the K1 polymer to structures expressed on host tissues. This is best illustrated by the observation that the K1 polysaccharide is identical to PSA moieties on the embryonic form of the neural cell adhesion molecule (N-CAM) (17). N-CAM is an integral membrane glycoprotein involved in homophilic neuronal cell adhesion, and its activity is modulated by alteration of the PSA content (14, 52). The structural identity between the K1 capsule and the polysialosyl chains on N-CAM suggests that immune tolerance may be important to the pathogenesis of disease caused by K1 organisms.

PSA is a unique biological structure that is found in a remarkably diverse range of biological contexts such as the surfaces of neurotropic bacteria, the vitelline envelopes of fish eggs, and the surfaces of vertebrate neurons (50, 68). Polysialylated N-CAM is also aberrantly expressed on human neuroblastomas and Wilms tumors (50, 51, 68). The PSA portion of N-CAM is therefore not only a developmentally regulated antigen but also an oncodevelopmental antigen in human kidney and neural tissues. As a result, studies to understand the synthesis of PSA are now of great interest to microbiologists, developmental biologists, neuroscientists, and oncologists.

BIOSYNTHESIS AND GENETICS OF THE K1 CAPSULE

The PSA capsule of *E. coli* K1 is a polymer that is composed of about 200 sialic acid residues and that terminates in a phosphodiester linkage to 1,2-diacylglycerol (19, 54). The lipid is postulated to anchor the polymer to the cell surface. Biosynthesis of the capsule requires the synthesis, activation, and polymerization of sialic acid subunits into a large polymer which must be transported through two lipid bilayers (4, 68). NeuNAc is synthesized within the cytoplasm in a condensation reaction between *N*-acetylmannosamine and phosphoenolpyruvate (60, 71). The monosaccharide is activated to a nucleotide derivative, CMP-NeuNAc, generating a high-energy sugar nucleotide glycoside intermediate (71). A membrane-associated sialyltransferase complex catalyzes the polymerization of NeuNAc from CMP-NeuNAc (68). The initiation of polymer synthesis occurs by the attachment of polysialosyl monomers to an endogenous acceptor in the membrane (68). The molecular nature of the acceptor is not known, although a 20-kDa sialylated membrane protein has been identified and is postulated to be the endogenous acceptor in *E. coli* K1 (76). Polymerization proceeds processively at the nonreducing terminus by the addition of activated NeuNAc molecules to the growing polysaccharide chain (65, 73). These reactions involve a lipid intermediate, NeuNAc-P-undecaprenol (68).

The proteins necessary for the synthesis and expression of the K1 capsule are chromosomally encoded by the 17-kb *kps* gene cluster located near *serA* on the *E. coli* genetic map (35). The *kps* gene cluster has been cloned, and the genetic organization, nucleotide sequence, and protein composition of the cluster have been determined (5, 39, 58, 59, 72). The gene cluster is functionally divided into three regions (6) (Fig. 1). The genes in regions 1 and 3 are conserved among *E. coli* isolates that synthesize serologically distinct capsules (5, 47, 56). Cells with mutations in region 3 of the cluster accumulate polysaccharide within their cytoplasms, while cells with mutations in region 1 were reported to accumulate polymer within their periplasmic spaces (31, 77). These observations led to the conclusion that region 3 encodes a transporter responsible for the export of polymer out

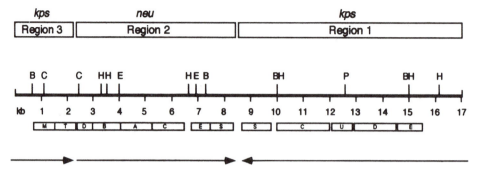

Figure 1. Genetic organization of the *kps* gene cluster of *E. coli* K1. The genes within the conserved regions 1 and 3 are designated *kps,* while the genes within the K1-specific region 2 are designated *neu.* Restriction endonuclease sites are indicated by the following abbreviations: B, *Bgl*II; C, *Cla*I; H, *Hind*III; E, *Eco*RI; BH, *Bam*HI; P, *Pst*I. The arrows below the map indicate the direction of transcription for each region.

of the cytoplasm, while region 1 is involved with the transport of polymer across the outer membrane. The recent report that certain mutations in region 1 also lead to the accumulation of cytoplasmic polysaccharide (7) underscores the complexity of polymer synthesis and export. Most likely, region 1 and 3 proteins function together in the export of capsular polysaccharide to the cell surface. This idea, however, does not imply that polymer translocation is a process temporally distinct from synthesis. Translocation of polymer most likely occurs concomitantly with its synthesis.

The six genes in region 2 contain information for the synthesis, activation, and polymerization of NeuNAc (2, 5, 71, 72). This region is unique for each *E. coli* serotype that synthesizes a chemically distinct capsular polysaccharide, and the number of region 2 genes correlates with the size and complexity of the polymer repeating unit (4, 5, 47). In general, cells harboring mutations in region 2 of the *kps* cluster do not synthesize polymer (2, 5, 72), although cells with a mutation in *neuE* are still capable of producing intracellular polymer (73). Two genes, *neuB* and *neuC,* are involved in sialic acid synthesis. Cells harboring mutations in either gene are defective in capsule synthesis but synthesize capsular polysaccharide when sialic acid is provided in the medium (59, 71, 72). The *neuA* gene encodes the 48.6-kDa CMP-NeuNAc synthetase, which catalyzes the activation of NeuNAc to the sugar nucleotide (79). The *neuA* gene product has been purified to homogeneity (70). The 47-kDa product of the *neuS* gene has been identified as the sialyltransferase required for polymerization of NeuNAc (66, 75). The function of *neuD,* the first gene in region 2, is not known, although the *neuD* gene product is not required for sialic acid synthesis or activation (2, 71). NeuD shows significant similarity to a family of bacterial acetyltransferases that includes LacA and CysE of *E. coli,* NodL of *Rhizobium leguminosarum* (13), and the Tn*2424* Cat protein (36). The proteins in the family and other acyl- and acetyltransferases that include the FirA (LpxD) and LpxA proteins share a novel region of homology that has been termed the "isoleucine patch" (11, 69).

STRUCTURE AND FUNCTION OF THE KpsMT TRANSPORTER

The mechanisms responsible for the initiation, polymerization, and termination of PSA synthesis as well as export of the polymer to the cell surface are not fully understood. The observation, however, that the polysaccharide is synthesized on the cytoplasmic surface of the cytoplasmic membrane (68) requires a mechanism for transporting the polymer across two membranes prior to attachment to the cell surface.

Region 3 of the *kps* gene cluster contains two genes, *kpsM* and *kpsT*. The *kpsM* gene product is a 29-kDa, highly hydrophobic, integral membrane protein, while *kpsT* encodes a 25-kDa hydrophilic peripheral inner membrane protein that contains a consensus ATP-binding site (38, 61). The characteristics of KpsM and KpsT and the phenotypes of mutants with region 3 mutations led to the proposal that KpsM and KpsT function together as an ATP-dependent system for transporting capsular polysaccharide across the inner membrane (38, 61). They belong to a large family of prokaryotic and eukaryotic membrane translocators involved in diverse biological processes (1, 3, 12, 15, 21). Most of the family members are made up of various membrane transport systems that are believed to move their respective substrates across membranes via energy obtained from ATP hydrolysis (1, 12, 21). They have been referred to as the ABC (ATP-binding cassette) superfamily of transport proteins (21), or traffic ATPases (12). The ABC transporters illustrate two simplifying principles of biology recently described by Koshland (29): the principle of redundancy and the principle of diversity. The principle of redundancy leads to similar systems in diverse organisms, while the principle of diversity suggests that each organism tailors the system to meet its specific needs (29).

The ABC transporters include three major groups: the periplasmic permeases of enteric bacteria, the eukaryotic transporters, and the bacterial exporters (15). In eukaryotes, several of these systems are of significant medical importance and include the cystic fibrosis transductance regulator (CFTR) (46) and the P glycoprotein, a mammalian drug resistance pump (20). More than 40 bacterial exporters that transport proteins, peptides, and nonprotein substrates such as polysaccharides have been identified (15). The KpsMT transporter is the prototypical exporter of capsular polysaccharides in bacteria. BexBA and CtrCD, proteins homologous to KpsMT, were identified in *H. influenzae* type b and *N. meningitidis* group B, respectively, and are postulated to play analogous roles in the transport of polysaccharide in both organisms (18, 30). Indeed, a major theme that has emerged from the study of capsule gene clusters in *E. coli, H. influenzae,* and *N. meningitidis* is that the organization of capsule determinants in gram-negative bacteria is remarkably similar (5, 18).

The members of the ABC transporter superfamily share a common organizational motif consisting of a hydrophobic membrane component and a hydrophilic, ATP-binding component (12, 21). It is believed that a functional transporter consists of two of each type of component (12, 21). The structures of KpsM and KpsT are consistent with the basic organizational model of the transport systems. They consist of a hydrophobic component situated in the membrane and a hydrophilic component containing the ATP-binding fold of Walker et al. (74) that is located on the cytoplasmic side of the membrane. In all instances, two membrane units must be present to form a functional transporter. In some systems this motif is expressed as separate polypeptides, while in others the hydrophobic and hydrophilic components are fused into larger multifunctional polypeptides (1, 12, 15, 21).

The results of mutational analysis indicate that cells lacking KpsM or KpsT are capable of polymer synthesis but are unable to transport the polymer across the cytoplasmic membrane (31, 37). Electron microscopy was used to examine the locations of polymers in cells carrying mutations in region 3 (31, 37, 40). These studies revealed characteristic electron-lucent zones within the cytoplasm, indicative of the intracellular accumulation of polysaccharide (37, 39, 40). Moreover, polymer accumulated against the inner membrane of cells carrying defined *kpsT* mutations (37), which is consistent with the view that that polysaccharide remains attached to the inner surface of the cytoplasmic membrane during synthesis.

The phenotypes of cells harboring mutations in region 3 include both the accumulation of intracellular polymer and a reduction in the level of endogenous sialyltransferase activity (37, 72). The endogenous sialyltransferase activity is a measure of the transfer of labeled sugar onto preexisting acceptors within the membranes, while the exogenous activity is a measure of the transfer of labeled sugar onto exogenously added PSA (68). The reason for this reduced level of activity is not apparent, since the exogenous activity in these cells is normal, indicating that the sialyltransferase enzyme encoded by the *neuS* gene is not affected. However, these observations do imply that transport defects in some way affect polymer synthesis. The current view of capsule expression and transport assumes coupling between the biosynthetic and transport machineries and posits that the membrane-bound components of the cluster form a heterooligomeric complex involved with both polymer synthesis and transport. One could envision that the loss of a component could prevent the proper assembly of the complex, causing pleiotropic effects on polysaccharide synthesis and transport. In this scenario, the loss of KpsM or KpsT would not only block polymer export but also result in an improperly assembled complex that is unable to carry out normal synthetic activities. My colleagues and I have observed, however, that the phenotypes of cells harboring the K44E mutant allele of KpsT (38) (see below) are the same as those of cells with a deletion of the gene (37). It appears, therefore, that the decrease in endogenous sialyltransferase activity is not due to the lack of a component but is more likely related to the inability of cells to transport polymer across the inner membrane. Synthesis and export of capsular polysaccharide likely occur at the same time, and it may be possible that a block in export could feed back on the synthesis pathway, perhaps affecting the nature or number of endogenous acceptors. In any event, an understanding of this particular phenotype will be possible only when the nature of the endogenous sialyl acceptor within the cytoplasmic membrane is determined.

The hydrophobicity profile of KpsM indicates a protein with at least six potential transmembrane helical domains (38). KpsM was expressed by using the T7 promoter-polymerase system (43). A radiolabeled protein of 25 kDa was observed, indicating that KpsM has a slightly faster mobility on sodium dodecyl sulfate (SDS)-polyacrylamide gels than expected from the calculated molecular mass (38). This is due to the hydrophobic nature of the protein and is consistent with the fact that it is an integral membrane protein. Radiolabeled KpsM was cross-linked in vivo with dithiobissuccinimydyl propionate (43). Cross-linked KpsM samples migrated at 44 kDa on SDS-polyacrylamide gels in hosts with and without the *kps* cluster (43). The size of the cross-linked product corresponds to the predicted size of a dimer of KpsM and is consistent with the "two-times-six" helix paradigm of membrane transporters (21).

To study the transmembrane organization of KpsM, Pigeon and I constructed KpsM–β-lactamase fusions (43). In this system, which was developed by Broome-Smith et al. (8), a truncated β-lactamase lacking its amino-terminal leader sequence acts as a topological reporter when fused to portions of a membrane protein. The β-lactamase that is exposed to the periplasm via a transmembrane domain confers resistance to high levels of ampicillin when single colonies are grown on ampicillin-containing plates (8). In contrast, strains in which the fusions are localized to the cytoplasm do not confer resistance to single colonies but can be detected by patch screening of isolates on ampicillin-containing plates (8). In the latter case, cells survive through the release of the hybrid protein by lysis of some cells within the patch. Pigeon and I isolated and characterized 35 distinct in-frame fusions of KpsM to the β-lactamase gene (*blaM*) (43). Our analysis of the KpsM-BlaM fusions provides evidence for six transmembrane-spanning regions (regions 1 to 6), resulting in three periplasmic loops (PI to PIII), two cytoplasmic loops (CI and CII), and both the amino- and carboxy-terminal domains facing the cytoplasm (Fig. 2). A region of the periplasmic loop connecting transmembrane regions 5 and 6 also appears to localize within the cytoplasmic membrane. Mutants in the periplasmic and cytoplasmic loops of KpsM were isolated by linker-insertion mutagenesis, and several regions of the protein essential for polymer transport were identified (Fig. 2). An insertion in the CI loop generates a protein that is nonfunctional and that is of particular interest. Comparison of the amino acid sequences of membrane components within the superfamily of ABC transporters reveals little primary sequence homology (12, 21). Only a short 20-amino-acid motif (the EAA loop) identified in bacterial transporters is conserved between the membrane domains of some ABC transporters (10). This region, which is positioned in a cytoplasmic loop, is postulated to interact with the ATP-binding subunit (10, 12, 21). The CI loop (Fig. 2) has homology to the conserved EAA motif, and my colleagues and I postulate that this region of the protein may be the site of interaction between KpsM and KpsT.

Computer-based structural models have been generated. In these models, secondary-structure predictions for the ATP-binding component of the ABC transporters were compared with known crystal structures of proteins that bind ATP such as adenylate cyclase, elongation factor EF-Tu, and p21^{ras} (23, 33). The main conclusion from that analysis was that these components are modular in design, consisting of an ATP-binding domain and a variable domain. According to this model, the ATP-binding domain, comprising a five-stranded beta sheet with inter-

Periplasm

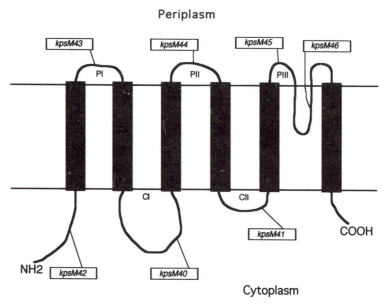

FIGURE 2. Proposed transmembrane topology of KpsM (258 amino acids) in the cytoplasmic membrane (43). Positions of the *kpsM* linker insertion mutations are indicated. Cells carrying the *kpsM42, kpsM40,* and *kpsM46* mutations exhibited a null phenotype, while the *kpsM41, kpsM43, kpsM44,* and *kpsM45* alleles did not affect protein function (43).

vening alpha helices, is interrupted by a variable "helical domain" (23, 33). The helical domain is postulated to interact with the membrane components of the ABC transporters (12, 21) and is important for the specific transport function of a particular system. Within this helical domain resides a short, conserved "linker peptide," a flexible region of the protein that is believed to function in transmembrane signaling (12, 21). The linker peptide is postulated to transmit conformational changes that occur in the ATP-binding domain to the helical domain, which in turn transmits the signal to the membrane component (12, 21). The ATP-binding fold and the linker peptide of the superfamily members are highly conserved, while the helical domain shows more variation (12, 21). Structural predictions for the KpsT protein best fit the model proposed by Mimura et al. (33) (Fig. 3).

The current model for the KpsMT transporter predicts that loops extending from KpsM (perhaps the CI loop) interact with the helical domain of KpsT, coupling the energy of ATP hydrolysis to the transport of PSA across the cytoplasmic membrane. According to this view, the binding of ATP and concomitant hydrolysis are presumed to induce a conformational change in the ATP-binding domain which is transmitted via the linker peptide to the transmembrane component and in some way mediates transport across the membrane. Support for the interaction between KpsT and ATP was derived both from structure-function analysis of KpsT mutants and from in vitro chemical modification studies with photolabeling analogs of ATP (37, 57). A mutant allele of *kpsT* (K44E), which contains a mutation in a highly conserved lysine residue within the ATP-binding consensus sequence of the pro-

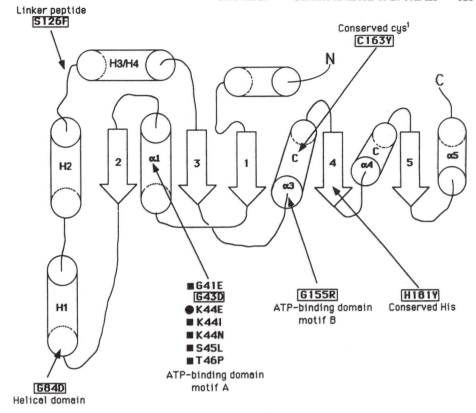

FIGURE 3. Proposed tertiary-structure model of the KpsT protein. The beta strands are represented by arrows, and the alpha helices are represented by cylinders. Each region of the protein is also indicated, and the locations of amino acid changes in mutant proteins are shown. Mutations were generated by hydroxylamine mutagenesis (shown boxed) and by site-directed (black circle) and saturation (black squares) mutagenesis by PCR (37). Note the locations of the two cysteines in alpha helices 3 and 4 of the ATP-binding domain.

tein, does not complement a strain with a chromosomal mutation in *kpsT* (38). Cells with the K44E mutant allele in the chromosome are acapsular and accumulate polysaccharide against the inner membrane (37). In addition, wild-type KpsT was photolabeled by the ATP analog 8-N$_3$[γ-^{32}P]ATP with good efficiency, while K44E was derivatized very poorly (37). Interestingly, efficient photoaffinity labeling of KpsT seems to require a membrane environment, but does not require any other *kps* gene products (37). My colleagues and I also performed saturation mutagenesis throughout motif A of the ATP-binding domain and isolated five additional mutants (G41E, K44I, K44N, S45L, and T46P) that produce nonfunctional proteins (Fig. 3). Each of the mutations changes a single conserved residue within the consensus ATP-binding domain of the protein. These results are consistent with the ABC transporter paradigm and indicate that ATP binding and, presumably, hydrolysis are important to KpsT function and the export of capsular polysaccharide.

Results obtained from chemical mutagenesis of KpsT are consistent with the structural model depicted in Fig. 3 and revealed characteristics particular to capsule transporters (37). For instance, mutation G84D resides in the helical domain and mutation S126F is located within the linker peptide (Fig. 3). Interestingly, a nonfunctional HisP (a KpsT homolog in the histidine permease of *Salmonella typhimurium*) mutant with a Ser-to-Phe mutation and nonfunctional CFTR proteins with Ser-to-Asn, -Lys, and -Ile mutations in the conserved Ser residue of the linker peptide have been reported (55). These KpsT mutants are still able to bind ATP, suggesting that the defects in these mutants may be due to alterations in the interaction between KpsT and KpsM. A highly conserved histidine residue at position 181 in KpsT was mutated to a tyrosine residue (Fig. 3). This histidine residue has been postulated to be important for events after ATP binding (55). Mutations in the histidine residue at this position in HisP yield mutant proteins that can bind ATP, but the proteins are unable to transport the substrate (55). The C163Y mutation (Fig. 3) is very interesting since KpsT has two cysteine residues, one at position 163 and the other at position 190 (38). These cysteines are conserved in the BexA and CtrD proteins, which are homologs of KpsT that are involved in capsular polysaccharide export in *H. influenzae* and *N. meningitidis*, respectively, but are not found in other members of the superfamily (37). Each cysteine residue is located in the ATP-binding fold, in α helices 3 and 4, and each is situated within the middle of its respective helix, on the hydrophobic side. These observations suggest that the two cysteines of KpsT are in proximity, allowing interaction with each other or some other *kps* component, and this interaction is required for polymer export. This conclusion is supported by the observation that the C163Y mutant can still bind ATP, indicating some other functional defect in the protein (37).

The *kpsM* and *kpsT* genes from the *E. coli* K5 serotype have also been cloned and sequenced (61). The *kpsM* gene products are essentially identical (38). In contrast, the *kpsT* gene products are 72% identical and 84% similar (38). The two proteins are very similar at their amino termini. The homology decreases, however, at their carboxy termini. In fact, the K5 protein is 5 amino acids longer than the K1 protein at the C terminus (38, 61). The 3' ends of the genes are located at the junction between region 3 and region 2 of the gene cluster, and the proximity of *kpsT* to the junction may explain the divergence in the C termini of the proteins. While this observation indicates that KpsT can tolerate changes in its carboxy terminus and still function, the two cysteine residues as well as the conserved histidine residue at position 182 are located within small "islands" of homology in the divergent part of the protein (37), suggesting that changes in these residues are not tolerated. These data support the idea that the cysteine residues are functionally important for the capsular polysacchride transport systems.

The KpsMT exporter of *E. coli* K1 functions in the transport of acidic capsular polysaccharides across the cytoplasmic membrane. The mechanism by which such long, hydrophilic molecules move across the lipid bilayer is an interesting and challenging problem. As research progresses, several key questions will need to be considered, including the following: (i) How is the polysaccharide recognized by the transport apparatus and directed into the pathway? (ii) How is the presumed ATPase activity used in polymer transport? (iii) If there is, indeed, a protein complex that bridges the two membranes, how is the complex assembled? From the results of studies to answer those questions, a detailed understanding of polysaccharide transport in *E. coli* will emerge, and this understanding will perhaps provide a direction for rational intervention in the disease process.

The biosynthesis of the PSA capsule of *E. coli* K1 is a complex process that involves the synthesis, activation, and polymerization of sialic acid subunits into a large polymer which must be transported through two lipid bilayers and anchored to the cell surface. The 17-kb *kps* gene cluster, which is divided into three functional regions, encodes the information necessary for polymer synthesis and expression at the cell surface. Most of the genes of the cluster have now been identified and the genetic organization has been determined. A major challenge for the future will be to determine the functions of the protein products of the *kps* cluster. The identification of interacting proteins is also essential to the eventual understanding of this complex system.

ACKNOWLEDGMENTS. I express my deepest gratitude to the members of my laboratory, Paula Annunziato, Joe Bliss, Marty Pavelka, Ron Pigeon, and Lori Wright, for reviewing the manuscript.

My work described here was supported by Public Health Service grant AI26655 from the National Institutes of Health and by a National Institutes of Health predoctoral training grant in microbial pathogenesis (5-T32-AI07362).

REFERENCES

1. **Ames, G. F.-L., C. S. Mimura, S. R. Holbrook, and V. Shyamala.** 1992. Traffic ATPases: a superfamily of transport proteins operating from *Escherichia coli* to humans. *Adv. Enzymol.* **65:**1–47.

2. **Annunziato, P., L. Wright, W. F. Vann, and R. P. Silver.** Submitted for publication.

3. **Blight, M. A., and I. B. Holland.** 1990. Structure and function of haemolysin B, P-glycoprotein, and other members of a novel family of membrane translocators. *Mol. Microbiol.* **4:**873–880.

4. **Boulnois, G. J., and K. Jann.** 1989. Bacterial polysaccharide capsule synthesis, export, and evolution of structural diversity. *Mol. Microbiol.* **3:**1819–1823.

5. **Boulnois, G. J., and I. S. Roberts.** 1990. Genetics of capsular polysaccharide production in bacteria. *Curr. Top. Microbiol. Immunol.* **150:**1–18.

6. **Boulnois, G. J., I. S. Roberts, R. Hodge, K. R. Hardy, K. Jann, and K. N. Timmis.** 1987. Analysis of K1 capsule biosynthesis genes of *Escherichia coli:* definition of three functional regions for capsule production. *Mol. Gen. Genet.* **208:**242–246.

7. **Bronner, D., V. Sieberth, C. Pazzani, I. S. Roberts, G. J. Boulnois, B. Jann, and K. Jann.** 1993. Expression of the capsular K5 polysaccharide of *Escherichia coli:* biochemical and electron microscopic analyses of mutants with defects in region 1 of the K5 gene cluster. *J. Bacteriol.* **175:**5984–5992.

8. **Broome-Smith, J. K., M. Tadayyon, and Y. Zhang.** 1990. β-Lactamase as a probe of membrane protein assembly and protein export. *Mol. Microbiol.* **4:**1637–1644.

9. **Cross, A. S.** 1990. The biologic significance of bacterial encapsulation. *Curr. Top. Microbiol. Immunol.* **150:**87–95.

10. **Dassa, E., and M. Hofnung.** 1985. Sequence of gene *malG* in *E. coli* K-12: homologies between integral membrane components from binding protein-dependent transport systems. *EMBO J.* **4:**2287–2293

11. **Dicker, I. B., and S. Seetharam.** 1992. What is known about the structure and function of the *Escherichia coli* protein FirA? *Mol. Microbiol.* **6:**817–823.

12. **Doige, C. A., and G. F.-L. Ames.** 1993. ATP-dependent transport systems in bacteria and humans: relevance to cystic fibrosis and multidrug resistance. *Annu. Rev. Microbiol.* **47:**291–319.

13. **Downie, J. A.** 1989. The *nodL* gene of Rhizobium leguminosarum is homologous to the acetyl transferases encoded by *lacA* and *cysE.* *Mol. Microbiol.* **3:**1649–1651.

14. **Edelman, G. M.** 1985. Cell adhesion and the molecular processes of morphogenesis. *Annu. Rev. Biochem.* **54**:135–169.

15. **Fath, M. J., and R. Kolter.** 1993. ABC-transporters: the bacterial exporters. *Microbiol. Rev.* **57**: 995–1017.

16. **Fearson, D. T., and K. F. Austen.** 1980. The alternative pathway of complement: a system for host resistance to microbial infection. *N. Engl. J. Med.* **303**:254–263.

17. **Finne, J.** 1982. Occurrence of unique polysialyl carbohydrate units in glycoprotein of developing brain. *J. Biol. Chem.* **257**:11966–11970.

18. **Frosch, M., U. Edwards, K. Bousset, B. Krausse, and C. Weisgerber.** 1991. Evidence for a common molecular origin of the capsule gene loci in gram-negative bacteria expressing group II capsular polysaccharides. *Mol. Microbiol.* **5**:1251–1263.

19. **Gotschlich, E. C., B. A. Fraser, O. Nishimura, J. B. Robbins, and T.-Y. Liu.** 1981. Lipids on capsular polysaccharides of gram-negative bacteria. *J. Biol. Chem.* **256**:8915–8921.

20. **Gross, R. J., J. Croop, and D. Housmen.** 1986. Mammalian multi-drug resistance gene: complete DNA sequence indicates strong homology to bacterial transport proteins. *Cell* **47**:371–380.

21. **Higgins, C. F.** 1992. ABC transporters: from microorganisms to man. *Annu. Rev. Cell Biol.* **8**: 67–113.

22. **Horowitz, M. A., and S. C. Silverstein.** 1980. Influence of the *Escherichia coli* capsule on complement fixation and on phagocytosis and complement killing by human phagocytes. *J. Clin. Invest.* **65**:82–94.

23. **Hyde, S. E., P. Emsley, M. J. Hartshorn, M. M. Mimmack, U. Gileadi, S. R. Pearce, M. P. Gallagher, D. R. Gill, R. E. Hubbard, and C. F. Higgins.** 1990. Structural model of ATP-binding proteins associated with cystic fibrosis, multidrug resistance and bacterial transport. *Nature* (London) **346**:362–368.

24. **Jann, B., and K. Jann.** 1990. Structure and biosynthesis of the capsular antigens of *Escherichia coli. Curr. Top. Microbiol. Immunol.* **150**:19–42.

25. **Jennings, H. J.** 1983. Capsular polysaccharides as human vaccines. *Adv. Carbohydr. Chem. Biochem.* **41**:155–208.

26. **Joiner, K. A.** 1988. Complement evasion by bacteria and parasites. *Annu. Rev. Microbiol.* **42**: 201–230.

27. **Kaijser, B., L. A. Hanson, V. Jodal, G. Linden-Johnson, and J. B. Robbins.** 1977. Frequency of *Escherichia coli* K antigen in urinary tract infection in children. *Lancet* **i**:664–666.

28. **Kasper, D. L., J. L. Winkelhake, W. D. Zollinger, B. L. Brandt, and M. S. Artenstein.** 1973. Immunochemical similarity between polysaccharide antigens of *Escherichia coli* O7:K1L:NM and group B *Neisseria meningitidis. J. Immunol.* **110**:262–268.

29. **Koshland, D. E., Jr.** 1993. The two-component pathway comes to eukaryotes. *Science* **262**:532.

30. **Kroll, J. S., B. Loynds, L. N. Brophy, and E. R. Moxon.** 1990. The *bex* locus in encapsulated *Haemophilus influenzae:* a chromosomal region involved in capsule polysaccharide export. *Mol. Microbiol.* **4**:1853–1862.

31. **Kroncke, K. D., G. Boulnois, I. Roberts, D. Bitter-Suermann, J. R. Golecki, B. Jann, and K. Jann.** 1990. Expression of the *Escherichia coli* K5 capsular antigen: immunoelectron microscopic and biochemical studies with recombinant *E. coli. J. Bacteriol.* **172**:1085–1091.

32. **McCracken, G. H., Jr., L. D. Sarff, M. P. Glode, S. G. Mize, M. S. Schiffer, J. B. Robbins, E. C. Gotschlich, I. Orskov, and F. Orskov.** 1974. Relation between *Escherichia coli* K1 polysaccharide antigen and clinical outcome in neonatal meningitis. *Lancet* **ii**:246–250.

33. **Mimura, C. S., S. R. Holbrook, and G. F.-L. Ames.** 1991. Structural model of the nucleotide-binding conserved component of periplasmic permeases. *Proc. Natl. Acad. Sci. USA* **88**:84–88.

34. **Moxon, E. R., and J. S. Kroll.** 1990. The role of bacterial polysaccharide capsules as virulence factors. *Curr. Top. Microbiol. Immunol.* **150**:65–85.

35. **Orskov, I., F. Orskov, B. Jann, and K. Jann.** 1977. Serology, chemistry, and genetics of O and K antigens of *Escherichia coli. Bacteriol. Rev.* **41**:667–710.

36. **Parent, R., and P. H. Roy.** 1992. The chloramphenicol acetyltransferase gene of Tn*2424:* a new breed of *cat. J. Bacteriol.* **174**:2891–2897.

37. **Pavelka, M. S., Jr., S. F. Hayes, and R. P. Silver.** Characterization of KpsT, the ATP-binding

component of the ABC-transporter involved with the export of capsular polysialic acid in *Escherichia coli* K1. *J. Biol. Chem.,* in press.

38. **Pavelka, M. S., Jr., L. F. Wright, and R. P. Silver.** 1991. Identification of two genes, *kpsM* and *kpsT,* in region 3 of the polysialic acid gene cluster of *Escherichia coli* K1. *J. Bacteriol.* **173:** 4603–4610.

39. **Pazzani, C., C. Rosenow, G. J. Boulnois, D. Bronner, K. Jann, and I. S. Roberts.** 1993. Molecular analysis of region 1 of the *Escherichia coli* K5 antigen gene cluster: a region encoding proteins involved in cell surface expression of capsular polysaccharide. *J. Bacteriol.* **175:**5978–5983.

40. **Pelkonen, S.** 1990. Capsular sialyl chains of *Escherichia coli* K1 mutants resistant to K1 phage. *Curr. Microbiol.* **21:**23–28.

41. **Peterson, P. K., and P. G. Quie.** 1981. Bacterial surface components and the pathogenesis of infectious diseases. *Annu. Rev. Med.* **32:**29–43.

42. **Philip, A. G. S.** 1985. *Neonatal Sepsis and Meningitis.* Hall Medical Publishers, Boston.

43. **Pigeon, R., and R. P. Silver.** Submitted for publication.

44. **Pluschke, G., J. Mayden, M. Achtman, and R. P. Levine.** 1983. Role of the capsule and the O antigen in resistance of O18:K1 *Escherichia coli* to complement-mediated killing. *Infect. Immun.* **42:**907–913.

45. **Pluschke, G., A. Mercer, B. Kusecek, A. Pohl, and M. Achtman.** 1983. The induction of bacteremia in newborn rats by *Escherichia coli* K1 is correlated with only certain O (lipopolysaccharide) antigens. *Infect. Immun.* **39:**599–608.

46. **Riordan, J. R., J. M. Rommens, B. Kerem, N. Alon, R. Rozmahel, Z. Grzelczak, J. Zielenski, S. Loc, N. Plavsic, J. Chou, M. L. Drumm, M. C. Iannuzzi, F. S. Collins, and L. Tsui.** 1989. Identification of the cystic fibrosis gene: cloning and characterization of complementary DNA. *Science* **245:** 1066–1073.

47. **Roberts, I. S., R. Mountford, R. Hodge, K. B. Jann, and G. J. Boulnois.** 1988. Common organization of gene clusters for the production of different capsular polysaccharides (K antigens) in *Escherichia coli. J. Bacteriol.* **170:**1305–1310.

48. **Robbins, J. B.** 1978. Vaccines for the prevention of encapsulated bacterial diseases. Current status, problems and prospects for the future. *Immunochemistry* **15:**839–854.

49. **Robbins, J. B., G. H. McCracken, Jr., E. C. Gotschlich, F. Ørskov, and L. A. Hanson.** 1974. *Escherichia coli* K1 capsular polysaccharide associated with neonatal meningitis. *N. Engl. J. Med.* **290:**1216–1220.

50. **Roth, J., U. Rutishauser, and F. A. Troy II (ed.).** 1993. *Polysialic Acid: from Microbes to Man.* Birkhauser Verlag, Basel.

51. **Roth, J., C. Zuber, P. Wagner, D. J. Taatjes, C. Weisgerber, P. U. Heits, C. Goridis, and D. Bitter-Suermann.** 1988. Reexpression of poly(sialic acid) units of the neural cell adhesion molecule in Wilms tumor. *Proc. Natl. Acad. Sci. USA* **85:**2999–3003.

52. **Rutishauer, U.** 1989. Polysialic acid as a regulator of cell interactions, p. 367–382. *In* R. U. Margolis and R. K. Margolis (ed.), *Neurobiology of Glycoconjugates.* Plenum Press, New York.

53. **Schauer, R.** 1982. Chemistry, metabolism, and biological functions of sialic acids. *Adv. Carbohydr. Chem.* **40:**131–194.

54. **Schmidt, A. M., and K. Jann.** 1982. Phospholipid substitution of capsular (K) polysaccharide antigens from *Escherichia coli* causing extraintestinal infections. *FEMS Microbiol. Lett.* **14:**69–74.

55. **Shayamala, V., V. Baichwal, E. Beall, and G. F.-L. Ames.** 1991. Structure-function analysis of the histidine permease and comparison with cystic fibrosis mutants. *J. Biol. Chem.* **266:**18714–18719.

56. **Silver, R. P., W. Aaronson, and W. F. Vann.** 1987. Translocation of capsular polysaccharides in pathogenic strains of *Escherichia coli* requires a 60-kilodalton periplasmic protein. *J. Bacteriol.* **169:**5489–5495.

57. **Silver, R. P., P. Annunziato, M. S. Pavelka, R. P. Pigeon, L. F. Wright, and D. E. Wunder.** 1993. Genetic and molecular analyses of the polysialic acid gene cluster of *Escherichia coli* K1, p. 59–71. *In* J. Roth, U. Rutishauser, and F. A. Troy II (ed.), *Polysialic Acid: from Microbes to Man.* Birkhauser Verlag, Basel.

58. **Silver, R. P., C. W. Finn, W. F. Vann, W. Aaronson, R. Schneerson, P. J. Kretschmer, and C. F. Garon.** 1981. Molecular cloning of the K1 capsular polysaccharide genes of *Escherichia coli. Nature* (London) **289:**696–698.

59. **Silver, R. P., W. F. Vann, and W. Aaronson.** 1984. Genetic and molecular analyses of *Escherichia coli* K1 antigen genes. *J. Bacteriol.* **157:**568–575.
60. **Silver, R. P., and E. R. Vimr.** 1990. Polysialic acid capsule of *Escherichia coli* K1, p. 39–60. *In* B. Iglewski and V. Clark (ed.), *The Bacteria*, vol. XI. *Molecular Basis of Bacterial Pathogenesis.* Academic Press, Inc., New York.
61. **Smith, A. N., G. J. Boulnois, and I. S. Roberts.** 1990. Molecular analysis of the *Escherichia coli* K5 *kps* locus: identification and characterization of an inner-membrane capsular polysaccharide transport system. *Mol. Microbiol.* **4:**1863–1869.
62. **Smith, H.** 1983. The elusive determinants of bacterial interference with non-specific host defences. *Philos. Trans. R. Soc. London Ser.* B **303:**99–113.
63. **Smith, H. W., and M. B. Huggins.** 1980. The association of the O18, K1 and H7 antigens and the ColV plasmid of a strain of *Escherichia coli* with its virulence and immunogenicity. *J. Gen. Microbiol.* **121:**387–400.
64. **Smith, T., and G. Bryant.** 1927. Studies on pathogenic B. coli from bovine sources. II. Mutations and their immunological significance. *J. Exp. Med.* **46:**133–140.
65. **Steenbergen, S. M., and E. R. Vimr.** 1990. Mechanism of polysialic chain elongation in *Escherichia coli* K1. *Mol. Microbiol.* **4:**603–611.
66. **Steenbergen, S. M., T. J. Wrona, and E. R. Vimr.** 1992. Functional analysis of the sialyltransferase complex in *Escherichia coli* K1 and K92. *J. Bacteriol.* **174:**1099–1108.
67. **Sussman, M.** 1985. *The Virulence of Escherichia coli.* Academic Press, London.
68. **Troy, F. A., II.** 1992. Polysialylation: from bacteria to brains. *Glycobiology* **2:**5–23.
69. **Vaara, M.** 1992. Eight bacterial proteins, including UDP-*N*-acetylglucosamine acyltransferase (LpxA) and three other transferases of *Escherichia coli,* consist of a six-residue periodicity theme. *FEMS Microbiol. Lett.* **97:**249–254.
70. **Vann, W. F., R. P. Silver, C. Abeijon, K. Chang, W. Aaronson, A. Sutton, C. W. Finn, W. Lindner, and M. Kotsatos.** 1987. Purification, properties, and genetic location of *Escherichia coli* cytidine 5′-monophosphate N-acetylneuraminic acid synthetase. *J. Biol. Chem.* **262:**17556–17562.
71. **Vann, W. F., G. Zapata, I. S. Roberts, G. J. Boulnois, and R. P. Silver.** 1993. Structure and function of enzymes in sialic acid metabolism in polysialic acid producing bacteria, p. 125–136. *In* J. Roth, U. Rutishauser, and F. A. Troy II (ed.), *Polysialic Acid: from Microbes to Man.* Birkhauser Verlag, Basel.
72. **Vimr, E. R., W. Aaronson, and R. P. Silver.** 1989. Genetic analysis of chromosomal mutations in the polysialic acid gene cluster of *Escherichia coli* K1. *J. Bacteriol.* **171:**1106–1117.
73. **Vimr, E. R., and S. Steenbergen.** 1993. Mechanisms of polysialic acid assembly in *Escherichia coli* K1, p. 73–92. *In* J. Roth, U. Rutishauser, and F. A. Troy II (ed.), *Polysialic Acid: from Microbes to Man.* Birkhauser Verlag, Basel.
74. **Walker, J. E., M. Saraste, M. J. Runswick, and N. J. Gay.** 1982. Distantly related sequences in the α and β-subunits of ATP synthase, myosin, kinases and other ATP-requiring enzymes and a common nucleotide binding fold. *EMBO J.* **8:**945–951.
75. **Weisgerber, C., A. Hansen, and M. Frosch.** 1991. Complete nucleotide and deduced amino acid sequence of CMP-NeuAc:poly-α-2,8-sialosyl sialyltransferase of *Escherichia coli* K1. *Glycobiology* **1:**357–363.
76. **Weisgerber, C., and F. A. Troy II.** 1990. Biosynthesis of the polysialic acid capsule in *Escherichia coli* K1. The endogenous acceptor of polysialic acid is a membrane protein of 20 kDa. *J. Biol. Chem.* **265:**1578–1587.
77. **Wunder, D. E., W. Aaronson, S. F. Hayes, J. M. Bliss, and R. P. Silver.** 1994. Nucleotide sequence and mutational analysis of the gene encoding KpsD: a periplasmic protein involved in transport of polysialic acid in *Escherichia coli* K1. *J. Bacteriol.* **176:**4025–4033.
78. **Wyle, F. A., M. S. Artenstein, B. L. Brandt, E. C. Tramont, D. L. Kasper, P. L. Altieri, S. L. Berman, and J. P. Lowenthal.** 1972. Immunologic response of man to group B meningococcal polysaccharide vaccines. *J. Infect. Dis.* **126:**514–522.
79. **Zapata, G., W. F. Vann, W. Aaronson, M. S. Lewis, and M. Moos.** 1989. Sequence of the cloned *Escherichia coli* K1 CMP-*N*-acetylneuraminic acid synthetase gene. *J. Biol. Chem.* **264:**14769–14774.

Molecular Genetics of Bacterial Pathogenesis
Edited by V. L. Miller, J. B. Kaper, D. A. Portnoy, and R. R. Isberg
© 1994 American Society for Microbiology, Washington, DC 20005

Chapter 21

Type III Capsular Polysaccharide of Group B Streptococci: Role in Virulence and the Molecular Basis of Capsule Expression

Craig E. Rubens

Group B streptococci (GBS) are weakly beta-hemolytic, facultatively anaerobic, gram-positive cocci which have emerged over the past 50 years as the most common cause of neonatal pneumonia, sepsis, and meningitis in the United States, western Europe, and some developing countries (6, 70, 77). Mortality in infants ranges from 10 to 20%, with significant neurological sequelae occurring in up to 50% of survivors (19, 69). In addition, it is a major cause of perinatal morbidity in pregnant women, and the number of infections in nonpregnant, immunocompromised adults has increased substantially (21). The mortality rate in the latter group is higher than that observed in infants, approaching 40%. The impact of perinatal infections alone on public health resources has been estimated at $700 million annually (34). The recognition of GBS as an important human pathogen has pressed the need for a better understanding of the pathogenesis of infections caused by this organism. This requires identification and characterization of the factors important for its virulence and their interactions with various host immune defense mechanisms.

CAPSULAR POLYSACCHARIDES OF GBS

GBS are primarily classified into serotypes on the basis of structurally and antigenically distinct capsular polysaccharide antigens composed of galactose, glucose, *N*-acetylglucosamine, and sialic acid (except type 6, which lacks *N*-acetylglucosamine) (13, 36). All isolates from human infections are encapsulated. Antibodies directed against the capsular polysaccharides provide protection by opsonization, enhancing phagocytic activity and killing of GBS by leukocytes in the presence of complement (1, 3, 4, 9, 16, 18, 31, 56, 60). A low level of capsule-specific maternal immunoglobulin G is considered a major risk factor for the development of subsequent infection by the neonate (7).

Craig E. Rubens • Division of Infectious Diseases, Children's Hospital and Medical Center, and Department of Pediatrics, University of Washington, Seattle, Washington 98105.

Although the primary serotypes Ia, Ib, II, and III are isolated equally from asymptomatic colonized infants and mothers, the predominant serotype isolated from infected infants is type III (for a review, see reference 6). About 90% of isolates from infants with meningitis are type III strains. The reason for the predominance of type III in ~70% of the GBS infections is as yet not clearly understood, but it has been suggested that the capsule might be an important virulence determinant. Indirect evidence that the type III capsule has a role in virulence was also suggested by initial laboratory studies with chemically derived mutants which failed to produce the capsule (76). These mutants were less virulent when compared with the wild-type strain in an animal model of GBS infection.

More direct evidence of the role of the type III capsule of GBS in virulence has been provided by deriving isogenic (differing from the wild-type strain by a single trait) acapsular mutants by transposon mutagenesis and comparing the mutant and wild-type strains in in vitro and in vivo models of virulence. The unencapsulated mutants failed to produce any recognizable capsule polymer on their surfaces (51, 53). In comparison with the wild-type parent strains, these mutants were susceptible to opsonophagocytosis in the presence of complement and peripheral blood leukocytes in vitro. The mechanism for this susceptibility has been investigated by Marques et al. (43), who demonstrated that the acapsular mutant bound significantly more complement factor C3 fragments than the parent strain. C3 fragments bound to the acapsular mutant were primarily in the active form, C3b, whereas the inactive form, C3bi, was predominantly bound to the surface of the parent strain. The degree of encapsulation influenced the amount of C3 bound as well, because an inverse correlation was shown between C3 binding and the amount of type III capsule on the bacterial surface (43). These data provided compelling evidence that the capsule inhibits C3 deposition and inactivates the alternative complement pathway, providing the organism a protective mechanism from phagocytic killing. How this polysaccharide prevents activation of complement and hinders binding of opsonic fragments has been elucidated by investigating the role of its side chain sialic acid moiety.

SIALIC ACID IS CRITICAL TO VIRULENCE FUNCTION OF TYPE III CAPSULE

The type III capsular polysaccharide is a high-molecular-weight polymer composed of repeating pentasaccharide subunits containing a trisaccharide backbone of galactose, glucose, and N-acetylglucosamine with a side chain of galactose and a terminal sialic acid moiety (Fig. 1). Earlier studies have demonstrated that the

```
        GLUCOSE-GLUCOSAMINE-GALACTOSE     GLUCOSE-GLUCOSAMINE-GALACTOSE
                      |                                 |
                  GALACTOSE                         GALACTOSE
Native                |                                            Asialo
                  SIALIC ACID
```

FIGURE 1. Repeating unit structure of native and asialo type III capsular polysaccharide.

presence of the terminal side chain sialic acid confers a critical conformational influence on the epitope which incited and recognized protective antibody. Alteration of sialic acid resulted in a backbone structure (asialo) which did not react with or incite protective antibodies (38). In epidemiological studies, a correlation was observed between newborns with antibody to the sialylated capsule and protection from infection, which is in contrast to the case for infants who lacked antibody or who had antibody to the desialylated polysaccharide (7, 38). Further studies with enzymatically treated type III GBS indirectly demonstrated that the sialic acid moiety of the type III capsular polysaccharide is critical for preventing opsonization by the alternative complement pathway in the absence of anticapsular antibody. Organisms treated with neuraminidase activated complement and were readily phagocytosed (17, 18).

Direct evidence for the role of sialic acid has been provided by investigating transposon mutants of type III GBS which produced a capsular polysaccharide lacking the terminal sialic acid moiety (72, 74) (asialo; Fig. 1). When compared with the parent strains, the asialo mutants were as susceptible to phagocytic killing in the presence of complement and peripheral blood leukocytes, as observed for the acapsular mutant. Marques et al. (43) demonstrated that the asialo mutant bound more of the active form of C3 (C3b) than the sialylated capsule of the parent strain but bound less C3b than the acapsular mutant. The asialo mutant also bound less C3bi than the parent strain. The fact that a significant amount of C3b bound to the type 14 pneumococcus, which produces a capsule with the same repeating unit structure as type III GBS but without sialic acid (asialo-like), provided further evidence that capsular sialic acid influenced C3 deposition and activation (32).

Together, the in vitro observations described above demonstrate that sialic acid is responsible for the virulence functions of the type III capsule by inhibiting binding of C3 to the organism and inhibiting the alternative complement pathway by promoting inactivation of C3b via sialic acid, therefore preventing phagocytic killing. The sialic acid-rich capsules of *Escherichia coli* K1 and group B *Neisseria meningitidis* regulate activation of complement similarly (17, 35, 67).

EFFECT OF CAPSULE MUTATIONS ON VIRULENCE IN VIVO

The acapsular and asialo mutants described above were significantly less virulent in animal models of GBS infection. Subcutaneous injection of the acapsular or asialo mutants into neonatal rats resulted in similar 50% lethal doses that were at least 2 orders of magnitude higher than those observed for the parent strain (51, 53, 74). These data confirm the above in vitro findings, further demonstrating that the capsule is an essential virulence determinant, and also confirm that sialic acid is critical to the function of the capsule in vivo.

Newborn infants, however, acquire GBS primarily by aspiration of infected amniotic fluid in utero or vaginal contents during descent through the birth canal. Once in the airways and alveoli, the organism replicates luxuriously and crosses the alveolar septal walls by invading the respiratory epithelium and capillary endothelium, resulting in bacteremia, sepsis, and dissemination to other organs, such as the central nervous system (23, 54, 55). In a more relevant model of neonatal GBS pneumonia and bacteremia, neonatal rats were inoculated with either the parent strain or an acapsular mutant by intratracheal injection (44). Immediately after injection and after 6 h, the animals were sacrificed and the lungs and spleens were removed for quantitative culture and histologic examination. For animals who received the acapsular mutant, fewer GBS were recovered per gram of lung, more bacteria were associated with resident alveolar macrophages, and the animals were significantly less bacteremic than animals who received the parent strain. These in vivo observations provide further compelling evidence that the capsule protects the organism from phagocytic clearance early during various stages in the pathogenesis of infection of a nonimmune host.

Previous observations have shown that the amount of capsule expressed by GBS may vary depending on the conditions of growth or the environment. Subpopulations of heavy and poorly encapsulated GBS can be subcultured from strains isolated from infected infants (25, 28, 29). Organisms passed on growth media become less encapsulated (phase shifting) than when they are passed in animals. The growth conditions in artificial media also influence the proportion of heavy and poorly encapsulated cells. In addition, the capsule attenuates, but does not prevent, invasion of epithelial and endothelial cells in vitro (23, 33). Similar observations have been reported for other invasive encapsulated bacteria such as *Haemophilus influenzae* (61). These observations suggest that the organism may regulate and alter capsule expression to suit the environmental conditions with which it is confronted.

Perhaps, on the mucosal surface, less capsule may promote adherence and invasion of the epithelium. In vitro the asialo and acapsular mutants were more adherent and significantly more invasive for respiratory epithelium than the parent strain (33). These observations suggest that the capsule hinders interaction with the epithelial cell surface and/or partially masks important bacterial surface ligands that recognize epithelial surface receptors (64). Lowering of the level of capsule production, however, renders the organism more susceptible to mucosal surface opsonins and phagocytic cells. This suggests that a dynamic balance between producing enough capsule to prevent bacterial clearance but not enough to prevent the bacterial cell-epithelial cell interaction may exist during the early stages of the infection. Once within the subepithelial tissue or in the bloodstream, an increase in the amount of capsule expressed would prevent rapid clearance by host defenses. An understanding of the regulation of capsule expression in response to environmental stimuli and the role of variable capsule synthesis during pathogenesis will require further definition of the molecular mechanisms responsible for capsule production, such as those described below.

GENETICS OF TYPE III CAPSULE BIOSYNTHESIS

Investigation into the molecular basis of capsular polysaccharide production by gram-positive organisms began with the early studies of transformation that led to the classic discovery of DNA as the transforming factor by Avery et al.

(5). Those studies were performed in *Streptococcus pneumoniae* as early as 1931 (14). In those studies, the capsule expressed by one serotype could be changed by transformation with genetic material from a heterologous serotype in vitro; acapsular strains were also transformed to a capsulated phenotype (26). Further insights into the genetics of pneumococcal capsule synthesis were provided by investigating spontaneous capsule mutants with characterized defects in synthetic functions. Using natural transformation and classic cross-over techniques, the investigators in those studies demonstrated that the capsule genes are closely linked. The fact that they were transferred together suggested that they are regionally clustered on the chromosome (10, 20, 47). The pneumococcal capsule genes and their organization have yet to be characterized.

The genetic characterization of complex polysaccharide production by gram-negative bacteria has proceeded more quickly because of advances in the molecular technology available for these organisms. Polysaccharide genes are usually clustered on the chromosome, as for *E. coli* and *Salmonella*, *Klebsiella*, and *H. influenzae*, or on megaplasmids, as shown for *Rhizobium* spp. (11, 12, 40, 42, 48, 49, 57, 58). The loci for nucleotide sugar precursor biosynthesis, the polymerization of activated sugars into oligosaccharide subunits, and subunit polymerization and transport are arranged in specific regions within these clusters. Regions containing the genes responsible for transport and modification are usually homologous within a given species. The differences in the types of capsules synthesized within a species are often due to different biosynthetic structural genes within a single region, as shown for the *E. coli*, *N. meningitidis*, and *H. influenzae* capsule genes (11, 22, 39, 49).

An understanding of the molecular biology and biochemistry of GBS capsule synthesis is developing. The derivation of GBS type III capsule mutants has provided the means for identifying the genes responsible for capsule biosynthesis. The DNAs flanking the transposon insertion sites from the acapsular and asialo mutants were cloned and used as probes of a type III genomic cosmid library cloned in *E. coli*. The transposon insertion sites for both types of mutants were mapped to a 30-kb region of the type III chromosome (Fig. 2) (41, 52). These data suggested that the GBS capsule genes may be clustered on the GBS chromosome, as observed for other encapsulated bacteria. Probes from this region were used

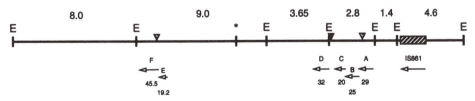

FIGURE 2. The 30-kb region of representative type III GBS chromosome. E, *Eco*RI sites; X, *Xba*I sites; *, *Eco*RI RFLP seen in some type III and all type Ia strains analyzed to date. Numbers above the map indicate the sizes of the *Eco*RI fragments (in kilobases). Inverted triangles are Tn916 or Tn916ΔE transposon insertion sites in type III GBS capsule mutants discussed in the text. Open arrows indicate open reading frames, with the orientation of transcription, the *cps* gene letter designation, and sizes (in kilodaltons) indicated.

in hybridization analyses with genomic DNAs from other type III strains and representative strains from the other GBS capsular serotypes. Southern hybridization analysis demonstrated that this region is highly conserved among all the serotypes (41).

Clues to the genes and biochemical pathways which encode for capsule synthesis in GBS have come from the results of studies with other encapsulated organisms. For example, in the generation of many polysaccharides, the monosaccharide precursors, either synthesized or scavenged from the environment, must be activated with a nucleotide by a condensation reaction prior to polymerization (63). In *E. coli* K1, sialic acid is synthesized de novo and then condensed with CTP to CMP-sialic acid through a synthetase encoded by *neuA* (68, 78), prior to its addition to a nascent polymer of sialic acid by sialyltransferase (see chapter 1). Similarly, investigators have begun to identify the genes in GBS involved in monosaccharide synthesis and activation. The transposon mutations in the type III asialo mutants resulted in the accumulation of free sialic acid and the production of an asialo surface polymer because of the interruption of a locus which encoded sialic acid synthetase activity (72). Nucleotide sequence analysis of this locus has identified the open reading frame, *cpsF*, which shares extensive homology with the *neuA* gene (78) from *E. coli* K1 (27). Those data have provided further support for prior observations that the mutants described above were defective in the biochemical reaction that activates sialic acid with a nucleotide phosphate (72).

Nucleotide sequence analysis of DNA 5′ to the *cpsF* gene identified another open reading frame designated *cpsE* in a different reading frame (Fig. 2). Best-fit analysis revealed significant homology with genes that code for the enzymes involved in acetylation reactions, e.g., *cat* (chloramphenicol acetyltransferase), *nodL* (acetylation of lipooligosaccharides in *Rhizobium* spp.), *cysE* (serine acetyltransferase), and *lacA* (lactose transacetylase) (50). The most striking homology was observed with *neuD* from *E. coli* K1, whose function is unknown. Mutations in *neuD* prevent capsule synthesis for reasons that are unclear, because sialic acid synthesis and the sialic acid synthetase activity in lysates of these *E. coli* capsule mutants remain intact (57). Common to all of the genes similar to *cpsE* was the presence of a consensus sequence which may be the active-site domain for their catalytic function (57). The homology with known acetylating enzymes and the presence of a common consensus domain suggest that the *cpsE* gene product is involved in an acetylation reaction during monosaccharide or oligosaccharide synthesis. The proximity to the *cpsF* gene may suggest that *cpsE* is involved in *N*-acetylneuraminic acid metabolism in GBS. Alternatively, the *cpsE* gene product may have a role in another pathway for polymer synthesis, e.g., metabolism of *N*-acetylglucosamine, which is the central monosaccharide in the oligosaccharide subunit. These data suggest that the genes for sialic acid metabolism in GBS may code for gene products with functions similar to those reported for *E. coli* and *N. meningitidis* (8, 45). Whether the genes responsible for the synthesis and activation of the other monosaccharides in the GBS capsule are similar has yet to be determined.

Once the monosaccharides are synthesized into nucleotide precursors, they are polymerized into an oligosaccharide or a polysaccharide, either directly or through linkage to a lipid intermediate. Group A streptococci do not appear to utilize a lipid intermediate in the production and secretion of their hyaluronic acid capsule (62). Lipid intermediates are often used in the production of complex polysaccharides, such as the O-polysaccharide of *Salmonella* lipopolysaccharide (75), and the capsules of *E. coli* (66), *Klebsiella* spp. (12), and *Rhizobium* spp. (48). Oligosaccharide subunits are assembled enzymatically on isoprenoid lipid phosphate intermediates by specific transmembrane glycosyltransferases. These subunits are then transported across the cytoplasmic membrane and polymerized by mechanisms which are poorly understood, but they appear to include ATP-binding transport proteins such as the gene products encoded by *bexA* (39), *kpsT* (57), and *exoP* (24).

The genes involved in the polymerization and transport of the oligosaccharide subunits of the type III capsule appear to be located within the 30-kb region shown in Fig. 2. Nucleotide sequence analysis of the DNA flanking the transposon insertion sites in the acapsular mutants identified four open reading frames, *cpsA*, *cpsB*, *cpsC*, and *cpsD*, the characteristics of which are shown in Fig. 2 (51). The transposon sites were mapped to *cpsA* and *cpsD*. No significant homology between *cpsA* and other known genes was found. *cpsB* is immediately distal but out of frame with the other three genes, and the function of its gene product is also unknown.

cpsC encodes a 20-kDa protein which shares homology with the *exoP* gene of *Rhizobium* spp. (50); the latter appears to be involved in the polymerization or transport of oligosaccharide subunits of the succinoglycan exopolysaccharide (24). Of note is that the carboxyl-terminal end of the *exoP* gene has a region which is homologous to putative nucleotide-binding sites observed in a superfamily of transporter proteins present in prokaryotes and eukaryotes (30). *cpsC* shares 50% identity in this same region. The proteins of both *exoP* and *cpsC* also contain a hydrophobic domain which may be a putative transmembrane domain. Whether these homologous regions imply significant similarities in structure and function for *cpsC* and *exoP* remains to be determined. The biochemical defects created by the transposon mutation in *cpsA* have not been characterized, but the similarity of *cpsC* to *exoP* suggests that the genes in this region may be involved in the polymerization or transport of oligosaccharide subunits to the bacterial surface.

Acapsular mutations as a result of transposon insertion very close to or just within the start of *cpsD* were also identified (Fig. 2) (51). This gene shared significant homology with *rfbP* and *exoY*, which are galactosyltransferase genes from the O-polysaccharide locus (*rfb*) of *Salmonella typhimurium* and the exopolysaccharide locus of *Rhizobium meliloti*, respectively (24, 37, 46). These galactosyltransferases catalyze the addition of the first monosaccharide, UDP-galactose, to a lipid intermediate (undecaprenol phosphate) in the synthesis of their oligosaccharide subunits. Mutations in *rfbP* or *exoY* interrupt oligosaccharide subunit synthesis and, therefore, prevent polysaccharide expression in *S. typhimurium* and *R. meliloti*, respectively. These data imply that *cpsD* may also function as a galactosyltransferase. Further evidence that this gene may encode a glycosyltransferase which associates with a lipid intermediate was provided by the identification of a consensus sequence for the binding of isoprenoid units in the lipid dolichol. This consensus binding site has been observed in several eukaryotic glycosyltransferases and the proteins encoded by *neuE* and *kpsM* in *E. coli,* which catalyze the synthesis and/or transport of glycoproteins or oligosaccharides through a lipid intermediate (2, 59, 65). The observation of a dolichol-binding sequence in *cpsD* was the first reported in a glycosyltransferase from a gram-positive organism.

Further evidence for the function of the *cpsD* gene product was provided by biochemical analysis of the acapsular *cpsD* mutation. Particulate membrane fractions of lysates of this mutant demonstrated a 50% reduction in the transfer of $[^{14}C]$UDP-galactose to a lipid carrier, undecaprenol phosphate, compared with that in lysates of the parent strain (51). These data suggest that the mutation causes a defect in galactosyltransferase activity, but that the mutation is either incomplete, which was doubtful since no polysaccharide was detected in the mutant, or that another gene in GBS expresses similar activity. Alternatively, some activity could be contributed by an endogenous epimerase, for example, the conversion of $[^{14}C]$UDP-galactose to glucose, which could be incorporated into a lipid-bound complex. In addition, those studies provided the first direct biochemical evidence that the GBS capsule is synthesized through a lipid intermediate.

These observations demonstrate that *cpsD* may reside within a region which codes for polymerization of the activated monosaccharide precursors into the pentasaccharide subunit. In addition, as observed for *cpsC, cpsE,* and *cpsF,* the gene product of *cpsD* shares significant homology with proteins involved in the synthesis of polysaccharides in gram-negative bacteria, suggesting that gram-positive organisms may utilize similar mechanisms for complex polymer synthesis. The evolutionary significance of these observations awaits further study.

SUMMARY

The type III capsular polysaccharide plays a major role in the virulence of GBS. Studies to date have provided compelling evidence that sialic acid is a critical sugar moiety in this complex polysaccharide for preventing phagocytic killing of the organism by inhibiting complement deposition and activation on the bacterial surface. The role of the capsule during an infection has been further defined by comparing isogenic capsule mutants with the wild-type strain in in vitro and in vivo models of GBS infection, which focus on specific stages in pathogenesis. These studies have demonstrated that the presence and/or amount of capsule synthesized influences the organism's association with mucosal epithelial cells and capillary endothelial cells of the lung. Although the capsule is required to prevent clearance by professional phagocytes on the mucosal surface, too much capsule on the surface may inhibit the organism's ability to invade epithelial and endothelial cells effectively. Once in the bloodstream and deeper tissues, GBS must protect themselves from opsonization by serum complement and killing by leukocytes, an environment in which the production of increased levels of a sialic acid-rich capsule may be paramount for the organism's survival.

The genes responsible for capsule synthesis have been localized so far to a specific region of the chromosome. This region appears to be highly conserved among all the serotypes. However, some minor differences which may reflect the minor structural differences of the various serotypes have been observed. Genes for monosaccharide synthesis and activation, oligosaccharide subunit polymerization, and potential subunit transport and polymerization appear to be clustered, as observed for capsule synthesis in *E. coli, N. meningitidis,* and *H. influenzae.* Perhaps, as shown for these organisms, the genes involved in the transport and regulation of the GBS capsule are highly conserved among the serotypes, with specific differences in genetic organization occurring between the specific structural genes according to the serotype. In addition, several of the GBS capsule genes share significant homology at the amino acid level with polysaccharide synthesis genes from gram-negative organisms. Although the evolutionary significance is unclear, these observations suggest that the mechanisms for polysaccharide production by gram-negative and gram-positive bacteria may be similar. Some differences must exist, however, since production of a carbohydrate polymer in gram-positive bacteria occurs across a single cytoplasmic membrane and much thicker peptidoglycan cell wall.

Molecular analysis of the genes involved in GBS capsule biosynthesis should provide an understanding of complex polysaccharide biosynthesis and transport in gram-positive organisms. Similar studies investigating capsule synthesis by *Streptococcus pneumoniae* (71) and *Streptococcus pyogenes* (15, 73) are being pursued. The results of these genetic studies will also provide the foundation for understanding how gram-positive organisms regulate expression of the capsule in response to the environment during the various stages in the pathogenesis of an infection. For group B streptococci, the severity of the infections in newborn infants and the increasing prevalence of infections in compromised adults underscore the importance of increasing our understanding of this important virulence factor and its role in pathogenesis.

REFERENCES

1. **Adamkin, D., B. A. Stitzel, J. Urmson, M. L. Farnett, E. Post, and R. Spitzer.** 1978. Activity of the alternative pathway of complement in the newborn infant. *J. Pediatr.* **93:**604–608.
2. **Albright, C. F., P. Orlean, and P. W. Robbins.** 1989. A 13-amino acid peptide in three yeast glycosyltransferases may be involved in dolichol recognition. *Proc. Natl. Acad. Sci. USA* **86:** 7366–7369.
3. **Anderson, D. C., M. S. Edwards, and C. J. Baker.** 1980. Luminol-enhanced chemiluminescence for evaluation of type III group B streptococcal opsonins in human sera. *J. Infect. Dis.* **141:** 370–381.
4. **Anthony, B. F.** 1976. Immunity to the group B streptococci: interaction of serum and macrophages with types Ia, Ib, and Ic. *J. Exp. Med.* **143:**1186–1198.
5. **Avery, O. T., C. M. MacLeod, and M. McCarty.** 1944. Studies on the chemical nature of the substance inducing transformation of pneumococcal types. Induction of transformation by a desoxyribonucleic acid fraction isolated from pneumococcus type III. *J. Exp. Med.* **79:**137–158.
6. **Baker, C. J., and M. S. Edwards.** 1990. Group B streptococcal infections, p. 742–811. *In* J. S. Remington and J. O. Klein (ed.), *Infectious Diseases of the Fetus and Newborn Infant.* The W. B. Saunders Co., Philadelphia.
7. **Baker, C. J., and D. L. Kasper.** 1976. Correlation of maternal antibody deficiency with susceptibility to neonatal group B streptococcal infection. *N. Engl. J. Med.* **294:**753–756.
8. **Balcklow, R. S., and L. Warren.** 1962. The biosynthesis of sialic acids by Neisseria meningitidis. *J. Biol. Chem.* **237:**3520–3526.

9. **Baltimore, R. S., D. L. Kasper, C. J. Baker, and D. K. Goroff.** 1977. Antigenic specificity of opsonophagocytic antibodies in rabbit antisera to group B streptococci. *J. Immunol.* **118:**673–678.

10. **Bernheimer, H. P., and I. E. Wermundsen.** 1972. Homology in capsular transformation fractions in Pneumococcus. *Mol. Gen. Genet.* **116:**68–83.

11. **Boulnois, G. J., and K. Jann.** 1989. Bacterial polysaccharide capsule synthesis, export and evolution of structural diversity. *Mol. Microbiol.* **3:**1819–1823.

12. **Clarke, B. R., and C. Whitfield.** 1992. Molecular cloning of the *rfb* region of *Klebsiella pneumoniae* serotype O1:K20: the *rfb* gene cluster is responsible for synthesis of the D-galactan I O polysaccharide. *J. Bacteriol.* **174:**4614–4621.

13. **d'Ascenzi, S., C. von Hunolstein, M. Dentini, G. Alfarone, V. Crescenzi, and G. Orefici.** 1992. Characterization of group B Streptococcus type capsular polysaccharides, p. 458–460. *In* G. Orefici (ed.), *New Perspectives on Streptococci and Streptococcal Infections*. Proceedings of the XI Lancefield International Symposium. Gustav Fisher Verlag, New York.

14. **Dawson, M. H., and R. H. P. Sia.** 1931. In vitro transformation of pneumococcal types. A technique for inducting transformation of pneumococcal types in vitro. *J. Exp. Med.* **54:**681–699.

15. **Dougherty, B. A., and I. van de Ryn.** 1992. Molecular characterization of a locus required for hyaluronic acid capsule production in group A streptococci. *J. Exp. Med.* **175:**1291–1299.

16. **Edwards, M. S., C. J. Baker, and D. L. Kasper.** 1979. Opsonic specificity of human antibody to the type III polysaccharide of group B Streptococcus. *J. Infect. Dis.* **140:**1004–1008.

17. **Edwards, M. S., D. L. Kasper, H. J. Jennings, C. J. Baker, and A. Nicholson-Weller.** 1982. Capsular sialic acid prevents activation of the alternative complement pathway by type III, group B streptococci. *J. Immunol.* **128:**1278–1283.

18. **Edwards, M. S., A. Nicholson-Weller, C. J. Baker, and D. L. Kasper.** 1980. The role of specific antibody in alternative pathway-mediated opsonophagocytosis of type III, group B Streptococcus. *J. Exp. Med.* **151:**1275–1287.

19. **Edwards, M. S., M. A. Rench, A. A. M. Haffer, M. A. Murphy, M. M. Desmond, and C. J. Baker.** 1985. Long-term sequelae of group B streptococcal meningitis in infants. *J. Pediatr.* **106:**717–722.

20. **Effrussi-Taylor, H.** 1951. Genetic aspects of transformation of pneumococci. *Cold Spring Harbor Symp. Quant. Biol.* **16:**445–456.

21. **Farley, M. M., R. C. Harvey, T. Stull, J. D. Smith, A. Schuchat, J. D. Wenger, and D. S. Stephens.** 1993. A population-based assessment of invasive disease due to group B streptococcus in nonpregnant adults. *N. Engl. J. Med.* **328:**1807–1811.

22. **Frosch, M., C. Weisgeber, and T. F. Meyer.** 1989. Molecular characterization and expression in Escherichia coli of the gene complex encoding the polysaccharide capsule of Neisseria meningitidis group B. *Proc. Natl. Acad. Sci. USA* **86:**1699–1673.

23. **Gibson, R. L., M. K. Lee, C. Soderland, E. Y. Chi, and C. E. Rubens.** 1993. Group B streptococci invade endothelial cells: type III capsular polysaccharide attenuates invasion. *Infect. Immun.* **61:**478–485.

24. **Glucksmann, M. A., T. L. Reuber, and G. C. Walker.** 1993. Genes needed for the modification, polymerization, export, and processing of succinoglycan by *Rhizobium meliloti:* a model for succinoglycan biosynthesis. *J. Bacteriol.* **175:**7045–7055.

25. **Gray, B. M., and D. G. Pritchard.** 1992. Phase variation in the pathogenesis of group B streptococcal infections, p. 452–454. *In* G. Orefici (ed.), *New Perspectives on Streptococci and Streptococcal Infections*. Proceedings of the XI Lancefield International Symposium. Gustav Fisher Verlag, New York.

26. **Griffith, F.** 1928. The significance of pneumococcal types. *J. Hyg.* **27:**119–159.

27. **Haft, R., M. Mebane, M. Wessels, and C. E. Rubens.** Unpublished data.

28. **Hakansson, S., A. Bergholm, S. E. Holm, B. Wagner, and M. Wagner.** 1988. Properties of high and low density subpopulations of group B streptococci: enhanced virulence of the low density variant. *Microb. Pathog.* **5:**345–355.

29. **Hakansson, S., S. E. Holm, and M. Wagner.** 1987. Density profile of group B streptococci, type III, and its possible relation to enhanced virulence. *J. Clin. Microbiol.* **25:**714–718.

30. **Higgins, C. F.** 1993. ABC transporters: from microorganisms to man. *Annu. Rev. Cell Biol.* **8:**67–113.

31. **Hill, H. R., A. O. Shigeoka, R. T. Hall, and V. G. Hemming.** 1979. Neonatal cellular and humoral immunity to group B streptococci. *Pediatrics* **S64:**787–794.

32. **Hostetter, M. K.** 1986. Serotypic variations among virulent pneumococci in deposition and degradation of covalently bound C3b: implications for phagocytosis and antibody production. *J. Infect. Dis.* **153:**682–692.

33. **Hulse, M. L., S. Smith, E. Y. Chi, A. Pham, and C. E. Rubens.** 1993. The effect of type III group B streptococcal capsular polysaccharide on invasion of respiratory epithelial cells. *Infect. Immun.* **61:**4835–4841.

34. **Institute of Medicine, National Academy of Sciences.** 1985. New vaccine development: establishing priorities, p. 242–439, appendix P. *In Diseases of Importance in the United States,* vol. 1. National Academy Press, Washington, D.C.

35. **Jarvis, G. A., and N. A. Vedros.** 1987. Sialic acid of group B *Neisseria meningitidis* regulates alternative complement pathway activation. *Infect. Immun.* **55:**174–180.

36. **Jelinkova, J., and J. Motlova.** 1985. The nomenclature of GBS. *Antibiot. Chemother.* **35:**49–52.

37. **Jiang, X. M., B. Neal, F. Santiago, S. J. Lee, L. K. Romana, and P. R. Reeves.** 1991. Structure and sequence of the rbf (O antigen) gene cluster of Salmonella serovar typhimurium (strains LT2). *Mol. Microbiol.* **5:**695–713.

38. **Kasper, D. L., C. J. Baker, R. S. Baltimore, J. H. Crabb, G. Schiffman, and H. J. Jennings.** 1979. Immunodeterminant specificity of human immunity to type III group B streptococcus. *J. Exp. Med.* **149:**327–339.

39. **Kroll, J. S., I. Hopkins, and E. R. Moxon.** 1988. Capsule loss in H. influenzae type b occurs by recombination-mediated disruption of a gene essential for polysaccharide export. *Cell* **53:**347–356.

40. **Kroll, J. S., B. M. Loynds, and E. R. Moxon.** 1991. The Haemophilus influenzae capsulation gene cluster: a compound transposon. *Mol. Microbiol.* **5:**1549–1560.

41. **Kuypers, J. M., L. M. Heggen, and C. E. Rubens.** 1989. Molecular analysis of a region of the group B Streptococcus chromosome involved in type III capsule expression. *Infect. Immun.* **57:** 3058–3065.

42. **Liu, D., A. M. Haase, L. Lindqvist, A. A. Lindberg, and P. R. Reeves.** 1993. Glycosyl transferases of O-antigen biosynthesis in *Salmonella enterica:* identification and characterization of transferase genes of groups B, C2, and E1. *J. Bacteriol.* **175:**3408–3413.

43. **Marques, M. B., D. L. Kasper, M. K. Pangburn, and M. R. Wessels.** 1992. Prevention of C3 deposition by capsular polysaccharide is a virulence mechanism of type III group B streptococci. *Infect. Immun.* **60:**3986–3993.

44. **Martin, T. R., J. T. Ruzinski, C. E. Rubens, E. Y. Chi, and C. B. Wilson.** 1992. The effect of type-specific polysaccharide capsule on the clearance of group B streptococci from the lungs of infant and adult rats. *J. Infect. Dis.* **165:**306–314.

45. **Merker, R. I., and F. A. Troy.** 1990. Biosynthesis of the polysialic acid capsule in Escherichia coli K1. Cold inactivation of sialic acid synthase regulates capsule expression below 20°C. *Glycobiology* **1:**93–100.

46. **Muller, P., M. Keller, W. M. Weng, J. Quandt, W. Arnold, and A. Puhler.** 1993. Genetic analysis of the Rhizobium meliloti exoYFQ operon: ExoY is homologous to sugar transferases and ExoQ represents a transmembrane protein. *Am. Phytopathol. Soc.* **6:**55–65.

47. **Ravin, A. W.** 1960. Linked mutations borne by deoxyribonucleic acid controlling the synthesis of capsular polysaccharide in pneumococcus. *Genetics* **45:**1387–1403.

48. **Reuber, T. L., and G. C. Walker.** 1993. Biosynthesis of succinoglycan, a symbiotically important exopolysaccharide of Rhizobium meliloti. *Cell* **74:**269–280.

49. **Roberts, I., R. Mountford, N. High, D. Bitter-Suermann, K. Jann, K. Timmis, and G. Boulnois.** 1986. Molecular cloning and analysis of genes for production of K5, K7, K12, and K92 capsular polysaccharides in *Escherichia coli. J. Bacteriol.* **168:**1228–1233.

50. **Rubens, C. E.** Unpublished data.

51. **Rubens, C. E., L. M. Heggen, R. F. Haft, and M. R. Wessels.** 1993. Identification of cpsD, a gene essential for type III capsule expression in group B streptococci. *Mol. Microbiol.* **8:**843–855.

52. **Rubens, C. E., J. M. Kuypers, L. M. Heggen, D. L. Kasper, and M. R. Wessels.** 1991. Molecular analysis of the group B streptococcal capsule genes, p. 179–183. *In* G. M. Dunny, P. P. Cleary, and

L. L. McKay (ed.), *Genetics and Molecular Biology of Streptococci, Lactococci, and Enterococci*. American Society for Microbiology, Washington, D.C.

53. **Rubens, C. E., M. R. Wessels, L. M. Heggen, and D. L. Kasper.** 1987. Transposon mutagenesis of group B streptococcal type III capsular polysaccharide: correlation of capsule expression with virulence. *Proc. Natl. Acad. Sci. USA* **84**:7208–7212.

54. **Rubens, C. E., H. V. Raff, J. C. Jackson, E. Y. Chi, J. T. Bielitzki, and S. L. Hillier.** 1991. Pathophysiology and histopathology of group B streptococcal sepsis in Macaca nemestrina primates induced after intraamniotic inoculation: evidence for bacterial cellular invasion. *J. Infect. Dis.* **164**:320–330.

55. **Rubens, C. E., S. Smith, M. Hulse, E. Y. Chi, and G. van Belle.** 1992. Respiratory epithelial cell invasion by group B streptococci. *Infect. Immun.* **60**:5157–5163.

56. **Shigeoka, A. O., R. T. Hall, V. G. Hemming, C. D. Allred, and H. R. Hill.** 1978. Role of antibody and complement in opsonization of group B streptococci. *Infect. Immun.* **21**:34–40.

57. **Silver, R. P., P. Annunziato, M. S. Pavelka, R. P. Pigeon, L. F. Wright, and D. E. Wunder.** 1993. Genetic and molecular analyses of the polysialic acid cluster of Escherichia coli K1, p. 59–71. *In* J. Roth, U. Rutishauser, and F. A. Troy II (ed.), *Polysialic Acid*. Birkhauser Verlag, Basel.

58. **Silver, R. P., C. W. Finn, W. F. Vann, W. Aaronson, R. Schneerson, P. J. Kretschmer, and C. F. Garon.** 1981. Molecular cloning of the K1 capsular polysaccharide genes of E. coli. *Nature* (London) **289**:696–698.

59. **Steenbergen, S. M., T. J. Wrona, and E. R. Vimr.** 1992. Functional analysis of the sialyltransferase complexes in *Escherichia coli* K1 and K92. *J. Bacteriol.* **174**:1099–1108.

60. **Stewardson-Krieger, P. B., K. Albrandt, T. Nevin, R. R. Kretschmer, and S. P. Gotoff.** 1977. Perinatal immunity to group B β-hemolytic Streptococcus type Ia. *J. Infect. Dis.* **136**:649–654.

61. **St. Geme, J. W., III, and S. Falkow.** 1990. *Haemophilus influenzae* adheres to and enters cultured human epithelial cells. *Infect. Immun.* **58**:4036–4044.

62. **Stoolmiller, A. C., and A. Dorfman.** 1969. The biosynthesis of hyaluronic acid by Streptococcus. *J. Biol. Chem.* **244**:236–246.

63. **Sutherland, I. W.** 1982. Biosynthesis of microbial exopolysaccharides. *Adv. Microb. Physiol.* **23**:79–150.

64. **Tamura, G., and C. Rubens.** Host-bacterial interactions in pathogenesis of group B streptococcal infection. *Curr. Opinions Infect. Dis.,* in press.

65. **Troy, F. A.** 1992. Polysialylation: from bacteria to brains. *Glycobiology* **2**:5–23.

66. **Troy, F. A., I. K. Vijay, and N. Tesche.** 1975. Role of undecaprenyl phosphate in synthesis of polymers containing sialic acid in Escherichia coli. *J. Biol. Chem.* **250**:156–163.

67. **VanDijk, W. C., H. A. Verbrugh, M. E. Tol, R. Peters, and J. Verhoef.** 1979. Role of *Escherichia coli* K capsular antigens during complement activation, C3 fixation and opsonization. *Infect. Immun.* **25**:603–609.

68. **Vann, W. F., R. P. Silver, C. Abeijon, K. Chang, W. Aaronson, A. Sutton, C. W. Finn, W. Linder, and M. Kotsatos.** 1987. Purification, properties, and genetic location of Escherichia coli cytidine 5′-monophosphate N-acetylneuraminic acid synthetase. *J. Biol. Chem.* **262**:17556–17562.

69. **Wald, E. R., I. Bergman, H. B. Taylor, D. Chiponis, C. Porter, and K. Kubek.** 1986. Long-term outcome of group B streptococcal meningitis. *Pediatrics* **77**:217–221.

70. **Walsh, J. A., and S. Hutchins.** 1989. Group B streptococcal disease: its importance in the developing world and prospect for prevention with vaccines. *Pediatr. Infect. Dis. J.* **8**:271–276.

71. **Watson, D. A., and D. M. Musher.** 1990. Interruption of capsule production in *Streptococcus pneumoniae* serotype 3 by insertion of transposon Tn916. *Infect. Immun.* **58**:3135–3138.

72. **Wessels, M. R., R. F. Haft, L. M. Heggen, and C. E. Rubens.** 1992. Identification of a genetic locus essential for capsule sialylation in type III group B streptococci. *Infect. Immun.* **60**:392–400.

73. **Wessels, M. R., A. E. Moses, J. M. Goldberg, and T. J. DiCesare.** 1991. Hyaluronic acid capsule is a virulence factor for mucoid group A streptococci. *Proc. Natl. Acad. Sci. USA* **88**:8317–8321.

74. **Wessels, M. R., C. E. Rubens, V.-J. Benedi, and D. L. Kasper.** 1989. Definition of a bacterial virulence factor: sialylation of the group B streptococcal capsule. *Proc. Natl. Acad. Sci. USA* **86**:8983–8987.

75. **Wright, A.** 1971. Mechanism of conversion of the *Salmonella* O antigen by bacteriophage e34. *J. Bacteriol.* **105**:927–936.

76. **Yeung, M. K., and S. J. Mattingly.** 1983. Isolation and characterization of type III group B streptococcal mutants defective in biosynthesis of the type-specific antigen. *Infect. Immun.* **42:**141–151.

77. **Zangwill, K. M., A. Schuchat, and J. Wenger.** 1990. Group B streptococcal disease in the United States: report from a multistate active surveillance system. *Morbid. Mortal. Weekly Rep.* **41:** 25–32.

78. **Zapata, G., W. F. Bann, W. Aaronson, M. S. Lewis, and M. Moos.** 1989. Sequence of the cloned Escherichia coli K1 CMP-*N*-acetylneuraminic acid synthetase gene. *J. Biol. Chem.* **264:** 14769–14774.

Molecular Genetics of Bacterial Pathogenesis
Edited by V. L. Miller, J. B. Kaper, D. A. Portnoy, and R. R. Isberg
© 1994 American Society for Microbiology, Washington, DC 20005

Chapter 22

Unorthodox Secretion by Gram-Negative Bacteria

Alison Ann Weiss

The cytoplasmic membrane can be considered the boundary between life and death. A cell with an intact membrane is alive, but pop the membrane and the scientist has entered the in vitro world. Chaos is kept at bay by the carefully controlled passage of compounds across the membrane. An extraordinary amount of the cellular machinery is directed at performing and regulating the traffic across the membrane. Gram-negative bacteria are double-challenged—unlike all other creatures on earth they possess two membranes separated by a compartment called the "periplasmic space." The outer membrane of pathogenic gram-negative bacteria is the surface where they interface with their host (a polite term for the molecular battlefield of the pathogen and the immune system). For these creatures, it is not enough to produce a virulence factor such as a toxin; they must also provide for its secretion. Moving their toxins and virulence factors through the inner membrane, across the periplasmic space, and out of the outer membrane must be comparable to Hannibal crossing the Alps with his elephants, but there is no doubt that the microbes succeed admirably in this respect. However, investigators may now regard this process in gram-negative bacteria as a therapeutic opportunity, since any cellular process that is essential for bacterial growth or pathogenesis that is not shared by the human host is a potential target for an antimicrobial agent. Knowledge of this bacterial process could allow humans to control the diseases caused by gram-negative bacteria. In this chapter, I will discuss the architecture of the gram-negative bacterial cell and several solutions to the gram-negative secretion dilemma.

GRAM-NEGATIVE BACTERIAL CELL STRUCTURE AND FUNCTION

If the membrane is the boundary of life and energy is required to keep entropy at bay, how is secretion fueled? Energy is available at the cytoplasmic membrane

Alison Ann Weiss • Molecular Genetics, Biochemistry, and Microbiology, 231 Bethesda Avenue, ML 524, University of Cincinnati, Cincinnati, Ohio 45267.

in the chemical form of ATP or from the proton motive force in the form of a pH and electrical gradient formed by the asymmetric movement of compounds across the membrane (5, 31). In contrast, the outer membrane appears to have few energy options. ATP is not thought to be available at this site, since phosphatases are present in the periplasm. Chemical and pH gradients are not thought to form since the outer membrane possesses porin proteins which allow free passage of molecules in the size range of a few thousand daltons or less (5, 18, 19). Since both of these arguments are indirect and logic dictates that there must be energy available at the outer membrane at least occasionally, this area needs to be experimentally revisited. Lack of understanding of the periplasm and the energy options available at the outer membrane is an important gap that must be filled before investigators can understand the secretion processes in gram-negative bacteria.

While it presents several unique challenges, the outer membrane does provide several benefits to gram-negative bacteria. Many compounds are excluded by the outer membrane. One important example is the beta-lactams and related antibiotics, which cannot reach their cell wall target as easily in gram-negative bacteria as they can in gram-negative bacteria (5). Gram-negative bacteria are often less susceptible to killing by complement (5, 25).

SECRETION

Gram-negative bacteria enjoy the protection of the outer membrane; however, it does not come without a price. Secretion across two membranes is more than twice the challenge faced by single-membrane cells. The well-characterized Sec secretion pathway, shared by all prokaryotes and eukaryotes, requires that the secreted protein have a secretion signal sequence, which is usually located on the N terminus of the protein (6, 18, 21, 31). This consists of positively charged amino acids followed by a string of hydrophobic amino acids. The signal is recognized by the Sec secretion machinery, and the protein is transported sequentially across the membrane as an unfolded strand. Folding occurs on the other side of the membrane. It must be obvious that this pathway deposits the protein in the periplasm and not on the outside of gram-negative bacterium. For a protein to be truly secreted by a gram-negative bacterium, it must either devise a Sec-independent mechanism of secretion (Fig. 1B) or, for a protein containing a secretion signal, evolve a second secretion mechanism, as shown on the right in Fig. 1A.

Pathogenic gram-negative bacteria have devised several mechanisms to accomplish this goal. One of the best-studied examples of a Sec-initiated postperiplasmic secretion pathway is the pullulanase secretion pathway of *Klebsiella oxytoca* (18, 19). The protein products of 16 genes have been shown to be required for the proper synthesis, maturation, and secretion of the enzyme pullulanase (Fig. 2). The out secretion pathway of *Erwinia chrysanthemi* shares 13 of these genes (for a review, see reference 18), suggesting that most of these genes encode information required for the shared secretion pathway rather than structural or maturation functions. Interestingly, several other postperiplasmic multigenic secretion pathways in gram-negative bacteria have a homolog to a single pullulanase

Sec-Mediated

FIGURE 1. Examples of protein secretion mechanisms in gram-negative bacteria. (A) Sec-mediated secretion results in deposition of the protein in the periplasmic space. For a protein containing a secretion signal to be exported past the outer membrane, a second secretion mechanism must be employed, as shown on the right. Examples of this are discussed in the text. (B) Sec-independent secretion can occur by a process that does not involve a periplasmic intermediate. Examples are the *Escherichia coli* hemolysin (see chapter 23) and the *Bordetella pertussis* adenylate cyclase toxin.

Not Sec-mediated

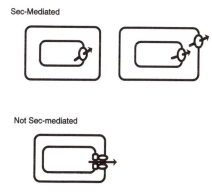

secretion gene, either *pulD* and *pulE,* suggesting that independent pathways of secretion converge at critical functions performed by either PulD or PulE and then branch off from the pullulanase pathway after that point (Fig. 2).

Interestingly, pathways employing the PulE pathway secrete complex multi-molecular structures, either DNA or protein. DNA is transferred in the case of the T-DNA of *Agrobacterium tumefaciens* (9) and the P conjugation pathway (10, 11, 14). Protein secretion pathways that use the PulE homologs include those that secrete pertussis toxin (29) and cholera toxin (20). The pathways requiring the PulD homolog but not the PulE homolog are involved in the secretion of virulence factors (Ysc or Mxi; Fig. 2), development of competence for DNA transformation (*Haemophilus influenzae*), and filamentous phage morphogenesis (1, 18).

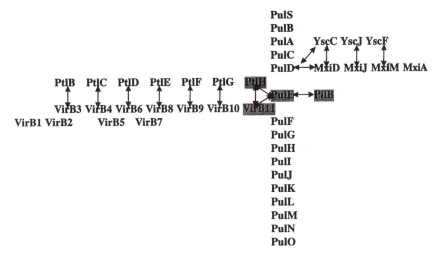

FIGURE 2. The scrabble family of secretion proteins of gram-negative bacteria. The proteins required for pullulanase (Pul) synthesis, maturation, and secretion are shown from top to bottom. The Ysc secretion pathway of *Yersinia* spp. and the Mxi pathway of *Shigella* spp. (1) have homologs only to the PulD protein (18). The Ptl secretion pathway for pertussis toxin (29) and the VirB operon required for T-DNA transfer by *A. tumefaciens* have homologs only to the PulE protein (18).

Bordetella pertussis has evolved numerous pathways for the secretion of its virulence factors. Sec-independent secretion occurs in the case of the adenylate cyclase toxin (see chapter 23) and filamentous hemagglutinin (32), and other examples of signal sequence-independent secretion in *B. pertussis* are under investigation (unpublished data). The rest of this discussion focuses on a Sec-dependent secretion mechanism followed by a postperiplasmic secretion mechanism, the pathway depicted in Fig. 1A.

Pertussis toxin is the most important virulence factor of *B. pertussis* (4, 16, 28). It is also the most complex bacterial toxin. It has the A-B toxin structure. The A component is responsible for causing the damage to the host, and the B component is responsible for delivering the toxin to the appropriate site in the body. The pertussis toxin A component is called S1 (for subunit 1). It is a single polypeptide which catalyzes the ADP-ribosylation of a family of mammalian GTP-binding proteins that regulate cellular communication (7). The pertussis toxin B subunit delivers the A portion to the target mammalian cell. It comprises five polypeptides that form a pentameric ring, composed of one copy each of subunits S2, S3, and S5 and two copies of subunit S4. The crystal structure of pertussis toxin has been solved recently. In spite of the lack of sequence similarity between the S2 through S5 subunits, they adopt a strikingly similar three-dimensional conformation (23). In addition, the B pentamer of pertussis toxin, which is composed of four different polypeptide subunits, closely resembles the B pentamers of other AB_5 toxins, including the heat-labile toxin and verotoxin 1, which are each composed of five identical polypeptide subunits. The subunits of these homopentamers do not share sequence homology with any of the pertussis toxin subunits (23).

SECRETION OF PERTUSSIS TOXIN

Each of the five pertussis toxin subunits is encoded by a separate gene, and each has an N-terminal secretion signal sequence that is proteolytically processed and absent from the mature toxin (16). This strongly suggests that pertussis toxin is secreted via the *Bordetella* equivalent of the Sec-mediated secretion pathway. As discussed above, at this point the job is only half done, since this will deliver the toxin only as far as the periplasmic space.

Recently, my colleagues and I have discovered an operon of seven genes, *ptlB* to *ptlH* (called *ptl* for pertussis toxin liberation), that is required for the secretion of pertussis toxin across the outer membrane (29). An eighth open reading frame, *ptlA*, appears to be the promoter (29). The entire *ptl* operon appears to share extensive homology with a secretion complex which is required to transfer DNA, not protein, out of the cell of a plant pathogen, *A. tumefaciens* (9, 27) (Fig. 2 and 3). PtlH homologs are present in several gram-negative postperiplasmic secretion systems, including PulE, which is needed for the secretion of pullulanase (18, 19), and the PilB and PilT proteins (18, 30), which are needed for pilus secretion by *Pseudomonas aeruginosa* (Fig. 3).

Why would the pertussis toxin secretion system resemble a DNA secretion apparatus more closely than the other protein secretion systems do? While the

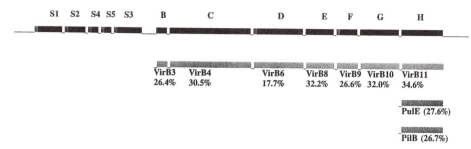

FIGURE 3. Comparison of the Ptl operon with the *A. tumefaciens* VirB operon. The 12-kb chromosomal region of the *B. pertussis*-encoding pertussis toxin structural genes (denoted S1 to S5) and the closely linked *ptl* secretion genes (denoted B to H) are shown on the top line. The proteins with a secretion signal are denoted by the shaded region at the start. The *virB* homologs and the percent identity with the *ptl* genes are shown below. *virB11*, *pulE*, and *pilB* are all homologous to *ptlH*, and the percent identities are given.

proteins of the VirB operon do indeed ultimately transfer a DNA molecule past both bacterial membranes, it is not transferred as naked DNA but, rather, is transferred as a protein-coated DNA complex (3). Pertussis toxin is also a protein complex. All of the evidence suggests that only the assembled complex is efficiently secreted (15, 17). In contrast, other proteins are transported as unfolded structures and fold after secretion (8, 18). Unfolded proteins present a slender (albeit long) structure to the secretion machinery, which is quite different from the case for pertussis toxin. The multisubunit complex of pertussis toxin may be better suited to be transported by a system evolved to handle a multisubunit protein-DNA complex.

In retrospect, it is clear why pertussis toxin must be secreted as an assembled complex. The individual subunits appear to have all of the information they need to self-assemble, since denatured toxin will refold, and the individual subunits produced in *Escherichia coli* by recombinant DNA techniques also refold (2). However, these are inefficient processes driven by the law of mass action. Unless the stoichiometry of the production of the subunits was carefully controlled, the bacterium would waste precious energy producing toxin subunits that failed to become incorporated into the assembled toxin. For a pathogenic bacterium, this is a more serious matter than simply metabolic waste. The toxin is functional only in the assembled state. The individual subunits are not only nonfunctional but also antigenic. Secretion of small amounts of the nonfunctional antigenic subunits would compromise the effectiveness of the holotoxin that did manage to assemble correctly. It is interesting to note that production of complex multisubunit toxin has not been described for the single-membrane gram-positive microorganisms. In this case, the outer membrane barrier may serve an important and necessary function: to block the secretion of the unassembled subunits. One of the functions of the Ptl secretion machinery must be to recognize a conformational epitope on the assembled toxin and permit it to be released from the cell while holding back the unassembled subunits.

The predicted localization of the Ptl proteins on the basis of the hydrophobic

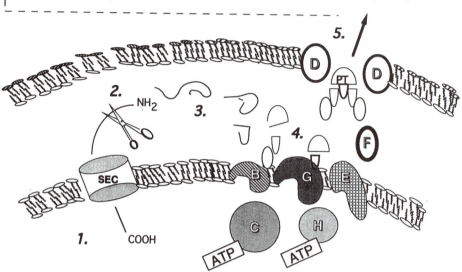

1. *Protein synthesis and sec-mediated secretion to periplasm*
2. *Processing by leader peptidase*
3. *Folding*
4. *Assembly of toxin*
5. *Secretion past outer membrane*

FIGURE 4. Steps required for proper pertussis toxin (PT) assembly and secretion. The proposed locations of the Ptl proteins B to G are based on the predicted amino acid sequence.

properties deduced from the predicted amino acid sequence is shown in Fig. 4. Only PtlF and PtlD have secretion signals and presumably would be the only periplasmic or cytoplasmic components. PtlD has considerable hydrophobic properties and is a candidate to be either inserted in the outer membrane (as shown in Fig. 4) or reinserted back into the inner membrane. PtlF is quite hydrophilic and could be soluble in the periplasm (as shown in Fig. 4). PtlC and PtlH are hydrophilic and would be predicted to be soluble in the cytoplasm. Both of these proteins, and their VirB counterparts, have a predicted ATP-binding cassette (9, 22, 29). By the dogma that ATP is available only in the cytoplasm, this seems to be a logical place for them, as depicted in Fig. 4. PtlB, PtlE, and PtlG are all quite hydrophobic on the N-terminal ends and quite hydrophilic on the C-terminal ends and could be integral membrane proteins with the C-terminal end either in the cytoplasm (as shown in Fig. 4) or in the periplasm. The predicted localization sites from the deduced amino acid sequences place most of the components in the inner membrane, in marked contrast to the simple model shown in Fig. 1, where the postperiplasmic secretion machinery is entirely in the outer membrane. This presents a bit of a paradox.

The extensive studies of the *Agrobacterium* VirB system give some resolution to this paradox and some clues to the mechanism of Ptl secretion. In an interesting study by Thorstenson et al. (26), immunoblotting of inner and outer membrane fractions generated by two independent techniques was used to localize a subset of the VirB components. Information was generated for VirB4 (PtlC homolog), VirB8 (PtlE homolog), VirB9 (PtlF homolog), VirB10 (PtlG homolog), and VirB11 (PtlH homolog). Different results were obtained if differential detergent solubilization or density gradient separation was used. Interestingly, all of the proteins with Ptl homologs, including the hydrophilic proteins VirB4 and VirB11, were found to be associated with both the inner and outer membrane fractions. From this the authors concluded that the distinction between the inner membrane and the outer membrane is lost at the sites where the secretion complex forms, and they proposed a model that transfer occurs via a gated protein channel that spans both membranes. This model would explain how these proteins can easily segregate into either the inner membrane or the outer membrane compartment.

However, this model does not fit well with the proposed periplasmic intermediate for pertussis toxin. It is hard to imagine that two secretion pathways that share so much homology would not share a mechanism. I would like to propose a model for pertussis toxin secretion (and, by analogy, T-DNA transfer) that is consistent with the experimental observations but quite different in its basic mechanism from that of Thorstenson et al. In this model, pertussis toxin subunits accumulate and assemble in the periplasm and secretion occurs by a membrane fusion mechanism (Fig. 5). Some support for this comes from data on the kinetics of secretion. Pertussis toxin is synthesized and accumulates in the cells late in the growth cycle, first appearing after about 20 h in culture. Secretion, however, is delayed until the late stationary phase, after about 60 h in culture (29). However, the Ptl operon is expressed throughout the growth cycle (12, 13) (unpublished data), suggesting that the toxin accumulates in association with the secretion machinery until a secretion signal is received in the late stationary phase. This mechanism resembles the quantal release seen in the neurotransmitter secretion pathway (24) more than the catalytic processes seen in other secretion pathways. This model can be extended to *A. tumefaciens* by proposing that T-DNA transfer starts

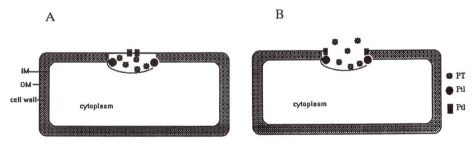

FIGURE 5. Proposed model of pertussis toxin (PT) secretion via a membrane fusion mechanism. For clarity, only two Ptl proteins are shown, but one could envision a secretion compartment formed by a complex of proteins on the inner membrane (IM) and outer membrane (OM). (A) Accumulation and assembly of pertussis toxin subunits occur in the periplasmic space at early times in the cell cycle. (B) Secretion occurs by a fusion mechanism in the late stationary phase.

with the protein-covered T-DNA associating with the VirB proteins on the inner side of the cytoplasmic membrane, which is in contrast to the case for pertussis toxin, which associates with the Ptl proteins on the outer side of the cytoplasmic membrane. For T-DNA transfer, a cytoplasmic membrane fusion step would generate a T-DNA-containing vesicle. This vesicle could bud off and fuse with the outer membrane, as proposed here for Ptl. This model also accounts for entry into the plant cell via a third membrane fusion. Thus, one mechanism can transfer pertussis toxin originating from the periplasm across the outer membrane as well as T-DNA originating from the cytoplasm across the outer membrane and the plant membrane.

Current efforts in the laboratory are aimed at testing the membrane fusion hypothesis and understanding how the seven Ptl subunits perform the secretion functions. Perhaps along the way investigators can devise a way to exploit this Achilles' heel as a target for therapy of infections caused by gram-negative bacteria.

ACKNOWLEDGMENTS. Stanley, a heart-felt thank you. I also acknowledge the artistic contributions of Gyeong Park and Louise Temple-Rosebrook.

REFERENCES

1. **Allaoui, A., P. J. Sansonetti, and C. Parsot.** 1993. MxiD, an outer membrane protein necessary for the secretion of the *Shigella flexneri* Ipa invasins. *Mol. Microbiol.* 7:59–68.
2. **Burnette, W. N., J. L. Arciniega, V. L. Mar, and D. L. Burns.** 1992. Properties of pertussis toxin B oligomer assembled in vitro from recombinant polypeptide produced by *Escherichia coli. Infect. Immun.* 60:2252–2256.
3. **Das, A.** 1998. *Agrobacterium tumefaciens virE* operon encodes a single-stranded DNA-binding protein. *Proc. Natl. Acad. Sci. USA* 85:2902–2913.
4. **Goodwin, M. S. M., and A. A. Weiss.** 1990. Adenylate cyclase toxin is critical for bacterial colonization and pertussis toxin is critical for lethal infection by *Bordetella pertussis* in infant mice. *Infect. Immun.* 58:3445–3447.
5. **Hancock, R. E. W.** 1991. Bacterial outer membranes: evolving concepts. *ASM News* 57:175–1182.
6. **Johansson, M., I. Nilsson, and G. von Heijne.** 1993. Positively charged amino acids placed next to a signal sequence block protein translocation more efficiently in *Escherichia coli* than in mammalian microsomes. *Mol. Gen. Genet.* 239:251–256.
7. **Katada, T., and M. Ui.** 1982. Direct modification of the membrane adenylate cyclase system by islet-activation protein due to ADP-ribosylation of a membrane protein. *Proc. Natl. Acad. Sci. USA* 79:3129–3133.
8. **Klauser, T., J. Pohlner, and T. F. Meyer.** 1993. The secretion pathway of IgA protease-type proteins in gram-negative bacteria. *BioEssays* 15:799–805.
9. **Kuldau, G. A., G. De Vos, J. Owen, G. McCaffrey, and P. Zambryski.** 1990. The *virB* operon of *Agrobacterium tumefaciens* pTiC58 encodes 11 open reading frames. *Mol. Gen. Genet.* 221:256–266.
10. **Lessl, M., D. Balzer, and E. Lanka.** 1992. Relationship of DNA-transfer-systems: essential factors of plasmids RP4, Ti and F share common sequences. *Nucleic Acids Res.* 20:6099–6100.
11. **Lessl, M., D. Balzer, W. Pansegrau, and E. Lanka.** 1992. Sequence similarities between the RP4 Tra2 and the Ti VirB region strongly support the conjugation model for T-DNA transfer. *J. Biol. Chem.* 267:20471–20480.
12. **Melton, A. R., and A. A. Weiss.** 1989. Environmental regulation of expression of virulence determinants in *Bordetella pertussis. J. Bacteriol.* 171:6206–6212.
13. **Melton, A. R., and A. A. Weiss.** 1993. Characterization of enviromental regulators of *Bordetella pertussis. Infect. Immun.* 61:807–815.

14. **Motallebi-Veshareh, M., D. Balzar, E. Lanka, G. Jagura-Burdzy, and C. M. Thomas.** 1992. Conjugative transfer functions of broad-host-range plasmid RK2 are coregulated with vegetative replication. *Mol. Microbiol.* **6:**907–920.

15. **Nencioni, L., M. Pizza, G. Volpini, M. T. de Magistris, F. Giovannoni, and R. Rappuoli.** 1991. Properties of the B oligomer of pertussis toxin. *Infect. Immun.* **59:**4732–4734.

16. **Nicosia, A., M. Perugini, C. Franzini, M. C. Casagli, M. G. Borri, G. Antoni, M. Almoni, P. Neri, G. Ratti, and R. Rappuoli.** 1986. Cloning and sequencing of the pertussis toxin genes: operon structure and gene duplication. *Proc. Natl. Acad. Sci. USA* **83:**4361–4635.

17. **Pizza, M., M. Bugnoli, R. Manetti, A. Covacci, and R. Rappuoli.** 1990. The subunit S1 important for pertussis toxin secretion. *J. Biol. Chem.* **265:**17759–17763.

18. **Pugsley, A. P.** 1993. The complete general secretory pathway in gram-negative bacteria. *Microbiol. Rev.* **57:**50–108.

19. **Pugsley, A. P., C. d'Enfert, I. Reyss, and M. G. Kornacker.** 1990. Genetics of extracellular protein secretion by gram-negative bacteria. *Annu. Rev. Genet.* **24:**67–90.

20. **Sandkvist, M., V. Morales, and M. Bagdasarian.** 1993. A protein required for secretion of cholera toxin through the outer membrane of *Vibrio cholerae. Gene* **123:**81–86.

21. **Schatz, P. J., and J. Beckwith.** 1990. Genetic analysis of protein export in *Escherichia coli. Annu. Rev. Genet.* **24:**215–248.

22. **Shirasn, K., Z. Koukolikova-Nicola, B. Hohn, and C. I. Kato.** 1994. An inner-membrane-associated virulence protein essential for T-DNA transfer from *Agrobacterium tumefaciens* to plants exhibits ATPase activity and similarities to conjugative transfer genes. *Mol. Microbiol.* **11:**581–588.

23. **Stein, P. E., A. Boodhoo, G. D. Armstrong, S. A. Cockle, M. H. Klein, and R. J. Read.** 1994. The crystal structure of pertussis toxin. *Structure* **2:**45–57.

24. **Stevens, C. F.** 1993. Quantal release of neurotransmitter and long-term potentiation. *Cell* **72:**55–63.

25. **Taylor, P. W.** 1992. Complement-mediated killing of susceptible gram-negative bacteria: an elusive mechanism. *Exp. Clin. Immunogenet.* **9:**48–56.

26. **Thorstenson, Y. R., G. A. Kuldau, and P. C. Zambryski.** 1993. Subcellular localization of seven VirB proteins of *Agrobacterium tumefaciens:* implications for the formation of a T-DNA transport structure. *J. Bacteriol.* **175:**5233–5241.

27. **Ward, J. E., Jr., E. M. Dale, P. J. Christie, E. W. Nester, and A. N. Binns.** 1990. Complementation analysis of *Agrobacterium tumefaciens* Ti plasmid *virB* genes by use of a *vir* promoter expression vector: *virB9, virB10,* and *virB11* are essential virulence genes. *J. Bacteriol.* **172:**5187–5199.

28. **Weiss, A. A., and E. L. Hewlett.** 1986. Virulence factors of *Bordetella pertussis. Annu. Rev. Microbiol.* **40:**661–686.

29. **Weiss, A. A., F. D. Johnson, and D. L. Burns.** 1993. Molecular characterization of an operon required for pertussis toxin secretion. *Proc. Natl. Acad. Sci. USA* **90:**2970–2974.

30. **Whitchurch, C. B., M. Hobbs, S. P. Livingston, V. Krishnapillai, and J. S. Mattick.** 1990. Characterization of a *Pseudomonas aeruginosa* twitching motility gene and evidence for a specialised protein export system widespread in eubacteria. *Gene* **101:**33–44.

31. **Wickner, W., A. J. M. Driessen, and F. U. Hartl.** 1991. The enzymology of protein translocation across the *Escherichia coli* plasma membrane. *Annu. Rev. Biochem.* **60:**101–124.

32. **Willems, R. J., L., C. Geuijen, H. G. J. van der Heide, G. Renauld, P. Bertin, W. M. R. van den Akker, C. Locht, and F. R. Mooi.** 1994. Mutational analysis of the *Bordetella pertussis fim/fha* gene cluster: identification of a gene with sequence similarities to haemolysin accessory genes involved in export of FHA. *Mol. Microbiol.* **11:**337–347.

Molecular Genetics of Bacterial Pathogenesis
Edited by V. L. Miller, J. B. Kaper, D. A. Portnoy, and R. R. Isberg
© 1994 American Society for Microbiology, Washington, DC 20005

Chapter 23

Holistic Perspective on the *Escherichia coli* Hemolysin

Rodney A. Welch

Bacterial pathogens that try to survive extracellularly in hosts are at combat with an army of efficient phagocytic warriors. In order to wage a war of survival, many pathogens have evolved thick carbohydrate shields that inhibit phagocytosis, some have become chameleons capable of hiding in an ever-changing camouflage that eludes the host radar-like processes, and yet others have turned to a no-holds-barred, cellular rumble complete with molecular guns, knives, and poisons. Members of a large group of gram-negative human and animal pathogens that have essentially extracellular lifestyles share the ability to produce a potent molecular weapon, an RTX exotoxin that is capable of punching holes in the host cell membranes. This chapter reviews the knowledge of the prototype of the RTX family of pore-forming toxins, the *Escherichia coli* hemolysin, and proposes that new efforts be made to study the significance and roles of these exotoxins, together with those of endotoxins, as an integrated synergistic system of multiple virulence factors. The need for this analysis arises from the belated acceptance that the presence of endotoxin in hemolysin preparations is not a purification problem that should be circumvented but is probably a biological reality that should be examined.

The fundamental approach taken by biochemists and cell biologists interested in understanding the role in pathogenesis of a bacterial virulence factor such as an exoprotein has been to purify the exoprotein and observe its activity on different substrates, cells, and hosts. Today, the pathogeneses of diphtheria, tetanus, and botulism are understood because of the logic and rigor of this approach. For the pathogenesis of cholera, a curious story can be told of how purification of the cholera toxin and the identification of its enzymatic activity led to the logical conclusion that it was the sole cause of the fluid imbalance seen in the ill host (26). However, genetic studies combined with human volunteer vaccine trials could not fulfill Falkow's version of molecular Koch's postulates (14). Isogenic mutants that lacked the cholera toxin gene did not cause as serious a disease as the parental

Rodney A. Welch • Department of Medical Microbiology and Immunology, University of Wisconsin-Madison, Madison, Wisconsin 53706.

strains, but they still caused a significant intestinal illness (32, 33, 36). Later, elegant work by the Kaper group (15, 55) demonstrated that there were two other toxins, Ace and Zot produced by *Vibrio cholerae,* that also contribute to enteric disease. Hence, one can see the success of using the logical combination of approaches involving biochemistry, cell biology, and molecular genetics. The inability to define satisfactorily a role of Shiga toxin in dysentery stands out as one of the greater frustrations in current molecular pathogenesis research. In a purified form, Shiga toxin is cytotoxic in vitro and is lethal to animals (54). It is produced in the largest amounts by the *Shigella* species and strains that cause the severest disease, yet, when null mutants were constructed and tested in monkeys, only modest amelioration of the disease signs were observed; this amelioration involved less blood in the stool and decreased damage to the local capillaries (18, 54). Thus, to cope with these results, animal models are blamed for their perpetual inapplicability to human pathogenesis research, and investigators must search for new methods to attack the question or to accept the fact that the Shiga toxin may not have a role in dysentery.

For the pathogens that are RTX toxin producers, the problems in the identification and proof of their significance in disease have some parallels with the problems with research done with the Shiga toxin, but there are additional issues. The diseases caused by RTX toxin producers are varied. Table 1 provides a list of the

TABLE 1
Selected features of toxins from the RTX toxin branch of the RTX exoprotein family

Genus and species	Toxin gene	Predicted gene product size (kb)	pI[a]	Cell and host specificity	Reference(s)
Escherichia coli	*hlyA*	110	6.1	Broad	17
Proteus vulgaris	*pvxA*	?	?	Broad	34, 58
Morganella morganii	*mmxA*	?	?	Broad	34, 58
Pasteurella haemolytica	*lktA*	102	5.8	Ruminant PMNs[b] or monocytes	12, 25, 37
Pasteurella haemolytica-like	*pllktA*	101.5	5.5	Bovine or porcine leukocytes	11
Actinobacillus pleuropneumoniae	*apxAI*	110	5.4	Broad	19
Actinobacillus pleuropneumoniae	*apxAII*	103	5.3	Porcine leukocytes[c]	13
Actinobacillus pleuropneumoniae	*apxAIII*	113	5.5	Porcine leukocytes[d]	28
Actinobacillus suis	*ashA*	102.5	5.6	Porcine leukocytes	8
Actinobacillus actinomycetemcomitans	*aaltA*	114	8.8	Human PMNs or leukocytes	35
Bordella pertussis	*cyaA*	177.5	4.4	Broad	20

[a] The predicted pI, based on the predicted amino acid sequence of the toxin gene product.
[b] PMNs, polymorphonuclear leukocytes.
[c] A weak hemolytic activity is observed.
[d] No hemolytic activity is observed.

known RTX toxins and some of their features. Perhaps the most remarkable feature is that, despite their unarguable homologous origins, they have different patterns of target cell and host cell specificity. The RTX-associated diseases include pertussis, juvenile periodontitis, pneumonia, urinary tract infections, and wound infections. The epidemiological evidence for an association between RTX toxin production and isolates that cause disease is strong for all of the organisms except *E. coli, Morganella morganii,* and *Proteus vulgaris* (53, 58, 62). Disease-causing isolates of these three species are not always hemolysin producers, and those that are phenotypically nonhemolytic are missing the genes for hemolysin (58, 61). The lack of a consistent positive epidemiological association does argue that the presence of hemolysin is not a significant selective factor in the colonization of the extraintestinal sites where these bacteria cause disease (1). Even though the hemolysin lacks the significant epidemiological association that Shiga toxin has with *Shigella* isolates that cause dysentery, it may provide a function during a disease such as cystitis or pyelonephritis that an unknown factor provides for nonhemolytic strains of *E. coli* (42, 62). Another consideration is the possibility that variations in the pathogenesis of a disease such as pyelonephritis exist and that there are undetected nuances in the host-parasite interaction that are hemolysin dependent. With the qualification of the epidemiological data, application of molecular Koch's postulates for the identification of the *E. coli* hemolysin as a virulence factor has been accomplished. The genetic constructions and rat peritonitis model work was performed in Falkow's laboratory by Dellinger, Minshew, Falkow, and I (60). The conclusion from that work was that the *E. coli* hemolysin contributes to the virulence of this organism; this conclusion was based on results obtained with isogenic strains and animal model data and has been supported by other workers (23, 51). Problems arise with the biochemical data that could lend support to our conclusion. With the possible exception of the *Bordetella pertussis* adenylate cyclase-hemolysin (24, 49), unambiguous purification of any of the RTX toxins has not been achieved. It is from the arguments surrounding the putative biochemical purity of the *E. coli* hemolysin that I will try to derive a holistic view of RTX toxin activity in pathogenesis. In the majority of gram-negative organisms that produce RTX toxins, the major exoprotein species produced is the RTX protein. A lot of effort has been devoted to the examination of the in vitro activities of RTX toxins against different cell types. The logic for this work is inescapable: it is a soluble toxin; therefore, it must have effects at distant host sites. The *beau ideal* for the biochemistry-minded investigator is that in order to understand the interplay within the whole system, the investigator must purify each component and study its activities in isolation, and the significance of that single factor can then be properly assessed. The two vexing problems for the cell biology and biochemical approach in assessing the significance of RTX toxins in pathogenesis are as follows. The cytolytic activities for the RTX hemolysins and leukotoxins are sufficiently labile to make their purification a major technical problem, and concomitantly, that which constitutes a purified RTX toxin preparation cannot be agreed upon.

WHAT IS THE *E. COLI* HEMOLYSIN?

In an endeavor to develop an answer to the question, "What is the *E. coli* hemolysin?," I will review the literature covering the determination of the size of the *E. coli* hemolysin, its composition, and the mode of its cytolytic activity. As implied above, the size and biochemical composition of the extracellular hemolytic complex are controversial. In order to critique the literature explicitly, I will use the term "hemolysin" or "hemolytic complex" to describe the molecule or molecules present in the cell-free culture medium that participate(s) in the lysis of an erythrocyte. I caution that the hemolysin has well-documented and more relevant cytotoxic activity toward cells other than erythrocytes (59) but that the convenient erythrocyte lytic assay used during purification of the hemolysin necessitates that I focus on that particular target cell in my discussion. The use of the term "HlyA" is reserved solely as the name for the structural protein that is part of the hemolytic complex. As I will shortly try to develop, it should not be implied that the hemolysin solely comprises HlyA molecules that have been activated by the intracellular HlyC protein (17, 27). There is little debate that the apparent size of active hemolysin found in culture supernatants exceeds the predicted 110,000 Da of a monomer of the HlyC-activated HlyA polypeptide (17). The sizes estimated independently by electron microscopy (46), gel exclusion chromatography (6, 21, 44, 46, 50, 56), ultrafiltration (10, 16), diffusion coefficient determination (46), and photon correlation spectroscopy (44) all exceed 300,000 Da. Gel exclusion chromatography and photon correlation spectroscopy methods indicate that the size of the hemolytic complex is heterogeneous, with a possible size range running from 300 to greater than 1,000 kDa (44). There are reports based on ultracentifugation sedimentation velocity behavior in either glycerol or sucrose gradients that indicate a uniformly smaller size for the active hemolysin (4, 52, 64). By this method, gradient fractions with approximate sizes of 100 to 150 kDa that possess hemolytic activity are subjected to sodium dodecyl sulfate (SDS)-polyacrylamide gel electrophoresis (PAGE) (4, 52). This results in the detection of the full-length 110-kDa HlyA polypeptide and antigenically related smaller polypepides as the predominant protein constituents of the hemolytic complex. This has led to the conclusion that the ultimate "active hemolytic complex" is the monomeric, HlyC-modified HlyA polypeptide. Perhaps underappreciated in the literature is the observation of Rennie and Arbuthnott (46) in 1974 that the presence of glycerol in the buffers used in size exclusion chromatography reduced the apparent size and heterogeneity of the hemolytic complex. The formal possibility exists that, once the ultracentrifugation gradient fractions are collected, they rapidly reassociate into a larger multimeric, active complex. The resolution to this issue awaits the direct demonstration of the molecular size of the lytic complex in a target cell membrane. Undoubtedly, adding to the confusion of the hemolysin size analysis is the knowledge that the hemolytic activity is labile because of the formation of nonfunctional, large molecular aggregates (21, 44). The solubilization of hemolysin aggregates by a chaotropic agent such as urea causes an increase in the specific lytic activity of the hemolysin (44). The urea treatment causes a decrease in the apparent size of the hemolysin, but not to the point at which it is the size of monomeric HlyA

(44). The most recent data which support a multimeric HlyA state in the hemolytic complex are from intragenic complementation experiments performed by Ludwig and colleagues (38). They constructed strains harboring compatible plasmids encoding two different, nonlytic HlyA mutant molecules which, when coexpressed in *E. coli,* secreted extracellularly active hemolysin (38). A fascinating observation made by those investigators is that the intragenic complementation occurs in vivo only when the mutants are coexpressed in the same cell. Mixtures of supernatants containing mutant forms of HlyA do not show hemolytic activity. The denaturation of the mixed supernatants by 6 M urea treatment also does not subsequently lead to hemolytic activity. This suggests that the formation of the HlyA multimer present in active hemolysin is dependent on a factor involved in the secretion or the addition of an unknown *E. coli* factor during the maturation of the large hemolysin complex.

The hemolysin is not only heterogeneous in size but also heterogeneous in charge. Hemolytically active fractions from Sephacryl S-1000 columns have been shown to separate by DEAE ion-exchange chromatography into two distinct fractions eluting at 0.3 and 0.4 M NaCl (6). The 0.4 M salt fraction material has 11-fold greater specific activity than the 0.3 M fraction (6). The estimated pI of 6.1 obtained on the basis of the predicted HlyA amino acid sequence does not match the experimentally derived pI of 4.0 to 4.6 for the hemolysin (46, 50). The discrepancy may be explained in part by the demonstration that the HlyC modification of HlyA makes it more negatively charged (41). Another possible contributor to the anionic state of the hemolysin is the presence of lipopolysaccharide (LPS). The two hemolytically active DEAE fractions isolated by Bohach and Snyder (6) were found to contain significant amounts of LPS. LPS cofractionated with the protein present in the hemolysin fractions. The LPS was further isolated by gas-liquid chromatography, and the 3-hydroxytetradecanoic acid methyl ester that is characteristic of LPS was identified by mass spectrum analysis. If their data are examined further with the assumption that all of the protein in their hemolytically active fractions is HlyA, the less active, less negatively charged fraction has, on a molar basis, an excess of 200 molecules of LPS for every molecule of HlyA. The more active, more negatively charged fraction has, by comparison, perhaps 25 to 33% less LPS. Ostolaza and colleagues (44) also demonstrated the presence of LPS-like, 2-keto-3-deoxyoctonate material in their purified preparations of hemolysin. In later work by Bohach and Snyder (7), LPS was isolated from their hemolysin-producing strain of *E. coli* and was used to prepare an anti-LPS monoclonal antibody. An anti-HlyA polypeptide-reactive monoclonal antibody (D12) was used in affinity purification to isolate the HlyA protein present in hemolysin. The HlyA protein that bound to the column was eluted, and hemolytic activity was corecovered. An enzyme-linked immunosorbent assay (ELISA) with the anti-LPS monoclonal antibody revealed that the LPS coeluted with the hemolytic activity (7). This provides strong evidence that hemolytically active material containing HlyA has LPS with it.

The perception that LPS may not be a simple contaminant of hemolysin preparations but a participant in the active extracellular hemolytic complex is given additional credence by the recent analysis of the effects of different mutations in

LPS biosynthesis on the synthesis, export, and activity of hemolysin. The first suggestion of this was from experiments examining the relative hemolytic activities of *Salmonella typhimurium rfa* mutant strains harboring plasmids encoding the *E. coli* hemolysin (9). Camprubi and associates (9) found decreases in relative hemolytic activities that ranged from 4- to 10-fold for different rough and deep rough mutants compared with that for a smooth strain of *S. typhimurium*. They did not determine at what stage in hemolysin production (transcription, translation, HlyC modification, secretion, or multimerization) the *rfa* mutations affected hemolysin activity levels. Wandersman and Letoffe (57) later found that spontaneous *galU* mutants that result in a deep rough LPS phenotype are not altered in their ability to synthesize HlyA but appear to be remarkably inefficient in secreting it free into the growth medium. Stanley and colleagues (52) found that another *E. coli* gene involved in LPS biosynthesis, *rfaP*, when interrupted by Tn*phoA*, does not result in an alteration in the synthesis or export of HlyA but results in a significant reduction in hemolysin activity. *rfaP* mutants are associated with a deep rough LPS phenotype because of their inability to modify the LPS inner core with phosphate or pyrophosphorylethanolamine (45). The reduction of hemolytic activity in the *rfaP* mutant is associated with differences in the migration distances of the hemolysin species in sucrose density gradients (52). The *rfaP* mutant-derived hemolysin appears to be much larger in size than the wild-type hemolysin. There does not appear to be any difference between the mutant and wild-type toxins on the basis of the migration distances of their HlyA-like species in SDS-PAGE. If the *rfaP* mutant hemolysin is treated with 4 M guanidine hydrochloride, its migration in the sucrose gradients mimics that of the wild-type hemolysin, and, significantly, some but not all of the hemolytic activity is regained when compared with the hemolytic activity of the wild-type hemolysin.

The putative association of LPS with the hemolysin complex may be reflected in some of its physical as well as functional characteristics. As described earlier in this chapter, the migration in the sucrose gradients of the wild-type hemolysin relative to molecular size markers suggests that the extracellular, cytolytically active hemolysin is monomeric HlyA (4, 52). That conclusion does not take into account the numerous independent observations that wild-type hemolysin is large and heterogeneous in size. One effect of the deep rough form of LPS produced by a *rfaP* mutant on hemolysin may be that it inhibits the dissociation of hemolysin into a smaller hemolysin species. In the wild-type cells, the smaller-molecular-size species are capable of reassociation into an active multimeric complex only in the context of their possible artifactual separation by the velocity sucrose gradient technique. The poor size resolution capabilities of the rate zonal centrifugation technique make it unlikely that a 110-kDa HlyA monomer can be resolved from, say, a 114.4-kDa molecule. That would be the predicted minimum size of a heterogeneous multimer that is made up of a simple, one-to-one LPS-HlyA hemolysin complex. The inability to recover 100% of the wild-type hemolytic activity in the presence of guanidine hydrochloride may be a very significant observation. It argues that the hemolysin produced in a *rfaP* mutant background remains functionally altered. The data on the treatment of aggregated, inactive hemolysin preparations with other chaotropic agents such as urea indicate that greater than 100%

recovery of hemolytic activity can occur (21). The potential effect of LPS on size, charge, and activity measurements may also be reflected in experiments in which hemolysin preparations are subjected to urea-PAGE. This technique does not lead to a discrete, stainable protein band in the gel (21). Therefore, despite the anionic nature of hemolysin, the HlyA molecules within it either are in a physical state that precludes their entry into such an electrophoretic gel or are within heterogeneously sized and heterogeneously charged complexes that are spread across a wide area of the gels. Nicaud et al. (41) published a photograph showing apparently monomeric HlyA in an electrophoresis gel run without SDS, but they gave no information about the sample preparation or gel buffer conditions. When hemolysin is subjected to electrophoresis in 4 to 15% native acrylamide electrophoresis gels without SDS and the gels are subjected to immunoblotting, it is apparent that a broad area at the top of the gel contains the HlyA antigen (40). This suggests that the presence of SDS is necessary to dissociate the large hemolysin complexes. It is possible that the interaction of HlyA with LPS mimics the physical and chemical interaction of LPS with the OmpF and OmpC porin proteins of *E. coli*. In those instances, boiling of purified porin preparations in 2% SDS dissociates some but not all LPS from OmpF (48). It is of particular significance that 4 M urea does not dissociate OmpF from LPS (48). This may explain the results described by Ostolaza and colleagues, who found that urea dissociates very large complexes into smaller ones, but that the hemolytic complex still had an apparent molecular size of greater than 300 kDa (44). This may also explain the failure to observe monomeric HlyA polypeptide on urea-PAGE because in the HlyA multimeric complex there is an unusual and strong association between LPS and HlyA, creating a large, heterogeneously sized and heterogeneously charged complex.

Among the three previously mentioned size estimates of hemolysin obtained on the basis of the velocity gradient technique, only in a subsequent report by Bhakdi and Martin (5) was there an attempt made to detect LPS in the hemolytically active fraction. The results of two different biological assays for the detection of LPS, pyrogenicity in rabbits and the *Limulus* assay, led to the conclusion that insignificant quantities of LPS (10% or less on a molar basis) were present in the purified hemolysin preparations. Both of those assays depended on the biological detection of LPS. One must give serious consideration to the possibility that the lipid A component responsible for the biological effects of LPS is masked by the physical association of LPS with HlyA. Attempts by those investigators (5) to detect fatty acids associated with hemolysin by gas-liquid chromatography were unsuccessful; however, they did not report the level of sensitivity of their particular method. These data are in obvious contrast to those from previously described studies (6, 44) in which LPS or substituents suggestive of LPS were found in hemolytic fractions. A big breakthrough in the study of the structure and function of the hemolysin that impinges on this discussion was the development of an in vitro method for acyl carrier protein-dependent, HlyC-dependent activation and radiolabeling of HlyA with [^{14}C]palmitate (27). Issartel et al. (27) demonstrated that the ^{14}C label is retained on the 110-kDa HlyA polypeptide after SDS-PAGE. This suggests but does not prove that the HlyC-dependent activation results in a covalent change in HlyA (48). At present there is no information about the final chemical structure of the HlyC modification of HlyA. It remains to be demonstrated that the ^{14}C label bound to HlyA remains in the form of a palmitate molecule.

HOW IS THE HEMOLYSIN TOXIC TO CELLS?

For more than 20 years, the hemolysin has been known to lyse erythrocytes through the formation of a membrane lesion by a one-hit, irreversible mechanism suggestive of a nonenzymatic activity (47, 50, 65). The original evidence that the hemolysin caused in erythrocytes a very rapid Ca^{2+} influx that is similar to the activity of the ionophore A23187 was published in 1983 by Jorgensen and colleagues (31). On the basis of results of kinetic experiments and dose-response experiments involving erythrocyte lysis, Jorgensen and members of her laboratory had actually proposed 3 years earlier that the hemolysin may lead to cellular lysis by a channel-like or an ionophore-like mechanism (30). Data which later supported their proposal came from osmotic protection experiments that indicated that the hemolysin produces discretely sized pores estimated to be 2 to 3 nm in diameter (4). Scanning electron micrographs from the Jorgensen laboratory showed that there is a dramatic change in the morphology of an erythrocyte within 5 min of exposure to a single hit of the hemolysin (30). The erythrocytes transform into echinocytes, in which gross cytoskeletal or membrane irregularities result in rounding of the cell and the formation of multiple teardrop-shaped projections from the cell surface. This cellular transformation is similar to that caused by the treatment of erthrocytes by a calcium ionophore (63). An example of the scanning electron micrographs showing hemolysin-treated erythrocytes is shown in Fig. 1. In the 1983 publication (31), the Jorgensen group tested the hypothesis that the hemolysin forms an ionophore-like membrane lesion. They measured the influx of radioactive inulin, sucrose, and phosphate into hemolysin-treated erthrocytes. In their experiments, they used low amounts of hemolysin and relatively high erythrocyte concentrations (30 to 35%) in order to reduce the likelihood that multiple hits of the hemolysin would occur on a single cell. During a short incubation time (8 to 12 min), during which fewer than 10% of the erythrocytes were lysed, there appeared to be a statistically insignificant amount of influx of the aforementioned radiolabeled molecules into the cells. These influx data were in clear contrast to the data obtained in an experiment of the influx of $^{45}Ca^{2+}$, in which approximately 50% of the initial extracellular $^{45}Ca^{2+}$ appeared to enter the cell. This led Jorgensen and colleagues to conclude that the hemolysin causes the membrane to become selectively permeable to the influx of cations. These conclusions differ in substance from those made by Bhakdi and colleagues (4), who later performed similar influx studies with radiolabeled molecules in which erythrocyte lysis was prevented by incubating the cell-hemolysin mixture in the presence of dextran 4. They observed under these conditions the progressive influx of larger and larger radiolabeled sugars into the hemolysin-treated erythrocytes. Those data led them to propose a model for hemolysin lytic action in which hemolysin creates

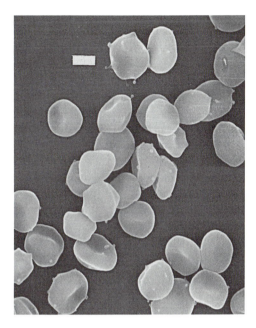

FIGURE 1. After a 5-min treatment with the *E. coli* hemolysin, sheep erythrocytes were fixed and visualized by scanning electron microscopy (30). The photograph shows the loss of the concave cellular structure typical of erythrocytes and the formation of multiple teardrop-shaped projections from the cell surface. Bar, 2 μm. The photograph was the gift of Sally Jorgensen, University of Minnesota.

a pore with a discrete diameter of approximately 2 to 3 nm. The static, presumably protein-lined pore dictates a constraint on the rate of diffusion of sugars into the erythrocyte on the basis of their sizes. Experiments in my laboratory indicate that there is a statistically significant variance in the influx of radiolabeled sugars into hemolysin-treated erythrocytes under either of the two experimental conditions just described (40). These experiments are problematic because they fundamentally require the diffusion of a sugar over a short time span of minutes into an intracellular space that is sludge-like because of the high concentrations of hemoglobin and other proteins. The results of these types of experiments indicate that it is extremely difficult to draw meaningful conclusions about the size of a hemolysin lesion in an erythrocyte. As is plainly evident in Fig. 1, during a short, 5-min time span, dramatic intracellular and membrane-cytoskeletal changes in the hemolysin-treated erythrocyte also occur. It may be too simple to assume that hemolysin lytic activity can be modeled as the formation of a fixed pore in a host membrane that results in lysis through osmotic swelling.

To further complicate acceptance of the static pore model of hemolysin, the apparent size of the hemolysin lesion in the erythrocyte membrane can be manipulated in osmotic protection experiments by varying the parameters of hemolysin concentration and the assay incubation time and temperature. For example, under sufficiently low hemolysin concentrations, in which approximately 75% of the available erythrocytes in a 2% suspension will lyse, even disaccharides with a diameter of approximately 0.9 nm osmotically protect the hemolysin-treated erythrocytes (40).

Results of experiments with artificial lipid bilayers lend strong support to the proposal that the hemolysin creates an ion-permeable channel that possesses cat-ion selectivity, as originally proposed by Jorgensen and colleagues and subse-quently supported by Benz et al. (2) and Menestrina et al. (39). The hemolysin pores detected by this method also appear to be smaller than those originally circumscribed by the osmotic protection experiments, with the diameter being closer to 1 nm. The cause or source of the larger hemolysin lesions in erythrocyte membranes that can be observed in osmotic protection experiments remains an enigma. A model that is most consistent with the available data would be one in which the hemolysin directly creates an ion channel that permits the rapid influx of Ca^{2+}. This is the direct toxic insult of hemolysin to a cell. The changes in cellular structure are the result of this injury, and the larger pores that are apparent after treatment with high toxin concentrations and for longer incubation times are either the result of an additional membrane multimerization process that creates a larger lesion or the consequence of secondary membrane damage caused by the initial toxic insult. I caution to add that it is also too simple for me to propose that the hemolysin acts identically to an ionophore such as A23187. In elegant experiments in which human neutrophils were treated with hemolysin, the data clearly show that it is a faster and more potent inducer of phosphoinositide-based signal transduction than A23187 (22). In addition, it appears that, in contrast to A23187, these sublytic effects by the hemolysin can be inhibited by pertussis toxin (22).

An interesting question that arises from the artificial lipid bilayer experiments is, what are the receptors for the *E. coli* hemolysin? The experiments performed by Benz et al. (2) and Menestrina et al. (39) used pure lipids without proteins or carbohydrates incorporated into the bilayer. The efficiency of hemolysin-depen-dent channel formation exhibited some dependence on what lipid was used. Al-though the two research groups differed in their opinions over the significance of these differences, these data, on the whole, have led to the suggestion that lipid molecules, alone in a nonspecific manner, represent the receptor for the hemoly-sin. However, it is also readily apparent from these experiments that relatively high concentrations of the hemolysin (in the low-nanomolar range) are required to observe channel formation. These experiments may not directly address the target cell specificity issue. This is because by a mass action phenomenon alone, rather than through a classic receptor ligand-facilitated interaction, a very small subset of hemolysin molecules may bind, insert, and form an ion channel in the artificial lipid. Although Ostolaza and colleagues (43) recently concluded that the lytic action of hemolysin is by a completely different, detergent-like activity, they confirmed that lipid vesicles made of a pure, single species of phospholipids are susceptible to leakage of preloaded fluorescent-labeled dextrans when treated with hemolysin. The nonspecific lipid receptor model for the hemolysin does seem consistent with the observation that the hemolysin affects many different cell types from many different hosts (for a review, see reference 59). The exception may be human platelets, which are refractory to the lytic action of hemolysin (3, cited in reference 29). The basis for that observation remains to be elucidated.

The suggestion that the *E. coli* hemolysin lacks a classic receptor-ligand inter-action that initiates the lytic process stands in contrast to the clear target cell and host cell specificities of the closely related RTX leukotoxins produced by *Pasteurella haemolytica* and *Actinobacillus actinomycetemcomitans* (Table 1) (for a review, see reference 59). The evolutionary events that must have led to the RTX toxin target cell specificity differences are intriguing. Did differences in the primary sequence of a common RTX ancestor toxin lead to the loss of the broad reactivity of a progenitor toxin or a gain in the broad reactivity of an originally narrowly reactive RTX toxin?

SUMMARY

In an earlier review of the RTX toxins (59), I proposed that greater attention should be paid to the sublytic effects of these toxins on nucleated cells. From further reflection on the *E. coli* hemolysin model system, it is evident that successful comprehension of the more subtle effects of the RTX toxins on the host requires a better understanding of the physical and biochemical natures of what has come to be called the *E. coli* hemolysin. A bias in pathogenesis research prescribes that the virulence-related activity of an exotoxin must be studied only within the context of a chromatographically pure protein. The motivation for covering the older literature in this chapter is to stimulate thought, experimentation, and discussion about the significance of the ever-present LPS molecules on the structure and activity of the *E. coli* hemolysin. Does LPS serve a true structural role in the hemolytic complex, or is it a contaminant of hemolysin preparations? Does LPS contribute to the numerous cytotoxic events attributed to sublytic concentrations of hemolysin? Even if it can be demonstrated unambiguously that HlyC-activated HlyA is solely responsible for the apparent membrane pore that directly results in host cell toxicity or host cell lysis, there remains the need to understand the synergies within the melange of weapons of an important gang of gram-negative warriors that are pitched into a jungle war against a much larger foe.

ACKNOWLEDGMENTS. The work in my laboratory described in this chapter was funded by Public Health Service grant AI20323 and the Romnes Fellowship from the University of Wisconsin.

I thank Margaret Bauer for critical comments on the manuscript and Irvin Snyder, Sally Jorgensen, and all of the people in my laboratory for helpful discussions. I thank Peg Riley for access to the Yale University libraries and encouragement. Lastly, I thank Stanley Falkow for insisting 14 years ago that I do a few experiments involving the *E. coli* hemolysin while I was writing a National Institutes of Health postdoctoral fellowship application on conjugative transposons.

REFERENCES

1. **Arthur, M., C. E. Johnson, R. H. Rubin, R. D. Arbeit, C. Campanelli, C. Kim, S. Steinbach, M. Agarwal, R. Wilkinson, and R. Goldstein.** 1989. Molecular epidemiology of adhesin and hemolysin virulence factors among uropathogenic *Escherichia coli*. *Infect. Immun.* **57:**303–313.
2. **Benz, R., A. Schmid, W. Wagner, and W. Goebel.** 1989. Pore formation by the *Escherichia coli* hemolysin: evidence for an association-dissociation equilibrium of the pore-forming aggregates. *Infect. Immun.* **57:**887–895.
3. **Bhakdi, S.** Unpublished data.
4. **Bhakdi, S., N. Mackman, J. M. Nicaud, and I. B. Holland.** 1986. *Escherichia coli* hemolysin may damage target cell membranes by generating transmembrane pores. *Infect. Immun.* **52:**63–69.

5. **Bhakdi, S., and E. Martin.** 1991. Superoxide generation by human neutrophils induced by low doses of *Escherichia coli* hemolysin. *Infect. Immun.* **59**:2955–2962.

6. **Bohach, G. A., and I. S. Snyder.** 1985. Chemical and immunological analysis of the complex structure of *Escherichia coli* α-hemolysin. *J. Bacteriol.* **164**:1071–1080.

7. **Bohach, G., and I. S. Snyder.** 1986. Composition of affinity-purified α-hemolysin of *Escherichia coli. Infect. Immun.* **53**:435–437.

8. **Burrows, L. L., and R. Y. C. Lo.** 1992. Molecular characterization of an RTX toxin determinant from *Actinobacillus suis. Infect. Immun.* **60**:2166–2173.

9. **Camprubi, S., J. Tomas, F. Munoa, C. Madrid, and A. Juarez.** 1990. Influence of lipopolysaccharide on external hemolytic activity of *Salmonella typhimurium* and *Klebsiella pneumoniae. Curr. Microbiol.* **20**:1–3.

10. **Cavalieri, S. J., and I. S. Snyder.** 1982. Cytotoxic activity of partially purified *Escherichia coli* alpha haemolysin. *J. Med. Microbiol.* **15**:11–21.

11. **Chang, Y., D. Ma, J. Shi, and M. M. Chengappa.** 1993. Molecular characterization of a leukotoxin gene from a *Pasteurella haemolytica*-like organism, encoding a new member of the RTX toxin family. *Infect. Immun.* **61**:2089–2095.

12. **Chang, Y.-F., R. Young, D. Post, and D. K. Struck.** 1987. Identification and characterization of the *Pasteurella haemolytica* leukotoxin. *Infect. Immun.* **55**:2348–2354.

13. **Chang, Y.-F., R. Young, and D. K. Struck.** 1989. Cloning and characterization of a hemolysin gene from *Actinobacillus (Haemophilus) pleuropneumoniae. DNA* **8**:635–647.

14. **Falkow, S.** 1988. Molecular Koch's postulates applied to microbial pathogenicity. *Rev. Infect. Dis.* **10**:S274–S276.

15. **Fasano, A., B. Baudry, D. Pumplin, S. Wasserman, B. D. Tall, J. Ketley, and J. B. Kaper.** 1991. *Vibrio cholerae* produces a second enterotoxin, which affects intestinal tight junctions. *Proc. Natl. Acad. Sci. USA* **88**:5242–5246.

16. **Felmlee, T., S. Pellett, E. Y. Lee, and R. A. Welch.** 1985. The *Escherichia coli* hemolysin is released extracellularly without cleavage of a signal peptide. *J. Bacteriol.* **163**:88–93.

17. **Felmlee, T., S. Pellett, and R. A. Welch.** 1985. The nucleotide sequence of an *Escherichia coli* chromosomal hemolysin. *J. Bacteriol.* **163**:94–105.

18. **Fontaine, A., J. Arondel, and P. J. Sansonetti.** 1988. Role of Shiga toxin in the pathogenesis of bacillary dysentery, studied by using a Tox⁻ mutant of *Shigella dysenteriae* 1. *Infect. Immun.* **56**:3099–3109.

19. **Frey, J., R. Meier, D. Gygi, and J. Nicolet.** 1991. Nucleotide sequence of the hemolysin I gene from *Actinobacillus pleuropneumoniae. Infect. Immun.* **59**:3026–3032.

20. **Glaser, P., D. Ladant, O. Sezer, F. Pichot, A. Ullman, and A. Danchin.** 1988. The calmodulin-sensitive adenylate cyclase of *Bordetella pertussis:* cloning and expression in *Escherichia coli. Mol. Microbiol.* **2**:19–30.

21. **Gonzalez-Carrero, M. I., J. Zabala, F. de la Cruz, and J. M. Ortiz.** 1985. Purification of a hemolysin from an overproducing *E. coli* strain. *Mol. Gen. Genet.* **109**:106–110.

22. **Grimminger, F., U. Sibelius, S. Bhakdi, N. Suttorp, and W. Seeger.** 1991. *Escherichia coli* hemolysin is a potent inductor of phosphoinositide hydrolysis and related metabolic responses in human neutrophils. *J. Clin. Invest.* **88**:1531–1539.

23. **Hacker, J. C., C. Hughes, H. Hof, and W. Goebel.** 1983. Cloned hemolysin genes from *Escherichia coli* that cause urinary tract infection determine different levels of toxicity in mice. *Infect. Immun.* **42**:57–63.

24. **Hewlett, E. L., V. M. Gordon, J. D. McCaffery, W. M. Sutherland, and M. C. Gray.** 1989. Adenylate cyclase toxin from *Bordetella pertussis:* identification of the holotoxin molecule. *J. Biol. Chem.* **264**:19379–19384.

25. **Highlander, S. K., M. Chidambaram, M. J. Engler, and G. M. Weinstock.** 1989. DNA sequence of the *Pasteurella haemolytica* leukotoxin gene cluster. *DNA* **8**:15–28.

26. **Holmgren, J., and A. Svennerholm.** 1977. Mechanism of disease and immunity in cholera: a review. *J. Infect. Dis.* **136S**:105–112.

27. **Issartel, J.-P., V. Koronakis, and C. Hughes.** 1991. Activation of *Escherichia coli* prohaemolysin to the mature toxin by acyl carrier protein-dependent fatty acylation. *Nature* (London) **351**:759–761.

28. **Jansen, R., J. Braire, E. M. Kamp, A. L. Gielkins, and M. A. Smits.** 1993. Cloning and characteriza-

tion of the *Actinobacillus pleuropneumoniae* RTX-toxin III (ApxIII) gene. *Infect. Immun.* **61:** 947–954.

29. **Jonas, D., B. Schultheis, C. Klas, P. H. Krammer, and S. Bhakdi.** 1993. Cytocidal effects of *Escherichia coli* hemolysin on human T lymphocytes. *Infect. Immun.* **61:**1715–1721.

30. **Jorgensen, S. E., R. F. Hammer, and G. K. Wu.** 1980. Effects of a single hit from the α-hemolysin produced by *Escherichia coli* on the morphology of sheep erythrocytes. *Infect. Immun.* **27:** 988–994.

31. **Jorgensen, S. E., P. F. Mulcahy, G. K. Wu, and C. F. Louis.** 1983. Calcium accumulation in human and sheep erythrocytes that is induced by *Escherichia coli* hemolysin. *Toxicon* **21:**717–727.

32. **Kaper, J. B., H. Lockman, M. M. Baldini, and M. M. Levine.** 1984. A recombinant live oral cholera vaccine. *Bio/Technology* **2:**345–349.

33. **Kaper, J. B., H. Lockman, M. M. Baldini, and M. M. Levine.** 1984. Recombinant nontoxigenic *Vibrio cholerae* strains as attenuated cholera vaccines candidates. *Nature* (London) **308:**655–658.

34. **Koronakis, V., M. Cross, B. Senior, E. Koranakis, and C. Hughes.** 1987. The secreted hemolysins of *Proteus mirabilis, Proteus vulgaris,* and *Morganella morganii* are genetically related to each other and to the α-hemolysin of *Escherichia coli. J. Bacteriol.* **169:**1509–1515.

35. **Kraig, E., T. Dailey, and D. Kolodrubetz.** 1990. Nucleotide sequence of the leukotoxin gene from *Actinobacillus actinomycetemcomitans:* homology to the α-hemolysin/leukotoxin gene family. *Infect. Immun.* **58:**920–929.

36. **Levine, M. M., J. B. Kaper, D. Herrington, G. Losonsky, J. G. Morris, M. L. Clements, R. E. Black, B. Tall, and R. Hall.** 1988. Volunteer studies of deletion mutants of *Vibrio cholerae* O1 prepared by recombinant techniques. *Infect. Immun.* **56:**161–167.

37. **Lo, R. Y. C., C. Strathdee, and P. Shewen.** 1987. Nucleotide sequence of the leukotoxin genes of *Pasteurella haemolytica* A1. *Infect. Immun.* **55:**1987–1996.

38. **Ludwig, A., R. Benz, and W. Goebel.** 1993. Oligomerization of *Escherichia coli* (HlyA) is involved in pore formation. *Mol. Gen. Genet.* **241:**89–96.

39. **Menestrina, G., N. Mackman, I. B. Holland, and S. Bhakdi.** 1987. *Escherichia coli* haemolysin forms voltage-dependent ion channels in lipid membranes. *Biochim. Biophys. Acta* **905:**109–117.

40. **Moayeri, M.** Unpublished data.

41. **Nicaud, J. M., N. Mackman, L. Gray, and I. B. Holland.** 1985. Characterization of HlyC and mechanism of activation and secretion of hemolysin from *E. coli* 2001. *FEBS Lett.* **187:**339–344.

42. **Opal, S. M., A. S. Cross, P. Gemski, and L. W. Lyhte.** 1990. Aerobactin and α-hemolysin as virulence determinants in *Escherichia coli* isolated from human blood, urine, and stool. *J. Infect. Dis.* **161:**794–796.

43. **Ostolaza, H., B. Bartolome, I. O. de Zarate, F. de la Cruz, and F. Goni.** 1993. Release of lipid vesicle contents by the bacterial protein toxin α-hemolysin. *Biochim. Biophys. Acta* **1147:**81–88.

44. **Ostolaza, H., B. Bartoleme, J. Serra, F. de la Cruz, and F. Goni.** 1991. α-Hemolysin from *E. coli:* purification and self-aggregation properties. *FEBS Lett.* **280:**195–198.

45. **Parker, C. T., A. W. Kloser, C. A. Schnaitman, M. A. Stein, S. Gottesman, and B. W. Gibson.** 1992. Role of the *rfaG* and *rfaP* genes in determination of the lipolysaccharide structure and cell surface properties of *Escherichia coli. J. Bacteriol.* **174:**2525–2538.

46. **Rennie, R. P., and J. P. Arbuthnott.** 1974. Partial characterization of *Escherichia coli* haemolysin. *J. Med. Microbiol.* **7:**179–188.

47. **Rennie, R. P., J. H. Freer, and J. P. Arbuthnott.** 1974. The kinetics of erythrocyte lysis by *Escherichia coli* haemolysin. *J. Med. Microbiol.* **7:**189–195.

48. **Rocque, W., R. Coughlin, and E. McGroarty.** 1987. Lipopolysaccharide tightly bound to porin monomers and trimers from *Escherichia coli* K-12. *J. Bacteriol.* **169:**4003–40i0.

49. **Rogel, A., J. Schultz, R. M. Brownlie, J. G. Coote, R. Parton, and E. Hanski.** 1989. *Bordetella pertussis* adenylate cyclase: purification and characterization of the toxic form of the enzyme. *EMBO J.* **8:**2755–2760.

50. **Short, E. C., Jr., and H. J. Kurtz.** 1971. Properties of the hemolytic activities of *Escherichia coli. Infect. Immun.* **3:**678–687.

51. **Smith, H. W., and M. B. Huggins.** 1985. The toxic role of alpha-haemolysin in the pathogenesis of experimental *Escherichia coli* infection in mice. *J. Gen. Microbiol.* **131:**395–403.

52. **Stanley, P. L. D., P. Diaz, M. J. A. Bailey, D. Gygi, A. Juarez, and C. Hughes.** 1993. Loss of

activity in the secreted form of *Escherichia coli* haemolysin caused by an *rfaP* lesion in core lipopolysaccharide assembly. *Mol. Microbiol.* **10**:781–787.

53. **Swihart, K. G., and R. A. Welch.** 1990. The HpmA hemolysin is more common than HlyA among *Proteus* isolates. *Infect. Immun.* **58**:1853–1860.

54. **Tesh, V. L., and A. D. O'Brien.** 1991. The pathogenic mechanisms of Shiga toxin and the Shiga-like toxins. *Mol. Microbiol.* **5**:1817–1822.

55. **Trucksis, M., J. E. Galen, J. Michalski, A. Fasano, and J. Kaper.** 1993. Accessory cholera entero-toxin (Ace), the third toxin of a *Vibrio cholerae* virulence cassette. *Proc. Natl. Acad. Sci. USA* **90**:5267–5271.

56. **Wagner, W., M. Kuhn, and W. Goebel.** 1988. Active and inactive forms of hemolysin (HlyA) from *Escherichia coli. Biol. Chem. Hoppe-Seyler* **369**:39–46.

57. **Wandersman, C., and S. Letoffe.** 1993. Involvement of lipopolysaccharide in the secretion of *Escherichia coli* α-hemolysin and *Erwinia chrysanthemi* protease. *Mol. Microbiol.* **7**:141–150.

58. **Welch, R. A.** 1987. Identification of two different hemolysin determinants in uropathogenic *Proteus* isolates. *Infect. Immun.* **55**:2183–2190.

59. **Welch, R. A.** 1991. Pore-forming cytolysins of gram-negative bacteria. *Mol. Microbiol.* **5**:521–528.

60. **Welch, R. A., E. P. Dellinger, B. Minshew, and S. Falkow.** 1981. Hemolysin contributes to virulence of extra-intestinal *Escherichia coli* infections. *Nature* (London) **294**:665–667.

61. **Welch, R. A., R. Hull, and S. Falkow.** 1983. Molecular cloning and physical characterization of a chromosomal hemolysin from *Escherichia coli. Infect. Immun.* **42**:178–186.

62. **Welch, R. A., S. Pellett, D. Robbins, W. Keane, G. Gekker, and P. Peterson.** 1989. Epidemiological observations involving the *Escherichia coli* hemolysin. *In* E. Kass and C. Svanborg-Eden (ed.), *Host-Parasite Interactions in Urinary Tract Infections.* University of Chicago Press, Chicago.

63. **White, J. G.** 1974. Effects of an ionophore, A23187, on the surface morphology of normal erythrocytes. *Am. J. Pathol.* **77**:507–518.

64. **Williams, P. H.** 1979. Determination of the molecular weight of *Escherichia coli* α-haemolysin. *FEMS Microbiol. Lett.* **5**:21–24.

65. **Zwadyk, P., and I. S. Snyder.** 1971. Purification and kinetic studies of the hemolysin from *Escherichia coli. Can. J. Microbiol.* **17**:741–745.

Molecular Genetics of Bacterial Pathogenesis
Edited by V. L. Miller, J. B. Kaper, D. A. Portnoy, and R. R. Isberg
© 1994 American Society for Microbiology, Washington, DC 20005

Chapter 24

Yops of the Pathogenic *Yersinia* spp.

James B. Bliska

During their interaction with host cells, the pathogenic *Yersinia* spp. export a set of proteins known as the *Yersinia* outer membrane proteins, or Yops. The Yops are encoded by 70- to 75-kb virulence plasmids that are highly conserved among the three pathogenic yersiniae, *Y. pestis, Y. enterocolitica,* and *Y. pseudotuberculosis.* Several Yops with essential roles in *Yersinia* pathogenesis are homologous to eukaryotic proteins involved in the transduction of extracellular stimuli. The yersiniae may express these Yops during infection of the reticuloendothelial system as part of a multifactorial strategy to subvert and/or exploit signal transduction in the host. This chapter reviews the work that led to the discovery of the Yops and discusses recent progress made in understanding the expression and function of these proteins.

YERSINIA PATHOGENESIS

The genus *Yersinia* (family *Enterobacteriaceae*) consists of seven species, of which *Y. pestis, Y. enterocolitica,* and *Y. pseudotuberculosis* are considered primary pathogens of humans and other animals. *Y. pestis* is endemic in rodent populations in many parts of the world and is the agent of bubonic plague, an acute febrile lymphadenitis (10). Fleas serve as vectors for transmitting the infection between rodent hosts and from rodents to humans and other animals. Inflammation in a lymph node draining the site of inoculation can lead to the formation of a bubo, a distinctive swelling that is the hallmark of bubonic plague. From infected lymph nodes the bacteria spread to the bloodstream and distal organs such as the liver, spleen, and lungs. Bacterial infection of the lungs results in pneumonic plague, a form of the disease that is transmissible by aerosolization and that is often fatal. Several plasmid-encoded virulence determinants unique to *Y. pestis* have been described, including the plasminogen activator/coagulase and pesticin, which are encoded by the 9.5-kb Pst plasmid (39, 60), and the fraction 1 capsule and murine exotoxin, which are encoded by the 100-kb Tox plasmid

---

James B. Bliska • Department of Molecular Genetics and Microbiology, State University of New York at Stony Brook, Stony Brook, New York 11794-5222.

(50). These factors and their roles in the virulence of *Y. pestis* have been reviewed previously (10).

Y. enterocolitica is responsible for a variety of human illnesses ranging in severity from mild gastroenteritis to acute terminal ileitis (12). Natural reservoirs for *Y. enterocolitica* include rodents and domestic animals such as pigs and sheep. Transmission of the infection commonly occurs by consumption of contaminated food or water. Following ingestion, the bacteria transit to the ileum, invade the intestinal mucosa, and replicate in the lamina propria. In normal hosts, systemic involvement is rare, but gastrointestinal disease commonly results from extensive bacterial replication in the ileal tissues. Most pathogenic *Y. enterocolitica* strains elaborate a chromosomally encoded heat-stable enterotoxin that appears to be an important factor in diarrheal disease in young rabbits and possibly other hosts (14).

Except for the presence of the Tox and Pst plasmids, *Y. pestis* and *Y. pseudotuberculosis* are closely related (>90%) at the DNA level. By comparison, *Y. pestis* and *Y. enterocolitica* are about 50% homologous (47). Nevertheless, the salient features of *Y. pseudotuberculosis* infection and disease are similar to those described above for *Y. enterocolitica*. Various rodents, farm animals, and birds are the normal reservoirs for *Y. pseudotuberculosis*. Rabbits, gerbils, and mice have often been used as animal infection models because they are particularly susceptible to the pathogenic *Yersinia* spp. In an electron microscopic study of infection in rabbits, the initial site of *Y. pseudotuberculosis* invasion occurred in the M cells overlying the ileal lymphoid follicles (Peyer's patches) (20). Following M-cell translocation, *Y. pseudotuberculosis* cells were taken up by macrophages within the Peyer's patches (20). In a similar study that used immunohistochemical detection, extracellular localization of *Y. enterocolitica* was observed in the lamina propria and few bacteria were taken up by polymorphonuclear leukocytes (PMNs) recruited to the site of infection (28). These observations suggest that the enteropathogenic *Yersinia* spp. undergo a transient intracellular phase at the initiation of the infection process and that the bulk of bacterial replication occurs extracellularly. In addition, although the enteropathogenic yersiniae and *Y. pestis* invade their hosts by different routes and cause diseases of very different severities, they share a common tropism for lymphoid tissues and a remarkable capacity to suppress the cellular immune response.

YERSINIA VIRULENCE PLASMIDS

The *yop* Regulon

The homologous ca. 70- to 75-kb virulence plasmids carried by the pathogenic yersiniae (1, 23) are generically designated as pYVs (for plasmids associated with *Yersinia* virulence) (46). Work that led to the discovery and characterization of the pYVs has been reviewed previously (47). The pYV-encoded *yop* regulon controls the synthesis and export of the Yops by the yersiniae in response to stimuli encountered during infection. Operons within the *yop* regulon encode the known Yops (see Table 1 below) as well as accessory proteins with regulatory and secretory functions. Yop synthesis is controlled primarily at the transcriptional level in response to temperature. Transcription of the *yop* regulon is increased at least 100-fold by a temperature shift from 25 to 37°C (13). Thermal induction of transcription is modulated by YmoA, a small histone-like protein, and LcrF (VirF), a transcriptional activator related to AraC, the arabinose operon regulator (13, 36). Under standard culture conditions at 37°C, the Yops are present in a membrane-bound form but are not exported (7). As discussed below, Yop export is controlled at an additional level, apparently in response to stimuli encountered at the host cell surface.

When *Yersinia* cells are cultured at 37°C in the presence of Mg^{2+} and the absence of millimolar concentrations of Ca^{2+} or nucleotides such as ATP, the Yop export pathway is activated, and transcription of the *yop* regulon is upregulated an additional two- to threefold (13, 63, 64). This phenomenon leads to the restriction of bacterial growth and is known historically as the Ca^{2+}-dependent phenotype or the low-Ca^{2+} response (LCR) (63, 64). Ca^{2+} is abundant in extracellular fluids, but the mammalian cell cytoplasm typically has low levels of free Ca^{2+}. Prior to the accumulation of evidence indicating that yersiniae do not enter the cytoplasms of eukaryotic cells, it was proposed that the LCR was designed to sense low cytoplasmic Ca^{2+} concentrations (63, 64). It is more likely that low Ca^{2+} concentrations artifactually activate Yop export by mimicking the stimuli encountered at the host cell surface. Although the physiological relevance of the LCR is in question, it has nevertheless facilitated a genetic analysis of the *yop* regulon. Yop export is constitutively activated at 37°C in the presence of Ca^{2+} in *lcrE* insertion mutants, which suggests that LcrE acts as a negative regulator of Yop export (63–65). In addition to LcrE (also known as YopN), other released regulatory proteins such as LcrV (V antigen) and LcrG appear to participate in this negative-feedback loop (13, 52, 63, 64).

lcr- and *ysc*-Encoded Yop Export Pathway

In addition to the regulatory proteins already mentioned, the Yop export process requires multiple gene products of several operons including the 13-cistron Yop secretion (*ysc*) operon, the multicistronic *lcrB* locus, and *lcrDR* (63, 64). A number of the *lcr* and *ysc* gene products are homologous to proteins involved in flagellar biosynthesis or *sec*-independent export of virulence determinants by other bacterial species (21, 42, 63, 64). The product of *yscC* shares significant homology with PulD, a protein required for the export of pullulanase by *Klebsiella pneumoniae* (42). At least eight pYV-encoded proteins are homologous to products of the *inv* locus, a protein export pathway involved in the entry of *Salmonella typhimurium* into mammalian cells (21). For example, *invC*, the homolog of *yscN*, encodes a protein homologous to the F_0F_1 ATPase family (15). These proteins could potentially use the energy of ATP hydrolysis to drive protein export (15). These findings lend support to the idea that a number pathogenic bacteria, including *Yersinia* and *Salmonella* spp., use a novel type of export pathway during infection to deliver effector molecules to host cells. Although the export machinery may be homologous, the effector proteins are likely to be uniquely tailored to the pathogenic strategy of the bacteria. In the case of *Salmonella* spp., the effector molecule(s), which has not been identified, presumably facilitates bacterial uptake by the induction of membrane ruffles (19, 44) (see chapters 14 and 16). In studies that use high-resolution low-voltage scanning electron microscopy, flagellar-like appendages have been observed on the surface of *S. typhimurium* prior to invasion into mammalian cells (24). Formation of these surface appendages required bacterial cell-host cell contact and the function of *invC* (24). These appendages may represent a supramolecular structure encoded by the *inv* locus that delivers effector proteins to the host cell (24). It is tempting to speculate that similar appendages will be observed when this microscopic technique is applied to the study of *Yersinia*-mammalian cell interactions.

DISCOVERY AND CHARACTERIZATION OF Yops, A HISTORICAL PERSPECTIVE

Expression of Yops in Enteropathogenic Yersiniae

In 1956, Burrows and Bacon (11) described the presence of two antigens, the V and W antigens, in extracts of virulent *Y. pestis* grown at 37°C. It was found subsequently that *Y. pseudotuberculosis* and *Y. enterocolitica* also expressed the V and W antigens under LCR conditions (for a review, see reference 47). By the late 1970s, it was established that virulence, Ca^{2+} dependency, and expression of the V and W antigens were associated with the presence of a plasmid in all three pathogenic yersiniae (12, 47). This correlation suggested the existence in *Yersinia* spp. of a common plasmid-mediated mechanism of pathogenicity. In 1980, Zink et al. (70) reported the involvement of a plasmid in the ability of *Y. enterocolitica* to evoke a positive Sereny test. In 1981, members of Stanley Falkow's laboratory investigated the relationship between the pYV in *Y. enterocolitica* and tissue cell pathology in culture (45). After precultivation of the bacteria at 25°C, strains harboring pYV and their cured derivatives were found to attach to and enter into HEp-2 cells grown in monolayers. In addition, by the end of the assay, the HEp-2 cell monolayers infected with pYV-containing bacteria became detached from the tissue culture flasks. The authors concluded that the pYV was not required for invasion of HEp-2 cells by *Y. enterocolitica,* but it most likely encoded a factor responsible for monolayer detachment. Furthermore, when outer membrane proteins of pYV-bearing and cured strains grown at 25 and 37°C were examined by sodium dodecyl sulfate (SDS)-polyacrylamide gel electrophoresis (PAGE), at least three additional proteins (molecular sizes, 50, 37, and 25 kDa) were detected in Triton X-100-insoluble outer membranes prepared from pYV-containing bacteria grown at 37°C (Fig. 1). These proteins were not expressed by pYV-containing bacteria grown at 25°C or plasmid-cured strains, regardless of the growth temperature. In addition, it was reported that the addition of 2.5 mM Ca^{2+} to the bacterial growth medium repressed the expression of the plasmid-associated polypeptides (this would suggest that the medium used [brain heart infusion] was sufficiently low in Ca^{2+} to induce the LCR) (45). Subsequent studies confirmed and extended these initial observations (for reviews, see references 12 and 47). In 1985, Bolin

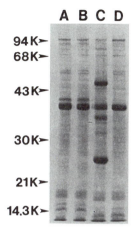

Figure 1. SDS-PAGE analysis of *Y. enterocolitica* outer membrane polypeptides. Lane A, strain 9576 grown at 25°C; lane B, strain 9576-c grown at 25°C; lane C, strain 9576 grown at 37°C; lane D, strain 9576-c grown at 37°C. From Portnoy et al. (45).

et al. (8) demonstrated that the pYV-associated outer membrane proteins were immunologically related in all three pathogenic yersiniae. The name YOPs, for yersiniae outer membrane proteins, was proposed (8) (the designation Yops is used here). Later work established the direct relationship between the expression of the Yops and the detachment of cultured mammalian cell monolayers in response to yersinial infection (54).

Expression of Yops in *Y. pestis*

In 1981, Straley and Brubaker (62) used one- and two-dimensional PAGE to examine differences in the expression and localization of the V and W antigens by *Y. pestis* and the enteropathogenic yersiniae. A number of additional polypeptides (later to be identified as the Yops) were detected in sucrose gradient-purified outer membranes from pYV-containing *Y. enterocolitica* and *Y. pseudotuberculosis* grown under LCR conditions. However, Yops were not immediately apparent in the outer membrane preparations of pYV-containing *Y. pestis,* regardless of the growth temperature (62).

This observation led initially to some doubt about the importance of the Yops in the pathogenicity of *Y. pestis*. However, the transfer of transposon-tagged pYV DNA to minicell-producing *Escherichia coli* strains and *Y. pseudotuberculosis* soon demonstrated that the *Y. pestis* pYV contained coding information for the Yops (48, 67). In addition, antibodies to Yops were detected in the sera of *Y. pestis*-infected animals (67). Finally, transposon-induced insertional mutations in several of the *yop* genes dramatically increased the 50% lethal doses of *Y. pestis* in mouse virulence assays (61). These observations indicated that the Yops were expressed in vivo by *Y. pestis* a J were important for pathogenicity. It was subsequently shown that the Yops were synthesized and exported by *Y. pestis* but then rapidly degraded by some factor encoded by the 9.5-kb Pst plasmid (57). Sodeinde et al. (59) found that deletion of the *pla* gene from this plasmid, which encodes the surface protease with plasminogen activator/coagulase activities, blocked the degradation of Yops.

This finding solved the mystery of the disappearing Yops, but the role of this phenomenon in the pathogenicity of *Y. pestis* remains to be clarified. Pla-mediated Yop proteolysis could be an artifact of in vitro culture, resulting from high local concentrations of Pla and the Yops together at the bacterial surface. Under the normal conditions of infection, delivery of the Yops directly to the host cell may minimize proteolysis. This concept is supported by the observation that Yop activity is detected in *pla*⁺ *Y. pestis* cells bound to cultured macrophages (25) and epithelial cells (54). High concentrations of the Pla substrate plasminogen in vivo could also competitively inhibit excessive Yop degradation during infection. As stated by Sodeinde et al. (59), "If this hypothesis proves correct, the primary role of *pla*-mediated Yop degradation in *Y. pestis* physiology has been to confuse researchers investigating the function of the 75-kb plasmid." Alternatively, in addition to its other activities (39, 60), Yop proteolysis by Pla could play a role in the pathogenesis of plague. There appears to be some specificity in proteolysis because several of the more soluble exported proteins, including YopM (see below), LcrV (the V antigen), and LcrE, escape degradation by Pla (63). The action of Pla on specific Yops could represent a posttranslational regulatory event if the resulting proteolytic fragments retain normal or enhanced activity. There is precedence for this in the case of YopH because a truncated form of this protein lacking the first 154 amino acids (see below) retains tyrosine phosphatase activity in vitro (26).

Yops as Released Proteins

In 1984, Heesemann et al. (30) characterized proteins released into the growth medium of pYV-containing *Y. enterocolitica*. On the basis of comparisons of electrophoretic mobilities, the authors surmised that the released proteins were related to the pYV-associated outer membrane proteins seen previously (45, 62). A later study established the identity of the released proteins as Yops and showed that their appearance in the growth medium did not occur by membrane blebbing but rather by a specific export process (29). When the N-terminal amino acid sequence of the released YopE was obtained and compared with that of the predicted product, it became apparent that Yops were released in intact form, i.e., without the removal of a signal sequence (17). Michiels et al. (43) provided evidence that the appearance of exported Yops in outer membrane fractions resulted either from their copurification with membranes or from their readsorption to the bacterial surface. Deletion analysis was used to localize the export recognition domains of YopH, YopE, and YopQ to their first 48, 98, and 76 N-terminal amino acids, respectively (41). Comparison of these regions failed to show any homology, suggesting that information for Yop export resided in the protein tertiary structure (41). More recently, it has been proposed that Yop-specific chaperone-like proteins provide specificity in the export process (66). On the basis of the results of those studies as well as analysis of deduced amino acid sequences, the Yops, despite their name, are not typical outer membrane proteins. Rather, they are exported proteins that are loosely associated with the outer membrane, possibly in the form of a multimeric structure.

ROLE OF Yops IN *YERSINIA* PATHOGENESIS

Yop Genes and Mutants

Most *yop* genes are found in monocistronic operons distributed around the pYV. *yopB* and *yopD* are found uncharacteristically at the end of the polycistronic operon *lcrGVHyopBD* (27). The majority of the genes encoding the known Yops have now been cloned by recombinant DNA methods and sequenced (Table 1). Four of these, *yopE, yopM, yopH*, and *ypkA*, have been shown to encode determinants essential for *Yersinia* virulence in animal models of acute disease (9, 22, 37, 54, 61). Activities were assigned to YopE and YopH in studies that analyzed the interaction of *Yersinia yop* mutants with cultured mammalian cells. The ability of *Yersinia* strains to induce rounding and detachment of cultured mammalian cells has been referred to as "cytotoxicity" (25, 54). YopE is known as the cytotoxin and is the primary factor involved in monolayer detachment (Table 1) (3, 54). The YopE protein present in secreted Yop preparations or lysates of bacterial cultures is not active in promoting cytotoxicity when added exogenously to the surfaces of cultured mammalian cells (25, 45, 55). A partial cytotoxic activity is reconstituted if these protein preparations are introduced into HeLa cells by the use of glass carrier beads (55). This indicates that YopE must enter the cytoplasm of the host cell to be active. A molecular function has not been assigned to YopE, but it may act to destabilize host cell actin microfilaments (55). Mutations in *yopD* (Table 1) appear to prevent YopE from functioning in a normal capacity (55). The defect

TABLE 1
Yops and other pYV-encoded proteins released by yersiniae

Protein name (mol mass [kDa])[a]	No. of amino acids[b]	Structural features	Identified function	Reference
YpkA (84)	732	Kinase homology	Serine/threonine kinase	22
YopH (51)	468	PTPase homology	PTPase	26 9
YopM (48)	367 (Ype)	GP1bα homology	Binds thrombin	38
YopB (44)	401	RTX toxin/IpaB homology	ND[c]	27
LcrV (41)	326 (Ype)	ND	Regulatory protein protective antigen	49
YopD (37)	306	Transmembrane region	Required for YopE function	27 55
LcrE (35)	293 (Ye)	ND	Regulatory protein	65
YopE (25)	219	ND	Cytotoxin	18
YopQ (20)	182 (Ye)	AraH homology	ND	43

[a] The names and apparent molecular masses of the proteins released from *Yersinia* spp. are given. Molecular masses are from Michiels et al. (43).

[b] Number of amino acids in each protein as predicted from the gene sequence. Values are for proteins encoded by *Y. pseudotuberculosis* unless indicated otherwise by the abbreviations in parentheses: Ype, *Y. pestis;* Ye, *Y. enterocolitica.*

[c] ND, not determined.

appears to occur after export, so it is conceivable that YopD is involved in the YopE entry process or the activity of YopE within the host cell (55). *Y. pseudotuberculosis yopE* insertion mutants express a residual level of cytotoxicity that is mediated by YopH (3, 54). The activity and function of this protein are described below.

Effect of Yops on Cellular Entry

Insertional or deletional inactivation of *yopH* or *yopE* in *Y. pseudotuberculosis* was reported to increase the efficiency of bacterial uptake by cultured macrophages (53, 54). Although the uptake pathway was not investigated in those studies, under the assay conditions used (minus complement or specific antibodies) (53, 54), uptake was most likely mediated by a bacterially encoded ligand. Two *Y. pseudotuberculosis* genes, chromosomal *inv* (34) and virulence plasmid-specific *yadA* (58), encode pathways of bacterial entry into cultured mammalian cells (3, 34, 68) (see chapter 15). These entry pathways involve coordinated movement of the actin cytoskeleton in response to signals generated by bacterial attachment to host cell integrin receptors (3, 34, 68). Thus, the negative effects of YopE and YopH on bacterial entry may be attributed to their activities against normal host cell signal transduction and/or cytoskeletal functions.

The available evidence is consistent with the idea that the Yops are exported in response to signals encountered at the host cell surface (31, 53, 54). Because of this, the ultimate fate of *Y. pseudotuberculosis* depends on the relative levels of *inv*, *yadA*, and *yop* regulon expression at the time of bacterial cell-host cell contact. In general, wild-type or *inv* mutant strains pregrown at 28°C enter cultured cells efficiently, or at least to detectable levels (3, 33, 68) (Fig. 2). In this case, there is sufficient time for entry to occur during a lag period at 37°C that is required to upregulate the *yop* regulon. Wild-type or *inv* mutants pregrown at 37°C generally enter cells at lower or undetectable levels (33) (Fig. 3). Under these conditions, cellular entry is less efficient because the Yops are much more rapidly exported and delivered to the host cell.

Homologies to Eukaryotic Signal Transduction Proteins

The predicted products of *yopM, ypkA,* and *yopH* have been found by sequence analysis to contain blocks of homology to eukaryotic proteins involved in the transduction of extracellular stimuli (see below and Table 1) (22, 26, 38). YopM contains homology to an external domain of the platelet-specific receptor GP1bα (38). GP1bα is a seven-transmembrane G protein-linked receptor for thrombin and the von Willebrand factor. Binding of these ligands to the receptor causes platelet aggregation and activation and the release of a number of inflammatory mediators. Initially, the YopM-homologous domain of GP1bα was thought to be directly involved in thrombin binding (38). Although this may not be the case, purified YopM nevertheless has thrombin-binding activity and competitively inhibits thrombin-induced platelet activation (37, 51). Thus, during infection, sequestration of thrombin by YopM may subvert host inflammatory responses mediated by activated platelets. A previously uncharacterized 82- to 84-kDa Yop was recently

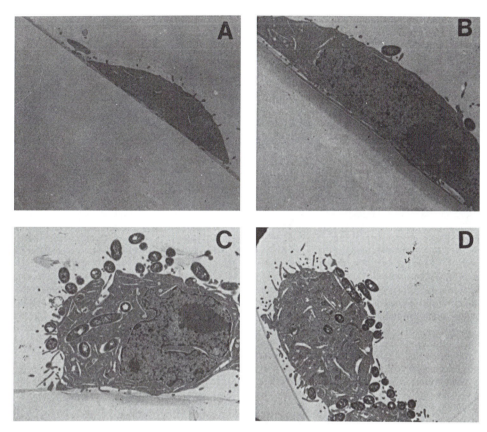

FIGURE 2. *Y. pseudotuberculosis inv* strains harboring the virulence plasmid and pregrown at 28°C are localized intracellularly in HEp-2 cells. Bacterial strains were grown at 28°C in L broth overnight and were added to monolayers of HEp-2 cells at 37°C. After 2 h for binding and entry, unbound bacteria were washed away and the monolayers were fixed, embedded, and sectioned for electron microscopy. (A and B) Virulence plasmid-cured *Y. pseudotuberculosis inv*. (C and D) Virulence plasmid-containing *Y. pseudotuberculosis inv*. From Isberg (33).

found to have extensive homology to the eukaryotic serine/threonine kinase family (22). This protein, designated YpkA, for *Yersinia* protein kinase A, undergoes autophosphorylation in vitro on serine (22). Because serine phosphorylation is a common protein modification in bacteria, YpkA could potentially play a regulatory role in *Yersinia* spp., but it is more likely that it functions in some capacity to undermine signal transduction in the host.

Tyrosine Phosphatase Activity of YopH

The *Y. pseudotuberculosis* YopH is a 468-amino-acid protein that migrates on SDS-polyacrylamide gels with an apparent molecular size of 51 kDa (Fig. 4). The *Y. enterocolitica* homolog of YopH is designated Yop51 (40). YopH and Yop51 are 99% homologous (462 of 468 shared amino acids). The nucleotide se-

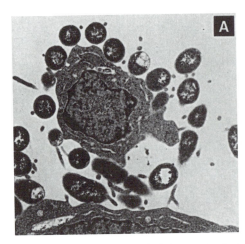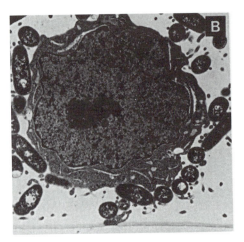

FIGURE 3. *Y. pseudotuberculosis inv* strains harboring the virulence plasmid and pregrown at 37°C do not enter HEp-2 cells. *Y. pseudotuberculosis inv* was grown overnight in L broth at 28°C. Freshly saturated cultures were diluted in the same medium and were aerated at 37°C for 3 h at this temperature before addition to monolayers of HEp-2 cells. After 2 h for binding and intracellular entry of the bacteria, the monolayers were washed and processed for thin-section electron microscopy. Displayed are two micrographs of *Y. pseudotuberculosis inv* infected in this fashion. From Isberg (33).

quence of *yopH* in *Y. pestis* (61) has not been reported. The sequence of the first 128 N-terminal residues of YopH is 42% identical to the sequence of YscM (LcrQ) (Fig. 5), a protein involved in the Yop regulatory and secretory pathway (see above) (42, 52). It is plausible that this domain of YopH originated as a tandem duplication of YscM. The carboxy-terminal 262 amino acids of Yop-51 and YopH are homologous to the catalytic domains of protein tyrosine phosphatases (PTPases) (26) (Fig. 5). Specific PTPase activity for phosphorylated peptides and the insulin receptor was present in recombinant Yop-51 expressed in *E. coli,* and this activity was lost upon conversion of the cysteine residue at position 403 in the catalytic site to alanine (the conversion is abbreviated as C403A) (26). In a comparative study of PTPase enzymatic properties, the k_{cat} value of Yop-51 was 25-fold higher than that of mammalian PTP1, indicating that the *Yersinia* enzyme is, by far, the most active PTPase known (69).

To determine if the PTPase activity of YopH was important for *Yersinia* pathogenesis, a mutant of *Y. pseudotuberculosis* lacking this activity was constructed and tested in a murine model of infection (6). For this purpose, allelic recombination was used to introduce the C403A codon substitution into the pYV of *Y. pseudotuberculosis* YPIII (6). The resulting strain expressed a form of YopH (YopHC403A) that differed significantly from the wild-type protein only in PTPase activity (Fig. 4) (6). Inactivation of YopH activity by the C403A mutation or by deletion of the gene had no obvious effects on bacterial growth or gene expression outside of the host (6, 9). The wild-type *Y. pseudotuberculosis* strain, the C403A mutant, and a plasmid-cured derivative were introduced into mice by intravenous injection, and bacterial growth in the spleen was determined. The C403A mutation

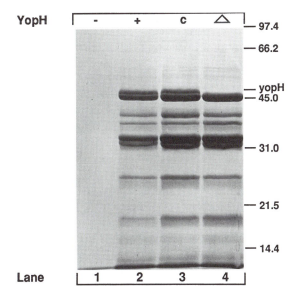

FIGURE 4. SDS-PAGE analysis of proteins released by *Y. pseudotuberculosis yopH* mutants. *Y. pseudotuberculosis* strains were grown at 37°C in a defined low-Ca^{2+} medium (6, 61). Proteins were isolated from the growth medium by precipitation with trichloroacetic acid and were analyzed by electrophoresis on an SDS-polyacrylamide gel (12%). The strains used are described in Bliska et al. (4). Lane 1, plasmid-cured *Y. pseudotuberculosis;* lane 2, wild-type *Y. pseudotuberculosis* (*yopH$^+$*); lane 3, *Y. pseudotuberculosis* catalytic YopH mutant (*yopHC403A*); lane 4, *Y. pseudotuberculosis* YopH deletion mutant (*yopH∆*). The positions of the molecular size markers (in kilodaltons) and YopH are indicated on the right.

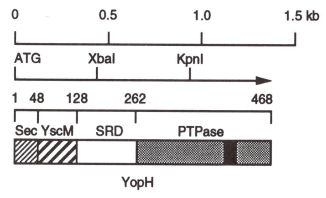

FIGURE 5. Structure of YopH. At the top is shown a restriction map of *yopH* and a corresponding scale in kilobases. At the bottom is shown a functional map of YopH with a scale corresponding to length in amino acids. The domains involved in various functions of the protein are indicated. Sec, N-terminal 48 residues sufficient for export (41); YscM, region of YopH that is homologous to YscM (LcrQ) and that corresponds to the first 128 amino acids and overlaps the export domain (42, 52): SRD; substrate recognition domain (residues 129 to 261), a region of the protein involved in phosphoprotein-binding activity (2, 4); PTPase, domain of YopH that is homologous to eukaryotic protein tyrosine phosphatase catalytic domains (26).

significantly reduced the virulence of *Y. pseudotuberculosis* in this assay, which measured the ability of the bacteria to maintain a systemic infection (6). To determine if the PTPase activity of YopH was also important for the establishment of infection by the oral route, three groups of three 8-week-old female Swiss mice were challenged orally with 7×10^8 CFU of either the wild-type strain, the C403A mutant, or the plasmid-cured derivative [IP2666(pIB1), IP2666(pIB1C403A), and IP2666c, respectively (6)]. After 5 days, bacterial growth in the spleen was quantitated. All three mice challenged with IP2666(pIB1) had high numbers of bacteria in their spleens (between 1×10^5 and 2×10^5 CFU), while the spleens of mice infected with IP2666c and IP2666(pIB1C403A) were not colonized. These results suggest that YopH PTPase activity is also important for the establishment of systemic infection by the oral route. A more careful study of the growth properties of these mutants in Peyer's patches and mesenteric lymph nodes at earlier time points is needed to clarify the role of YopH during infection.

PTPases are a large family of enzymes that act in concert with protein tyrosine kinases (PTKases) to facilitate signal transduction and regulate cell division in eukaryotic cells (16, 32). There are reports of tyrosine phosphorylated proteins in pathogenic bacteria (for a review, see reference 5), but in general, this protein modification is absent from most prokaryotes including the yersiniae. It has not been ruled out that YopH PTPase activity is important in *Yersinia* physiology, but as discussed below, several lines of evidence suggest that the primary function of YopH involves entery into host cells and upsetting of the normal balance in PTPase and PTKase activities. Several tyrosine-phosphorylated host proteins are rapidly dephosphorylated (within 15 min) in cultured murine macrophages infected by wild-type *Y. pseudotuberculosis* (4). These proteins were not dephosphorylated in cells infected by the C403A mutant. Although these data do not prove that YopH directly dephosphorylated these proteins, they indicate that PTPase activity is somehow responsible for altering normal host tyrosine phosphorylation.

YopHC403A binds specifically to several tyrosine-phosphorylated proteins present in detergent lysates of murine macrophages (4). This interaction may reflect the affinity of YopHC403A for a class of proteins that are the normal substrates of YopH during infection. Although the exact regions of YopH responsible for this putative substrate-binding activity have not been defined, sequences found between the borders of the YscM (LcrQ) homology domain and the PTPase homology domain (amino acids 129 to 261) appear to be involved (2). This region of YopH is referred to as the "substrate recognition domain" (Fig. 5). To understand the molecular function of YopH within host cells, it will be important to study this region of the bacterial PTPase and to identify eukaryotic proteins that bind to it.

OVERVIEW AND CONCLUSION

A microbial pathogen must be transmitted to a susceptible host, find a suitable niche, replicate, and disseminate to successfully complete its infectious life cycle (see the introduction to this volume by Stanley Falkow). The three pathogenic

Yersinia spp. share a common tropism for the lymphoid tissues of their hosts but have evolved two distinct modes of infection. The enteropathogenic yersiniae cause chronic gastrointestinal infections, are excreted for extended periods of time, and are transmitted to new hosts by the fecal-oral route. *Y. pestis* causes an acute, systemic infection to facilitate ingestion of the bacteria by the flea vector during a blood meal. Death of the host, which typically follows the onset of bacteremia, prompts the infected flea to search for a new host. During a subsequent blood meal, the infection is transmitted to the susceptible host by subcutaneous or intradermal inoculation. The relatively small numbers of genotypic and phenotypic differences that exist between the enteropathogens and *Y. pestis* are probably responsible for these distinct pathogenic lifestyles (10).

The ability of *Yersinia* spp. to replicate in the host reticuloendothelial system involves a mechanism of pathogenicity that is common to all medically significant members of this genus. This process is encoded by the 70- to 75-kb pYV that is highly conserved in *Y. pestis*, *Y. enterocolitica*, and *Y. pseudotuberculosis* (47). *Yersinia* strains cured of pYV retain the ability to enter host tissues (and mammalian cells in culture), but they lack the capacity to replicate to any significant degree. Because virulent pYV-containing *Yersinia* strains are able to overcome the normal inflammatory response and replicate extracellularly in host tissues, pYV-encoded elements appear to suppress the cell-mediated immune response.

Investigators are just beginning to appreciate the complexity of this virulence apparatus. The system appears to consist of a regulatory hierarchy that is coupled to the export and assembly of a multiprotein complex on the bacterial cell surface. Many of the individual components of this process such as transcriptional activators (i.e., LcrF or VirF), export machinery (i.e., the YscN ATPase), and end products or effector proteins (i.e., the PTPase YopH) have been identified and characterized (63, 64). It is also clear that a number of other bacteria pathogenic for plants or animals encode a homologous system of virulence protein export (21, 63). Domain-swapping experiments currently under way should provide important insights into the structures and functions of many of these individual components. Many similarities also exist between these virulence protein export pathways and the process of synthesis and assembly of flagellar organelles in bacteria. It should be instructive to examine the genetic, molecular, and biochemical strategies used in the study of flagellar biosynthesis when investigating the mechanism of virulence protein export.

During their pathogenic life cycles, the yersiniae transit from environmental niches into the tissues of the host. To conserve energy and to avoid inappropriate virulence gene expression, Yop synthesis and export are regulated in response to stimuli encountered during infection of the mammalian host. The primary stimulus for synthesis of the Yops and the export machinery is growth at 37°C, the physiological temperature of the mammalian host (13). Upon binding to a host cell surface, a second signal is received and the Yops are rapidly (probably within 5 to 15 min) exported and delivered to the mammalian cell. The speed of this process may be critical, since host cell responses to bacterial infection can occur equally rapidly. For example, the antibacterial machinery of phagocytic cells is activated within several minutes of microbial binding. Major issues that remain to be addressed in this regard are the nature of the stimuli for Yop export and the mechanism of Yop delivery into the host cell. Bacterial attachment mediated either by invasin or by YadA is necessary and sufficient to allow the export and delivery of the Yops in tissue culture assays (3, 54). It is likely that both mechanisms of adherence involve bacterial attachment to host cell integrin receptors (35), either directly in the case of invasin (34) or via extracellular matrix ligands in the case of YadA (3, 68). One could speculate from these data that bacterial attachment to integrins is necessary for Yop export and delivery. However, neither invasin nor YadA is necessary for lethal infections of the rodent host by *Y. pseudotuberculosis* or *Y. pestis* (56). Thus, additional mechanisms of productive attachment must be expressed or available in the in vivo environment. One possibility is that the Yops are exported when opsonized yersiniae are bound to phagocyte receptors, such as the complement integrin receptor CR3. Alternatively, one of the additional adherence factors known to be expressed by the yersiniae (33) (see chapter 15) might fulfill this role in vivo. To gain further insight into the nature of the stimulus for Yop export, it will be necessary to investigate the roles of these receptor-ligand interactions in *Yersinia* pathogenesis.

Signal transduction is critically involved in a number of cellular host responses to microbial infection. From the homologies that exist between eukaryotic proteins and several of the Yops (i.e., YopH, YopM, and YpkA), it is reasonable to conclude that these proteins function in some manner to subvert signal transduction in the host during infection. Host phosphotyrosine signaling pathways would appear to be the primary targets of YopH (4, 6). There is already some evidence that YopH acts negatively on the transduction of signals involved in bacterial entry into mammalian cells (3, 53). Other Yops may function synergistically with YopH by interfering with host cell functions involving actin microfilaments (YopE), serine or threonine phosphorylation (YpkA), and G-protein-linked receptor stimulation (YopM) (22, 37, 55). The outcome of this multifactorial process is the ability of the yersiniae to suppress a cellular immune response while interfacing with a number of different host cell types during infection.

ACKNOWLEDGMENT. I thank Daniel A. Portnoy for reviewing the manuscript.

REFERENCES

1. **Ben Gurion, R., and A. Shafferman.** 1981. Essential virulence determinants of different *Yersinia* species are carried on a common plasmid. *Plasmid* 5:183–187.
2. **Bliska, J. B.** Unpublished data.
3. **Bliska, J. B., M. C. Copass, and S. Falkow.** 1993. The *Yersinia pseudotuberculosis* adhesin YadA mediates intimate bacterial attachment to and entry into HEp-2 cells. *Infect. Immun.* 61:3914–3921.
4. **Bliska, J. B., J. C. Clemens, J. E. Dixon, and S. Falkow.** 1992. The Yersinia tyrosine phosphatase: specificity of a bacterial virulence determinant for phosphoproteins in the J774A.1 macrophage. *J. Exp. Med.* 176:1625–1630.
5. **Bliska, J. B., J. E. Galan, and S. Falkow.** 1993. Signal transduction in the mammalian cell during bacterial attachment and entry. *Cell* 73:903–920.
6. **Bliska, J. B., K. Guan, J. E. Dixon, and S. Falkow.** 1991. Tyrosine phosphate hydrolysis of host proteins by an essential *Yersinia* virulence determinant. *Proc. Natl. Acad. Sci. USA* 88:1187–1191.
7. **Bolin, I., A. Forsberg, L. Norlander, M. Skurnik, and H. Wolf-Watz.** 1988. Identification and

mapping of the temperature-inducible, plasmid-encoded proteins of *Yersinia* spp. *Infect. Immun.* **56**:343–348.

8. **Bolin, I., D. A. Portnoy, and H. Wolf-Watz.** 1985. Expression of the temperature-inducible outer membrane proteins of yersiniae. *Infect. Immun.* **48**:234–240.

9. **Bolin, I., and H. Wolf-Watz.** 1988. The plasmid-encoded Yop2b protein of Yersinia pseudotuberculosis is a virulence determinant regulated by calcium and temperature at the level of transcription. *Mol. Microbiol.* **2**:237–245.

10. **Brubaker, R. R.** 1991. Factors promoting acute and chronic diseases caused by yersiniae. *Clin. Microbiol. Rev.* **4**:309–324.

11. **Burrows, T. W., and G. A. Bacon.** 1956. The basis of virulence in *Pasteurella pestis:* an antigen determining virulence. *Br. J. Exp. Pathol.* **37**:481–493.

12. **Cornelis, G., Y. Laroche, G. Balligand, M. P. Sory, and G. Wauters.** 1987. Yersinia enterocolitica, a primary model for bacterial invasiveness. *Rev. Infect. Dis.* **9**:64–87.

13. **Cornelis, G. R., T. Biot, C. Lambert de Rouvroit, T. Michiels, B. Mulder, C. Sluiters, M.-P. Sory, M. Van Bouchaute, and J.-C. Vanooteghem.** 1989. The Yersinia yop regulon. *Mol. Microbiol.* **3:** 1455–1459.

14. **Delor, I., and G. R. Cornelis.** 1992. Role of *Yersinia enterocolitica* Yst toxin in experimental infection of young rabbits. *Infect. Immun.* **60**:4269–4277.

15. **Eichelberg, K., C. C. Ginochio, and J. E. Galan.** Molecular and functional characterization of the *Salmonella typhimurium* invasion genes *invB* and *invC:* homology of InvC to the F_0F_1 ATPase family of proteins. *J. Bacteriol.,* in press.

16. **Fischer, E. H., H. Charbonneau, and N. Tonks.** 1991. Protein tyrosine phosphatases: a diverse family of intracellular and transmembrane enzymes. *Science* **253**:401–406.

17. **Forsberg, A., I. Bolin, L. Norlander, and H. Wolf-Watz.** 1987. Molecular cloning and expression of calcium-regulated, plasmid-encoded proteins of *Yersinia pseudotuberculosis*. *Microb. Pathog.* **2**:123–137.

18. **Forsberg, A., and H. Wolf-Watz.** 1988. The virulence protein Yop5 of Yersinia pseudotuberculosis is regulated at transcriptional level by plasmid-pIB1-encoded trans-acting elements controlled by temperature and calcium. *Mol. Microbiol.* **2**:121–133.

19. **Francis, C. L., T. A. Ryan, B. D. Jones, S. J. Smith, and S. Falkow.** 1993. Ruffles induced by *Salmonella* and other stimuli direct macropinocytosis of bacteria. *Nature* (London) **364**:639–642.

20. **Fujimura, Y., T. Kihara, and H. Mine.** 1992. Membranous cells as a portal of Yersinia pseudotuberculosis entry into rabbit ileum. *J. Clin. Electron Microsc.* **25**:35–45.

21. **Galan, J. E., and C. C. Ginocchio.** The molecular genetic basis of *Salmonella* entry into mammalian cells. *Biochem. Trans.,* in press.

22. **Galyov, E. E., S. Hakansson, A. Forsberg, and H. Wolf-Watz.** 1993. A secreted protein kinase of *Yersinia pseudotuberculosis* is an indispensable virulence determinant. *Nature* (London) **361:** 730–732.

23. **Gemski, P., J. R. Lazere, T. Casey, and J. A. Wohlhieter.** 1980. Presence of a virulence-associated plasmid in *Yersinia pseudotuberculosis. Infect. Immun.* **28**:1044–1047.

24. **Ginocchio, C. C., S. B. Olmsted, C. L. Wells, and J. E. Galan.** 1994. Contact with epithelial cells induces the formation of surface appendages on Salmonella typhimurium. *Cell* **76**:717–724.

25. **Goguen, J. D., W. S. Walker, T. P. Hatch, and J. Yother.** 1986. Plasmid-determined cytotoxicity in *Yersinia pestis* and *Yersinia pseudotuberculosis. Infect. Immun.* **51**:788–794.

26. **Guan, K., and J. E. Dixon.** 1990. Protein tyrosine phosphatase activity of an essential virulence determinant in *Yersinia. Science* **249**:553–556.

27. **Hakansson, S., T. Bergman, J. Vanooteghem, G. Cornelis, and H. Wolf-Watz.** 1993. YopB and YopD constitute a novel class of *Yersinia* Yop proteins. *Infect. Immun.* **61**:71–80.

28. **Hanski, C., U. Kutschka, H. P. Schmoranzer, M. Naumann, A. Stallmach, H. Hahn, and H. Menge.** 1989. Immunohistochemical and electron microscopic study of interaction of *Yersinia enterocolitica* serotype O8 with intestinal mucosa during experimental enteritis. *Infect. Immun.* **57**:673–678.

29. **Heesemann, J., B. Algermissen, and R. Laufs.** 1986. Immunochemical analysis of plasmid-encoded proteins released by enteropathogenic *Yersinia* sp. grown in calcium-deficient media. *Infect. Immun.* **54**:561–567.

30. **Heesemann, J., B. Algermissen, and R. Laufs.** 1984. Genetically manipulated virulence of *Yersinia enterocolitica. Infect. Immun.* **46:**105–110.

31. **Heesemann, J., and R. Laufs.** 1985. Double immunofluorescence microscopic technique for accurate differentiation of extracellularly and intracellularly located bacteria in cell culture. *J. Clin. Microbiol.* **22:**168–175.

32. **Hunter, T.** 1989. Protein modification: phosphorylation on tyrosine residues. *Curr. Opin. Cell Biol.* **1:**1168–1181.

33. **Isberg, R. R.** 1989. Determinants for thermoinducible cell binding and plasmid-encoded cellular entry detected in the absence of the *Yersinia pseudotuberculosis* invasin protein. *Infect. Immun.* **57:**1998–2005.

34. **Isberg, R. R.** 1991. Discrimination between intracellular uptake and surface adhesion of bacterial pathogens. *Science* **252:**934–938.

35. **Juliano, R. J., and S. Haskill.** 1993. Signal transduction from the extracellular matrix. *J. Cell Biol.* **120:**577–585.

36. **Lambert de Rouvroit, C., C. Sluiters, and G. R. Cornelis.** 1992. Role of the transcriptional activator, VirF, and temperature in the expression of the pYV plasmid genes of *Yersinia enterocolitica. Mol. Microbiol.* **6:**395–409.

37. **Leung, K. Y., B. S. Reisner, and S. C. Straley.** 1990. YopM inhibits platelet aggregation and is necessary for virulence of *Yersinia pestis* in mice. *Infect. Immun.* **58:**3262–3271.

38. **Leung, K. Y., and S. C. Straley.** 1989. The *yopM* gene of *Yersinia pestis* encodes a released protein having homology with the human platelet surface protein GPIbα. *J. Bacteriol.* **171:**4623–4632.

39. **McDonough, K. A., A. M. Barnes, T. J. Quan, J. Montenieri, and S. Falkow.** 1993. Mutation in the pla gene of *Yersinia pestis* alters the course of the plague bacillus-flea (Siphonaptera: Ceratophyllidae) interaction. *J. Med. Entomol.* **30:**772–780.

40. **Michiels, T., and G. Cornelis.** 1988. Nucleotide sequence and transcription analysis of yop51 from *Yersinia enterocolitica* W22703. *Microb. Pathog.* **5:**449–459.

41. **Michiels, T., and G. R. Cornelis.** 1991. Secretion of hybrid proteins by the *Yersinia* Yop export system. *J. Bacteriol.* **173:**1677–1685.

42. **Michiels, T., J.-C. Vanooteghem, C. Lambert de Rouvoit, B. China, A. Gustin, P. Boudry, and G. R. Cornelis.** 1991. Analysis of *virC*, an operon involved in the secretion of Yop proteins by *Yersinia enterocolitica. J. Bacteriol.* **173:**4994–5009.

43. **Michiels, T., P. Wattiau, R. Brasseur, J.-C. Ruysschaert, and G. Cornelis.** 1990. Secretion of Yop proteins by yersiniae. *Infect. Immun.* **58:**2840–2849.

44. **Pace, J., M. J. Hayman, and J. E. Galan.** 1993. Signal transduction and invasion of epithelial cells by *Salmonella typhimurium. Cell* **72:**505–514.

45. **Portnoy, D., S. L. Moseley, and S. Falkow.** 1981. Characterization of plasmids and plasmid-associated determinants of *Yersinia enterocolitica* pathogenesis. *Infect. Immun.* **31:**775–782.

46. **Portnoy, D. A., and S. Falkow.** 1981. Virulence-associated plasmids from *Yersinia enterocolitica* and *Yersinia pestis. J. Bacteriol.* **148:**877–883.

47. **Portnoy, D. A., and R. J. Martinez.** 1985. Role of a plasmid in the pathogenicity of Yersinia species. *Curr. Top. Microbiol. Immunol.* **118:**29–51.

48. **Portnoy, D. A., H. Wolf-Watz, I. Bolin, A. B. Beeder, and S. Falkow.** 1984. Characterization of common virulence plasmids in *Yersinia* species and their role in the expression of outer membrane proteins. *Infect. Immun.* **43:**108–114.

49. **Price, S. B., K. Y. Leung, S. S. Barve, and S. C. Straley.** 1989. Molecular analysis of *lcrGVH*, the V antigen operon of *Yersinia pestis. J. Bacteriol.* **171:**5646–5653.

50. **Protsenko, O. A., P. I. Anisimov, O. T. Mosarov, N. P. Donnov, Y. A. Popov, and A. M. Kokushkin.** 1983. Detection and characterization of *Yersinia pestis* plasmids determining pesticin 1, fraction 1 antigen and mouse toxin synthesis. *Genetika* **19:**1081–1090.

51. **Reisner, B. S., and S. C. Straley.** 1992. *Yersinia pestis* YopM: thrombin binding and overexpression. *Infect. Immun.* **60:**5242–5252.

52. **Rimpilainen, M., A. Forsberg, and W. H. Wolf.** 1992. A novel protein, LcrQ, involved in the low-calcium response of *Yersinia pseudotuberculosis* shows extensive homology to YopH. *Infect. Immun.* **174:**3355–3363.

53. **Rosqvist, R., I. Bolin, and H. Wolf-Watz.** 1988. Inhibition of phagocytosis in *Yersinia pseudotuber-*

culosis: a virulence plasmid-encoded ability involving the Yop2b protein. *Infect. Immun.* **56:** 2139–2143.

54. **Rosqvist, R., A. Forsberg, M. Rimpilainen, T. Bergman, and H. Wolf-Watz.** 1990. The cytotoxic protein YopE of *Yersinia* obstructs the primary host defence. *Mol. Microbiol.* **4:**657–667.

55. **Rosqvist, R., A. Forsberg, and W. H. Wolf.** 1992. Intracellular targeting of the *Yersinia* YopE cytotoxin in mammalian cells induces actin microfilament disruption. *Infect. Immun.* **59:** 4562–4569.

56. **Rosqvist, R., M. Skurnik, and H. Wolf-Watz.** 1988. Increased virulence of *Yersinia pseudotuberculosis* by two independent mutations. *Nature* (London) **334:**522–524.

57. **Sample, A. K., J. M. Fowler, and R. R. Brubaker.** 1987. Modulation of the low calcium response in *Yersinia pestis* via plasmid-plasmid interaction. *Microb. Pathog.* **2:**443–453.

58. **Skurnik, M., and H. Wolf-Watz.** 1989. Analysis of the *yopA* gene encoding the Yop1 virulence determinants of Yersinia spp. *Mol. Microbiol.* **3:**517–529.

59. **Sodeinde, O. A., A. Sample, R. R. Brubaker, and J. D. Goguen.** 1988. Plasminogen activator/ coagulase gene in *Yersinia pestis* is responsible for degradation of plasmid-encoded outer membrane proteins. *Infect. Immun.* **56:**2749–2752.

60. **Sodeinde, O. A., Y. V. Subrahmanyam, K. Stark, T. Quan, Y. Bao, and J. D. Goguen.** 1993. A surface protease and the invasive character of plague. *Science* **258:**1004–1007.

61. **Straley, S. C., and W. S. Bowmer.** 1986. Virulence genes regulated at the transcriptional level by Ca^{2+} in *Yersinia pestis* include structural genes for outer-membrane proteins. *Infect. Immun.* **51:** 445–454.

62. **Straley, S. C., and R. R. Brubaker.** 1981. Cytoplasmic and membrane proteins of yersiniae cultivated under conditions simulating mammalian intracellular environment. *Proc. Natl. Acad. Sci. USA* **78:**1224–1228.

63. **Straley, S. C., G. V. Plano, E. Skrzypek, and J. B. Bliska.** 1993. Yops of *Yersinia* spp. pathogenic for humans. *Infect. Immun.* **61:**3105–3110.

64. **Straley, S. C., G. V. Plano, E. Skrzypek, P. L. Haddix, and K. A. Fields.** 1993. Regulation by Ca^{2+} in the *Yersinia* low-Ca^{2+} response. *Mol. Microbiol.* **8:**1005–1010.

65. **Viitanen, A.-M., P. Toivanen, and M. Skurnik.** 1990. The *lcrE* gene is part of an operon in the *lcr* region of *Yersinia enterocolitica* O:3. *J. Bacteriol.* **172:**3152–3162.

66. **Wattiau, P., and G. R. Cornelis.** 1993. SycE, a chaperone-like protein of *Yersinia enterocolitica* involved in the secretion of YopE. *Mol. Microbiol.* **8:**123–131.

67. **Wolf-Watz, H., D. A. Portnoy, I. Bolin, and S. Falkow.** 1985. Transfer of the virulence plasmid of *Yersinia pestis* to *Yersinia pseudotuberculosis*. *Infect. Immun.* **48:**241–243.

68. **Yang, Y., and R. Isberg.** 1993. Cellular internalization in the absence of invasin expression is promoted by the *Yersinia pseudotuberculosis yadA* product. *Infect. Immun.* **61:**3907–3913.

69. **Zhang, Z. Y., J. C. Clemens, H. L. Schubert, J. A. Stuckey, M. W. F. Fischer, D. M. Hume, M. A. Saper, and J. E. Dixon.** 1992. Expression, purification and physiochemical characterization of a recombinant *Yersinia* tyrosine phosphatase. *J. Biol. Chem.* **267:**23759–23766.

70. **Zink, D. L., J. C. Feeley, J. G. Wells, C. Vanderzant, J. C. Vickery, W. D. Roof, and G. A. O'Donovan.** 1980. Plasmid-mediated tissue invasiveness in *Yersinia enterocolitica*. *Nature* (London) **283:**224–226.

Molecular Genetics of Bacterial Pathogenesis
Edited by V. L. Miller, J. B. Kaper, D. A. Portnoy, and R. R. Isberg
© 1994 American Society for Microbiology, Washington, DC 20005

Chapter 25

Role of Flagella in *Campylobacter* Pathogenesis

Patricia Guerry

The thermophilic campylobacters *Campylobacter jejuni* and *Campylobacter coli* together constitute one of the major causes of bacterial diarrhea worldwide. In the United States it has been estimated that there are 2.5 million cases of campylobacter diarrhea annually (43), and the estimates for developing nations are higher (44). Despite their prevalence as human pathogens, surprisingly little is understood about the pathogenesis of these obligate microaerophiles. The first and only virulence determinant identified in *Campylobacter* spp. is the single polar flagellum, an organelle which mediates a characteristic rapid, darting motility so unique that it can be used to identify the organism in stool samples (8). Early work with spontaneous nonmotile variants (33) or chemically induced mutants with motility defects (31) indicated that motility was required for colonization in mouse models, most likely in order to mediate penetration of the thick mucous lining of the intestinal tract (22). Moreover, data indicate that antibodies to flagellin correlate with the development of protection against disease (27). Considerable data on *Campylobacter* flagella have been published over the last few years, and this chapter attempts to summarize the current understanding of the structure and function of these organelles as well as their roles in pathogenesis.

FLAGELLAR STRUCTURE

Most of the molecular biological studies on *Campylobacter* flagella have been done on two strains, *C. coli* VC167 (1–5, 13–16, 23, 25, 26) and *C. jejuni* 81116 (12, 35–38, 46, 47), although flagellin genes have also been cloned from several other strains (11, 19). The flagella of both VC167 (13–15) and 81116 (36, 38, 46) are complex structures composed of two highly related flagellin subunits. These flagellins are the products of two flagellin genes, *flaA* and *flaB*, which are located adjacent to one another on the chromosomes of both VC167 and 81116. Virtually all *Campylobacter* strains have been shown to contain two flagellin genes (3). The

Patricia Guerry • Enteric Diseases Program, Naval Medical Research Institute, Bethesda, Maryland 20889.

two genes, which in VC167 and 81116 are >90% homologous to each other, are under the control of distinctive and independent promoters: the *flaA* gene is controlled by a σ^{28} promoter (13–15, 26, 36, 38, 46), typical of flagellin genes of the members of the family *Enterobacteriaceae* and *Bacillus* spp. (17), and *flaB* is controlled by a σ^{54} promoter (4, 13, 14), a class of promoter that has been found to control nitrogen-regulated genes in the enteric organisms (6), the pilin genes of *Pseudomonas aeruginosa* (18) and *Neisseria gonorrhoeae* (29), and the temporally regulated flagellin genes of *Caulobacter crescentus* (30). Under normal laboratory growth conditions in VC167, the *flaA* gene is expressed at much higher levels than *flaB*, but both genes are expressed concomitantly in the same cell (13, 14). In VC167 approximately 70% of the primary amino acid differences between the FlaA and FlaB flagellins occur in the amino- and carboxy-terminal domains, regions which are normally highly conserved among flagellins, since they are involved in flagellin export and polymerization into the filament (14). Sequence differences in these regions suggest that the two flagellins may have different structural roles in the assembled filament. Indeed, in VC167 the *flaB* gene product is apparently interspersed among the *flaA* subunits throughout the length of the flagellar filament (13, 14), an organization which is distinct from those of other complex filaments in which individual flagellins are assembled sequentially into distinct regions of the filament (10, 21, 42).

Since two highly homologous, tandem genes would be expected to be unstable, there must be a strong selective advantage in maintaining this genomic arrangement. Genetic studies in strain VC167 have indicated that both subunits are required for full motility (13, 14). Site-specific insertional inactivation of the *flaB* gene results in a full-length flagellar filament with slightly reduced motility compared with that of a wild-type filament (13, 14). Mutation of the *flaA* gene, however, results in the production of a short, truncated filament which is composed exclusively of the FlaB protein and which imparts greatly reduced motility (13, 14). Thus, it appears that a filament composed of both subunits is slightly more motile than one composed of the FlaA flagellin only. The FlaB flagellin has been shown to be environmentally regulated by studies in which the σ^{54} promoter was fused to a chloramphenicol acetyltransferase reporter gene (4). Variations in growth temperature, pH, and inorganic ion contents led to increases in expression of *flaB* up to 2.6-fold over the basal level of expression in standard medium (4), suggesting that campylobacters can modulate the compositions of their flagellar filaments and their motilities in direct response to environmental signals. It may be that other, as yet unidentified signals can increase *flaB* expression still more, although it remains to be determined if the FlaB flagellin is structurally capable of being incorporated into a fully functional, simple flagellar filament.

Another reason for maintaining two flagellar genes is to ensure motility with a backup or reserve copy. Alm et al. (5) observed that *flaA flaB*[+] mutants, in which the *flaA* gene had been insertionally inactivated with a kanamycin resistance cassette, frequently segregated off fully motile revertants. In the absence of selective pressure for kanamycin resistance, these motile revertants all represented simple, intragenomic recombinational events which resulted in a hybrid flagellin gene containing the 5' end of *flaA* and the 3' end of *flaB* and the elimination of the kanamycin resistance gene and other intervening flagellin information. However, when selection for kanamycin resistance was maintained, we observed motile revertants which resulted from more complex genomic recombinational events involving additional duplication of flagellin information. Kanamycin-resistant, motile revertants were never obtained in the presence of DNase, indicating that these recombinational events were due to autolysis of cells and natural transformation-mediated (45), intergenomic recombinational events (5). Such a mechanism also offers a way of rapidly introducing genetic variability among flagellin genes.

PHASE AND ANTIGENIC VARIATION

Campylobacter flagella undergo two types of variation which have been termed "phase" (9) and "antigenic" (16) variations. Phase variation refers to the ability of some campylobacter strains to turn the expression of flagella on and off at high frequencies. This is best demonstrated by use of a semisolid pour plate method in which mixed colonial types (motile and nonmotile) are readily apparent in many cultures of *Campylobacter* spp. One can purify cultures which are enriched for each phenotype, but the cells retain the ability to revert to the alternate phenotype at measurable frequencies. When a culture of nonmotile-phase variants was fed to rabbits, only motile forms were recovered from subsequent stool cultures, again suggesting a role for motility in colonization (9). Similarly, when a mixed culture of nonmotile and motile bacteria was fed to human volunteers, only motile forms were recovered in stool samples (7). This phenomenon has been shown to involve transcriptional regulation of flagellin gene expression (35), but the molecular basis remains to be elucidated. It should be mentioned that it has also not been determined if phase variation affects expression of other genes in addition to those for flagellin.

Another better-characterized variation of campylobacter flagellin is antigenic variation. Numerous workers have observed that flagellin preparations from campylobacters often contained doublets or ran as very broad bands on sodium dodecyl sulfate (SDS)-polyacrylamide gels (16, 24, 33, 35) or as multiple species in isoelectric focusing gels (32). Harris et al. (16) studied this phenomenon in strains of the Lior 8 serogroup. Flagellin preparations from several strains of both *C. jejuni* and *C. coli* from this serogroup displayed a doublet pattern on SDS-polyacrylamide gels, and the antigenic variation in one of these strains, *C. coli* UVC167, has been intensively studied. The doublets in UVC167 represent flagellins with distinct apparent molecular weights and distinct antigenicities, subsequently called antigenic type 1 (T1; apparent M_r, 61,500) and antigenic type 2 (T2; apparent M_r, 59,500). When cells were grown in antisera specific for T1, a population that produced T2 flagellin was selected. Growth of T2 cells in an anti-T2 sera allowed for the back-selection of cells the produced T1 flagellin. Logan et al. (23) fed rabbits T1 and T2 cells and monitored the antigenic specificities of the flagella in campylobacters subsequently shed in fecal material. Rabbits fed T2 cells shed T2 cells exclusively during the course of the infection; rabbits fed T1 cells shed T1 cells on day 1 postfeeding, but by day 3 postfeeding exclusively shed T2 cells. These data suggest that T2 cells are preferred in nonimmune animals, although they do not distinguish between the selection of a few cells in the population which may have already been expressing T2 flagella from the active induction of expression of cells with alternate antigenic specificities in vivo. Comparison of the DNA sequences of the *flaA* and *flaB* genes from a T1 and a T2 background revealed only minor changes in the predicted primary amino acid sequences between the two antigenic backgrounds, and these changes are insufficient to account for the observed differences in apparent molecular weight and antigenicity (14). Using variants of UVC167 fixed in the T1 or T2 antigenic state, Alm et al. (2) showed that flagellin antigenicity and apparent molecular weight were independent of the primary amino acid sequence but, instead, were dependent on the host background in which the flagellin genes were expressed. For example, when flagellin genes were transferred from the T1 strain into the T2 background, they ran at the apparent molecular weight of T2 flagellins (59,500) in SDS-polyacrylamide gels rather than the apparent molecular weight of 61,500 typical of T1 and reacted with T2-specific antisera only. These data indicate that the antigenic differences between T1 and T2 are due to differences in posttranslational modifications to the flagellins rather than changes in primary structure (2). The chemical nature of the modifications and the molecular basis for the alternate expression are under investigation.

ANALYSIS OF FLAGELLIN MUTANTS IN IN VIVO AND IN VITRO MODELS

As mentioned above, early studies of Newell (33) and Morooka et al. (31) suggested a requirement of flagella for *C. jejuni* to colonize mice. While interpretation of the data from those early studies, which did not employ isogenic mutations, is subject to some reservations, their overall observations have been confirmed. Pavlovskis et al. (41) showed that a *flaA flaB* nonmotile mutant of strain VC167 (15) was unable to colonize rabbits and, subsequently, that a *flaA flaB*$^+$ mutant was also unable to colonize rabbits (5). Similarly, Wassenaar et al. (47) also showed that nonflagellated mutants of strain 81116 are unable to colonize chickens.

The lack of a suitable nonprimate disease model has precluded examination of nonflagellated isogenic mutants in a pathogenesis model, but several groups have examined the role of flagella in the invasion and adherence of eukaryotic cells in vitro. Some strains of *C. jejuni* have been shown to invade eukaroytic cells at low levels by using a microfilament-dependent invasion system (20). Van der Zeijst and colleagues (38, 46) studied the ability of flagellin mutants of strain 81116 to invade INT407 cells. Their data indicted that *flaA*$^+$ *flaB* mutants invade INT407 cells at levels comparable to that of the wild type, but that *flaA flaB* mutants and *flaA flaB*$^+$ mutants invade INT407 cells at 0.06 to 1% the invasion levels of the wild type when the bacteria are not centrifuged to bring them into contact with the monolayer. If a centrifugation step is added, the level of invasion of the mutants rises approximately 30-fold, but remains significantly lower than that of the wild type (38, 46). Studies with independently isolated flagellin mutants of strain 81116 by Grant et al. (12) and an assay in which the bacteria were centrifuged onto the monolayer showed that *flaA flaB* mutants invaded INT407 cells at 2.0 to 2.5% the wild-type levels and that a *flaA flaB*$^+$ mutant invaded

INT407 cells at 18% of the level of the wild type. Grant et al. (12) also demonstrated an inability of both *flaA flaB* and *flaA flaB$^+$* mutants to translocate across Caco-2 cell monolayers. More recently, a microfilament-independent, microtubule-dependent invasion system has been described for some strains of *Campylobacter* spp. (39). Studies on the role of flagella in these strains are in progress, but preliminary data indicate that motility is also required for invasion in this system (48). Since several groups had previously suggested that flagella may function as an adhesin (28, 34), Grant et al. (12) also examined the abilities of these mutants to adhere to INT407 cells, but found no significant difference among wild-type and flagellin mutants in their adherence abilities. Collectively, the studies with 81116 flagellin mutants have been interpreted to indicate that either the FlaA flagellin (but not the FlaB flagellin) or the ability to be fully motile is necessary for efficient uptake into eukaryotic cells. Clearly, further studies are needed to distinguish these possibilities and others (such as the role of chemotaxis in invasion) and to further elucidate the role of flagella in pathogenesis.

SUMMARY

Research on *Campylobacter* flagella has become, in a sense, a model system in which to study several aspects of the biology of these unusual and challenging microaerophilic bacteria. Flagellin genes were among the first campylobacter genes to be cloned and mutated by site-specific mutagenesis protocols (26). The subsequent manipulation of these genes not only has led to greater understanding about the structure of an unusual flagellar filament but also has increased understanding of other aspects of the biology of these organisms. For example, studies of flagellin have enhanced understanding of one of the major serotyping schemes of *Campylobacter* spp. (1, 3) and the potential role of natural transformation in the antigenic diversity of campylobacters in general (5) and was led to the first diagnostic test for *C. jejuni* and *C. coli* based on PCR (40). Finally, an understanding of the mechanisms by which campylobacters regulate expression of flagellin genes and genes encoding alternate posttranslational modifications on the flagellins will contribute further to understanding of the molecular pathogenesis of these important pathogens.

ACKNOWLEDGMENTS. Research in my laboratory was supported by U.S. Naval Medical Research and Development Command, Research Work Unit no. 61102A3M161102BS13AK.111.

I thank Trevor J. Trust and Richard A. Alm for comments on the manuscript.

REFERENCES

1. **Alm, R. A., P. Guerry, M. E. Power, H. Lior, and T. J. Trust.** 1991. Analysis of the role of flagella in the heat-labile Lior serotyping scheme of thermophilic campylobacters by mutant allele exchange. *J. Clin. Microbiol.* **29**:2438–2445.
2. **Alm, R. A., P. Guerry, M. E. Power, and T. J. Trust.** 1992. Variation in antigenicity and molecular weight of *Campylobacter coli* VC167 flagellin in different genetic backgrounds. *J. Bacteriol.* **174:** 4230–4238.
3. **Alm, R. A., P. Guerry, and T. J. Trust.** 1993. Distribution and polymorphism of the flagellin genes from isolates of *Campylobacter coli* and *Campylobacter jejuni*. *J. Bacteriol.* **175**:3051–3057.

4. **Alm, R. A., P. Guerry, and T. J. Trust.** 1993. The *Campylobacter* sigma^{54} flagellin promoter is subject to environmental regulation. *J. Bacteriol.* **175:**4448–4455.

5. **Alm, R. A., P. Guerry, and T. J. Trust.** 1993. Significance of duplicated flagellin genes in *Campylobacter. J. Mol. Biol.* **230:**359–363.

6. **Ames, G. F.-L., and K. Nikaido.** 1985. Nitrogen regulation in *Salmonella typhimurium.* Identification of an *ntrC* protein binding site and definition of the concensus. *EMBO J.* **4:**539–547.

7. **Black, R. E., M. M. Levine, M. L. Clements, T. P. Hughes, and M. J. Blaser.** 1988. Experimental *Campylobacter jejuni* infections in humans. *J. Infect. Dis.* **157:**472–479.

8. **Butzler, J. P., and M. B. Skirrow.** 1979. *Campylobacter* enteritis. *Clin. Gastroenterol.* **8:**737–765.

9. **Caldwell, M. B., P. Guerry, E. C. Lee, and J. P. Burans.** 1985. Reversible expression of flagella in *Campylobacter jejuni. Infect. Immun.* **50:**941–943.

10. **Ely, B., and L. Shapiro.** 1984. Regulation of cell differentiation in *Caulobacter crescentus.* Microbial development, p. 1–26. *In* R. Losick and L. Shapiro (ed.), *Microbial Development.* Cold Spring Harbor Laboratory, Cold Spring Harbor, N.Y.

11. **Fischer, S. H., and I. Nachamkin.** 1991. Common and variable domains of the flagellin gene, flaA, in *Campylobacter jejuni. Mol. Microbiol.* **5:**1151–1158.

12. **Grant, C. C. R., M. E. Konkel, W. Cieplak, Jr., and L. S. Tompkins.** 1993. Role of flagella in adherence, internalization, and translocation of *Campylobacter jejuni* in nonpolarized and polarized epithelial cell cultures. *Infect. Immun.* **61:**1764–1771.

13. **Guerry, P., R. A. Alm, M. E. Power, S. M. Logan, and T. J. Trust.** 1991. The role of two flagellin genes in *Campylobacter* motility. *J. Bacteriol.* **173:**4757–4764.

14. **Guerry, P., R. A. Alm, M. E. Power, and T. J. Trust.** 1992. Molecular and structural analysis of *Campylobacter* flagellin, p. 267–281. *In* I. Nachamkin, M. J. Blaser, and L. S. Tompkins (ed.), *Campylobacter jejuni: Current Status and Future Trends.* American Society for Microbiology, Washington, D.C.

15. **Guerry, P., S. M. Logan, S. Thornton, and T. J. Trust.** 1990. Genomic organization and expression of *Campylobacter* flagellin genes. *J. Bacteriol.* **172:**1853–1860.

16. **Harris, L. A., S. M. Logan, P. Guerry, and T. J. Trust.** 1987. Antigenic variation of *Campylobacter* flagella. *J. Bacteriol.* **169:**5066–5071.

17. **Helmann, J. D., and M. J. Chamberlain.** 1987. DNA sequence analysis suggests that expression of flagellar and chemotaxis genes in *Escherichia coli* and *Salmonella typhimurium* is controlled by an alternative sigma factor. *Proc. Natl. Acad. Sci. USA* **84:**6422–6424.

18. **Ishimoto, K., and S. Lory.** 1989. Formation of pilin in *Pseudomonas aeruginosa* requires the RpoN subunit of RNA polymerase. *Proc. Natl. Acad. Sci. USA* **86:**1954–1957.

19. **Khawaja, R., K. Neote, H. L. Bingham, J. L. Penner, and V. L. Chan.** 1992. Cloning and sequence analysis of the flagellin gene of *Campylobacter jejuni* TGH9011. *Curr. Microbiol.* **24:**213–221.

20. **Konkel, J. E., and L. A. Joens.** 1989. Adhesion to and invasion of HEp-2 cells by *Campylobacter* spp. *Infect. Immun.* **57:**2984–2990.

21. **Kostrzynska, M., J. D. Betts, J. W. Austin, and T. J. Trust.** 1991. Identification, characterization, and spatial location of two flagellin species in *Helicobacter pylori* flagella. *J. Bacteriol.* **173:** 937–946.

22. **Lee, A., J. L. O'Rourke, P. J. Barrington, and T. J. Trust.** 1986. Mucus colonization by *Campylobacter jejuni:* a mouse cecal model. *Infect. Immun.* **51:**536–546.

23. **Logan, S. M., P. Guerry, D. M. Rollins, D. H. Burr, and T. J. Trust.** 1989. In vivo antigenic variation of *Campylobacter* flagellin. *Infect. Immun.* **57:**2583–2585.

24. **Logan, S. M., L. A. Harris, and T. J. Trust.** 1987. Isolation and characterization of *Campylobacter* flagellins. *J. Bacteriol.* **169:**5072–5077.

25. **Logan, S. M., and T. J. Trust.** 1986. Location of epitopes on *Campylobacter jejuni* flagellin. *J. Bacteriol.* **168:**739–745.

26. **Logan, S. M., T. J. Trust, and P. Guerry.** 1989. Evidence for posttranslational modification and gene duplication of *Campylobacter* flagellin. *J. Bacteriol.* **171:**3031–3038.

27. **Martin, P. M., J. Mathiot, J. Ipero, M. Kirimat, A. J. Georges, and M.-C. Georges-Courbot.** 1989. Immune response to *Campylobacter jejuni* and *Campylobacter coli* in a cohort of children from birth to 2 years of age. *Infect. Immun.* **57:**2542–2546.

28. **McSweegan, E., and R. I. Walker.** 1986. Identification and characterization of two *Campylobacter jejuni* adhesins for cellular and mucous substrates. *Infect. Immun.* **53:**141–148.

29. **Meyer, T. F., E. Billyard, R. Haas, S. Storzbach, and M. So.** 1984. Pilus genes of *Neisseria gonorrhoeae:* chromosomal organization and DNA sequence. *Proc. Natl. Acad. Sci. USA* **81:**6110–6114.

30. **Minnich, S. A., and A. Newton.** 1987. Promoter mapping and cell cycle regulation of flagellin gene transcription in *Caulobacter crescentus. Proc. Natl. Acad. Sci. USA* **81:**1142–1146.

31. **Morooka, T., A. Umeda, and K. Amako.** 1985. Motility as an intestinal colonization factor for *Campylobacter jejuni. J. Gen. Microbiol.* **131:**1973–1980.

32. **Nachamkin, I., and X. H. Yang.** 1988. Isoelectric focusing of *Campylobacter jejuni* flagellin: microheterogeneity and restricted antigenicity of charged species with a monoclonal antibody. *FEMS Microbiol. Lett.* **49:**235–238.

33. **Newell, D. G.** 1986. Monoclonal antibodies directed against the flagella of *Campylobacter jejuni:* production, characterization and lack of effect on colonization of infant mice. *J. Hyg.* **96:**131–141.

34. **Newell, D. G., H. McBride, and J. M. Dolby.** 1985. Investigations on the role of flagella in the colonization of infant mice with *Campylobacter jejuni* and attachment of *Campylobacter jejuni* to human epithelial cell lines. *J. Hyg.* **95:**217–227.

35. **Nuijten, P. J. M., N. M. C. Bleumink-Pluym, W. Gaastra, and B. A. M. van der Zeijst.** 1989. Flagellin expression in *Campylobacter jejuni* is regulated at the transcriptional level. *Infect. Immun.* **57:**1084–1088.

36. **Nuijten, P. J. M., F. J. A. M. van Asten, W. Gaastra, and B. A. M. van der Zeijst.** 1991. Structural and functional analysis of two *Campylobacter jejuni* flagellin genes. *J. Biol. Chem.* **265:** 17798–17804.

37. **Nuijten, P. J. M., B. A. M. van der Zeijst, and D. G. Newell.** 1991. Localization of immunogenic regions of the flagellin proteins of *Campylobacter jejuni* 81116. *Infect. Immun.* **59:**1100–1105.

38. **Nuijten, P. J. M., T. M. Wassenaar, D. G. Newell, and B. A. M. van der Zeijst.** 1992. Molecular characterization and analysis of *Campylobacter jejuni* flagellin genes and proteins, p. 282–296. *In* I. Nachamkin, M. J. Blaser, and L. S. Tompkins (ed.), *Campylobacter jejuni: Current Status and Future Trends*. American Society for Microbiology, Washington, D.C.

39. **Oelschlaeger, T. A., P. Guerry, and D. J. Kopecko.** 1993. Unusual microtubule-dependent endocytosis mechanisms triggered by *Campylobacter jejuni* and *Citrobacter freundii. Proc. Natl. Acad. Sci. USA* **90:**6884–6888.

40. **Oyofo, B. A., S. A. Thornton, D. H. Burr, T. J. Trust, and P. Guerry.** 1992. Specific detection of *Campylobacter jejuni* and *Campylobacter coli* by using the polymerase chain reaction. *J. Clin. Microbiol.* **30:**2613–2619.

41. **Pavlovskis, O. R., D. M. Rollins, R. L. Haberberger, Jr., A. E. Green, L. Habash, S. Stroko, and R. I. Walker.** 1991. Significance of flagella in colonization resistance of rabbits immunized with *Campylobacter* spp. *Infect. Immun.* **59:**2259–2264.

42. **Pleier, E., and R. Schmitt.** 1991. Expression of two *Rhizobium meliloti* flagellin genes and their contribution to the complex filament structure. *J. Bacteriol.* **173:**2077–2085.

43. **Tauxe, R. V.** 1992. Epidemiology of *Campylobacter jejuni* infections in the United States and other industrialized nations, p. 9–19. *In* I. Nachamkin, M. J. Blaser, and L. S. Tompkins (ed.), *Campylobacter jejuni: Current Status and Future Trends*. American Society for Microbiology, Washington, D.C.

44. **Taylor, D. N.** 1992. Campylobacter infections in developing countries, p. 20–30. *In* I. Nachamkin, M. J. Blaser, and L. S. Tompkins (ed.), *Campylobacter jejuni: Current Status and Future Trends*. American Society for Microbiology, Washington, D.C.

45. **Wang, Y., and D. E. Taylor.** 1990. Natural transformation in *Campylobacter* species. *J. Bacteriol.* **172:**949–955.

46. **Wassenaar, T. M., N. M. C. Bleumink-Pluym, and B. A. M. van der Zeijst.** 1991. Inactivation of *Campylobacter jejuni* flagellin genes by homologous recombination demonstrates that *flaA* but not *flaB* is required for invasion. *EMBO J.* **10:**2055–2061.

47. **Wassenaar, T. M., B. A. M. van der Zeijst, R. Ayling, and D. N. Newell.** 1993. Colonization of chicks by motility mutants of *Campylobacter jejuni* demonstrates the importance of flagellin-A expression. *J. Gen. Microbiol.* **139:**1171–1175.

48. **Yao, R., H. Niu, T. J. Trust, and P. Guerry.** Unpublished data.

Molecular Genetics of Bacterial Pathogenesis
Edited by V. L. Miller, J. B. Kaper, D. A. Portnoy, and R. R. Isberg
© 1994 American Society for Microbiology, Washington, DC 20005

Chapter 26

Molecular Biology and Role in Disease of the Verotoxins (Shiga-Like Toxins) of *Escherichia coli*

Jim Brunton

Shiga toxin was discovered in 1903 by Conradi, and Van Heynigen and Gladstone succeeded in substantially purifying it in 1953 (for a review, see reference 40). The paralysis produced by the toxin in experimental animals was found to be caused by thrombotic occlusion of the microvasculature of the brain and spinal cord (4). A direct cytotoxic activity on tissue culture cells was later demonstrated. Further progress required full purification of the toxin, which was achieved in the early 1980s (40). The cytotoxicity was shown to be due to catalytic inactivation of the 60S ribosomal subunit of host cells, with the resulting inhibition of protein synthesis (46). Later, the cellular receptor for the toxin was found to be the neutral glycolipid globotriaosylceramide (Gb3) (31).

The recognition that some strains of *Escherichia coli* produced cytotoxins immunologically similar to Shiga toxin and that these strains, particularly serotype O157:H7, were the cause of epidemic and sporadic cases of hemmorrhagic colitis (HC) (50) and childhood hemolytic uremic syndrome (HUS) (26) was responsible for a tremendous resurgence in interest in the Shiga toxin family in the 1980s and 1990s (17). H. Williams Smith (62) demonstrated bacteriophage-mediated verotoxin (VT) conversion, facilitating cloning of the VT1 operon. A variant of the family was found to be associated with edema disease of swine (61). Karmali et al. (26) postulated that VT acts on endothelial cells to produce thrombotic microvasculopathy characteristic of pig edema disease, HC, and HUS. This chapter discusses the current understanding of this family of toxins, with an emphasis on recent advances in structural analysis, and in the understanding of receptor binding and the role of the toxins in pathogenesis.

NOMENCLATURE

The toxins are called Shiga-like toxins (SLTs) by American workers because of the sequence similarity to the prototype Shiga toxin of *Shigella dysenteriae* 1.

Jim Brunton • Departments of Microbiology and Medicine, The Toronto Hospital, and Samuel Lunenfeld Research Institute at Mount Sinai Hospital, Toronto, Ontario, Canada M5G 2C4.

TABLE 1
SLT and VT nomenclatures and biological properties

Nomenclature		Producing organisms (prototype strain)	Cross-neutralized by antiserum to:	Receptor	Disease
SLT	VT				
Shiga	Shiga	*S. dysenteriae* 1 (60R)	VT1	Gb3	Human HUS, HC, diarrhea
SLT-I	VT1	*E. coli* serotypes (O26:H11 strain H30)	Shiga, not VT2	Gb3	Human HUS, HC, diarrhea
SLT-II	VT2	*E. coli* serotypes (C600 [933W])[a]	VT2, VT2e, VT2c *not* with Shiga, VT1[b]	Gb3	Human HUS, HC, diarrhea
SLT-IIc[c]	VT2c[c]	*E. coli* serotypes (O157:H⁻ strain E32511[a] [M. Karmali[b]]	VT2, VT2e	Gb3	Human HUS, HC, diarrhea
SLT-IIe[d]	VT2e[d]	*E. coli* serotypes (O139 strain S1191)	VT2, VT2c	Gb4	Pig edema disease

[a] The most common serotypes producing SLT or VT associated with human disease are O157:H7, O157:H⁻, O26: H11, O111:H⁻, and H8. Other important serotypes have been O91:H⁻, O113:H21, O103:H2, O128:H⁻?, and O145: H⁻. For a more complete list see Karmali (23).

[b] VT2 related toxins are not neutralized by antiserum raised against Shiga toxin or VT1.

[c] SLT-IIc has been called SLT-IIvha or SLT-IIvhb in the past. VT2c was often called VT2 in the past. M. Karmali's strain E32511 is cured of an SLTII-specifying bacteriophage.

[d] SLT-IIe was called SLT-IIv and SLT-IIvp. VT2e has been called VT2v and VTe.

British and Canadian workers prefer the name ''verotoxins'' because of the work of Konowalchuk et al. (28), who demonstrated that they are cytotoxic to Vero cells. Recently, a consensus was reached that the terms VT and SLT should be interchangeable; an agreement on the identification of type toxins and numbering was also secured. Table 1 lists the family of toxins and their two sets of nomenclatures, together with their salient biological properties (25).

GENERAL FEATURES OF THE FAMILY

The toxins all consist of a single toxic 31-kDa A subunit noncovalently linked to a pentamer of 7-kDa B subunits. The B oligomer mediates binding to the eukaryotic cell surface receptor glycolipid Gb3, which is a prerequisite for internalization. Sandvig and colleagues (40, 52) have elegantly demonstrated that Shiga toxin is internalized in coated vesicles. Routing to the trans-Golgi apparatus appears to be a prerequisite for cellular intoxication. Recently, it has been shown that the toxin may undergo retrograde transport to the endoplasmic reticulum (51). Low vesicular pH is not necessary for A-subunit translocation. Calcium ions are necessary for cellular intoxication, which can be blocked by inhibitors of calcium transport (40). The A subunit is activated by proteolytic cleavage and reduction, yielding a 27-kDa toxic A_1 fragment and a 4-kDa A_2 fragment which is noncovalently associated with the B pentamer. Enzymatic inactivation of the 60S ribosomal subunit caused by the 27-kDa A_1 fragment leads to irreversible inhibition of protein synthesis and cell death.

The salient features of the VTs are listed in Table 1. The family can be divided into two subgroups on the basis of their neutralization and sequence similarities. Shiga toxin and VT1 are virtually identical and are cross-neutralizable. VT2 and the variants VT2e and VT2c share more sequence similarity with each other and are cross-neutralizable by antiserum to VT2; they are not neutralized at all by Shiga toxin antiserum. VT1, VT2, and VT2c bind to Gb3 and are associated with HC and HUS in humans. VT2e binds primarily to Gb4 and is associated with pig edema disease (5, 23, 24, 40).

The genes for VT1 and VT2 are located on the genomes of lysogenic lambdoid bacteriophages (5, 49). Shiga toxin and VTe are encoded on chromosomal genes and do not appear to be associated with phages (40). The genes are organized as bicistronic operons. The A cistron is proximal to the promoter and precedes the B cistron by 12 to 15 nucleotides. Lee (29) and Habib and Jackson (18) have shown that the B cistron of VT1 or Shiga toxin is translated more efficiently than the A cistron, probably because of a more efficient ribosomal binding site and the secondary structure, which places the ribosomal binding site of the B cistron in the loop of a stem-loop structure. The expression of Shiga toxin and VT1 is negatively regulated by the FUR/Fe^{2+} system, and a ferric uptake repressor (FUR)-binding sequence is found to overlap the -35 and part of the -10 sequences of the promoter (5, 6). In contrast, expression of VT2 and VT2e is not regulated by iron. Their promoters have been mapped and have been shown to have similar sequences; the sequences of the promoters differ significantly from the *E. coli* promoter consensus sequence (55).

LOW-LEVEL SLT PRODUCTION

VT-producing *E. coli* strains associated with HC and HUS produce relatively large amounts of toxin, which is found in the periplasmic space and the culture supernatant (23, 26). During the 1970s and 1980s, diverse bacterial strains, e.g., *Vibrio cholerae*, *Shigella* sp. other than *S. dysenteriae*, *Salmonella* sp., *Campylobacter jejuni*, *Vibrio parahaemolyticus*, classical enteropathogenic *E. coli* serotypes, enteroinvasive *E. coli*, and laboratory strains such as *E. coli* K-12 were found on bioassay to produce minute amounts of SLT. These strains were referred to as "low-level SLT producers," and it was speculated that SLT might play a role in the pathogenesis of the associated disease (39). Attempts to identify the putative genes by hybridization and cloning have failed. Two studies have shown that the activity is not neutralized by Shiga toxin antiserum; the concept of low-level SLT production should be laid to rest (5, 13).

SHIGA TOXIN FAMILY INACTIVATES RIBOSOMES BY AN *N*-GLYCOSIDASE ACTIVITY

Reisbig et al. (46) showed that the A subunit of Shiga toxin catalytically inactivates the eukaryotic 60S ribosomal subunit in a cell-free protein synthesis system.

This activity is similar to that of the ribosome-inactivating proteins (RIPs) such as α-momorcharin and ricin, which are found in many plants and their seeds (19). The mechanisms of action of Shiga toxin and the RIPs were a mystery until Endo and colleagues (11, 12) demonstrated that they all possess N-glycosidase activity. They specifically remove the adenine base at position 4324 (A4324) of eukaryotic 28S RNA, leaving the phosphoribose backbone intact. The region surrounding A4324 (in E. coli 23S RNA, the homologous residue is A2660) is highly conserved in E. coli, yeast, chloroplast, plant, and animal RNAs (19). It forms a stem-loop structure, with A4324 at the center of the loop. In E. coli, it has been shown that A2660 is an important site for binding the peptide chain elongation factors EF-Tu and EF-G (35). This accounts for the finding that Shiga toxin and ricin alter ribosomes to prevent binding of EF-1 and EF-2 (the homologs of ET-Tu and EF-G, respectively), irreversibly inhibiting protein synthesis (43). This region of the ribosome is also targeted by the fungal toxin α-sarcin, which cleaves the phosphodiester backbone of eukaryotic 28S RNA between residues 4325 and 4326.

MODELS OF THE N-GLYCOSIDASE REACTION

The amino acid sequences of the RIPs and Shiga toxin family are highly conserved in the region of the active site. The residues corresponding to Glu at position 167 (Glu-167), Arg-170, Tyr-114, and Tyr-77 of Shiga toxin are invariably present. The crystal structures of ricin, α-momorcharin, and Shiga toxin show that the spatial organization of the active-site residues is also highly conserved. Substitution of Asp for Glu-167 of Shiga toxin produces a 1,000-fold reduction in activity (21). Studies on the ricin A chain showed that individual substitution of the residues described above results in a dramatic reduction in activity (45). Crystallographic analyses of substrate analog complexes with ricin and α-momorcharin have recently been reported, and models for the N-glycosidase reaction have been proposed (37, 47). The guanidium group of the Arg-170 homolog is thought to form a hydrogen bond with the N-3 of adenine, while the amide hydrogen of the residue 78 homolog forms a hydrogen bond with N-1, drawing electrons from the N-glycoside bond into the adenine ring. This leads to the formation of an oxycarbonium ion in the ribose which is thought to be stabilized by the carboxylate anion of Glu-167. Cleavage of the glycosidic bond is completed by attack by an active-site water molecule. Residues corresponding to Tyr-77 and Tyr-114 of Shiga toxin likely play an important role in substrate binding by intercalating on either side of the purine ring. Hydrogen bonds to the enzyme also specifically anchor the adenine in the active site. These models are probably directly relevant to Shiga toxin, since the A-subunit fold is very similar to that of the ricin A chain, particularly with respect to the active-site residues (15). An interesting difference is that access to substrate by the active site appears to be blocked by Met-260, which is located in the A_2 fragment. This probably accounts for the observation that the inhibitory activity of the A subunit on cell-free protein synthesis is significantly increased by removal of the A_2 fragment (46). Cleavage probably occurs through the action of bacterial or cellular proteases after Arg-251; this is followed by reduction of the disulfide bond between Cys-242 and Cys-261 (16).

INTERACTION OF A AND B SUBUNITS

The A subunit of Shiga toxin lies on the side of the B-subunit pentamer, nearest to the C-terminal end of the B-subunit α helices. It interacts with the B pentamer through a hydrophobic α helix (residues 282 to 287 of the A_2 region) which extends half of the 2.0-nm length of the pore in the B pentamer. This pore is lined by the hydrophobic side chains of the B-subunit α helices (15, 54). The A subunit also interacts with the B subunit via a four-stranded mixed β sheet composed of residues of both the A_2 and A_1 fragments (15).

B-SUBUNIT PENTAMER

Each B subunit is composed of six antiparallel β strands forming a closed β barrel capped by a single α helix between strands 3 and 4 (15, 38, 54). The fold has been noted in a number of proteins that bind to oligosaccharides or oligonucleotides and has been called the "oligomer-binding fold" (38). The fold is almost identical to that of the B subunit of the cholera toxin (CT) family and to subunits S_2, S_3, S_4, and S_5 of pertussis toxin (PT), although CT and PT are larger and do not share significant sequence similarities with the Shiga toxin family. Differences between the molecules are found in the lengths and conformations of the loops between the strands and the helices. The S_2 and S_3 subunits of PT also have significant extensions at the amino and carboxyl termini (53). Monomers of the B pentamer of the Shiga toxin family assemble to form six-stranded antiparallel β sheets between adjacent monomers. The hydrophobic α helices line the central pore of the pentamer (15, 54).

PUTATIVE RECEPTOR-BINDING POCKET OF THE B SUBUNIT

Investigators have recently found that the phenylalanine at position 30 plays a critical role in receptor binding. Purified Phe-30-Ala mutant B subunit has a 30-fold lower affinity than the wild type for Gb3. Mutant holotoxin has 100,000-fold lower cytotoxic activity on Vero cell monolayers. The mutant B subunit has been crystallized, and difference maps comparing it with the wild-type B subunit of VT1 show that the structures are identical with the exception of the substituted side chain. Phe-30 forms part of a potential carbohydrate-binding domain located on the face of the B pentamer opposite the A subunit. In addition to Phe-30, the residues forming the pocket include Asp-17 and Thr-19, whose side groups would be expected to form a hydrogen bond to the sugar (7). Although the results strongly suggest that the phenyl ring of Phe-30 plays a major role in binding to the Galα1-4Gal receptor probably by interacting with a hydrophobic face of one of the sugars, the structure of the putative complex remains to be determined. The proposed location is compatible with the results of Jackson et al. (21a), who reported that a double substitution, Asp-16–His and Asp-17–His, resulted in ablation of VT1 binding to receptor analog; this is probably due to steric hindrance or local alteration in conformation. Maloney and Lingwood have reported significant amino acid sequence similarity between the B subunit of VTs and the interferon α receptor and the B lymphocyte differentiation antigen CD19. The latter two have now been shown to bind the Gb3 which appears to modulate their function (8, 33a).

RECEPTOR BINDING OF THE B SUBUNIT

All members of the Shiga toxin family bind to globoseries glycolipids: Shiga toxin, VT1, VT2, and VT2c bind to Gb3, while VT2e binds primarily to Gb4 (5, 10, 24, 40). High-affinity binding depends on the multivalent presentation of the carbohydrate, as would be provided by glycolipids in a membrane. Lindberg et al. (31) showed that free galabiose did not inhibit toxin binding to cell monolayers, while bovine serum albumin substituted with galabiose did. The potency of inhibition was much greater when 18 to 25 mol of galabiose was substituted per mol of bovine serum albumin than when 7 mol was substituted. Recently, Toone's group has been able to study monovalent binding of synthetic Galα1-4Galβ1-4Gluc to purified VT1 B subunit by microcalorimetry. They have shown that the K_d is ≈ 1 to 2 mM; this is in contrast to the K_d of $\approx 10^{-8}$ M for VT binding to glycolipids (57). Whether this difference can be completely accounted for by multivalent sugar binding remains to be seen.

IS Gb3 THE PHYSIOLOGICAL RECEPTOR?

Cohen et al. (8) reported that Daudi lymphoma cells selected for resistance to VT were deficient in Gb3 and were also resistant to the growth inhibitory effects of alpha 2 interferon. The insusceptibility to both was due to the reduced level of binding to the cell surface because of the absence of receptor glycolipid. Waddell et al. (60) showed that susceptibilities of the cells could be restored by fusing Gb3 containing liposomes to the cell membrane. Strong support for the physiological relevance of Gb3 as a receptor also comes from studies showing the Gb3 expression is developmentally regulated in the rabbit small intestine and that the intestinal fluid response correlates with the Gb3 content (22, 36).

ALTERED BINDING RELATED TO CARBOHYDRATE SPECIFICITY

The B subunits of VT2 and VT2e differ at only 11 positions, yet VT2e binds primarily to Gb4, while VT2 binds only to Gb3 (10, 58). The physiological significance of these in vitro binding studies is supported by the finding that VT-resistant cells reconstituted with Gb3 are killed by VT1 and, to a lesser extent, by VTe. Cells reconstituted with Gb4 are unaffected by VT1 but are killed by VT2e (3). Site-directed mutagenesis of VT2e in which residues were substituted for those found in VT2 produced a VT2e mutant, a Glu for Gln at position 64 (Gln-64–Glu) and Lys-66–Gln, which had significantly decreased levels of binding to Gb4 and which had no cytotoxic effect on resistant cells reconstituted with Gb4 (3, 58). The residues Glu-64 and Lys-66 are found in β strand 6, which lies beside the cleft between adjacent B subunits (54). It remains to be seen whether the alteration of binding specificity is due to a conformational effect on the postulated Gb3-binding pocket or whether these residues interact with the terminal GalNAc residue of Gb4.

ROLE OF CERAMIDE IN TOXIN BINDING

There is evidence that, in the context of the cell membrane, the ceramide portion of Gb3 has important effects on the receptor properties of the Galα1-4Gal group. Lingwood et al. (44) showed that digalactosyl diglyceride, which has the same Galα1-4Gal sugar as Gb3, has no receptor activity for VTs. Lingwood's group (44) has studied the roles that the fatty acyl chains of Gb3 play. Homogeneous semisynthetic Gb3 species were prepared by saponifying defined fatty acids with deacylated (lyso) Gb3. The affinity of VT1 for Gb3 isoforms was found to be influenced by the fatty acyl chain length and by its level of saturation. Of considerable interest was the finding that a mixture of Gb3 isoforms with various acyl chains gave the highest binding affinity (44). In addition, the system has been used to explore the binding specificities of different members of the VT family. Gb3 with $C_{20:0}$ and $C_{22:1}$ fatty acyl chains had the greatest capacity to bind VT1. In contrast, $C_{18:0}$ and $C_{18:1}$ homologs had the greatest capacity to bind VT2c in vitro. The physiological significance of the in vitro findings described above was confirmed by examining the relative abilities of the Gb3 isoforms to restore toxin susceptibility to resistant cell lines (27). The basis for these findings may be related to the ability of different Gb3 isoforms to present multivalent sugar-binding sites in the optimal orientation and position at the membrane surface. It is also possible that different fatty acyl groups affect the conformations of individual receptor epitopes on the sugar. These are plausible explanations for the lack of an absolute correlation between quantitative Gb3 content and the susceptibilities of cells to members of the VT family which has been noted in cell culture systems and animal models (20, 31, 32).

ROLE IN HUMAN DISEASE

The importance of VT-producing *E. coli* to human health was emphasized by the recent fast-food-associated outbreak in the U.S. Pacific Northwest. Of 477 patients with recognized *E. coli* O157:H7 infections, 144 patients required hospitalization; there were 30 cases of HUS, with 21 patients requiring dialysis. Analysis of this and many other outbreaks and sporadic cases has shown that VT-producing *E. coli* strains which produce VT1, VT2, and/or VT2c are associated with a spectrum of disease ranging from mild diarrhea, to HC, to HUS (17, 23, 24, 26). The primary reservoir of these organisms is the bovine intestinal tract, particularly dairy calves and cows (17). VT-producing *E. coli* strains produce HC in cattle, although the most virulent strain for humans (O157:H7) is carried asymptomatically by cattle. The vehicles for transmission to humans are undercooked meat, unpasteurized milk, apple cider, and contaminated lake and potable water. The frequency of human-to-human transmission suggests that the infectious inoculum may approximate that of *Shigella* spp. (i.e., 10^2) (12).

There is no satisfactory animal model for HC and HUS, and the severity of the disease precludes study of experimental infections in human volunteers. Therefore, understanding of the pathogenesis comes from a combination of pathological studies on human material in combination with animal model and endothelial cell studies (24, 40).

Most VT-producing *E. coli* strains adhere to the epithelium of the terminal ileum, cecum, and colon by the attaching and effacing (AE) lesion (see chapter 12). Studies in gnotobiotic pigs have shown that VT production is not necessary for *E. coli* O157:H7 to produce diarrhea, while AE lesion formation is necessary and sufficient to produce diarrhea (59). Experimental infection of pigs and cattle causes diarrhea but not HC. Although VTs have been shown to be cytotoxic to primary human colonic epithelial cells in tissue culture, it is likely that the signaling, cytoskeletal rearrangement, effacement of the microvilli, and intimate adherence associated with AE lesion formation cause diarrhea in humans. Furthermore, the same processes could increase the level of absorption of toxin either by transcytosis through epithelial cells or by passage between cells through leaky tight junctions, although at the moment there is no experimental evidence to support this notion.

Mobassaleh et al. (36) studied the effects of Shiga toxin on rabbit jejunum and found that cells at the bottom of the crypts lack Gb3 and are not affected by the toxin. In contrast, villus cells are rich in Gb3. Toxin affects these cells, reducing sodium reabsorption and leading to net fluid accumulation in the intestinal lumen because of continued secretion from the crypts (22). Whether a similar process plays a role in human diarrhea is not clear.

HEMORRHAGIC COLITIS

HC was first defined in 1983 in the course of an investigation of cases of bloody diarrhea associated with the consumption of undercooked hamburger (50). *E. coli* O157:H7, a previously rarely isolated serotype, was found in the stools of affected individuals and was isolated from the meat implicated in outbreaks. Fully developed HC is characterized by intense abdominal pain and bloody diarrhea. In contrast to shigellosis, fever is usually absent, although the blood leukocyte count can be considerably elevated. Disease occurs primarily in the cecum and ascending colon (50). There is edema and hemorrhage of the lamina propria, together with superficial ulceration and pseudomembrane formation. Thrombosis of the microvasculature of the lamina propria with necrosis of the superficial colonic mucosa and preservation of the crypts is characteristic. These changes are also found in ischemic colitis and suggest that microvascular thrombosis plays an important role in the pathogenesis of HC (40). The pathology is remarkable for the lack of inflammatory cells in and around the microvasculature, suggesting that the thrombosis is not due to localized inflammation or cytokine release. This microvascular thrombosis without cellular infiltrate is also characteristic of pig edema disease, HUS, and thrombotic thrombocytopenic purpura (24, 40). The best evidence that VTs may play a role in microvascular thrombosis in the human colon is that the changes are found only with VT-producing *E. coli* infections; they are not associated with enteropathogenic *E. coli* strains which produce AE lesions but not toxin. In a rhesus monkey model, a wild-type *S. dysenteriae* strain produced microvascular thrombosis, whereas a Shiga toxin deletion mutant did not (13). While interpretation of this result is complicated by the marked inflammation produced by *S. dysenteriae*, it strongly suggests that the toxin exacerbates the process by causing microvascular thrombosis. Cecal edema and hemorrhagic cecitis have been produced in rabbits by intravenous or intraperitoneal injection of purified VT1, VT2, or VT2c (1, 48).

PIG EDEMA DISEASE

Edema disease occurs in weanling pigs as a result of intestinal infection with strains of *E. coli* which produce VTe, usually of serotypes O138, O139, and O141. It is characterized by loss of appetite, neurological abnormalities, and edema of the gut and eyelids. All symptoms are due to microvascular thrombosis in the relevant tissues. The disease can be reproduced by intravenous injection of VTe, and it is presumed that VTe is absorbed from the gut during the course of intestinal infection. Boyd et al. (3) showed that injection of a VTe mutant which does not bind to Gb4 produced central nervous system and subcutaneous pathologies but spared the gut. This suggests that the Gb4 receptors on the endothelial cells of the gut are of primary importance (3).

HEMOLYTIC UREMIC SYNDROME

HUS consists of a prodrome of bloody diarrhea. This is followed about 7 days later by thrombocytopenia, microangiopathic hemolytic anemia, and microvascular thrombosis in the renal glomerulus and, sometimes, other organs, including the central nervous system. The microvascular thrombosis is not accompanied by an inflammatory infiltrate. It is thought that this is due to the direct intravascular aggregation of platelets; activation of the coagulation cascade and fibrin production are thought to be relatively minor, secondary processes (34). To date, there is no satisfactory animal model of HUS (23, 24, 26). However, several studies have yielded important information on the in vivo toxicity of VTs to the microvasculature. Richardson et al. (48) injected ^{125}I-labeled VT1 into rabbits by the intravenous route. The toxin disappeared rapidly from the blood and accumulated primarily in the cecum and central nervous system. Immunofluorescence studies localized the toxin to the endothelium of postcapillary venules of the cecum and central nervous system. This localization reflected the pathological findings of microvascular thrombosis (without inflammatory cell infiltrates) in the same vessels and organs. This vascular damage is the presumed cause of cecal edema, diarrhea, and paralysis seen after injection of toxin. However, toxin accumulation in the microvasculature of the colon and small bowel did not lead to thrombosis, suggesting that receptor binding does not inevitably lead to intoxication (48). The implication of that work is that the microvascular beds of different organs differ in their susceptibilities to toxin-induced thrombosis because of their differential abilities to bind toxin. The model did not produce the renal glomerular thrombosis, thrombocytopenia, or hemolytic anemia characteristic of human HUS (48). The pathogenesis of these important features of HUS in humans is not fully understood. It has been proposed that the endothelial damage characteristic of HUS is caused by systemic absorption of VT produced by *E. coli* adherent to the colonic mucosa, but circulating VT has never been demonstrated in humans (24). There is good evidence for the absorption of toxin in gnotobiotic piglets, which were found to develop microvascular thrombosis of the central nervous system during the course of intestinal infection with *E. coli* O157:H7 (14). As noted above, in HUS, there is evidence of platelet activation and consumption without disseminated intravascular coagulation. It has been postulated that unusually large multimers of von Willebrand's factor found in the sera of patients with HUS may be produced by the damaged endothelial cells. Unusually large multimers of von Willebrand's factor have been shown to bind to the platelet surface glycoproteins GPIb-IX and GPIIb-IIIa, resulting in platelet aggregation in areas of high fluid shear stress such as the renal microcirculation (34).

Recently, Obrig et al. (42) have shown that primary cultures of renal endothelial cells are rich in Gb3 and are very susceptible to VT. So far, however, there is no convincing evidence that there are VT receptors on the vascular surfaces of glomerular endothelial cells in human kidney sections. In contrast, human umbilical vein endothelial cells have little Gb3 unless they are stimulated with tumor necrosis factor alpha or bacterial lipopolysaccharide (41, 42). This suggests that expression of Gb3 may be controlled differently in different cell lines and tissues (42). Tesh and Samuel (56) and other investigators have shown that VT2 induces tumor necrosis factor alpha production by human monocytes. This cytokine is known to induce procoagulant activity and leukocyte cytoadherence molecules in endothelial cells. Under normal conditions, endothelial cells express proteins on their surfaces which prevent coagulation or lyse a clot in case one forms. Obrig's group (33) looked at the effect of VT on tissue plasminogen activator and tissue plasminogen activator inhibitor but found no clear evidence for changes toward a net procoagulant effect.

HUMAN IMMUNITY TO VT

Neutralizing antibody responses are easily produced in animals by parenteral injection of toxoid or the native B subunit of VT (2). Antibody responses to Shiga toxin in patients recovering from *S. dysenteriae* infections are poor. Similar results have been obtained in patients recovering from HUS and HC (30). This may be due to the inability of minute amounts of toxin to elicit a response as occurs in tetanus. Also, the main response might be mucosal rather than systemic. In the case of VTs, however, there is also the intriguing finding that human B cells are rich in Gb3 and that VTs kill immunoglobulin A (IgA)- and IgG-committed B cells, leaving only IgM-committed cells to produce antibody (9). No VT2 antibodies have been demonstrated to date. However, a nonimmunoglobulin inhibitor of the VT2 group has been discovered in human serum. It is not clear what effect this would have on the development of VT2-associated pathology.

CONCLUSION

Epidemiological and pathological evidence strongly implicates VTs in the pathogenesis of HC and HUS. However, many features of HUS remain unexplained. Ultimately, production of specific immunity to VT in large populations in which there is a high attack rate may be the only method of proving a direct role of VTs in the pathogenesis of these conditions. It is hoped that the recent advances in structure-function correlation will lead to an improved understanding of the pathogenesis of HC and HUS in the near future.

REFERENCES

1. **Barrett, T., M. Potter, and I. Wachsmuth.** 1989 Continuous peritoneal infusion of Shiga-like toxin II (SLT-II) as a model of SLT-II-induced diseases. *J. Infect. Dis.* **159:**774–777.
2. **Boyd, B., S. Richardson, and J. Gariepy.** 1991. Serological responses to the B subunit of Shiga-like toxin 1 and its peptide fragments indicate that the B subunit is a vaccine candidate to counter the action of the toxin. *Infect. Immun.* **59:**750–757.
3. **Boyd, B., G. Tyrrell, M. Malone, C. Gyles, J. Brunton, and C. Lingwood.** 1993. Alteration of the glycolipid binding specificity of the pig edema toxin from globotetraosyl to globotriaosyl ceramide alters in vivo tissue targetting and results in verotoxin 1-like disease in pigs. *J. Exp. Med.* **177:**1745–1753.
4. **Bridgewater, F., R. Morgan, K. Rowson, and G. Wright.** 1955. The neurotoxin of *Shigella shigae*. Morphological and functional lesions produced in the central nervous system of rabbits. *Br. J. Exp. Pathol.* **36:**447–453.
5. **Brunton, J.** 1990. The Shiga toxin family: molecular nature and possible role in disease, p. 377–398. *In* B. Iglewski (ed.), *The Bacteria,* vol. XI. Academic Press, Inc., New York.
6. **Calderwood, S., and J. Mekalanos.** 1987. Iron regulation of Shiga-like toxin in *Escherichia coli* is mediated by the *fur* locus. *J. Bacteriol.* **169:**4759–4764.
7. **Clark, C., D. Bast, R. Agha, A. Sharp, R. Read, and J. Brunton.** Unpublished data.
8. **Cohen, A., G. Hannigan, B. Williams, and C. Lingwood.** 1987. Roles of globotriosyl- and galabiosylceramide in Verotoxin binding and high affinity interferon receptor. *J. Biol. Chem.* **262:**17088–17091.
9. **Cohen, A., V. Madrid-Marina, Z. Estrov, M. Freedman, C. A. Lingwood, and H.-M. Dosch.** 1990. Expression of glycolipid receptors to Shiga-like toxin on human B lymphocytes: a mechanism for failure of long-lived antibody response to dysenteric disease. *Int. Immunol.* **2:**1–8.
10. **DeGrandis, S., H. Law, J. Brunton, C. Gyles, and C. Lingwood.** 1989. Globotetraosyl ceramide is recognized by the pig edema disease toxin. *J. Biol. Chem.* **264:**12520–12525.
11. **Endo, Y., and K. Tsurugi.** 1988. The RNA N-glycosidase activity of ricin A chain. The characteristics of the enzymatic activity of ricin A chain with ribosomes and with rRNA. *J. Biol. Chem.* **263:**8735–8739.
12. **Endo, Y., K. Tsurugi, Y. Takeda, T. Ogasawara, and K. Igarashi.** 1988. Site of action of verotoxin 2 (VT2) from *Escherichia coli* O157:H7 on eukaryotic ribosomes. *Eur. J. Biochem.* **171:**45–50.
13. **Fontaine, A., A. Josette, and P. J. Sansonetti.** 1988. Role of Shiga toxin in the pathogenesis of bacillary dysentery studied by using a *tox* mutant of *Shigella dysenteriae* 1. *Infect. Immun.* **56:**3099–3109.
14. **Francis, D., R. Moxley, and C. Androas.** 1989. Edema disease-like lesions in gnotobiotic piglets infected with *Escherichia coli* serotype O157:H7. *Infect. Immun.* **57:**1339–1342.

15. **Frazer, M., M. Chernai, Y. Kozlov, and M. James.** 1994. Crystal structure of the holotoxin from *Shigella dysenteriae* at 2.5 A resolution. *Nat. Struct. Biol.* **1**:59–64.

16. **Gordon, V., and S. Leppla.** 1994. Proteolytic activation of bacterial toxins: role of bacterial and host cell proteases. *Infect. Immun.* **62**:333–340.

17. **Griffin, P., and P. Tauxe.** 1991. The epidemiology of infections caused by *Escherichia coli* O157: H7, other enterohemorrhagic *E. coli* and the associated hemolytic syndrome. *Epidemiol. Rev.* **13**: 60–98.

18. **Habib, N. F., and M. P. Jackson.** 1993. Roles of a ribosome-binding site and mRNA secondary structure in differential expression of Shiga toxin genes. *J. Bacteriol.* **175**:597–603.

19. **Hartley, M., G. Legname, R. Osborn, Z. Chen, and J. M. Lord.** 1991. Single-chain ribosome inactivating proteins from plants depurinate *Escherichia coli* 23S ribosomal RNA. *FEBS Lett.* **290**:65–68.

20. **Head, S., M. Karmali, and C. Lingwood.** 1991. Preparation of VT1 and VT2 hybrid toxins from their purified dissociated subunits. *J. Biol. Chem.* **266**:3617–3621.

21. **Hovde, C. J., S. Calderwood, J. Mekalanos, and R. J. Collier.** 1988. Evidence that glutamic acid 167 is an active site residue of Shiga-like toxin I. *Proc. Natl. Acad. Sci. USA* **85**:2568–2572.

21a. **Jackson, M. P., E. Wadolkowski, D. Weinstein, R. Holmes, and A. O'Brien.** 1990. Functional analysis of the Shiga toxin and Shiga-like toxin type II variant binding subunits by site-directed mutagenesis. *J. Bacteriol.* **172**:653–658.

22. **Kandel, G., A. Donohue-Rolfe, M. Donowitz, and G. Keusch.** 1991. Pathogenesis of Shigella diarrhea. XVI. Selective targeting of Shiga toxin to villus cells of the rabbit jejunum explains the effect of the toxin on intestinal electrolyte transport. *J. Clin. Invest.* **84**:1509–1517.

23. **Karmali, M.** 1989. Infection by verocytotoxin-producing *Escherichia coli*. *Clin. Microbiol. Rev.* **2**:15–38.

24. **Karmali, M.** 1992. The association of verocytotoxins and the classical hemolytic uremic syndrome, p. 199–212. *In* K. Kaplan, R. Trompeter, and J. Moake (ed.), *Hemolytic Uremic Syndrome and Thrombotic Thrombocytopenic Purpura.* Marcel Dekker, Inc., New York.

25. **Karmali, M., A. O'Brien, and S. Scotland.** Personal communication.

26. **Karmali, M., M. Petric, C. Lim, R. Cheung, and G. Arbus.** 1985. The association between idiopathic hemolytic uremic syndrome and infection by verotoxin-producing *Escherichia coli*. *J. Infect. Dis.* **151**:775–782.

27. **Kiarash, A., B. Boyd, and C. Lingwood.** 1994. Glycosphingolipid receptor function is modified by fatty acid content: Verotoxin 1 and Verotoxin 2c preferentially recognize different globotriaosyl ceramide fatty acid homologues. *J. Biol. Chem.* **269**:11138–11146.

28. **Konowalchuk, J., J. Speirs, and S. Stavric.** 1977. Vero response to a cytotoxin of *Escherichia coli*. *Infect. Immun.* **18**:775–779.

29. **Lee, S. L.** 1990. M.Sc. thesis. University of Toronto, Toronto, Ontario, Canada.

30. **Levine, M., J. McWeen, G. Losonsky, M. Reymann, I. Harari, J. Brown, D. Taylor, A. Donohue-Rolfe, D. Cohen, M. Bennish, Y. L. Lim, and R. Arnon.** 1992. Antibodies to Shiga holotoxin and to two synthetic peptides of the B subunit in sera of patients with *Shigella dysenteriae* 1 dysentery. *J. Clin. Microbiol.* **30**:1636–1641.

31. **Lindberg, A. A., J. Brown, N. Stromberg, M. Westling-Ryd, M. Schultz, and K. A. Karlsson.** 1987. Identification of the carbohydrate receptor for Shiga toxin produced by *Shigella dysenteriae* type 1. *J. Biol. Chem.* **262**:1779–1785.

32. **Lindgren, S., J. Samuel, C. Schmitt, and A. O'Brien.** 1994. The specific activities of Shiga-like toxin type II (SLT-II) and SLT-II-related toxins of enterohemorrhagic *Escherichia coli* differ when measured by Vero cell cytotoxicity but not by mouse lethality. *Infect. Immun.* **62**:623–631.

33. **Louise, C., T. Daniel, L. Barley-Maloney, and T. Obrig.** 1993. Shiga toxin-related vascular disease: LPS/endotoxin and TNFα effects on fibrinolysis factor synthesis in human renal and umbilical endothelial cells, abstr. B-94, p. 42. *Abstr. 93rd Gen. Meet. Am. Soc. Microbiol. 1993.* American Society for Microbiology, Washington, D.C.

33a. **Maloney, M. D., and C. A. Lingwood.** CD19 has a potential CD77 (globotriaosyl ceramide)-binding site with sequence similarity to verotoxin B subunits: implications of molecular mimicry for B cell adhesion and enterohemorrhagic *Escherichia coli* pathogenesis. *J. Exp. Med.*, in press.

34. **Moake, J. L.** 1994. Haemolytic-uraemic syndrome: basic science. *Lancet* **343**:393–397.

35. **Moazed, D., J. Robertson, and H. Noller.** 1988. Interaction of elongation factors EF-G and EF-Tu with a conserved loop in 23S RNA. *Nature* (London) **324:**362–364.

36. **Mobassaleh, M., A. Donohue-Rolfe, M. Jacewicz, R. Grand, and G. Keusch.** 1988. Pathogenesis of Shigella diarrhea: evidence for a developmentally regulated glycolipid receptor for Shigella toxin involved in the fluid secretory response of rabbit small intestine. *J. Infect. Dis.* **157:**1023–1031.

37. **Monzingo, A., and J. D. Robertus.** 1992. X-ray analysis of substrate analogs in the ricin A chain active site. *J. Mol. Biol.* **227:**1136–1145.

38. **Muzzin, A. G.** 1993. OB (oligonucleotide/oligosaccharide binding) fold: common structural and functional solution for non-homologous sequences. *EMBO J.* **12:**861–867.

39. **O'Brien, A., and R. K. Holmes.** 1987. Shiga and Shiga-like toxins. *Microbiol. Rev.* **51:**206–220.

40. **O'Brien, A., V. Tesh, A. Donohue-Rolfe, M. Jackson, S. Olsnes, K. Sandvig, A. Lindberg, and G. T. Keusch.** 1992. Shiga toxin: biochemistry, genetics mode of action and role in pathogenesis. *Curr. Top. Microbiol. Immunol.* **180:**65–94.

41. **Obrig, T., P. Del Vecchio, J. Brown, T. Moran, B. Rowland, K. Judge, and S. W. Rothman.** 1988. Direct cytotoxic action of Shiga toxin on human vascular endothelial cells. *Infect. Immun.* **56:**2373–2378.

42. **Obrig, T., C. Louise, C. Lingwood, B. Boyd, L. Barley Maloney, and T. Daniel.** 1993. Endothelial heterogeneity in Shiga toxin receptors and responses. *J. Biol. Chem.* **268:**15484–15488.

43. **Obrig, T. G., T. Moran, and J. E. Brown.** 1987. The mode of action of Shiga toxin on peptide elongation of eukaryotic protein synthesis. *Biochem. J.* **244:**287–294.

44. **Pellizzari, A., H. Pang, and C. Lingwood.** 1992. Binding of verocytotoxin 1 to its receptor is influenced by differences in receptor fatty acid content. *Biochemistry* **31:**1363–1370.

45. **Ready, M., P. Kim, and J. D. Robertus.** 1991. Site directed mutagenesis of ricin A chain and implications for the mechanism of action. *Proteins* **10:**270–278.

46. **Reisbig, R., S. Olsnes, and K. Eiklid.** 1981. The cytotoxic activity of Shigella toxin: evidence for catalytic inactivation of the 60S ribosomal subunit. *J. Biol. Chem.* **256:**8739–8744.

47. **Ren, P., Y. Wang, Y. Dong, and D. I. Stuart.** 1994. The *N*-glycosidase mechanism of ribosome-inactivating proteins implied by crystal structures of α momorchain. *Structure* **2:**7–15.

48. **Richardson, S., T. Rotman, V. Jay, C. Smith, L. Becker, M. Petric, N. Olivieri, and M. Karmali.** 1992. Experimental verocytotoxemia in rabbits. *Infect. Immun.* **60:**4154–4167.

49. **Rietra, P., G. Willshaw, H. Smith, A. Field, S. Scotland, and B. Rowe.** 1989. Comparison of verocytotoxin-encoding phages from *Escherichia coli* of human and bovine origin. *J. Gen. Microbiol.* **135:**2307–2318.

50. **Riley, L., R. Remis, S. Helgerson, H. McGee, J. Wells, B. Davis, R. Herbert, E. Olcott, L. Johnson, N. Hagrett, P. Blake, and M. Cohen.** 1983. Hemorrhagic colitis associated with a rare *Escherichia coli* serotype. *N. Engl. J. Med.* **308:**681–685.

51. **Sandvig, K., O. Garred, K. Prydz, J. Kozlor, S. Hansen, and B. van Deurs.** 1992. Retrograde transport of endocytosed Shiga toxin to the endoplasmic reticulum. *Nature* (London) **358:**510–512.

52. **Sandvig, K., K. Prydz, M. Ryd, and B. Van Deurs.** 1991. Endocytosis and intracellular transport of the glycolipid-binding ligand Shiga toxin in polarized MDCK cells. *J. Cell Biol.* **113:**553–562.

53. **Stein, P., A. Boodhoo, G. Armstrong, C. Cockle, M. Klein, and R. J. Read.** 1994. The crystal structure of pertussis toxin. *Structure* **2:**45–57.

54. **Stein, P., A. Boodhoo, G. Tyrrell, J. Brunton, and R. J. Read.** 1992. Crystal structure of the cell-binding B oligomer of verotoxin-1 from *E. coli*. *Nature* (London) **355:**748–750.

55. **Sung, L. M., M. Jackson, A. O'Brien, and R. Holmes.** 1990. Analysis of transcription of the Shiga-like toxin type II and Shiga-like toxin type II variant operons of *Escherichia coli*. *J. Bacteriol.* **172:**6386–6395.

56. **Tesh, V. L., and J. E. Samuel.** 1993. Immunomodulatory properties of purified Shiga toxin, abstr. B-93, p. 42. *Abstr. 93rd Gen. Meet. Am. Soc. Microbiol. 1993.* American Society for Microbiology, Washington, D.C.

57. **Toone, E., and P. St. Hilaire.** Unpublished data.

58. **Tyrrell, G., K. Ramotar, B. Toye, B. Boyd, C. Lingwood, and J. Brunton.** 1992. Alteration of the carbohydrate binding specificity of verotoxins from gal α 1–4 gal to gal NAC β 1–3 galα 1–4 gal and vice versa by site-directed mutagenesis of the binding subunit. *Proc. Natl. Acad. Sci. USA* **89:**524–528.

59. **Tzipori, S., H. Karch, I. Wachsmuth, R. Robbins-Browne, A. O'Brien, H. Lior, M. Cohen, J. Smithers, and M. Levine.** 1987. Role of a 60-megadalton plasmid and Shiga-like toxins in the pathogenesis of infection caused by enterohemorrhagic *Escherichia coli* O157:H7 in gnotobiotic piglets. *Infect. Immun.* **55:**3117–3125.

60. **Waddell, T., A. Cohen, and C. Lingwood.** 1990. Induction of verotoxin sensitivity in receptor deficient cell lines using the receptor glycoliid globotriaosyl ceramide. *Proc. Natl. Acad. Sci. USA* **87:**7898–7901.

61. **Weinstein, D., M. Jackson, L. Perera, R. Holmes, and A. O'Brien.** 1988. Cloning and sequencing of a shiga-like toxin II variant from an *Escherichia coli* strain responsible for edema disease of swine. *J. Bacteriol.* **170:**4223–4230.

62. **Williams Smith, H., P. Green, and Z. Parsell.** 1983. Verocell toxins in *Escherichia coli* and related bacteria: transfer by phage and conjugation and toxic action in laboratory animals, chickens and pigs. *J. Gen. Microbiol.* **129:**3121–3137.

Part 5. Virulence Gene Regulation

Although environmental influences on the production of virulence factors have long been noted, it is only recently, with the application of molecular genetic approaches to studying virulence, that the mechanisms of environmental control have been elucidated. Results of those studies indicated that virulence gene expression in response to environmental signals occurs in many complex and interesting ways, some of which are described in the chapters that follow. It has also become clear that the expression of individual virulence genes is coordinated, frequently by a central regulatory factor. Stanley Falkow and Alison Weiss (while a graduate student) made some of the earliest observations in this area. Alison developed methods for manipulating *Bordetella pertussis* genetically and generated a bank of Tn5 insertion mutants; these mutants were screened for the production of a variety of virulence-associated factors such as filamentous hemagglutinin, pertussis toxin, and hemolysin (4). While most mutants were defective for the production of a single factor, two were defective for the production of all the virulence factors examined (3). To explain this phenotype, Weiss and Falkow proposed that the transposon had inactivated a positive effector (Vir) required for the expression of virulence genes. In addition, they proposed that modulation of virulence gene expression by environmental factors (temperature, Mg^{2+} ions, nicotinic acid) is mediated by this positive effector. Several people in Stan's lab followed up on Alison's observations (Scott Stibitz, Craig Roy, Jeff Miller), and their work showed that the *vir* locus (renamed *bvg*) encodes two regulatory proteins, BvgA and BvgS, that are related to the family of two-component signal transduction systems (see chapter 27 by S. Stibitz and J. Miller). These two-component regulatory systems have received a great deal of attention lately and are involved in regulating functions as diverse as chemotaxis and the porin compositions of bacterial outer membranes.

Several other interesting virulence gene regulation "stories" grew out of projects initiated in Stan's lab. David Low spent much of his time while in Stan's lab cloning genes encoding adhesins from *Escherichia coli*. One of these adhesins, Pap (pyelonephritis-associated pili), has been the focus of research in his own laboratory. He has shown that expression of the *pap* operon is subject to multiple regulatory systems, including DNA methylation patterns and DNA-binding proteins such as CRP and PapB. Studies by Jorge Crosa on the regulation of iron uptake by the fish pathogen *Vibrio anguillarum* grew out of molecular epidemiological studies of naturally occurring virulent and avirulent strains. He recognized early on that virulence was associated with the presence of a large plasmid, and he has been studying this plasmid ever since. Research from his lab has demonstrated that expression of the iron uptake system of *V. anguillarum* requires the activities of two positive regulators, AngR and TAF, both of which are encoded on the 65-kb plasmid pJM1. Negative control is exerted by Fur (chromosomally encoded) and a plasmid-encoded antisense RNA.

Early rationales for the regulation of virulence gene expression focused on the energy savings to the bacteria of expressing products only when the bacteria are in an environment where the product is needed (e.g., inside a host). However, recent studies have indicated that the ability to regulate gene expression throughout the course of an infection is also important for virulence. This is clearly illustrated in the case of the regulatory gene *phoP* from *Salmonella* species (1, 2). In this case, both *phoP*-negative and *phoP* constitutive mutants are avirulent. Since PhoP acts as both a positive and a negative regulator, either the loss of PhoP or the overproduction of PhoP results in inappropriate expression of a number of virulence-associated genes. A similar type of story is emerging for the bacterial ureases (see chapter 29 by C. Collins). The genes encoding urease have now been cloned and sequenced from a variety of bacterial species. Despite considerable similarity between the ureases themselves, regulation of expression of the urease genes varies from species to species. This variation is seen both in terms of the regulatory proteins involved and in the environmental signals that they recognize. Carleen Collins speculates that the mode of regulation reflects both the niche occupied by the particular bacterial species in question and the role of urease for the successful occupation of that niche.

So it is clear that investigators have come a long way in identifying regulatory systems and that understanding of regulatory systems has moved beyond identification of a simple on-off switch. Some hints of the complexity of these regulatory loops can be seen in the chapters that follow.

Virginia L. Miller

REFERENCES

1. **Miller, S. I., A. M. Kukral, and J. J. Mekalanos.** 1989. A two-component regulatory system (*phoP phoQ*) controls *Salmonella typhimurium* virulence. *Proc. Natl. Acad. Sci. USA* **86:**5054–5058.
2. **Miller, S. I., and J. J. Mekalanos.** 1990. Constitutive expression of the PhoP regulon attenuates *Salmonella* virulence and survival within macrophages. *J. Bacteriol.* **172:**2485–2490.
3. **Weiss, A. A., and S. Falkow.** 1984. Genetic analysis of phase change in *Bordetella pertussis. Infect. Immun.* **43:**263–269.
4. **Weiss, A. A., E. L. Hewlett, G. A. Myers, and S. Falkow.** 1983. Tn*5*-induced mutations affecting virulence factors of *Bordetella pertussis. Infect. Immun.* **42:**33–41.

Molecular Genetics of Bacterial Pathogenesis
Edited by V. L. Miller, J. B. Kaper, D. A. Portnoy, and R. R. Isberg
© 1994 American Society for Microbiology, Washington, DC 20005

Chapter 27

Coordinate Regulation of Virulence in *Bordetella pertussis* Mediated by the *vir* (*bvg*) Locus

Scott Stibitz and Jeff F. Miller

Understand and things are all one;
If you don't understand, there are myriad distinctions,
  a thousand differences.
When you don't understand, things are all one;
Understand, and there are myriad distinctions,
  a thousand differences.

—Wumen (62)

The bacterial agent that causes the human disease pertussis or whooping cough was isolated in 1906 by Bordet and Gengou (11) but did not receive its current designation, *Bordetella pertussis,* until 1952 (38). At that time it was grouped together with two other respiratory agents, the human pathogen *Bordetella parapertussis* and the veterinary pathogen *Bordetella bronchiseptica.* Many molecular genetic data available today support this grouping. Two types of regulation of virulence potential have been recognized in *B. pertussis*: phase variation and antigenic modulation.

Leslie and Gardner (28) reported in 1931 that *B. pertussis* could grow in any of four different phases (phases I to IV). These phases were distinguished by colony morphology, hemolysis, and antigenic profile. Subsequent reports differed on the exact number and nature of the phases of *B. pertussis* (23). However, the basic finding that one could isolate spontaneous derivatives which no longer expressed specific antigens or virulence proteins, were avirulent, and were stable on passage has stood the test of time.

The realization that the virulence potential of this organism is clearly regulated in response to changes in environmental conditions is attributed to Lacey (26) in 1960, a phenomenon he termed "modulation." By culturing *B. pertussis* under a

Scott Stibitz • Division of Bacterial Products, Center for Biologics Evaluation and Research, Food and Drug Administration, Bethesda, Maryland 20892.   *Jeff F. Miller* • Department of Microbiology and Immunology, University of California, Los Angeles, California 90024.

wide variety of conditions of temperature and with culture additives, he showed that it could grow in two different modes, which he termed the X (xanthic) mode and the C (cyanic) mode, that differed in terms of the presentation of specific surface components. Lacey distinguished between these two forms using colonial characteristics, including hemolysis, and antisera raised against the different forms. We now know a great deal more about specific protein molecules or virulence factors, such as pertussis toxin, adenylate cyclase toxin, filamentous hemagglutinin, pertactin, and fimbriae, which were probably detected by Lacey in those experiments. Some of the culture additives Lacey used to induce the change from the X mode to the C mode, such as $MgSO_4$, are commonly used today to study this and related phenomena.

As discussed in more detail below, an accumulation of evidence over recent years strongly suggests that both of these forms of regulation are based in the *vir* (now *bvg*) locus that is the focus of this chapter.

BEGINNING OF THE MOLECULAR GENETIC ERA AND DISCOVERY OF THE *vir* (*bvg*) LOCUS

By 1983 the biochemical characterization of *B. pertussis* had defined several virulence-associated proteins. Pertussis toxin, perceived to be one of the most important, had been purified, characterized biochemically, and shown to be an ADP-ribosylating toxin with host GTP-binding proteins as cellular targets (24, 55). Filamentous hemagglutinin (FHA), believed to be important in the specific adherence that *B. pertussis* demonstrates to the ciliated epithelium, had also been purified (15). It was also known that *B. pertussis* made fimbriae (10) and a novel toxin, the extracytoplasmic adenylate cyclase toxin (14).

Despite this biochemical characterization, the contribution of these virulence-associated factors to the pathogenic capabilities of *B. pertussis* had not been assessed experimentally, and little was known about the genetic makeup of this organism. The stage was thus set for the work of Weiss et al. (58), who isolated insertion mutations in *B. pertussis* using the transposon Tn5. That study was pivotal for at least two reasons. First, several of these mutations affected specific virulence traits such as pertussis toxin, FHA, and adenylate cyclase/hemolysin, and thus defined the genetic loci required for their elaboration. It is now known that, with the exception of the insertions affecting FHA, these mutations defined the structural genes for these factors (53, 58). These mutant strains were then tested for their ability to cause disease in an animal model in order to directly evaluate, for the first time, the contribution of specific molecules to the virulence of this pathogen (59). Second, two of these Tn5 insertion mutations defined a new locus called the *vir* locus. Mutations in this locus abolished the synthesis of all of the known virulence-associated factors, suggesting that the *vir* locus is involved in the coordinate regulation of virulence.

An additional chapter in the history of *vir*-mediated regulation concerns the discovery of *vir*-repressed genes, or *vrg*'s. These were discovered by Tn*phoA* mutagenesis and screening to identify genes for secreted proteins which were regulated in a reciprocal fashion to *vir*-activated genes or *vag*'s (25). In other words, under conditions in which *vag* expression is reduced, i.e., in a *bvg* mutant or under modulating conditions, *vrg* expression is increased, and vice versa. In the same study, a number of new *vag*'s were identified, the functions of which are under study.

Cloning of the *vir* locus demonstrated that the *vir* and *fha* loci are tightly linked. Plasmids which contained both loci were capable of directing the expression of FHA in *Escherichia coli,* and this expression was regulated by the same environmental conditions that regulated its expression in *B. pertussis* (53). On the basis of this expression, mapping of the *vir* locus by Tn5 mutagenesis in *E. coli* indicated that it was approximately 5 kb in size. The structural gene for FHA was found to be contiguous with the *vir* locus and to be transcribed away from *vir* (53). DNA sequence analysis revealed the direction of transcription of the *vir* locus and showed it to be divergent with respect to *fha* (5). At this time the *vir* locus was renamed the *bvg* locus (for *Bordetella vir* gene) to distinguish it from other loci termed *vir* in other organisms (5).

The original DNA sequence predicted the presence of three *bvg*-encoded polypeptides, two of which shared significant sequence similarity with a large family of bacterial regulatory proteins commonly called "two-component" systems (5, 40, 51). This term refers to a common mechanism for translating changes in environmental conditions to changes in gene expression. The two components are a "sensor" or "transmitter" protein, which can span the gram-negative inner membrane, and a "regulator" or "receiver" protein, which is cytoplasmic and is often a DNA-binding regulatory protein. From studies on many such systems, it can be seen that the receiver receives input from the transmitter via phosphorylation at a highly conserved aspartate residue. This phosphate is donated by the transmitter after autophosphorylation at a highly conserved histidine residue. The C-terminal kinase domain of the transmitter and the N-terminal phosphoacceptor domain of the receiver are the regions which are conserved between systems. The domains presumably responsible for sensing environmental conditions (the N terminus of the transmitter protein) and for effecting changes in gene expression (the C terminus of the receiver protein) have diverged widely, presumably as a result of different evolutionary pressures.

Further studies showed that other key structural features of two-component systems were embodied in the *vir* locus. One such feature was the transmembrane nature of the transmitter protein. This was shown by examining the alkaline phosphatase activities of fusions of the *E. coli phoA* gene to the *bvg* genes at various locations (54). Use of these fusion proteins to generate antisera which could detect the native proteins in bacterial cell extracts led to a reevaluation of the DNA sequence such that only two *bvg*-encoded proteins were seen to be encoded (54). The current interpretation of these results is shown in Fig. 1, where it can be seen that the *bvg* locus encodes BvgA, a typical "receiver" protein, and BvgS, a more complex protein containing a transmitter domain. As in other two-component systems, the transmitter protein has its N-terminal sequences in the periplasm, and the cytoplasmic C-terminal portion contains the transmitter histidine kinase homology. The presence of a receiver domain at the C-terminal end of BvgS is

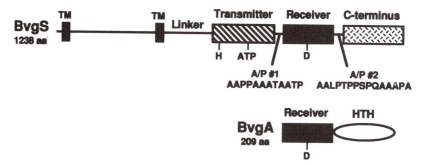

FIGURE 1. Features of the BvgA and BvgS proteins. The portion of BvgS between the two hydrophobic transmembrane sequences (TM) is located in the periplasm, while the remainder of BvgS and BvgA are cytoplasmic. The conserved histidine (H) and ATP-binding motif present in transmitter modules are indicated along with the conserved aspartate residues (D) found in receiver modules. The BvgS linker is the site at which signal-insensitive mutations map. The alanine- and proline-rich sequences (A/P #1, A/P #2) connecting the transmitter, receiver, and C-terminal domains are likely to be conformationally flexible, allowing interdomain interactions. The C-terminal region of BvgA contains a helix-turn-helix (HTH) motif. aa, amino acids.

somewhat unusual, but it is a feature shared with FrzE of *Myxococcus xanthus,* ArcB of *E. coli,* VirA of *Agrobacterium tumefaciens,* and LemA of *Pseudomonas syringae* (13, 20, 22, 32).

The mechanistic features of the *bvg* locus which are conserved among other two-component systems, such as intermolecular signaling via phosphorylation and DNA binding of the receiver protein, have also now been well documented (42, 56). Both of these features will be discussed in greater detail in later sections of this chapter.

PHASE VARIATION

As described above, the observation that stable derivatives of *B. pertussis,* which had lost the ability to synthesize most of their virulence-associated characteristics, could arise spontaneously was made some time ago. In most cases the capacity of avirulent-phase derivatives to regain this ability by reverting to the virulent phase has not been reported. However, in a genetic analysis of phase variation, Weiss and Falkow (57) isolated a series of phase variants of the strain Tohama III derived one from the other in a sequential fashion. The frequency of switching between the virulent and the avirulent phases that they observed was relatively high, about 1 in 10^3 (57). In an effort to determine the mechanism for phase variation, the difference between the virulent and avirulent phases of strains belonging to this series was mapped and determined at the DNA sequence level. It was found that the change resulting in the avirulent phenotype was a frameshift mutation caused by the addition of a G residue within a run of six G residues near the end of *bvgS*. Reversion to the virulent phenotype was associated with reversion of this mutation (51). Another analysis of phase variation was undertaken by using avirulent phase variants of *B. bronchiseptica.* In this case it was found that these variants arose through the occurrence of small spontaneous deletions in the *bvg* locus of this organism (36). This change was thus nonrevertable.

The biological relevance of these observations is difficult to assess but would appear questionable. Nonreverting deletions in the *bvg* locus would seem to represent a dead end for organisms experiencing them. The high rate of phase variation caused by frameshift mutations is now known to be associated with a mutator phenotype in the strains in which it was observed (50). On the other hand, it is intriguing that variation of fimbrial expression in *B. pertussis* has been shown to occur by a somewhat similar mechanism. In this case the context for insertion and deletion of single bases is a run of 13 C residues in the promoter region of the serotype 3 fimbrial subunit gene (60). Fimbrial phase variation can be observed to occur at a frequency of approximately 1 in 10^3 colonies examined with the Fim2 antibody (50). This high rate of phase variation together with the striking sequence features strongly suggests that fimbrial phase variation represents a conserved mechanism for introducing variability, which is of biological relevance. Whether the same can be said of the frameshifting seen in the *bvg* locus remains to be determined.

MODULATION

In contrast to phase variation, the biological relevance of modulation, the other phenomenon mediated through the *bvg* locus, seems assured. It is now widely held that the appropriate expression of specific virulence factors under the appropriate conditions (e.g., in the appropriate locations in the host) is an important aspect of a bacterium's adaptation to the pathogenic lifestyle. The demonstration that this regulatory phenomenon in *B. pertussis* was also mediated by the *bvg* locus was less straightforward than for phase variation. Although the sequence similarity observed to two-component systems was suggestive of such a role for *bvg*, this did not constitute experimental proof.

The first indication that environmental responsiveness was in fact encoded by the *bvg* locus was the observation that plasmids containing *bvg* and *fha* reconstituted expression of FHA-immunoreactive material in *E. coli* in a manner which was dependent on the integrity of the *bvg* locus and which was responsive to the same signals which function in *B. pertussis* (53). This initial observation was examined more quantitatively by using gene fusions of β-galactosidase to *fha*, which resulted in the clear demonstration that the *bvg* locus acted in *trans* (34). Further studies in *E. coli* showed that the activation of *fha-lacZ* could be achieved by the overexpression of BvgA alone, but that such activation was unresponsive to environmental signals (43).

All of these results clearly showed that the *bvg* locus is a key player in the environmentally modulated regulation of *B. pertussis* virulence genes but failed to implicate this locus in the direct sensing of environmental signals. The strongest evidence which supported such a role was the isolation of a particular class of mutations which, in contrast to the previously examined mutations, did not abolish *bvg* activation of regulated genes, but rendered such activation unresponsive to modulating signals (25, 33). These mutations were ultimately shown to affect the *bvg* locus, specifically, the *bvgS* gene at positions encoding amino acid residues just C terminal to the transmembrane region, and therefore cytoplasmic (33). Because these mutations appear to affect only the ability of the BvgS protein to respond to environmental signals, and not its ability to function in gene activation, these mutations provide strong evidence that BvgS is in fact the environmental sensor that responds to modulating signals.

MECHANISMS OF BvgAS ACTION

Some mechanistic features of BvgAS action can be predicted as a consequence of membership in the family of two-component systems. These predictions are primarily related to the central mechanism associated with this family, i.e., signaling between proteins by phosphorylation of the receiver at a conserved aspartate residue by the transmitter after autophosphorylation at a conserved histidine residue (40). Some predictions have recently been fulfilled through the demonstration in vitro that a purified, truncated 'BvgS protein is capable of autophosphorylation and of phosphorylating, in turn, purified BvgA. A mutation introducing a conservative amino acid change at the conserved histidine in the transmitter domain of BvgS abolished autophosphorylation (56). Other predictions are that the level of phosphorylated BvgA in vivo is affected by environmental conditions and that phosphorylation of BvgA results in a change in its ability to bind specific DNA sites and/or to promote the expression of *bvg*-regulated genes, but these have not yet been tested experimentally.

One of the aspects of BvgAS action that may be unique, or at least that remains to be elucidated for other two-component systems, concerns the role of the "extra" receiver domain present at the C terminus of BvgS. The majority of two-component systems are able to function without such a domain, but BvgAS function is dependent on it, as shown by a variety of mutations which affect this domain and which completely eliminate BvgAS activity. Such mutations include deletions, in-frame linker insertions, and site-directed mutations which destroy the predicted site of phosphorylation in this domain (33, 48, 56). Biochemically, this domain appears to be involved in the transfer of phosphate from BvgS to BvgA, in that the 'BvgS protein, in which the phosphoacceptor site has been destroyed by mutation, is incapable of phosphorylating BvgA (56). A model for the Bvg phosphorylation cascade is shown in Fig. 2. In contrast, deletion of the receiver domain of the VirA protein of *A. tumefaciens,* a transmitter protein with the same overall configuration as BvgS, resulted in increased levels of expression of a regulated *virB-lacZ* fusion, suggesting that in this protein the receiver domain has an inhibitory function (13). It is thus unclear how well conclusions regarding the function of this domain can be extrapolated from one system to another.

Another intriguing mechanistic question addresses how environmental signals which are presumed to be extracellular are translated into intracellular changes in phosphorylation. A simple model would suggest that the periplasmic domain is involved in sensing these signals and that changes in this domain are somehow

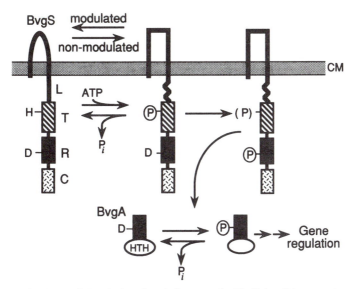

FIGURE 2. Model for the BvgS-BvgA phosphorylation cascade. The linker (L), transmitter (T), receiver (R), and C-terminal (C) domains of BvgS are indicated. It is proposed that BvgS autophosphorylates at the conserved histidine (H) of the transmitter. Phosphorylation of the conserved aspartate (D) in the BvgS receiver is then required for phosphotransfer to the receiver of BvgA. Phosphorylated BvgA is then able to activate or repress regulated promoters. HTH, helix-turn-helix motif; CM, cytoplasmic membrane. Adapted from reference 56.

propagated to the cytoplasmic domains of this protein. Such a model appears incorrect in that, to date, no mutations which specifically affect sensory transduction and which map to the periplasmic domain have been described. In contrast, the modulation-insensitive mutations described above map to a cytoplasmic location, in the "linker" region between the second transmembrane sequence and the transmitter domain (33). In the VirAG system of *A. tumefaciens,* deletion of the entire periplasmic region of VirA failed to destroy environmentally responsive regulation mediated by the truncated protein (13). One exception to this observation is the response of VirA to monosaccharides such as galactose, which is in fact dependent on the periplasmic domain of VirA. This particular sensing pathway is known to require the presence of a periplasmic galactose-binding protein, ChvE (13). In contrast to these systems, the periplasmic domain of BvgS does not appear dispensable because in-frame linker insertion mutations in this region abolish BvgAS activity (33, 48). Deletion of the periplasmic region of BvgS also abolishes activity, unless combined with one of the BvgS constitutive mutations described earlier. In this case, BvgAS activity is restored but it is not environmentally responsive (33). It may be that, as with monosaccharide sensing by VirA, BvgS-mediated environmental responsiveness also requires an as yet unidentified periplasmic protein. Of unknown significance to the mechanism of BvgS action is the possibility that BvgS may be active as a dimer or a higher multimer. The primary support for such a notion is genetic and twofold: some periplasmic in-frame linker insertion mutations appear to be *trans* dominant, negative, and intra-cistronic complementation between *bvgS* alleles is possible in some combinations (48). Both of these findings are consistent with a multimeric configuration for BvgS, but they have yet to be corroborated by biochemical experiments.

cis-ACTING SEQUENCES IMPLICATED IN *bvg*-RESPONSIVE REGULATION

The sensing of environmental signals by the *bvg* locus and the transduction of these signals into changes in expression of multiple virulence genes is only one component of *bvg*-mediated regulation. For a complete understanding, the sites present near the promoters of *bvg*-responsive genes which affect the regulation of these genes must also be defined. These will be discussed in turn.

fha and *bvg*

The *fha* and *bvg* operons are distinguished from other *bvg*-regulated genes by their ability to respond to the presence of BvgA and BvgS in *E. coli* (34, 53). As mentioned earlier, *E. coli* lysogens carrying bacteriophage lambda derivatives with *fhaB-lacZ* transcriptional fusions show a several hundred-fold increase in β-galactosidase production in the presence of *bvgAS* supplied in *trans* on a pBR322 vector, with transcription initiating at the same site as it initiates in *B. pertussis*. Transcription from the *bvg* promoter is also increased by BvgAS in *E. coli,* demonstrating the presence of a positive autoregulatory circuit which functions in *B. pertussis* as well (44, 47). BvgAS-mediated transcriptional activation of *fhaB-lacZ* and *bvgA-lacZ* fusions in *E. coli* is downregulated by the same environmental signals that modulate virulence gene expression in *B. pertussis* (34, 44).

Primer extension and ribonuclease protection studies have identified several transcripts that initiate in the 425-bp intergenic region separating the *fhaB* and *bvgA* open reading frames. Transcription of the *fhaB* locus initiates at a site located 70 bp upstream from the translation initiation codon (47). Transcription of the *bvgAS* operon occurs from at least three promoters located 93 bp (*bvgP1*), 143 bp (*bvgP2*), and 271 bp (*bvgP3*) upstream of the *bvgA* structural gene (44, 47). In the Bvg⁻ phase, which results from null mutations in *bvg* or from the presence of modulating environmental signals, only *bvgP2* is active. This promoter is presumably responsible for maintaining basal levels of BvgA and BvgS. In the Bvg⁺ phase, the *bvgP2* promoter is no longer expressed; however, transcription from the *fhaB, bvgP1, bvgP3,* and *bvgP4* promoters is activated. The *bvgP2* promoter is relatively weak in relation to *bvgP1*, and this may account for the increased levels of expression of *bvgAS* in the Bvg⁺ phase. The *bvgP4* promoter produces an antisense transcript that initiates 19 bp upstream from *bvgA;* however, its role in regulating *bvgAS* expression is unknown.

Deletion analysis, gel retardation, and DNase I protection studies have identified *cis*-acting sequences and defined specific BvgA-DNA interactions involved

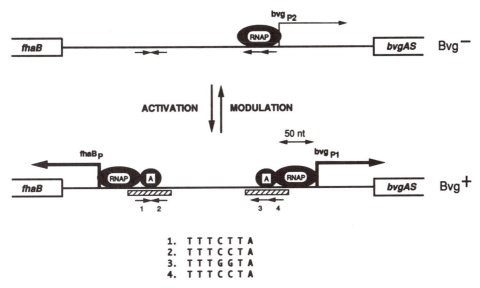

FIGURE 3. Model for the regulation of *bvgAS* and *fhaB* (44). Transcription of the *bvgAS* operon initiates in the Bvg⁻ phase from the *bvgP2* promoter. In the Bvg⁺ phase, BvgA (A) is able to direct RNA polymerase (RNAP) to activate transcription of the *bvgAS* and *fhaB* operons at the *bvgP1* and *fhaB* promoters. Relative levels of transcription are represented by arrow thickness. Locations of inverted and direct repeats in the *bvgAS* and *fhaB* intergenic regions are illustrated by small numbered arrows, and the sequences of these repeats are given at the bottom of the figure. Sites of BvgA-dependent protection from DNase I digestion are indicated by the hatched rectangles (42).

in *fhaB* and *bvgAS* gene regulation (42). As shown in Fig. 3, two major sites of BvgA-dependent protection from DNase I digestion are located between *bvgA* and *fhaB*. In both cases the protected sequences contain two copies of a heptanucleotide repeat, TTT(C/G)NTA, which is present in an inverted orientation upstream of the *fhaB* promoter and in direct orientation upstream from the *bvgP1* promoter. Deletion analysis demonstrated that the repeated sequences are required for *trans*-activation by *bvgAS*. In addition, the presence of an oligonucleotide containing the inverted repeat upstream of *fhaB* on a high-copy-number plasmid inhibits *bvgAS* activation of an *fhaB-lacZ* fusion, most likely as a result of the sequence-specific titration of BvgA (35). These results suggest that the heptanucleotide repeats serve as specific recognition sites for BvgA and are required for gene regulation.

The BvgA-binding sites positioned upstream from the *bvgP1* promoter are located between the − 10 and − 35 sequences of *bvgP2*. This arrangement is probably responsible for the phase-dependent alteration in promoters that initiate transcription of *bvg*. The model shown in Fig. 3 suggests that, during a transition from the Bvg⁻ to the Bvg⁺ phase, BvgA becomes activated and binds to specific DNA sequences. The result is transcriptional activation of the *fhaB* and *bvgP1* promoters and simultaneous repression of *bvgP2*. BvgA therefore appears to act as both an activator and a repressor of transcription. Presumably, BvgS-dependent phosphorylation of BvgA is related to its ability to associate with DNA, to interact with RNA polymerase, or both.

ptx

The pertussis toxin operon encoding the five subunits of this toxin was one of the first *B. pertussis* operons for which the DNA sequence was determined (31, 39). At that time a putative promoter sequence was noted, with expected homology to the canonical −10 and −35 regions of *E. coli* promoters, but with a spacing between these regions of 21 bp in comparison with the canonical spacing of 17 bp. Deletion analysis of the region upstream of the *ptx* promoter indicated that signals necessary for *bvg* responsiveness lay within 171 bp of the transcription start site. From positions −157 to −117 is a tandem duplication of a 20- to 21-bp sequence (16 of 21 bases match). Because of their spacings relative to each other, the repeated sequences are predicted to be aligned to the same face of the DNA helix. Deletions extending into one of these sequences reduced *bvg* responsiveness and promoter strength by approximately half, and deletion into the remaining repeat removed *bvg* responsiveness entirely. Complete loss of promoter activity, however, was achieved only by deletions which extended into the putative −35 region (19). More recent analysis suggests that the spacing between the repeated sequences and the −35 region of the promoter may be critical (18). It should be noted that this sequence in the *ptx* promoter region is not homologous to the BvgA-binding site identified in the *fha* and *bvg* promoter regions. Such BvgA-binding sites are also not apparent elsewhere in the *ptx* promoter region, and in vitro experiments have repeatedly failed to demonstrate binding of BvgA to this region (18, 42).

cya

The genes in the *cya* operon encoding the adenylate cyclase toxin/hemolysin of *B. pertussis* are arranged in the order *cyaABDE*. The *cyaA* gene is the structural gene for the toxin, and the *cyaBDE* genes are required for proper secretion of this toxin (16). In addition, upstream of *cyaA* and in a transcriptionally divergent orientation is the *cyaC* gene, which is required for activation of the toxin (6). Expression of the *cyaA* gene is driven by a *bvg*-responsive promoter 115 bp upstream of the start codon for this gene, but expression of the downstream *cyaBDE* genes is driven by a non-*bvg*-responsive promoter (27). Deletion analysis of the *cyaA* promoter suggests that sequences required for *bvg* responsiveness are encoded within 569 bp upstream of the start of transcription for the *cyaA* promoter (17). Analysis of the DNA sequence of this region has failed to reveal the presence of BvgA-binding sites homologous to those in the *fha* and *bvg* promoter regions (42). The best fit for sites similar to the repeats near the *ptx* promoter is found from positions −114 to −157 relative to the start of transcription, where 20 of 44 bases match the 41-bp sequence defined near the *ptx* promoter (21). A role for this sequence in the regulation of *cya* expression has not been definitively established.

fim Loci

B. pertussis has generally been considered to be capable of synthesizing fimbriae of two different serotypes. The structural genes for the fimbrial subunit, *fim-2* and *fim-3*, have been cloned and sequenced (29, 37). A third fimbrial subunit gene, *fimX*, has also been cloned and sequenced and appears to be expressed at a low level (41). Other genes required for the export and assembly of the fimbrial subunits have been found to be encoded within the *fha* operon and are thus *bvg* regulated (30, 61). The genes for the fimbrial subunits themselves have been found both to be *bvg* responsive and to undergo a phase variation of their own. A comparison of the DNA sequences of the promoter regions of the *fim-2*, *fim-3*, and *fimX* genes and of *fim-3* phase variants suggests a common mechanism for phase variation and regulation of *fim* gene expression. In these promoters, a putative activator-binding sequence is separated from the −10 region of the *fim* promoter by a stretch of C residues (15 for *fim-2*, 13 for *fim-3*). As with the *ptx* locus, spacing appears critical, because fimbrial phase variants can arise by the deletion of a single C residue and the poor expression of *fimX* may be attributed to its short (eight-residue) C stretch (60). As with the *ptx* and *cya* loci, no BvgA-binding sites can be discerned in the DNA sequences of these promoter regions. Some sequence similarity is seen between a consensus *fim* promoter sequence and the *ptx* promoter, but this homology does not extend to the direct repeats that are apparently involved in the regulation of *ptx* expression (19, 60).

vir-Repressed Genes

To date the DNA sequences of the promoter regions of five *vrg*'s identified by Tn*phoA* mutagenesis have been reported (7, 8). Four of these five *vrg*'s contain a 32-bp sequence within the coding region that is similar to a consensus sequence, with 22 of the 32 consensus nucleotides being the same in at least three of the four sites (8). A 6-bp in-frame linker insertion in this sequence in one of these genes, *vrg-6*, abolished responsiveness to modulation (7). Analogously, isolation of a *vrg-6* repressor-resistant mutant after *mutD* mutagenesis and screening identified a nucleotide within this same sequence, which, when changed from T to A, abolished modulation of the *vrg-6* gene, resulting in constitutive expression (8). Northern blots suggested that the regulation seen for the *vrg-6* gene is at the transcriptional level and that replacement of the *vrg-6* promoter with the non-*bvg*-regulated *asd* promoter resulted in a construct which was regulated normally, thus supporting the hypothesis that the sites required for this regulation lie within the coding region of this gene (8). A 62-bp fragment containing the consensus regulatory sequence from *vrg-6* was found to bind to a band corresponding to a 34-kDa protein in Southwestern experiments; under modulating conditions this binding was greatly reduced (8). These experiments suggest a model for regulation of at least some *vrg*'s in *B. pertussis* by a repressor protein which is itself encoded by a *bvg*-activated gene.

bvg REGULONS

The term ''regulon'' refers to a group of genes which are coregulated and, thus, at some level share a common regulatory mechanism. Thus, the virulence factors of *B. pertussis* which require the *bvg* locus for their expression may be said to constitute a regulon. However, at what level a common *bvg*-mediated mechanism may be operating remains a mystery. For the *bvg* and *fha* loci, a straightforward mechanism for activation involving stimulation of transcription via binding of BvgA to a defined operator sequence is apparent (42). For the other *bvg*-activated loci such as *ptx, cya,* and *fim,* such a mechanism is not supported by the currently available data. The promoter regions of these genes do not display BvgA-binding sites and do not bind BvgA, and their expression cannot be reconstituted in *E. coli* under the control of *bvg* (17, 18, 34, 42, 60). Two general models can be put forward to explain these differences. In one, the *bvg* locus alone is seen to be sufficient for mediating regulation, and one must therefore postulate that the way in which BvgA acts at loci such as *ptx, cya,* and *fim* is mechanistically different than the way in which it acts at *fha* and *bvg.* In the other type of model, the existence of accessory regulatory factors is postulated. These can be envisaged to act together with BvgA at *bvg*-responsive loci or to be themselves turned on by the *bvg* locus either by direct activation or at the level of synthesis (as in a regulatory cascade).

Currently available data which address this issue are somewhat confusing and contradictory. In contrast to earlier reports, a recent study suggests that, under the appropriate conditions of plasmid context and DNA topology, the *bvg* locus is in fact sufficient to *trans*-activate *ptx* expression in *E. coli* (46). However, under these conditions, environmental responsiveness is not seen. Biochemical evidence for the existence of an accessory regulator has been reported. A 23-kDa protein which bound DNA fragments from the *ptx* and *cya* promoter regions and which appeared to be regulated by *bvg* was detected (21). A recent study reports the cloning and sequencing of a gene encoding a 23- to 24-kDa protein from *B. pertussis* which is a non-sequence-specific DNA-binding protein. The predicted protein shows significant sequence identity to the eukaryotic histone H1 (54.7%) and to the H1c protein of *Chlamydia trachomatis* (52.7%) but not to other prokaryotic histone-like proteins and it does not appear to be regulated by *bvg* (45). At a genetic level, *B. pertussis* mutants which display a phenotype consistent with mutations identifying accessory regulators (Fha$^+$, Ptx$^-$, Cya$^-$) have been isolated, but to date these have been shown to map only to *rpoA,* the gene for the alpha subunit of RNA polymerase, or to the *bvg* locus itself (12, 49, 52). Clearly, additional research is required to find the missing factors, if they exist, or to define the different mechanisms of action that the *bvg* locus displays at different regulated loci, if they do not.

Although phenotypic modulation by environmental signals has sometimes been viewed as an "on-off" switch for *Bordetella* virulence factors, it should now be clear that this is an incomplete picture. BvgAS-mediated signal transduction results in a biphasic transition, with the Bvg$^-$ phase expressing prominent and unique characteristics of its own (Fig. 4). This was recognized at an antigenic level early in *B. pertussis* research and is now supported by studies at the molecular genetic level (25, 26). Thus, the characteristics unique to the Bvg$^-$ phase are seen to be encoded by *vrg*'s (25). At least one of these, *vrg-6,* appears to be required for full persistence in a mouse model of respiratory infection, suggesting

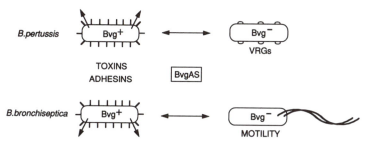

FIGURE 4. BvgAS-mediated biphasic alteration of *B. pertussis* and *B. bronchiseptica*. In the Bvg$^+$ phase, both species express a variety of adhesins and toxins encoded by BvgAS-activated genes and operons. In the Bvg$^-$ phase, which results from modulation or null mutations in the *bvg* locus, BvgAS-activated virulence factors are no longer expressed and *vrg* loci (*B. pertussis*) and flagellar genes (*B. bronchiseptica*) are induced.

that *vrg*'s may also be virulence factors (9). Another dramatic example of *vrg*'s is the motility apparatus of *B. bronchiseptica,* which is synthesized only in the Bvg$^-$ phase (2, 4). It is tempting to speculate that motility could play a role in allowing *B. bronchiseptica* to colonize mucosal surfaces.

As with the *vag*'s, a complete picture of the factors involved in regulation of *vrg*'s has not yet emerged, but it has been studied in the greatest depth in *B. bronchiseptica*. In this organism, both motility and siderophore production are controlled by *bvgAS* in a manner reciprocal to that in which *bvg*-activated genes are controlled (1, 4). Two motility loci in *B. bronchiseptica, flaA* (the structural gene for flagellin) and *frlAB* (the motility master regulatory locus, similar to *flhCD* of *E. coli*), have been characterized (2). Transcription of both *flaA* and *frlAB* is negatively regulated by BvgAS in the Bvg$^+$ phase, and genetic studies have identified a regulatory cascade that connects the BvgAS system to flagellar synthesis. According to a recent model, BvgAS repression of *frlAB* transcription is relieved in the Bvg$^-$ phase (2). The *frl* gene products function as transcriptional activators that control the expression of several loci, including the structural gene for an alternative sigma factor, sigma F. Sigma F then activates late genes which include the flagellin locus. *frlAB* is the first-characterized example of an auxiliary regulatory locus within the *bvg* regulon. Although *B. pertussis* appears to be nonmotile, it does encode a *frl* locus that is functional when expressed in *B. bronchiseptica* (3). The role of *frl* in *B. pertussis* is unknown.

The little more than 10 years since the discovery of the *bvg* locus has seen a tremendous increase in the knowledge and understanding of the regulation of virulence potential in *Bordetella* species. However, this understanding has been primarily at a mechanistic level. What is perhaps the most interesting question is one of the most difficult to address. Why and under what conditions has it been advantageous to these bacterial pathogens to be able to radically alter their phenotypic makeups, to play down their overtly pathogenic nature, and to present a different face to their environment? Several possibilities for the role of this alternate personality, the Bvg$^-$ phase, have been suggested, including the establishment of infection, modulation of disease, persistence of the organism by downregulation of antigens, transmission, or survival in microenvironments in the host or in environmental reservoirs, but experimental support for any of these has been lacking. With hope, the next 10 years will lead to a greater understanding of the nature of these alternate states, the role of BvgAS-mediated signal transduction in general, and the pathogenic strategies of members of the genus *Bordetella*.

REFERENCES

1. **Agiato Foster, L. A., P. C. Giardina, M. Wang, B. J. Akerley, J. F. Miller, and D. W. Dyer.** 1994. Siderophore biosynthesis in *Bordetella bronchiseptica* is controlled by the *bvg* regulon, abstr. D-178, p. 127. *Abstr. 94th Gen. Meet. Am. Soc. Microbiol. 1994.* American Society for Microbiology, Washington, D.C.

2. **Akerley, B. J., and J. F. Miller.** 1993. Flagellin transcription in *Bordetella bronchiseptica* is regulated by the BvgAS virulence control system. *J. Bacteriol.* **175:**3468–3479.

3. **Akerley, B. J., and J. F. Miller.** Unpublished data.

4. **Akerley, B. J., D. M. Monack, S. Falkow, and J. F. Miller.** 1992. The *bvgAS* locus negatively controls motility and synthesis of flagella in *Bordetella bronchiseptica*. *J. Bacteriol.* **174:**980–990.

5. **Arico, B., J. F. Miller, C. Roy, S. Stibitz, D. Monack, S. Falkow, R. Gross, and R. Rappuoli.** 1989. Sequences required for expression of *Bordetella pertussis* virulence factors share homology with prokaryotic signal transduction proteins. *Proc. Natl. Acad. Sci. USA* **86:**6671–6675.

6. **Barry, E. M., A. A. Weiss, I. E. Ehrmann, M. C. Gray, E. L. Hewlett, and M. S. Goodwin.** 1991. *Bordetella pertussis* adenylate cyclase toxin and hemolytic activities require a second gene, *cyaC*, for activation. *J. Bacteriol.* **173:**720–726.

7. **Beattie, D. T., S. Knapp, and J. J. Mekalanos.** 1990. Evidence that modulation requires sequences downstream of the promoters of two *vir*-repressed genes of *Bordetella pertussis*. *J. Bacteriol.* **172:**6997–7004.

8. **Beattie, D. T., M. J. Mahan, and J. J. Mekalanos.** 1993. Repressor binding to a regulatory site in the DNA coding sequence is sufficient to confer transcriptional regulation of the *vir*-repressed genes (*vrg* genes) in *Bordetella pertussis*. *J. Bacteriol.* **175:**519–527.

9. **Beattie, D. T., R. Shahin, and J. J. Mekalanos.** 1992. A *vir*-repressed gene of *Bordetella pertussis* is required for virulence. *Infect. Immun.* **60:**571–577.

10. **Blom, J., G. A. Hansen, and F. M. Poulsen.** 1983. Morphology of cells and hemagglutinogens of *Bordetella* species: resolution of substructural units in fimbriae of *Bordetella pertussis*. *Infect Immun.* **42:**308–317.

11. **Bordet, J., and O. Gengou.** 1906. Le microbe de la Coqueluche. *Ann. Inst. Pasteur* **20:**731–741.

12. **Carbonetti, N. H., N. Khelef, N. Guiso, and R. Gross.** 1993. A phase variant of *Bordetella pertussis* with a mutation in a new locus involved in the regulation of pertussis toxin and adenylate cyclase toxin expression. *J. Bacteriol.* **175:**6679–6688.

13. **Chang, C.-H., and S. C. Winans.** 1992. Functional roles assigned to the periplasmic, linker, and receiver domains of the *Agrobacterium tumefaciens* VirA protein. *J. Bacteriol.* **174:**7033–7039.

14. **Confer, D. L., and J. W. Eaton.** 1982. Phagocyte impotence caused by an invasive bacterial adenylate cyclase. *Science* **217:**948–950.

15. **Cowell, J. L., Y. Sato, H. Sato, B. An der Lan, and C. R. Manclark.** 1982. Separation, purification, and properties of the filamentous hemagglutinin and the leukocytosis promoting factor-hemagglutinin from *Bordetella pertussis*, p. 371–379. *In* J. B. Robbins, J. C. Hill, and J. C. Sadoff (ed.), *Bacterial Vaccines*. Thieme-Stratton Inc., New York.

16. **Glaser, P., H. Sakamoto, J. Bellalou, A. Ullmann, and A. Danchin.** 1988. Secretion of cyclolysin, the calmodulin-sensitive adenylate cyclase-haemolysin bifunctional protein of *Bordetella pertussis*. *EMBO J.* **7:**3997–4004.

17. **Goyard, S., and A. Ullmann.** 1991. Analysis of *Bordetella pertussis cya* operon regulation by use of *cya-lac* fusions. *FEMS Microbiol. Lett.* **77:**251–256.

18. **Gross, R., N. H. Carbonetti, R. Rossi, and R. Rappuoli.** 1992. Functional analysis of the pertussis toxin promoter. *Res. Microbiol.* **143:**671–681.

19. **Gross, R., and R. Rappuoli.** 1988. Positive regulation of pertussis toxin expression. *Proc. Natl. Acad. Sci. USA* **85:**3913–3917.

20. **Hrabak, E. M., and D. K. Willis.** 1992. The *lemA* gene required for pathogenicity of *Pseudomonas syringae* pv. syringae on bean is a member of a family of two-component regulators. *J. Bacteriol.* **174:**3011–3020.

21. **Huh, Y. J., and A. A. Weiss.** 1991. A 23-kilodalton protein, distinct from BvgA, expresses by virulent *Bordetella pertussis* binds to the promoter region of *vir*-regulated genes. *Infect. Immun.* **59:**2389–2395.

22. **Iuchi, S., Z. Matsuda, T. Fujiwara, and E. C. C. Lin.** 1990. The *arcB* gene of *Escherichia coli* encodes a sensor-regulator protein for anaerobic repression of the *arc* modulon. *Mol. Microbiol.* **4:**715–727.

23. **Kasuga, T., Y. Nakase, and K. Ukishima.** 1953. Studies on *Haemophilus pertussis*. Part I. Antigen structure of *H. pertussis* and its phases. *Kitasato Arch. Exp. Med.* **26:**121–134.

24. **Katada, T., and M. Ui.** 1982. Direct modification of the membrane adenylate cyclase system by islet-activating protein due to ADP-ribosylation of a membrane protein. *Proc. Natl. Acad. Sci. USA* **79:**3129–3133.

25. **Knapp, S., and J. J. Mekalanos.** 1988. Two *trans*-acting regulatory genes (*vir* and *mod*) control antigenic modulation in *Bordetella pertussis*. *J. Bacteriol.* **170:**5059–5066.

26. **Lacey, B. W.** 1960. Antigenic modulation of *Bordetella pertussis*. *J. Hyg.* **31:**423–434.

27. **Laoide, B. M., and A. Ullmann.** 1990. Virulence dependent and independent regulation of the *Bordetella pertussis cya* operon. *EMBO J.* **9:**999–1005.

28. **Leslie, P. H., and A. D. Gardner.** 1931. The phases of *Haemophilus pertussis*. *J. Hyg.* **31:**423–434.

29. **Livey, I., C. J. Duggleby, and A. Robinson.** 1987. Cloning and nucleotide sequence analysis of the serotype 2 fimbrial subunit gene of *Bordetella pertussis*. *Mol. Microbiol.* **1:**203–209.

30. **Locht, C., M.-C. Geoffroy, and G. Renauld.** 1992. Common accessory genes for the *Bordetella pertussis* filamentous hemagglutinin and fimbriae share sequence similarities with the *papC* and *papD* gene families. *EMBO J.* **11:**3175–3183.

31. **Locht, C., and J. M. Keith.** 1986. Pertussis toxin gene: nucleotide sequence and genetic organization. *Science* **232:**1258–1264.

32. **McCleary, W. R., and D. R. Zusman.** 1990. FrzE of *Myxococcus xanthus* is homologous to both CheA and CheY of *Salmonella typimurium*. *Proc. Natl. Acad. Sci. USA* **87:**5898–5902.

33. **Miller, J. F., S. A. Johnson, W. J. Black, D. T. Beattie, J. J. Mekalanos, and S. Falkow.** 1992. Constitutive sensory transduction mutations in the *Bordetella pertussis bvgS* gene. *J. Bacteriol.* **174:**970–979.

34. **Miller, J. F., C. R. Roy, and S. Falkow.** 1989. Analysis of *Bordetella pertussis* virulence gene regulation by use of transcriptional fusions in *Escherichia coli*. *J. Bacteriol.* **171:**6345–6348.

35. **Miller, J. F., C. R. Roy, and S. Falkow.** 1991. Regulation of *fhaB, bvg,* and *ptx* transcription in *E. coli:* a comparative analysis, p. 217–224. *In* C. R. Manclark (ed.), *Proceedings of the Sixth International Symposium on Pertussis.* U.S. Department of Health and Human Services, Bethesda, Md.

36. **Monack, D. M., B. Arico, R. Rappuoli, and S. Falkow.** 1989. Phase variants of *Bordetella bronchiseptica* arise by spontaneous deletions in the *vir* locus. *Mol. Microbiol.* **3:**1719–1728.

37. **Mooi, F. R., A. ter Avest, and H. G. J. van der Heide.** 1990. Structure of the *Bordetella pertussis* gene coding for the serotype 3 fimbrial subunit. *FEMS Microbiol. Lett.* **55:**285–289.

38. **Moreno-López, M.** 1952. El género *Bordetella*. *Microbiol. Esp.* **5:**177–181.

39. **Nicosia, A., M. Perugini, C. Franzini, M. C. Casagli, M. G. Borri, G. Antoni, M. Almoni, P. Neri, G. Ratti, and R. Rappuoli.** 1986. Cloning and sequencing of the pertussis toxin genes: operon structure and gene duplication. *Proc. Natl. Acad. Sci. USA* **83:**4631–4635.

40. **Parkinson, J. S., and E. C. Kofoid.** 1992. Communication modules in bacterial signaling proteins. *Annu. Rev. Genet.* **26:**71–112.

41. **Pedroni, P., B. Riboli, F. de Ferra, G. Grandi, S. Toma, B. Arico, and R. Rappuoli.** 1988. Cloning of a novel pilin-like gene from *Bordetella pertussis:* homology to the *fim2* gene. *Mol. Microbiol.* **2:**539–543.

42. **Roy, C. R., and S. Falkow.** 1991. Identification of *Bordetella pertussis* regulatory sequences re-

quired for transcriptional activation of the *fhaB* gene and autoregulation of the *bvgAS* operon. *J. Bacteriol.* **173**:2385–2392.

43. **Roy, C. R., J. F. Miller, and S. Falkow.** 1989. The *bvgA* gene of *Bordetella pertussis* encodes a transcriptional activator required for coordinate regulation of several virulence genes. *J. Bacteriol.* **171**:6338–6344.

44. **Roy, C. R., J. F. Miller, and S. Falkow.** 1990. Autogenous regulation of the *bvgABC* operon of the bacterial pathogen *Bordetella pertussis. Proc. Natl. Acad. Sci. USA* **87**:3763–3767.

45. **Scarlato, V.** Unpublished data.

46. **Scarlato, V., B. Arico, and R. Rappuoli.** 1993. DNA topology affects transcriptional regulation of the pertussis toxin gene of *Bordetella pertussis* in *Escherichia coli* and in vitro. *J. Bacteriol.* **175**: 4764–4771.

47. **Scarlato, V., A. Prugnola, B. Arico, and R. Rappuoli.** 1990. Positive transcriptional feedback at the *bvg* locus controls expression of virulence factors in *Bordetella pertussis. Proc. Natl. Acad. Sci. USA* **87**:6753–6757.

48. **Stibitz, S.** Submitted for publication.

49. **Stibitz, S.** Submitted for publication.

50. **Stibitz, S.** Unpublished data.

51. **Stibitz, S., W. Aaronson, D. Monack, and S. Falkow.** 1989. Phase-variation in *Bordetella pertussis* by frameshift mutation in a gene for a novel two-component system. *Nature* (London) **338**:226–229.

52. **Stibitz, S., and N. Carbonetti.** Unpublished data.

53. **Stibitz, S., A. A. Weiss, and S. Falkow.** 1988. Genetic analysis of a region of the *Bordetella pertussis* chromosome encoding filamentous hemagglutinin and the pleiotropic regulatory locus *vir. J. Bacteriol.* **170**:2904–2913.

54. **Stibitz, S., and M.-S. Yang.** 1991. Subcellular localization and immunological detection of proteins encoded by the *vir* locus of *Bordetella pertussis. J. Bacteriol.* **173**:4288–4296.

55. **Tamura, M., K. Nogimori, S. Murai, M. Yajima, K. Ito, T. Katada, M. Ui, and S. Ishii.** 1982. Subunit structure of islet-activating protein, pertussis toxin, in conformity with the A-B model. *Biochemistry* **21**:5516–5522.

56. **Uhl, M. A., and J. F. Miller.** 1994. Autophosphorylation and phosphotransfer in the *Bordetella pertussis* BvgAS signal transduction cascade. *Proc. Natl. Acad. Sci. USA* **91**:1163–1167.

57. **Weiss, A. A., and S. Falkow.** 1984. Genetic analysis of phase change in *Bordetella pertussis. Infect. Immun.* **43**:263–269.

58. **Weiss, A. A., E. L. Hewlett, G. A. Meyers, and S. Falkow.** 1983. Tn*5*-induced mutations affecting virulence factors of *Bordetella pertussis. Infect. Immun.* **42**:33–41.

59. **Weiss, A. A., E. L. Hewlett, G. A. Myers, and S. Falkow.** 1984. Pertussis toxin and extracytoplasmic adenylate cyclase as virulence factors of *Bordetella pertussis. J. Infect. Dis.* **150**:219–222.

60. **Willems, R., A. Paul, H. G. van der Heide, A. R. ter Avest, and F. R. Mooi.** 1990. Fimbrial phase variation in *Bordetella pertussis:* a novel mechanism for transcriptional regulation. *EMBO J.* **9**: 2803–2809.

61. **Willems, R. J. L, H. G. J. van der Heide, and F. R. Mooi.** 1992. Characterization of a *Bordetella pertussis* fimbrial gene cluster which is located directly downstream of the filamentous haemagglutinin gene. *Mol. Microbiol.* **6**:2661–2671.

62. **Wumen.** 1228. Wumen's verse on the 16th koan of the Wumenguan "Putting on a formal vestment at the sound of a bell." (*In* T. Cleary [translation and commentary]. 1993. *No Barrier: Unlocking the Zen Koan.* Bantam Books, New York.)

Molecular Genetics of Bacterial Pathogenesis
Edited by V. L. Miller, J. B. Kaper, D. A. Portnoy, and R. R. Isberg
© 1994 American Society for Microbiology, Washington, DC 20005

Chapter 28

Methylation-Dependent and Lrp-Dependent Fimbrial Gene Regulation in *Escherichia coli*

David A. Low

REGULATION OF FIMBRIA-ADHESIN GENE EXPRESSION

Phenomenon of Phase Variation

The majority of bacterial pathogens and opportunists colonize the mucosal surfaces of their hosts by expressing specific adhesins that bind to host target receptors. For example, *Escherichia coli* strains collectively express a multitude of adhesins that enable this species to infect different host tissues including the oropharynx, the small and large intestines, and the urinary tract. Analysis of the *E. coli* isolates that express these adhesins has indicated that a limited number of clonally derived strains have evolved (50), and each one is capable of expressing a variety of adhesin determinants, usually from one to four. Many of these adhesins are physically associated with fibrous appendages known as fimbriae or pili, which appear to form a bridge between the bacterium and its target cell, both of which bear net negative charges. In Pap (pyelonephritis-associated pili) and Pap-related fimbriae, the PapG adhesin is located at the fimbrial tip and facilitates binding to either globotriaosylceramide (class I), globoside (class II), or the Forssman glycolipid (class III) (37, 39). In addition, the PapE protein, a major component of the shaft of the tip fibrillum, appears to have binding specificity for fibronectin (57). Thus, the fimbria-adhesin structure is complex and multifunctional.

Many *E. coli* urinary tract isolates, in particular those that cause kidney infections, contain multiple *pap* operons, each of which is capable of expressing and assembling cell surface fimbriae. For example, *E. coli* J96 contains a *pap* operon as well as a related sequence designated *prs,* for *pap*-related sequence (38). The *pap* sequence encodes a class I G adhesin, whereas *prs* encodes a class III G adhesin. The G adhesins can be assembled into heterologous fimbria-adhesin complexes on the basis of complementation data (38). Isolate J96 also contains a type 1 fimbrial operon (*fim*), which codes for a mannose-binding adhesin, also located at the type 1 fimbrial tip (28). Interestingly, all of these three fimbrial operons are

David A. Low • Division of Cell Biology and Immunology, Department of Pathology, Room 5B310MC, University of Utah Medical School, Salt Lake City, Utah 84132.

regulated by phase variation, defined as the ability to switch between on and off expression states. Thus, individual bacteria within a population will express different combinations of fimbria-adhesin complexes.

The rates of phase variation measured for the *pap-17* operon of *E. coli* C1212 in M9 minimal medium containing glycerol as the carbon source are about 10^{-2} per cell per generation for on to off and 10^{-4} per cell per generation for off to on (10). The approximately 100-fold higher switch rate to off results in a small fraction of P-fimbriated cells in the bacterial population. In contrast, the *pap-21* operon, also present in strain C1212, has about a 10-fold higher rate of transition from off to on than *pap-17* (10^{-3} per cell per generation), resulting in a larger fraction of Pap-21-fimbriated cells in the bacterial population (10). Although a comprehensive survey of *pap*$^+$ uropathogenic *E. coli* has not been reported, preliminary evidence suggests that the *pap* operons in many uropathogenic isolates behave similarly to *pap-17* from isolate C1212, with a low off-to-on switch rate (unpublished data). This probably accounts for the observation that, upon primary subculture of uropathogenic isolates, many are found to lack adhesin expression, even though further subculture may yield adhesin-positive bacteria (43).

Initial studies on the regulation of type 1 fimbriae in *E. coli* indicated that the rate of switching from on to off (about 1×10^{-3} per cell per generation) was similar to the off-to-on rate (3×10^{-3} per cell per generation) (16). Recent work, however, has shown that the *E. coli* strain (CSH50) used to study *fim* regulation contains an IS*1* insertion in the *fimE* regulatory gene (8). Since FimE appears to facilitate the conversion of on to off, this insertion results in an artifactually low on-to-off switch rate (40). Current estimated switch rates for *fim* lacking the IS*1* insertion are very similar to those measured for *pap* phase variation, with an approximately 100-fold higher rate of switching to the off state (21). Thus, both Pap and type 1 fimbrial expression appears to be biased toward the off state, with only a low fraction of cells in a bacterial population expressing these appendages.

Recent analysis of the expression of F1845 fimbriae, which bind to the Dr blood antigen (3, 4), and S fimbriae, which recognize sialyl-containing receptors (27), has shown that these fimbriae are also under phase variation control (52). Measurement of fimbrial on-to-off and off-to-on switching frequencies showed that the rates of switching to the off state were 66- and 142-fold higher than the rate of switching to the on state for the *sfa* (codes for S fimbriae) and *daa* (codes for F1845 fimbriae) operons, respectively. The off-to-on and on-to-off transition rates were similar (within fivefold) to the rates reported for type 1 and Pap fimbriae. Taken together, these results show that the expression of a number of different fimbriae in *E. coli* is under phase variation control and that in all cases the switch bias is toward the off expression state.

Possible Biological Roles of Phase Variation

What is the biological role of fimbrial phase variation and why is the bias set toward the off state? Although neither of these questions has directly been addressed yet, I present some hypotheses below. Recent work by Bloch et al. (5) has shown that type 1 fimbriae are required for colonization of the rat oropharynx and facilitate the transmission of *E. coli* from one animal to the other. In contrast, expression of type 1 fimbriae may be detrimental to *E. coli* in the intestine (41) and other tissues such as the renal pelvis since binding to leukocytes is facilitated by the mannose-specific adhesin associated with these fimbriae (48). In fact, bacteria isolated from these tissues do not express type 1 fimbriae (41). In this example, phase variation would allow *E. coli* to colonize different host environments through the generation of different fimbrial expression states.

As discussed above, the majority of pathogenic *E. coli* are capable of expressing multiple fimbrial types on their cell surfaces. Since these fimbriae are usually antigenically distinct and bind to different tissue receptors, this maximizes the potential of the *E. coli* population to interact with the diverse environments of its hosts and to avoid detection by the host's immune system. If all possible fimbrial types were expressed constitutively by bacteria, this would elicit immune responses to the fimbriae which might eliminate all members of the monotypic population. In addition, this would necessitate a major investment in resources for the bacterium since fimbrial expression is metabolically costly. Finally, if all bacteria in a population expressed fimbriae, this could inhibit their ability to move from one tissue to another since even if the bacteria were motile, they and their progeny would be tethered to tissue sites. These hypotheses need to be tested by altering the regulation of fimbrial genes in animal pathogens such as *Salmonella typhimurium* and determining the consequences of these regulatory changes to colonization, tissue localization, immune response, and disease in natural hosts.

Environmental Control of Phase Variation

Phase variation initially appeared to be a stochastic process, unaffected by environmental conditions (15). However, work on the *pap* operon showed that the carbon source greatly influenced the switch rate from off to on via the cyclic AMP (cAMP)-cAMP receptor protein (CRP) regulon (defined as a group of operons controlled by a single regulatory molecule (10). This switching rate is 35-fold higher in cells growing on glycerol as the sole carbon source than in glucose-grown cells. This difference in switching rates appears to be due to binding of cAMP-CRP near the *papI* promoter, which stimulates the transcription and the subsequent expression of PapI, which, in turn, is a critical factor in promoting *pap* pilin transcription (2, 22).

Recent work shows that type 1 fimbrial expression is environmentally regulated by aliphatic amino acids. Ian Blomfield, Barry Eisenstein, and colleagues (21) have shown that leucine and alanine stimulate both *fimB*-promoted switching (bidirectional) as well as *fimE*-promoted switching (on to off). For example, *fimE*-promoted switching from on to off was increased about 16-fold in the presence of aliphatic amino acids, which accounted for virtually all of the effect of defined rich medium on fimbrial switching compared with the effect of defined minimal medium (21). Thus, in both *pap* and *fim* regulation, nutrient-rich environments which have high levels of glucose and amino acids shut down expression of Pap and type 1 fimbriae, respectively, whereas nutrient-poor conditions enhance the expression of these fimbriae. Presumably, the responses of *pap* and *fim* to different signals reflect the environment in which the expression of each fimbrial type is favored; however, this has not been tested. Although the intestinal environment has a large amount of nutrients in transit, it is likely that the microenvironment of *E. coli* is nutrient poor, at least some of the time, because of competition from the abundant endogenous flora. However, very little is known about the relative nutrient levels and environmental signals in the small and large intestines or about how these levels fluctuate with time and nutrient input. Although this is a difficult area of study, it will need to be explored to gain a full understanding of the biological significance of different fimbrial regulatory mechanisms.

PHASE VARIATION REGULATORY MECHANISMS

Central Role of the Lrp Regulon

The regulatory mechanisms of type 1 and Pap fimbrial phase variation are very different. The type 1 fimbrial switch is regulated by inversion of a small DNA fragment containing the pilin promoter. The *fimB* and *fimE* genes, which share significant sequence similarities with bacteriophage lambda integrase (35), are likely to be site-specific recombinases that mediate the inversion event. The *fimB* gene promotes inversion in both directions, whereas *fimE* promotes only the on-to-off inversion. Integration host factor is required for this inversion, whereas the histone-like protein H-NS (PilG) appears to act as an inhibitor of promoter inversion (17, 34). In addition, recent results indicate that the leucine-responsive regulatory protein (Lrp) bidirectionally stimulates *fim* inversion (7). It is not known if these factors directly affect the inversion event or regulate components such as FimB and FimE that are necessary for inversion (or both).

The *pap* phase variation system, in contrast, employs an epigenetic mechanism involving DNA methylation patterns. In this case there are no DNA rearrangements and the *pap* regulatory DNA sequences present in phase on and phase off cells are identical (9). Instead, the methylation states of two DNA GATC sites (GATC-I and GATC-II), located upstream of the *papBA* fimbrial promoter, govern the binding of the Lrp and PapI regulatory proteins to DNA regions overlapping these GATC sites (Fig. 1). Binding of these proteins near the GATC-I and GATC-II sites prevents their methylation by deoxyadenosine methylase (Dam).

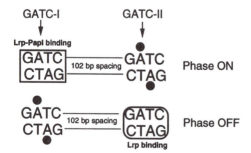

FIGURE 1. Methylation patterns of Pap phase on and phase off cells. The square at the upper left represents binding of Lrp-PapI around the GATC-I site, whereas the oval at the lower right represents binding of Lrp near the GATC-I site. Black dots indicate methyl groups covalently attached to adenosine residues by Dam.

The gene encoding Lrp was first identified in 1976 as a regulator of branched-chain amino acid transport (1). It was subsequently independently identified by different groups studying amino acid biosynthesis and degradation, peptide transport, and fimbrial gene expression (36, 45). Lrp is a relatively abundant (about 3,000 molecules per cell) basic DNA-binding protein of 19 kDa which appears to exist in solution as a dimer (55). On the basis of the binding of Lrp to a number of DNA target sites, a consensus sequence, TTTATTCNNNAT, where "N" is any deoxyribonucleotide, was proposed by Rex et al. (46). However, further comparison with *ilvIH* (56) and *pap* (unpublished data) Lrp-binding sites indicates that there are very few sequence similarities among them. These results suggest the possibility that Lrp recognizes DNA structure, such as DNA bend sites. Indeed, Lrp has been shown to bend DNA by about 50° when bound to a single site and to 135° when bound to two adjacent sites (55).

Lrp is a regulatory protein that exerts global control over as many as 50 genes in *E. coli* (14). In some cases, such as *ilvIH*, *fim* switching, and *pap* pilin transcription, Lrp is a positive regulator. In other cases, such as *sdaA*, *livJ*, and *livKHMGF*, Lrp acts negatively. Moreover, many of the genes in the Lrp regulon are responsive to the aliphatic amino acids leucine and alanine. For example, leucine turns off the transcription of *ilvH*, enhances Lrp-mediated *fim* switching, and turns on the transcription of *sda* (14). In other cases such as *pap*, leucine does not have any effect on transcription (13). Although the mechanism by which leucine acts is not known, it appears to interact with Lrp since mutations in *lrp* which yield a leucine-nonresponsive phenotype have been isolated (44). These mutations map to the carboxyl side of Lrp, whereas the DNA-binding domain appears to be within the amino-terminal side (44).

Roles of Lrp, PapI, and Dam in Pap Fimbrial Phase Variation

Lrp is required for the methylation protection of both of the *pap* GATC sites, whereas PapI, an 8-kDa regulatory protein, is required only for the protection of GATC-I. In the absence of PapI, Lrp binds to multiple sites flanking GATC-II, which overlaps the *papBA* pilin promoter, which was determined on the basis of DNase I footprint analysis (42). My colleagues and I hypothesize that this occurs in phase off cells since Lrp is required for the methylation protection of GATC-II in vivo (11) and GATC-II is nonmethylated in phase off cells (Fig. 1). Upon the addition of PapI, additional sites flanking GATC-I are protected by DNase I cleavage, forming the phase on methylation pattern (Fig. 1).

Analysis of mutations within *pap* regulatory DNA that allow transcription of *pap* in the absence of PapI indicates that PapI does not interact with *pap* DNA. Instead, PapI appears to bind to Lrp and increase its affinity for sequences near GATC-I (33). Binding of Lrp-PapI to GATC-I does not occur if GATC-I is fully methylated, which is characteristic of phase off cells. Therefore, cells are locked in the off state until DNA replication, which generates a hemimethylated GATC-I site. On the basis of the results of in vitro DNA binding analyses, Lrp-PapI binds to hemimethylated GATC-I sites, although with a lower affinity than to nonmethylated sites (42). Maintenance of this binding of Lrp-PapI to GATC-I through a second round of cell division would yield *pap* DNA with a fully nonmethylated GATC-I site, which is characteristic of phase on cells (9, 12). A prediction of this model that needs to be tested is that DNA replication is essential for phase switching.

As shown in Fig. 1, analysis of phase on cells showed that GATC-I is nonmethylated, whereas GATC-II is fully methylated (9). Presumably, this occurs as a result of binding of Lrp-PapI to GATC-I and the subsequent blockage of Dam methylation, although this has not been directly shown. In contrast, the addition of Lrp and PapI to nonmethylated *pap* DNA in vitro results in binding to both the GATC-I and GATC-II regions, which does not correlate with the phase on methylation pattern in which GATC-II is not protected from methylation. One possible reason for this discrepancy is that in these in vitro experiments the DNA used contained a nonmethylated GATC-II site, whereas this GATC site is methylated in phase on DNA. Methylation of GATC-II appears to weaken the binding of Lrp to this DNA region on the basis of the results of DNase I footprint analysis (42). Moreover, mutational analysis has recently shown that methylation of GATC-II is necessary for *pap* phase variation (12). A mutant of *E. coli* containing a GCTC-II site which is nonmethylatable displayed a phase off phenotype, even though binding of Lrp to this site was not disrupted. In addition, a mutant containing GCTC-I, which normally displays a locked on phenotype, is phase off in the absence of Dam. Together these results showed that methylation of the GATC-II site is required for the transition to the phase on state. Methylation of GATC-II might aid in the binding of Lrp-PapI to GATC-I by reducing the affinity of Lrp for GATC-II. As discussed above, binding of Lrp to GATC-II is weakened by methylation of this site. On the basis of the results of in vivo methylation analysis, once binding of Lrp-PapI to GATC-I occurs, the GATC-II site is unbound and is free to be methylated. Since the GATC-II site overlaps the −35 and −10 RNA polymerase recognition domain (9), it is likely that Lrp must disengage from this site prior to transcriptional activation.

In summary, Lrp, PapI, and Dam methylase play important roles in the *pap* phase switch. The characteristic methylation pattern of phase on cells (GATC-I nonmethylated, GATC-II methylated) is essential for transcriptional activation and appears to be maintained by binding of the Lrp-PapI complex near GATC-I (42). As outlined above, in phase on cells methylation of GATC-II is essential for *pap* transcription. In contrast, in phase off cells, GATC-I is methylated, preventing binding of Lrp-PapI near GATC-I and blocking transcription activation. Thus, the methylation states of both the GATC-I and GATC-II sites are critical in the regulation of the *pap* phase switch.

Roles of PapB, CRP, and H-NS Proteins in *pap* Fimbrial Phase Variation

As discussed above, the *pap* phase switch is influenced by the carbon source, which reflects the nutritional status of the immediate environment. This response appears to be due to the binding of CRP at a site about 105 bp upstream of the *papI* transcription start site (22). Mutation of this site so that cAMP-CRP cannot bind to the site blocks both *papI* and *papBA* transcription (19). In addition, PapB,

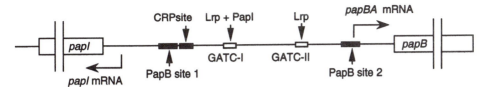

FIGURE 2. Organization of the regulatory region of the *pap* operon. Binding domains for PapB (18), Lrp and Lrp-PapI (12, 42), and CRP (22) as well as the 27-bp inverted repeats containing GATC-I and GATC-II are shown as boxes. Also shown are the transcription start sites of the *papI* and *papBA* transcripts. PapB site 2 overlaps the −10 RNA polymerase recognition site at the *papBAp* promoter.

a 12-kDa regulatory protein encoded by the *papBA* mRNA, binds near cAMP-CRP at site 1 and also appears to play a role in stimulating *papI* transcription (18) (Fig. 2). Thus, both CRP and PapB appear to regulate *papI* transcription. Under conditions with high levels of cAMP (glycerol minimal medium, for example), PapI levels will increase and the probability of transition to the phase on state will increase concomitantly, since PapI is a limiting factor in the phase switch. Under these conditions, the level of PapB will also rise as a result of *papBA* transcription, generating even higher levels of PapI by stimulating *papI* transcription. This feedback loop is limited by a second PapB-binding site near the *papBA* promoter (site 2), which has a lower affinity for PapB than site 1 (18). It appears that binding of PapB to site 2 inhibits transcription initiated at the *papBA* promoter, thereby indirectly regulating the levels of PapI.

The results discussed above indicate that CRP controls the *pap* switch indirectly because of its control over PapI levels. In addition, Uhlin and coworkers (19) have presented evidence suggesting that CRP may directly stimulate transcription at the *papBA* promoter. In that experiment, a multicopy plasmid containing a *papBA-lacZ* operon fusion was constructed; in that fusion, both *papI* and *papB* contained mutations that blocked their activities. Introduction of a plasmid expressing PapI under *lacUV5* promoter control, which is cAMP-CRP independent, stimulated transcription about sevenfold but did not stimulate transcription in a *cya* mutant background (cAMP negative). These results suggest that cAMP-CRP stimulates *papBA* transcription independently of any affects on *papI* transcription. Because the CRP-binding site is about 50 bp from the GATC-I site, it is possible that binding of cAMP-CRP might increase the affinity of Lrp-PapI for this region or modulate the spatial orientation of Lrp-PapI by its DNA-bending attribute. These hypotheses need to be tested.

In addition to Lrp and CRP, a third global regulatory protein, H-NS, has been reported to regulate *pap* transcription by an unknown mechanism denoted "silencing." The introduction of *hns* mutations inhibits *pap* thermoregulation, in which *pap* pilin transcription is greatly reduced at temperatures below 26°C (24). In addition, it has been reported that although *pap* transcription in CRP⁻ and PapB⁻ mutant isolates is greatly reduced, introduction of an *hns* mutation (the *drdX* allele) into these strains restores *pap* transcription (19). On the basis of these results, it has been proposed that CRP and PapB act as "antirepressors" which in some way modulate the repressor activity of H-NS (19). If this hypothesis is correct, it would constitute a new role for CRP which does not involve any "positive" activity such as increasing the level of binding of RNA polymerase through direct contact. Moreover, it suggests the possibility that Lrp and PapI also might act solely as antirepressors which activate *pap* transcription by counteracting H-NS repression. A preliminary analysis by my colleagues and I (51) indicates that this is probably not the case since H-NS is not essential for Pap phase variation. Further studies need to be carried out to determine what role, if any, H-NS might play in *pap* gene regulation.

Lrp AND DNA METHYLATION CONTROL OTHER FIMBRIAL SWITCHES IN *E. COLI* AND *S. TYPHIMURIUM*

Lrp regulates a wide variety of fimbrial operons in *E. coli* (52, 53) and at least one fimbrial operon (*pef*) in *S. typhimurium* (20, 32). What is striking is the level of diversity regarding the types of regulation which Lrp carries out. Although many of the operons contain conserved GATC box sites with 102- to 103-bp spacing which serve as Lrp-binding domains, operons such as *fan* do not (Fig. 3). All of the operons that display phase variation except *fim* (encoding type 1 fimbriae) contain GATC box motifs as well as a *papI* homolog, and some of these operons are regulated by a methylation-dependent mechanism similar to that of the *pap* system (53). In the case of *fim*, Lrp probably binds near the *fim* switch and regulates type 1 fimbrial phase variation in a methylation-independent manner distinct from that of the *pap* system (6).

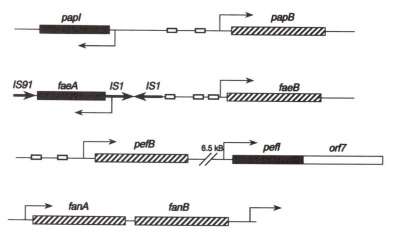

FIGURE 3. Regulatory regions of four Lrp-regulated *E. coli* fimbrial operons. The organization of the *pap* (10), *fae* (29), *pef* (20), and *fan* (30, 49) fimbrial regulatory regions are shown. Genes sharing sequence similarities with *papI* are denoted by dark shading, whereas genes with sequence similarities to *papB* are denoted by diagonal stripes. DNA sequences with similarity to the *pap* GATC-I box (CGATCTTTTAT) and GATC-II box (AAGATCGT) (53) are depicted as small open boxes. Transcription start sites are shown by arrows. The insertion sequences within *fae* are not drawn to scale. In the *pef* operon, ORF 7 overlaps the 3' end of the *pefI* gene.

The *fae* operon (coding for K88 fimbriae) contains GATC box motifs but does not display phase variation (29) (Fig. 3). Also, Lrp appears to inhibit *fae* transcription, in contrast to its role as a positive regulator in the other fimbrial operons analyzed. Unlike *pap, sfa, daa, prs,* and *pef,* the *fae* regulatory region contains an additional GATC box which is an inverted GATC box I (29) (Fig. 3). This third box is located near the *fae* promoter, suggesting the possibility that, in the presence of FaeA (the PapI homolog), Lrp-FaeA binds at this box and inhibits *pap* transcription. In addition, *fae* contains an IS*1* sequence upstream of the *fae* fimbrial promoter; that sequence appears to contribute to transcriptional control (29).

Recently, Robert Kadner and colleagues sequenced a large region from the 60-MDa plasmid of *S. typhimurium* and identified *pef* (plasmid-encoded fimbriae) DNA sequences with similarity to *pap* (20). Notably, *pef* contains conserved GATC boxes with the same spacing as *pap* as well as both *papI* (Fig. 3) and *papB* homologs. However, the *papI* gene is located almost 7 kb distant from the GATC box region (20). Although Pef fimbrial expression is dependent on Lrp (32), it is not known if Pef is subject to methylation-dependent phase variation. Comparison of PefI with other fimbrial PapI-like genes shows that Pef shares amino acid sequences with DaaF (3) and FaeA (29) that are not present in the other PapI homologs (Fig. 4, boxed regions). These results suggest the possibility that horizontal transfer between these genera may have occurred.

Currently, very little is known about structure-function relationships in the PapI homologs. Certain groups of amino acids such as YQ[A or V]R-YL are highly conserved, suggesting that they may be important for a function common to all proteins such as an Lrp-binding domain (Fig. 4). Although PapI has been shown to complement an *sfaC* mutant (23) and DaaF complemented a *papI* mutant (52), it is not yet clear if these regulatory proteins have identical activities with their own operons compared with their activities with other operons in the *pap* family.

CONCLUSIONS

The regulation of *pap* gene expression is complex and involves a variety of factors including Lrp, Dam, CRP, PapI, PapB, and possibly H-NS. In addition, RimJ, the N-terminal acetylase of ribosomal protein S5, may play a role in thermoregulation (58). As a member of the Lrp regulon, Pap fimbrial regulation is tied to cellular metabolism, although it is not leucine responsive (Fig. 5). Under the CRP regulon, Pap fimbrial expression is shut down when rich carbon sources such as glucose are present. Thus, Lrp and CRP constitute "modulons," defined as a family of target operons under the control of multiple global regulatory proteins (31). In addition, Pap is one of 30 to 40 genes in *E. coli* that contain nonmethylated GATC sites (47), placing it under the control of Dam. Notably, nonmethylated GATC sites appear to overlap CRP-binding sites in the *mtl, cdd, flh,* and *gut* operons, suggesting the possibility that these operons may, like *pap,* be regulated by DNA methylation patterns (54) (Fig. 5). Placement of genes under methylation control may function as a type of cellular memory since methylation patterns

```
      0         10        20        30        40        50        60        70        80        90
cons  ....mkneilefln.Rhnggk--------Taeiªeal.v.dYQaRyYLllªekeg.vqrsplrrGmat.Wflkge.qagqscsstt*
PapI  M.SEYMKNEILEFLN-RHDGGK--------TAEIAEALAVTDYQARYYLLLLEKAGMVQRSPLRRGMATYWFLKGEKQAGQSCSSTT*
SfaC  ....MQNEIMGFLS-RHNVGK--------TAEIAEALAVTDYQARYYLLLLLEKEGMVQRSPLRRGMATYWTLKGEMQAGQSCSSTT*
PrsI  MASEYMKNEILEFLN-RHNGGK--------TAEIAEALAVTDYQARYYLLLLLEKEGMVQRSPLRRGMAT-WFLKGEMQAGQNCSSTT*
DaaF  ....MKINKLT-LNªRKNDILSYFGEINAªCRTSEVAEHI GVSAYQARHYLQCLEKEGKIKRSPVRRGASTªWEISSIPP*
PefI  M.SESIVTKIISIVQªRQNMDD------GAªVKTRDIAªAªGISIYQVRLYLEQLHDVGVLEKVNAGKGVPGªWRIIª*
FaeA  ....MKEHILEYINSRGSCT--------IREIAªATGISAYQARYYLMTLDREKKIRRTPLRQGARTªWGVLREK*
```

FIGURE 4. Comparative amino acid analysis of PapI-related DNA sequences. The amino acid sequences of PapI (10), SfaC (25), PrsI (39), DaaF (3), PefI (21), and FaeA (29) are shown. Dashes represent gaps introduced for alignment purposes. On the top row, cons denotes total consensus at a given amino acid position among all six regulatory proteins. Amino acids that are conserved among all six PapI-like proteins are shown in boldface type. Lowercase letters indicate that at least three sequences contain the same amino acid at a single position. Boxed regions show identities between either PefI and DaaF or PefI and FaeA or identities among PefI, DaaF, and FaeA that are not present in PapI, SfaC, or PrsI sequences.

FIGURE 5. Regulatory networks affecting the *pap* operon. Large circles represent the operons controlled within each regulatory network or regulon, and numbers denote the various classes of regulatory control. The operons *pap*, *daa*, and *sfa* are members of class 2 since they are controlled by Lrp, CRP, and DNA methylation patterns. The *fan* operon belongs to class 3 since it is not regulated by Dam. The *mtl*, *cdd*, *flh*, and *gut* operons (54) are possible members of class 1, although it is not known if any of these operons are regulated by Lrp or Dam. Finally, my colleagues and I (26) recently identified a gene located at 77 min on the *E. coli* chromosome; the gene contains a nonmethylated GATC site dependent on the presence of Lrp, possibly a member of class 4.

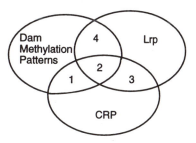

are heritable. It seems likely that by linking Pap fimbrial expression to multiple regulatory systems that respond to extracellular signals, cells optimize their responses to intra- and extra-host environments. Future research should provide a more detailed view of how these responses are coordinated and what role(s) they might play in pathogenesis.

ACKNOWLEDGMENTS. I am grateful to Marjan van der Woude, Linda Kaltenbach, and Bruce Braaten for help in preparing some of the figures presented here and for helpful discussions. I also thank my colleagues S. Bilge, I. Blomfield, J. Calvo, B. Eisenstein, F. de Graaf, J. Hacker, T. Huisman, R. Isaacson, R. Kadner, R. Matthews, S. Moseley, P. Orndorff, and B. E. Uhlin for graciously communicating their unpublished results and work in press. I also express my gratitude to the National Institutes of Health (grant 2 RO1 AI23348) and National Science Foundation (grant MCB-9305166) for continuing support. Lastly, I thank Stan Falkow for my initiation into bacterial pathogenesis and for a chance to walk a mile in the bacterium's shoes.

REFERENCES

1. **Anderson, J. J., S. C. Quay, and D. L. Oxender.** 1976. Mapping of two loci affecting the regulation of branched-chain amino acid transport in *Escherichia coli* K-12. *J. Bacteriol.* **126**:80–90.
2. **Baåga, M., M. Goransson, S. Normark, and B. E. Uhlin.** 1985. Transcriptional activation of a *pap* pilus virulence operon from uropathogenic *Escherichia coli*. *EMBO J.* **4**:3887–3893.
3. **Bilge, S. S., J. M. M. Apostol, J. J. Fullner, and S. L. Moseley.** 1993. Transcriptional organization of the F1845 fimbrial adhesin determinant of *Escherichia coli*. *Mol. Microbiol.* **7**:993–1006.
4. **Bilge, S. S., C. R. Clausen, W. Lau, and S. L. Moseley.** 1989. Molecular characterization of a fimbrial adhesin, F1845, mediating diffuse adherence of diarrhea-associated *Escherichia coli* to HEp-2 cells. *J. Bacteriol.* **171**:4281–4289.
5. **Bloch, C. A., B. A. D. Stocker, and P. E. Orndorff.** 1992. A key role for type 1 pili in enterobacterial communicability. *Mol. Microbiol.* **6**:697–701.
6. **Blomfield, I.** Personal communication.
7. **Blomfield, I. C., P. J. Calie, K. J. Eberhardt, M. S. McClain, and B. I. Eisenstein.** 1993. Lrp stimulates phase variation of type 1 fimbriation in *Escherichia coli* K-12. *J. Bacteriol.* **175**:27–36.
8. **Blomfield, I. C., M. S. McClain, J. A. Princ, P. J. Calie, and B. I. Eisenstein.** 1991. Type 1 fimbriation and *fimE* mutants of *Escherichia coli* K-12. *J. Bacteriol.* **173**:5298–5307.
9. **Blyn, L. B., B. A. Braaten, and D. A. Low.** 1990. Regulation of *pap* pilin phase variation by a mechanism involving differential Dam methylation states. *EMBO J.* **9**:4045–4054.
10. **Blyn, L. B., B. A. Braaten, C. A. White-Ziegler, D. H. Rolfson, and D. A. Low.** 1989. Phase-variation of pyelonephritis-associated pili in *Escherichia coli*: evidence for transcriptional regulation. *EMBO J.* **8**:613–620.
11. **Braaten, B. A., L. B. Blyn, B. S. Skinner, and D. A. Low.** 1991. Evidence for a methylation-

blocking factor (*mbf*) locus involved in *pap* pilus expression and phase variation in *Escherichia coli. J. Bacteriol.* **173:**1789–1800.

12. **Braaten, B. A., X. Nou, L. S. Kaltenbach, and D. A. Low.** 1994. Methylation patterns in *pap* regulatory DNA control the pyelonephritis-associated pili phase variation in *Escherichia coli. Cell* **76:**577–588.

13. **Braaten, B. A., J. V. Platko, M. W. van der Woude, B. H. Simons, F. K. de Graaf, J. M. Calvo, and D. A. Low.** 1992. Leucine-responsive regulatory protein controls the expression of both the *pap* and *fan* pili operons in *Escherichia coli. Proc. Natl. Acad. Sci. USA* **89:**4250–4254.

14. **Calvo, J. M., and R. G. Matthews.** The leucine-responsive regulatory protein (Lrp), a global regulator of metabolism in *Escherichia coli. Microbiol. Rev.,* in press.

15. **Eisenstein, B. I.** 1981. Phase variation of type 1 fimbriae in *Escherichia coli* is under transcriptional control. *Science* **214:**347–349.

16. **Eisenstein, B. I.** 1982. Operon fusion of the phase variation switch. A virulence factor in *Escherichia coli. Infection* **10:**112–115.

17. **Eisenstein, B. I., D. S. Sweet, V. Vaughn, and D. I. Friedman.** 1987. Integration host factor is required for the DNA inversion that controls phase variation in *Escherichia coli. Proc. Natl. Acad. Sci. USA* **84:**6506–6510.

18. **Forsman, K., M. Gorannson, and B. E. Uhlin.** 1989. Autoregulation and multiple DNA interactions by a transcriptional regulatory protein in *E. coli* pili biogenesis. *EMBO J.* **8:**1271–1277.

19. **Forsman, K., B. Sonden, M. Goransson, and B. E. Uhlin.** 1992. Antirepression function in *Escherichia coli* for the cAMP-cAMP receptor protein transcriptional activator. *Proc. Natl. Acad. Sci. USA* **89:**9880–9884.

20. **Friedrich, M. J., N. E. Kinsey, J. Vila, and R. J. Kadner.** 1993. Nucleotide sequence of a 13.9 kb segment of the 90 kb virulence plasmid of *Salmonella typhimurium:* the presence of fimbrial biosynthetic genes. *Mol. Microbiol.* **8:**543–558.

21. **Gally, D. L., J. A. Bogan, B. I. Eisenstein, and I. C. Blomfield.** 1993. Environmental regulation of the *fim* switch controlling type 1 fimbrial phase variation in *Escherichia coli* K-12: effects of temperature and media. *J. Bacteriol.* **175:**6186–6193.

22. **Gorannson, M., K. Forsman, P. Nilsson, and B. E. Uhlin.** 1989. Upstream activating sequences that are shared by two divergently transcribed operons mediate cAMP-CRP regulation of a pilus-adhesin is *Escherichia coli. Mol. Microbiol.* **3:**1557–1565.

23. **Gorannson, M., K. Forsman, and B. E. Uhlin.** 1988. Functional and structural homology among regulatory cistrons of pili-adhesin determinants in *Escherichia coli. Mol. Gen. Genet.* **212:**412–417.

24. **Gorannson, M., B. Sonden, P. Nilsson, B. Dagberg, K. Forsman, K. Emanuelsson, and B. E. Uhlin.** 1990. Transcriptional silencing and thermoregulation of gene expression in *Escherichia coli. Nature* (London) **344:**682–685.

25. **Hacker, J.** 1990. Genetic determinants coding for fimbriae and adhesins of extraintestinal *Escherichia coli. Curr. Top. Microbiol. Immunol.* **151:**1–27.

26. **Hale, W. B., M. van der Woude, and D. A. Low.** Analysis of nonmethylated GATC sites in the *Escherichia coli* chromosome and identification of sites that are differentially methylated in response to environmental stimuli. *J. Bacteriol.,* in press.

27. **Hanisch, F. G., J. Hacker, and H. Schroten.** 1993. Specificity of S fimbriae on recombinant *Escherichia coli:* preferential binding to gangliosides expressing NeuGcα(2-3)Gal and NeuAcα(2-8)NeuAc. *Infect. Immun.* **61:**2108–2115.

28. **Hanson, M. S., and C. Brinton, Jr.** 1988. Identification and characterization of the *E. coli* type-1 pilus tip adhesion protein. *Nature* (London) **332:**265–268.

29. **Huisman, T. T., D. Bakker, P. Klaasen, and F. K. de Graaf.** 1994. Leucine-responsive regulatory protein, IS1 insertions, and the negative regulator FaeA control the expression of the fae (K88) operon in *Escherichia coli. Mol. Microbiol.* **11:**525–536.

30. **Inoue, O. J., J. H. Lee, and R. E. Isaacson.** 1993. Transcriptional organization of the *Escherichia coli* pilus adhesin K99. *Mol. Microbiol.* **10:**607–613.

31. **Iuchi, S., and E. C. C. Lin.** 1988. ArcA (dye), a global regulatory gene in *Escherichia coli* mediating repression of enzymes in areobic pathways. *Proc. Natl. Acad. Sci. USA* **85:**1888–1892.

32. **Kadner, R.** Personal communication.

33. **Kaltenbach, I.** Unpublished data.

34. Kawula, T. H., and P. E. Orndorff. 1991. Rapid site-specific DNA inversion in *Escherichia coli* mutants lacking the histone-like protein H-NS. *J. Bacteriol.* **173:**4116–4123.

35. Klemm, P. 1986. Two regulatory genes fim genes, *fimB* and *fimE,* control the phase variation of type 1 fimbriae in *Escherichia coli. EMBO J.* **5:**1389–1393.

36. Lin, R. T., R. D'Ari, and E. B. Newman. 1990. The leucine regulon of *E. coli* K-12: a mutation in *rblA* alters expression of L-leucine-dependent metabolic operons. *J. Bacteriol.* **172:**4529–4535.

37. Lindberg, F., B. Lund, L. Johansson, and S. Normark. 1987. Localization of the receptor-binding protein adhesin at the tip of the bacterial pilus. *Nature* (London) **328:**84–87.

38. Lund, B., B. I. Marklund, B.-I. Stromberg, F. Lindberg, K. A. Karlsson, and S. Normark. 1988. Uropathogenic *Escherichia coli* can express serologically identical pili of different receptor binding specificities. *Mol. Microbiol.* **2:**225–263.

39. Marklund, B. I., J. M. Tennent, E. Garcia, A. Hamers, M. Baåga, F. Lindberg, W. Gaastra, and S. Normark. 1992. Horizontal gene transfer of the *Escherichia coli pap* and *prs* operons as a mechanism for the development of tissue-specific adhesive properties. *Mol. Microbiol.* **6:** 2225–2242.

40. McClain, M. S., I. C. Blomfield, and B. I. Eisenstein. 1991. Roles of *fimB* and *fimE* in site-specific DNA inversion associated with phase variation of type 1 fimbriae in *Escherichia coli. J. Bacteriol.* **173:**5308–5314.

41. Nagy, B., H. Moon, R. Isaacson, C. C. To, and C. C. Brinton. 1978. Immunization of suckling pigs against enteric enterotoxigenic *Escherichia coli* infection by vaccinating dams with purified pili. *Infect. Immun.* **21:**269–274.

42. Nou, X., B. Skinner, B. Braaten, L. Blyn, D. Hirsh, and D. Low. 1993. Regulation of pyelonephritis-associated pili phase variation in *Escherichia coli:* binding of the PapI and Lrp regulatory proteins is controlled by DNA methylation. *Mol. Microbiol.* **7:**545–553.

43. O'Hanley, P., D. Low, I. Romero, D. Lark, K. Vosti, S. Falkow, and G. Schoolnik. 1985. Gal-Gal binding and hemolysin phenotypes and genotypes associated with uropathogenic *Escherichia coli. N. Engl. J. Med.* **313:**414–447.

44. Platko, J. V., and J. M. Calvo. 1993. Mutations affecting the ability of *Escherichia coli* Lrp to bind DNA, activate transcription, or respond to leucine. *J. Bacteriol.* **175:**1110–1117.

45. Platko, J. V., D. A. Wilins, and J. M. Calvo. 1990. The *ilvIH* operon of *Escherichia coli* is positively regulated. *J. Bacteriol.* **172:**4563–4570.

46. Rex, J. H., B. D. Aronson, and R. L. Somerville. 1991. The *tdh* and *serA* operons of *Escherichia coli:* mutational analysis of the regulatory elements of leucine-responsive genes. *J. Bacteriol.* **173:** 5944–5993.

47. Rinquist, S., and C. L. Smith. 1992. The *Escherichia coli* chromosome contains specific, unmethylated *dam* and *dcm* sites. *Proc. Natl. Acad. Sci. USA* **89:**4539–4543.

48. Rodriguez-Ortega, M., I. Ofek, and N. Sharon. 1987. Membrane glycoproteins of human polymorphonuclear leukocytes that act as receptors for mannose-specific *Escherichia coli. Infect. Immun.* **55:**968–973.

49. Roosendaal, E., M. Boots, and F. K. de Graaf. 1987. Two novel genes, *fanA* and *fanB,* involved in the biogenesis of K99 fimbriae. *Nucleic Acids Res.* **15:**5973–5984.

50. Selander, R. K., T. K. Korhonen, V. Vaisanen-Rhen, P. H. Williams, P. E. Pattison, and D. A. Caugant. 1986. Genetic relationships and clonal structure of strains of *Escherichia coli* causing neonatal septicemia and meningitis. *Infect. Immun.* **52:**213–222.

51. van der Woude, M., L. Kaltenbach, and D. A. Low. Unpublished data.

52. van der Woude, M., and D. A. Low. 1994. Leucine-responsive regulatory protein and deoxyadenosine methylase control the phase variation and expression of the *sfa* and *daa* pili operons in *Escherichia coli. Mol. Microbiol.* **11:**605–618.

53. van der Woude, M. J., B. A. Braaten, and D. A. Low. 1992. Evidence for global regulatory control of pilus expression in *Escherichia coli* by Lrp and DNA methylation: model building based on analysis of *pap. Mol. Microbiol.* **6:**2429–2435.

54. Wang, M. X., and G. M. Church. 1992. A whole genome approach to in vivo DNA-protein interactions in *E. coli. Nature* (London) **360:**606–610.

55. Wang, Q., and J. M. Calvo. 1993. Lrp, a global regulatory protein of *Escherichia coli,* binds cooperatively to multiple sites and activates transcription of *ilvIH. J. Mol. Biol.* **229:**306–318.

56. **Wang, Q., M. Sacco, E. Ricca, C. T. Lago, M. DeFelice, and J. M. Calvo.** 1993. Organization of Lrp binding sites upstream of *ilvIH* in *Salmonella typhimurium. Mol. Microbiol.* **7:**883–891.

57. **Westerlund, B., I. van Die, C. Kramer, P. Kuusela, H. Holthofer, A.-M. Tarkkhanen, R. Virkola, N. Riegman, H. Bergmans, W. Hoekstra, and T. K. Korhonen.** 1991. Multifunctional nature of P fimbriae of uropathogenic *Escherichia coli:* mutations in *fsoE* and *fsoF* influence fimbrial binding to renal tubuli and immobilized fibronectin. *Mol. Microbiol.* **5:**2965–2975.

58. **White-Ziegler, C. A., and D. A. Low.** 1992. Thermoregulation of the *pap* operon: evidence for the involvement of RimJ, the N-terminal acetylase of ribosomal protein S5. *J. Bacteriol.* **174:**7003–7012.

Molecular Genetics of Bacterial Pathogenesis
Edited by V. L. Miller, J. B. Kaper, D. A. Portnoy, and R. R. Isberg
© 1994 American Society for Microbiology, Washington, DC 20005

Chapter 29

The Tao of Urease

Carleen M. Collins

Urease catalyzes the hydrolysis of urea to carbonic acid and two molecules of ammonia, and thus generates the preferred nitrogen source of many microorganisms (34, 62). In addition to its role in nitrogen utilization and assimilation, urease is a virulence determinant when expressed by certain bacterial pathogens. Urease plays a significant role in bacterial infections of both the urinary tract (5, 32) and the gastroduodenal region (4). Urinary tract infections (UTIs) with a urease-producing pathogen are more difficult to clear and are associated with greater tissue damage than if a nonurease producer is the infecting agent (26, 40, 41). For *Helicobacter pylori,* a gastroduodenal pathogen, urease is necessary for survival, and urease-negative *H. pylori* mutants are not able to colonize in an animal model system (22). Most, if not all, of the increase in the level of injury and the persistence of infection associated with urease is due to the production of ammonia. Therefore, urease is primarily a metabolic enzyme, but it can facilitate and enhance the virulence of a pathogen when expressed in the appropriate environment. Accordingly, I consider this enzyme an opportunistic virulence determinant. Although opportunistic in this sense, urease is a potent virulence determinant, and production of the enzyme can result in significant injury to an infected individual.

Only in recent years have bacterial ureases and the genes that encode them been characterized. A primary reason for the current interest in urease is the association of the enzyme with bacterial pathogenesis. This chapter describes the role of bacterial urease in infection, the structure of urease, and the genetic organization and regulation of urease genes.

UREASE AS A VIRULENCE FACTOR IN URINARY TRACT INFECTIONS

When urease is produced by uropathogens in an infected urinary tract, the hydrolysis of urea increases the ammonia concentration in the urine, and there is a subsequent elevation of the urine pH (49). The elevated urine pH can lead to a number of effects, the most damaging of which is the production of a urinary

Carleen M. Collins • Department of Microbiology and Immunology, University of Miami School of Medicine, Miami, Florida 33101.

Ascending urinary tract infection with urease-producing bacteria:

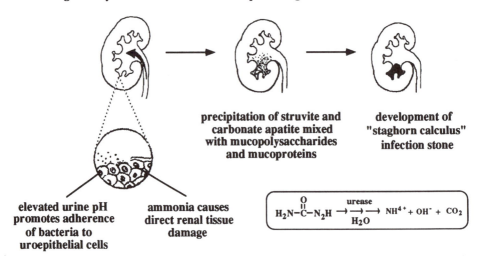

precipitation of struvite and
carbonate apatite mixed
with mucopolysaccharides
and mucoproteins

development of
"staghorn calculus"
infection stone

elevated urine pH
promotes adherence
of bacteria to
uroepithelial cells

ammonia causes
direct renal tissue
damage

$$H_2N-\overset{\overset{\text{O}}{\|}}{C}-N_2H \xrightarrow[H_2O]{\text{urease}} NH^{4+} + OH^- + CO_2$$

FIGURE 1. Role of urease in the formation of struvite stones. Adapted from Griffith and Osborne (33). Reprinted with permission of S. Karger AG, Basel.

stone or calculus (Fig. 1) formed by the precipitation of magnesium ammonium phosphate salts (struvite) (32). Stone formation can cause urinary tract obstruction and interfere with voiding, thereby making it more difficult to clear the infecting organism from the urinary tract (49, 64). Stones can also harbor the infecting bacteria in a protected site, which in many instances reduces the effectiveness of antibiotic treatment (23). Struvite precipitates can be found in the renal pelvis and the bladder and can be found encrusted on urinary catheters.

Prolonged survival of ureolytic bacteria in the urinary tract is seen even in the absence of stone formation. Recent studies have examined the virulences of isogenic pairs of urease-positive or urease-negative strains of *Proteus mirabilis* (40, 41) and *Staphylococcus saprophyticus* (26) in rodent animal model systems. Those studies indicate that UTIs with urease-producing organisms are increased in duration; have greater numbers of bacteria colonizing the bladder, urine, and kidney; and result in more damage to the kidney tissue. In addition to investigations with urease-negative mutants, the role of urease in UTIs has been defined by treating infections in rodents with urease inhibitors, such as acetohydroxamic acid (31, 67, 88) or flurofamide (59). Treatment of UTIs with urease inhibitors decreases bacterial invasion of the kidneys, the severity of the kidney lesions, and the incidence of bacteremia resulting from infection.

The exact mechanism responsible for the increased level of survival of the pathogen and the damage associated with urease can only be inferred. Increased bacterial colonization of the urinary tract might result from the generation of an easily assimilated nitrogen source and/or from alkalinization of the urine to a more favorable pH for growth (57). Ammonia can inactivate the fourth component of complement, and thus might prevent the immune system from functioning efficiently (1). Tissue damage associated with the infection can result either from direct damage of the renal cells by ammonia (51) or, possibly, by an indirect mechanism such as those associated with *H. pylori* infection (see below).

Frequently isolated ureolytic uropathogens include *P. mirabilis, Klebsiella pneumoniae, Providencia stuartii,* and *S. saprophyticus* (51). These organisms are associated with both community- and hospital-acquired infections. Patients with chronic indwelling catheters are most at risk for infection by urease-producing organisms, and the infections can be debilitating because of obstruction of the urinary catheter and the risk of subsequent bacteremia (16).

UREASE AS A FACTOR IN DEVELOPMENT OF CHRONIC GASTRITIS AND PEPTIC ULCERS

Urease is produced in abundance by *H. pylori,* a gram-negative curved or spiral bacterium that lives in the mucous layer overlaying the gastric epithelium (4). The organism can colonize the stomach for years or decades, causing chronic inflammation that may result in chronic active gastritis and gastric and duodenal ulcers. There is epidemiological evidence linking *H. pylori* with gastric carcinoma (71, 72).

Urease is a crucial virulence determinant in these infections; ammonia resulting from the hydrolysis of urea presumably buffers the surrounding acidic gastric environment, resulting in a pH that is more favorable for *H. pylori* growth and survival (56). Urease has been shown to be essential for colonization of *H. pylori* in a gnotobiotic piglet model of infection (22). Ammonia production is believed to injure the gastric mucosa both by direct action on mucosal cells (79, 84) and by altering the surroundings of the gastric epithelium such that the normal passage of hydrogen ions is blocked and there is an increase in hydrogen ion back-diffusion (36). *H. pylori*-infected gastric mucosa is characterized by a predominant neutrophil accumulation, and ammonia is thought to enhance a neutrophil-dependent cytocidal activity (86). In addition, urease itself has been shown to possess chemotactic activity for leukocytes, resulting in leukocyte-dependent damage of the surrounding area (54).

OTHER EXAMPLES OF UREASE-PRODUCING PATHOGENS

Urease has been implicated in the pathogenesis of reactive arthritis, a sterile inflammatory disease that can follow gastrointestinal or urogenital infection by *Yersinia enterocolitica* and other gram-negative organisms. The β subunit of urease is thought to be a target for the synovial T-cell response of patients with reactive arthritis (74). In addition, the β subunit of *Y. enterocolitica* urease can induce arthritis after intra-articular injection into rats (58, 83). The exact mechanism by which urease facilitates the development of this disease is unclear at this time.

An interesting urease-producing organism is the urogenital pathogen *Ureaplasma urealyticum. U. urealyticum* inhabits the human lower genital tract and is associated with the development of nongonococcal urethritis and chorioamnionitis. Both clinical and experimental studies indicate that *U. urealyticum* infection of the urinary tract can result in the development of struvite stones (30, 73, 87). It is not known if urease plays a role in *U. urealyticum* infections at sites other than the urinary tract. The urease of *U. urealyticum* is unique in that it is essential for the survival of the bacterium (44). The action of urease is thought to increase the intracellular ammonium ion concentration to produce a transmembrane potential, which in turn drives the synthesis of ATP (76).

HETEROMERIC SUBUNIT STRUCTURE OF BACTERIAL UREASES

Urease has been purified from a number of pathogenic bacterial species including *Klebsiella aerogenes* (90), *P. mirabilis* (6), *P. stuartii* (66), *H. pylori* (21, 38), and *U. urealyticum* (85, 89). With the exception of *H. pylori* urease, the bacterial ureases studied thus far are composed of three distinct subunits, termed α, β, and γ. The subunits differ in size between the bacterial species: α subunits vary between 60 and 72 kDa, β subunits vary between 10 and 14 kDa, and γ subunits vary between 8 and 11 kDa. The purified enzyme from *K. aerogenes* is the most extensively studied of the bacterial ureases. In *K. aerogenes* the three subunits associate in an apparent $(\alpha_1\beta_2\gamma_2)_2$ stoichiometry to form a mature enzyme of 211 kDa (90). All bacterial ureases except the *H. pylori* urease are cytoplasmic enzymes (14, 42, 63). *H. pylori* urease differs from other bacterial ureases in that it is composed of only two subunits, with molecular masses of 62 and 27 kDa, and it is associated with the bacterial cell membrane (21, 35, 38).

The primary amino acid sequences of each of the bacterial urease subunits display high degrees of similarity to one another. When the various proteins are compared, the α subunit from *K. aerogenes* urease and the 62-kDa subunit from *H. pylori* urease align with one another, while the γ and β subunits from *K. aerogenes* urease and the 27-kDa subunit from *H. pylori* urease align with each other. Therefore, although the enzymes differ in the number of subunits, the amino acid compositions of urease are highly conserved between these species.

UREASE GENE CLUSTERS OF THE FAMILY *ENTEROBACTERIACEAE*

Each of the urease gene clusters of *P. mirabilis* (43), *K. aerogenes* (65), *P. stuartii* (the plasmid-encoded locus) (19), and *Escherichia coli* (18) contains seven genes necessary for the synthesis of enzymatically active urease. The relative positions of these genes are similar in each species. The seven genes are located immediately adjacent to each other (Fig. 2). *ureA, ureB,* and *ureC* encode the structural subunits γ, β, and α, respectively. The structural genes are highly conserved among the various urease gene clusters, and their DNA sequences are at least 60% identical.

ureD, ureE, ureF, and *ureG* encode urease accessory polypeptides. The amino acid sequences of the accessory polypeptides are not as conserved between the species as are the sequences of the structural polypeptides (Fig. 3). In *K. aerogenes* the *ureD, ureE, ureF,* and *ureG* gene products have molecular sizes

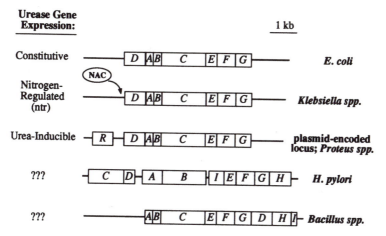

Urease Gene Expression:

1 kb

Constitutive — D A B C E F G — *E. coli*

NAC

Nitrogen-Regulated (ntr) — D A B C E F G — *Klebsiella spp.*

Urea-Inducible — R D A B C E F G — plasmid-encoded locus; *Proteus spp.*

??? — C D A B I E F G H — *H. pylori*

??? — A B C E F G D H I — *Bacillus spp.*

FIGURE 2. Maps of various bacterial urease gene clusters. Shown are the relative sizes and orientations of the chromosomal *E. coli* urease genes (10), *K. pneumoniae* and *K. aerogenes* urease genes (11, 65), the plasmid-encoded and *P. mirabilis* urease genes (19, 20, 43), *H. pylori* urease genes (45), and *Bacillus* sp. strain TB-90 urease genes (52). Open boxes indicate the positions of each gene (*ure*). The oval represents Nac, the nitrogen assimilation control protein (2, 3). The arrow points to the Nac-binding site, located approximately 90 nucleotides upstream of *ureD* in *K. pneumoniae* (11, 28).

of 29.8, 17.6, 25.2, and 21.9 kDa, respectively (65). The exact functions of the *ureD, ureE, ureF,* and *ureG* gene products are not known, although evidence indicates that each is involved in the incorporation of nickel by urease apoenzyme (48). All ureases studied thus far require nickel for enzymatic activity, and nickel ions are thought to have an essential role in the catalytic mechanism of the enzyme (17, 34).

FIGURE 3. Similarity between the amino acid sequences of UreA (structural subunit) and UreD (accessory polypeptide) from various bacterial species. A dendrogram representation of the clustering relationships between the amino acid sequences is shown. The horizontal branch lengths are proportional to the similarities between the sequences. A pairwise alignment of the sequences was performed by the algorithm of Needleman and Wunsch (68) by using a re-scaled version of the scoring matrix of Dayhoff et al. (13), as found in the Genetics Computer Group Sequence Analysis Software package, version 7.0.2 (15). The similarity scores obtained were used to create a clustering order by the unweighted pair-group method by using arithmetic averages (81). For *H. pylori* the first 101 amino acids of UreA were analyzed. *H. pylori* UreH is thought to be equivalent to UreD of the other clusters.

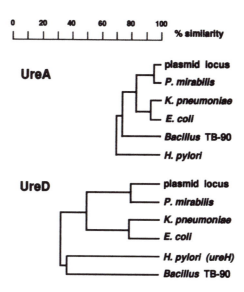

0   20   40   60   80   100   % similarity

UreA
— plasmid locus
— *P. mirabilis*
— *K. pneumoniae*
— *E. coli*
— *Bacillus* TB-90
— *H. pylori*

UreD
— plasmid locus
— *P. mirabilis*
— *K. pneumoniae*
— *E. coli*
— *H. pylori* (ureH)
— *Bacillus* TB-90

UreE produced by *K. aerogenes* is a cytoplasmic protein capable of binding nickel (48). It has been suggested that UreE binds intracellular nickel and acts as a nickel donor during assembly of urease. Purified urease preparations from strains with a mutation in *ureE* contain approximately 50% of the wild-type levels of incorporated Ni (47).

The urease produced by strains containing mutations in either *ureD*, *ureF*, or *ureG* has less than 10% of the Ni content of wild-type urease (47). Acquisition of nickel by urease is energy dependent (46). UreG contains a nucleotide-binding motif and therefore might be involved in the energy-dependent incorporation of Ni into urease (47). It is interesting to note that the primary amino acid sequence of UreG has identity with HypB, a member of the hydrogenase system of *E. coli* (47, 55). Similar to urease, *E. coli* hydrogenases are Ni metalloenzymes (75). HypB binds GTP and is required for nickel incorporation into hydrogenases (55). These data suggest that HypB and UreG might perform similar functions in nickel metallocenter biosynthesis for the respective enzymes. However, a mutation in *hypB* cannot be complemented by *K. pneumoniae ureG* (55). There are few clues as to the function of the *ureD* and *ureF* gene products. In *Klebsiella* species both are apparently made at levels lower than the other gene products and are difficult to visualize on polyacrylamide gels after in vitro transcription and translation experiments (7, 47).

UREASE GENE CLUSTER OF *H. PYLORI*

The genetic organization of *H. pylori* urease genes differs from that found in members of the family *Enterobacteriaceae* (Fig. 2). The complete DNA sequence of the *H. pylori* urease gene cluster, which is composed of nine genes (*ureC, ureD, ureA, ureB, ureI, ureE, ureF, ureG,* and *ureH*) has been determined (12, 38, 45). *H. pylori* urease contains two structural genes: *ureA*, which encodes the 27-kDa subunit, and *ureB*, which encodes the 62-kDa subunit (21, 38, 45). *ureA* of *H. pylori* is likely to have resulted from a fusion of ancestral genes represented by *ureA* and *ureB* of the *Enterobacteriaceae*. The 5' half of *H. pylori ureA* has identity with *ureA* found in the *Enterobacteriaceae*, and the 3' half of *H. pylori ureA* has identity with *ureB* from the *Enterobacteriaceae*. *H. pylori ureB* corresponds to the *ureC* found in *Enterobacteriaceae* gene clusters.

It is difficult to assess the exact role of the other seven *H. pylori* urease genes. In most studies, *H. pylori* urease genes have been expressed either in *Campylobacter jejuni* or in *E. coli*. *C. jejuni* can be transformed to a urease-positive phenotype by the introduction of *H. pylori ureC, ureD, ureA*, and *ureB* genes. The *H. pylori ureE, ureI, ureF, ureG*, and *ureH* genes are not required for production of the enzyme in *C. jejuni* (45). In contrast, *ureA, ureB, ureE, ureF, ureG*, and *ureH*, but not *ureC* or *ureD*, must be present for the synthesis of active urease in *E. coli* (12, 38). *H. pylori ureE, ureF*, and *ureG* display some identity with *ureE, ureF*, and *ureG* of the *Enterobacteriaceae* urease gene clusters and may encode proteins with similar functions. There is also a limited degree of identity between the predicted amino acid sequence of the *ureH* gene product from *H. pylori* and the *ureD* gene product from the *Enterobacteriaceae*, indicating that these two genes may represent homologs of each other.

UREASE GENES OF GRAM-POSITIVE ORGANISMS

Urease genes have been isolated from two gram-positive organisms: *S. saprophyticus* (27) and the thermophilic *Bacillus* sp. strain TB-90 (52). While the genetic organization of the staphylococcal cluster has not been determined, the complete DNA sequence of the *Bacillus* urease locus is known. The *Bacillus* urease gene cluster differs from the previously described clusters (Fig. 2). Similar to the enteric organisms, three structural genes (*ureA, ureB,* and *ureC*) are present. *ureD*, which encodes an accessory polypeptide, is located downstream of *ureG* in the *Bacillus* species, which is similar to its location in *H. pylori* (as mentioned above, the presumed *ureD* homolog in *H. pylori* is *ureH*) and different from the position in *Klebsiella* or *Proteus* species. The *Bacillus* species urease gene cluster appears to code for more accessory polypeptides than are found in *Klebsiella* and *Proteus* species urease gene clusters. The functions of these additional accessory polypeptides are not known at this time.

UREA-INDUCIBLE UREASE GENE EXPRESSION: POSITIVE REGULATION BY UreR

Urease in *P. mirabilis* (42), *P. vulgaris* (61, 82), and *P. stuartii* (20, 63) is produced only when urea is present in the growth medium. The urease genes of *P. mirabilis* are found on the chromosome, while those of *P. stuartii* are encoded by large plasmids (82 to 230 kb) (29, 60). The plasmid-borne urease genes have been found in some *E. coli* and *Salmonella* isolates, and this urease locus has been termed the "plasmid-encoded urease gene cluster" (8, 20).

Expression of the plasmid-encoded urease genes appears to be regulated by a transcriptional activator termed UreR (20). The plasmid-encoded UreR is a 34.1-kDa protein (Table 1) that contains a helix-turn-helix motif and that shows significant amino acid similarity to the AraC family of transcriptional activators. The

TABLE 1
Urease gene transcriptional activators

UreR	Nac
34.1 kDa	32.9 kDa
AraC family	LysR family
Effector: urea	Effector: nitrogen starvation
Activates transcription at *ureR* and *ureD* promoters	Activates transcription at *ureD* promoter; *nac* transcription is dependent on NtrA and NtrC
Maps within urease locus	Maps outside urease locus
Found in *P. mirabilis* and plasmid-encoded gene cluster	Controls expression of *ure, hut,* and *put* in *Klebsiella* species

plasmid-encoded *ureR* is found 414 bp upstream of *ureD*, and the two genes are divergently transcribed (Fig. 2). Studies with *ureR-lacZYA* and *ureD-lacZYA* transcriptional fusions indicate that UreR is a positive regulator that activates transcription at the *ureD* and *ureR* promoters (18, 20). Thus, UreR regulates its own synthesis at the transcriptional level. Additional studies with *ureR-lacZYA* and *ureD-lacZYA* transcriptional fusions indicate that the *ureR* and *ureD* promoter regions reside in the approximate middle of the 414-bp fragment and that these promoters overlap (18). The chromosomal *P. mirabilis* urease gene cluster also contains *ureR*. *P. mirabilis* UreR is a 33.4-kDa protein with 70% deduced amino acid identity to the plasmid-encoded UreR (69). Future studies should indicate if the plasmid-encoded UreR and *P. mirabilis* UreR function in a similar manner.

It is not known at this time if UreR interacts directly with the effector molecule urea. However, it has been clearly established that the induction of urease genes in *P. mirabilis* occurs in specific response to urea. Nicholson et al. (70) showed that structural analogs of urea were unable to induce transcription from a *ureA-lacZYA* translational fusion. If UreR binds urea, it might occur via the amino-terminal portion of the protein, since with other members of the AraC family the effector molecule interacts with the amino-terminal region of the protein (25).

NITROGEN-REGULATED UREASE GENE EXPRESSION: POSITIVE REGULATION BY Nac

In *K. pneumoniae* and *K. aerogenes* synthesis of urease is induced when the nitrogen source of the growth medium is limiting. This suggests that urease gene expression in these species is controlled by the global nitrogen regulation system (ntr) (24). Mutations in two integral components of the ntr cascade, *ntrA* and *ntrC*, eliminate urease expression in these *Klebsiella* species (11, 50). *ntrA* codes for an alternative sigma factor, σ^{54}, and *ntrC* codes for a regulatory protein that can activate transcription at σ^{54}-dependent promoters (53). *K. aerogenes* contains an additional member of the nitrogen regulation system termed *nac* (nitrogen assimilation control) (2, 3) (Table 1). *K. aerogenes nac* mutants exhibit substantially less urease activity than the wild-type *K. aerogenes* strain (the decrease in urease activity is from 1.5- to 16-fold, depending upon the mutant) (50). Nac is a 32-kDa protein that is similar to the LysR family of transcriptional activators (2, 78) (Table 1). *nac* maps to a site near the *his* operon in *K. aerogenes* and is transcribed from an *ntrA*- and *ntrC*-dependent promoter (3, 78).

A nitrogen-regulated promoter that controls the transcription of at least two *K. pneumoniae* genes, *ureD* and *ureA*, has been localized to a 158-bp region upstream of *ureD* (Fig. 2) (11). This promoter requires Nac for high levels of activity. In the histidine utilization operon (*hut*), Nac activates transcription from a σ^{70}-dependent promoter (50). There is a sequence closely resembling the consensus $\sigma^{70} - 10$ binding site in the promoter region of *K. pneumoniae ureD* (11), and Nac has been shown to bind immediately upstream of this putative $- 10$ sequence (28). Recent studies on the effects of site-directed mutations in this region demonstrate that these sequences are involved in the transcription of *ureD* (9). It is believed at this time that Nac activates transcription at a σ^{70} promoter upstream of *ureD*.

Urease genes have been reported to be nitrogen-regulated in *Pseudomonas aeruginosa* (39). Similar to the *Klebsiella* species, a mutation in *P. aeruginosa rpoN* (*ntrA*) results in a urease-negative phenotype (91). Although some investigators believe that *H. pylori* urease genes are constitutively expressed, Cussac et al. (12) suggested that *H. pylori* urease also might be nitrogen regulated. More data are needed to confirm or disprove this assertion.

CONSTITUTIVE EXPRESSION OF UREASE GENES

In some organisms urease gene expression is not regulated. Constitutive expression of urease activity is seen with *Morganella morganii* (77, 80), *S. saprophyticus* (18), and rare *E. coli* isolates that contain a chromosomal urease locus (8). Analysis of the chromosomal *E. coli* locus indicates that it contains only seven of the previously described urease genes, *ureD*, *ureA*, *ureB*, *ureC*, *ureE*, *ureF*, and *ureG* (Fig. 2) (18). Therefore, in contrast to the inducible gene clusters, no additional gene products are required for the constitutive expression of urease in *E. coli*. Information on the complete genetic organization of *M. morganii*, *S. saprophyticus*, and *Y. enterocolitica* is not currently available.

Why does urease gene regulation differ between various bacterial species? In general, species that contain constitutive gene clusters, such as *E. coli* and *M. morganii*, are less ureolytic than species with regulated clusters, such as *K. pneumoniae* and *P. mirabilis* (7, 80). This suggests that it is not advantageous for a bacterium to produce high levels of urease in a constitutive manner. Why there are two distinct mechanisms of urease gene regulation, i.e., urea inducible and nitrogen regulated, is not clear. It must result from a difference in the evolution of the various urease-producing species.

ACKNOWLEDGMENTS. I am grateful to Stanley Falkow for the inspiration, advice, support, and friendship he has extended to me over the past decade. I thank Sarah D'Orazio for assistance with the figures in this chapter.

My laboratory is supported by Public Health Service grant AI 27907 from the National Institutes of Health.

REFERENCES

1. **Beeson, P. B., and D. Rowley.** 1959. The anticomplementary effect of kidney tissue. Its association with ammonia production. *J. Exp. Med.* **110:**695–698.
2. **Bender, R. A.** 1991. The role of the NAC protein in the nitrogen regulation of *Klebsiella aerogenes*. *Mol. Microbiol.* **5:**2575–2580.
3. **Bender, R. A., P. M. Snyder, R. Bueno, M. Quinto, and B. Magasanik.** 1983. Nitrogen regulation system of *Klebsiella aerogenes:* the *nac* gene. *J. Bacteriol.* **156:**444–446.
4. **Blaser, M. J.** 1992. *Helicobacter pylori:* its role in disease. *Clin. Infect. Dis.* **15:**386–393.
5. **Braude, A. I., and J. Siemienski.** 1960. Role of bacterial urease in experimental pyelonephritis. *J. Bacteriol.* **80:**171–179.
6. **Breitenbach, J. M., and R. P. Hausinger.** 1988. *Proteus mirabilis* urease: partial purification and inhibition by boric acid and boronic acids. *Biochem. J.* **250:**917–920.
7. **Collins, C. M.** Unpublished data.
8. **Collins, C. M., and S. Falkow.** 1990. Genetic analysis of *Escherichia coli* urease genes: evidence for two distinct loci. *J. Bacteriol.* **172:**7138–7144.

9. **Collins, C. M., and D. Gutman.** Unpublished data.
10. **Collins, C. M., and D. M. Gutman.** 1992. Insertional inactivation of an *Escherichia coli* urease gene by IS*3411*. *J. Bacteriol.* **174:**883–888.
11. **Collins, C. M., D. M. Gutman, and H. Laman.** 1993. Identification of a nitrogen-regulated promoter controlling expression of *Klebsiella pneumoniae* urease genes. *Mol. Microbiol.* **8:**187–198.
12. **Cussac, V., R. L. Ferrero, and A. Labigne.** 1992. Expression of *Helicobacter pylori* urease genes in *Escherichia coli* grown under nitrogen-limiting conditions. *J. Bacteriol.* **174:**2466–2473.
13. **Dayhoff, M. O., R. M. Schwartz, and B. C. Orcutt.** 1978. Matrices for detecting distant relationships, p. 353–358. *In* M. O. Dayhoff (ed.), *The Atlas of Protein Sequence and Structure.* National Biomedical Research Foundation, Washington, D.C.
14. **Delisle, G. J.** 1977. Multiple forms of urease in cytoplasmic fractions of *Ureaplasma urealyticum.* *J. Bacteriol.* **130:**1390–1392.
15. **Devereux, J., P. Haeberli, and O. Smithies.** 1984. A comprehensive set of sequence analysis programs for the VAX. *Nucleic Acids Res.* **12:**387–395.
16. **DeVivo, M. J., P. R. Fine, G. R. Cutter, and H. M. Maetz.** 1984. The risk of renal calculi in spinal cord injury patients. *J. Urol.* **131:**857–860.
17. **Dixon, N. E., R. L. Blakeley, and B. Zerner.** 1980. Jack bean urease (EC 3.5.1.5). III. The involvement of active-site nickel ion in inhibition by β-mercaptoethanol, phosphoramidate, and fluoride. *Can. J. Biochem.* **58:**481–488.
18. **D'Orazio, S. E. F., and C. M. Collins.** Unpublished data.
19. **D'Orazio, S. E. F., and C. M. Collins.** 1993. Characterization of a plasmid-encoded urease gene cluster found among members of the family *Enterobacteriaceae.* *J. Bacteriol.* **175:**1860–1864.
20. **D'Orazio, S. E. F., and C. M. Collins.** 1993. The plasmid-encoded urease gene cluster of the *Enterobacteriaceae* is positively regulated by UreR, a member of the AraC family of transcriptional activators. *J. Bacteriol.* **175:**3459–3467.
21. **Dunn, B. E., G. P. Campbell, G. I. Perez-Perez, and M. J. Blaser.** 1990. Purification and characterization of urease from *Helicobacter pylori.* *J. Biol. Chem.* **265:**9464–9469.
22. **Eaton, K., C. L. Brooks, D. R. Morgan, and S. Krakowka.** 1991. Essential role of urease in pathogenesis of gastritis induced by *Helicobacter pylori* in gnotobiotic piglets. *Infect. Immun.* **59:**2470–2475.
23. **Fowler, J. E. J.** 1984. Bacteriology of branched renal calculi and accompanying urinary tract infection. *J. Urol.* **131:**213–215.
24. **Friedrich, B., and B. Magasanik.** 1977. Urease of *Klebsiella aerogenes:* control of its synthesis by glutamine synthetase. *J. Bacteriol.* **131:**446–452.
25. **Gallegos, M.-T., C. Michan, and J. L. Ramos.** 1993. The XylS/AraC family of regulators. *Nucleic Acids Res.* **21:**807–810.
26. **Gatermann, S., J. John, and R. Marre.** 1988. *Staphylococcus saprophyticus* urease: characterization and contribution to uropathogenicity in unobstructed urinary tract infection of rats. *Infect. Immun.* **57:**110–116.
27. **Gatermann, S., and R. Marre.** 1989. Cloning and expression of *Staphylococcus saprophyticus* urease gene sequences in *Staphylococcus carnosus* and contribution of the enzyme to virulence. *Infect. Immun.* **57:**2998–3002.
28. **Goss, T. G., and R. A. Bender.** Personal communication.
29. **Grant, R. B., J. L. Penner, J. N. Hennesey, and B. J. Jackowski.** 1981. Transferable urease activity in *Providencia stuartii.* *J. Clin. Microbiol.* **13:**561–565.
30. **Grenbro, L., H. Hedelin, and S. Petterson.** 1988. Urinary infection stones caused by *Ureaplasma urealyticum:* a review. *Scand. J. Infect. Dis.* **53:**46–49.
31. **Griffith, D. P., F. Khonsari, J. H. Skurnick, K. E. James, and V. A. C. S. Group.** 1988. A randomized trial of acetohydroxamic acid for the treatment and prevention of infection-induced urinary stones in spinal cord injury patients. *J. Urol.* **140:**318–324.
32. **Griffith, D. P., D. M. Musher, and C. Itin.** 1976. Urease: the primary cause of infection-induced urinary stones. *Invest. Urol.* **13:**346–350.
33. **Griffith, D. P., and C. A. Osborne.** 1987. Infection (urease) stones. *Min. Electrol. Metab.* **13:**278.
34. **Hausinger, R. P.** 1987. Nickel utilization by microorganisms. *Microbiol. Rev.* **51:**22–24.

35. **Hawtin, P. R., A. R. Stacey, and D. G. Newell.** 1990. Investigation of the structure and location of the urease of *Helicobacter pylori* using monoclonal antibodies. *J. Gen. Microbiol.* **136:**1995–2000.

36. **Hazell, S. L., and A. Lee.** 1986. *Campylobacter pyloridis,* urease, hydrogen ion back diffusion, and gastric ulcers. *Lancet* **ii:**15–17.

37. **Hu, L., and H. L. T. Mobley.** 1990. Purification and N-terminal analysis of urease from *Helicobacter pylori. Infect. Immun.* **58:**992–998.

38. **Hu, L.-T., P. A. Foxall, R. Russell, and H. L. T. Mobley.** 1992. Purification of recombinant *Helicobacter pylori* urease apoenzyme encoded by *ureA* and *ureB. Infect. Immun.* **60:**2657–2666.

39. **Janssen, D. B., W. J. A. Habets, J. T. Mamgg, and C. van der Drift.** 1982. Nitrogen control in *Pseudomonas aeruginosa:* mutants affected in the synthesis of glutamine synthetase, urease, and NADP-dependent glutamate dehydrogenase. *J. Bacteriol.* **151:**22–28.

40. **Johnson, D. E., R. G. Russell, C. V. Lockatell, J. C. Zulty, J. W. Warren, and H. L. T. Mobley.** 1993. Contribution of *Proteus mirabilis* urease to persistence, urolithiasis, and acute pyelonephritis in a mouse model of ascending urinary tract infection. *Infect. Immun.* **61:**2748–2754.

41. **Jones, B. D., C. V. Lockatell, D. E. Johnson, J. W. Warren, and H. L. T. Mobley.** 1990. Construction of a urease-negative mutant of *Proteus mirabilis:* analysis of virulence in a mouse model of ascending urinary tract infection. *Infect. Immun.* **58:**1120–1123.

42. **Jones, B. D., and H. L. T. Mobley.** 1988. *Proteus mirabilis* urease: genetic organization, regulation, and expression of structural genes. *J. Bacteriol.* **170:**3342–3349.

43. **Jones, B. D., and H. L. T. Mobley.** 1989. *Proteus mirabilis* urease: nucleotide sequence determination and comparison with jack bean urease. *J. Bacteriol.* **171:**6414–6422.

44. **Kenny, G. E., and F. D. Cartwright.** 1977. Effect of urea concentration on growth of *Ureaplasma urealyticum* (T-strain mycoplasma). *J. Bacteriol.* **132:**144–150.

45. **Labigne, A., V. Cussac, and P. Courcoux.** 1991. Shuttle cloning and nucleotide sequences of *Helicobacter pylori* genes responsible for urease activity. *J. Bacteriol.* **173:**1920–1931.

46. **Lee, M. H., S. B. Mulrooney, and R. P. Hausinger.** 1990. Purification, characterization, and in vivo recombination of *Klebsiella aerogenes* urease apoenzyme. *J. Bacteriol.* **172:**4427–4431.

47. **Lee, M. H., S. B. Mulrooney, M. J. Renner, Y. Markowicz, and R. P. Hausinger.** 1992. *Klebsiella aerogenes* urease gene cluster: sequence of *ureD* and demonstration that four accessory genes (*ureD, ureE, ureF, ureG*) are involved in nickel metallocenter biosynthesis. *J. Bacteriol.* **174:**4324–4330.

48. **Lee, M. H., H. S. Prankratz, S. Wang, R. A. Scott, M. G. Finnegan, M. K. Johnson, J. A. Ippolito, D. W. Christianson, and R. P. Hausinger.** 1993. Purification and characterization of *Klebsiella aerogenes* UreE protein: a nickel binding protein that functions in urease metallo center assembly. *Prot. Sci.* **2:**1042–1052.

49. **Lerner, S. P., M. J. Gleeson, and D. P. Griffith.** 1989. Infection stones. *J. Urol.* **141:**753–758.

50. **Macaluso, A., E. A. Best, and R. A. Bender.** 1990. Role of the *nac* gene product in the nitrogen regulation of some NTR-regulated operons of *Klebsiella aerogenes. J. Bacteriol.* **172:**7249–7255.

51. **MacLaren, D. M., and P. G. H. Peerbooms.** 1986. Urinary infections by urea splitting microorganisms. *In* A. W. Brumfit (ed.), *Microbial Diseases in Nephrology.* John Wiley & Sons, New York.

52. **Maeda, M., M. Hidaka, A. Nakamura, H. Masaki, and T. Uozumi.** 1994. Cloning, sequencing, and expression of the thermophilic *Bacillus* sp. strain TB-90 urease gene complex in *Escherichia coli. J. Bacteriol.* **176:**432–442.

53. **Magasanik, B., and F. C. Neidhardt.** 1987. Regulation of carbon and nitrogen utilization, p. 1318–1325. *In* F. C. Neidhardt, J. L. Ingraham, K. B. Low, B. Magasanik, M. Schaechter, and H. E. Umbarger (ed.), *Escherichia coli and Salmonella typhimurium: Cellular and Molecular Biology.* American Society for Microbiology, Washington D.C.

54. **Mai, U. E. H., G. I. Perez-Perez, J. B. Allen, S. M. Wahl, M. J. Blaser, and P. D. Smith.** 1992. Surface proteins from *Helicobacter pylori* exhibit chemotactic activity for human leukocytes and are present in gastric mucosa. *J. Exp. Med.* **175:**517–525.

55. **Maier, T., A. Jacobi, M. Sauter, and A. Hock.** 1993. The product of the *hypB* gene, which is required for nickel incorporation into hydrogenases, is a novel guanine nucleotide-binding protein. *J. Bacteriol.* **175:**630–635.

56. **Marshall, B. J., L. Barret, C. Prakesh, R. McCallum, and R. Guerrant.** 1988. Protection of *Campy-*

lobacter pyloridis but not *Campylobacter jejuni* against acid susceptibility by urea, p. 402–403. *In* B. Kaijser and E. Falsen (ed.), *Campylobacter IV*. University of Goteborg, Goteborg, Sweden.

57. **McLean, R. J. C., J. C. Nickel, K.-J. Cheng, and J. W. Costerton.** 1988. The ecology and pathogenicity of urease-producing bacteria in the urinary tract. *Crit. Rev. Microbiol.* **16**:37–79.

58. **Mertz, A., S. Batsford, E. Curschellas, M. Kist, and K. Gondolf.** 1991. Cationic *Yersinia* antigen-induced chronic allergic arthritis in rats. *J. Clin. Invest.* **88**:632–642.

59. **Millner, O. E. J., J. A. Andersen, M. E. Appler, C. E. Benjamin, J. G. Edwards, D. T. Humphrey, and E. M. Shearer.** 1982. Flurofamide: a potential inhibitor of bacterial urease with a potential clinical utility in the treatment of infection induced urinary stones. *J. Urol.* **127**:346–350.

60. **Mobley, H. L. T., G. R. Chippendale, M. H. Fraiman, J. H. Tenney, and J. W. Warren.** 1985. Variable phenotypes of *Providencia stuartii* due to plasmid-encoded traits. *J. Clin. Microbiol.* **22**: 851–853.

61. **Mobley, H. L. T., G. R. Chippendale, K. G. Swihart, and R. A. Welch.** 1991. Cytotoxicity of the HmpA hemolysis and urease of *Proteus mirabilis* and *Proteus vulgaris* against cultured human renal proximal tubular epithelial cells. *Infect. Immun.* **59**:2036–2042.

62. **Mobley, H. L. T., and R. P. Hausinger.** 1989. Microbial ureases: significance, regulation, and molecular characterization. *Microbiol. Rev.* **53**:85–108.

63. **Mobley, H. L. T., B. D. Jones, and A. E. Jerse.** 1986. Cloning of urease gene sequences from *Providencia stuartii*. *Infect. Immun.* **54**:161–169.

64. **Mobley, H. L. T., and J. W. Warren.** 1987. Urease-positive bacteriuria and obstruction of long-term urinary catheters. *J. Clin. Microbiol.* **25**:2216–2217.

65. **Mulrooney, S. B., and R. P. Hausinger.** 1990. Sequence of the *Klebsiella aerogenes* urease genes and evidence for accessory proteins facilitating nickel incorporation. *J. Bacteriol.* **172**:5837–5843.

66. **Mulrooney, S. B., M. J. Lynch, H. L. T. Mobley, and R. P. Hausinger.** 1988. Purification, characterization, and genetic organization of recombinant *Providencia stuartii* urease expressed by *Escherichia coli*. *J. Bacteriol.* **170**:2202–2207.

67. **Musher, D. M., D. P. Griffith, D. Yawn, and R. D. Rossen.** 1975. Role of urease in pyelonephritis resulting from urinary tract infection with *Proteus*. *J. Infect. Dis.* **131**:177–181.

68. **Needleman, S. B., and C. D. Wunsch.** 1970. A general method applicable to the search for similarity in the amino acid sequence of two proteins. *J. Mol. Biol.* **48**:443–453.

69. **Nicholson, E. B., E. A. Concaugh, P. A. Foxall, M. D. Island, and H. L. T. Mobley.** 1993. *Proteus mirabilis* urease: transcriptional regulation by UreR. *J. Bacteriol.* **175**:465–473.

70. **Nicholson, E. B., E. A. Concaugh, and H. L. T. Mobley.** 1991. *Proteus mirabilis* urease: use of a *ureA-lacZ* fusion demonstrates that induction is highly specific for urea. *J. Bacteriol.* **59**: 3360–3365.

71. **Nomura, A., G. N. Stemmermann, P.-H. Chyou, I. Kato, G. I. Perez-Perez, and M. J. Blaser.** 1991. *Helicobacter pylori* infection and gastric carcinoma among Japanese Americans in Hawaii. *N. Engl. J. Med.* **325**:1132–1136.

72. **Parsonnet, J., G. D. Friedman, D. P. Vandersteen, Y. Chang, J. H. Vogelman, N. Orentreich, and R. K. Sibley.** 1991. *Helicobacter pylori* infection and the risk of gastric carcinoma. *N. Engl. J. Med.* **325**:1121–1131.

73. **Pickering, W. J., and D. F. Birch.** 1989. Bacteriologic and serologic findings in experimental pyelonephritis caused by *Ureaplasma urealyticum*. *Infect. Immun.* **57**:1235–1239.

74. **Probst, P., E. Hermann, K.-H. Meyer zum Buschenfelde, and B. Fleischer.** 1993. Identification of the *Yersinia enterocolitica* urease β subunit as a target antigen for human synovial T lymphocytes in reactive arthritis. *Infect. Immun.* **61**:4507–4509.

75. **Przybyla, A. E., J. Robbins, N. Menon, and H. D. J. Peck.** 1992. Structure/function relationships among the nickel-containing hydrogenases. *FEMS Microbiol. Rev.* **88**:109–136.

76. **Romano, N., R. La Litica, and D. R. Alesi.** 1986. Energy production in *Ureaplasma urealyticum*. *Pediatr. Infect. Dis. J.* **5**:S308–S312.

77. **Rosenstein, I., J. M. T. Hamilton-Miller, and W. Brumfit.** 1980. The effect of acetohydroxamic acids on the induction of bacterial ureases. *Invest. Urol.* **18**:112–114.

78. **Schwada, A., and R. A. Bender.** 1993. The *nac* (nitrogen assimilation control) gene from *Klebsiella aerogenes*. *J. Bacteriol.* **175**:2107–2115.

79. **Segal, E. D., J. Shon, and L. S. Tompkins.** 1992. Characterization of *Helicobacter pylori* urease mutants. *Infect. Immun.* **60:**1883–1889.

80. **Senior, B. W.** 1983. *Proteus morganii* is less frequently associated with urinary tract infections than *Proteus mirabilis*—an explanation. *J. Med. Microbiol.* **16:**317–322.

81. **Sheath, P. H. A., and R. R. Sokal.** 1973. *Numerical Taxonomy.* W. H. Freeman & Co., San Francisco.

82. **Sizer, I. W.** 1941. Temperature activation of the urease-urea system using urease of *Proteus vulgaris. J. Bacteriol.* **41:**511–527.

83. **Skurnik, M., S. Batsford, A. Mertz, E. Schiltz, and P. Toivanen.** 1993. The putative arthritogenic cationic 19-kilodalton antigen of *Yersinia enterocolitica* is a urease β subunit. *Infect. Immun.* **61:** 2498–2504.

84. **Smoot, D. T., H. L. T. Mobley, G. R. Chippendale, J. F. Lewison, and J. H. Resau.** 1990. *Helicobacter pylori* urease activity is toxic to human gastric epithelial cells. *Infect. Immun.* **58:**1992–1994.

85. **Stemke, G. W., J. A. Robertson, and M. Nhan.** 1987. Purification of urease from *Ureaplasma urealyticum. Can. J. Microbiol.* **33:**857–862.

86. **Suzuki, M., S. Miura, M. Suematsu, D. Fukumura, I. Kurose, H. Suzuki, A. Kai, Y. Kudoh, M. Ohashi, and M. Tsuchiya.** 1992. *Helicobacter pylori*-associated ammonia production enhances neutrophil-dependent gastric mucosal cell injury. *Am. J. Physiol.* **263:**G719–G725.

87. **Takebe, S., A. Numata, and K. Kobashi.** 1984. Stone formation by *Ureaplasma urealyticum* in human urine and its prevention by urease inhibitors. *J. Clin. Microbiol.* **20:**869–873.

88. **Takeuchi, H., K. Kobashi, and O. Yoshida.** 1980. Prevention of infected urinary stones in rats by urease inhibitor. *Invest. Urol.* **18:**102–105.

89. **Thirkell, D., A. D. Myles, B. L. Precious, J. S. Frost, J. C. Woodall, M. G. Burdon, and W. C. Russell.** 1989. The urease of *Ureaplasma urealyticum. J. Gen. Microbiol.* **135:**315–323.

90. **Todd, M. J., and R. P. Hausinger.** 1987. Purification and characterization of the nickel-containing multicomponent urease fron *Klebsiella aerogenes. J. Biol. Chem.* **262:**5963–5967.

91. **Totten, P. A., J. C. Lara, and S. Lory.** 1990. The *rpoN* gene product of *Pseudomonas aeruginosa* is required for expression of diverse genes, including the flagellin gene. *J. Bacteriol.* **172:**389–396.

Molecular Genetics of Bacterial Pathogenesis
Edited by V. L. Miller, J. B. Kaper, D. A. Portnoy, and R. R. Isberg
© 1994 American Society for Microbiology, Washington, DC 20005

Chapter 30

pJM1 Plasmid-Mediated Iron Transport-Virulence System of *Vibrio anguillarum*

Jorge H. Crosa

Vibrio anguillarum causes vibriosis, one of the most devastating diseases affecting salmonid fishes, which results in a hemorrhagic septicemia and death caused by hypoxia and the dysfunction of various organs (15, 16).

The disease caused by this organism has remarkable similarities to invasive septicemic diseases in humans and is thus an interesting model to study bacterial pathogenicity. Since the bacterium is the actual pathogen for this vertebrate, the study of this system leads to valid assessments of the host-parasite relationship.

My laboratory has been concerned with understanding the mechanisms of pathogenesis of this disease by elucidating the genetics and molecular biology of *V. anguillarum* (14–18, 56). In many strains of this bacterium, we identified a 65-kb plasmid, pJM1, which encodes an essential component of its virulence repertoire. In the past few years we have elucidated the nature of this component as an iron uptake system that consists of the 348-Da siderophore anguibactin and the iron transport proteins FatA to FatD (1–3, 14, 17, 62). Either elimination of the pJM1 plasmid or mutation of the iron uptake system leads to a dramatic decrease in virulence (14, 50, 53, 54).

Experimental infections in which a virulent strain carrying the wild-type plasmid was coinfected with a nonvirulent strain were also performed. In one of these experiments, the nonvirulent strain harbors the pJM1 plasmid with a mutation in the anguibactin biosynthetic pathway; i.e., it still retained expression of the FatA outer membrane protein, the receptor for ferric-anguibactin complexes. In another experiment, however, the nonvirulent strain had a plasmid with a deletion of the FatA receptor gene (67). In the first experiment, both strains, virulent and avirulent, were recovered after infection, suggesting that the wild-type strain produced enough anguibactin siderophore to cross-feed the receptor-proficient, siderophore-deficient mutant strain. Only the wild type could be recovered from mixed infections with the wild-type strain and the FatA receptor-deficient avirulent strain. Therefore, it is clear that the siderophore anguibactin itself is an important virulence factor in *V. anguillarum* infections of salmonid fishes.

Jorge H. Crosa • Department of Molecular Microbiology and Immunology L-220, Oregon Health Sciences University, 3181 S.W. Sam Jackson Park Road, Portland, Oregon 97201.

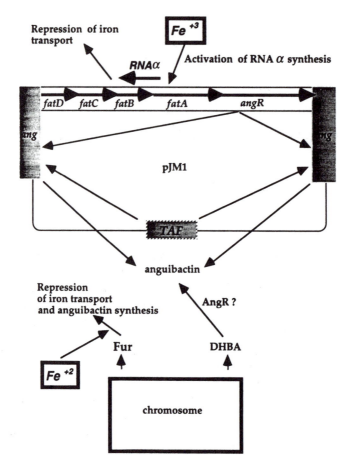

FIGURE 1. Schematic representation of the pJM1 plasmid-mediated system of *V. anguillarum* and the chromosome contribution to its expression. *ang,* regions containing anguibactin biosynthetic genes; *fatA* to *fatD,* iron transport genes (*fatA* is an 86-kDa outer membrane protein receptor, *fatB* is a ca. 40-kDa periplasmic protein, and *fatC* and *fatD* are ca. 40-kDa cytoplasmic membrane proteins); *angR,* a gene encoding the regulatory (and possibly biosynthetic) protein AngR; TAF, region encoding a *trans*-acting transcriptional regulator; RNAα, antisense RNA that controls the expression of iron transport genes; Fur, chromosomally mediated protein that, in conjunction with iron, inhibits expression of the pJM1 iron transport and siderophore production systems; DHBA, 2,3-dihydroxybenzoic acid, which is a chromosomally mediated compound that is a precursor of the otherwise pJM1 plasmid-mediated siderophore anguibactin.

Mutagenesis experiments of the pJM1 plasmid with Tn*3*-HoHo1 led to the identification of six genetic units and identified the iron uptake region to approximately 25 kb of pJM1 DNA (50). Analysis of the mutants generated by this procedure indicated that genetic units I, IV, V, and VI harbored genes for the biosynthesis of anguibactin, while genetic unit III consisted exclusively of the gene for the 100-kDa positive regulator AngR. Another positive regulatory factor, TAF, is encoded in a region noncontiguous to the iron uptake region (Fig 1). Both

TAF and AngR are required for anguibactin production. Their action occurs by enhancement of the transcription of anguibactin biosynthetic genes (50).

Genetic unit II consisted of four genes: *fatA,* which encodes the 86-kDa outer membrane protein which is the anguibactin receptor, the periplasmic protein *fatB,* and the hydrophobic cytoplasmic membrane proteins *fatC* and *fatD.* These four proteins were sufficient to ensure the actual transport of iron into the cell cytosol, as long as ferric anguibactin was provided as a source of ferric iron (2, 31, 61). The existence of a chromosomal iron transport system in some strains of *V. anguillarum* isolated in various geographical regions was also reported (32). It is possible that it may be a component of virulence for these strains.

BIOSYNTHESIS OF ANGUIBACTIN

My colleagues and I carried out experiments to determine the biosynthetic pathway of anguibactin biosynthesis. Previous X-ray diffraction studies (30) identified the presence of a diphenolic group in the molecule of anguibactin, while genetic and biochemical studies demonstrated that a plasmidless derivative of *V. anguillarum* 775 does not produce anguibactin, although it still produces 2,3-dihydroxybenzoic acid (DHBA) (1). The fact that DHBA is produced in the plasmidless strain supports the notion that enzymes for its synthesis are encoded in the chromosome rather than the pJM1 plasmid. Our present hypothesis is that the chromosomally encoded DHBA is used by plasmid-encoded enzymes as one of the building blocks to synthesize anguibactin. To prove this hypothesis, we decided to construct mutants deficient in the biosynthesis of DHBA and assess whether this mutation in turn affected anguibactin production.

To perform this task, we started by cloning the genes involved in the biosynthesis of DHBA in *V. anguillarum.* The cloned genes could subsequently be disrupted by introducing a fragment carrying an antibiotic resistance gene. The disrupted gene could then be recombined into a wild-type gene by allele replacement (marker exchange) to generate the desired mutation in the *V. anguillarum* chromosome.

In *Escherichia coli* and other bacteria, *aroC* is an important gene in the biosynthesis of DHBA (16, 40). Consequently, we transduced a *V. anguillarum* cosmid library of chromosomal DNA to an *aroC* mutant of *E. coli* and identified a clone that could actually complement this lesion (12). By following this approach, we were able to clone the chromosomal *aroC* gene encoding chorismate synthase from *V. anguillarum* 775 by complementation of the *E. coli aroC* mutant AB2849 with a cosmid gene bank of the *V. anguillarum* plasmidless strain H775-3. In that same work, we determined the nucleotide sequence of this gene and found an open reading frame that corresponds to a protein of 372 amino acids with a calculated mass of 40,417 Da; this correlated with the size of the *V. anguillarum aroC* product detected in vitro. The homology of the *V. anguillarum aroC* gene to the *aroC* genes of *E. coli* and *Salmonella typhi* is 68% at the nucleotide level and 78% at the protein level. The expression of the *aroC* transcript is not regulated by iron, as determined by Northern blot hybridization analysis. After insertion of a resistance gene cassette within the cloned *aroC* gene, an *aroC* mutant of *V. anguillarum* was generated by allelic exchange. This mutant was deficient in the production of DHBA. Further analysis of this mutant demonstrated that this strain was also deficient in the production of anguibactin. The deficiency could be complemented by introducing a clone containing the wild-type *aroC* gene. Therefore, these results showed that the chromosomally encoded DHBA is indeed a precursor of the pJM1 plasmid-mediated siderophore anguibactin (12).

Tomalsky, Actis, and I (55) have also recently identified a sequence in the pJM1 plasmid iron uptake region that shares homology with domains of the histidine decarboxylase found in *Morganella morganii, Klebsiella planticola,* and *Enterobacter aerogenes.* Since the product of the action of this enzyme on histidine is histamine and the latter is also a component of the anguibactin molecule, we believe that this finding is of the utmost importance in our quest to dissect the biosynthetic genes for anguibactin, especially because mutations in this region lead to a cessation of anguibactin biosynthesis and to a concomitant attenuation of virulence (50).

REPRESSION BY IRON OF THE EXPRESSION OF THE GENE FOR THE FatA OUTER MEMBRANE PROTEIN, THE RECEPTOR FOR FERRIC ANGUIBACTIN IN *V. ANGUILLARUM*

In *E. coli* and most other bacteria, iron forms a ferrous complex with the Fur protein, which acts as a repressor by binding to the Fur-binding sequence found at the promoter region of iron-regulated genes (5–7, 10, 19–22, 26–28, 33, 34, 41, 46, 47, 63, 66). The *V. anguillarum fur* gene is highly related (95% identity) to those of *Vibrio vulnificus* and *Vibrio cholerae* but shows only 76% relatedness to those of *Yersinia* spp. and *E. coli* (58). By different approaches, such as marker exchange for *V. cholerae* (33), *V. vulnificus* (34), and *Yersinia pestis* (46, 47), and by using selection at high Mn^{2+} concentrations (28) for *Pseudomonas aeruginosa* (41) and *V. anguillarum* (58), it was possible to demonstrate that mutations in the *fur*-like genes in these bacteria lead to an impairment of iron regulation for specific iron-regulated systems. To gain information on the structure-function relationships in Fur proteins, my colleagues and I (65) examined two of these mutants, 9 and 11, in more detail.

The mutations could be complemented by a recombinant clone carrying the wild-type *fur* gene. The regions in the *V. anguillarum* chromosome carrying the mutated *fur* gene in both derivatives were cloned in a pBluescript vector after PCR. Sequencing of the mutated genes demonstrated that, in the case of mutant 9, a single nucleotide was changed: a T was replaced by a C in nucleotide 332. This change resulted in the change of the amino acid at position 104 from D to G. The mutated Fur protein migrated faster on sodium dodecyl sulfate-polyacrylamide gel electrophoresis than the wild-type Fur protein. The other mutant, mutant 11, which is apparently null, showed a change in the initiation codon from AUG to GUG, which is more rarely used in initiation. There were no other changes within the open reading frame and upstream region. In the case of mutant 9, it is clear that this mutation occurs in a highly conserved region among different *fur* genes and it has been predicted that it may disrupt the α-helix structure of the Fur protein (13). This protein alteration, which leads to a faster-migrating form, occurs not only when the mutated chromosomal *fur* gene is expressed in the *V. anguillarum* cell environment but also as a cloned gene in an *E. coli* background, suggesting that the mutation results in a change that is intrinsically related to the protein encoded by the mutated gene. Furthermore, the fact that this mutant can be complemented by the wild-type gene strongly suggests that this is the only mutation accounting for the Fur⁻ phenotype of this mutant. This observation was confirmed by restoring the original nucleotide by site-directed mutagenesis by using a PCR-generated *fur* mutant DNA fragment cloned in pBluescript. The protein expressed by the new derivative was indistinguishable from the wild-type Fur protein, both from the standpoint of gel migration as well as from the standpoint of an iron-regulated phenotype for the β-galactosidase gene in the *E. coli* reporter strain RRJC1. It was of interest that there was a selection for bacteria carrying the mutant Fur protein with the altered mobility when bacteria were grown under conditions of iron limitation, while there was a reversion to the wild-type mobility under iron-rich conditions, suggesting a pressure for the existence of a wild-type Fur protein at high iron concentrations, possibly because this metal, in high concentrations, could be deleterious to the cell, for instance, by catalyzing the production of free radicals. The mutation in derivative 9 is located in the carboxy-terminal half of the Fur protein, which is involved in iron binding, and it may likely lead to a conformational change in the protein. Bacteria carrying either the wild-type or mutant conformation may be selected for by the environmental growth conditions. Understanding of the function-structure relationships leading to the interaction of the Fur protein with iron, DNA, and other parameters will likely be increased by analysis of these and other mutants that we are generating.

In order to further analyze the contribution of this Fur-like product to the regulation of the anguibactin-mediated iron transport system, we introduced a plasmid harboring the *fatA* gene (as well as the rest of the iron transport region of pJM1, i.e., *fatB, fatC,* and *fatD*) in both the *fur* mutant strain of *E. coli* and its isogenic derivative carrying a plasmid with the *E. coli fur* gene. By using a probe complementary to *fatA* mRNA in RNase protection assays, we determined that synthesis of *fatA* mRNA was not very effectively shut off by the *E. coli* Fur protein. These results, as well as the low degree of homology shared by the *E. coli* and *V. anguillarum* Fur proteins (61), strongly suggest that the *E. coli* Fur product may not have a strong affinity for the *fatA* sequences that bind the *V. anguillarum* Fur protein. However, it was puzzling that even though *fatA* mRNA was still synthesized under iron-rich conditions, especially in the *fur* mutant strain, no synthesis of the FatA protein occurred. It was obvious from these results that another mechanism(s) not involving Fur must act under iron-rich conditions to repress the synthesis of the FatA protein.

Indeed, we have recently identified a pJM1 plasmid-mediated repressor, encoded in the cloned pJM1 fragment carrying the iron transport region *fatABCD*, which showed inhibitory activity in the synthesis of FatA. Several derivatives carrying portions of this region were tested for the presence of the inhibitory effect. Inhibition occurred only when the intergenic region between *fatA* and *fatB* was part of the derivative. To further investigate the nature of the repressor, we inserted the Ω fragment, which carries transcription and translation termination signals, in one of these derivatives with a cloned 2.4-kb *Hind*III fragment of *fatB-fatA* intergenic DNA. Only insertions of Ω at the right *Hind*III site (within a region encoding the untranslated *fatA* mRNA) abolished inhibition, while no changes in inhibition resulted from the Ω insertion either at the left *Hind*III site (within the *fatA* mRNA region containing the open reading fame) or in the *Pvu*I site within this fragment. These results and the fact that the direction of transcription of the *fatA* mRNA is toward the right (toward the right *Hind*III site) suggested that the repressor was likely an antisense RNA transcribed opposite from the direction of transcription of the *fatA* mRNA. An RNA probe that could detect antisense RNA was then used in RNase protection experiments with RNA obtained from *V. anguillarum* cells harboring various derivatives and grown under both iron-rich and iron-limiting conditions (44). An antisense RNA, designated RNAα, was synthesized only under iron-rich conditions in the case of wild-type strains of *V. anguillarum,* while it was constitutively synthesized in the case of strains containing the cloned 2.4-kb *Hind*III fragment since the portion of DNA encoding RNAα in this construct had lost the wild-type promoter during the cloning procedure and was now under the control of the tetracycline resistance promoter from the pBR322 vector. We identified three possible sites for the initiation of RNAα synthesis in the wild-type strains by using primer extension experiments and also determined that RNAα is 600 nucleotides long by using Northern blots. These results are consistent with the fact that RNAα is encoded in a region spanning the intergenic region between *fatB* and *fatA*. Thus, RNAα must be complementary to the 5'-untranslated end of *fatA* mRNA and to the 3'-translated end of *fatB* mRNA. Under iron-rich conditions, there is no detectable *fatA* mRNA in *V. anguillarum*; therefore, it is likely that a Fur-like product must act first by inhibiting transcription initiation of the *fatA* gene. It is also possible that under these conditions RNAα may also contribute to this repression. However, our results fit with a model in which RNAα must act mostly posttranscriptionally, preventing the translation of already started *fatA* transcripts, and thus is a fine-tuning control of iron regulation in *V. anguillarum* (44, 61).

We also identified within RNAα the region essential for its inhibitory capacity. The results indicated that about 30% of the 5' end of RNAα can be deleted without an appreciable change in its inhibitory activity (44, 61).

POSITIVE REGULATION OF ANGUIBACTIN PRODUCTION IN *V. ANGUILLARUM*

The combined action of the positive regulators TAF and AngR leads to the biosynthesis of anguibactin (43). Recently, my colleagues and I have concentrated on the analysis of the *angR* gene and have obtained a clearer picture of the role that it plays in anguibactin biosynthesis.

The *V. anguillarum* 775 *angR* gene (*angR$_{775}$*) encodes a regulatory protein,

AngR, of 1,048 amino acids (23, 43, 57) that has two helix-turn-helix motifs typical of prokaryotic DNA-binding proteins (8, 13, 48). We also identified leucine zippers just upstream of each one of these two helix-turn-helix motifs (51). In eukaryotic regulators, a leucine zipper is often followed by a basic DNA-binding region to form the so-called bZIP (11, 36). The presence of leucine zippers in prokaryotic proteins other than AngR was also reported recently (25). In eukaryotic systems, these proteins interact with another regulatory protein molecule and bind DNA (4, 8, 24, 36, 39, 42, 48, 59, 60). It is possible that one or both of the leucine zipper–helix-turn-helix domains found in AngR might play a role in protein and DNA recognition and that one of them may be involved in the synergistic action found between AngR and the regulator TAF (43, 57). We were able to generate mutations in the leucine zipper and helix-turn-helix domains of AngR by using site-directed mutagenesis (50). We determined anguibactin production by these mutants after transfer to an appropriate strain containing a plasmid with the cloned iron uptake region, mutated by transposon insertion in the *angR* gene, and another plasmid that provided the TAF factor. Anguibactin production was measured by the ability of the strain to grow in a medium containing the iron chelator ethylenedi-amine-di(*o*-hydroxyphenyl)acetic acid. We also sequenced each of the mutant derivatives as well as the *angR*$_{531A}$ gene, which is found on plasmids harbored by strains of *V. anguillarum* that produce high levels of anguibactin (52, 57). Our results showed that the only difference between *angR*$_{531A}$ (high anguibactin producer) and *angR*$_{775}$ (the wild-type gene) was a substitution of His (in AngR$_{775}$) for Asn (in AngR$_{531A}$) at amino acid 267. It was of interest that the substituted amino acid is located between the first leucine zipper and the helix-turn-helix motif (50). Substitution of the His with Leu or Gln generated AngR derivatives that produced lower levels of anguibactin than those produced by the wild-type *V. anguillarum,* confirming the importance of this amino acid for AngR activity. Since the mutation occurs at a site between a leucine zipper and a helix-turn-helix motif, it is possible that it affects the regulatory role of AngR by modifying protein-protein or DNA-protein interactions involving AngR. Some of the mutations con-sisted of substitutions of the original amino acid for Pro, which is known to disrupt helix structures (13), while other mutations resulted in substitutions of one or more Leu or Val residues, producing a disruption of the leucine zipper. All of these mutations affected the function of AngR.

AngR also possesses a specific domain, IIYTSGSTGLPKG, which shares homology with similar sequences found in several proteins of three groups of ATP-utilizing enzymes (29, 35, 45, 49). These proteins are acid-thiol ligases and are responsible for the activation reactions for the biosynthesis of enterobactin and for the synthetases for tyrocidine, gramicidine S, and penicillin, also known as the firefly luciferase family: the gramicidine S and tyrocidine synthetases from *Bacillus brevis,* the 57-kDa polypeptide of the 4-chlorobenzoate dehalogenase from *Pseudomonas* spp., the coumarate coenzyme A:ligase from *Petrosinum crispum,* the luciferase from *Photinus pyralis,* the D-alanine-activating enzyme from *Lactobacillus casei,* and the DHBA-AMP ligase (EntE) from *E. coli.* EntE catalyzes the activation of DHBA, an essential step in the biosynthetic pathway of the diphenolic siderophore enterobactin. Since the molecule of anguibactin also possesses a diphenolic ring and it is synthesized from the chromosomally encoded DHBA by plasmid pJM1-encoded enzymes, it is possible that AngR, in addition to its regulatory functions, could have a role as an activating enzyme in the biosynthesis of anguibactin. It was of interest that a clone expressing AngR can replace EntE, leading to the complementation of the *entE* mutation in *E. coli,* indicating that, in addition to its regulatory role, AngR also has an enzymatic function related to that of DHBA-AMP ligase, which may play an important role in the biosynthesis of anguibactin. Therefore, it is possible that AngR possesses distinct domains that are related to either biosynthetic or regulatory functions. Further analysis of site-specific mutations in the AngR protein will allow us to ascribe specific domains to these functions.

SUMMARY

Iron, an essential nutrient for all living cells, is very abundant on the earth's crust; however, in aerobic environments, it is found as a precipitate consisting of complex ferric hydroxides. In biological systems, both the ferric as well as the ferrous forms of this metal are normally found as complexes, mostly with high-affinity iron-binding proteins (7, 9, 16, 35, 38, 64, 66). In mammals, most of the iron in biological fluids is in complexes with either transferrin or lactoferrin or is in erythrocytes as part of hemoglobin. The consequence of this free iron shortage is that microorganisms attempting to establish an infection will depend heavily on their ability to use the complexed iron. Bacteria evolved clever strategies to accomplish this task. One of them was the production of a low-molecular-weight compound, a siderophore, with a very high affinity for iron. This molecule is capable of extracting iron from the complexed state in the vertebrate host by forming a very strong siderophore-iron complex (in the ferric form). This complex can in turn be recognized by an outer membrane protein receptor on the bacterial cell and can be internalized. Another mechanism of using the bound iron implicates receptors on the bacterial outer membrane that can recognize heme or complexes of either transferrin or lactoferrin with iron (37).

The pJM1 plasmid in *V. anguillarum* possesses the genetic determinants for an iron uptake system that is tightly regulated by both positive and negative factors and that is an important factor in the virulence repertoire of this bacterium. TAF and AngR are essential as positive enhancers of the expression of anguibactin biosynthesis, at the transcription level, while a chromosomally mediated Fur protein as well as a plasmid-mediated antisense RNAα play roles in the negative control of the expression of the iron transport system. It is remarkable that one of the positive-acting factors, AngR, also shares homology in a motif with members of the firefly luciferase family, of which the DHBA-AMP ligase (EntE) from *E. coli* is a member, and that AngR can replace EntE in *E. coli entE* mutants and could therefore play a role in the biosynthesis of anguibactin. Whether the regulatory activity of AngR can be separated from the EntE-like activity necessitates further analysis by site-directed mutagenesis, which is under investigation.

ACKNOWLEDGMENT. The research reported here was supported by grant AI19018 from the National Institutes of Health.

REFERENCES

1. **Actis, L. A., W. Fish, J. H. Crosa, K. Kellerman, S. Ellenberger, F. Hauser, and J. Sanders-Loher.** 1986. Characterization of anguibactin, a novel siderophore from *Vibrio anguillarum* 775(pJM1). *J. Bacteriol.* **167:**57–65.

2. **Actis, L. A., S. Potter, and J. H. Crosa.** 1985. Iron-regulated outer membrane protein OM2 of *Vibrio anguillarum* is encoded by virulence plasmid pJM1. *J. Bacteriol.* **161:**736–742.

3. **Actis, L. A., M. E. Tolmasky, D. Farrell, and J. H. Crosa.** 1988. Genetic and molecular characterization of essential components of the *Vibrio anguillarum* plasmid-mediated iron transport system. *J. Biol. Chem.* **263:**2853–2860.

4. **Agre, P., P. Johnson, and S. McKnight.** 1989. Cognate DNA binding specificity retained after leucine zipper exchange between GCN4 and C/EBP. *Science* **246:**922–925.

5. **Baggs, A., and J. B. Neilands.** 1987. Molecular mechanism of regulation of siderophore-mediated iron assimilation. *Microbiol. Rev.* **51:**509–518.

6. **Baggs, A., and J. B. Neilands.** 1987. Ferric uptake regulation protein acts as a repressor, employing iron(II) as a cofactor to bind the operator of an iron transport operon in *Escherichia coli. Biochemistry* **26:**5471–5477.

7. **Braun, V., and K. Hantke.** 1991. Genetics of bacterial iron transport, p. 107–138. *In* G. Winkelmann (ed.), *Handbook of Microbial Iron Chelates.* CRC Press, Inc., Boca Raton, Fla.

8. **Brennan, R. G., L. H. Weaver, and B. W. Matthews.** 1986. Use of protein sequence and structure to infer distant evolutionary relationships. *Chem. Scr.* **26B:**251–255.

9. **Bullen, J. J., H. J. Rogers, and E. Griffiths.** 1978. Role of iron in bacterial infections. *Curr. Top. Microbiol. Immunol.* **80:**1–35.

10. **Calderwood, S. B., and J. J. Mekalanos.** 1988. Confirmation of the Fur operator site by insertion of a synthetic oligonucleotide into an operator fusion plasmid. *J. Bacteriol.* **170:**1015–1017.

11. **Chakerian, A., V. Tesmer, S. Manly, J. Brackett, M. Lynch, J. Hoh, and K. Matthews.** 1991. Evidence for a leucine zipper motif in lactose repressor protein. *J. Biol. Chem.* **266:**1371–1374.

12. **Chen, Q., L. Actis, M. Tolmasky, and J. H. Crosa.** 1994. Chromosome-mediated 2,3-dihydroxybenzoic acid is a precursor in the biosynthesis of the plasmid-mediated siderophore anguibactin in *Vibrio anguillarum. J. Bacteriol.* **176:**4226–4234.

13. **Chow, P. Y., and C. D. Fasman.** 1978. Empirical predictions of protein conformation. *Annu. Rev. Biochem.* **47:**251–276.

14. **Crosa, J. H.** 1980. A plasmid associated with virulence in the marine fish pathogen *Vibrio anguillarum* specifies an iron-sequestering system. *Nature* (London) **284:**566–568.

15. **Crosa, J. H.** 1984. The relationship of plasmid mediated iron transport and bacterial virulence. *Annu. Rev. Microbiol.* **38:**69–89.

16. **Crosa, J. H.** 1989. Genetics and molecular biology of siderophore-mediated iron transport in bacteria. *Microbiol. Rev.* **53:**517–530.

17. **Crosa, J. H., L. Hodges, and M. Schiewe.** 1980. Curing of a plasmid is correlated with an attenuation of virulence in the marine fish pathogen *Vibrio anguillarum. Infect. Immun.* **27:**897–902.

18. **Crosa, J. H., L. Hodges, M. H. Schiewe, and S. Falkow.** 1977. Evidence for plasmid contribution to the virulence of the fish pathogen *Vibrio anguillarum. Infect. Immun.* **18:**509–513.

19. **de Lorenzo, V., F. Giovannini, M. Herrero, and J. B. Neilands.** 1988. Metal ion regulation of gene expression. Fur repressor-operator interaction at the promoter region of the aerobactin system of pColV-K30. *J. Mol. Biol.* **203:**875–884.

20. **de Lorenzo, V., F. Giovannini, M. Herrero, and J. B. Neilands.** 1988. Fur (ferric uptake regulation) protein and CAP (catabolite-activator protein) modulate transcription of fur gene in *Escherichia coli. Eur. J. Biochem.* **173:**537–546.

21. **de Lorenzo, V., S. Wee, M. Herrero, and J. B. Neilands.** 1987. Operator sequences of the aerobactin operon of plasmid ColV-K30 binding the ferric uptake regulation (*fur*) repressor. *J. Bacteriol.* **169:**2624–2630.

22. **Ernst, J. F., R. L. Bennett, and L. I. Rothfield.** 1978. Constitutive expression of the iron entero-

chelin and ferrichrome uptake systems in a mutant strain of *Salmonella typhimurium*. *J. Bacteriol.* **135**:928–934.

23. Farrell, D., P. Mikesell, L. Actis, and J. H. Crosa. 1990. A regulatory gene, *angR*, of the iron uptake system of *Vibrio anguillarum*: similarity with phage P22 *cro* and regulation by iron. *Gene* **86**:45–51.

24. Fuma, S., Y. Fujishima, N. Corbell, C. D'Souza, M. Nakano, P. Zuber, and K. Yamane. 1993. Nucleotide sequence of 5′ portion of srfA that contains the region required for competence establishment in Bacillus subtilis. *Nucleic Acids Res.* **21**:93–97.

25. Geraldo, R., C. Nieto, M. Tresguerres, and R. Diaz. 1989. Bacterial zipper. *Nature* (London) **342**: 66.

26. Goldberg, M. B., S. A. Boyko, and S. B. Calderwood. 1990. Transcriptional regulation by iron of a *Vibrio cholerae* virulence gene and homology of the gene to the *Escherichia coli* Fur system. *J. Bacteriol.* **172**:6863–6870.

27. Hantke, K. 1984. Cloning of the repressor protein gene of iron regulated system in *E. coli* K-12. *Mol. Gen. Genet.* **197**:337–341.

28. Hantke, K. 1987. Selection procedure for deregulated iron transport mutants (*fur*) in *E. coli* K-12: *fur* not only affects iron metabolism. *Mol. Gen. Genet.* **210**:135–139.

29. Heaton, M., and F. Neuhaus. 1992. Biosynthesis of D-alanyl-lipoteichoic acid: cloning, nucleotide sequence, and expression of the *Lactobacillus casei* gene for the D-alanine-activating enzyme. *J. Bacteriol.* **174**:4707–4717.

30. Jalal, M., D. Hossain, J. van der Helm, J. Sanders-Loerh, L. A. Actis, and J. H. Crosa. 1989. Structure of anguibactin, a unique plasmid-related bacterial siderophore from the fish pathogen *Vibrio anguillarum*. *J. Am. Chem. Soc.* **111**:292–296.

31. Koster, W. L., L. A. Actis, L. Waldbeser, M. E. Tolmasky, and J. H. Crosa. 1991. Molecular characterization of the iron transport system mediated by the pJM1 plasmid in *Vibrio anguillarum* 775. *J. Biol. Chem.* **266**:23829–23833.

32. Lemos, M. L., P. C. Salinas, A. E. Toranzo, J. L. Barja, and J. H. Crosa. 1988. Chromosome-mediated iron uptake system in pathogenic strains of *Vibrio anguillarum*. *J. Bacteriol.* **170**: 1920–1925.

33. Litwin, C. M., S. A. Boyko, and S. B. Calderwood. 1992. Cloning, sequencing, and transcriptional regulation of the *Vibrio cholera fur* gene. *J. Bacteriol.* **174**:1897–1903.

34. Litwin, C. M., and S. B. Calderwood. 1993. Cloning and genetic analysis of the *Vibrio vulnificus fur* gene and construction of a *fur* mutant by in vivo marker exchange. *J. Bacteriol.* **168**:228–236.

35. Matzanke, B. F., G. I. Muller, E. Bill, and A. X. Trautwein. 1989. *Eur. J. Biochem.* **183**:371–379.

36. Maxon, M., J. Wigboldus, N. Brot, and H. Weissbach. 1990. Structure-function studies on *Escherichia coli* MetR protein, a putative prokaryotic leucine zipper protein. *Proc. Natl. Acad. Sci. USA* **87**:7076–7079.

37. Nau Cornelissen, C., G. D. Biswas, J. Tsai, D. K. Paruchuri, S. A. Thompson, and P. F. Sparling. 1992. Gonococcal transferrin-binding protein 1 is required for transferrin utilization and is homologous to TonB-dependent outer membrane receptors. *J. Bacteriol.* **174**:5788–5797.

38. Neilands, J. B. 1981. Microbial iron compounds. *Annu. Rev. Biochem.* **50**:715–731.

39. O'Shea, E., R. Rutkowski, and P. Kim. 1989. Evidence that the leucine zipper is a coiled coil. *Science* **243**:538–542.

40. Pollack, J. R., B. N. Ames, and J. B. Neilands. 1970. Iron transport in *Salmonella typhimurium*: mutants blocked in the biosynthesis of enterobactin. *J. Bacteriol.* **104**:635–639.

41. Prince, R. W., D. C. Cox, and M. L. Vasil. 1993. Coordinate regulation of siderophore and endotoxin A production: molecular cloning and sequencing of the *Pseudomonas aeruginosa fur* gene. *J. Bacteriol.* **175**:2589–2598.

42. Pu, W., and K. Struhl. 1991. The leucine zipper symmetrically positions the adjacent basic regions for specific DNA binding. *Proc. Natl. Acad. Sci. USA* **88**:6901–6905.

43. Salinas, P., M. E. Tolmasky, and J. H. Crosa. 1989. Regulation of the iron uptake system in *Vibrio anguillarum*: evidence for a cooperative affect between two transcriptional activators. *Proc. Natl. Acad. Sci. USA* **86**:3529–3533.

44. Salinas, P., L. Waldbeser, and J. H. Crosa. 1993. Regulation of the expression of bacterial iron transport genes: possible role of an antisense RNA as a repressor. *Gene* **123**:33–38.

45. **Scholten, J. D., K. Chang, P. Babbit, H. Charest, M. Sylvestre, and D. Dunaway-Mariano.** 1991. Novel enzymatic hydrolytic dehalogenation of a chlorinated aromatic. *Science* **253:**182–185.

46. **Staggs, T. M., and R. D. Perry.** 1991. Identification and cloning of a *fur* regulatory gene in *Yersinia pestis. J. Bacteriol.* **173:**417–425.

47. **Staggs, T. M., and R. D. Perry.** 1992. Fur regulation in *Yersinia* species. *Mol. Microbiol.* **6:** 2507–2516.

48. **Talanian, R., C. McKnight, and P. Kim.** 1990. Sequence-specific DNA binding by a short peptide dimer. *Science* **249:**769–771.

49. **Toh, H.** 1991. Sequence analysis of firefly luciferase family reveals a conservative sequence motif. *Protein Sequence Data Anal.* **4:**111–117.

50. **Tolmasky, M. E., L. A. Actis, and J. H. Crosa.** 1988. Genetic analysis of the iron uptake region of the *Vibrio anguillarum* plasmid pJM1: molecular cloning of genetic determinants encoding a novel *trans* activator of siderophore biosynthesis. *J. Bacteriol.* **170:**1913–1919.

51. **Tolmasky, M. E., L. A. Actis, and J. H. Crosa.** 1993. A single aminoacid change in AngR, a protein encoded by pJM1-like virulence plasmids, results in hyperproduction of anguibactin. *Infect. Immun.* **61:**3228–3233.

52. **Tolmasky, M. E., L. A. Actis, A. Toranzo, J. Barja, and J. H. Crosa.** 1985. Plasmids mediating iron uptake in *Vibrio anguillarum* strains isolated from turbot in Spain. *J. Gen. Microbiol.* **131:** 1989–1997.

53. **Tolmasky, M. E., and J. H. Crosa.** 1984. Molecular cloning and expression of genetic determinants for the iron uptake system mediated by the *Vibrio anguillarum* plasmid pJM1. *J. Bacteriol.* **160:** 860–866.

54. **Tolmasky, M. E., and J. H. Crosa.** 1991. Regulation of plasmid-mediated iron transport and virulence in *Vibrio anguillarum. Bio. Metals* **4:**33–35.

55. **Tolmasky, M. E., L. A. Actis, and J. H. Crosa.** Submitted for publication.

56. **Tolmasky, M. E., A. Gammie, and J. H. Crosa.** 1992. Characterization of the *recA* gene of *Vibrio anguillarum. Gene* **110:**41–48.

57. **Tolmasky, M. E., P. C. Salinas, L. A. Actis, and J. H. Crosa.** 1988. Increased production of the siderophore anguibactin mediated by pJM1-like plasmids in *Vibrio anguillarum. Infect. Immun.* **56:**1608–1614.

58. **Tolmasky, M. E., A. M. Wertheimer, L. A. Actis, and J. H. Crosa.** 1994. Characterization of the *Vibrio anguillarum fur* gene: role in regulation of expression of the FatA outer membrane protein and catechol. *J. Bacteriol.* **176:**213–220.

59. **Turner, R., and R. Tijan.** 1989. Leucine repeats and an adjacent DNA binding domain mediate the formation of functional cFos-cJun heterodimers. *Science* **243:**1689–1694.

60. **Vinson, C., P. Sigler, and L. McKnight.** 1989. Scissor-grip model for DNA recognition by a family of leucine zipper proteins. *Science* **249:**911–916.

61. **Waldbeser, L., M. E. Tolmasky, L. A. Actis, and J. H. Crosa.** 1993. Mechanisms for negative regulation by iron of the *fatA* outer membrane protein gene expression in *Vibrio anguillarum* 775. *J. Biol. Chem.* **268:**10433–10439.

62. **Walter, M., S. Potter, and J. H. Crosa.** 1983. Iron uptake system mediated by *Vibrio anguillarum* plasmid pJM1. *J. Bacteriol.* **156:**880–887.

63. **Wee, S., J. B. Neilands, M. L. Bittner, B. C. Hemming, B. L. Haymore, and R. Seetharam.** 1988. Expression, isolation, and properties of Fur (ferric uptake regulation) protein of *Escherichia coli* K-2. *Bio. Metals* **1:**62–68.

64. **Weinberg, E. D.** 1984. Iron withholding: a defense against infection and neoplasia. *Physiol. Rev.* **64:**65–100.

65. **Wertheimer, A., M. Tolmasky, L. Actis, and J. H. Crosa.** Structural and functional analysis of mutant Fur proteins with impaired regulatory function. *J. Bacteriol.,* in press.

66. **Williams, P. H.** 1979. Novel iron uptake system specified by ColV plasmids: an important component in the virulence of invasive strains of *Escherichia coli. Infect. Immun.* **26:**925–932.

67. **Wolf, M., and J. H. Crosa.** 1986. Evidence for the role of a siderophore in promoting *Vibrio anguillarum* infections. *J. Gen. Microbiol.* **132:**2949–2952.

Part 6. Other Aspects of Bacterial Pathogenesis

This part of the book illustrates several themes that emerged during Stanley's career. The first is the unusual breadth of scope of his research interests. This breadth is a result of a questioning and open mind that perceives intriguing questions frequently missed by others. In addition, he is undaunted by potential technical difficulties and can usually persuade *someone* to give it a try. This led David Relman down the path of unculturable pathogens. In his chapter (chapter 34), David describes the use of PCR to identify microbial pathogens directly from infected tissue—a process that does not require the ability to cultivate the microbe. This technique is based on the assumption that these microbes have sufficiently conserved 16S rRNA that degenerate primers can be used for the amplification of a portion of the 16S rDNA. Subsequent cloning and sequencing of the amplified fragment allows a determination of phylogenetic relationships. Although this approach is not without its own limitations, it has been successfully used to identify the causative agent of several diseases such as bacillary angiomatosis and Whipple's disease.

The second theme is that fundamental questions in the field of bacterial pathogenesis often touch upon questions of importance for the fields of bacterial physiology, genetics, evolution, and ecology. An example of this is the work of Don LeBlanc on sucrose metabolism by mutans streptococci, the major etiological agent of dental caries (chapter 31). This research deals with aspects of sugar metabolism and transport. Many of the virulence traits of these bacteria are attributable to the transport and/or metabolism of sucrose; these virulence traits include the ability to adhere to enamel, the ability to produce organic acids, and the ability to compete effectively during periods of low carbohydrate availability. Another example is the work Pam Small has been doing on the acid tolerance of *Shigella* spp.; very few of these bacteria are needed to establish an infection, and the ability to survive passage through the acidic environment of the stomach is hypothesized to facilitate this. In the course of her work she discovered that there may be overlaps between the response of *Shigella* spp. to a pH stress and to other stresses; *katF* (*rpoS*), originally characterized as a sigma factor required for the expression of genes induced in the stationary phase, was also implicated in acid tolerance.

The third theme that can be gleaned from this part of the volume, and Stanley's career as a whole, is that basic studies on bacterial pathogenesis *can* lead to practical applications. Some of this was clearly illustrated in an earlier chapter

(chapter 5) on molecular epidemiology by L. Tompkins. It is also beautifully illustrated in the chapter by Gordon Dougan (chapter 33). In that chapter he describes
how knowledge of attenuated *Salmonella typhimurium* and the host response to
attenuated strains can be exploited to develop better oral vaccines. He also describes studies of immunogens, such as cholera toxin subunit B, which are effective at stimulating mucosal immunity, as models for the future development of
nonliving mucosal vaccines.

It is evident from the chapters in this part and from the book as a whole that
Stanley Falkow's interests have touched many aspects of bacterial genetics and
pathogenesis, and undoubtedly will continue to do so. The underlying goal, however, even when working on such fundamental questions as plasmid replication
and transfer, has been to understand the interaction of pathogens with their environment—the human host.

Virginia L. Miller

Molecular Genetics of Bacterial Pathogenesis
Edited by V. L. Miller, J. B. Kaper, D. A. Portnoy, and R. R. Isberg
© 1994 American Society for Microbiology, Washington, DC 20005

Chapter 31

Role of Sucrose Metabolism in the Cariogenicity of the Mutans Streptococci

Donald J. LeBlanc

Dental caries constitutes the most common, and very likely the most expensive, of human diseases (19). Yet, a microbial etiology of caries was clearly established only 34 years ago (8, 12). Among the 200 to 300 bacterial species that have been identified in dental plaque, only a few are considered cariogenic. Some cariogenic bacteria, such as the lactobacilli, are associated with the development of specific types of carious lesions (19). However, the group of streptococci collectively referred to as the mutans streptococci (MS) generally are considered the major etiological agents of this disease. Members of the MS are found in human or animal dental plaque, are able to ferment mannitol and sorbitol, produce extracellular glucans from sucrose, and, with the exception of *Streptococcus ferus,* are cariogenic in animal models. Several properties of these microorganisms are thought to contribute to their virulence, including the ability to adhere to enamel surfaces, to produce large amounts of organic acids from dietary carbohydrates (acidogenicity), to compete effectively in the oral cavity during periods of low carbohydrate availability, and to exhibit high degrees of tolerance for acid environments (aciduricity). Two members of the MS, *Streptococcus mutans* and *Streptococcus sobrinus,* are the most common human cariogenic pathogens (19, 39). The frequency of isolation of either *S. mutans* or *S. sobrinus* from humans has been dependent, to a large extent, on the geographical locations of the subjects under study, although *S. mutans* generally has predominated. For the purposes of discussion in this chapter, the term MS includes only these two genospecies.

Historically, two factors have been associated consistently with the development of human caries: the presence of MS and the presence of fermentable carbohydrates in the diet (19, 39). By far the most common carbohydrate connected with caries formation has been sucrose. It has been demonstrated among several populations that shortly after the introduction of sucrose into the diet there is a marked increase in the incidence of caries. Furthermore, restriction of sucrose in the diet has led to significant declines in decay rates and to concomitant declines

Donald J. LeBlanc • Department of Microbiology, University of Texas Health Science Center at San Antonio, 7703 Floyd Curl Drive, San Antonio, Texas 78284.

in the levels of MS in plaque. Many of the suspected virulence traits of the MS have been linked to the expression by these bacteria of multiple mechanisms for the metabolism of sucrose.

MECHANISMS OF SUCROSE METABOLISM BY MS

Several enzymes or enzyme systems in the MS are associated with the metabolism of sucrose (Fig. 1), and most may be involved in the virulence of these oral pathogens. All MS have glucosyltransferases (GTFs) responsible for the synthesis of extracellular homopolymers of glucose which may be water soluble or insoluble, depending on the proportion of α-1,3 linkages present. To date, three different GTF genes have been cloned and characterized from *S. mutans* (14), and three such genes also have been cloned from *S. sobrinus* (14), although four different GTF enzymes have been described in the latter species (43). The formation of such glucans has been shown to be a virulence factor in the development of smooth surface caries by *S. sobrinus* (19) and *S. mutans* (23), at least in animals. *S. mutans* also exhibits fructosyltransferase (FTF) activity, which catalyzes the synthesis of fructans from sucrose (39), and a genetic determinant encoding FTF has been cloned in *Escherichia coli* from *S. mutans* (29). Evidence that FTF activity also may contribute to the virulence of *S. mutans* has been published (32). *S. sobrinus* exhibits very little if any FTF activity (39). GTFs and FTFs have very high K_ms for sucrose and require high concentrations of substrate for their activity (39). Under these conditions, the free fructose derived from the activity of GTF and the free glucose derived from the activity of FTF may serve as carbon and energy sources and may contribute to the formation of acid by the MS. *S. mutans,* but not *S. sobrinus,* synthesizes large amounts of intracellular glycogen-like polysaccharides in the presence of high carbohydrate-to-nitrogen ratios (39). These storage compounds also may serve as sources of carbon and energy during periods of carbohydrate deficiency.

Less than 10% of the sucrose available to MS is converted to hexose polymers (39). Most of the sucrose is transported into the cell and serves directly as a source of carbon and energy for these bacteria. In 1979, two groups described a sucrose-specific phosphoenolpyruvate-dependent phosphotransferase system (PTS) in the MS (33, 37). Because of the low apparent K_m (70 μM) of this PTS for sucrose, it was suggested that the MS could function as effective scavengers of low levels of sucrose from the environment (37). In this way these organisms would have a competitive advantage in the oral cavity during periods characterized by low levels of environmental sucrose. All MS strains tested have possessed this high-affinity PTS (33, 37). A second enzyme activity, sucrose 6-phosphate hydrolase (S6PH; Fig. 1), also was described; this enzyme was able to split the sucrose 6-phosphate product of the PTS to glucose 6-phosphate and fructose (37, 38). The apparent K_m for the sucrose phosphate of this enzyme was ~300 μM, but it also had activity on sucrose, with an apparent K_m of ~40 mM (38), the latter being similar to the previously reported K_m for intracellular invertase (sucrose hydrolytic) activity (13, 40). A mutant of *S. sobrinus* that was completely devoid of S6PH activity and

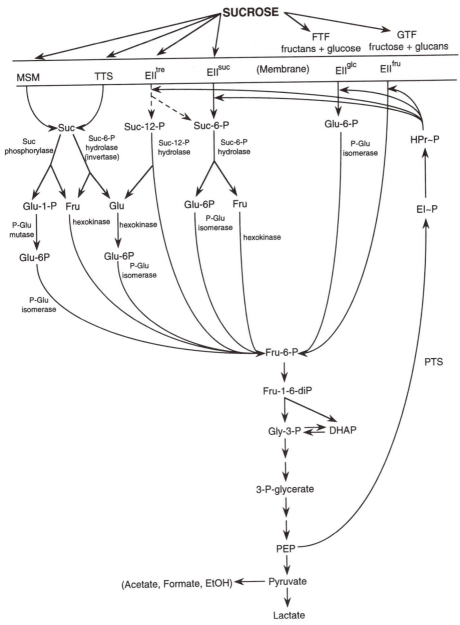

FIGURE 1. Metabolism of sucrose by MS. Abbreviations: FTF, fructosyltransferase; GTF, glucosyl-transferase; MSM, membrane-associated transport proteins of the *msm* operon (28); TTS, very low affinity, non-PTS, third (sucrose) transport system (35); EII^{tre}, EII^{suc}, EII^{glc}, and EII^{fru}, trehalose-, sucrose-, glucose-, and fructose-specific enzyme IIs, respectively, of the PTS; Suc, sucrose; Suc-6-P, sucrose phosphorylated in the C-6 position of the glucose moiety; Suc-12-P, sucrose phosphorylated in the C-6 position of the fructose moiety; Glu, glucose; Glu-1-P and Glu-6-P, glucose phosphorylated in the C-1 and C-6 positions, respectively; Fru, fructose; Fru-6-P and Fru-1,6-diP, fructose phosphory-lated in the C-6 position and in both the C-1 and C-6 carbon positions, respectively; P-Glu, phosphoglu-cose; Gly-3-P, glyceraldehyde-3-phosphate; DHAP, dihydroxyacetone phosphate; PEP, phosphoenol-pyruvate; PTS, PEP-dependent phosphotransferase system; EI~P and HPr~P, phosphorylated EI (enzyme I) and HPr (heat-stable protein), respectively, which are cytoplasmic phosphocarrier proteins of the PTS providing phosphate groups for all sugar-specific EIIs; EtOH, ethanol.

that had lost 80% of its sucrose hydrolytic activity was described (38). These results would suggest that the MS, or at least *S. sobrinus,* possesses a single S6PH enzyme, but may have two or more hydrolytic enzymes able to hydrolyze unphosphorylated sucrose. The genetic determinants encoding sucrose-specific PTS activity, i.e., EII^{suc} (Fig. 1), designated *scrA* (31), and S6PH, designated *scrB* (10, 20), have been cloned from *S. mutans*. More recently, *scrA* and *scrB* also were cloned from *S. sobrinus* (3). The nucleotide base sequences of the cloned DNA encoding both genes from each MS strain were determined (4, 30, 31). The two genes were adjacent to each other on the respective MS genomes and were transcribed in opposite orientations (4, 30). The amino acid sequences of the proteins predicted by the respective *scrA* genes shared only 45% identity, whereas the corresponding SPH proteins predicted by the *scrB* genes shared ~70% amino acid identity (4). Loss of S6PH activity by *S. mutans* because of gene-specific mutagenesis by allelic replacement resulted in the loss of viability of the mutant when it was propagated in the presence of sucrose (20). This was consistent with the results of St. Martin and Wittenberger (38) obtained with chemically induced mutants of *S. sobrinus*. Both sets of results also imply that there is but one S6PH activity in each of these strains.

Soon after its discovery, it became clear that the high-affinity sucrose-specific PTS was neither the major nor the only sucrose transport system associated with the utilization of sucrose as a source of carbon or energy, or both. Slee and Tanzer (34) showed that the sucrose-specific PTS activity was repressed during growth in the presence of high concentrations of sucrose, but that it became derepressed when the cells reached the stationary phase. Those investigators also reported increased sucrose-specific PTS activity at low pH during slow growth on low concentrations of sucrose (2 mM). Ellwood and Hamilton (6) reported that slow growth of *S. mutans* in continuous culture and under conditions of limiting sucrose resulted in the expression of PTS activity, but that growth at faster rates and/or with higher sucrose concentrations required a second transport system for sucrose, since PTS activity was repressed under these conditions (6). Subsequently, Slee and Tanzer (35) assayed for sucrose uptake by mutagenized isolates of *S. mutans* and *S. sobrinus* harvested at different stages of growth in batch culture. The results of those experiments were consistent with the presence of at least three transport systems for sucrose in these MS species: the high-affinity sucrose-specific PTS described above (K_m, ~60 to 70 µM), a second PTS system with much lower affinity for sucrose (K_m, ~240 to 250 µM), and a third, non-PTS transport system (TTS; Fig. 1) with a very low affinity for sucrose (K_m, 3 mM). Recently, Poy and Jacobson (25) studied sucrose transport in a mutant of *S. mutans* containing an insertionally inactivated *scrA* gene. Their transport results provided evidence for a second PTS with a high K_m, 330 µM for sucrose, very similar to the K_m of the low-affinity sucrose PTS described by Slee and Tanzer (35). The results of inhibition studies were consistent with the interpretation that this low-affinity sucrose PTS activity was actually due to a high-affinity trehalose-specific PTS (EII^{tre}; Fig. 1) which recognizes sucrose as a substrate. The very low affinity TTS for sucrose suggested by the results of Slee and Tanzer (35), and probably equivalent to the non-PTS sucrose transport system active at fast growth rates in a chemostat described by Ellwood and Hamilton (6), has yet to be identified.

Russell and collaborators (28) have described a binding protein-dependent sugar transport system in *S. mutans* Ingbritt that is associated with the transport and/or metabolism of several sugars. The genes associated with this system are located contiguously on an 11-kb region of the *S. mutans* chromosome and are cotranscribed as a single operon, termed the *msm* operon. The system apparently is responsible for the transport of melibiose, raffinose, and isomaltotriose. Of particular interest here is the involvement of the system in the metabolism of not only melibiose and isomaltosaccharides but also sucrose. More specifically, the raffinose that is transported by Msm proteins is first cleaved by an α-galactosidase to galactose and sucrose. Further metabolism of the sucrose is facilitated by the product of the Msm system gene, *gtfA,* which is a sucrose phosphorylase that catalyzes the splitting and phosphorylation of sucrose, resulting in the production of glucose 1-phosphate and fructose (Fig. 1). More recently, Tao et al. (41) reported the ability of sucrose to inhibit melibiose transport by *S. mutans,* suggesting that sucrose may be a substrate for transport by the Msm system (Fig. 1). Since *S. sobrinus* does not grow at the expense of either melibiose or raffinose (19), it would not be expected to have a complete Msm system. However, *S. sobrinus* may contain a gene equivalent to *gtfA.*

HIGH-AFFINITY SUCROSE-SPECIFIC PTS OF MS

Kuramitsu and associates cloned and obtained the nucleotide sequences of *scrA* (31) and *scrB* (30) from *S. mutans* GS5. A recombinant plasmid, pMH613, containing a 6.6-kb *Eco*RI fragment from the GS5 strain and encoding both *scrA* and *scrB* was obtained from Kuramitsu. The 6.6-kb *Eco*RI fragment, as well as subfragments internal to each of the structural genes, was purified and used as a hybridization probe against Southern blots of chromosomal DNA from *S. mutans* GS5, *S. mutans* 10449, and *S. sobrinus* 6715 digested with a variety of restriction endonucleases. The results were consistent with the presence in *S. mutans* 10449 of *scrA* and *scrB* genes that were identical in structure and arrangement on the chromosome to their counterparts in *S. mutans* GS5 (2). It was necessary to lower the hybridization stringencies considerably in order to detect any homology between the GS5-derived DNA probes and genomic DNA from *S. sobrinus* 6715 in the Southern blots. DNA fragments internal to the GS5 *scrA* and *scrB* genes each hybridized under nonstringent conditions (requiring less than 50% identity) to a 14-kb *Eco*RI fragment from *S. sobrinus* 6715 genomic DNA. A bacteriophage lambda library was constructed with 12- to 16-kb *Eco*RI fragments from the 6715 chromosome, and plaque lifts were hybridized to a combined GS5 *scrA-scrB* probe. Numerous hybridizing plaques were obtained, but all were false positives (2). Subsequently, two *Hin*dIII fragments were cloned from the 6715 chromosome that hybridized to fragments internal to either *scrA* or *scrB* from GS5. These two fragments, 2.7 kb (*scrA*) and 1.2 kb (*scrB*) in size, hybridized under stringent conditions to the same 14-kb *Eco*RI fragment from the 6715 chromosome as the

GS5 DNA probes, but not to any of the DNA from the bacteriophage lambda library. It was suspected that the 14-kb *Eco*RI fragment contained sequences that were unclonable in *E. coli*. Subsequently, a new λgt10 genomic library was constructed from a *Sau*3A partial digest of *S. sobrinus* 6715 genomic DNA. The two *Hin*dIII fragments described above were used as probes, under stringent conditions, to isolate from the library clones containing both *scrA* and *scrB* (3). Three such clones, of 4.2, 5.5, and 6.6 kb, all contained a common intact 4.2-kb region of the *S. sobrinus* chromosome. The smallest fragment was subcloned onto the low-copy-number vector pGB2 (5), and the recombinant molecule was maintained stably by the *E. coli* clone. This clone expressed sucrose hydrolytic activity on MacConkey agar base supplemented with raffinose (27) or sucrose. Deletion analyses also narrowed the amount of *S. sobrinus* DNA required to express sucrose hydrolytic activity to a 3.5-kb segment. Since previous work (26) had shown that PTS components from gram-positive and gram-negative bacteria either do not complement each other or do so very poorly, it was felt that analysis of the *scrA* gene product would best be accomplished in a gram-positive host, preferably a streptococcal species. *Lactococcus lactis* LM0230 (formerly *Streptococcus lactis*) was derived from a strain able to metabolize sucrose via EII^{suc} and S6PH activities (21). This strain was cured of all resident plasmids, as well as the ability to grow at the expense of sucrose. When the cloned *S. mutans* GS5-derived 6.6-kb *Eco*RI fragment containing *scrA* and *scrB* was subcloned onto the shuttle vector pDL278 (16) and transferred to *L. lactis* LM0230 by electroporation (24), transformants were able to grow at the expense of sucrose, albeit with a generation time of approximately 12 h (2). However, when the 4.2-kb fragment from *S. sobrinus* 6715 was subcloned onto pDL278 and transferred to strain LM0230, these transformants were able to grow at the expense of sucrose with a generation time comparable to that seen with glucose as the carbon and energy source, i.e., ~70 min. These transformants were assayed for sucrose-specific enzyme activities and were shown to express both EII^{suc} and S6PH activities (3). Restriction endonuclease maps of the *scrA* and *scrB* regions of the *S. mutans* GS5 and *S. sobrinus* 6715 chromosomes showed, in keeping with the hybridization results as well as the physiological results described above, that they were quite different.

The nucleotide base sequence of the cloned 4.2-kb fragment of *S. sobrinus* DNA that permitted the growth of *L. lactis* LM0230 on sucrose and that expressed EII^{suc} and S6PH activities was determined (4). The fragment contained two open reading frames (ORFs) transcribed divergently with their respective initiation codons 192 bp apart. The ORF corresponding to *scrB* was 1,437 nucleotides long and predicted a hydrophilic protein of 54,501 Da. This protein exhibited ~70% amino acid identity with that predicted by *scrB* of *S. mutans* GS5. The ORF corresponding to *scrA* had no stop codon in the 4.2-kb fragment and predicted a protein of approximately the same size as EIII-dependent sugar-specific EIIs. The previously cloned 2.7-kb *Hin*dIII fragment from the *S. sobrinus* chromosome, which was used as the *S. sobrinus scrA* probe, extended beyond the 4.2-kb fragment and was also sequenced. The results indicated that intact *scrA* contained 1,896 bases and actually encoded an EIII-independent EII^{suc} enzyme of 66,529 Da. The segment of the gene missing in the 4.2-kb DNA fragment was equivalent to the EIII domain of this protein, shared significant homology with EIII^{glc} of gram-negative bacteria, and probably was replaced by the EIII^{glc} of *L. lactis* LM0230 recombinant clones containing the 4.2-kb *S. sobrinus*-derived DNA fragment. The predicted amino acid sequence of intact *S. sobrinus* EII^{suc} exhibited only 45% identity with that predicted by *scrA* of *S. mutans,* indicating considerable divergence between the two MS genospecies with respect to this sucrose catabolic enzyme. The transcription initiation sites for *scrA* and *scrB* of *S. sobrinus* were determined by primer extension analyses, and the putative promoter regions of these two genes were partially overlapping.

NEW TOOLS AND FUTURE PROSPECTS

The 4.2-kb fragment of *S. sobrinus* DNA that provided *L. lactis* LM0230 with the capacity to grow at the expense of sucrose with a generation time of just over 1 h encoded both EII^{suc} and S6PH activities. However, the nucleotide sequence of this cloned DNA revealed the presence of a complete *scrB* (S6PH) gene, but only two of three domains of the putative *scrA* (EII^{suc}) gene. Additional sequence analysis of a cloned 2.7-kb *Hin*dIII fragment from *S. sobrinus* showed that the *scrA* gene was indeed composed of three domains and that the missing domain in the 4.2-kb cloned fragment corresponded to an EIII component. The interpretation of these results has been that the missing EIII domain was being provided in *L. lactis* by another EIII, possibly EIII^{glc}. Unfortunately, the only clones obtained in the λgt10 library with *scrA*-specific DNA ended at the same location in the *scrA* gene (3). Since all of these clones were obtained from a *Sau*3A partial digest of *S. sobrinus* genomic DNA, it is possible that the site for this enzyme at the EII-EIII junction is a preferred site and that a cloned fragment with a complete *scrA-scrB* gene region will never be obtained from such a library. However, the 6.6-kb *Eco*RI fragment from *S. mutans* GS5 contains a complete *scrA-scrB* gene region. When this was subcloned on pDL278 and transferred to *L. lactis* LM0230, growth at the expense of sucrose was very poor, with a generation time of more than 12 h. The initial assumption was that because the *E. coli* clone had been around for some time and also had been manipulated without selection for function, one or more mutations in the fragment led to poor expression of the *scrA* and/or *scrB* genes. One possibility that may account for all of the results obtained with the cloned *S. sobrinus* and *S. mutans* DNAs is that of inhibition of the EII^{suc} by the complementing *L. lactis* EIII^{glc} or some other EIII factor. This also might imply modulation of *scrA*-encoded EII^{suc} by unrelated EIII factors in *S. sobrinus* or *S. mutans,* or both. It should be possible to obtain answers to these questions by cloning a complete *scrA-scrB* region from *S. sobrinus* and again from *S. mutans* and by establishing whether all or only a portion of the respective *scrA* genes with a functional *scrB* gene will provide *L. lactis* with the ability to grow at a reasonable rate at the expense of sucrose.

An extremely useful feature of recombinant DNA technology is the ability to manipulate cloned genetic determinants from a particular organism, to replace the wild-type determinant with the altered one, and then to test the effects of such manipulation on the phenotype associated with that determinant. Such a process requires an ability to introduce DNA into the organism under study. Unlike *S. mutans*, many strains of which are naturally transformable, there has been no mechanism available to reintroduce recombinant DNA molecules into *S. sobrinus*. Numerous attempts to establish competence in *S. sobrinus*, either natural or artificially induced, have been unsuccessful. Recently, a mobilization gene, *mob*, was discovered on the oral streptococcal plasmid pVA380-1 (15). The product of this gene facilitated the mobilization, by the conjugative streptococcal plasmid pAMβ1, of a pVA380-1 derivative in *S. sobrinus* 6715. In addition, the mobilizing plasmid, pAMβ1, was lost from the transconjugants at a very high frequency under nonselective conditions. These features of the transfer system, plus the observed loss of stability of pVA380-1 under nonselective conditions following deletion of its minus origin (16), indicated that it may hold promise as a tool for such genetic manipulations in *S. sobrinus* as allelic replacement and transposon delivery. Thus, a new *E. coli*-*Streptococcus* shuttle vector is being constructed with the following components: (i) the pVA380-1 plus origin, *rep* gene, *mob* gene, RSa, and a partially deleted minus origin; (ii) the replication origin of a low-copy-number *E. coli* plasmid, for greater stability in *E. coli* of cloned *S. sobrinus* DNA; (iii) a *lacZ*-containing multiple cloning site; and (iv) either the *aphA3* (42) or *aad-9* (17) gene, each from *Enterococcus faecalis* and encoding constitutive kanamycin or spectinomycin resistance, respectively, in gram-positive and gram-negative bacterial hosts. The availability of such a vector will facilitate studies on the role of each of the different sucrose metabolic enzyme systems of *S. sobrinus* in its virulence with respect to dental caries.

On the basis of results described on page 468, Slee and Tanzer (35) concluded that *S. mutans* and *S. sobrinus* possess at least three transport systems for sucrose, a high-affinity and a low-affinity PTS and a very low affinity non-PTS transport system (TTS). The genetic determinants encoding the presumed sucrose-specific, high-affinity PTS, i.e., the *scrA* and *scrB* regions, of *S. mutans* (10, 20, 30, 31) and *S. sobrinus* (3, 4) have been cloned and sequenced. Very little information on the actual role of *scrA* in the metabolism of sucrose by these MS species is available. Transposon Mu dE insertions into *scrA* of *S. mutans* abolished the PEP-dependent transport of sucrose at concentrations of 25 μM (high-affinity sucrose PTS), but not 100 mM (low-affinity sucrose PTS [31]). One of these mutants was used by Poy and Jacobson (25) to provide evidence that the activity of an EII^tre (Fig. 1) was responsible for the low-affinity sucrose PTS activity. Ellwood and Hamilton (6) correlated the ability of *S. mutans* Ingbritt to transport sucrose via PTS and non-PTS mechanisms at different growth rates under conditions of limiting sucrose. They were able to measure PTS activity only at very slow growth rates, suggesting that at faster growth rates a non-PTS transport mechanism was responsible for sucrose transport. Furthermore, under conditions conducive to PTS activity, the *S. mutans* cells were heterofermentative, producing only small amounts of lactic acid (see Fig. 1 for other fermentation products from pyruvate), whereas at the faster growth rates, when sucrose transport apparently occurred via a non-PTS mechanism, the cells exhibited homolactic fermentation of sucrose. In light of the results of Hillman (11), which clearly implied a role of lactic acid production in caries formation by *S. mutans,* the results of Ellwood and Hamilton (6) suggest that the PTS may play no direct role in caries induction, but rather may function only as a scavenger of sucrose during periods of sucrose limitation.

Growth of bacteria in batch culture represents a constantly changing system with respect to concentration of the carbon and energy source, actual growth rates, and the general physiological status of the cells. In the natural environment, bacteria generally grow at much slower rates than can be maintained for any reasonable amount of time in batch culture. Estimates of mean generation times for the bacterial components of dental plaque vary between 3 and 14 h (9, 22, 36). Only in continuous culture, e.g., in a chemostat, can such conditions be established and maintained for essentially unlimited periods of time. Furthermore, generation times at specific substrate concentrations (limiting versus excess) or even at different limiting concentrations can be regulated by varying the dilution rates (7). Clearly, continuous culture techniques offer many advantages for studying the role of different enzyme systems in the metabolism of a given substrate. Thus, future studies on the role of the *scrA*-encoded sucrose-specific PTS activity in the metabolism of sucrose by the two species of MS will combine the molecular technique of allelic replacement, now feasible in both *S. mutans* and *S. sobrinus,* with the physiologically relevant technique of continuous culture. Results of comparative studies of the wild-type versus *scrA*-deficient MS mutants should provide valuable information on the role of the high-affinity sucrose-specific (*scrA*-encoded) PTS in the metabolism of sucrose, acid production, and the types of fermentation end products that contribute to the acid pH in both *S. mutans* and *S. sobrinus.* On the basis of K_m determinations for sucrose-specific PTS activities in the wild-type and mutant strains, it should be possible to confirm the presence of a second sucrose-active PTS as well as its affinity for the substrate. Evidence for modulation of sucrose-specific PTS activity by other EIIIs, e.g., EIII^{glc}, also may be obtained from the results of such studies.

Whether sucrose enters the cell as a phosphorylated intermediate, via the activity of one or more PTS-associated EIIs, or as unalterred sucrose, e.g., via a TTS (31) or an Msm system (41), the disaccharide must be split into two monosaccharides for further metabolism (Fig. 1). Inactivation of the S6PH activity by chemical mutagenesis of *S. sobrinus* (13) or by allelic replacement of the *scrB* gene of *S. mutans* (20) resulted in the loss of viability of each strain when grown in the presence of sucrose, presumably because of the accumulation of toxic sucrose phosphate. If there is a second PTS able to transport sucrose into the cell as sucrose phosphate, is there also a second sucrose phosphate hydrolase (SPH) activity? Why, then, is sucrose toxic to *scrB*-defective mutants? It may be that the second SPH does not recognize sucrose phosphate when phosphorylated in the C-6 position of glucose, because the fructose moiety of sucrose is phosphorylated by this EII (Fig. 1). Chassy and Porter (1) purified to homogeneity the S6PH of *S. sobrinus* and reported a K_m for sucrose phosphate of 210 μM and a K_m for sucrose of 120 mM. Lunsford and Macrina (20) purified from an *E. coli* clone the S6PH protein encoded by *scrB* of *S. mutans* GS5. This enzyme had a K_m for sucrose phosphate of 80 μM and a K_m for sucrose of 180 mM. It has been suggested that intracellular invertase (Fig. 1) activity is actually due to S6PH. However, St.

Martin and Wittenberger (38) also showed that an S6PH mutant of *S. sobrinus* had no S6PH activity but retained 20% of its normal sucrose hydrolytic activity. The results of continuous culture experiments and assays for enzymatic activities conducted with *scrA-scrB* double mutants should help to elucidate the role of S6PH in the metabolism of sucrose by *S. mutans* and *S. sobrinus*. They also should provide data in support of the presence in these species of a second SPH activity, as well as a true invertase, if they exist.

Finally, the availability of gene transfer systems and appropriate vector molecules will permit comparative analyses of the regulation of sucrose metabolism in the MS. For instance, the synthesis of EII^{suc} is thought to be inducible on the basis of the presence of detectable levels of activity when *S. sobrinus* is grown in the presence of sucrose, but not when it is grown on other carbohydrates (38). On the other hand, Lodge and Jacobson (18) reported increased sucrose PTS activity in cells that had been starved for sucrose relative to the activity detected in mid-exponential-phase cells growing in the presence of sucrose. The latter investigators suggested that the differences in PTS activity were not due to increased levels of PTS protein, but rather to differences in the activities of preexisting proteins. An attempt to elucidate the regulation of *scrA* and the activity of its product, EII^{suc}, might involve the construction of a mutant that will retain an intact *scrA* gene and an intact *scrB* gene in its chromosome, each under the control of the *scrA-scrB* promoter region, but that will also contain a reporter gene, e.g., β-galactosidase or *cat*, expression of which will be under the control of the *scrA* promoter. Transcriptional and translational control of *scrA* by growth in the presence of different carbohydrates could be monitored by measuring the level of β-galactosidase or chloramphenicol acetyltransferase activity, whereas the regulation of EII^{suc} activity could be monitored by assays for PEP-dependent sucrose phosphorylation. Thus, it should be possible to distinguish between regulation at the level of enzyme synthesis versus enzyme activity.

SUMMARY

MS are generally considered the major etiological agents of dental caries. Many of the suspected virulence traits of the MS, such as the ability to adhere to enamel surfaces, to produce large amounts of organic acids at the expense of dietary carbohydrates, and to compete effectively in the oral cavity during periods of low carbohydrate availability, are attributable to the presence of multiple enzymes and enzyme systems associated with the transport and/or metabolism of sucrose. There are three to four very low affinity GTF enzymes that catalyze the production of water-soluble and water-insoluble glucans in the presence of high concentrations of sucrose. The free fructose produced from sucrose by the GTF activity may serve as a carbon and an energy source and may assist in the production of acid. Similarly, the glucose moiety remaining after the action of *S. mutans*-encoded FTF activity also may serve as a source of carbon and energy. The glucose and fructose polymers synthesized from sucrose, via the activities of GTF and FTF, respectively, contribute to the attachment of the MS to enamel surfaces and to caries production in animal models. *S. mutans* is able to synthesize and store intracellular polysaccharides from sucrose. *S. mutans* and *S. sobrinus* each produce at least three different enzymes or enzyme systems directly associated with the transport of sucrose into the cell. These include a high-affinity sucrose-specific PTS, a lower-affinity PTS, the primary substrate of which may be trehalose, and a very low affinity ATP or proton motive force-dependent transport system. The two PTS proteins transport sucrose as sucrose phosphate, which is then hydrolyzed by sucrose phosphate hydrolase, only one of which has been described in members of each of the two species. The non-PTS system would transport sucrose unchanged into the cell. Hydrolysis of the sucrose molecule might be catalyzed by a sucrose phosphate hydrolase, a true invertase, or the product of *S. mutans* *gtfA* or its equivalent in *S. sobrinus*. The interactions of the various enzymatic activities for the synthesis of extracellular and intracellular polymers from sucrose, sucrose transport, and sucrose hydrolysis under different physiological conditions have yet to be elucidated. It is now possible to combine genetic, molecular, and physiological approaches to initiate studies on the interactions of these sucrose metabolic enzymes and to assess their precise roles in the virulence of MS.

ACKNOWLEDGMENT. Work conducted in my laboratory was supported in part by Public Health Service grant DE08915 from the National Institute for Dental Research.

REFERENCES

1. **Chassy, B. M., and E. V. Porter.** 1982. Sucrose-6-phosphate hydrolase from *Streptococcus mutans*. *Methods Enzymol.* **90**:556–559.
2. **Chen, Y. M.** 1993. Genetic analysis of *scrA* and *scrB* from *Streptococcus sobrinus* strain 6715. Ph.D. thesis. University of Texas Health Science Center at San Antonio, San Antonio.
3. **Chen, Y. M., and D. J. LeBlanc.** 1992. Genetic analysis of *scrA* and *scrB* from *Streptococcus sobrinus* 6715. *Infect. Immun.* **60**:3739–3746.
4. **Chen, Y.-Y. M., L. N. Lee, and D. J. LeBlanc.** 1993. Sequence analysis of *scrA* and *scrB* from *Streptococcus sobrinus* 6715. *Infect. Immun.* **61**:2602–2610.
5. **Churchward, G., D. Belin, and Y. Nagamine.** 1984. A pSC101-derived plasmid which shows no sequence homology to other commonly used cloning vectors. *Gene* **31**:165–171.
6. **Ellwood, D. C., and I. R. Hamilton.** 1982. Properties of *Streptococcus mutans* Ingbritt growing on limiting sucrose in a chemostat: repression of the phosphoenolpyruvate phosphotransferase transport system. *Infect. Immun.* **36**:576–581.
7. **Ellwood, D. C., J. R. Hunter, and V. M. C. Longyear.** 1974. Growth of *Streptococcus mutans* in a chemostat. *Arch. Oral Biol.* **19**:659–664.
8. **Fitzgerald, R. J., and P. H. Keyes.** 1960. Demonstration of the etiologic role of streptococci in experimental caries in the hamster. *J. Am. Dent. Assoc.* **61**:9–19.
9. **Gibbons, R. J.** 1964. Bacteriology of caries. *J. Dent. Res.* **46**(Suppl.):1021–1028.
10. **Hayakawa, M., H. Aoki, and H. K. Kuramitsu.** 1986. Isolation and characterization of the sucrose-6-phosphate hydrolase gene from *Streptococcus mutans*. *Infect. Immun.* **53**:582–586.
11. **Hillman, J. D.** 1978. Lactate dehydrogenase mutants of *Streptococcus mutans;* isolation and preliminary characterization. *Infect. Immun.* **21**:206–212.
12. **Keyes, P. H.** 1960. The infectious and transmissible nature of experimental dental caries. Findings and implications. *Arch. Oral Biol.* **1**:304–320.
13. **Kuramitsu, H. K.** 1973. Characterization of invertase activity from cariogenic *Streptococcus mutans*. *J. Bacteriol.* **115**:1003–1010.
14. **Kuramitsu, H. K., T. Shiroza, S. Sato, and M. Hayakawa.** 1987. Genetic analysis of *Streptococcus*

mutans glucosyltransferases, p. 209–211. *In* J. J. Ferretti and R. Curtiss III (ed.), *Streptococcal Genetics*. American Society for Microbiology, Washington, D.C.

15. **LeBlanc, D. J., Y.-Y. M. Chen, and L. N. Lee.** 1993. Identification and characterization of a mobilization gene in the streptococcal plasmid, pVA380-1. *Plasmid* **30:**296–302.

16. **LeBlanc, D. J., L. N. Lee, and A. Abu-Al-Jaibat.** 1992. Molecular, genetic, and functional analysis of the basic replicon of pVA380-1, a plasmid of oral streptococcal origin. *Plasmid* **28:**130–145.

17. **LeBlanc, D. J., L. N. Lee, and J. M. Inamine.** 1991. Cloning and nucleotide base sequence analysis of a spectinomycin adenyltransferase AAD(9) determinant from *Enterococcus faecalis*. *Antimicrob. Agents Chemother.* **35:**1804–1810.

18. **Lodge, J., and G. R. Jacobson.** 1988. Starvation-induced stimulation of sugar uptake in *Streptococcus mutans* is due to an effect on the activities of preexisting proteins of the phosphotransferase system. *Infect. Immun.* **56:**2594–2600.

19. **Loesche, W. J.,** 1986. Role of *Streptococcus mutans* in human dental decay. *Microbiol. Rev.* **50:** 353–380.

20. **Lunsford, R. D., and F. L. Macrina.** 1986. Molecular cloning and characterization of *scrB*, the structural gene for the *Streptococcus mutans* phosphoenolpyruvate-dependent sucrose phosphotransferase system sucrose-6-phosphate hydrolase. *J. Bacteriol.* **166:**426–434.

21. **McKay, L. L., K. A. Baldwin, and J. D. Efstathiou.** 1976. Transductional evidence for plasmid linkage of lactose metabolism in *Streptococcus lactis* C2. *Appl. Environ. Microbiol.* **32:**45–52.

22. **Mikx, F. H. M., and M. Svanberg.** 1978. Considerations about microbial interactions in relation to modification of the microflora of dental plaque, p. 109–118. *In* B. G. Bibby and R. J. Shern (ed.), *Proceedings, Methods of Caries Prediction*, special supplement to *Microbial Abstracts*. Information Retrieval, Inc., Washington, D.C.

23. **Munro, C., S. M. Michalek, and F. L. Macrina.** 1991. Cariogenicity of *Streptococcus mutans* V403 glucosyltransferase and fructosyltransferase mutants constructed by allelic exchange. *Infect. Immun.* **59:**2316–2323.

24. **Powell, I. B., M. G. Achen, A. J. Hillier, and B. E. Davidson.** 1988. A simple and rapid method for genetic transformation of lactic streptococci by electroporation. *Appl. Environ. Microbiol.* **54:** 655–660.

25. **Poy, F., and G. R. Jacobson.** 1990. Evidence that a low-affinity sucrose phosphotransferase activity in *Streptococcus mutans* GS-5 is a high-affinity trehalose uptake system. *Infect. Immun.* **58:** 1479–1480.

26. **Reizer, J., M. H. Saier, Jr., J. Deutscher, F. Grenier, J. Thompson, and W. Henstenberg.** 1988. The phosphoenolpyruvate:sugar phosphotransferase system in gram-positive bacteria: properties, mechanism, and regulation. *Crit. Rev. Microbiol.* **15:**297–338.

27. **Robeson, J. P., R. G. Barletta, and R. Curtiss III.** 1983. Expression of a *Streptococcus mutans* glucosyltransferase gene in *Escherichia coli*. *J. Bacteriol.* **153:**211–221.

28. **Russell, R. R. B., J. Aduse-Opoku, I. C. Sutcliffe, L. Tao, and J. J. Ferretti.** 1992. A binding protein-dependent transport system in *Streptococcus mutans* responsible for multiple sugar metabolism. *J. Biol. Chem.* **267:**4631–4637.

29. **Sato, S., and H. K. Kuramitsu.** 1986. Isolation and characterization of a fructosyltransferase gene from *Streptococcus mutans* GS5. *Infect. Immun.* **52:**166–170.

30. **Sato, Y., and H. K. Kuramitsu.** 1988. Sequence analysis of the *Steptococcus mutans scrB* gene. *Infect. Immun.* **56:**1956–1960.

31. **Sato, Y., F. Poy, G. R. Jacobson, and H. K. Kuramitsu.** 1989. Characterization and sequence analysis of the *scrA* gene encoding enzyme II^{scr} of the *Streptococcus mutans* phosphoenolpyruvate-dependent sucrose phosphotransferase system. *J. Bacteriol.* **171:**263–271.

32. **Schroeder, V. A., S. M. Michalek, and F. L Macrina.** 1989. Biochemical characterization and evaluation of virulence of a fructosyltransferase-deficient mutant of *Streptococcus mutans* V403. *Infect. Immun.* **57:**3560–3569.

33. **Slee, A. M., and J. M. Tanzer.** 1979. Phosphoenolpyruvate-dependent sucrose phosphotransferase activity in *Streptococcus mutans* NCTC 10449. *Infect. Immun.* **24:**821–828.

34. **Slee, A. M., and J. M. Tanzer.** 1980. Effect of growth conditions on sucrose phosphotransferase activity of *Streptococcus mutans*. *Infect. Immun.* **27:**922–927.

35. **Slee, A. M., and J. M. Tanzer.** 1982. Sucrose transport by *Streptococcus mutans:* evidence for multiple transport systems. *Biochim. Biophys. Acta* **692:**415–424.

36. **Socransky, S. S., A. D. Manganielli, D. Propas, V. Orum, and J. van Houte.** 1977. Bacteriological studies of developing supragingival dental plaque. *J. Periodontal Res.* **12:**90–106.

37. **St. Martin, E. J., and C. L. Wittenberger.** 1979. Characterization of a phosphoenolpyruvate-dependent sucrose phosphotransferase system in *Streptococcus mutans. Infect. Immun.* **24:**865–868.

38. **St. Martin, E. J., and C. L. Wittenberger.** 1979. Regulation and function of sucrose 6-phosphate hydrolase in *Streptococcus mutans. Infect. Immun.* **26:**487–491.

39. **Tanzer, J. M.** 1992. Microbiology of dental caries, p. 377–424. *In* J. Slots and M. Taubman (ed.), *Contemporary Oral Microbiology and Immunology.* Mosby Year Book, St. Louis.

40. **Tanzer, J. M., A. T. Brown, and M. F. McInerney.** 1973. Identification, preliminary characterization, and evidence for regulation of invertase in *Streptococcus mutans. J. Bacteriol.* **116:**192–202.

41. **Tao, L., I. C. Sutcliffe, R. R. B. Russell, and J. J. Ferretti.** 1993. Transport of sugars, including sucrose, by the *msm* transport system of *Streptococcus mutans. J. Dent. Res.* **72:**1386–1390.

42. **Trieu-Cuot, P., and P. Courvalin.** 1983. Nucleotide sequence of the *Streptococcus faecalis* plasmid gene encoding the 3′-5′-aminoglycoside phosphotransferase type III. *Gene* **23:**331–341.

43. **Yamashita, Y., N. Hanada, and T. Takehara.** 1989. Purification of a fourth glucosyltransferase from *Streptococcus sobrinus. J. Bacteriol.* **171:**6265–6270.

Molecular Genetics of Bacterial Pathogenesis
Edited by V. L. Miller, J. B. Kaper, D. A. Portnoy, and R. R. Isberg
© 1994 American Society for Microbiology, Washington, DC 20005

Chapter 32

How Many Bacteria Does It Take To Cause Diarrhea and Why?

Pamela L. C. Small

Diarrheal disease is one of the most common diseases of humans. Although diarrhea is more of an annoyance than a serious health threat to healthy adults, it poses a serious risk to infants and young children throughout the developing world. The causative agents of infectious diarrhea and dysentery can be viral, bacterial, or protozoal. All of these pathogens share a common route of infection, the oral-fecal route, and yet the infectious dose required to cause diarrhea varies between species by several orders of magnitude (Table 1). Whereas 10^8 *Vibrio cholerae* are required to cause cholera (7), as few as 10 *Shigella dysenteriae* are able to do so (13, 14). The specific determinants of the infective dose are largely unknown, although both host and pathogen factors are clearly important.

In order to discuss the determinants of infective dose, it is first necessary to examine how infective dose is defined and what experimental methods are used in its determination. The infective dose for a pathogen usually refers to the lowest inoculum necessary to produce disease in a susceptible host. The term is somewhat misleading, in that "disease" rather than "establishment of infection" is implied by the term in most studies. Determination of infective dose is usually based on volunteer studies, although evidence from epidemiological studies can make a significant contribution toward arriving at a meaningful definition.

VOLUNTEER STUDIES

For obvious reasons, volunteer studies have served as the "gold standard" for determining infective dose, and these are the studies most often referenced. Such studies have the advantage of being prospective studies in which all experimental conditions can be well defined and controlled. Subjects are healthy adults, the inoculum is administered in a defined medium and is carefully quantitated, a well-characterized strain is usually used for the inoculum, and the volunteer's symptoms are closely monitored.

There are, however, some serious limitations to volunteer studies which may

Pamela L. C. Small • Rocky Mountain Laboratories, Hamilton, Montana 59840.

TABLE 1
Infective doses of enteric pathogens required to cause disease on the basis of volunteer studies

Pathogen	Infective dose (CFU)	Reference(s)
Campylobacter jejuni	500–800	4, 5, 42
Escherichia coli		
Enteropathogenic	$10^8–10^{10a}$	10, 12, 34, 43
Enterotoxigenic	10^{6a}	36
Enteroinvasive	$10^{6a}–10^{10}$	12
Rotavirus	$1–10^a$	8, 26, 45
Salmonella typhi	$10^5–10^9$	6, 27
Salmonella spp., not *Salmonella typhi*	$10^5–10^{10}$	6
Shigella dysenteriae, Shigella flexneri, and *Shigella sonnei*	10–200	13, 14
Vibrio cholerae	$10^{4a}–10^8$	7

[a] The inoculum was administered with or following the ingestion of sodium bicarbonate.

make it difficult to use the results of some of the studies as a basis for making valid inferences about the infective dose in a natural infection. Ideally, subjects in a volunteer study should be representative of the population in which diarrheal disease occurs. However, some enteric pathogens such as rotavirus and entero-pathogenic *Escherichia coli* (EPEC) cause diarrhea primarily in infants and small children (8, 10, 41, 43, 45). Clearly, it is not possible to use children or infants in volunteer studies, and so adults must be used, even though some of these patho-gens have a limited ability to cause disease in adults. For example, in one study an inoculum of 10^8 EPEC produced symptoms in only one of five volunteers (34). Yet, the ability of EPEC to spread readily from person to person in nursery or day-care environments (30, 35) suggests that the infective dose in infants may be several orders of magnitude lower than that shown in volunteer studies with adults.

A second problem with the selection of subjects for these studies is that the immune status of the subjects with respect to the organism being tested is not always known. The high frequency of diarrheal disease in many populations means that many people will have recently been infected with an enteric pathogen. To ensure that the volunteer is in fact a susceptible host, wherever possible volunteers need to be screened for antibodies to the pathogen being tested. This is not always done.

A third problem common to volunteer studies is that, at most, a few and often only a single strain of a pathogen can be tested. It is clear at least with some pathogens such as *Campylobacter jejuni* that there are strains which differ consid-erably in their virulences and infective doses (4, 5). Particularly in older studies of infective dose, the isolates used may not have been well characterized with respect to virulence traits and may not have been wisely chosen. Blaser and New-man (6), in an excellent review of human salmonellosis, discuss this issue at length. For example, *Salmonella pullorum,* a chicken-adapted *Salmonella* species with little ability to produce disease in humans, was used in a study on the infective dose for salmonellosis (6). Not surprisingly, a large inoculum was required to produce disease.

Although volunteer studies often have well-defined parameters, all studies do not share the same parameters, making comparison between studies difficult. For example, in volunteer studies with rotavirus, volunteers were given sodium bicarbonate before ingestion of the inoculum (26, 45) because low pH is known to inactivate the virus. Results of studies in which stomach acidity is neutralized with sodium bicarbonate cannot be realistically compared with results of studies which do not employ this treatment. Despite these limitations, volunteer studies have contributed many useful data. They are most valid when it is possible to use subjects who are representative of the population capable of being infected and when the inoculum is administered without neutralization of gastric acidity.

EPIDEMIOLOGICAL STUDIES

Another approach to studying infective dose is the use of epidemiological data. Most of these studies are retrospective; they work backward from the cases of disease. By identifying a source of infection and reconstructing a model of transmission, inferences about infective dose can sometimes be drawn. Retrospective studies can be particularly instructive when an outbreak of diarrhea can be traced to the consumption of a common food or beverage. If unconsumed portions of the suspected contaminated food or beverage are available, it may be possible to isolate the pathogen from the source and calculate the concentration of organisms in the infected food at the time of ingestion. This information combined with details about the amount of contaminated food ingested by affected individuals can lead to an estimate of what the infective dose must have been.

Determination of the route of transmission can also be helpful in making broad statements about infective dose. Pathogens which are transmitted almost exclusively through the ingestion of common food are believed to require a higher infective dose to produce disease than organisms capable of spreading directly from person to person. This is because food-borne pathogens have often had the opportunity to replicate to high concentrations in foods before consumption. Direct person-to-person spread is typical of enteric pathogens, such as *Shigella* species, which have a low infective dose. With these pathogens, secondary spread within a family is common (46).

The fact that most infectious infant diarrhea is transmitted from person to person rather than via a common source (30) argues for a low infective dose for these pathogens. In this case, however, it is important not to ignore the role that the naive immune status of nonnursing infants may play in determining the infectious dose for diarrheal disease.

Although epidemiological studies lack the solid quantitative data and controlled environment characteristic of volunteer studies, their strength is that they describe the natural history of the disease. In most cases neither epidemiological investigations nor volunteer studies alone can provide a realistic determination of infective dose. However, with data from both types of studies it is possible to determine a realistic infective dose for many enteric pathogens. This is particularly true for pathogens that produce disease with either a very high or a very low infective dose.

ENTERIC PATHOGENS CAPABLE OF PRODUCING DISEASE WITH A LOW INFECTIVE DOSE

Both volunteer studies and epidemiological data concur in assigning a low infective dose to *Shigella* (13, 14, 38, 46) and *C. jejuni* (4, 5) diarrhea. Evidence is particularly strong in the case of *Shigella* species, in which numerous volunteer studies with three different *Shigella* species have shown that the infective dose for shigellosis is between 10 and 500 organisms (13, 14). Although there are common-source food-borne outbreaks of shigellosis, direct person-to-person spread is the most common form of transmission (14). Secondary spread is high, and hand washing is extremely effective in decreasing shigellosis when outbreaks occur (31, 38, 41, 46). The ability and frequency with which *Shigella* species infect researchers working with the organism is legend; the majority of people who do extensive research on this organism have been infected with their own strains (16).

Volunteer studies with *C. jejuni* report that, at least with some isolates, disease can be caused by as few as 500 to 800 organisms (5, 42). The data are complicated by the fact that different isolates of *C. jejuni* vary considerably in virulence (4, 5, 18), and there is clearly strain-to-strain variability in *C. jejuni* with respect to infective dose (5). In one volunteer study with *C. jejuni,* an inoculum of 10^8 organisms resulted in infection but no disease in five subjects. Disease did occur in two of four subjects who ingested sodium bicarbonate along with the bacterial inoculum (4). In the same study with another isolate, 800 organisms were sufficient to cause disease. Although person-to-person spread has not been documented as often with *C. jejuni* as with *Shigella* species, epidemiological studies clearly show that direct transmission from animals, particularly cats and poultry, to humans is probably extremely common (3).

Rotavirus is one of the most common causative agents of diarrheal disease in infants throughout the world. It is not a common cause of diarrheal disease in adults. Although several volunteer studies on the infective dose of rotavirus have been conducted, the subjects for those studies were, of necessity, adults. Those studies place the infective dose of rotavirus at about 10 focus-forming units (8, 26, 45), but the fact that subjects were treated with sodium bicarbonate to neutralize gastric acidity makes it difficult to evaluate those studies. However, epidemiological studies on rotavirus support a low infective dose for infants (8, 41).

ENTERIC PATHOGENS THAT REQUIRE A HIGH INFECTIVE DOSE TO PRODUCE DISEASE

At the other end of the spectrum are organisms such as *V. cholerae* and enterotoxigenic *E. coli* (ETEC); evidence from both volunteer studies and epidemiological data supports a high infective dose for these pathogens to produce disease. In volunteer studies with these pathogens, more than 10^8 organisms were required to cause diarrhea (7, 36). Epidemiological data support a high infective dose as well, since direct person-to-person transmission of disease caused by these organisms occurs infrequently (30, 36).

ENTERIC PATHOGENS CHARACTERIZED BY AN INTERMEDIATE INFECTIVE DOSE TO PRODUCE DISEASE

Of all the bacterial pathogens known to cause diarrheal disease, evidence on the infective dose of *Salmonella* species is perhaps the most difficult to evaluate. This is partially due to a lack of consistency between volunteer studies and epidemiological data as well as to the fact that several *Salmonella* species are capable of producing enteric disease in humans. The host specificity of *Salmonella* species differs greatly from species to species. Thus, whereas *Salmonella typhi* is able to cause disease only in humans and primates and *S. pullorum* causes disease almost exclusively in poultry, other *Salmonella* species have a less restricted host range. For example, *Salmonella cholerae-suis* and *Salmonella typhimurium* are host adapted to pigs and mice, respectively, yet they cause disease in humans as well.

All volunteer studies of the infective doses of species of *Salmonella* that exist report that at least 10^5 organisms are required to produce disease (6, 27). It might be expected that since *S. typhi* is a human-adapted pathogen the infective dose for typhoid fever would be lower than that for other salmonella infections; however, this does not appear to be the case. Studies on the infective dose required for salmonellosis have been critically evaluated by Blaser and Newman (6), who argue, using several lines of evidence, that the infectious dose of *S. typhi* must be considerably lower than that determined in volunteer studies. They suggest that the isolates of *S. typhi* used in those studies were not fully virulent. In addition, epidemiological evidence based on the inferred concentration of infectious agents in food or water associated with common-source outbreaks of salmonellosis often suggest an infective dose much lower than 10^5 organisms (6). The evidence taken together suggests that, at least in some cases, the infective dose of *S. typhi* may be lower than that reported in volunteer studies, but it is probably considerably higher than that for *Shigella* species or *C. jejuni*.

DETERMINANTS OF INFECTIVE DOSE: ROLE OF THE GASTRIC BARRIER

It is clear from the above discussion that the inocula of enteric pathogens required to produce diarrheal disease differ by several orders of magnitude. The specific factors involved in determining infective dose are largely unidentified. All enteric pathogens follow a similar path through the body and encounter similar innate barriers to infection when administered via the oral-fecal route. An early and rather formidable barrier is the highly acidic environment in the normal stomach through which enteric pathogens must pass before gaining access to the intestines. Many studies on the bacterial flora of the gastrointestinal tracts of healthy and achlorhydric people show that whereas the normal fasting stomach with a pH of less than 3.0 is essentially sterile (*Helicobacter* species were not known at the time of many of the studies), the stomach contents of subjects with gastric achlorhydria contains a large and varied flora (1, 11, 23, 24). The gastric contents of people with a gastric pH of greater than 3.0 contains microbial flora representative of the normal oral flora. Research has also shown that although gastric juice contains a number of electrolytes and enzymes, hydrochloric acid is the only constituent of gastric secretions that plays a significant role in preventing bacterial colonization of the stomach (19, 22–24). Epidemiological studies have also established a close link between achlorhydria and the presence of enteric infections (22–24, 28, 29, 39). Although infants attain a normal low gastric pH within the first few weeks of life (15), severe protein-calorie deficiency can cause hypochlorhydria in infants and young children (2). This may contribute to the high level of diarrheal disease found among infants and young children in developing countries (2, 29).

The most dramatic evidence for the relationship between the acidic environment of the stomach and infective dose comes from volunteer studies. By using human volunteers, the infective dose of *V. cholerae* was lowered from 10^8 to 10^4 bacteria by feeding *V. cholerae* along with 2 g of sodium bicarbonate (7). In volunteer studies with enteroinvasive *E. coli* (EIEC) and *C. jejuni,* sodium bicarbonate has also been effectively used to lower the infective dose and enhance infection (4, 12). Volunteer studies with EPEC, ETEC, and rotavirus use pretreatment with sodium bicarbonate as part of the standard protocol in studies on infective dose (26, 34, 36, 45). In animal studies, pretreatment with sodium bicarbonate is also used to establish infection (9). Finally, studies on colonization by orally administered vaccine strains show that colonization of the intestines is markedly enhanced by feeding oral vaccine strains with sodium bicarbonate. This has become a standard protocol for administering oral vaccines (13).

The research cited above strongly supports a role for the relationship between low gastric pH and infective dose. It is interesting, then, to consider what is known about the differential abilities of enteric pathogens to survive low pH. The prediction would be that organisms which have a low infective dose would be better able to withstand low pH than organisms which have a much higher one. However, the results of studies on the ability of *Shigella* species to survive at a pH of less than 3.0 are inconsistent. Whereas several investigators found *Shigella* species to be more acid resistant than other enteric pathogens (21, 25, 40), other investigators found that, at least under some conditions, *Shigella flexneri* and *Shigella sonnei* were not able to survive under the extreme acid conditions found in the normal stomach (19, 40). A possible explanation for the discrepancies between those studies may lie in the experimental methods used. Assays conducted in buffered water or buffered low-phosphate medium indicate that *Shigella* species are susceptible to acid. When *Shigella* species were tested in acidified milk, Luria broth, or postprandial gastric juice adjusted to pH 1.5, they were able to survive for several hours. In most early studies, the acid resistance of a single isolate of a species was examined.

A recent study by Gorden and Small (25) compared the abilities of *Shigella* species, *Salmonella* species, and *E. coli* to survive for 2 h in acidified Luria broth acidified with HCl to pH 2.5. In that study, 11 of 12 isolates of three *Shigella* species as well as 11 of 15 isolates of *E. coli* tested were found to be able to survive at pH 2.5 for 2 h. None of 12 clinical isolates of five different *Salmonella* species tested, including two recent clinical isolates of *S. typhi*, were able to survive for 15 min at less than pH 3.0. Foster and Hall (20) have described an inducible acid tolerance response in *S. typhimurium*. When *S. typhimurium* is exposed to a moderately low pH (5.9) and subsequently exposed to pH 3.3, a significant portion of the inoculum survives the treatment. If the pH of the ingested food was slightly acidic, this acid tolerance response could be an important determinant of infective doses in circumstances in which *Salmonella*-contaminated food was the source of infection. Aside from this circumstance, inducible acid tolerance is unlikely to play a role in the survival of bacteria during passage through the stomach because of the rapidity with which organisms are dashed into an acid bath upon ingestion. It is of interest that although *E. coli* isolates have been found to be heterogeneous with respect to acid resistance, all *Salmonella* isolates tested were found to be uniformly susceptible (25). Studies on the in vitro acid susceptibility of *C. jejuni* have not been reported.

GENETIC DETERMINANTS OF ACID RESISTANCE IN *SHIGELLA* SPECIES

The in vitro studies of acid resistance described above demonstrate that enteric pathogens differ widely in their abilities to survive in an environment as acidic as the contents of the normal stomach (pH of less than 3.0). Despite this, until recently little was known about the specific genes required for the survival of enteric bacteria at such low pH. A genetic analysis of acid resistance in *S. flexneri* and *E. coli* by Small et al. (44) has shown that the ability to survive low pH is a characteristic of stationary-phase organisms and that expression of this phenotype requires an alternative sigma factor, *rpoS* (*katF*). Mutants defective or lacking *rpoS* are extremely acid susceptible, and acid resistance can be restored to such mutants by complementation of a cloned *rpoS* gene. It is known that *rpoS* regulates the expression of over 30 proteins in stationary-phase *E. coli* (37), but the proteins required for acid resistance are unknown.

Although *rpoS* is required for the expression of acid resistance in *Shigella* species, the acid susceptibility of *Salmonella* species is apparently not due to a lack of this gene. Fang and colleagues (17) have shown that a functional *rpoS* homolog is present in *S. typhimurium*. Therefore, it is unlikely that the susceptibility of *S. typhimurium* to low pH is due to a defect in *rpoS*. It is more likely that *Salmonella* species are deficient in one or more of the genes regulated by *rpoS*.

It is tempting to draw conclusions about the determinants of infective dose for disease from the studies on in vitro acid resistance described above. However, conclusive evidence for the role of acid resistance as a determinant of infective dose requires the testing of acid-susceptible mutants in an in vivo model. The fact that *Shigella* species cause disease only in humans and primates makes in vivo testing difficult. However, it is possible, using an animal model, to determine whether acid-susceptible mutants are able to survive passage through the low-pH environment of the stomach. My colleagues and I have done preliminary experiments in our laboratory with mice to address the question of whether acid susceptibility in vitro correlates with the ability to survive passage through the gastrointestinal tracts of mice. Mice were orally coinfected with a 1:1 ratio of an acid-susceptible *S. flexneri*::Tn5 strain and the wild-type parental *S. flexneri* strain. After 30 min the mice were sacrificed, the pH of their stomachs was determined, and the bacterial contents of the stomachs was plated. Only the acid-resistant parental *S. flexneri* was recovered from the gastric contents of mice which had a gastric pH of 3.0 or less. A similar result was obtained in a coinfection study with an acid-susceptible *rpoS*-deficient *E. coli* HB101 strain and strain HB101 containing a cloned *rpoS* gene from *S. flexneri* (unpublished data). Although that experiment did not address the issue of infective dose for disease, since *S. flexneri* does not cause dysentery in mice, it does show a correlation between in vitro acid resistance and the ability to survive in an acidic stomach.

OTHER POTENTIAL DETERMINANTS OF INFECTIVE DOSE

Although the evidence presented so far supports a primary role for acid resistance in determining infective dose for disease, acid resistance is almost certainly not the only determinant. Specific virulence determinants also are undoubtedly involved. For example, when enteric pathogens are classified according to whether they reside primarily in an extracellular or an intracellular compartment in the host, a pattern emerges. Most of the enteric pathogens which require a very large inoculum to cause disease, such as *V. cholerae* and ETEC, are extracellular pathogens whose mode of virulence involves adherence and toxin secretion. Although the infective dose of *V. cholerae* for disease can be lowered from 10^8 to 10^4 by the administration of sodium bicarbonate (7), an inoculum of 10^4 is still several orders of magnitude greater than that required for *Shigella* species to produce disease. Conversely, those pathogens characterized by a very low infective dose for disease, such as *Shigella* species and rotavirus, are most often intracellular organisms with the ability to replicate and spread within cells. Theoretically, a single organism with this lifestyle might be capable of causing disease. For example, although rotavirus, like *V. cholerae*, is extremely susceptible to acid, in the presence of sodium bicarbonate, the infective dose of rotavirus for disease in volunteer and animal studies is 1 to 10 focus-forming units (45). As mentioned previously, the infective dose for *Shigella* species, organisms that enter, replicate within, and spread from cell to cell, is between 10 and 200 organisms (14). Interestingly, the infective dose of EIEC for disease appears to be several log units greater than that of *Shigella* species, even though genetic determinants for entry and cell-to-cell spread are homologous and functionally interchangeable with those in *Shigella* species (12). The fact that EIEC isolates are heterogeneous with respect to acid resistance may explain this discrepancy (25), although there may also be differences in virulence determinants between EIEC and *Shigella* species which have yet to be discovered.

According to the rationale presented above, facultative intracellular pathogens with limited ability for intracellular replication or spread would be expected to be intermediate with respect to the infective dose required to produce disease. This would appear to be the case with *S. typhimurium*, an organism which enters intestinal cells but probably does not replicate extensively within intestinal cells and does not spread from cell to cell. It is not clear how *C. jejuni* fits into this model. All *C. jejuni* isolates apparently have the ability to enter cells (18, 33), but the lack of a good animal model for *C. jejuni* disease has made it difficult to determine the relevancy of the invasive phenotype to virulence. Invasive mutants have been constructed, but data on the infective doses of these mutants required to cause disease have not yet been reported (32).

SUMMARY

Even though all pathogens capable of causing diarrhea are transmitted via the oral-fecal route, there are large differences between enteric pathogens in the minimal inoculum required for disease. The abilities of organisms to survive passage through the normally acidic stomach is probably an important factor for determining infective dose, but other, more specific determinants such as the ability to survive and replicate within cells are likely to play significant roles as well.

REFERENCES

1. **Belitsos, P. C., J. K. Greenson, J. H. Yardley, J. R. Sisler, and J. G. Bartlett.** 1991. Association of gastric hypoacidity with opportunistic enteric infections in patients with AIDS. *J. Infect. Dis.* **166:**277–284.

2. **Black, R. E., K. H. Brown, and S. Becker.** 1984. Malnutrition is a determining factor in diarrheal duration, but not incidence, among young children in a longitudinal study in rural Bangladesh. *Am. J. Clin. Nutr.* **37:**87–94.

3. **Black, R. E., G. L. De Romaña, K. H. Brown, N. Bravo, O. G. Bazalar, and H. C. Kanashiro.** 1988. Incidence and etiology of infantile diarrhea and major routes of transmission in Huascar, Peru. *Am. J. Epidemiol.* **129:**785–799.

4. **Black, R. E., M. M. Levine, M. L. Clements, T. P. Hughes, and M. J. Blaser.** 1988. Experimental *Campylobacter jejuni* infection in humans. *J. Infect. Dis.* **157:**472–479.

5. **Black, R. E., D. Perlman, M. L. Clements, M. M. Levine, and M. J. Blaser.** 1992. Human volunteer studies with *Campylobacter jejuni,* p. 207–215. *In* I. Nuchamkin, M. J. Blaser, and L. Tompkins (ed.), *Campylobacter jejuni.* American Society for Microbiology, Washington, D.C.

6. **Blaser, M. J., and L. S. Newman.** 1982. A review of human salmonellosis. I. Infective dose. *Rev. Infect. Dis.* **4:**1096–1106.

7. **Cash, R. A., S. I. Music, J. P. Libonati, M. J. Snyder, R. P. Wenzel, and R. B. Hornick.** 1974. Response of man to infection with *Vibrio cholerae.* I. Clinical, serologic, and bacteriologic responses to a known inoculum. *J. Infect. Dis.* **129:**45–52.

8. **Christensen, M. L.** 1989. Human viral gastroenteritis. *Clin. Microbiol. Rev.* **2:**51–89.

9. **Cray, W. C., Jr., E. Tokunaga, and N. F. Pierce.** 1983. Successful colonization and immunization of adult rabbits by oral inoculation with *Vibrio cholerae* O1. *Infect. Immun.* **41:**735–741.

10. **Donnenberg, M. S., C. O. Tacker, S. P. James, G. Losonsky, J. P. Nataro, S. S. Wasserman, J. B. Kaper, and M. M. Levine.** 1993. Role of the *eaeA* gene in experimental enteropathogenic *Escherichia coli* infection. *J. Clin. Invest.* **92:**1412–1417.

11. **Drasar, B. S., M. Shiner, and G. M. McLeod.** 1969. Studies on the intestinal flora. *Gastroenterology* **56:**71–79.

12. **Dupont, H. L., S. B. Formal, R. B. Hornick, M. J. Synder, J. P. Libonati, D. G. Sheahan, E. H. LaBrec, and J. P. Kalas.** 1971. Pathogenesis of *Escherichia coli* diarrhea. *N. Engl. J. Med.* **285:** 1–9.

13. **Dupont, H. L., R. B. Hornick, M. J. Snyder, J. P. Libonati, S. B. Formal, and E. J. Gangarosa.** 1972. Immunity in shigellosis. II. Protection induced by oral live vaccine or primary infection. *J. Infect. Dis.* **125:**12–16.

14. **Dupont, H. L., M. M. Levine, R. B. Hornick, and S. B. Formal.** 1989. Inoculum size in shigellosis and implications for expected mode of transmission. *J. Infect. Dis.* **159:**1126–1128.

15. **Euler, A. R., W. J. Bryne, P. J. Meis, R. D. Leake, and M. E. Ament.** 1979. Basal and pentagastrin-stimulated acid secretion in newborn human infants. *Pediatr. Res.* **13:**36–37.

16. **Falkow, S.** Personal communication.

17. **Fang, F. C., J. Libby, N. A. Buchmeier, P. C. Loewen, J. Seitala, J. Harwood, and D. G. Guiney.** 1993. The alternative sigma factor KatF (rpos) regulates *Salmonella* virulence. *Proc. Natl. Acad. Sci. USA* **90:**3511–3515.

18. **Fauchere, J. L., A. Rosenau, M. Veron, E. N. Moyen, S. Richard, and A. Pfister.** 1986. Association with HeLa cells of *Campylobacter jejuni* and *Campylobacter coli* isolated from human feces. *Infect. Immun.* **54:**283–287.

19. **Felsen, J., and A. G. Osofsky.** 1939. Gastric barrier in bacillary dysentery. *Arch. Intern. Med.* **63:** 64–70.

20. **Foster, J. W., and H. K. Hall.** 1990. Adaptive acidification tolerance response of *Salmonella typhimurium.* *J. Bacteriol.* **172:**771–778.

21. **Garrod, L. P.** 1937. The susceptibility of different bacteria to destruction in the stomach. *J. Pathol. Bacteriol.* **45:**473–474.

22. **Giannella, R. A., S. A. Broitman, and N. Zamcheck.** 1972. Gastric acid barrier to ingested microorganisms in man: studies *in vivo* and *in vitro.* *Gut* **13:**251–256.

23. **Giannella, R. A., S. A. Broitman, and N. Zamcheck.** 1973. Influence of gastric acidity on bacterial and parasitic enteric infections. *Ann. Intern. Med.* **78:**271–276.

24. **Gitelson, S.** 1971. Gastrectomy, achlorhydria and cholera. *Isr. J. Med. Sci.* **7:**663–667.

25. **Gorden, J., and P. L. C. Small.** 1993. Acid resistance in enteric bacteria. *Infect. Immun.* **61:** 364–367.

26. **Graham, D. Y., G. R. Dufour, and M. K. Estes.** 1987. Minimal infective dose of rotavirus. *Arch. Virol.* **92:**261–271.

27. **Hornick, R. B., S. E. Greisman, T. E. Woodward, H. L. DuPont, A. T. Dawkins, and M. J. Snyder.** 1970. Typhoid fever: pathogenesis and immunologic control. *N. Engl. J. Med.* **283:**686–691.

28. **Howden, C. W., and R. H. Hunt.** 1985. Gastric secretion in patients with typhoid. *Gut* **26:** 1387–1388. (Letter.)

29. **Howden, C. W., and R. H. Hunt.** 1987. Relationship between gastric secretion and infection. *Gut* **28:**96–107.

30. **Jose, M. V., and J. R. Bobadilla.** 1994. Epidemiological model of diarrhoeal diseases and its application in prevention and control. *Vaccine* **12:**109–116.

31. **Khan, M. U.** 1982. Interruption of shigellosis by hand washing. *Trans. R. Soc. Trop. Med. Hyg.* **76:**164–168.

32. **Konkel, M.** Personal communication.

33. **Konkel, M. E., and L. A. Joens.** 1989. Adhesion to and invasion of HEp-2 cells by *Campylobacter* spp. *Infect. Immun.* **57:**2984–2990.

34. **Levine, M. M., E. J. Bergquist, D. R. Nalin, D. H. Waterman, R. B. Hornick, C. R. Young, and S. Sotman.** 1978. *Escherichia coli* strains that cause diarrhoea but do not produce heat-labile or heat-stable enterotoxins and are non-invasive. *Lancet* **ii:**1119–1122.

35. **Levine, M. M., and R. Edelman.** 1984. Enteropathogenic *Escherichia coli* of classic serotypes associated with infant diarrhea: epidemiology and pathogenesis. *Epidemiol. Rev.* **6:**31–51.

36. **Levine, M. M., D. R. Nalin, D. L. Hoover, E. J. Bergquist, R. B. Hornick, and C. R. Young.** 1979. Immunity to enterotoxigenic *Escherichia coli.* *Infect. Immun.* **23:**729–736.

37. **McCann, M. P., J. P. Kidwell, and A. Matin.** 1991. The putative sigma factor *katF* has a central role in development of starvation-mediated general resistance in *Escherichia coli.* *J. Bacteriol.* **173:**4188–4194.

38. **Mosley, W. H., B. Adams, and E. D. Lyman.** 1962. Epidemiologic and sociologic features of a large urban outbreak of shigellosis. *JAMA* **182:**1307–1311.

39. **Nalin, D. R., R. J. Levine, M. M. Levine, D. Hoover, E. Bergquist, J. McLaughlin, J. Libonati, J. Alam, and R. B. Hornick.** 1978. Cholera, non-vibrio cholera, and stomach acid. *Lancet* **ii:** 856–859.

40. **Peterson, W. L., P. A. Mackowiak, C. C. Barnett, M. Marling-Cason, and M. L. Haley.** 1989. The human gastric bactericidal barrier: mechanisms of action, relative antibacterial activity, and dietary influences. *J. Infect. Dis.* **159:**977–983.

41. **Pickering, L. K., D. G. Evans, H. L. DuPont, J. J. Vollet III, and D. J. Evans.** 1981. Diarrhea caused by *Shigella,* rotavirus, and *Giardia* in day-care centers: prospective study. *J. Pediatr.* **99:** 51–56.

42. **Robinson, D. A.** 1981. Infective dose of *Campylobacter jejuni* in milk. *Br. Med. J.* **282:**1004.

43. **Savarino, S. J.** 1993. Diarrhoeal disease: current concepts and future challenges. *Trans. R. Soc. Trop. Med. Hyg.* **87:**49–53.

44. **Small, P., D. Blankenhorn, D. Welty, E. Zinser, and J. L. Slonczewski.** 1994. Acid and base resistance in *Escherichia coli* and *Shigella flexneri:* role of *rpoS* and growth pH. *J. Bacteriol.* **176:** 1729–1737.

45. **Ward, R. L., D. I. Bernstein, E. C. Young, J. R. Sherwood, D. R. Knowlton, and G. M. Schiff.** 1986. Human rotavirus studies in volunteers: determination of infectious dose and serological response to infection. *J. Infect. Dis.* **154:**871–880.

46. **Wharton, M., R. A. Spiegel, J. M. Horan, R. V. Tauxe, J. G. Wells, N. Barg, J. Herndon, R. A. Meriwether, J. N. MacCormack, and R. H. Levine.** 1990. A large outbreak of antibiotic-resistant shigellosis at a mass gathering. *J. Infect. Dis.* **162:**1324–1328.

Molecular Genetics of Bacterial Pathogenesis
Edited by V. L. Miller, J. B. Kaper, D. A. Portnoy, and R. R. Isberg
© 1994 American Society for Microbiology, Washington, DC 20005

Chapter 33

Genetics as a Route toward Mucosal Vaccine Development

Gordon Dougan

The development of safe and effective vaccines is an ethical and cost-effective approach to controlling disease in populations, a fact that has been clearly illustrated by the eradication of smallpox and the control of other infectious diseases such as polio. Following the invention of the early vaccines, progress in vaccine development was generally sporadic and was dominated by an empirical approach to vaccine design. With the emergence of genetic engineering and related technologies in the early 1970s, there was tremendous optimism that vaccinology would be one of the fields that could benefit most from the new scientific approaches. The ability to manipulate genomes and individual genes and to express antigens in heterologous hosts offered new opportunities for handling proteins and other antigens derived from fastidious infectious agents. Indeed, the first reports of the expression of foreign antigens from pathogens, for example, the VP1 protein of foot-and-mouth disease virus, in heterologous hosts such as *Escherichia coli* was widely publicized in the media. However, enthusiasm was quickly tempered when many individual antigens expressed as soluble proteins in foreign organisms turned out to be ineffective vaccines in target species for a variety of reasons (4). Very rapidly, the mood changed from wild optimism to much more pessimism.

The earlier disappointments might not have been so painful if consideration had been given to the scientific data then in existence of how vaccines actually worked. It was clear that there was much to learn about vaccine-related issues such as immune recognition, antigen delivery, and protective immune mechanisms. The advances in vaccinology that have occurred in the past several years have been significant, although they have been more limited in nature that was originally generally predicted. One of the first contributions to vaccine development that was mediated through recombinant DNA technology was the development of a vaccine for use in pigs against enterotoxigenic *E. coli* infection. Research in Stanley Falkow's laboratory involving the cloning of the genetic determinants for the *E. coli* adhesins (42, 55) and toxins, in particular the heat-labile toxin (LT) (56),

Gordon Dougan • Department of Biochemistry, Imperial College of Science, Technology and Medicine, London SW7 2AZ, United Kingdom.

was the foundation which made possible the development of this vaccine. Indeed, the early work on LT generated what was probably the first genetically engineered toxoid when the LT B subunit was expressed separately from the A subunit for the first time. This approach provided a general option for other toxin-mediated diseases including cholera, pertussis, and tetanus (see chapters 4, 12, and 27). In spite of the early setbacks, new vaccines against several human diseases, including hepatitis B, now exist, and more are in various phases of clinical trial.

A goal in vaccinology is to obtain fully antigenically defined vaccines that can be delivered via mucosal routes, thus avoiding the requirement for injections with needles. Indeed, this is one of the aims of the Children's Vaccine Initiative sponsored by the World Health Organization (63). A major problem with the development of oral vaccines is that most antigens are poorly immunogenic when presented via mucosal surfaces. The reasons for this are only partially understood. Presentation of antigens directly at mucosal surfaces is generally considered to be the most effective means of inducing local responses, although improved methods for inducing mucosal responses to systemically delivered antigens are under intensive investigation. One problem associated with this mode of delivery is simply getting intact antigens at reasonable concentrations to these protected surfaces. Antigens delivered orally or even intranasally are rapidly subjected to dilution and degradation effects. A second problem, which is even more poorly understood, is that mucosal presentation of some antigens induces a tolerance effect on the immune system when the same antigen is subsequently administered systemically. This phenomenon may be controlled by mechanisms which have evolved to limit potentially damaging immune responses to dietary antigens. Thus, the immune system may be able to partially discriminate between dietary antigens and antigenic components of pathogens.

The immune system is also organized into compartments associated with different tissues of the body. Immune tissues and cells present at mucosal surfaces constitute the mucosal immune system. In general terms, the mucosal and systemic immune systems are regulated independently, although there are clearly in existence important mechanisms for extensive cross-talk between the two. Further mucosal immune systems operating at different mucosal surfaces act to a large extent independently, although again, there is evidence for the existence of a common mucosal immune system (2, 38).

It is against this background that new oral vaccines must be designed. Progress may be made in several areas, including (i) the development of improved particulate delivery systems designed to offer protection to antigens against degradation, (ii) the delivery of antigens systemically in a formulation capable of inducing mucosal immunity, (iii) the design of antigens which are improved mucosal immunogens, and (iv) the development of oral adjuvants. In this chapter, I consider recent advances in areas covered by the last two points.

PROPERTIES OF MUCOSAL IMMUNOGENS

The majority of antigens are poor mucosal immunogens. That is, they are ineffective at inducing significant local and/or systemic immune responses when delivered via mucosal surfaces. Immune recognition of antigens can often be achieved if high levels of antigens are repeatedly administered to a particular mucosal surface. However, this is a poor route for inducing practical, protective immunity. There are few examples of truly effective mucosal immunogens. Many pathogens that are capable of entering the body and causing infections via mucosal surfaces can be considered mucosal immunogens. Examples are enteric bacterial pathogens including *Vibrio cholerae, Salmonella* species, and viruses such as poliovirus. Vaccines which are based on attenuated variants of these pathogens have been designed, and the methodologies being applied in this area are considered in the next section. There are very few examples of pure, defined nonliving antigens that can induce immune responses when delivered in microgram quantities to mucosal surfaces. The few examples that have been defined normally originate from pathogens. A good example is the cholera enterotoxin. Current work with nonliving antigens will be considered later in the chapter.

MUCOSAL VACCINES BASED ON LIVE ORGANISMS: GENERAL CONSIDERATIONS

In practical terms, potentially the quickest route toward the development of mucosal vaccines is the use of attenuated variants of mucosal pathogens. Indeed, this approach has been used to develop several vaccines that are in general clinical use. Examples include the Sabin polio vaccine and the oral typhoid vaccine based on *Salmonella typhi* Ty21a (24). Both of these vaccines are based upon attenuated strains generated by using relatively empirical approaches. The Sabin polio vaccines were generated by passaging virus outside of the human host (41), and Ty21a was generated by using random chemical mutagenesis (24). Other live vaccines that are now generally administered parenterally, including vaccines against *Mycobacterium bovis* BCG and vaccinia virus, were or can be used as mucosal vaccines. Thus, in spite of the problems normally associated with vaccination with live vaccines, live vaccines can find acceptance for clinical use. It is arguable if vaccines in which the genetic basis of attenuation remains undefined should now be acceptable for registration as novel vaccines. With the rapidly improving knowledge of the genetic basis of virulence, it is possible to construct attenuated variants of many pathogens. It should be easier to control the quality of and reproducibly manufacture a vaccine strain harboring two or more independently attenuating, fully defined lesions. The likelihood of reversion to virulence can be effectively eliminated. However, in relation to virulence, the definition of an attenuating lesion requires proper analysis. For example, vaccinia virus is attenuated in healthy individuals, yet it can cause lethal viremia in immunocompromised individuals such as those suffering from AIDS. In the modern world, it is important that attenuated lesions are truly attenuating, even in the absence of a fully functional immune system, if they are to be incorporated into vaccines for use in human populations. In conclusion, any new live vaccine strain should be constructed by using fully defined lesions.

LIVE BACTERIAL VACCINES: APPLICATION OF GENETICS

Over the past few years many genes that are required for the survival of pathogens in vivo, and that thus contribute to the infection process, have been identified. If these key genes are mutated and incorporated back into the host pathogen by using crosses that result in replacement of the functional wild-type gene with the mutation, the strain constructed may show a degree of attenuation in a particular assay system. A number of examples of attenuated variants of different pathogens are described in this volume. As mentioned above, attenuation can be measured by various in vitro or in vivo techniques. For the development of effective vaccines, it is important to consider the consequences of the attenuating lesion on the level of morbidity and immunity resulting from infections by the attenuated variant. By selecting different combinations of mutations and strain backgrounds, it is possible to build up a series of strains that exhibit different degrees of attenuation and/or immunizing potency in a particular assay system.

An example of such a comparison is shown in Table 1 for *Salmonella typhimurium* harboring different, defined mutations assayed in BALB/c mice. A number of points can be gleaned from the information displayed in Table 1. First of all, the wild-type bacteria are highly virulent in this system. Administered parenterally, fewer than 10 organisms will establish a systemic infection that reproducibly leads to death within a few days. The introduction of different defined mutations into this background results in strains with different potentials of causing infection. These differences also have major effects on the immune response to *S. typhimurium* monitored in these animals following infection. *purE* mutations have a moderate attenuating effect on this strain of *Salmonella* in BALB/c mice (Fig. 1B). The animals can more readily control the infection, but scattered deaths still occur in

TABLE 1
Effects of level of attenuation on immunogenicity of *S. typhimurium* SL1344 (wild type) defined mutants in BALB/c mice[a]

Mutation	50% Lethal dose (CFU/ml)		Immune response	
	Oral route	Intravenous route	Antibody	Cellular[b]
Wild type	10^5	<1	Death	Death
purE	10^7	10^3	—[c]	—[c]
ompR[d]		10^6	+	+
aroA[d]		10^6	+	+
purA		10^8	+	±
aroA purA		10^9	+	−

[a] The mutations were all in the *S. typhimurium* SL1344 background. Wild type refers to the parent strain. All mutations were introduced by using precise genetic methods. Antibody production was monitored by an enzyme-linked immunosorbent assay against whole, inactivated SL1344 bacteria. Data are based on work published previously (44, 45).

[b] The cellular response was measured by using a number of parameters but, in particular, refers to the ability of immunized animals to control an intravenous challenge with *L. monocytogenes* 2 weeks after vaccination with a single intravenous dose of the attenuated SL1344 derivative.

[c] —, abscesses and scattered deaths.

[d] Animals were protected against salmonellosis by using a single oral dose of 10^9 bacteria.

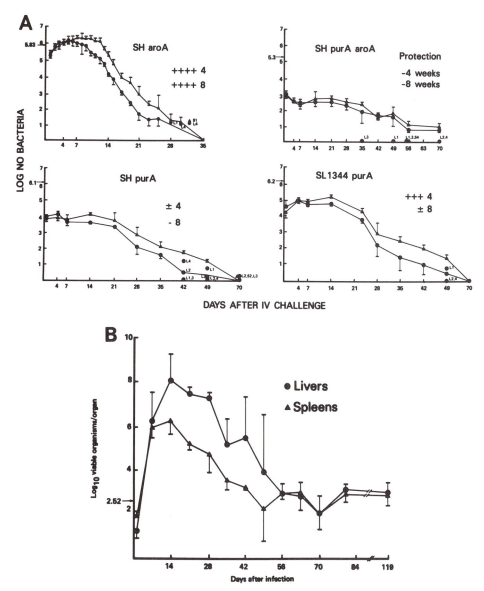

FIGURE 1. Persistence and immunogenicity of *S. typhimurium* SH and SL1344 derivatives following intravenous (IV) inoculation into BALB/c mice. The horizontal arrow on the left axis indicates the dose of vaccine administered. Levels of bacteria in the livers (●) and spleens (▲) were measured as described previously (44). L1 to L4 and S1 to S4 refer to the number of organs (liver [L] or spleen [S]) cleared of bacteria at a particular sampling point. Protection was monitored at 4 or 8 weeks as indicated. Each plus sign represents a log unit of protection against virulent challenge with parenterally administered virulent *S. typhimurium*. (A) Results for *S. typhimurium* SH *aroA*, SH *purA*, and SH *aroA purA*, as indicated. (B) Results for *S. typhimurium* SH *purE*.

the animals and survivors have often controlled the infection at the expense of abscess formation (44). The data obtained by using this assay system would suggest that *purE* mutations are not suitable for use in vaccine construction. The incorporation of *aro* mutations into these strains results in a higher level of attenuation (Fig. 1A). Animals are no longer killed following the administration of bacteria via the oral route, and the bacteria are extremely attenuated when administered parenterally. Interestingly, if the persistence of *aro* mutants is carefully monitored in vivo following either oral or parenteral administration, there is evidence for limited replication of these strains in vivo (37). *aro* mutants are extremely potent oral vaccines in this assay system. Animals can be effectively protected against salmonellosis with a single oral dose and display potent mucosal, humoral, and cellular immune responses to salmonellae (45). Indeed, *aro*-vaccinated mice display a period of nonspecific immunity to non-*Salmonella* species, such as *Listeria monocytogenes,* for 2 to 3 weeks after vaccination. This nonspecific protection may be a consequence of the generally activated immune cells by the vaccine strain (45).

When *purA* mutations are introduced into the same strain background, they display a higher level of attenuation than *aro* mutants. An important point here is that different mutations in the purine biosynthetic pathway have vastly different effects on the attenuated phenotype, whereas a number of different *aro* mutations give a similar level of attenuation. Strains harboring *purA* mutations are able to induce humoral and systemic responses in these mice, but they are not effective oral vaccines against salmonellosis. Furthermore, although T-cell proliferative responses to salmonellae are positive, the animals do not display nonspecific immunity when they are injected with high levels of live vaccine (45). Also, *purA* mutations display a different persistence pattern in vivo. Another important property of *purA* mutations is that the level of persistence differs significantly if the same mutation is incorporated into different strain backgrounds (Fig. 1). Also in this respect there are different consequences for the induction of nonspecific immunity but not oral protective immunity. The combination of *purA* and *aro* mutations into the same strain results in a very highly attenuated strain which is an ineffective oral or parenteral vaccine. Furthermore, the strains display an unusual persistence pattern in which the animals become chronically infected with the strain, develop a humoral response, but do not become immune.

Analysis like that described above for salmonellae can provide important clues for the design of practical live vaccines. Not only is the choice of individual attenuating mutations important but the effects of combining different mutations and the selection of strain background are also critical considerations. This is of extreme importance when considering the design of human vaccines, in which small changes in virulence may have significant effects on reactogenicity. There are very few examples in which a careful comparison of the effect on the immunogenic potential of the same strain harboring selected individual mutations has been made. In the case of *S. typhimurium* mutations, the pattern of immunogenicity has proved to be very similar when defined strains have been evaluated in other hosts, including humans. *Salmonella* strains harboring single or double *aro* mutations have been shown to be effective single-dose oral vaccines in mice and cattle

(19, 25, 60). *S. typhi aro purA* strains were more attenuated than similar strains harboring double *aro* mutations when they were evaluated in volunteers (31, 60, 61). Also, strain background was found to have a significant effect on morbidity in humans (60, 61).

Genetic analysis of the genes required for *Salmonella* virulence has uncovered an array of different genes covering a wide range of functions. Genes associated with biosynthetic or metabolic processes were described above. Another attenuating lesion that might fall into this category is *hemA,* which is required for hemin biosynthesis (1). Several groups have shown that mutations affecting genes involved in regulating the expression of gene families can attenuate salmonellae. Examples of regulatory mutations include *ompR* (16), *phoP* (40), and *cyal crp* (14). Another gene class is made up of genes that are regulated by stresses, such as *htrA* (28). Some of these mutations have been evaluated in live vaccination experiments. *cya* and *crp* mutations have been incorporated into *S. typhi* strains and tested in volunteers. *htrA* mutations, in combination with *aro* mutations, will shortly undergo a similar evaluation (9).

The construction of candidate human vaccine strains requires the use of precise gene replacement systems in order to ensure that any introduced mutation is fully characterized. Often, this will require the cloning and sequencing of the target gene in order to introduce fully sequenced deletion mutations into each gene (11). Such mutations can readily be monitored in vaccine lots by using techniques such as PCR. Work on defined *Salmonella* vaccines is currently one of the most advanced in terms of human testing. Work on other pathogens is described elsewhere in this volume. There are other potential targets to which similar technology might be applied. Already, *aro* mutants of *Shigella* spp. are being evaluated (30), and similar technology, probably with other mutations, might be applied to other enteric pathogens for which vaccines are not currently available. The use of live vaccines for other pathogens such as *Neisseria gonorrhoeae* might also be considered, although the high recombinogenic potential of this organisms will serve as a deterrent to vaccine developers. Certainly, this organism can be attenuated, as monitored in some model systems (6).

LIVE VACCINES AS CARRIERS OF HETEROLOGOUS ANTIGENS

Since the early use of conjugative crosses to exchange genetic material between different bacterial species, the prospect of using these hybrids as a route toward the development of multivalent vaccines has been an attractive one. Perhaps one of the first examples of this type of work was that conducted by using crosses between *Shigella* spp. and either *E. coli* or *Salmonella* spp. This type of approach was simplified by the invention of recombinant DNA technology. The ability to move genes between organisms meant that normal species barriers could be broken, even between bacteria, viruses, and parasites. A number of different bacteria have been considered for use as antigen delivery systems, including members of the genera *Salmonella* (7, 13, 15) and *Yersinia* (3, 46). For a variety of reasons, the potential of salmonellae has received the most attention. *Salmonella* spp. are genetically quite closely related to *E. coli,* and systems for the genetic manipulation of genes in *Salmonella* spp. are well advanced. *Salmonella* spp. can infect the host across mucosal surfaces, making possible the development of mucosal delivery systems. Furthermore, some *Salmonella* strains and the vaccines derived from them are potent immunogens, being able to induce secretory, humoral, and cellular responses in target species. *Salmonella* spp. are able to invade eukaryotic cells and persist for periods of time within immune cells (23). Thus, salmonellae can be considered an active targeting system for immune cells.

There are a number of *Salmonella* vaccine strains available for experimental use, particularly in the murine system, which differ in their immunogenic potentials. In theory, any of these strains could be selected for antigen delivery. Most work to date with strains with defined mutations has been conducted by using either *aro* mutants (11, 25, 39, 50, 53, 62) or *cya crp* mutants (14). Strains of *Salmonella* harboring these mutations are able to induce protective immunity against salmonellosis. The choice of carrier may, to some extent, depend on the type of immune response required against the carried antigen. For example, less invasive strains of salmonellae may induce local responses in the absence of substantial systemic responses. There are now many reports in the literature describing the use of *Salmonella* strains to deliver heterologous antigens derived from a wide variety of pathogens (7, 11, 12). In addition, the coexpression of cytokine genes in this system has also been used in an attempt to modulate the responses (5). In some instances protection has been demonstrated against the pathogen from which the antigen was derived.

The expression of heterologous antigens in bacteria can be problematic. Some foreign antigens, particularly those derived from eukaryotic organisms or viruses, do not fold correctly in bacteria or are toxic when expressed at high levels. Thus, there are substantial limitations to the potential use of salmonellae as an antigen delivery system. If the heterologous antigens are expressed from recombinant plasmids, selective pressures can lead to segregation of the plasmid from the bacterial population or the accumulation of mutations in the heterologous gene. Genetic instability has been a major problem encountered with the use of live *Salmonella* vaccines for the expression of heterologous antigens. These vaccine strains may undergo a short period of replication within the host, and during this period unstable plasmids may be rapidly lost. The result is that the majority of vaccinating bacteria that reach the host tissues fail to express the heterologous protein. This leads to poor immune responses to the carried antigen. Early experiments utilized promoters that were used to express recombinant proteins in fermenters. During batch culture growth, protein expression is induced by the addition of gratuitous inducers. An example is the use of the *lac* or *tac* promoters. Although these promoters are highly suitable for this purpose, such inducers cannot be added to bacteria resident within the host. In order to use these promoters in vaccine strains, expression must be constitutive (in the absence of repressor), and hence, this increases the likelihood of plasmid instability.

A number of approaches have been taken to overcome these instability problems. One is to introduce the heterologous gene onto the *Salmonella* chromosome by using a homologous recombination system (27, 58). This means that only one or a few copies of the heterologous gene are present in the vaccine strain genome and expression levels are normally low. Nevertheless, chromosome integration has been used successfully to increase stability, and these strains have been used to induce immune responses to some heterologous proteins such as the pertactin protein of *Bordetella pertussis* (57). A second approach is to use a plasmid tagged with an essential gene which has been deleted from the chromosome of the vaccine strain. This system has been successfully established by using the *asd* gene (43). An alternative but compatible approach is to drive expression of the heterologous antigen from promoters which are regulated by environmental cues likely to be encountered within the host tissues. In this way vaccine lots can be prepared by using growth conditions in which the heterologous gene expression is low yet the gene is turned on when the appropriate conditions are encountered by the vaccine strain in the host.

An example of the latter approach is the use of the *nirB* promoter to construct a single-dose oral vaccine against tetanus (10). Tetanus neurotoxin is a 150,000-Da polypeptide which is responsible for the clinical symptoms associated with tetanus disease. The protein is one of the most toxic substances known to humans. In spite of the high toxicity of the holotoxin, a 50,000-Da nontoxic polypeptide fragment, known as fragment C, can be cleaved from the C-terminal end of tetanus toxin. In addition to being nontoxic, fragment C is highly immunogenic when administered parenterally with adjuvant to animals and can induce protection against tetanus toxin challenge (22). Recombinant fragment C has been expressed at high levels in *E. coli* by using a partially synthetic gene driven by the *tac* promoter (21). The *tac* expression system was initially used to construct a *Salmonella aro* vaccine strain which constitutively expressed fragment C at high levels. Administration of this vaccine strain orally to mice was able to induce some protection against tetanus and complete protection against salmonellosis (10). However, protection against tetanus was variable because of the instability of the vaccine strain. The plasmid encoding the fragment C gene rapidly segregated from the salmonellae present in the tissues of vaccinated animals. The plasmid could be stabilized by incorporating the *lac* repressor gene onto the plasmid, but in this case, the mice failed to seroconvert to tetanus toxin fragment C. Thus, an alternative approach was required if a practical vaccine strain was to become available.

The *nirB* promoter is regulated by cues from the environment, including the levels of oxygen (48, 49). The promoter is highly repressed in the presence of normal levels of oxygen but is induced in anaerobic conditions. A *Salmonella aro* vaccine strain was constructed in which fragment C expression was driven from the *nirB* promoter. Fragment C expression levels were low when the strain was grown with aeration but were high during anaerobic growth (10). Oral administration of this vaccine strain to mice reproducibly induced excellent protection against tetanus and salmonellosis even when a single dose was used. Furthermore, the plasmid encoding fragment C was found to be highly stable in the salmonellae recovered from the tissues of vaccinated animals. Thus, the use of so-called in vivo-inducible promoters may be an excellent practical method for live vaccine development. Other in vivo-inducible promoters currently under evaluation include promoters associated with the heat shock response, such as the *groE* or *htrA* promoters, or promoters regulated by osmolarity.

NONLIVING MUCOSAL ANTIGENS

Most purified proteins are poorly immunogenic when administered to mucosal surfaces. The reasons for this were briefly described above. Since administration to mucosal surfaces is an established method for stimulating local immune responses, the search for effective mucosal immunogens could be a valuable pursuit. If investigators could identify a number of molecules that are effective mucosal immunogens, it might be possible to identify common characteristics that are associated with this class of antigen. These properties might then be incorporated into poor mucosal immunogens in order to improve their immunogenic properties at mucosal surfaces. Perhaps the best-characterized example of an effective mucosal immunogen is cholera enterotoxin (CT) and its relatives such as the *E. coli* LT. Both of these molecules and their B-subunit components are highly immunogenic when delivered to mucosal surfaces either orally or intranasally (20, 26). As little as a few micrograms can be effective in some animal species. The immunogenicity of CT subunit (CT-B) B in humans has been exploited for the development of an oral cholera vaccine which has undergone extensive evaluation in volunteers and in the field (59). This vaccine is composed of CT-B mixed with inactivated *V. cholerae* whole cells.

The oral immunogenicity of CT-B has been carefully studied in volunteers. Oral administration of the CT-B–inactivated *V. cholerae* vaccine to volunteers induces a systemic anti-CT-B response which can be detected a few days after vaccination. The immune serum contains significant levels of both immunoglobulin G (IgG) and IgA class anti-CT-B antibodies (33). The levels of antibodies can be boosted by administration of further doses of the vaccine even several years later (59). The appearance of anti-CT-B antibodies in the serum is associated with strong T-cell responses as well as production of local antibodies (32). Studies by the ELISPOT assay (32) in peripheral blood indicate that B cells that produce antibodies to CT-B enter the bloodstream soon after oral vaccination, reproducibly peaking on day 7 after primary exposure and day 5 after secondary exposure to antigen. These cells may be involved in the establishment of the common mucosal response. After administration of the second dose of vaccine, as many as 25% of the circulating IgA-producing B cells may be producing anti-CT-B antibodies (32, 33). Thus, this is a very potent immune response to a mucosally delivered antigen.

CT and CT-B are potent mucosal antigens. Studies are under way to determine the characteristics of the protein that contributes to this property. Early studies suggest that the ability to bind to eukaryotic target cells and resist the degradative actions of proteases may be important factors.

CT AS A MUCOSAL ADJUVANT

As well as being a potent mucosal immunogen in its own right, CT has been shown to be able to act as a mucosal adjuvant to coadministered antigens. Mucosal administration of simple mixtures of CT and bystander antigens, which are poor mucosal immunogens, can stimulate local and systemic responses to the bystander (35). This phenomenon has been studied extensively in murine systems by using both oral and intranasal administration of vaccine (29, 32, 34). Some studies have suggested that CT-B may also possess the adjuvanting property. This has been clearly demonstrated by using both CT-B and LT-B when antigens are directly coupled to the carrier by genetic or chemical linkages (34, 64). The situation is more controversial for uncoupled antigens. The picture has been clouded because many workers have used commercially purchased CT-B purified from *V. cholerae,* which may be contaminated with low levels of active CT (47). Studies with highly purified recombinant molecules are required. One study has already suggested that at least some LT molecules that lack ADP-ribosyltransferase activity are poor mucosal adjuvants, suggesting that the activity associated with CT toxicity may also be required for adjuvanticity (36). Colleagues and I (18) have examined the adjuvant properties of a number of CT mutants that still bind to cells but that lack the toxic ADP-ribosyltransferase activity. Results of studies with fragment C of tetanus toxin as a bystander antigen suggest that some of these mutants retain significant activity as mucosal adjuvants. Such studies may open up the testing in humans of novel nontoxic, nonliving mucosal vaccines.

OTHER MUCOSAL IMMUNOGENS

There are a few other examples of other pure antigens that can induce significant immune responses when administered to mucosal surfaces. Most of these are proteins purified from pathogens, and in some cases they have been used to induce protection against challenge with virulent organisms. Colleagues and I have examined the properties of proteins derived from *B. pertussis,* including fimbrial hemagglutinin (FHA), pertactin-P.69 (8), and pertussis toxin (Ptx). Both FHA and pertactin-P.69 can be used to mucosally immunize mice against challenge with *B. pertussis* (52, 54). Both of these proteins can induce local and some systemic responses after intranasal immunization. Interestingly, the preparations of both of these proteins used in the immunizations retain their abilities to bind to eukaryotic cells, a factor which may contribute to their mucosal immunogenicities.

Colleagues and I (17) have studied this further using a nontoxic derivative of Ptx (Ptx-crm 9K/129G). This protein still folds to take up the conformation of the holotoxin and can bind to eukaryotic cells but lacks the toxic ADP-ribosyltransferase activity normally associated with pertussis toxin. The protein has undergone extensive characterization as a component of an acellular, parenterally administered whooping cough vaccine (51). Intranasal administration of a few micrograms of the Ptx-crm to BALB/c mice induces both local secretory and systemic anti-Ptx antibody responses. Subcutaneous administration of the same protein induces systemic but not secretory antibody production. Gentle formaldehyde treatment was used to destroy the cell-binding activity of the protein but retain holotoxin formation and immunogenicity upon parenteral administration. In contrast to the untreated Ptx-crm toxoid, intranasal administration of the formaldehyde-treated

TABLE 2
Influence of route of vaccination and formaldehyde treatment on secretory immune response to pertussis toxoid Ptx-crm in BALB/c mice[a]

Immunogen	Route of immunization	Individual mouse	Anti-Ptx IgA titers[b]	Mean specific IgA levels
Ptx-crm	i.n.	A	0.38	0.355
		B	0.38	
		C	0.16	
		D	0.50	
	s.c.	A	0	
		B	0	
Ptx-crm + formaldehyde	i.n.	A	0	
		B	0	
		C	0	
		D	0	
	s.c.	A	0	
		B	0	

[a] Levels of antibody were measured by an enzyme-linked immunosorbent assay against Ptx-crm. Ptx-crm + formaldehyde is material treated with formaldehyde to remove the cell-binding activity but retain the holotoxin formation ability. Letters indicate titers in individual mice. i.n., intranasal; s.c., subcutaneous.

[b] Specific anti-Ptx IgA titers in nasal lavage fluid following intranasal or subcutaneous immunization of mice. Each anti-Ptx IgA value has been corrected for total IgA in each individual lavage.

toxoid failed to induce any local or systemic anti-Ptx-crm antibody responses (Table 2), indicating that cell binding may play a significant role in the mucosal immunization activity of this toxoid.

CONCLUSIONS

A great deal of progress has been made in recent years in the development of mucosal vaccines, and investigators have been able to increase their understanding of how mucosal vaccines induce immune responses. In the future a combination of approaches may lead to more significant advances, particularly in the production of mucosal vaccines of practical use. The use of genetics has made a major contribution to this field, as have basic studies on virulence and immunology. It is expected that the field will provide enormous practical benefits, especially in developing countries, where mucosal vaccines might prove to be extremely valuable.

ACKNOWLEDGMENTS. I thank Stanley Falkow for introducing me to medical microbiology and the concept of applying molecular techniques to solving medical problems.

This work was supported by grants from the British Medical Research Council, the World Health Organization, and the Wellcome Trust.

REFERENCES

1. **Benjamin, W. H., P. Hall, and D. E. Briles.** 1991. A *hemA* mutation renders *Salmonella typhimurium* avirulent in mice, yet capable of eliciting protection against intra-venous infection with *S. typhimurium. Microb. Pathog.* **11:**289–296.

2. **Bienenstock, J., M. McDermott, D. Befus, and L. M. O'Neil.** 1978. A common mucosal immunological system involving the bronchus, breast and bowel. *Adv. Exp. Med. Biol.* **107**:53–69.

3. **Bowe, F., P. O'Goara, D. Maskell, M. Cafferkey, and G. Dougan.** 1989. In vivo properties of *Yersinia enterocolitica aroA* mutants. *Infect. Immun.* **57**:3234–3236.

4. **Brown, F., G. Dougan, and D. Snary.** 1984. Genetically engineered vaccines: problems and promises. *Microbiol. Sci.* **1**:123–126.

5. **Carrier, M. J., S. N. Chatfield, G. Dougan, V. T. A. Nowicka, D. O'Callaghan, J. E. Beesley, S. Milano, E. Cillais, and F. Y. Liew.** 1992. Expression of human interleukin-1 B in *Salmonella typhimurium:* a model system for the delivery of recombinant therapeutic proteins in vivo. *J. Immunol.* **148**:1176–1181.

6. **Chamberlain, L., R. Strugnell, G. Dougan, C. E. Hormaeche, and R. De Marco De Hormaeche.** 1993. *Neisseria gonorrhoeae* strain MS11 harbouring a mutation in gene *aroA* is attenuated and immunogenic. *Microb. Pathog.* **15**:51–63.

7. **Charles, I., and G. Dougan.** 1990. Gene expression and the development of live enteric vaccines. *TIBtech.* **8**:117–121.

8. **Charles, I., G. Dougan, D. Pickard, S. N. Chatfield, M. Smith, P. Novotny, P. M. Morrissey, and N. F. Fairweather.** 1988. Molecular cloning and characterisation of P. 69, a protective outer membrane protein from *Bordetella pertussis. Proc. Natl. Acad. Sci. USA* **89**:3554–3558.

9. **Chatfield, S.** Personal communication.

10. **Chatfield, S. N., I. G. Charles, A. J. Makoff, M. D. Oxer, G. Dougan, D. Pickard, D. Slater, and N. F. Fairweather.** 1992. Use of the *nirB* promoter to direct the stable expression of heterologous antigens in *Salmonella* oral vaccine strains: development of a single-dose oral tetanus vaccine. *Bio/Technology* **10**:888–892.

11. **Chatfield, S. N., N. F. Fairwether, I. Charles, D. Pickard, M. M. Levine, D. Hone, M. Posada, R. A. Strugnell, and G. Dougan.** 1992. Construction of a genetically defined *Salmonella typhi* Ty2 *aroA, aroC* mutant for the engineering of a candidate oral typhoid-tetanus vaccine. *Vaccine* **10**: 53–60.

12. **Clements, J. D., F. L. Lyon, K. L. Lowe, A. L. Farrand, and S. El-Morshidy.** 1986. Oral immunization of mice with attenuated *Salmonella enteritidis* containing a recombinant plasmid which codes for production of the B subunit of heat-labile enterotoxin of *Escherichia coli. Infect. Immun.* **53**: 685–693.

13. **Curtiss, R., III, S. M. Kelly, P. A. Gulig, and K. Nakayama.** 1989. Selective delivery of antigens by recombinant bacteria. *Curr. Top. Microbiol. Immunol.* **146**:35–54.

14. **Curtiss, R. R., and S. M. Kelly.** 1987. *Salmonella typhimurium* deletion mutants lacking adenylate cyclase and cyclase receptor protein are avirulent and immunogenic. *Infect. Immun.* **55**:3035–3043.

15. **Dallas, W. S., D. M. Gill, and S. Falkow.** 1979. Cistrons encoding *Escherichia coli* heat-labile toxin. *J. Bacteriol.* **139**:850–858.

16. **Dorman, C. J., S. N. Chatfield, C. H. Higgins, C. Hayward, and G. Dougan.** 1989. Characterization of porin and *ompR* mutants of a virulent strain of *Salmonella typhimurium: omp* mutants are attenuated in vivo. *Infect. Immun.* **57**:2136–2140.

17. **Douce, G., M. Pizza, I. Cropley, M. Roberts, R. Rappuoli, and G. Dougan.** Submitted for publication.

18. **Douce, G., M. Pizza, R. Rappuoli, and G. Dougan.** Submitted for publication.

19. **Dougan, G., S. N. Chatfield, D. Pickard, D. O'Callaghan, and D. Maskell.** 1988. Construction and characterisation of *Salmonella* vaccine strains harbouring mutations in two different *aro* genes. *J. Infect. Dis.* **158**:1329–1335.

20. **Elson, C. J., and W. Ealding.** 1984. Generalised systemic and mucosal immunity in mice after mucosal stimulation with cholera toxin. *J. Immunol.* **132**:2736–2743.

21. **Fairweather, N. F., S. N. Chatfield, A. J. Makoff, R. A. Strugnell, J. Bester, D. J. Maskell, and G. Dougan.** 1990. Oral vaccination against tetanus using a live, rationally attenuated *Salmonella* carrier. *Infect. Immun.* **58**:1323–1329.

22. **Fairweather, N. F., V. Lyness, and D. Maskell.** 1987. Immunization of mice against tetanus with fragments of tetanus toxin synthesized in *Escherichia coli. Infect. Immun.* **55**:2541–2545.

23. **Fields, P. I., R. V. Swanson, C. G. Heidaris, and F. Heffron.** 1986. Mutants of *Salmonella typhimu-*

rium that cannot survive inside macrophages are avirulent. *Proc. Natl. Acad. Sci. USA* **83**: 5189–5193.

24. **Germanier, R., and E. Furer.** 1975. Isolation and characterisation of a *galE* mutant Ty21a of *Salmonella typhi:* a candidate strain for a live oral typhoid vaccine. *J. Infect. Dis.* **131**:553–558.

25. **Hoiseth, S. K., and B. A. D. Stocker.** 1981. Aromatic-dependent *Salmonella typhimurium* are non-virulent and effective as live vaccines. *Nature* (London) **291**:238–239.

26. **Holmgren, J., and C. Czerkinsky.** 1992. Cholera as a model for research on mucosal immunity and development of oral vaccines. *Curr. Opin. Immunol.* **4**:387–392.

27. **Hone, D., S. Attridge, L. Van den Bosch, and J. Hackett.** 1988. A chromosomal integration system for stabilisation of heterologous genes in *Salmonella* based vaccine strains. *Microb. Pathog.* **5**: 407–418.

28. **Johnson, K. J., I. G. Charles, I. A. Miller, D. Pickard, P. O'Goara, G. Costa, T. Ali, and C. E. Hormaeche.** 1991. The role of a stress-response protein in *Salmonella typhimurium* virulence. *Mol. Microbiol.* **5**:401–407.

29. **Jones, P. W., G. Dougan, C. Hayward, N. MacKenzie, P. Collins, and S. N. Chatfield.** 1990. Oral vaccination of calves against experimental salmonellosis using a double *aro* mutant of *Salmonella typhimurium. Vaccine* **9**:29–34.

30. **Karnell, A., P. D. Cam, N. Verma, and A. A. Lindberg.** 1993. *aroD* deletion attenuates *Shigella flexneri* strain 2457T and makes it a safe and efficatious oral vaccine in monkeys. *Vaccine* **11**: 830–836.

31. **Levine, M. M., D. Herrington, J. R. Murphy, J. G. Morriss, G. Losonsky, B. Tall, A. Lindberg, S. Svenson, S. Baqar, M. F. Edwards, and B. A. D. Stocker.** 1987. Safety, infectivity, immunogenicity and in vivo stability of two attenuated auxotrophic mutant strains of *Salmonella typhi,* 541Ty and 543Ty as live oral vaccines in humans. *J. Clin. Invest.* **79**:888–902.

32. **Lewis, D. J. M., L. R. R. Castello-Branco, P. Novotny, G. Dougan, and G. E. Griffin.** 1993. Circulating cellular immune response to oral immunisation of humans with cholera toxin B-subunit. *Vaccine* **11**:119–121.

33. **Lewis, D. J. M., P. Novotny, G. Dougan, and G. E. Griffin.** 1991. The early cellular and humoral immune response to primary and booster oral immunisation with cholera toxin B-subunit in humans. *Eur. J. Immunol.* **21**:2087–2094.

34. **Lipscombe, M., I. G. Charles, M. Roberts, G. Dougan, J. Tite, and N. F. Fairweather.** 1991. Intranasal immunisation using B subunit of the *Escherichia coli* heat labile toxin fused to an epitope of the *Bordetella pertussis* P.69 antigen. *Mol. Microbiol.* **5**:1385–1392.

35. **Lycke, N., and J. Holmgren.** 1986. Strong adjuvant properties of cholera toxin on gut mucosal immune responses to orally presented antigens. *Immunology* **59**:301–308.

36. **Lycke, N., T. Tsuji, and J. Holmgren.** 1992. The adjuvant effect of *Vibrio cholerae* and *Escherichia coli* heat-labile enterotoxins are linked to their ADP-ribosyl-transferase activity. *Eur. J. Immunol.* **22**:2277–2281.

37. **Maskell, D. J., K. J. Sweeney, D. O'Callaghan, C. E. Hormaeche, F. Y. Liew, and G. Dougan.** 1987. *Salmonella typhimurium aroA* mutants as carriers of the *Escherichia coli* heat-labile enterotoxin B subunit to the murine secretory and systemic immune systems. *Microb. Pathog.* **2**:211–221.

38. **McGhee, J. R., J. Mestecky, M. T. Dertzbaugh, J. H. Eldridge, M. Hirasawa, and H. Kiyono.** 1992. The mucosal immune system: from fundamental concepts to vaccine development. *Vaccine* **10**: 75–81.

39. **Miller, I. A., S. N. Chatfield, G. Dougan, L. DeSilva, H. Joysey, and C. H. Hormaeche.** 1989. Bacteriophage P22 as a vehicle for transducing cosmid gene banks between smooth strains of *Salmonella typhimurium:* use in identifying a role for *aroD* in attenuating virulent Salmonella strains. *Mol. Gen. Genet.* **215**:312–316.

40. **Miller, S. I., A. M. Kukral, and J. J. Mekalanos.** 1989. A two-component regulatory system (*phoP phoQ*) controls *Salmonella typhimurium* virulence in mice. *Proc. Natl. Acad. Sci. USA* **86**:5054–5058.

41. **Minor, P.** 1992. The molecular biology of polio vaccines. *J. Gen. Virol.* **73**:3065–3077.

42. **Morrissey, P., and G. Dougan.** 1986. Cloning and characterisation of the *E. coli* 987P adhesion fimbriae determinant. *Gene* **43**:79–84.

43. **Nakayama, K., S. M. Kelley, and R. Curtiss III.** 1988. Construction of an *asd* expression-cloning

vector: stable maintenance and high level expression of cloned genes in salmonella based vaccine strains. *Bio/Technology* **6**:693–697.

44. O'Callaghan, D., D. Maskell, F. Y. Liew, C. S. F. Easmon, and G. Dougan. 1988. Characterization of aromatic-dependent and purine-dependent *Salmonella typhimurium:* studies on attenuation, persistence, and ability to induce protective immunity in BALB/c mice. *Infect. Immun.* **56:** 419–423.

45. O'Callaghan, D., D. Maskell, J. Tite, and G. Dougan. 1990. Immune responses in BALB/c mice following immunisation with aromatic compound or purine dependent *Salmonella typhimurium* strains. *Immunology* **69**:184–189.

46. O'Gaora, P., M., Roberts, F. Bowe, C. Hormaeche, R. DeMarco de Hormaeche, M. Cafferkey, J. Tite, and G. Dougan. 1990. *Yersinia enterocolitica aroA* mutants as carriers of the B subunit of the *Escherichia coli* heat-labile enterotoxin to the murine immune system. *Microb. Pathog.* **9:** 105–116.

47. O'Hagan, D. T. 1990. Novel non-replicating antigen delivery systems. *Curr. Opin. Infect. Dis.* **3:** 393–401.

48. Oxer, M. D., C. M. Bentley, J. G. Doyle, T. C. Peakman, I. G. Charles, and A. J. Makoff. 1991. High level heterologous expression in *E. coli* using the anaerobically-activated nirB promoter. *Nucleic Acids Res.* **19**:1889–1897.

49. Peakman, T., J. Crouzet, J. F. Jayaux, S. Busby, S. Mohan, N. Harborne, J. Wooton, R. Nicholson, and J. Cole. 1990. Nucleotide sequence, organisation and structural analysis of the products of genes in the *nirB-cysB* region of the *Escherichia coli* chromosome. *Eur. J. Biochem.* **191**:315–324.

50. Poirier, T. P., M. Kehoe, and E. H. Beachey. 1988. Protective immunity evoked by oral administration of attenuated *aroA Salmonella typhimurium* expressing streptococcal M protein. *J. Exp. Med.* **168**:25–36.

51. Rappuoli, R., A. Podda, M. Pizza, A. Covacci, A. Bartolini, M. Magistris, and L. Nencioni. 1992. Progress towards the development of new vaccines against whooping cough. *Vaccine* **10:** 1027–1032.

52. Roberts, M., I. Cropley, S. N. Chatfield, and G. Dougan. 1993. Protection of mice against respiratory *Bordetella pertussis* infection by intranasal immunisation with P.69 and FHA. *Vaccine* **11:** 866–872.

53. Sadoff, J. C, R. W. Ballou, L. S. Baron, W. R. Marjaran, R. N. Brey, W. T. Hockmeyer, J. Young, J. J. Cryz, J. Ou, G. H. Lowell, and J. D. Chulay. 1988. Oral *Salmonella typhimurium* vaccine expressing circumsporozoite protein protects against malaria. *Science* **240**:236–240.

54. Shahin, R. D., D. F. Amsbaugh, and M. F. Leef. 1992. Mucosal immunization with filamentous hemagglutinin protects against *Bordetella pertussis* respiratory tract infection. *Infect. Immun.* **60:** 1482–1488.

55. Shipley, P., G. Dougan, and S. Falkow. 1981. Identification and cloning of the genetic determinant that encodes the K88ac adherence antigen. *J. Bacteriol.* **145**:920–925.

56. So, M., W. S. Dallas, and S. Falkow. 1978. Characterization of an *Escherichia coli* plasmid encoding for the synthesis of heat-labile toxin: molecular cloning of the toxin determinant. *J. Bacteriol.* **139**:850–858.

57. Strugnell, R., G. Dougan, S. N. Chatfield, I. C. Charles, N. F. Fairweather, J. Tite, L. Jing Li, J. Beesely, and M. Roberts. 1992. Characterization of a *Salmonella typhimurium aro* vaccine strain expressing the P.69 antigen of *Bordetella pertussis. Infect. Immun.* **60**:3994–4002.

58. Strugnell, R., D. Maskell, N. F. Fairweather, D. Pickard, A. Cockayne, C. Penn, and G. Dougan. 1990. Stable expression of foreign antigens from the chromosome of *Salmonella typhimurium* vaccine strains. *Gene* **88**:57–63.

59. Svennerholm, A.-M., and J. Holmgren. 1986. Oral combined B subunit-whole cell cholera vaccine, p. 33–43. *In* J. Holmgren, A. Lindberg, and R. Mollby (ed.), *Development of Vaccines and Drugs against Diarrhoea.* 11th Nobel Conference, Stockholm, 1985. Student Lituratur, Lund, Sweden.

60. Tacket, C. O., D. M. Hone, R. Curtiss III, S. M. Kelly, G. Losonsky, L. Guers, A. M. Harris, R. Edelman, and M. M. Levine. 1992. Comparison of the safety and immunogenicity of ΔaroC, ΔaroA and Δcya Δcrp *Salmonella typhi* strains in adult volunteers. *Infect. Immun.* **60**:536–544.

61. Tacket, C. O., D. M. Hone, G. A. Losonsky, L. Guers, R. Adelman, and M. M. Levine. 1992.

Clinical acceptability and immunogenicity of CVD908 *Salmonella typhi* vaccine strain. *Vaccine* **10:**443–446.

62. **Tite, J. P., S. M. Russell, G. Dougan, D. O'Callaghan, I. Jones, G. Brownlee, and F. Y. Liew.** 1988. Antiviral immunity induced by recombinant nucleoprotein of influenzae A virus. I. Characteristics and cross-reactivity of T-cell responses. *J. Immunol.* **141:**3980–3987.

63. **World Health Organization.** *Ten Years of Progress 1984–1993. The Programme for Vaccine Development.* World Health Organization, Geneva.

64. **Wu, H.-Y., and M. W. Russell.** 1993. Induction of mucosal immunity by intranasal application of a streptococcal surface protein antigen with the cholerae toxin B subunit. *Infect. Immun.* **61:** 314–322.

Molecular Genetics of Bacterial Pathogenesis
Edited by V. L. Miller, J. B. Kaper, D. A. Portnoy, and R. R. Isberg
© 1994 American Society for Microbiology, Washington, DC 20005

Chapter 34

Phylogenetic Diversity of Microbial Pathogens

David A. Relman

In 1949, Winogradsky remarked that readily cultivated bacteria in natural microbial communities "draw importance to themselves, whereas the other forms, being less docile, or even resistant, escape attention" (54). Winogradsky and other microbial ecologists realized their limited understanding of microbial growth requirements and the biases of in vitro growth enrichment methods. This comment proved prophetic, but to a far greater degree than might have been anticipated a half century ago. Culture methods remain significantly inadequate in duplicating the physiological needs of microorganisms and, in particular, environmental microorganisms. With the advent of sequence-based (molecular) phylogeny has come the ability to detect and analyze the compositions of microbial communities without reliance on in vitro microbial cultivation. Early results from studies on a variety of environmental microbial habitats suggest that the predominant species in aquatic and terrestrial locations are those that have remained uncultivated and, thus, previously unidentified and uncharacterized (3, 17, 18, 29, 46, 55). To what extent are these findings relevant to the microbial flora in mammals?

Pathogenic microorganisms often elicit host tissue pathology and clinical manifestations as they propagate in privileged anatomic sites. In this sense they "draw attention" to themselves, despite the failure of some of these organisms to propagate on artificial laboratory media or within alternative hosts. With the use of various refined stains (e.g., silver or periodic acid-Schiff), some of these microorganisms can be visualized in direct proximity to damaged areas of tissue. Pathogenicity might be assumed on the basis of "guilt by association," especially if clinical response or relapse can be correlated with antimicrobial therapy and the disappearance or reappearance of the visualized microorganisms. Some human commensal microorganisms occasionally draw attention to themselves when host defenses become compromised, and these microorganisms penetrate into privileged sites and induce pathophysiology. Nonetheless, morphology is not terribly useful for phylogeny. And, despite intensive efforts, at least some of these pathogens cannot

David A. Relman • Departments of Medicine and of Microbiology and Immunology, Stanford University School of Medicine, Stanford, California 94305, and Department of Veterans Affairs Medical Center, Palo Alto, California 94304.

be purified or propagated in the laboratory. Whipple's disease, bacillary angiomatosis, and cat scratch disease are examples of diseases with relatively distinct pathologies that have been associated with visible but uncultivated bacilli for as long as 87 years (59). The presumed causative microorganisms resisted identification by traditional methods until recently, when an alternative approach was brought to bear on this problem.

INFERENCE OF ORGANISMAL EVOLUTIONARY RELATIONSHIPS FROM rRNA SEQUENCES

Not only are phenotypic characters notoriously unreliable for microbial identification and classification (33, 49) but they are also essentially unavailable for uncultivated and unpurified microorganisms. In contrast, genotypic characters offer a greater wealth of phylogenetic information that is more consistent, selectively neutral, and "clock-like" (63). Some molecules are thought to reflect more accurately the evolutionary history of the organism. Among these molecules, rRNA has become the most widely used and, consequently, the most useful. In particular, analysis of the small-subunit (ss) or 16S (in prokaryotes) rRNA has facilitated reconstructions of organismal phylogeny, leading to new insights concerning the organization of and relationships among all known extant species on earth (15, 32, 33, 63, 64). Comparative rRNA analysis depends upon selection of an appropriate group of homologous sequences, rigorous sequence alignment with preservation of conserved secondary structures, and use of appropriate phylogenetic inference methods (23, 25). Using ss rRNA sequence analysis, nearly all known extant species can be grouped into one of three domains, the *Bacteria* (Fig. 1), the *Archaea,* and the *Eukarya* (64). Interestingly, the *Archaea* are more closely related to the *Eukarya* than they are to the other prokaryotic domain—the *Bacteria.* Both appear to be of thermophilic origin (34).

Several features make ss rRNA molecules useful and reliable evolutionary clocks. First, they are ubiquitous among living cells, and they do not seem to be transferred horizontally among mixed populations. Of course, a truly accurate picture of the relationships between pathogenic microorganisms should also take into account their known propensity to share blocks of virulence-associated genes by lateral transfer (mediated by plasmids, bacteriophages, transposable elements, and natural transformation). Second, the function of the molecule is essential, and as a result, its secondary and tertiary structures are highly conserved. This feature ensures consistent patterns of evolutionary change among nearly all ss rRNAs. Conserved secondary structures allow more accurate sequence alignments. Third, these molecules are large enough to contain sufficient information content for comparisons among diverse organisms. Frequencies of sequence change vary between different regions of the molecule. More volatile regions provide greater degrees of phylogenetic specificity among closely related organisms.

Evolutionary relationships within traditional taxonomic groups of cultivated pathogenic microorganisms have been redefined with 16S rRNA sequence analysis. The bacterial groups studied by this method include members of the proteo-

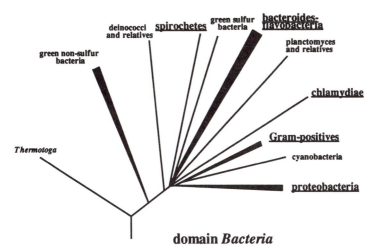

FIGURE 1. Evolutionary tree of the domain *Bacteria* on the basis of comparative analysis of 16S rRNA sequences. A revised organization of this domain was published recently (34). The divisions that contain organisms known to be pathogenic for humans are underlined. Line segment length is proportional to the number of fixed point mutations in the rRNA, and thus reflects evolutionary distance.

bacteria division—the genus *Yersiniae* (24) (gamma-subdivision); the genus *Vibrio* (12) (gamma-subdivision); the families *Pasteurellaceae* (9, 10) (gamma-subdivision), *Neisseriaceae* (8) (beta-subdivision), and *Rickettsiaceae* (57, 58) (alpha-subdivision); the genus *Mycobacterium* within the gram-positive division (38, 48); and the pathogenic spirochetes within the spirochete division (37). Some of these traditional taxonomic units, such as the *Rickettsiaceae,* include phylogenetically diverse organisms. For example, although the genera *Rickettsia* and *Ehrlichia* are alpha-proteobacteria, *Coxiella burnetii* is a gamma-proteobacterium and is closely related to *Legionella* species (57, 58). In some cases, the evolutionary origins of important phenotypic characters are revealed by this type of analysis and are found to be monophyletic and distributed vertically. However, some phenotypes appear to have arisen independently on numerous occasions, e.g., photosynthesis (see below). One of the challenges in molecular phylogenetic analysis will be the inference of more complex phenotypic relationships from sequence-based analysis.

One of the limitations in 16S rRNA-based phylogenetic analysis is in the delineation of species boundaries. Although members of different, traditionally defined species within some genera contain distinct gene sequences, this does not always hold true (16). On the other hand, genus boundaries are usually more easily established. In general, by examining multiple gene sequences, one acquires a more accurate representation of organismal history. Some of the more useful molecules, besides rRNA, include RNA polymerases, elongation factor G, and proton-translocating ATPases (33). Other molecules behave as reliable evolutionary clocks for only limited groups of organisms, e.g., the citrate synthase gene for the rickettsiae (40).

AN APPROACH FOR IDENTIFICATION OF MICROORGANISMS
DIRECTLY FROM HOST TISSUE

A variety of methods have been used to generate rRNA sequences. Some of the earliest studies relied upon RNase T_1 oligonucleotide catalogs, a cumbersome approach. Lane et al. (28) later described reverse transcriptase sequencing of 16S rRNA by using bulk cellular RNA and synthetic oligodeoxynucleotide primers designed from universally conserved 16S rRNA sequences. Following the advent of nucleic acid amplification techniques, rapid methods for 16S rRNA and ribosomal DNA (rDNA) sequencing were developed, again relying upon scattered regions of 16S rRNA sequence that are conserved among large phylogenetic groups (e.g., all three or separate individual domains) (5, 30, 56, 62). Boddinghaus et al. (4) proposed the use of genus-specific mycobacterial 16S rRNA primers for PCR amplification of 16S ribosomal complementary DNA (rcDNA) and species-specific mycobacterial detection (4). All of these methods were applied to purified or cultivated cells.

In 1990, a series of investigations illustrated the feasibility of directly studying the compositions of natural microbial communities from environmental samples (18, 55). Libraries of cloned 16S rRNA genes were created from 16S rcDNA or PCR-amplified rDNA without in vitro cultivation. With analogous objectives, colleagues and I became interested in the identification of microbial pathogens directly from infected human tissue, i.e., the culture-independent study of pathogenic microorganisms in a disease-associated "natural" habitat (41). For the purpose of identifying previously uncharacterized members of the domain *Bacteria,* it was assumed that these organisms would share 16S rRNA sequences conserved in nearly all previously studied members of the domain (see Fig. 1 in reference 41) (28, 63). At the same time, these "broad-range" sequences would distinguish *Bacteria* from all members of the domain *Eukarya,* including, and in particular, humans. A crude digest of infected or diseased tissue would serve as target in a PCR with primers designed from conserved broad-range bacterial sequences. With the sequence of the amplified partial or complete 16S rDNA fragments, one could then infer the phylogenetic relationships of the putative causative bacterial agent. A more detailed description of this experimental approach has been published elsewhere (42, 47).

The experimental approach described above was initially applied to an infection of humans associated with angioproliferative pathology, known as bacillary angiomatosis (BA) (6, 50). This disease had been attributed to pleomorphic bacilli that were consistently detected in these tissues with a silver stain but that had not been reproducibly cultivated as of 1990. A single predominant 16S rDNA sequence was identified in tissues from four independent patients with BA. This sequence suggested the presence of a previously uncharacterized alpha-proteobacterium closely related to *Rochalimaea quintana* (see below) (44). In subsequent studies, the same approach revealed the nature of the putative agent of Whipple's disease, a chronic systemic infection of the gastrointestinal tract, its lymphatic drainage, and occasionally the heart, central nervous system, and other visceral sites (45, 61). Whipple's disease was first described in 1907 and has been associated with a visible but uncultivated bacillus since then (11, 59). Phylogenetic analysis suggested that the Whipple bacillus is a member of the gram-positive division of *Bacteria* and a previously unidentified actinomycete. Colleagues and I have proposed a genus and species name for this organism solely on the basis of its specific 16S rRNA sequence, bacillary morphology, and the unusual disease attributed to it (45).

Each of these two disease investigations illustrates the usefulness of culture-independent approaches for microbial identification. The results also provide an opportunity to speculate about the nature and diversity of microbial pathogens (see below). At the same time, some of the technical difficulties associated with this work raise other important issues concerning species definitions, the sterility of privileged anatomic sites, and proof of disease causation. For example, when 16S rDNA microheterogeneity (<1%) is observed in amplified molecules from a single tissue, how does one sort through the various possible explanations? Necrosis or chronic inflammation, as well as tissue fixation, may lead to damaged DNA (e.g., nucleotide substitutions) (22, 35). Many organisms contain multiple *rrn* operons with variant sequences (for an extreme example, see reference 31). *Taq* polymerase has an intrinsic nucleotide incorporation error rate of approximately 0.15% (13). Microheterogeneity may also reflect the presence of multiple strains or species. But because there are not yet enough sequence data from which to define species boundaries, the interpretation of amplified sequence is problematic. The possibility of PCR-generated chimeric molecules from samples with mixed infections further complicates the picture (36). Given the extreme sensitivity of PCR, further thought will be required in order to understand the meaning of the presence of amplified microbial ss rDNA in the absence of tissue pathology. And new postulates (14) will be needed in order to prove that the organism whose existence is inferred from an amplified sequence is, in fact, the true cause of the disease in question. A key approach may involve in situ hybridization with specific 16S rRNA oligonucleotide probes (1, 7, 19).

BACILLARY ANGIOMATOSIS AND THE ALPHA-PROTEOBACTERIA

Analysis of a complete 16S rRNA gene sequence amplified from tissue from patients with BA confirmed that the dominant organism associated with this form of pathology is a member of the alpha subdivision of the proteobacteria (formerly purple bacteria) (43, 44). Proteobacteria in general are found in diverse habitats (54). This subdivision contains a number of cultivated and characterized bacteria that are pathogenic for humans, including the genera *Rickettsia*, *Ehrlichia*, *Bartonella*, and *Rochalimaea*. The agents now associated with BA include *Rochalimaea henselae* (initially identified by amplified 16S rDNA) and *R. quintana* (27, 39). Interestingly, the *Rochalimaea* species are most closely related (98.5% sequence similarity) to *Bartonella bacilliformis* (43), a sandfly-borne, epierythrocytic pathogen in humans and the only other known bacterial cause of angioproliferative pathology. In this instance, molecular phylogeny predicts the monophyletic inheritance of an important virulence-associated phenotype. Cats are a reservoir for *R. henselae*, and cat fleas may act as a vector (26).

The alpha-proteobacteria tend to be intra- or epicellular endosymbionts of protozoa, insects, and other invertebrates. Many are obligate intracellular parasites. *Wolbachia pipientis* is an insect gonad tissue endosymbiont that is associated with cytoplasmic incompatibility and sex ratio distortion in a wide variety of insect genera. Some alpha-proteobacteria establish commensal relationships with plants but can occasionally cause disease (e.g., *Agrobacterium* and *Rhizobium* species). When these organisms find themselves in a vertebrate or human host, they frequently cause disease. *Anaplasma marginale* is an intracellular agent that propagates within ticks but that can act as an intraerythrocytic pathogen of cattle and other ruminants. These shared microbial features have suggested to some investigators that the genera *Rickettsia*, *Ehrlichia*, and *Rochalimaea* may have arisen as plant endosymbionts (58). More interesting, however, is the close relationship between the alpha-proteobacteria and mitochondria. Analysis of rRNA sequences from the latter place the origins of this organelle within the deeper branchings of the alpha-proteobacteria (21). In fact, it seems likely that mitochondria are the descendents of an endosymbiotic alpha-proteobacterium (or alpha-proteobacteria) that evolved to become an integral part of the eukaryotic genome. Likewise, chloroplasts may be the descendants of separate endosymbiotic lineages within the cyanobacteria division of *Bacteria* (52). Gray (21) has hypothesized that other organelles may also be of endosymbiotic origin. Could phylogenetically dispersed phototrophy reflect the polyphyletic establishment of intimate endosymbiotic relationships by ancient intracellular *Bacteria* or *Archaea?*

Thus, it may not be surprising that a number of recently identified uncultivated or fastidious human microbial pathogens are related to endosymbiotic alpha-proteobacteria. The intimate relationship between this group of organisms and eukaryotic host cells has probably led to nutritional dependencies that will be difficult to duplicate in a laboratory setting. By examining arthropod tissue directly, culture-independent methods will likely reveal a multitude of microorganisms—perhaps alpha-proteobacteria—with pathogenic potential for humans. As an example, a novel typhus-like organism has been detected in cat fleas by direct amplification of common but variable rickettsial gene sequences (2). This alpha-proteobacterium, the ELB agent, is harbored by opossums and may be the cause of a typhus-like illness in humans in southern California (60).

WHIPPLE'S DISEASE AND THE ACTINOMYCETES

The phylogeny of the Whipple's disease bacillus reveals a second theme that may be common to previously uncultivated microbial pathogens. The bacterial subdivision known as "high G + C gram-positive bacteria," or actinomycetes, comprises a large number of common soil and water commensal or saprophytic microorganisms. Some have adapted to and reside on the skin or mucous membranes of animals. Most of these microorganisms rarely behave as pathogens, if ever. Examples of actinomycetes include members of the genera *Arthrobacter*, *Streptomyces*, and *Terrabacter*, which are common soil organisms but which are usually nonpathogenic, and the genera *Mycobacterium*, *Rhodococcus*, *Nocardia*,

Actinomyces, Micrococcus, and *Dermatophilus,* which are common soil and water commensal organisms as well as colonizers of animals and which are occasionally pathogens. The Whipple's disease bacillus is most closely related to the genera *Dermatophilus, Arthrobacter,* and *Terrabacter,* but it is not closely related (<92.5% sequence similarity) to any of these (45). Nonetheless, these 16S rRNA-based phylogenetic relationships and the wide geographical distribution of Whipple's disease suggest that the Whipple's disease bacillus may be an environmental (soil and/or water) commensal organism to which humans are commonly exposed. An analysis of Whipple's disease patient occupations is consistent with this hypothesis; i.e., farmers are disproportionately represented among this population (11).

If the Whipple's disease bacillus is an environmental organism, then why is classical Whipple's disease so rare? Either its distribution in nature is limited or only human hosts with a subtle (and unrecognized) immune defect are susceptible to disease. If the organism is common, then one would suspect that other infected individuals develop less well recognized forms of pathology or else remain asymptomatic. It is interesting that two other actinomycetes, *Mycobacterium avium* complex and *Rhodococcus equi,* are both associated with a Whipple's disease-like syndrome in human immunodeficiency virus (HIV)-infected hosts (51, 53). One population in which one might expect to find the Whipple's disease bacillus is HIV-infected individuals with chronic unexplained diarrhea.

On the basis of a culture-independent analysis of bacterial diversity within soil, it is not surprising that previously unidentified human microbial pathogens would be found in this environment. Liesack and Stackebrandt (29) studied 86 16S rRNA gene fragments randomly cloned from a pool of molecules amplified directly from extracted soil DNA. The inferred organisms belonged to a variety of groups in the domain *Bacteria,* including the alpha-proteobacteria, the *Planctomyces-Chlamydia,* the gram-positive bacteria, and a previously undefined group. None of these sequences corresponded to known cultivated organisms. Gold (20) speculates that there may be a vast diversity of life below the surface of the earth at depths of up to 10 km comparable in mass and volume to the biomass on the earth's surface. Thus, it is clear that the terrestrial as well as the aquatic natural environments are replete with previously uncharacterized microorganisms (see below). Because humans have frequent and sometimes intimate contact with these environments, any of these organisms might be found as transient or permanent members of the human mucosal or skin commensal microflora and might occasionally cause disease.

CHARACTERIZATION OF MICROBIAL COMMUNITIES

As environmental microbiologists discover unexplored diversity in natural microbial communities, there is a clear message for medical microbiologists and for those who study microbial pathogenesis. The complex and intricate coadaptations of microorganisms with higher-order hosts may often hide their most critical biological features to the outside observer—even their very existence! Studies of microbial diversity in hot spring and ocean habitats based on the directly cloning or amplification of 16S rRNA genes reveal a preponderance of organisms that have not been cultivated (17, 18, 46, 55). These results are quite similar to those from investigations of bacterial diversity in soil described above (29). Perhaps it should come as no surprise that in extreme, nonphysiological environments, e.g., anaerobic, high-temperature hot springs, the vast majority of microorganisms enjoy growth conditions that would be difficult to duplicate in the laboratory, and hence remain uncultivated. Fuhrman et al. (17) and Barns et al. (3) have each discovered novel groups of *Archaea* in these environments that are most closely related to deeply branching, thermophilic lineages. What about the more familiar environments of the mammalian host? What fraction of the microorganisms that inhabit the human body have been cultivated or identified?

Clinical microbiology laboratories and the traditional methods used for microbial cultivation and identification are certainly biased in favor of the growth conditions and physiological properties displayed by known human commensal and pathogenic microorganisms. In addition, the clinical imperative generated by an infected and deathly ill patient motivates a more intense effort to detect and identify pathogenic microorganisms than might be brought to bear on environmental microbial communities. Therefore, a greater proportion of the transient or permanent microflora in humans, as well as the pathogenic microorganisms that intrude from the external environment to cause disease, may already have been identified. Nonetheless, one might envision an experiment similar to that of Barns et al. (3) in order to determine the extent to which cultivation yields an accurate representation of microbial diversity in a well-characterized human anatomic niche, the subgingival crevice (Fig. 2). One would compare the diversity revealed by traditional culture-based methods to that revealed by comparative analysis of directly amplified 16S rDNA sequences. To end with a quote from a modern-day Winogradsky

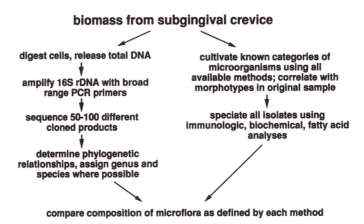

FIGURE 2. Proposed analysis of the human commensal microbial flora found in the subgingival crevice. This experiment is designed to address the question, to what extent does cultivation yield an accurate representation of microbial diversity in this human anatomic niche?

(Stanley Falkow), "Stop talking about it, and just do the experiment." This experiment is in progress.

ACKNOWLEDGMENTS. Stanley Falkow was instrumental in conceiving this work on the identification of previously uncharacterized pathogens. The idea for this work may have arisen as Stanley sat next to me at the weekly Stanford Infectious Diseases clinical conference. Between intermittent discrete bodily noises, Stanley leaned over, after looking at a photomicrograph of the bacteria associated with bacillary angiomatosis, and pronounced that we should certainly be able to identify these organisms by using rRNA sequences, without relying upon culture.

Stanley has been that rare person who combines great scientific vision, bench smarts, and clinical insight. Stanley is one of the most astute infectious disease "clinicians" I know. He has certainly had a profound influence on my intellectual and personal development. Finally, in deference to Stanley's language usage sensitivities, I have tried to use the term "uncultivated" rather than "uncultured."

Support for this work was provided by the Lucille P. Markey Charitable Trust (D.A.R. is a Lucille P. Markey Biomedical Scholar) and the SmithKline-Stanford PMGM Director's Research Fund.

REFERENCES

1. **Amann, R. I., L. Krumholz, and D. A. Stahl.** 1990. Fluorescent-oligonucleotide probing of whole cells for determinative, phylogenetic, and environmental studies in microbiology. *J. Bacteriol.* **172:**762–770.

2. **Azad, A. F., J. J. Sacci, W. M. Nelson, G. A. Dasch, E. T. Schmidtmann, and M. Carl.** 1992. Genetic characterization and transovarial transmission of a typhus-like rickettsia found in cat fleas. *Proc. Natl. Acad. Sci. USA* **89:**43–46.

3. **Barns, S. M., R. E. Fundyga, M. W. Jeffries, and N. R. Pace.** 1994. Remarkable archaeal diversity detected in a Yellowstone National Park hot spring environment. *Proc. Natl. Acad. Sci. USA* **91:** 1609–1613.

4. **Boddinghaus, B., T. Rogall, T. Flohr, H. Blocker, and E. C. Bottger.** 1990. Detection and identification of mycobacteria by amplification of rRNA. *J. Clin. Microbiol.* **28:**1751–1759.

5. **Chen, K., H. Neimark, P. Rumore, and C. R. Steinman.** 1989. Broad range DNA probes for detecting and amplifying eubacterial nucleic acids. *FEMS Microbiol. Lett.* **48:**19–24.

6. **Cockerell, C. J., and P. E. LeBoit.** 1990. Bacillary angiomatosis: a newly characterized, pseudoneoplastic, infectious, cutaneous vascular disorder. *J. Am. Acad. Dermatol.* **22:**501–512.

7. **DeLong, E. F., G. S. Wickham, and N. R. Pace.** 1989. Phylogenetic stains: ribosomal RNA-based probes for the identification of single cells. *Science* **243:**1360–1363. (Erratum, **245:**1312.)

8. **Dewhirst, F. E., C. K. Chen, B. J. Paster, and J. J. Zambon.** 1993. Phylogeny of species in the family *Neisseriaceae* isolated from human dental plaque and description of *Kingella orale* sp. nov. *Int. J. Syst. Bacteriol.* **43:**490–499.

9. **Dewhirst, F. E., B. J. Paster, I. Olsen, and G. J. Fraser.** 1992. Phylogeny of 54 representative strains of species in the family *Pasteurellaceae* as determined by comparison of 16S rRNA sequences. *J. Bacteriol.* **174:**2002–2013.

10. **Dewhirst, F. E., B. J. Paster, I. Olsen, and G. J. Fraser.** 1993. Phylogeny of the Pasteurellaceae as determined by comparison of 16S ribosomal ribonucleic acid sequences. *Int. J. Med. Microbiol. Virol. Parasitol. Infect. Dis.* **279:**35–44.

11. **Dobbins, W. O.** 1987. *Whipple's Disease.* Charles C Thomas, Springfield, Ill.

12. **Dorsch, M., D. Lane, and E. Stackebrandt.** 1992. Toward a phylogeny of the genus *Vibrio* based on 16S rRNA sequences. *Int. J. Syst. Bacteriol.* **42:**58–63.

13. **Ennis, P. D., J. Zemmour, R. D. Salter, and P. Parham.** 1990. Rapid cloning of HLA-A,B cDNA by using the polymerase chain reaction: frequency and nature of errors produced in amplification. *Proc. Natl. Acad. Sci. USA* **87:**2833–2837.

14. **Evans, A. S.** 1976. Causation and disease: the Henle-Koch postulates revisited. *Yale J. Biol. Med.* **49:**175–195.

15. **Fox, G. E., E. Stackebrandt, R. B. Hespell, J. Gibson, J. Maniloff, T. A. Dyer, R. S. Wolfe, W. E. Balch, R. S. Tanner, L. J. Magrum, L. B. Zablen, R. Blakemore, R. Gupta, L. Bonen, B. J.**

Lewis, D. A. Stahl, K. R. Luehrsen, K. N. Chen, and C. R. Woese. 1980. The phylogeny of prokaryotes. *Science* **209**:457–463.

16. Fox, G. E., J. D. Wisotzkey, and P. J. Jurtshuk. 1992. How close is close: 16S rRNA sequence identity may not be sufficient to guarantee species identity. *Int. J. Syst. Bacteriol.* **42**:166–170.

17. Fuhrman, J. A., K. McCallum, and A. A. Davis. 1992. Novel major archaebacterial group from marine plankton. *Nature* (London) **356**:148–149.

18. Giovannoni, S. J., T. B. Britschgi, C. L. Moyer, and K. G. Field. 1990. Genetic diversity in Sargasso Sea bacterioplankton. *Nature* (London) **345**:60–63.

19. Giovannoni, S. J., E. F. DeLong, G. J. Olsen, and N. R. Pace. 1988. Phylogenetic group-specific oligodeoxynucleotide probes for identification of single microbial cells *J. Bacteriol.* **170**:720–726. (Erratum, **170**:2418.)

20. Gold, T. 1992. The deep, hot biosphere. *Proc. Natl. Acad. Sci. USA* **89**:6045–6049.

21. Gray, M. W. 1992. The endosymbiont hypothesis revisited. *Int. Rev. Cytol.* **141**:233–357.

22. Greer, C. E., S. L. Peterson, N. B. Kiviat, and M. M. Manos. 1991. PCR amplification from paraffin-embedded tissues: effects of fixative and fixation time. *Am. J. Clin. Pathol.* **95**:117–124.

23. Gutell, R. R., B. Weiser, C. R. Woese, and H. F. Noller. 1985. Comparative anatomy of 16-S-like ribosomal RNA. *Prog. Nucleic Acids Res. Mol. Biol.* **32**:155–216.

24. Ibrahim, A., B. M. Goebel, W. Liesack, M. Griffiths, and E. Stackebrandt. 1993. The phylogeny of the genus Yersinia based on 16S rDNA sequences. *FEMS Microbiol. Lett.* **114**:173–177.

25. James, B. D., G. J. Olsen, and N. R. Pace. 1989. Phylogenetic comparative analysis of RNA secondary structure. *Methods Enzymol.* **180**:227–239.

26. Koehler, J. E., C. A. Glaser, and J. W. Tappero. 1994. *Rochalimaea henselae* infection. A new zoonosis with the domestic cat as reservoir. *JAMA* **271**:531–535.

27. Koehler, J. E., F. D. Quinn, T. G. Berger, P. E. LeBoit, and J. W. Tappero. 1992. Isolation of Rochalimaea species from cutaneous and osseous lesions of bacillary angiomatosis. *N. Engl. J. Med.* **327**:1625–1631.

28. Lane, D. J., B. Pace, G. J. Olsen, D. A. Stahl, M. L. Sogin, and N. R. Pace. 1985. Rapid determination of 16S ribosomal RNA sequences for phylogenetic analyses. *Proc. Natl. Acad. Sci. USA* **82**:6955–6959.

29. Liesack, W., and E. Stackebrandt. 1992. Occurrence of novel groups of the domain *Bacteria* as revealed by analysis of genetic material isolated from an Australian terrestrial environment. *J. Bacteriol.* **174**:5072–5078.

30. Medlin, L., H. J. Elwood, S. Stickel, and M. L. Sogin. 1988. The characterization of enzymatically amplified eukaryotic 16S-like rRNA-coding regions. *Gene* **71**:491–499.

31. Mylvaganam, S., and P. P. Dennis. 1992. Sequence heterogeneity between the two genes encoding 16S rRNA from the halophilic archaebacterium Haloarcula marismortui. *Genetics* **130**:399–410.

32. Olsen, G. J., D. J. Lane, S. J. Giovannoni, N. R. Pace, and D. A. Stahl. 1986. Microbial ecology and evolution: a ribosomal RNA approach. *Annu. Rev. Microbiol.* **40**:337–365.

33. Olsen, G. J., and C. R. Woese. 1993. Ribosomal RNA: a key to phylogeny. *FASEB J.* **7**:113–231.

34. Olsen, G. J., C. R. Woese, and R. Overbeek. 1994. The winds of (evolutionary) change: breathing new life into microbiology. *J. Bacteriol.* **176**:1–6.

35. Paabo, S., R. G. Higuchi, and A. C. Wilson. 1989. Ancient DNA and the polymerase chain reaction. The emerging field of molecular archaeology. *J. Biol. Chem.* **264**:9709–9712.

36. Paabo, S., D. M. Irwin, and A. C. Wilson. 1990. DNA damage promotes jumping between templates during enzymatic amplification. *J. Biol. Chem.* **265**:4718–4721.

37. Paster, B. J., F. E. Dewhirst, W. G. Weisburg, L. A. Tordoff, G. J. Fraser, R. B. Hespell, T. B. Stanton, L. Zablen, L. Mandelco, and C. R. Woese. 1991. Phylogenetic analysis of the spirochetes. *J. Bacteriol.* **173**:6101–6109.

38. Pitulle, C., M. Dorsch, J. Kazda, J. Wolters, and E. Stackebrandt. 1992. Phylogeny of rapidly growing members of the genus *Mycobacterium*. *Int. J. Syst. Bacteriol.* **42**:337–343.

39. Regnery, R. L., B. E. Anderson, J. E. Clarridge III, M. C. Rodriguez-Barradas, D. C. Jones, and J. H. Carr. 1992. Characterization of a novel *Rochalimaea* species, *R. henselae* sp. nov., isolated from blood of a febrile, human immunodeficiency virus-positive patient. *J. Clin. Microbiol.* **30**:265–274.

40. Regnery, R. L., C. L. Spruill, and B. D. Plikaytis. 1991. Genotypic identification of rickettsiae

and estimation of intraspecies sequence divergence for portions of two rickettsial genes. *J. Bacteriol.* **173:**1576–1589.

41. **Relman, D. A.** 1993. The identification of uncultured microbial pathogens. *J. Infect. Dis.* **168:**1–8.

42. **Relman, D. A.** 1993. Universal bacterial 16S rDNA amplification and sequencing, p. 489–495. *In* D. H. Persing, T. F. Smith, F. C. Tenover, and T. J. White (ed.), *Diagnostic Molecular Microbiology: Principles and Applications.* American Society for Microbiology, Washington, D.C.

43. **Relman, D. A., P. W. Lepp, K. N. Sadler, and T. M. Schmidt.** 1992. Phylogenetic relationships among the agent of bacillary angiomatosis, *Bartonella bacilliformis,* and other alpha-proteobacteria. *Mol. Microbiol.* **6:**1801–1807.

44. **Relman, D. A., J. S. Loutit, T. M. Schmidt, S. Falkow, and L. S. Tompkins.** 1990. The agent of bacillary angiomatosis. An approach to the identification of uncultured pathogens. *N. Engl. J. Med.* **323:**1573–1580.

45. **Relman, D. A., T. M. Schmidt, R. P. MacDermott, and S. Falkow.** 1992. Identification of the uncultured bacillus of Whipple's disease. *N. Engl. J. Med.* **327:**293–301.

46. **Schmidt, T. M., E. F. DeLong, and N. R. Pace.** 1991. Analysis of a marine picoplankton community by 16S rRNA gene cloning and sequencing. *J. Bacteriol.* **173:**4371–4378.

47. **Schmidt, T. M., and D. A. Relman.** 1994. Phylogenetic identification of uncultured pathogens using ribosomal RNA sequences. *Methods Enzymol.* **235:**205–222.

48. **Stahl, D. A., and J. W. Urbance.** 1990. The division between fast- and slow-growing species corresponds to natural relationships among the mycobacteria. *J. Bacteriol.* **172:**116–124.

49. **Stanier, R. Y., and C. B. vanNiel.** 1941. The main outlines of bacterial classification. *J. Bacteriol.* **42:**437–466.

50. **Stoler, M. H., T. A. Bonfiglio, R. T. Steigbigel, and M. Pereira.** 1983. An atypical subcutaneous infection associated with acquired immune deficiency syndrome. *Am. J. Clin. Pathol.* **80:**714–718.

51. **Strom, R. L., and R. P. Gruninger.** 1983. AIDS with *Mycobacterium avium-intracellulare* lesions resembling those of Whipple's disease. *N. Engl. J. Med.* **309:**1323–1324. (Letter.)

52. **Turner, S., W. T. Burger, S. J. Giovannoni, L. R. Mur, and N. R. Pace.** 1989. The relationship of a prochlorophyte Prochlorothrix hollandica to green chloroplasts. *Nature* (London) **337:**380–382.

53. **Wang, H. H., D. Tollerud, D. Danar, P. Hanff, K. Gottesdiener, and S. Rosen.** 1986. Another Whipple-like disease in AIDS? *N. Engl. J. Med.* **314:**1577–1578. (Letter.)

54. **Ward, D. M., M. M. Bateson, R. Weller, and A. L. Ruff-Roberts.** 1992. Ribosomal RNA analysis of microorganisms as they occur in nature. *Adv. Microb. Ecol.* **12:**219–286.

55. **Ward, D. M., R. Weller, and M. M. Bateson.** 1990. 16S rRNA sequences reveal numerous uncultured microorganisms in a natural community. *Nature* (London) **345:**63–65.

56. **Weisburg, W. G., S. M. Barns, D. A. Pelletier, and D. J. Lane.** 1991. 16S ribosomal DNA amplification for phylogenetic study. *J. Bacteriol.* **173:**697–703.

57. **Weisburg, W. G., M. E. Dobson, J. E. Samuel, G. A. Dasch, L. P. Mallavia, O. Baca, L. Mandelco, J. E. Sechrest, E. Weiss, and C. R. Woese.** 1989. Phylogenetic diversity of the rickettsiae. *J. Bacteriol.* **171:**4202–4206.

58. **Weisburg, W. G., C. R. Woese, M. E. Dobson, and E. Weiss.** 1985. A common origin of rickettsiae and certain plant pathogens. *Science* **230:**556–558.

59. **Whipple, G. H.** 1907. A hitherto undescribed disease characterized anatomically by deposits of fat and fatty acids in the intestinal and mesenteric lymphatic tissues. *Johns Hopkins Hosp. Bull.* **18:**382–391.

60. **Williams, S. G., J. J. Sacci, M. E. Schriefer, E. M. Andersen, K. K. Fujioka, F. J. Sorvillo, A. R. Barr, and A. F. Azad.** 1992. Typhus and typhuslike rickettsiae associated with opossums and their fleas in Los Angeles County, California. *J. Clin. Microbiol.* **30:**1758–1762.

61. **Wilson, K. H., R. Blitchington, R. Frothingham, and J. A. Wilson.** 1991. Phylogeny of the Whipple's-disease-associated bacterium. *Lancet* **338:**474–475.

62. **Wilson, K. H., R. B. Blitchington, and R. C. Greene.** 1990. Amplification of bacterial 16S ribosomal DNA with polymerase chain reaction. *J. Clin. Microbiol.* **28:**1942–1946.

63. **Woese, C. R.** 1987. Bacterial evolution. *Microbiol. Rev.* **51:**221–271.

64. **Woese, C. R., O. Kandler, and M. L. Wheelis.** 1990. Towards a natural system of organisms: proposal for the domains Archaea, Bacteria, and Eucarya. *Proc. Natl. Acad. Sci. USA* **87:** 4576–4579.

Index